Lecture Notes in Computer Science 13256

More information about this series at https://link.springer.com/bookseries/558

Armando J. Pinho · Petia Georgieva ·
Luís F. Teixeira · Joan Andreu Sánchez (Eds.)

Pattern Recognition and Image Analysis

10th Iberian Conference, IbPRIA 2022
Aveiro, Portugal, May 4–6, 2022
Proceedings

Springer

Editors
Armando J. Pinho ⓘ
University of Aveiro
Aveiro, Portugal

Petia Georgieva ⓘ
University of Aveiro
Aveiro, Portugal

Luís F. Teixeira ⓘ
University of Porto
Porto, Portugal

Joan Andreu Sánchez ⓘ
Universitat Politècnica de València
Valencia, Spain

ISSN 0302-9743   ISSN 1611-3349 (electronic)
Lecture Notes in Computer Science
ISBN 978-3-031-04880-7   ISBN 978-3-031-04881-4 (eBook)
https://doi.org/10.1007/978-3-031-04881-4

This Springer imprint is published by the registered company Springer Nature Switzerland AG
The registered company address is: Gewerbestrasse 11, 6330 Cham, Switzerland

Preface

When, in July 2019, during the 9th edition of IbPRIA in Madrid, the city of Aveiro was announced as the 2021 location for the 10th edition, we were all far from imagining what was to come: in March 2020, the world almost came to a halt. Like many other activities, scientific conferences also had to adapt to the new reality and became remote. The 10th edition of IbPRIA was scheduled to occur in June 2021. However, after an illusive return to a certain normality by the end of 2020, soon it was realized that having a face to face conference would be highly improbable. Therefore, in January 2021, in agreement with both supporting associations, AERFAI and APRP, the conference was postponed to May 2022. Instead of keeping the original date and running the conference remotely, the main reason for this decision was our strong belief that one of the key aims of IbPRIA is to give researchers the opportunity to be together for three days, fostering new collaborations. This is also why the conference is single track, and tutorials and lunches are included in the registration. There is no excuse for not expanding our network of fellow researchers!

So, it is indeed a great pleasure to be here, in the beautiful city of Aveiro, receiving all of you for what we believe will be a very rewarding event. After nine editions, always occurring in odd years, namely in Andraxt (2003), Estoril (2005), Girona (2007), Póvoa de Varzim (2009), Las Palmas de Gran Canaria (2011), Madeira (2013), Santiago de Compostela (2015), Faro (2017), and Madrid (2019), this edition happens in an even year, almost three years after the previous one. However, the odd-year tradition will be resumed with IbPRIA 2023, in Spain!

IbPRIA is an international conference co-organized by the Portuguese APRP (Associação Portuguesa de Reconhecimento de Padrões) and Spanish AERFAI (Asociación Española de Reconocimiento de Formas y Análisis de Imágenes) chapters of IAPR (International Association for Pattern Recognition). For this edition, we received 72 full paper submissions, from which we selected 26 to be presented in oral sessions. Also, 28 papers were accepted for poster presentation. There were submissions from authors in 15 countries, showing that, although most of the contributions come traditionally from Portugal and Spain, there is a good level of interest in IbPRIA from researchers in other countries. On average, each paper received three reviews, mostly from members of the Program Committee. The final decision of acceptance was made by the editors of this volume, leaving out a number of very interesting papers that could not be accommodated in the final program.

Of course, a successful event depends on the invaluable effort of many people, including authors, reviewers, chairs, and members of the conference committees. A special thanks to the invited speakers, Bob Fisher, Isabel Trancoso, and Battista Biggio, and tutorial presenters Hermann Ney and Gregory Rogez. A final word to the outstanding members of the local committee, the real force behind the organization of IbPRIA 2022. Thanks!

May 2022

Armando J. Pinho
Petia Georgieva
Luís F. Teixeira
Joan Andreu Sánchez

Organization

IbPRIA 2022 was co-organized by the Spanish AERFAI and the Portuguese APRP chapters of IAPR (International Association for Pattern Recognition), and locally organized by IEETA (Institute of Electronics and Informatics Engineering of Aveiro), University of Aveiro, Portugal.

General Co-chair APRP

Luís F. Teixeira   University of Porto, Portugal

General Co-chair AERFAI

Joan Andreu Sánchez   Polytechnic University of Valencia, Spain

Local Chair

Armando J. Pinho   University of Aveiro, Portugal

Local Committee

Diogo Pratas   University of Aveiro, Portugal
Raquel Sebastião   University of Aveiro, Portugal
Samuel Silva   University of Aveiro, Portugal
Sónia Gouveia   University of Aveiro, Portugal
Susana Brás   University of Aveiro, Portugal

Invited Speakers

Robert Fisher   University of Edinburgh, UK
Isabel Trancoso   Instituto Superior Técnico, University of Lisbon,
                                          Portugal
Battista Biggio   University of Cagliari, Italy

Program Chairs

Catarina Silva   University of Coimbra, Portugal
Hélder Oliveira   INESC TEC, University of Porto, Portugal
Petia Georgieva   University of Aveiro, Portugal
Manuel J. Marín   Universidad de Córdoba, Spain

Tutorial Chairs

Luís Alexandre   University of Beira Interior, Portugal
Antonio Pertusa   University of Alicante, Portugal

Program Committee

Abhijit Das	Griffith University, Australia
Adrian Perez-Suay	University of Valencia, Spain
Alessia Saggese	University of Salerno, Italy
Ana Mendonça	University of Porto, Portugal
António Cunha	University of Trás-os-Montes and Alto Douro, Portugal
António J. R. Neves	University of Aveiro, Portugal
Antonio-Javier Gallego	University of Alicante, Spain
Antonio-José Sánchez-Salmerón	Polytechnic University of Valencia, Spain
Arsénio Reis	University of Trás-os-Montes and Alto Douro, Portugal
Bilge Gunsel	Istanbul Technical University, Turkey
Billy Peralta	Universidad Andres Bello, Chile
Carlo Sansone	University of Naples Federico II, Italy
David Menotti	UFPR, Brazil
Diego Sebastián Comas	UNMDP, Argentina
Diogo Pratas	University of Aveiro, Portugal
Enrique Vidal	Polytechnic University of Valencia, Spain
Fernando Monteiro	Polytechnic Institute of Bragança, Portugal
Filiberto Pla	Universitat Jaume I, Spain
Francisco Casacuberta	Polytechnic University of Valencia, Spain
Francisco Manuel Castro	University of Malaga, Spain
Helio Lopes	PUC-Rio, Brazil
Hugo Proença	Univeristy of Beira Interior, Portugal
Ignacio Ponzoni	Planta Piloto de Ingeniería Química, UNS, and CONICET, Argentina
Jacques Facon	Universidade Federal do Epirito Santo, Sao Mateus, Brazil
Jaime Cardoso	University of Porto, Portugal
Javier Hernandez-Ortega	Universidad Autonoma de Madrid, Spain
Jesus Ariel Carrasco-Ochoa	INAOE, Mexico
João Carlos Neves	Instituto de Telecomunicações, Portugal
João M. F. Rodrigues	Universidade do Algarve, Portugal
Jordi Vitria	CVC, Spain
Jorge Calvo-Zaragoza	Unversity of Alicante, Portugal
Jorge S. Marques	Instituto Superior Técnico, Portugal

Jose Miguel Benedi	Polytechnic University of Valencia, Spain
Juan Valentín Lorenzo-Ginori	Universidad Central "Marta Abreu" de Las Villas, Cuba
Julian Fierrez	Universidad Autonoma de Madrid, Spain
Kalman Palagyi	University of Szeged, Hungary
Larbi Boubchir	University of Paris 8, France
Laurent Heutte	Université de Rouen, France
Lev Goldfarb	UNB, Canada
Luis-Carlos González-Gurrola	Universidad Autonoma de Chihuahua, Mexico
Marcelo Fiori	Universidad de la República, Uruguay
Marcos A. Levano	Universidad Catolica de Temuco, Chile
Mario Bruno	Universidad de Playa Ancha, Chile
Martin Kampel	Vienna University of Technology, Austria
Paulo Correia	Instituto Superior Técnico, Portugal
Pedro Latorre Carmona	Universidad de Burgos, Spain
Pedro Cardoso	Universidade do Algarve, Portugal
Pedro Real Jurado	University of Seville, Spain
Rafael Medina-Carnicer	University of Córdoba, Spain
Ramón A. Mollineda Cárdenas	Universitat Jaume I, Spain
Raquel Sebastião	University of Aveiro, Portugal
Rebeca Marfil	University of Malaga, Spain
Roberto Alejo	Tecnológico Nacional de México, Toluca, Mexico
Ruben Tolosana	Universidad Autonoma de Madrid, Spain
Ruben Vera-Rodriguez	Universidad Autonoma de Madrid, Spain
Samuel Silva	University of Aveiro, Portugal
Sónia Gouveia	University of Aveiro, Portugal
Susana Brás	University of Aveiro, Portugal
V. Javier Traver	Universitat Jaume I, Spain
Ventzeslav Valev	Institute of Mathematics and Informatics, Bulgarian Academy of Sciences, Bulgaria
Vicente Garcia	Universidad Autónoma de Ciudad Juárez, Mexico
Vitaly Kober	CICESE, Mexico
Vitomir Struc	University of Ljubljana, Slovenia
Vítor Filipe	University of Trás-os-Montes and Alto Douro, Portugal

Sponsoring Institutions

AERFAI – Asociación Española de Reconocimiento de Formas y Análisis de Imágenes
APRP – Associação Portuguesa de Reconhecimento de Padrões
FCT – Fundação para a Ciência e a Tecnologia, Portugal
UA – University of Aveiro, Portugal

Plenary Talks

The Vision Subsystems in the TrimBot2020 Gardening Robot

Robert Fisher

School of Informatics, University of Edinburgh, UK

Abstract. The TrimBot2020 gardening robot was designed to work outdoors in varying lighting conditions. The robot also needed to perform various tasks: model the garden, navigate without collisions, itentify and servo the vehicle to trimming locations, and identify and servo the cutters to trimming targets. Each of these tasks led to a different computer vision approach. This talk will give an overview of each of the systems, which included 2D intensity, 2D-to-3D, and 3D point cloud processing, using both traditional and deepnet algorithms.

Speech as Personal Identifiable Information

Isabel Trancoso

IST, University of Lisbon, Portugal

Abstract. Speech is the most natural and immediate form of communication. It is ubiquitous. The tremendous progress in language technologies that we have witnessed in the past few years has led to the use of speech as input/output modality in a panoplia of applications which have been mostly reserved for text until recently. Many of these applications run on cloud-based platforms that provide remote access to powerful models in what is commonly known as Machine Learning as a Service (MLaaS), enabling the automation of time-consuming tasks (such as transcribing speech), and help users to perform everyday tasks (e.g. voice-based virtual assistants). When a biometric signal such as speech is sent to a remote server for processing, however, this input signal can be used to determine information about the user, including his/her preferences, personality traits, mood, health, political opinions, among other data such as gender, age range, height, accent, education level, etc. Although there is a growing society awareness about user data protection (the GDPR in Europe is such an example), most users of such remote servers are unaware of the amount of information that can be extracted from a handful of their sentences - in particular, about their health status. In fact, the potential of speech as a biomarker for health has been realised for diseases affecting respiratory organs, such as Obstructive Sleep Apnea or COVID-19, for mood disorders such as Depression, and Bipolar Disease, and neurodegenerative diseases such as Parkinson's or Alzheimer's disease. The potential for mining this type of information from speech is however largely unknown. The current state of the art in speaker recognition is also largely unknown. Many research studies with humans involve speech recordings. In the past, such recordings were stored, claiming that all user information is anonymised, but given that recent challenges in speaker recognition involve corpora of around 6,000 speakers, this anonymity may nowadays be questionable. Users are also generally unaware of the potential misuse of their speech data for voice cloning from a handful of utterances. This enormous potential can be used for impersonation attacks that may be harmfully used for defamation, misinformation or incrimination. Moreover, this potential also raises crucial spoofing concerns for automatic speaker verification systems. Machine Learning as a Service has been used to process different types of signal, namely using cryptographic techniques such as homomorphic encryption, secure multiparty computation, distance preserving hashing techniques, or differential privacy, among others. Their use for remote speech processing, however, raises many problems, namely in terms of computational and transmission costs, and their use may also imply some performance degradation. The discussion of all these issues thus requires joining forces of different research communities - speech, cryptographic, usability - but also the legal and the policy and non-legal governance communities. Their different taxonomy is probably the first obstacle to conquer. The GDPR contains few norms that have direct applicability to inferred data, requiring an effort of extensive interpretation of many of its norms, with adaptations, to guarantee the effective protection of people's rights in an era where speech must be legally regarded as PII (Personal Identifiable Information).

Machine Learning Security: Attacks and Defenses

Battista Biggio

Department of Electrical and Electronic Engineering,
University of Cagliari, Italy

Abstract. In this talk, I will briefly review some recent advancements in the area of machine learning security, including attacks and defenses, with a critical focus on the main factors which are hindering progress in this field. These include the lack of an underlying, systematic and scalable framework to properly evaluate machine-learning models under adversarial and out-of-distribution scenarios, along with suitable tools for easing their debugging. The latter may be helpful to unveil flaws in the evaluation process, as well as the presence of potential dataset biases and spurious features learned during training. I will finally report concrete examples of what our laboratory has been recently working on to enable a first step towards overcoming these limitations, in the context of different applications, including malware detection.

Invited Tutorials

Speech Recognition and Machine Translation: From Bayes Decision Theory to Machine Learning and Deep Neural Networks

Hermann Ney

RWTH Aachen University, Germany

Abstract. The last 40 years have seen a dramatic progress in machine learning and statistical methods for speech and language processing like speech recognition, handwriting recognition and machine translation. Many of the key statistical concepts had originally been developed for speech recognition. Examples of such key concepts are the Bayes decision rule for minimum error rate and sequence-to-sequence processing using approaches like the alignment mechanism based on hidden Markov models and the attention mechanism based on neural networks.

Recently the accuracy of speech recognition, handwriting recognition machine translation could be improved significantly by the use of artificial neural networks and specific architectures, such as deep feed-forward multi-layer perceptrons and recurrent neural networks, attention and transformer architectures. We will discuss these approaches in detail and how they form part of the probabilistic approach.

Human 3D Sensing from Monocular Visual Data Using Classification Techniques

Gregory Rogez

NAVER LABS Europe, Grenoble, France

Abstract. In this tutorial, I will explain how the complex and severely ill-posed problem of 3D human pose estimation from monocular images can be tackled as a detection problem using standard object classifiers. I will review the classification-based techniques that were proposed over the past 15 years to handle different levels of the human body including full-body, upper body, face and hands. I will discuss advantages and drawbacks of classification approaches and present in detail some solutions involving training data synthesis, CNN architectures, distillation and transformers.

Contents

Pattern Recognition and Machine Learning

Computer Vision

Other Applications

Document Analysis

Test Sample Selection for Handwriting Recognition Through Language Modeling

Adrian Rosello, Eric Ayllon, Jose J. Valero-Mas, and Jorge Calvo-Zaragoza[✉]

Department of Software and Computing Systems, University of Alicante,
Alicante, Spain
{arp129,eap56}@alu.ua.es, {jjvalero,jorge.calvo}@ua.es

Abstract. When it comes the need to automatically recognize hand-written information, there may be a scenario in which the capture device (such as the camera of a mobile phone) retrieves several samples—a burst—of the same content. Unlike general classification scenarios, combining different inputs is not straightforward for state-of-the-art handwriting recognition approaches based on neural end-to-end formulations. Moreover, since not all images within the burst may depict enough quality for contributing in the overall recognition task, it is hence necessary to select the appropriate subset. In this work, we propose a pilot study which addresses this scenario in which there exists a burst of pictures of the same manuscript to transcribe. Within this context, we present a straightforward strategy to select a single candidate out of the elements of the burst which achieves the best possible transcription. Our hypothesis is that the best source image is that whose transcription most likely matches a Language Model (LM) estimated on the application domain—in this work implemented by means of n-grams. Our strategy, therefore, processes all the images of the burst and selects the transcription with the highest *a priori* probability according to the LM. Our experiments recreate this scenario in two typical benchmarks of Handwritten Text Recognition (HTR) and Handwritten Music Recognition (HMR), validating the goodness of our proposal and leaving room for promising future work.

Keywords: Handwriting recognition · Language modeling ·
Convolutional recurrent neural network

1 Introduction

Handwriting recognition is the research field which devises computational methods for automatically interpreting handwritten data from sources such as documents, images or touchscreens, among others [3]. It represents a long-standing,

This paper is part of the project I+D+i PID2020-118447RA-I00, funded by MCIN/AEI/10.13039/501100011033. The third author is supported by grant APOSTD/2020/256 from "Programa I+D+i de la Generalitat Valenciana".

yet challenging, area with a wide range of applications as, for instance, signature recognition and verification [5] or writer identification [13].

One of the most considered use cases in handwriting recognition is that of automatically extracting information from document images. Within this broad field, two specific research endeavours are widely considered in the literature: Handwritten Text Recognition (HTR) [14] and Handwritten Music Recognition (HMR) [2]. Note that these two fields constitute key processes in cultural heritage preservation tasks as they allow for the automatic digitization and transcription of historical documents only existing as physical manuscripts [7].

Generally, handwritten recognition frameworks perform an initial set of pre-processes to segment the given manuscript into a set of atomic units—most commonly words or lines—which are eventually recognized. While these stages have been commonly performed using computer vision heuristics or classic machine learning approaches, modern formulations mainly consider the use deep neural networks due to their competitive performance [4].

Traditionally, handwriting recognition is performed on single samples, i.e., there is only one picture of the manuscript to digitize. However, in a real-world scenario, it might be possible to have more than a sole image of the same content to be transcribed, either because the recording device performs a burst of image captures (e.g., mobile devices) or because the aforementioned segmentation pre-process provides several atomic units of the same region of the document. Note that, while the aggregation of the individual recognition results from multiple inputs improves the overall performance in conventional classification scenarios [1], the consolidated approaches for handwriting recognition based on neural sequence labeling formulations do not depict such a straightforward integration, up to the point of being currently unfeasible.

In this work, we perform a pilot study to adequately tackle this scenario in which a group of available images are known to represent the same sequence to transcribe. More precisely, we assume the commented case of a burst of images retrieved from a given capture device, particularly a mobile phone in a regular daily use. While most of those images may depict slight differences among them (contrast variations, color alterations, etc.), in some extreme cases they may be remarkably slanted or even non-recognizable. Hence, since each picture of the burst is expected to depict a different recognition hypothesis depending on the quality of the capture, the problem relies in how to obtain the best overall transcription.

Our premise is that an accurate estimate may be selected from the individual hypotheses retrieved the burst of images attending to their likelihoods based on a statistical language model. This proposal is validated on two handwritten corpora of text and music scores, showing that the use of a language model is capable of outperforming the baseline considered. Besides, it is also shown that the proper configuration of this model remarkably determines the overall success of the proposed framework.

The rest of the work is structured as follows: Sect. 2 introduces the proposed recognition scheme; Sect. 3 describes the experimentation considered for assessing the proposal; Sect. 4 presents and discusses the results obtained; finally, Sect. 5 concludes the work and poses future lines to address.

2   Methodology

Due to their relevance in the related literature, our proposal builds upon recognition models that work at the "line" level[1] using neural end-to-end approaches. Within this formulation, given an image of a single line, the goal is retrieving the series of symbols that appear therein. Thus, our problem defined as a *sequence labeling* task [8], where each input must be associated to a sequence of symbols from a predefined alphabet, without any further input/output alignment.

In our particular scenario, the recognition model is provided with a set of images from the same piece of the manuscript to be recognized, namely a burst of pictures. Within this framework, given that the considered end-to-end recognition scheme hinders the integration of the individual predictions, we resort to a *sample selection* approach: the best estimation may be obtained by assessing the *correctness* of the individual hypotheses provided by the burst of pictures, following a specific policy. Note that, since the scenario considers the recognition of images from manuscripts, such correctness can be measured using a certain Language Model (LM) estimated from a set of reference data.

Figure 1 shows the proposed scheme for addressing the presented scenario. During the train stage, a set of reference data is used for training both the recognition framework—aided by data augmentation processes—and an LM. At inference, the L-size burst of images is processed with the trained model and the resulting individual hypotheses are scored using the LM for eventually selecting the optimal one using a pre-defined policy.

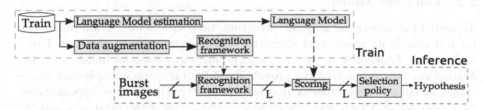

Fig. 1. Description of the recognition methodology proposed. Once the recognition framework and Language Model have been trained, a set of L hypotheses is drawn out of the burst of images and scored for eventually selecting the optimal one using a pre-defined policy.

The rest of the section introduces the particular recognition and LM frameworks considered in this work.

2.1   Neural End-to-end Recognition Framework

Formally, let $\mathcal{T} = \{(x_i, \mathbf{z}_i) : x_i \in \mathcal{X}, \ \mathbf{z}_i \in \mathcal{Z}\}_{i=1}^{|\mathcal{T}|}$ represent a set of data, where sample x_i is drawn from the image space $\mathcal{X}$ and $\mathbf{z}_i = (z_{i1}, z_{i2}, \ldots, z_{iM_i})$ corresponds to its transcript in terms of predefined set of symbol symbols. Note that

---

[1] In the case of music documents, the analogous unit is a staff.

$\mathcal{Z} = \Sigma^*$, where Σ represents the vocabulary. The per-line recognition task can be formalized as learning the underlying function $h : \mathcal{X} \to \mathcal{Z}$.

Given its good results in end-to-end approaches, we consider a Recurrent Neural Network (CRNN) [15] to approximate this function as $\hat{h}\,(\cdot)$. (CRNN) constitutes a particular neural architecture formed by an initial block of *convolutional* layers, which aim at learning the adequate set of features for the task, followed by another group of *recurrent* stages, which model the temporal dependencies of the elements from the initial feature-learning block.

To attain a proper end-to-end scheme, the CRNN is trained using the Connectionist Temporal Classification (CTC) algorithm [9], which allows optimizing the weights of the neural network using unsegmented sequential data. In our case, this means that, for a given image $x_i \in \mathcal{X}$, we only have its associated sequence of characters $\mathbf{z}_i \in \mathcal{Z}$ as its expected output, without any correspondence at pixel level or similar input-output alignment. Due to its particular training procedure, CTC requires the inclusion of an additional *"blank"* symbol within the Σ vocabulary, i.e., $\Sigma' = \Sigma \cup \{blank\}$.

The output of the CTC can be seen as a *posteriorgram*, i.e., the probability of each $\sigma \in \Sigma'$ to be located in each frame of the input image. Most commonly, the actual sequence prediction can be obtained out of this posteriorgram by using a *greedy* approach, which keeps the most probable symbol per step, merges repeated symbols in consecutive frames, and eventually removes the *blank* tokens.

2.2 Language Model

To model the underlying language structure, we resort to an n-gram probabilistic LM due to its conceptual simplicity and adequateness to the task. Given a sequence of tokens $\mathbf{w} = (w_1, \dots, w_{M-1})$, an n-gram model builds upon the Markovian assumption that the probability of finding symbol w_M is only conditioned by the previous n tokens. In other words, $P(w_M|\mathbf{w})$ may be approximated as $P(w_M|\mathbf{w}) \approx P(w_M|w_{M-n}, w_{M-n+1}, \dots w_{M-1}) = P(w_M|w_{M-n} : w_{M-1})$, where n represents the order of the n-gram model (unigram when $n = 1$, bigram when $n = 2$, etc.).

Based on this n-gram premise, the likelihood of a certain sequence $\mathbf{z} = (z_1, \dots, z_{|\mathbf{z}|})$ predicted by the CRNN recognition model of being a possibly correct sequence is computed as:

$$P\,(\mathbf{z}) \approx \prod_{m=1}^{|\mathbf{z}|} P(z_m|z_{m-n} : z_{m-1}) \tag{1}$$

In this work we consider as scoring function the so-called *perplexity* of the sequence $\mathbf{z}$, $PP(\mathbf{z})$, which is computed as:

$$\mathrm{PP}(\mathbf{z}) = P(\mathbf{z})^{\frac{1}{|\mathbf{z}|}} \approx \sqrt[|\mathbf{z}|]{\prod_{m=1}^{|\mathbf{z}|} \frac{1}{P(z_m|z_{m-n} : z_{m-1})}} \tag{2}$$

Note that, since this metric reflects the average branching factor in predicting a forthcoming word, lower scores denote better language representations.

As a final remark, while we acknowledge the simplicity of the approach compared to other more sophisticated, the goal of this work is stating whether language modeling processes may retrieve a proper best estimate in the considered scenario: a burst of images representing the same sequence. Hence, the search of the most appropriate model for the task may be deemed as a future work to explore.

3 Experimental Setup

This section introduces the set of corpora, neural architectures, evaluation protocol, and best hypothesis selection policies devised.

3.1 Corpora

We considered two different corpora of handwritten data to assess the goodness of our proposal:

- *Washington* set [6]: collection of 21 handwritten historical letters from George Washington annotated at the character level and segmented at both line and word levels. In this work we consider the former case—manuscripts segmented at the line level—, which comprises 656 images with 83 possible tokens. Figure 2b shows an example of those images.
- *Capitan* corpus [2]: set of sacred music manuscripts from the 17th century written in mensural notation which comprises 704 images of isolated monophonic staves with 320 different symbols. An example of a particular staff from this corpus is depicted in Fig. 2a.

(a) Staff section of the *Capitan* corpus.

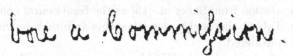

(b) Text line of the *Washington* set (with the message "bore a Commission.").

Fig. 2. Examples of the handwritten corpora considered.

In terms of data partitioning, each corpus is divided intro three fixed partitions of train, validation, and test data comprising each one 70%, 10%, and 20% of the total amount of data, respectively.

Train Data Augmentation. To improve the recognition capabilities of the neural model, we perform some controlled distortion on the train data partitions. These operations comprise the contrast modifications, rotation alterations, and erosion/dilation processes, as they are reported to enhance the robustness of the recognition model [12,16].

Burst Data Scenario. To recreate the considered scenario of a burst of L images representing the same target sequence, we derive a set of additional distorted images out of each sample in the test data partition. In these experiments the size of the burst is set to $L = 7$, being one of them the original image without any type of distortion.

In this regard, we consider two types of procedures for creating the data in the burst which differ on the severity of the distortion:

– **Mild procedure**: This first case creates new artificial samples directly using the *Train data augmentation* process previously introduced.
– **Hard procedure**: This second scenario first considers a *mild distortion* with the previous *Mild procedure* for then severely distorting the sample by blurring and/or cropping the sample.

In a real-world scenario, most images in the burst are expected to be *mildly* distorted, whereas a minor—and maybe none—fraction of them may depict a severe or hard distortion. To simulate this condition we consider that all images in the burst are distorted with the *Mild* procedure but, with a probability of $P_B \in [0,1]$—parameter of our laboratory experiments—, each of these data undergo a *Hard* distortion process. In our experimentation we consider the set of values $P_B = [0, 10\%, 20\%, 30\%]$ to assess the influence of this parameter in the overall recognition performance of the framework.

Finally, Fig. 3 shows an example of *Mild* and *Hard* distortions on two examples of the corpora considered. The original case is included for reference purposes.

3.2   Neural Architectures

Due to its reported good results in end-to-end recognition tasks, this work takes the CRNN configuration from Shi et al. [15] as the base neural configuration for the experiments. Table 1 thoroughly describes this architecture with the particular parameters considered.

The model was trained using the CTC backpropagation method using ADAM optimization [10], a fixed learning rate of 10^{-3}, and a batch size of 16 samples. A maximum of 100 training epochs was set, keeping the network weights of the best validation result. All images were scaled at the input of the model to a height of 64 pixels, forcing the maintenance of the aspect ratio for its width, and keeping their respective original color space.

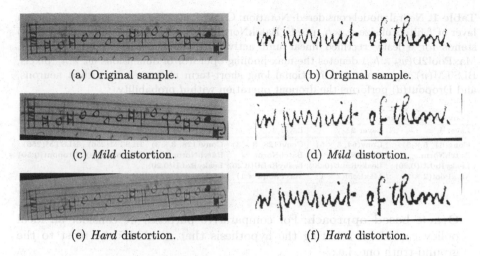

(a) Original sample.   (b) Original sample.

(c) *Mild* distortion.   (d) *Mild* distortion.

(e) *Hard* distortion.   (f) *Hard* distortion.

Fig. 3. Examples of the distortions considered, together with their respective original versions, for simulating the burst of data. Left column provides an example for the Capitan corpus whereas the right one shows the same procedures for the Washington set.

3.3   Best Hypothesis Selection Policies

This section introduces the different policies considered for selecting the most likely hypothesis from those given by the CRNN recognition framework when processing the images of the burst. Note that the proposed and compared strategies in this section do not aim at selecting the *best* image in the burst but the most suitable hypothesis from those obtained by performing the recognition process to the pictures in that set.

For the sake of clarity, let $\mathbf{x} = \{x_1, \dots, x_L\}$ represent a burst of L pictures corresponding to sequence $\mathbf{z} = (z_1, \dots, z_M)$ of length M. Additionally, let $\mathcal{H}(\mathbf{x}) = \left\{ \hat{h}(x_1), \dots, \hat{h}(x_L) \right\}$ denote the set of L hypotheses associated to burst $\mathbf{x}$ and let $\mathbf{z}' \in \mathcal{H}(\mathbf{x})$ denote the selected hypothesis.

We now present the actual policies assessed and compared in the work for selecting $\mathbf{z}'$:

- **LM directed**: This case represents the actual proposal of the work and it selects the hypothesis that minimizes the perplexity score $\mathrm{PP} : \mathcal{Z} \to \mathbb{R}$ of the considered n-gram model. Mathematically, this may be expressed as:

$$\mathbf{z}' = \operatorname*{argmin}_{\hat{h} \in \mathcal{H}(\mathbf{x})} \mathrm{PP}\left(\hat{h}(x_i)\right) \tag{3}$$

In addition, we assess the relation between the order of the model (parameter n) and the overall recognition performance of the posed framework considering values $n \in [2, 3, 4]$.

Table 1. Neural model considered. Notation: $\mathrm{Conv}(f, w_c \times h_c)$ represents a convolution layer of f filters of size $w_c \times h_c$ pixels, BatchNorm normalizes the batch, LeakyReLU(α) stands for a leaky rectified linear unit activation with negative slope value of α, MaxPool2D($w_p \times h_p$) denotes the max-pooling operator of dimensions $w_p \times h_p$ pixels, BLSTM(n) represents a bidirectional long short-term memory unit with n neurons, and Dropout(d) performs the dropout operation with d probability.

Convolutional stage				Recurrent stage	
Layer 1	Layer 2	Layer 3	Layer 4	Layer 5	Layer 6
Conv(64, 5 × 5)	Conv(64, 5 × 5)	Conv(128, 3 × 3)	Conv(128, 3 × 3)	BLSTM(256)	BLSTM(256)
BatchNorm	BatchNorm	BatchNorm	BatchNorm	Dropout(0.50)	Dropout(0.50)
LeakyReLU(0.20)	LeakyReLU(0.20)	LeakyReLU(0.20)	LeakyReLU(0.20)		
MaxPool(2 × 2)	MaxPool(2 × 1)	MaxPool(2 × 1)	MaxPool(2 × 1)		

- **Oracle-based approach**: For comparative purposes we consider an ideal policy capable of selecting the hypothesis that resembles the most to the ground-truth one, i.e.:

$$\mathbf{z}' = \underset{\hat{h} \in \mathcal{H}(\mathbf{x})}{\mathrm{argmin}} \ \mathrm{ED}\left(\hat{h}, \mathbf{z}\right) \qquad (4)$$

where $\mathrm{ED} : \mathcal{Z} \times \mathcal{Z} \to \mathbb{N}_0$ represents the string Edit distance [11]. Note that this approach is considered for stating the performance bound which would be achieved if the optimal hypothesis was selected.

- **Random approach**: In this policy we resort to a random selection of one of the sequences in set $\mathcal{H}(\mathbf{x})$. This case represents the baseline of the scenario posed as it models the cases in which there exists no method for selecting an optical hypothesis out of those from the burst. While more sophisticated policies could be devised (e.g., attending to the overall quality of the image or selecting the hypothesis with the maximum mutual agreement), we considered that such cases could be further explored in future lines to tackle provided that this work constitutes a pilot case of study meant to assess whether the proposed framework might be of interest in terms of research.

3.4   Evaluation Protocol

We consider the Symbol Error Rate (SER) to measure the performance of the proposal. Considering a set of test data $\mathcal{S} \subset \mathcal{X} \times \mathcal{Z}$, this metric is defined as the average number of elementary editing operations (insertions, deletions, or substitutions) necessary to match the sequence predicted by the model with the ground truth sequence and normalized by the length of the latter. Mathematically, this is represented as follows:

$$\mathrm{SER}\ (\%) = \frac{\sum_{i=1}^{|\mathcal{S}|} \mathrm{ED}\left(\mathbf{z}_i, \hat{h}(x_i)\right)}{\sum_{i=1}^{|\mathcal{S}|} |\mathbf{z}_i|} \qquad (5)$$

where $\hat{h}(x_i)$ and $\mathbf{z}_i$ respectively represent the estimated and ground-truth sequences for the i-th sample.

4   Results

The results obtained for the two corpora considered with the different hypothesis selection policies are reported in Table 2. Note that each row within the corpora represents the different burst data scenarios obtained when varying the parameter P_B related to the probability of having a severely distorted image within the group.

Table 2. Results obtained in terms of Sequence Error Rate (%)—SER—with the different hypothesis selection policies devised for the two handwritten data collections considered. Each row represents the different burst data scenarios obtained when varying the probability parameter P_B.

Corpus	Random	LM directed			Oracle-based
		2	3	4	
Washington					
0	7.18	6.34	4.66	4.72	3.62
10	7.20	6.13	4.66	4.66	3.58
20	12.81	16.25	4.68	4.66	3.61
30	13.89	12.20	4.64	4.70	3.71
Capitan					
0	13.33	12.77	12.71	12.71	9.99
10	13.22	12.95	12.71	12.71	9.92
20	19.14	12.74	12.71	12.71	10.05
30	19.06	12.83	12.71	12.71	10.18

On a broad sense, it may be observed that the baseline (Random choice) reports the lowest performance of the policies devised for all studied cases. Furthermore, these error rates progressively deteriorate as the number of samples with severe distortions increases. Oppositely, the proposed LM approach allows alleviating this phenomenon, retrieving figures close to those of the oracle case depending on the configuration of the actual model. We now analyze the results for each of the particular corpora.

Concerning the Washington dataset (text data), we observe that the LM plays a fundamental role in the performance of the recognition strategy. While with 2-grams, the trend is similar (yet slightly better) to the baseline, the use of 3-grams and 4-grams reports a remarkable improvement in the figures. In the simplest case ($P_B = 0$), the results with respect to the baseline improve from a SER of 7.18% to 4.72%, while in the most extreme case ($P_B = 30\%$), it is reduced from 13.89% to 4.70%. As expected, if we always chose the best option (Oracle-based) we would get the best recognition rates, although the best LM-driven approach remains just around 1% higher in terms of SER.

When addressing the Capitan corpus (music data) some slight differences are observed compared to the previous case of text data. More precisely, the use of higher orders of n-gram causes less impact on this occasion: 2-grams are already sufficiently stable and, in fact, 3-grams and 4-grams do not report any differences. In all cases, the best LM strategy maintains a SER figure of 12.71%, reducing this error from 13.33% and 19.06% for the simplest ($P_B = 0\%$) and most severe ($P_B = 30\%$) distortion scenarios, respectively. Furthermore, the results show a greater margin of improvement with respect to the oracle case than in the Washington set.

Despite the slight differences, the figures report some general trends regardless of the specific corpus. On the one hand, the LM-directed strategy can be considered successful as it yields competitive recognition results even when the amount of severely distorted images in the burst increases. On the other hand, there is still room for improvement in the hypothesis selection policy—as the Oracle-based approach is always better. However, the scale of the possible improvement does depend on the intrinsic difficulty of the manuscript to be recognized.

5 Conclusions

In this work, we present a straightforward strategy to choose the best hypothesis over a set of images from different captures of the same manuscript content. Our strategy is guided by a n-gram Language Model (LM), which is capable of scoring the different hypotheses based on their prior probability for that domain.

In our experiments, carried out on two corpora of handwritten text and music, it is observed that the proposed strategy is effective, and that a proper LM configuration is a key factor in the success of the task. In all cases, our strategy noticeably outperforms a random selection policy. However, we also observe that there is room for improvement, as our LM-based proposal does not always choose the most appropriate hypothesis.

As future work, we consider two main lines of research. The first, backed by the results obtained in this work, is to evaluate and propose better selection strategies. This can be addressed by improving the LM itself (for instance, with more sophisticated models based on neural networks or attention mechanisms) or developing more sophisticated selection strategies. The second avenue for further research is to study the possible aggregation of different hypotheses, as long as they are considered to report and appropriate transcription, instead of resorting to a selection policy which neglects the possible synergies when using several elements from the burst.

References

1. Calvo-Zaragoza, J., Rico-Juan, J.R., Gallego, A.-J.: Ensemble classification from deep predictions with test data augmentation. Soft. Comput. **24**(2), 1423–1433 (2019). https://doi.org/10.1007/s00500-019-03976-7

2. Calvo-Zaragoza, J., Toselli, A.H., Vidal, E.: Handwritten music recognition for mensural notation: Formulation, data and baseline results. In: 14th IAPR International Conference on Document Analysis and Recognition, ICDAR 2017, Kyoto, Japan, November 9–15, 2017, pp. 1081–1086. IEEE (2017)
3. Doermann, D., Tombre, K.: Handbook of Document Image Processing and Recognition. Springer, London (2014). https://doi.org/10.1007/978-0-85729-859-1
4. Dutta, K., Krishnan, P., Mathew, M., Jawahar, C.: Improving cnn-rnn hybrid networks for handwriting recognition. In: 16th International Conference on Frontiers in Handwriting Recognition, pp. 80–85. IEEE (2018)
5. Faundez-Zanuy, M.: On-line signature recognition based on VQ-DTW. Pattern Recogn. **40**(3), 981–992 (2007)
6. Fischer, A., Keller, A., Frinken, V., Bunke, H.: Lexicon-free handwritten word spotting using character HMMs. Pattern Recogn. Lett. **33**(7), 934–942 (2012)
7. Granell, E., Martinez-Hinarejos, C.D.: Multimodal crowdsourcing for transcribing handwritten documents. IEEE/ACM Trans. Audio Speech Lang. Process. **25**(2), 409–419 (2016)
8. Graves, A.: Supervised sequence labelling. In: Supervised Sequence Labelling with Recurrent Neural Networks, vol. 385, pp. 5–13. Springer, Heidelberg (2012). https://doi.org/10.1007/978-3-642-24797-2_2
9. Graves, A., Fernández, S., Gomez, F., Schmidhuber, J.: Connectionist temporal classification: labelling unsegmented sequence data with recurrent neural networks. In: Proceedings of the 23rd International Conference on Machine Learning. ICML 2006, New York, NY, USA, pp. 369–376. ACM (2006)
10. Kingma, D.P., Ba, J.: Adam: a method for stochastic optimization. In: 3rd International Conference on Learning Representations. San Diego, USA (2015)
11. Levenshtein, V.I., et al.: Binary codes capable of correcting deletions, insertions, and reversals. In: Soviet Physics Doklady, vol. 10, pp. 707–710. Soviet Union (1966)
12. López-Gutiérrez, J.C., Valero-Mas, J.J., Castellanos, F.J., Calvo-Zaragoza, J.: Data augmentation for end-to-end optical music recognition. In: Barney Smith, E.H., Pal, U. (eds.) ICDAR 2021. LNCS, vol. 12916, pp. 59–73. Springer, Cham (2021). https://doi.org/10.1007/978-3-030-86198-8_5
13. Rehman, A., Naz, S., Razzak, M.I.: Writer identification using machine learning approaches: a comprehensive review. Multimedia Tools Appl. **78**(8), 10889–10931 (2018). https://doi.org/10.1007/s11042-018-6577-1
14. Sánchez, J., Romero, V., Toselli, A.H., Villegas, M., Vidal, E.: A set of benchmarks for handwritten text recognition on historical documents. Pattern Recognit. **94**, 122–134 (2019)
15. Shi, B., Bai, X., Yao, C.: An end-to-end trainable neural network for image-based sequence recognition and its application to scene text recognition. IEEE Trans. Pattern Anal. Mach. Intell. **39**(11), 2298–2304 (2017)
16. Wigington, C., Stewart, S., Davis, B., Barrett, B., Price, B., Cohen, S.: Data augmentation for recognition of handwritten words and lines using a CNN-LSTM network. In: 2017 14th IAPR International Conference on Document Analysis and Recognition (ICDAR), vol. 1, pp. 639–645. IEEE (2017)

Classification of Untranscribed Handwritten Notarial Documents by Textual Contents

Juan José Flores[1] , Jose Ramón Prieto[1(✉)] , David Garrido[2] ,
Carlos Alonso[3] , and Enrique Vidal[1]

[1] PRHLT Research Center, Universitat Politècnica de València, Valencia, Spain
{juafloar,joprfon,evidal}@prhlt.upv.es
[2] HUM313 Research Group, Universidad de Cádiz, Cadiz, Spain
[3] tranSkriptorium IA, Valencia, Spain

Abstract. Huge amounts of digital page images of important manuscripts are preserved in archives worldwide. The amounts are so large that it is generally unfeasible for archivists to adequately tag most of the documents with the required metadata so as to allow proper organization of the archives and effective exploration by scholars and the general public. The class or "typology" of a document is perhaps the most important tag to be included in the metadata. The technical problem is one of automatic classification of documents, each consisting of a set of untranscribed handwritten text images, by the textual contents of the images. The approach considered is based on "probabilistic indexing", a relatively novel technology which allows to effectively represent the intrinsic word-level uncertainty exhibited by handwritten text images. We assess the performance of this approach on a large collection of complex notarial manuscripts from the *Spanish Archivo Histórico Provincial de Cádiz*, with promising results.

Keywords: Content-based image retrieval · Document classification · Historical manuscripts

1 Introduction

Content-based classification of manuscripts is an important task that is generally performed by expert archivists. Unfortunately, however, many manuscript collections are so vast that it is not possible to have the huge number of archive experts that would be needed to perform this task.

Current approaches for textual-content-based manuscript classification require the handwritten images to be first transcribed into text – but achieving sufficiently accurate transcripts are generally unfeasible for large sets of historical manuscripts. We propose a new approach to perform automatically this classification task which does not rely on any explicit image transcripts.

© Springer Nature Switzerland AG 2022
A. J. Pinho et al. (Eds.): IbPRIA 2022, LNCS 13256, pp. 14–26, 2022.
https://doi.org/10.1007/978-3-031-04881-4_2

Hereafter, bundles or folders of manuscript images are called "image bundles" or just "bundles" or "books". A bundle may contain several "files", also called "acts" or just "image documents". The task consists of classifying a given image document, that may range from a few to tens of handwritten text images, into a predefined set of classes or "types". Classes are associated with the topic or (semantic) content conveyed by the text written in the images of the document.

This task is different from other related tasks which, are often called with similar names, such as "content-based image classification", applied to single, natural scene (not text) images, and "image document classification", where classification is based on visual appearance or page layout. See [12] for a more detailed discussion on these differences, as well as references to previous publications dealing with related problems, but mainly aimed at printed text.

Our task is comparable to the time-honoured and well known task of *content-based document classification*, were the data are plain text documents. Popular examples of this traditional task, are *Twenty News Groups, Reuters, WebKB*, etc. [1,9,11]. The task here considered (textual-content-based handwritten text image document classification), is similar, except for a severe difference: our data are sets of digital images of handwritten text rather than file of (electronic) plain text. The currently accepted wisdom to approach our task would be to split the process into two sequential stages. First, a handwritten text recognition (HTR) system is used to transcribe the images into text and, second, content-based document classification methods, such as those referred to above, can be applied to the resulting text documents.

This approach might work to some extent for simple manuscripts, where HTR can provide over 90% word recognition accuracy [18]. But it is not an option for large historical collections, where the best available HTR systems can only provide word recognition accuracies as low as 40–60% [4,15,18]. This is the case of the collection which motivates this work, which encompasses millions of handwritten notarial files from the Spanish Archivo Histórico Provincial de Cádiz. A small subset of these manuscripts was considered in the Carabela project [4] and the average word recognition accuracy achieved was below 65% [15], dropping to 46% or less when conditions are closer to real-world usage [4]. Clearly, for these kinds of manuscript collections, the aforementioned two-stage idea would not work and more holistic approaches are needed.

In previous works [4,12], we have proposed an approach which strongly relies on the so-called *probabilistic indexing* (PrIx) technology, recently developed to deal with the intrinsic word-level *uncertainty* generally exhibited by handwritten text and, more so, by historical handwritten text images [3,10,14,20,21]. This technology was primarily developed to allow search and retrieval of textual information in large untranscribed manuscript collections [3,4,19].

In our proposal, PrIx provides the probability distribution of words which are likely written in the images, from which statistical expectations of *word* and *document frequencies* are estimated. These estimates are then used to compute well-known text features such as *Information Gain* and Tf·Idf [11], which are in turn considered inputs to a Multilayer Perceptron classifier [12].

In this paper, we consolidate this approach and, as mentioned above, apply it to a new collection of handwritten notarial documents from the Archivo Provincial de Cádiz. In contrast with [4,12], where the underlying class structure was very limited (just three rather artificial classes), here the classes correspond to real typologies, such as *power of attorney*, *lease*, *will*, etc. Our results clearly show the capabilities of the proposed approach, which achieves classification accuracy as high as 90–97%, depending on the specific set of manuscripts considered.

2   Probabilistic Indexing of Handwritten Text Images

The Probabilistic Indexing (PrIx) framework was proposed to deal with the intrinsic word-level uncertainty generally exhibited by handwritten text in images and, in particular, images of historical manuscripts. It draws from ideas and concepts previously developed for keyword spotting, both in speech signals and text images. However, rather than caring for "key" words, any element in an image which is likely enough to be interpreted as a word is detected and stored, along with its *relevance probability* (RP) and its location in the image. These text elements are referred to as *"pseudo-word spots"*.

Following [14,21], the image-region word RP is denoted as $P(R = 1 \mid X = x, V = v)$, but for the sake of conciseness, the random variable names will be omitted and, for $R = 1$, we will simply write R. As discussed in [22], this RP can be simply approximated as:

$$P(R \mid x, v) = \sum_{b \sqsubseteq x} P(R, b \mid x, v) \approx \max_{b \sqsubseteq x} P(v \mid x, b) \qquad (1)$$

where b is a small, word-sized image sub-region or Bounding Box (BB), and with $b \sqsubseteq x$ we mean the set of all BBs contained in x. $P(v \mid x, b)$ is just the posterior probability needed to "recognize" the BB image (x, b). Therefore, assuming the computational complexity entailed by (1) is algorithmically managed, any sufficiently accurate isolated word classifier can be used to obtain $P(R \mid x, v)$.

This word-level indexing approach has proved to be very robust, and it has been used to very successfully index several large iconic manuscript collections, such as the French CHANCERY collection [3], the BENTHAM PAPERS [19], and the Spanish CARABELA collection considered in this paper, among others.[1]

3   Plain Text Document Classification

If a text document is given in some electronic form, its words can be trivially identified as discrete, unique elements, and then the whole field of *text analytics* [1,11] is available to approach many document processing problems, including *document classification* (DC). Most DC methods assume a document representation model known as *vector model* or *bag of words* (BOW) [1,6,11]. In this

---

[1] See: http://transcriptorium.eu/demots/KWSdemos.

model, the order of words in the text is ignored, and a document is represented as a *feature vector* (also called "word embedding") indexed by V. Let $\mathcal{D}$ be a set of documents, $D \in \mathcal{D}$ a document, and $\vec{D} \in \mathbb{R}^N$ its BOW representation, where $N \stackrel{\text{def}}{=} |V|$. For each word $v \in V$, $D_v \in \mathbb{R}$ is the value of the v-th feature of $\vec{D}$.

Each document is assumed to belong to a unique class c out of a finite number of classes, C. The task is to predict the best class for any given document, D. Among many pattern recognition approaches suitable for this task, from those studied in [12] the Multi-Layer Perceptron (MLP) was the one most promising.

3.1 Feature Selection

Not all the words are equally helpful to predict the class of a document D. Thus, a classical first step in DC is to determine a "good" vocabulary, V_n, of reasonable size $n < N$. One of the best ways to determine V_n is to compute the *information gain* (IG) of each word in V and retain in V_n only the n words with highest IG.

Using the notation of [12], let t_v be the value of a boolean random variable that is *True* iff, for some random D, the word v appears in D. So, $P(t_v)$ is the probability that $\exists D \in \mathcal{D}$ such that v is used in D, and $P(\bar{t}_v) = 1 - P(t_v)$ is the probability that *no* document uses v. The IG of a word v is then defined as:

$$\text{IG}(v) = -\sum_{c \in C} P(c) \log P(c)$$
$$+ P(t_v) \sum_{c \in C} P(c \mid t_v) \log p(c \mid t_v)$$
$$+ P(\bar{t}_v) \sum_{c \in C} P(c \mid \bar{t}_v) \log P(c \mid \bar{t}_v) \tag{2}$$

where $P(c)$ is de prior probability of class c, $P(c \mid t_v)$ is the conditional probability that a document belongs to class c, given that it contains the word v, and $P(c \mid \bar{t}_v)$ is the conditional probability that a document belongs to class c, given that it does *not* contain v. Note that the first addend of Eq. (2) does not depend on v and can be ignored to rank all $v \in V$ in decreasing order of $\text{IG}(v)$.

To estimate the relevant probabilities in Eq. 2, let $f(t_v) \leq M \stackrel{\text{def}}{=} |\mathcal{D}|$ be the number of documents in $\mathcal{D}$ which contain v and $f(\bar{t}_v) = M - f(t_v)$ the number of those which do *not* contain v. Let $M_c \leq M$ be the number of documents of class c, $f(c, t_v)$ the number of these documents which contain v and $f(c, \bar{t}_v) = M_c - f(c, t_v)$ the number of those that do *not* contain v. Then, the relevant probabilities used in Eq. (2) can be estimated as follows:

$$P(t_v) = \frac{f(t_v)}{M} \qquad\qquad P(\bar{t}_v) = \frac{M - f(t_v)}{M} \tag{3}$$
$$P(c \mid t_v) = \frac{f(c, t_v)}{f(t_v)} \qquad\qquad P(c \mid \bar{t}_v) = \frac{M_c - f(c, t_v)}{M - f(t_v)} \tag{4}$$

3.2 Feature Extraction

Using information gain, a vocabulary V_n of size $n \leq N$ can be defined by selecting the n words with highest IG. By attaching a (real-valued) feature to each $v \in V_n$, a document D can be represente by a n-dimensional feature vector $\vec{D} \in \mathbb{R}^n$.

The value D_v of each feature v is typically related with the frequency $f(v, D)$ of v in D. However, absolute word frequencies can dramatically vary with the size of the documents and normalized frequencies are generally preferred. Let $f(D) = \sum_{v \in V_n} f(v, D)$ be the total (or "running") number of words in D. The normalized frequency of $v \in V_n$, often called *term frequency* and denoted $\mathrm{Tf}(v, D)$, is the ratio $f(v, D) / f(D)$, which is is a max-likelihood estimate of the conditional probability of word v, given a document D, $P(v \mid D)$.

While Tf adequately deals with document size variability, it has been argued that better DC accuracy can be achieved by further weighting each feature with a factor that reflects its "importance" to predict the class of a document. Of course, IG could be used for this purpose, but the so-called *inverse document frequency* (Idf) [2,8,17] is argued to be preferable. Idf is defined as $\log(M / f(t_v))$, which, according to Eq. (3), can be written as $-\log P(t_v)$.

Putting it all together, a document D is represented by a feature vector $\vec{D}$. The value of each feature, D_v, is computed as the Tf-Idf of D and v; i.e., $\mathrm{Tf}(v, D)$, weighted by $\mathrm{Idf}(t)$:

$$
\begin{aligned}
D_v &= \mathrm{Tf} \cdot \mathrm{Idf}(v, D) &&= \mathrm{Tf}(v, D) \cdot \mathrm{Idf}(v) \\
&= P(v \mid D) \log \frac{1}{P(t_v)} &&= \frac{f(v, D)}{f(D)} \log \frac{M}{f(t_v)}
\end{aligned}
\tag{5}
$$

4 Textual-Content-Based Classification of Sets of Images

The primary aim of PrIx is to allow fast and accurate search for textual information in large image collections. However, the information provided by PrIx can be useful for many other text analytics applications which need to rely on incomplete and/or imprecise textual contents of the images. In particular, PrIx results can be used to estimate all the text features discussed in Sect. 3, which are needed for image document classification.

4.1 Estimating Text Features from Image PrIx's

Since R is a binary random variable, theRP $P(R \mid x, v)$ can be properly seen as the statistical expectation that v is written in x. As discussed in [12], the sum of RPs for all the pseudo-words indexed in an image region x is the statistical expectation of the number of words written in x. Following this estimation principle, all the text features discussed in Sect. 3, which are needed for image document classification can be easily estimated.

Let $n(x)$ be the total (or "running") number of words written in an image region x and $n(X)$ the running words in an image document X encompassing several pages (i.e., $f(D)$, see Sect. 3.2). Let $n(v, X)$ be the frequency of a specific (pseudo-)word v in a document X. And let $m(v, \mathcal{X})$ be the number of documents in a collection, $\mathcal{X}$, which contain the (pseudo-)word v. As explained in [12], the expected values of these counts are:

$$E[n(x)] = \sum_v P(R \mid x, v) \tag{6}$$

$$E[n(X)] = \sum_{x \subseteq X} \sum_v P(R \mid x, v) \tag{7}$$

$$E[n(v, X)] = \sum_{x \subseteq X} P(R \mid x, v) \tag{8}$$

$$E[m(v, \mathcal{X})] = \sum_{X \subseteq \mathcal{X}} \max_{x \in X} P(R \mid x, v) \tag{9}$$

4.2   Estimating Information Gain and Tf·Idf of Sets of Text Images

Using the statistical expectations of document and word frequencies of Eqs. (6–9), IG and TfIdf can be strightforwardly estimated for a collection of text images. According to the notation used previously, a document D in Sect. 3 becomes a set of text images or *image document*, X. Also, the set of all documents $\mathcal{D}$ becomes the text image collection $\mathcal{X}$, and we will denote $\mathcal{X}_c$ the subset of image documents of class c. Thus $M \stackrel{\text{def}}{=} |\mathcal{X}|$ is now the total number of image documents and $M_c \stackrel{\text{def}}{=} |\mathcal{X}_c|$ the number of them which belong to class c.

The document frequencies needed to compute the IG of a word, v are summarized in Eqs. (3–4). Now the number of image documents that contain the word v, $f(t_v) \equiv m(v, \mathcal{X})$, is directly estimated using Eq. (9), and the number of image documents of class c which contain v, $f(c, t_v)$, is also estimated as in Eq. (9) changing $\mathcal{X}$ with $\mathcal{X}_c$.

On the other hand, the frequencies needed to compute the Tf·Idf document vector features are summarized in Eq. (5). In addition to $f(t_v) \equiv m(v, \mathcal{X})$, we need the total number of running words in a document D, $f(D)$, and the number of times the word v appears in D, $f(v, D)$. Clearly, $f(D) \equiv n(X)$ and $f(v, D) \equiv n(v, X)$, which can be directly estimated using Eq. (7) and (8), respectively.

4.3   Image Document Classification

Using the Tf·Idf vector representation $\vec{X}$ of an image document $X \in \mathcal{X}$, optimal prediction of the class of X is achieved under the minimum-error risk statistical framework as:

$$c^*(X) = \operatorname*{argmax}_{c \in \{1,\dots,C\}} P(c \mid \vec{X}) \tag{10}$$

The posterior $P(c \mid \vec{X})$ can be computed following several well-known approaches, some of which are discussed and tested in [12]. Following the results reported in that paper, the Multi-Layer Perceptron (MLP) was adopted for the present work. The output of all the MLP architectures considered is a softmax layer with C units and training is performed by backpropagation using the cross-entropy loss. Under these conditions, it is straightforward that the outputs for an input $\vec{X}$ approach $P(c \mid \vec{X})$, $1 \leq c \leq C$. Thus Eq. (10) directly applies.

Three MLP configurations with different numbers of layers have been considered. In all the cases, every layer except the last one is followed by batch normalization and ReLU activation functions [7]. The basic configuration is a plain C-class perceptron where the input is totally connected to each of the C neurons of the output layer (hence no hidden layers are used). For the sake of simplifying the terminology, here we consider such a model as a "0-hidden-layers MLP" and refer to it as MLP-0. The next configuration, MLP-1, is a proper MLP including one hidden layer with 128 neurons. The hidden layer was expected to do some kind of intra-document clustering, hopefully improving the classification ability of the last layer. Finally, we have also considered a deeper model, MLP-2, with two hidden layers and 128 neurons in each layer. Adding more hidden layers did not provide further improvements.

5   Dataset and Experimental Settings

The dataset considered in this work is a small part of a huge manuscript collection of manuscripts held by the Spanish Archivo Histórico Provincial de Cádiz (AHPC). In this section, we provide details of the dataset and of the settings adopted for the experiments discussed in Sect. 6.

5.1   A Handwritten Notarial Document Dataset

The AHPC (Provincial Historical Archive of Cádiz) was established in 1931, with the main purpose of collecting and guarding notarial documentation that was more than one hundred years old. Its functions and objectives include the preservation of provincial documentary heritage and to offer a service to researchers that allows the use and consultation of these documentary sources.

The notarial manuscripts considered in the present work come from a very large collection of 16 849 bundles or "notarial protocol books", with an average of 250 notarial acts or files and about 800 pages per book. Among these books, 50 were included in the collection compiled in the Carabela project [4].[2] From these 50 books, for the present work we selected two notarial protocol books, JMBD_4949 and JMBD_4950, dated 1723–1724, to be manually tagged with GT annotations. Figure 1 shows examples of page images of these two books.

The selected books were manually divided into sequential sections, each corresponding to a notarial act. A first section of about 50 pages, which contains a kind of table of contents of the book, was also identified but not used in the

---

[2] In http://prhlt-carabela.prhlt.upv.es/carabela the images of this collection and a PrIx-based search interface are available.

Fig. 1. Example of corpus pages from books JMDB_4949 and JMBD_4950.

present experiments. It is worth noting that each notarial act can contain from one to dozens of pages, and separating these acts is not straightforward. In future works, we plan to develop methods to also perform this task automatically, but for the present work we take the manual segmentation as given.

During the segmentation and labeling of the two notarial protocol books, the experts found a total of 558 notarial acts belonging to 38 different types or classes. However, for most classes, only very few acts were available. To allow the classification results to be sufficiently reliable, only those classes having at least *five* acts in each book were taken into account. This way, *five classes* were retained as sufficiently representative and 419 acts (i.e., documents) were finally selected: 220 in JMBD_4949 and 199 in JMBD_4950. So, in total, 139 acts (25%) were set aside, which amounted to 1 321 page images (including the long tables of contents mentioned above), out of the 3 186 pages of both books.

The five types (*classes*) we are finally left with are: *Power of Attorney* (P, from Spanish "Poder"), *Letter of Payment* (CP, "Carta de Pago"), *Debenture* (O, "Obligación"), *Lease* (A, "Arrendamiento") and *Will* (T, "Testamento"). Details of this dataset are shown in Table 1. The machine learning task consists in training a model to classify each document into one of these $C = 5$ classes.

Table 1. Number of documents and page images for JMBD_4949 and JMBD_4950: per class, per document & class, and totals.

Classes	JMBD_4949						JMBD_4950					
	P	CP	O	A	T	Total	P	CP	O	A	T	Total
Number of documents	141	35	21	12	11	**220**	100	39	23	19	18	**199**
Avgerage pages per doc.	3.6	4.5	4.2	4.6	6.0	**4.0**	3.7	4.8	5.4	5.2	10.0	**4.8**
Min-max pages per doc.	2–46	2–28	2–20	2–16	4–10	**2–46**	2–56	2–30	2–32	2–14	4–48	**2–56**
Total pages	514	158	90	56	66	**884**	370	188	124	100	179	**961**

5.2 Empirical Settings

PrIx vocabularies typically contain huge amounts of pseudo-word hypotheses. However, many of these hypotheses have low relevance probability and most of the low-probability pseudo-words are not real words. Therefore, as a first step, the huge PrIx vocabulary was pruned out avoiding entries with less than three characters, as well as pseudo-words v with too low estimated document frequency; namely, $E[m(v, \mathcal{X})] < 1.0$. This resulted in a vocabulary V of 559 012 pseudo-words for the two books considered in this work. Secondly, to retain the most relevant features as discussed in Sect. 3.1, (pseudo-)words were sorted by decreasing values of IG and the first n entries of the sorted list were selected to define a BOW vocabulary V_n. Exponentially increasing values of n from 8 up to 16 384 were considered in the experiments. Finally, a Tf·Idf n-dimensional vector was calculated for each document, $D \equiv X \in \mathcal{X}$. For experimental simplicity, $\mathrm{Tf} \cdot \mathrm{Idf}(v, D)$ was estimated just once all for all $v \in V$, using the normalized factor $f(D) \equiv E[n(X)]$ computed for all $v \in V$, rather than just $v \in V_n$.

For MLP classification, document vectors were normalized by subtracting the mean and dividing by the standard deviation, resulting in zero-mean and unit-variance input vectors. The parameters of each MLP architecture were initialized following [5] and trained according to the cross-entropy loss for 100 epochs using the SGD optimizer [16] with a learning rate of 0.01. This configuration has been used for all the experiments presented in Sect. 6.

The same leaving-one-out training and testing experiments were carried out for each book. In each experiment 10 runs were made with different initialization seeds, ending up with the average results for all runs. This amounts to $10\,M$ leaving-one-out executions for each experiment, where M is the total number of documents in each book (see Table 1).

The source code and data used in the experiments presented in this paper are publicly available.[3]

6   Experiments and Results

The empirical work has focused on MLP classification of documents (handwritten notarial *acts*) of two books, JMBD_4949 and JMBD_4950. For each book separately, we classify its documents (groups of handwritten page images) into the five classes established in Sect. 5.1.

Classification error rates are presented in Fig. 2 for 12 increasing values of n, the number of (pseudo-)words selected with maximum Information Gain.

---

[3] https://github.com/PRHLT/docClasifIbPRIA22.

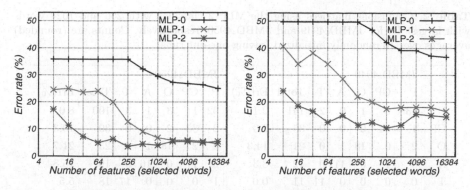

Fig. 2. Leaving-one-out classification error rate for three classifiers for JMBD_4949 (left), and JMBD_4950 (right). 95% confidence intervals (not shown for clarity) are all smaller than ±6.0% and smaller than ±3.0% for all the errors below than 6%.

Taking into account to the number of documents per each class (see Table 1), the error rates of a naive clasifier based only on estimated prior probability per class would be 35.9% and 49.7% for JMBD_4949 and JMBD_4950, respectively.

For JMBD_4949, best results are obtained using MLP-2, achieving a 3.6% error rate with a vocabulary of $n = 256$ words. For this model, accuracy remains good if the vocabulary size is increased, but it degrades significantly for lower values of n. The MLP-1 model cannot achieve the same accuracy as MLP-2 for $n = 256$ or less words, although it achieves good accuracy for larger n, reaching a 4.5% error rate for $n = 16\,384$ words. The plain perceptron classifier (MLP-0) fails to overcome the accuracy of the naive prior-based classifier for vocabulary sizes $n \leq 256$ words. Its lowest error rate is 25.0% for $n = 16\,384$.

Results for JMBD_4950 are also presented in Fig. 2 (right). Since it departs from a much higher prior-based naive classification error rate (49.7%), it is not surprise that all the accuracies for this book are significantly worse than those achieved for JMBD_4949. Best results are also obtained using MLP-2, with 10.6% error rate with a vocabulary of 1 024 words. For this model, accuracy degrades either if $n < 256$ or $n > 2\,048$. The accuracy of MLP-1 is lower than that of MLP-2, the best result being a 16.5% error rate with $n = 16\,384$ words Finally, for the plain perceptron (MLP-0), we see that it again does not achieve an accuracy good enough to be taken into account.

Model complexity, in terms of numbers of parameters to train, grows with the number of features, n as:

$$\text{MLP-0: } 5\,n + 5 \qquad \text{MLP-1: } 128\,n + 773 \qquad \text{MLP-2: } 128\,n + 17\,285$$

For all $n > 64$, the least complex model is MLP-0, followed by MLP-1 and MLP-2. For $n = 2\,048$, MLP-0, MLP-1 and MLP-2 have 10 245, 262 917 and 279 429 parameters, respectively. Therefore, despite the complexity of the model, MLP-2 is the best choice for the task considered in this work.

Table 2 shows the average confusion matrix and the specific error rate per class, using the best model (MLP-2) with the best vocabulary (n) for each book.

Table 2. Confusions matrices using the MLP-2 classifier with 256 and 1024 words with largest IG for JMBD_4949 and JMBD_4950, respectively. Counts are (rounded) averages over 10 randomly initialized leaving-one-out runs.

	JMBD_4949							JMBD_4950						
	P	CP	O	A	T	Total	Err(%)	P	CP	O	A	T	Total	Err(%)
P	138	0	1	0	2	141	2.1	92	1	3	1	3	100	8.0
CP	1	34	0	0	0	35	2.9	2	35	2	0	0	39	10.3
O	2	0	18	1	0	21	14.3	2	2	18	1	0	23	21.7
A	0	1	0	11	0	12	8.3	2	1	0	16	0	19	15.8
T	0	0	0	0	11	11	0.0	1	0	0	0	17	18	5.5
All	141	35	19	12	13	220	3.6	99	39	23	18	20	199	10.6

7 Conclusion

We have presented and showcased an approach that is able to perform textual-content-based document classification directly on multi-page documents of untranscribed handwritten text images. Our method uses rather traditional techniques for plaintext document classification, estimating the required word frequencies from image probabilistic indexes. This overcomes the need to explicitly transcribe manuscripts, which is generally unfeasible for large collections.

The present work successfully extends previous studies, but its scope is still fairly limited: only 419 document samples of five classes. Nevertheless, the experimental results achieved so far clearly support the capabilities of the proposed approach to model the textual contents of text images and to accurately discriminate content-defined classes of image documents. In future studies we plan to further extend the present work by taking into consideration all the document samples and classes available. Using the two bundles JMDB4949 and JMBD_4950 together will allow us to roughly double the number of classes with enough documents per class to allow reasonable class modeling. Furthermore, in order to approach practical situations, experiments will also assess the model capability of rejecting test documents of classes not seen in training.

In our opinion, probabilistic indexing opens new avenues for research in textual-content-based image document classification. In a current study we capitalize on the observation that fairly accurate classification can be achieved with relatively small vocabularies, down to 64 words in the task considered in this paper. In this direction, we are exploring the use of information gain and/or other by-products of the proposed MLP classifiiers to derive a small set of words that semantically describes the contents of each image document. A preliminary work in this direction is described in [13]. It aimas at automatic or semi-automatic creation of metadata which promises to be extremely useful for scholars and the general public searching for historical information in archived manuscripts.

Finally, in future works, we plan to explore other classification methods such as recurrent models that can take into account the sequential regularities exhibited by textual contents in succesive page images of formal documents.

Acknowledgments. Work partially supported by the research grants: Ministerio de Ciencia Innovación y Universidades "DocTIUM" (RTI2018-095645-B-C22), Generalitat Valenciana under project DeepPattern (PROMETEO/2019/121) and PID2020-116813RB-I00a funded by MCIN/AEI/ 10.13039/501100011033. The second author's work was partially supported by the Universitat Politècnica de València under grant FPI-I/SP20190010.

References

1. Aggarwal, C.C., Zhai, C.: Mining text data. Springer, Boston (2012). https://doi.org/10.1007/978-1-4614-3223-4
2. Aizawa, A.: An information-theoretic perspective of TF-IDF measures. Inf. Proc. Manag. **39**(1), 45–65 (2003)
3. Bluche, T., et al.: Preparatory KWS experiments for large-scale indexing of a vast medieval manuscript collection in the HIMANIS project. In: 14th IAPR International Conference on Document Analysis and Recognition (ICDAR), vol. 01, pp. 311–316, November 2017
4. Vidal, E., et al.: The Carabela project and manuscript collection: large-scale probabilistic indexing and content-based classification. In: 16th ICFHR, September 2020
5. Glorot, X., Bengio, Y.: Understanding the difficulty of training deep feedforward neural networks. J. Mach. Learn. Res. **9**, 249–256 (2010)
6. Ikonomakis, M., Kotsiantis, S., Tampakas, V.: Text classification using machine learning techniques. WSEAS Trans. Comput. **4**(8), 966–974 (2005)
7. Ioffe, S., Szegedy, C.: Batch Normalization: Accelerating Deep Network Training by Reducing Internal Covariate Shift (2015)
8. Joachims, T.: A probabilistic analysis of the Rocchio algorithm with TFIDF for text categorization. Technical report, Carnegie-mellon univ pittsburgh pa dept of computer science (1996)
9. Khan, A., Baharudin, B., Lee, L.H., Khan, K.: A review of machine learning algorithms for text-documents classification. J. Adv. Inf. Technol. **1**(1), 4–20 (2010)
10. Lang, E., Puigcerver, J., Toselli, A.H., Vidal, E.: Probabilistic indexing and search for information extraction on handwritten German parish records. In: 2018 16th International Conference on Frontiers in Handwriting Recognition (ICFHR), pp. 44–49, August 2018
11. Manning, C.D., Raghavan, P., Schtze, H.: Introduction to Information Retrieval. Cambridge University Press, New York (2008)
12. Prieto, J.R., Bosch, V., Vidal, E., Alonso, C., Orcero, M.C., Marquez, L.: Textual-content-based classification of bundles of untranscribed manuscript images. In: 2020 25th International Conference on Pattern Recognition (ICPR), pp. 3162–3169. IEEE (2021)
13. Prieto, J.R., Vidal, E., Sánchez, J.A., Alonso, C., Garrido, D.: Extracting descriptive words from untranscribed handwritten images. In: Proceedings of the 2022 Iberian Conference on Pattern Recognition and Image Analysis (IbPRIA) (2022)
14. Puigcerver, J.: A Probabilistic Formulation of Keyword Spotting. Ph.D. thesis, Univ. Politècnica de València (2018)

15. Romero, V., Toselli, A.H., Vidal, E., Sánchez, J.A., Alonso, C., Marqués, L.: Modern vs diplomatic transcripts for historical handwritten text recognition. In: Cristani, M., Prati, A., Lanz, O., Messelodi, S., Sebe, N. (eds.) ICIAP 2019. LNCS, vol. 11808, pp. 103–114. Springer, Cham (2019). https://doi.org/10.1007/978-3-030-30754-7_11
16. Ruder, S.: An overview of gradient descent optimization algorithms **14**, 2–3 (2017)
17. Salton, G., Buckley, C.: Term-weighting approaches in automatic text retrieval. Inf. Proc. Manag. **24**(5), 513/523 (1988)
18. Sánchez, J.A., Romero, V., Toselli, A.H., Villegas, M., Vidal, E.: A set of benchmarks for handwritten text recognition on historical documents. Pattern Recogn. **94**, 122–134 (2019)
19. Toselli, A., Romero, V., Vidal, E., Sánchez, J.: Making two vast historical manuscript collections searchable and extracting meaningful textual features through large-scale probabilistic indexing. In: 15th IAPR International Conference on Document Analysis and Recognition (ICDAR) (2019)
20. Toselli, A.H., Vidal, E., Puigcerver, J., Noya-García, E.: Probabilistic multi-word spotting in handwritten text images. Pattern Anal. Appl. **22**(1), 23–32 (2018). https://doi.org/10.1007/s10044-018-0742-z
21. Toselli, A.H., Vidal, E., Romero, V., Frinken, V.: HMM word graph based keyword spotting in handwritten document images. Inf. Sci. **370–371**, 497–518 (2016)
22. Vidal, E., Toselli, A.H., Puigcerver, J.: A probabilistic framework for lexicon-based keyword spotting in handwritten text images. Technical report, UPV (2017)

Incremental Vocabularies in Machine Translation Through Aligned Embedding Projections

Salvador Carrión[✉] and Francisco Casacuberta[✉]

Universitat Politècnica de València, Camí de Vera, 46022 València, Spain
{salcarpo,fcn}@prhlt.upv.es

Abstract. The vocabulary of a neural machine translation (NMT) model is often one of its most critical components since it defines the numerical inputs that the model will receive. Because of this, replacing or modifying a model's vocabulary usually involves re-training the model to adjust its weights to the new embeddings.

In this work, we study the properties that pre-trained embeddings must have in order to use them to extend the vocabulary of pre-trained NMT models in a zero-shot fashion.

Our work shows that extending vocabularies for pre-trained NMT models to perform zero-shot translation is possible, but this requires the use of aligned, high-quality embeddings adapted to the model's domain.

Keywords: Neural machine translation · Zero-shot translation · Continual learning

1 Introduction

Machine translation is an open vocabulary problem. There are many approaches to this problem, such as using vocabularies of characters, subwords, or even bytes.

Even though these approaches solve many of the open-vocabulary problems, rare or unknown *words* are still problematic because there is little to no information about them in the dataset. As a result, the model cannot learn good representations about them. For example, if we create a character-based vocabulary of lower and upper case letters but our training set only contains lower case letters, our model will not be able to learn any information about uppercase letters despite being in its vocabulary. Similarly, if they appear just a couple of times, the model will also not be able to learn enough information about how to use them.

As there are always constraints when training a model (e.g., amount of data or computational resources), we need a method to update these vocabularies after our model has been trained, but without the need to re-train the whole

Supported by Pattern Recognition and Human Language Technology Center (PRHLT).

A. J. Pinho et al. (Eds.): IbPRIA 2022, LNCS 13256, pp. 27–40, 2022.
https://doi.org/10.1007/978-3-031-04881-4_3

model again or without the risk of shifting its domain or weight distribution due to the initialization and re-training of the new entries.

Most approaches to extend a model's vocabulary either replace the old embeddings with new ones (randomly initialized) or add new entries (randomly initialized) on top of the old ones. However, the problem with these approaches is that they require the fine-tuning of the model, thing that is not possible when there is no data, or when there is so little data that its fine-tuning might hurt the performance of the model by shifting its weight distribution too much from the previous one.

In this work, we first study the properties that word embeddings must have in order to use them in a zero-shot scenario. We do this by studying the convergence of different word embeddings from several domains and qualities, projected to a common latent space (e.g., GloVe, FastText and custom ones). Then, we use FastText as a workaround to generate aligned embeddings for unknown words and extend the vocabulary of a model in a zero-shot manner. Finally, we adapt these embeddings to the target domain, achieving a significant boost in performance for our zero-shot models.

The contributions of this work are two-fold:

- First, we show that word embeddings from different domains do not appear to converge regardless of their quality. Hence, this suggests that latent space alignment is a requirement for zero-shot translation.
- Then, we show that it is possible to improve the performance of an NMT model in a zero-shot scenario by extending its vocabulary with aligned embeddings. However, to achieve significant gains in performance, these embeddings must be adapted to the target domain.

2   Related Work

In recent years, many approaches have been proposed to learn good word representations from large corpora. Among them, we can highlight Word2Vec [18], which was one of the first successful neural models to learn high-quality word vectors from vast amounts of data using neural networks; GloVe [21], which achieved state-of-the-art performance by learning through a matrix factorization technique the co-occurrence probability ratio between words as vector differences; and FastText [5], which extends Word2Vec to represent words as the sum of n-gram vectors.

Despite the success of these approaches, static word embeddings could not capture the meaning of a word based on its context. To deal with problem[1], researchers began to focus their efforts on contextualized word embeddings such as ELMo [22], a deep bidirectional language model, or BERT [9], a language model that make use of the encoder part of Transformer architecture [27].

The zero-shot problem in machine translation has been widely studied with relative success. For example, GNMT [12] showed that zero-shot machine translation was possible through multilingual systems; Ha et al. [11] presented strategies

---

[1] This is typically seen in *homograph* words.

with which to improve multilingual systems to tackle zero-shot scenarios better; and Zhang et al. [31] improved the zero-shot performance by using random online back-translation to enforce the translation of unseen training language pairs.

In contrast, the problem of lifelong learning in neural machine translation has not received enough attention, and it was not until 2020 that it was introduced as a task in the WMT [3] conference. Following this research line, Xu et al. [29] proposed a meta-learning method that exploits knowledge from past domains to generate improved embeddings for a new domain; Liu et al. [15] learned corpus-dependent features by sequentially updating sentence encoders (previously initialized with the help of corpus-independent features) using Boolean operations of conceptor matrices; and Qi et al. [24] showed that pre-trained embeddings could be effective in low-resource scenarios.

3 Models

3.1 Tokenization

Most tasks related to natural language require some form of pre-processing, being tokenization one of the most common techniques. There are many tokenization strategies, such as the ones based on bytes [10,30], characters [8,19,28], subwords [14,25], or words [7,26]. However, given the purpose of our work, we considered that word-based tokenization would be enough.

Specifically, we used the Moses tokenizer [13] because it is a reversible word-based tokenizer. That is, it can transform a tokenized text into its original version, and it also considers punctuation marks and nuances about the language.

3.2 Transformer Architecture

Nowadays, most machine translation systems are based on encoder-decoder neural architectures. However, amongst these Seq-to-Seq architectures [26], the Transformer [27] is the one that has obtained the latest state-of-the-art results.

The Transformer model has the advantage that it is based entirely on the concept of *attention* [4,16] to draw global dependencies between the input and output. Because of this, the Transformer does not need to make use of any recurrent layers to deal with temporal sequences, and it can process its sequences in parallel, obtaining significant performance improvements.

To accomplish this, the Transformer uses linear and normalization layers, residual connections, attention mechanisms, and mask-based tricks to encode the temporal information of its sequences.

3.3 Projecting Vectors into Different Latent Spaces

One of the requirements to extend the vocabulary of a pre-trained model with new entries is that the dimensionality of the new entries matches the dimensionality of the old ones.

If both entries have the same dimensionality, there is no problem, but if they do not, we need to transform them to have the same dimensionality (i.e., vector compression). There are many ways to project a set of vectors into a lower-dimensional latent space. We decided to explore the most common techniques, such as the projection of embeddings through PCA and Autodercoders (linear and non-linear), and their reconstruction error.

It is essential to highlight that although two sets of vectors might have the same dimensionality, their vectors might be projected into different regions of the same latent space. Namely, they could show similar distances amongst their respective word-pairs, but at the same time, they could reside in different locations.

Principal Component Analysis. Principal Component Analysis (PCA) is a method commonly used for dimensionality reduction. This method allows us to project an n-dimensional data point from a given dataset into another vector that resides in another n-dimensional space but with lower dimensionality while preserving as much of the data's variation as possible.

We will use this method to transform the embeddings from their original dimension to a lower dimension to match the one that our models need.

Autoencoders. Autoencoders are a type of Artificial Neural Network used to learn specific representations from unlabeled data. That is, they can encode a vector in a given n-dimensional space into another n-dimensional space while preserving as much of its information as possible.

These models usually are made from two separated blocks: i) the encoder, which learns how to encode the input data into a latent space; ii) and the decoder, which is in charge of transforming the encoded input data into the original input data.

In this work, we will use linear autoencoders (which in theory should be equivalent to PCA) and non-linear autoencoders. This comparison aims to explore potential non-linear representations for the embeddings.

4   Experimental Setup

4.1   Datasets

The data used in this work comes mainly from the WMT tasks [1,2]. In Table 1 we find all datasets used in this work.

Table 1. Datasets partitions

Dataset	Train	Val	Test
Europarl v7 (DE-EN)	2M/100K	5000	5000
Multi30K (DE-EN)	29K	1014	1000

The values in this table indicate the number of sentences.

These datasets contain parallel sentences extracted from different sources and domains:

- **Europarl** contains parallel sentences extracted from the European Parliament website
- **Multi30k** is a small dataset from WMT 2016 multimodal task, also known as Flickr30k

4.2 Training Details

As we were interested in exploring techniques to expand the vocabularies of a pre-trained model, we did not use subword-based tokenizers such as BPE [25] or Unigram [14]. Instead, we pre-processed our datasets using Moses [13] to create a word-based vocabulary (one per language).

For each dataset, we created word-based vocabularies with 250, 500, 1000, 2000, 4000, 8000, and 16000 words each. Then, we encode the Multi30K (de-en), Europarl-100k (de-en), and Europarl-2M (de-en) datasets using these vocabularies.

In addition to this, we use AutoNMT [6] as our sequence modeling toolkit. All the experimentation was done using a small version of the standard Transformer with 4.1 to 25M parameters depending on the vocabulary size (See Table 2). The reason for this was to speed up our research as the focus of this work was not to achieve state-of-the-art performance but to extend the vocabulary of pre-trained models in a zero-shot manner.[2]

Table 2. NMT models hyperparameters

Model	Parameters	Hyperparameters
Transformer small	4.1M (S)—25.0M (L)	3 layers/8 heads/256 dim/512 ffnn

Common hyperparamters: Loss function: CrossEntropy (without label smoothing). Optimizer: Adam. Batch size: 4096 tokens/batch, Clip-norm: 1.0. Maximum epochs: 50–100 epochs with early stopping (patience = 10). Beam search with 5 beams.

In order to make a fair comparison for all the models studied here, we use the same base transformer with similar (or equal) training hyper-parameters as long as the model was able to train correctly.

---

[2] All models were trained using 2x NVIDIA GP102 (TITAN XP) - 12 GB each.

4.3   Evaluation Metrics

Automatic metrics compute the quality of a model by comparing its output with a reference translation written by a human.

Due to the sensitivity of most metric systems to their hyper-parameters and implementation, we used SacreBLEU [23] which produces shareable, comparable, and reproducible BLEU scores. Similarly, we also evaluated our models with BERTScore [32], but no additional insights were gained from this.

- **BiLingual Evaluation Understudy (BLEU)** [20]: Computes a similarity score between the machine translation and one or several reference translations, based on the n-gram precision and a penalty for short translations.
- **BERTScore** [32]: Using pre-trained contextual embeddings from BERT, it matches words in candidate and reference sentences by cosine similarity

5   Experimentation

5.1   Projecting Pre-trained Embeddings

The typical approach when using pre-trained embeddings is to initialize the model's embedding layers, freeze them, train the model, and finally, fine-tune these layers using a small learning rate. Although effective, we have to consider that the dimensions of the embedding layers of our model and the ones from the pre-trained embeddings must match. This is usually not a problem when training models from scratch as we can leave the original embedding dimension or simply add a linear layer to transform one dimension into another, but for pre-trained models, this could become a non-trivial task.

When adding new embeddings for a pre-trained model, we need to consider a few things. For instance, if we can afford to replace all the embeddings, we can simply drop the old embeddings, add the new ones (plus a linear layer to match dimensions, if needed), and then fine-tune the model. The advantage of this approach is that all vectors share the same latent space. However, if we need to extend an existing embedding layer, not only the dimensions of the old and the new embeddings must match, but the region of the latent space in which their vectors are expected. That is, let say that in a embedding A, the word *dog* is at position $(1,0)$ and the word *cat* at $(1,1)$. But in the embedding B, *dog* is at $(-1, 0)$ and *cat* at $(-1, -1)$. In both cases, their respective distance is 1, and their embedding dimension is 2. However, both embeddings are in different locations of the same latent space. Because of that, it is expected that the new embeddings have to be aligned with the previous ones to use them in a zero-shot environment.

In Fig. 1a we have the reconstruction errors (R^2) for the embeddings of a translation task, corresponding to a Transformer trained on the Multi30k (de-en) datasets. Similarly, in Fig. 1b we have the reconstruction errors (R^2) for the GloVe [21] embeddings, which are high-quality embeddings from modeling co-occurrences probabilities through a matrix factorization process.

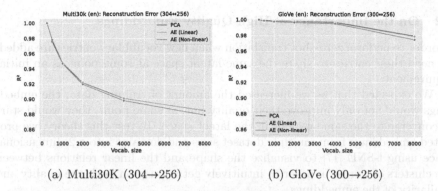

(a) Multi30K (304→256)   (b) GloVe (300→256)

Fig. 1. Reconstruction errors of the compressed embeddings using Linear and No-linear methods. Comparing both images, we can see that the higher the quality of the embeddings, the lower the reconstruction error. Furthermore, non-linear methods do not seem to be needed, as the embeddings seem to basically hold linear relations amongst them

In both figures, we see that PCA and autoencoders performed remarkably similarly. For the linear projections, PCA had a perfect overlap with the linear autoencoder, as expected, but the former significantly faster[3]. In contrast, the non-linear autoencoder performed slightly worse than these two, probably due to its additional complexity. From these observations, it seems that there is no need for non-linear methods when it comes to projecting embeddings to other latent spaces due to the linear relations of their vectors.

Concerning the vocabulary size (number of embeddings), as we increase it, the reconstruction error also increases regardless of the model. We believe that this is because the entries in our vocabulary are sorted by frequency, forming a long-tail distribution. Hence, as we increase the vocabulary size, we add less and less frequent entries to our vocabulary. As a result, their embedding vectors will not have the same average quality as the previous ones due to the lack of samples (frequency), so they will be noisy. Because of this, and given that noise cannot be compressed, it is to be expected that some of their information is lost, given that both PCA and Autoencoders perform lossy compression.

In the case of the GloVe embeddings, we see that the reconstruction errors remained pretty low regardless of the vocabulary size. In the same line as the previous explanation, we think that this is because the GloVe embeddings have more quality than the Multi30k embeddings because the GloVe embeddings resulted from a technique specifically designed to learn high-quality embeddings, while the Multi30K embeddings were the byproduct of a translation model trained on a very small dataset.

---

[3] We compare these linear methods despite their theoretical equivalence as a sanity check, and because we cannot use PCA when the dimensionality of the vectors is greater than the number of samples.

5.2   On the Importance of High-Quality Embeddings

In order to perform zero-shot translation when new vocabulary entries are added, we need these entries to share the same latent space at some point as an initial requirement.

We expected that as we increased the amount of training data, the embeddings would not only improve their quality but, at some point, they would start to converge to the same regions in the latent space. To test this theory, we projected embeddings from multiple dataset sizes and domains into a 2-dimensional space using t-SNE [17] to visualize the shape and the linear relations between the clusters of words, in order to intuitively get a grasp on both the quality and similarity of the embeddings.

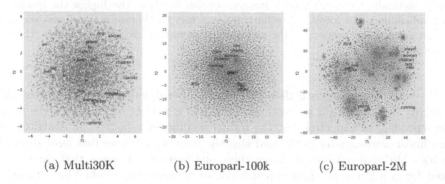

(a) Multi30K   (b) Europarl-100k   (c) Europarl-2M

Fig. 2. t-SNE projections for the 8000-word embedding layer (en) of a transformer trained on Multi30k, Europarl-100k and Europarl-2M.

In Fig. 2 we see that it was not until we used a dataset with 2M sentences (See Fig. 3a) that we started to see clusters of similar words. However, when we re-trained the model, we ended up with a completely different set of high-quality embeddings (different clusters) that would not have allowed zero-shot translation.

Finally, we wanted further explore the non-convergence problem by adding two well-known high quality embeddings: GloVe [21] and FastText [5]. From the t-SNE projections [17] in Fig. 3 we can see that although these high-quality embeddings group similar words together (i.e., *man, woman, children*), their clusters are completely different as well as the distance between non-related words (i.e., distance between *bird* and *man*).

From these results, it is clear to us that in order to perform zero-shot translation, the new embeddings have to be perfectly aligned with the previous ones. It is essential to highlight that we ended up with different projections each time we trained a t-SNE projection (figures). However, these projections were remarkably similar for the same embedding set (e.g., GloVe-GloVe, FastText-FastText, etc.), but very different between different embedding sets (e.g., GloVe-FastText,

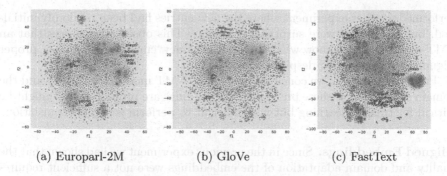

(a) Europarl-2M   (b) GloVe   (c) FastText

Fig. 3. t-SNE projections for the Europarl, Glove and FastText embeddings

Europarl-FastText). Because of this, we consider that our findings are sound and not the artifact of a specific projection.

5.3   Zero-shot Translation

To study whether it was a strong requirement that the new embeddings had to be aligned with previous ones in order to perform zero-shot translation, we designed three experiments in which we studied vocabulary expansion using:

- Unaligned Embeddings
- Aligned Embeddings (out-of-domain)
- Aligned Embeddings (in-domain)

For each experiment, we trained six translation models with vocabularies of 250, 500, 1000, 2000, 4000, and 8000 words. We then extended these vocabularies to 16K words using pre-trained embeddings, but without fine-tuning (zero-shot). Finally, we evaluated the performance of each model as a zero-shot task before and after the vocabulary expansion.

Unaligned Embeddings. In this experiment, we trained six translation models (with randomly initialized embeddings) on the Europarl-100k (de-en) dataset. Once the models were trained, we extended their vocabularies up to 16K (without fine-tuning) using the following pre-trained embeddings:

- Multi30K (Low-quality/out-of-domain)
- Europarl-100k (Low-quality/in-domain)
- GloVe (High-quality/out-of-domain)

However, the results in all cases were worse than the base model, although not by much, with a decrease of 0.25 to 1.0 pts of BLEU. This happened regardless of the quality of the embeddings (GloVe vs. Multi30k/Europarl-100k) and their domain (Europarl-100k/in-domain vs. GloVe/out-of-domain). Similarly, we performed another experiment where the new entries had been randomly initialized, but the results were surprisingly similar. This observation shows that an NMT model does not know what to do with the new embeddings without proper alignment or re-training despite their quality.

From these results, we conclude that for an NMT model, the quality and the domain adaptation of these pre-trained embeddings are properties that might be critical for transfer learning but not sufficient to perform zero-shot translation.

Aligned Embeddings. Since in the previous experiment we had shown that the quality and domain adaptation of the embeddings were not a sufficient requirement for their use in zero-shot settings, this time we wanted to study whether, as we previously hypothesized, the alignment of the embeddings was a critical requirement for extending the vocabularies of an NMT model in a zero-shot setting.

To this end, we designed a twofold experiment in which we studied the use of aligned embeddings, but with in-domain and out-of-domain embeddings.

For the experiment with the out-of-domain aligned embeddings, we decided to use FastText to initialize the embeddings of our models[4]. After initializing the embeddings layers and freezing them, we trained six NMT models with vocabularies from 250 to 8K words on the Europarl-100k (de-en) dataset, similar to previous experiments. Finally, we extended the vocabularies of these models up to 16K words. As the previous and new embeddings had been generated using FastText (and not modified), both embeddings were already aligned, so there was no need to use any alignment method.

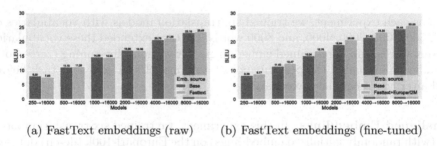

(a) FastText embeddings (raw)   (b) FastText embeddings (fine-tuned)

Fig. 4. BLEU scores from using aligned embeddings: FastText (raw) and FastText (fine-tuned). Adapting the aligned embeddings to the model's domain seem to be particularly relevant.

Interestingly, we can see in Figs. 5 and 4a that when we trained our models using the FastText embeddings (frozen), we got a wide range of results depending

---

[4] The reason why we switched from GloVe to FastText is that the latter had multilingual support and in addition to this, it allowed us to generate embeddings for unknown words.

on the original vocabulary size. When we extended the vocabulary from 250 to 16K, we lost 0.37 pts of BLEU. Then, the increases or decreases in performance were marginal after expanding the 500, 1K, and 2K vocabularies up to 16K words. In contrast, after we extended the 4K and 8K vocabularies to 16K, we obtained small but consistent performance increases of up to +0.53 pts of BLEU.

These results seem to indicate that even though the embeddings were aligned, an NMT model still needs enough words to learn to generalize to other but similar words. However, we wondered how domain adaptation could affect these results, so we repeated the previous experiment, but this time, fine-tuning the FastText embeddings to the Europarl domain.

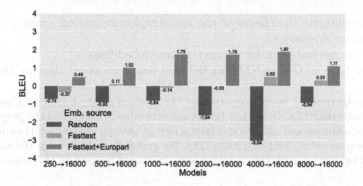

Fig. 5. BLEU scores differences from using Random (baseline) and FastText (raw and fine-tuned) aligned embeddings. Adapting the aligned embeddings to the model's domain seem to be particularly relevant but the model still requires an initial medium vocabulary to learn to generalize to unseen embeddings

Surprisingly, this fine-tuning turned out to be much more important than we initially expected as we got significant improvements in all our models, with a peak improvement of +1.92 pts of BLEU (See Fig. 4b and Fig. 5).

These results appear to indicate that if we want to expand the vocabulary of an NMT model in a zero-shot fashion, both the alignment and domain adaptation of the embeddings are strong requirements for zero-shot translation.

6   Conclusions

In this paper, we have studied the properties required to extend the vocabularies of an NMT model in a zero-shot fashion.

First, we have shown that high-quality embeddings have low-construction errors when projected into lower-dimensional spaces. However, despite the large amounts of data needed to obtain high-quality embeddings, these embeddings do not seem to converge to the same regions of the latent feature space. Hence, embedding alignment is required. Next, we have shown that extending the vocabulary of an NMT model in a zero-shot fashion is not only possible, but it is also a simple and effective way to improve the performance of an NMT model without re-training.

Finally, our work suggests that to expand the vocabulary of an NMT model and perform zero-shot translation, we must use high-quality embeddings adapted to the domain and, above all, properly aligned with the previous embeddings.

7   Future Work

Many adaptations and experiments have been left for the future due to a lack of time. Some of the ideas that we would have like to study are:

- Can we mitigate the effects of the catastrophic forgetting problem through vocabulary extensions?
- Does this method work for context-aware embeddings?
- Can we have an Online NMT model that learns its vocabulary on the fly?

Acknowledgements. Work supported by the Horizon 2020 - European Commission (H2020) under the SELENE project (grant agreement no 871467), and the project Deep learning for adaptive and multimodal interaction in pattern recognition (DeepPattern) (grant agreement PROMETEO/2019/121). We gratefully acknowledge the support of NVIDIA Corporation with the donation of a GPU used for part of this research.

References

1. ACL: Ninth workshop: Statistical machine translation (2014)
2. ACL: First conference: Machine translation (wmt16) (2016)
3. ACL: Fifth conference on machine translation (wmt20) (2020)
4. Bahdanau, D., Cho, K., Bengio, Y.: Neural machine translation by jointly learning to align and translate. CoRR abs/1409.0473 (2015)
5. Bojanowski, P., Grave, E., Joulin, A., Mikolov, T.: Enriching word vectors with subword information. CoRR abs/1607.04606 (2016)
6. Carrión, S., Casacuberta, F.: Autonmt: a framework to streamline the research of seq2seq models (2022). https://github.com/salvacarrion/autonmt/
7. Cho, K., et al.: Learning phrase representations using RNN encoder-decoder for statistical machine translation. In: Proceedings of the 2014 Conference on EMNLP, pp. 1724–1734 (2014)
8. Conneau, A., Schwenk, H., Barrault, L., LeCun, Y.: Very deep convolutional networks for natural language processing. CoRR abs/1606.01781 (2016)
9. Devlin, J., Chang, M., Lee, K., Toutanova, K.: BERT: pre-training of deep bidirectional transformers for language understanding. CoRR abs/1810.04805 (2018)
10. Gillick, D., Brunk, C., Vinyals, O., Subramanya, A.: Multilingual language processing from bytes. CoRR abs/1512.00103 (2015). http://arxiv.org/abs/1512.00103
11. Ha, T., Niehues, J., Waibel, A.H.: Effective strategies in zero-shot neural machine translation. CoRR abs/1711.07893 (2017). http://arxiv.org/abs/1711.07893
12. Johnson, M., et al.: Google's multilingual neural machine translation system: enabling zero-shot translation. CoRR abs/1611.04558 (2016). http://arxiv.org/abs/1611.04558

13. Koehn, P., et al.: Moses: open source toolkit for statistical machine translation. In: Proceedings of the 45th Annual Meeting of the ACL on Interactive Poster and Demonstration Sessions, ACL 2007, pp. 177–180 (2007)
14. Kudo, T.: Subword regularization: improving neural network translation models with multiple subword candidates. In: Proceedings of the 56th Annual Meeting of the ACL (Volume 1: Long Papers), pp. 66–75 (2018)
15. Liu, T., Ungar, L., Sedoc, J.: Continual learning for sentence representations using conceptors. ArXiv abs/1904.09187 (2019)
16. Luong, T., Pham, H., Manning, C.D.: Effective approaches to attention-based neural machine translation. In: Proceedings of the 2015 Conference on EMNLP, pp. 1412–1421 (2015)
17. van der Maaten, L., Hinton, G.: Visualizing data using t-SNE. J. Mach. Learn. Res. **9**, 2579–2605 (2008)
18. Mikolov, T., Chen, K., Corrado, G., Dean, J.: Efficient estimation of word representations in vector space. In: Bengio, Y., LeCun, Y. (eds.) 1st International Conference on Learning Representations, ICLR 2013, Scottsdale, Arizona, USA, May 2–4, 2013, Workshop Track Proceedings (2013)
19. Neubig, G., Watanabe, T., Mori, S., Kawahara, T.: Substring-based machine translation. Mach. Transl. **27**(2), 139–166 (2013)
20. Papineni, K., Roukos, S., Ward, T., Zhu, W.J.: Bleu: a method for automatic evaluation of machine translation. In: Proceedings of the 40th Annual Meeting on ACL. ACL 2002, pp. 311–318 (2002)
21. Pennington, J., Socher, R., Manning, C.: GloVe: global vectors for word representation, pp. 1532–1543
22. Peters, M.E., et al.: Deep contextualized word representations. CoRR abs/1802.05365 (2018)
23. Post, M.: A call for clarity in reporting BLEU scores. In: Proceedings of the Third Conference on Machine Translation: Research Papers, pp. 186–191 (2018)
24. Qi, Y., Sachan, D., Felix, M., Padmanabhan, S., Neubig, G.: When and why are pre-trained word embeddings useful for neural machine translation? In: Proceedings of the 2018 Conference of the North American Chapter of the Association for Computational Linguistics: Human Language Technologies, Volume 2 (Short Papers), June 2018
25. Sennrich, R., Haddow, B., Birch, A.: Neural machine translation of rare words with subword units. In: Proceedings of the 54th Annual Meeting of the ACL (Volume 1: Long Papers), pp. 1715–1725 (2016)
26. Sutskever, I., Vinyals, O., Le, Q.V.: Sequence to sequence learning with neural networks. In: Ghahramani, Z., Welling, M., Cortes, C., Lawrence, N., Weinberger, K.Q. (eds.) NIPS, vol. 27 (2014)
27. Vaswani, A., et al.: Attention is all you need. In: Proceedings of the 31st NeurIPS. NIPS 2017, pp. 6000–6010 (2017)
28. Vilar, D., Peter, J.T., Ney, H.: Can we translate letters? In: Proceedings of the Second WMT. StatMT 2007, pp. 33–39 (2007)
29. Xu, H., Liu, B., Shu, L., Yu, P.S.: Lifelong domain word embedding via meta-learning. In: Proceedings of the Twenty-Seventh International Joint Conference on Artificial Intelligence, IJCAI 2018, pp. 4510–4516, July 2018
30. Xue, L., et al.: Byt5: towards a token-free future with pre-trained byte-to-byte models. CoRR abs/2105.13626 (2021). https://arxiv.org/abs/2105.13626

31. Zhang, B., Williams, P., Titov, I., Sennrich, R.: Improving massively multilingual neural machine translation and zero-shot translation. CoRR abs/2004.11867 (2020). https://arxiv.org/abs/2004.11867
32. Zhang, T., Kishore, V., Wu, F., Weinberger, K.Q., Artzi, Y.: Bertscore: evaluating text generation with BERT. CoRR abs/1904.09675 (2019)

An Interactive Machine Translation Framework for Modernizing the Language of Historical Documents

Miguel Domingo[✉] and Francisco Casacuberta

PRHLT Research Center, Universitat Politècnica de València, Valencia, Spain
{midobal,fcn}@prhlt.upv.es

Abstract. In order to make historical documents accessible to a broader audience, language modernization generates a new version of a given document in the modern version of its original language. However, while they succeed in their goal, these modernizations are far from perfect. Thus, in order to help scholars generate error-free modernizations when the quality is essential, we propose an interactive language modernization framework based on interactive machine translation. In this work, we deployed two successful interactive protocols into language modernization. We evaluated our proposal on a simulated environment observing significant reductions of the human effort.

Keywords: Interactive machine translation · Language modernization · Spelling normalization · Historical documents

1 Introduction

Despite being an important part of our cultural heritage, historical documents are mostly limited to scholars. This is due to the nature of human language, which evolves with the passage of time, and the linguistic properties of these documents: the lack of spelling conventions made orthography to change depending on the author and time period. With the aim of making historical documents more accessible to non-experts, language modernization aims to automatically generate a new version of a given document written in the modern version of its original language.

However, language modernization is not error free. While it succeeds in helping non-experts to understand the content of a historical document, there are times in which error-free modernized versions are needed. For example, scholars manually modernize the language of classic literature in order to make works understandable to contemporary readers [33]. In order to help scholars to generate these error-free versions, we propose to apply the interactive machine translation (IMT) framework into language modernization.

IMT proposes a collaborative framework in which a human and a translation system work together to produce the final translation. In this work, we propose to integrate two successful IMT protocols into language modernization. Our contributions are as follow:

- Integration of prefix-based and segment-based IMT into language modernization.
- Experimentation over three datasets from different languages and time periods.

A. J. Pinho et al. (Eds.): IbPRIA 2022, LNCS 13256, pp. 41–53, 2022.
https://doi.org/10.1007/978-3-031-04881-4_4

2 Related Work

Despite being manually applied to literature for centuries (e.g., *The Bible* has been adapted and translated for generations in order to preserve and transmit its contents [15]), automatic language modernization is a young research field. One of the first related works was a shared task for translating historical text to contemporary language [41]. While the task was focused on normalizing the document's spelling, they also approached language modernization using a set of rules. Domingo et al. [9] proposed a language modernization approach based on statistical machine translation (SMT). Domingo and Casacuberta [7] proposed a neural machine translation (NMT) approach. Sen et al. [35] augmented the training data by extracting pairs of phrases and adding them as new training sentences. Domingo and Casacuberta [8] proposed a method to profit from modern documents to enrich the neural models and conducted a user study. Lastly, Peng et al. [26] proposed a method for generating modernized summaries of historical documents.

IMT was introduced during the *TransType* project [11] and was further developed during *TransType2* [3]. New contributions to IMT include developing new generations of the suffix [43]; and profiting from the use of the mouse [34]. Marie et al. [22] introduced a touch-based interaction to iteratively improve translation quality. Lastly, Domingo et al. [10] introduced a segment-based protocol that broke the left-to-right limitation.

With the rise of NMT, the interactive framework was deployed into the neural systems [18,29], adding online learning techniques [28]; and reinforcement and imitation learning [20].

3 Language Modernization Approaches

Language modernization relies on machine translation (MT) which, given a source sentence x_1^J, aims to find the most likely translation $\hat{y}_1^{\hat{I}}$ [6]:

$$\hat{y}_1^{\hat{I}} = \arg\max_{y_1^I} Pr(y_1^I \mid x_1^J) \tag{1}$$

3.1 SMT Approach

This approach relies on SMT, which has been the prevailing approach to compute Eq. (1) for years. SMT, uses models that rely on a log-linear combination of different models [24]: namely, phrase-based alignment models, reordering models and language models; among others [45].

The SMT, modernization approach tackles language modernization as a conventional translation task: given a parallel corpora—where for each original sentence of a document its modernized version is also available—an SMT, system is trained by considering the language of the original document as the source language and its modernized version as the target language.

3.2 NMT Approaches

These approaches are based on NMT, which models Eq. (1) with a neural network. While other architectures are possible, its most frequent architecture is based on an encoder-decoder, featuring recurrent networks [2,39], convolutional networks [13] or attention mechanisms [44]. The source sentence is projected into a distributed representation at the encoding state. Then, the decoder generates at the decoding step its most likely translation—word by word—using a beam search method [39]. The model parameters are typically estimated jointly—via stochastic gradient descent [32]—on large parallel corpora. Finally, at decoding time, the system obtains the most likely translation by means of a beam search method.

Like the previous approach, the NMT, approaches tackle language modernization as a conventional translation task but using NMT, instead of SMT. Moreover, due to NMT, needing larger quantities of parallel training data than we have available (the scarce availability of parallel training data is a frequent problem for historical data [5]), we followed Domingo and Casacuberta's [8] proposal for enriching the neural models with synthetic data: Given a monolingual corpus, we apply feature decay algorithm (FDA) to filter it and obtain a more relevant subset. Then, following a backtranslation approach [36], we train an inverse SMT system—using the modernized version of the training dataset as source, and the original version as target. After that, we translate the monolingual data with this system, obtaining a new version of the documents which, together with the original modern documents, conform the synthetic parallel data. Following that, we train an NMT modernization system with the synthetic corpus. Finally, we fine-tune the system by training a few more steps using the original training data.

We made use of two different NMT modernization approaches, whose difference is the architecture of the neural systems:

- NMT$_{LSTM}$: this approach uses a recurrent neural network (RNN) [16] architecture with long short-term memory (LSTM) [16] cells.
- NMT$_{Transformer}$: this approach uses a Transformer [44] architecture.

4 Interactive Machine Translation

In this work, we made use of two different IMT protocols: prefix-based and segment-based.

4.1 Prefix-Based IMT

In this protocol, the system proposes an initial translation y_1^I of length I. Then, the user reviews it and corrects the leftmost wrong word y_i. Inherently, this correction validates all the words that precede this correction, forming a validated prefix $\tilde{y}_1^i$, that includes the corrected word $\tilde{y}_i$. Immediately, the system reacts to this user feedback ($f = \tilde{y}_1^i$), generating a suffix $\hat{y}_{i+1}^I$ that completes $\tilde{y}_1^i$ to obtain a new translation of $x_1^J : \hat{y}_i^I = \tilde{y}_1^i\,\hat{y}_{i+1}^I$. This process is repeated until the user accepts the system's complete suggestion.

The suffix generation was formalized by Barrachina et al. [3] as follows:

$$\hat{y}_{i+1}^I = \arg\max_{I, y_{i+1}^I} Pr(\tilde{y}_1^i \, y_{i+1}^I \mid x_1^J) \qquad (2)$$

This equation is very similar to Eq. (1): at each iteration, the process consists in a regular search in the translations space but constrained by the prefix $\tilde{y}_1^i$.

Similarly, Peris et al. [29] formalized the neural equivalent as follows:

$$p(\hat{y}_{i'} \mid \hat{y}_1^{i'-1}, x_1^J, f = \tilde{y}_1^i; \Theta) = \begin{cases} \delta(\hat{y}_{i'}, \tilde{y}_{i'}), & \text{if } i' \leq i \\ \bar{\mathbf{y}}_{i'}^\top \mathbf{p}_{i'} & \text{otherwise} \end{cases} \qquad (3)$$

where x_1^J is the source sentence; $\tilde{y}_1^i$ is the validated prefix together with the corrected word; Θ are the models parameters; $\bar{\mathbf{y}}_{i'}^\top$ is the one hot codification of the word i'; $\mathbf{p}_{i'}$ contains the probability distribution produced by the model at time-step i; and $\delta(\cdot, \cdot)$ is the Kronecker delta.

This is equivalent to a forced decoding strategy and can be seen as generating the most probable suffix given a validated prefix, which fits into the statistical framework deployed by Barrachina et al. [3].

4.2   Segment-Based IMT

This protocol extends the human—computer collaboration from the prefix-based protocol. Now, besides correcting a word, the user can validate segments (sequences of words) and combine consecutive segments to create a larger one.

Like with the previous protocol, the process starts with the system suggesting an initial translation. Then, the user reviews it and validates those sequences of words which they consider to be correct. After that, they can delete words between validated segments to create a larger segment. Finally, they make a word correction.

These three actions constitute the user feedback, which Domingo et al. [10] formalized as: $\tilde{\mathbf{f}}_1^N = \tilde{\mathbf{f}}_1, \ldots, \tilde{\mathbf{f}}_N$; where $\tilde{\mathbf{f}}_1, \ldots, \tilde{\mathbf{f}}_N$ is the sequence of N correct segments validated by the user in an interaction. Each segment is defined as a sequence of one or more target words, so each action taken by the user modifies the feedback in a different way. Therefore, the user can:

1. Validate a new segment, inserting a new segment $\tilde{\mathbf{f}}_i$ in $\tilde{\mathbf{f}}_1^N$.
2. Merge two consecutive segments $\tilde{\mathbf{f}}_i$, $\tilde{\mathbf{f}}_{i+1}$ into a new one.
3. Introduce a word correction. This is introduced as a new one-word validated segment, $\tilde{\mathbf{f}}_i$, which is inserted in $\tilde{\mathbf{f}}_1^N$.

The first two actions are optional (an iteration may not have new segments to validate) while the last action is mandatory: it triggers the system to react to the user feedback, starting a new iteration of the process.

The system's reaction to the user's feedback results in a sequence of new translation segments $\hat{\mathbf{h}}_0^{N+1} = \hat{\mathbf{h}}_0, \ldots, \hat{\mathbf{h}}_{N+1}$. That means, an $\hat{\mathbf{h}}_i$ for each pair of validated segments $\tilde{\mathbf{f}}_i$, $\tilde{\mathbf{f}}_{i+1}$, being $1 \leq i \leq N$; plus one more at the beginning of the hypothesis, $\hat{\mathbf{h}}_0$; and another at the end of the hypothesis, $\hat{\mathbf{h}}_{N+1}$. The new translation of x_1^J is obtained by alternating validated and non-validated segments: $\hat{y}_1^I = \hat{h}_0, \tilde{f}_1, \ldots, \tilde{f}_N, \hat{h}_{N+1}$. The goal is to obtain the best sequence of translation segments, given the user's feedback and the source sentence:

$$\hat{h}_0^{N+1} = \arg\max_{h_0^{N+1}} Pr(h_0, \tilde{f}_1, \ldots, \tilde{f}_N, h_{N+1} \mid x_1^J) \tag{4}$$

This equation is very similar to Eq. (2). The difference is that, now, the search is performed in the space of possible substrings of the translations of x_1^J, constrained by the sequence of segments $\tilde{f}_1, \ldots, \tilde{f}_N$, instead of being limited to the space of suffixes constrained by $\tilde{y}_1^i$.

Similarly, Peris et al. [29] formalized the neural equivalent of this protocol as follows:

$$p(y_{i_n+i'} \mid y_1^{i_n+i'-1}, x_1^J, f_1^N; \Theta) = y_{i_n+i'}^\top \mathbf{P}_{i_n+i'} \tag{5}$$

where $f_1^N = f_1, \ldots, f_N$ is the feedback signal and $f_1, \ldots, f_N$ are a sequence of non-overlapping segments validated by the user; each alternative hypothesis y (partially) has the form $y = \ldots, f_n, h_n, f_{n+i}, \ldots$; g_n is the non-validated segment; $1 \le i' \le \hat{l}_n$; and l_n is the size of this non-validated segment and is computed as follows:

$$\hat{l}_n = \arg\max_{0 \le l_n \le L} \frac{1}{l_N + 1} \sum_{i'=i_n+1}^{i_n+l_n+1} \log p(y_{i'} \mid y_1^{i'-1}, x_1^J; \Theta) \tag{6}$$

5 Experimental Framework

In this section, we present the details of our experimental session: user simulation, systems, corpora and metrics.

5.1 User Simulation

In this work, we conducted an evaluation with simulated users due to the time and economic costs of conducting frequent human evaluations during the development stage. These users had as goal to generate the modernizations from the reference.

Prefix-Based Simulation. At each iteration, the user compares the system's hypothesis with the reference and looks for the leftmost different word. When they find it, they make a correction, validating a new prefix in the process. This correction has an associated cost of one mouse action and one word stroke. The system, then, reacts to the user feedback and generates a new suffix that completes the prefix to conform a new modernization hypothesis. This process is repeated until the hypothesis and the reference are the same.

This simulation was conducted using Domingo et al.'s [10] updated version of Barrachina et al.'s [3] software[1] for the SMT systems, and *NMT-Keras* [27]'s interactive branch for the NMT systems.

---

[1] https://github.com/midobal/pb-imt.

Segment-Based Simulation. Like Domingo et al. [10], in this simulation we assumed that validated word segments must be in the same order as in the reference. Therefore, segments that need to be reordered are not validated. Moreover, for the sake of simplicity and without loss of generality, we assumed that the user always corrects the leftmost wrong word.

At each iteration, the user validates segments by computing the longest common subsequence [1] between the system's hypothesis and the reference. This has an associated cost of one action for each one-word segment and two actions for each multi-word segment. Then, the user checks if any pair of consecutive validated segments should be merged into a single segment (i.e., they appear one consecutively in the reference but are separated by some words in the hypothesis). If so, they merge them, increasing mouse actions in one where there is a single word between them or two otherwise. Finally, they correct the leftmost wrong word (like in the prefix-based simulation). The system, then, reacts to this feedback generating a new hypothesis. This process is repeated until the hypothesis and the reference are the same.

This simulation was conducted using Domingo et al.'s [10] software[2] for the SMT systems, and *NMT-Keras*'s [27] interactive branch for the NMT systems.

5.2 Systems

SMT systems were trained with *Moses* [19], following the standard procedure: we estimated a 5-gram language model—smoothed with the improved KneserNey method—using *SRILM* [38], and optimized the weights of the log-linear model with MERT [23]. SMT systems were used both for the SMT modernization approach and for generating synthetic data to enrich the neural systems (see Sect. 3.2).

We built NMT systems using *NMT-Keras* [27]. We used long short-term memory units [14], with all model dimensions set to 512 for the RNN architecture. We trained the system using Adam [17] with a fixed learning rate of 0.0002 and a batch size of 60. We applied label smoothing of 0.1 [40]. At inference time, we used beam search with a beam size of 6. In order to reduce vocabulary, we applied joint byte pair encoding (BPE) [12] to all corpora, using 32, 000 merge operations.

For the Transformer architecture [44], we used 6 layers; Transformer, with all dimensions set to 512 except for the hidden Transformer feed-forward (which was set to 2048); 8 heads of Transformer self-attention; 2 batches of words in a sequence to run the generator on in parallel; a dropout of 0.1; Adam [17], using an Adam beta2 of 0.998, a learning rate of 2 and Noam learning rate decay with 8000 warm up steps; label smoothing of 0.1 [40]; beam search with a beam size of 6; and joint BPE applied to all corpora, using 32, 000 merge operations.

5.3 Corpora

We made use of the following corpora in our experimental session:

---

[2] https://github.com/midobal/sb-imt.

Dutch Bible [41]: A collection of different versions of the Dutch Bible. Among others, it contains a version from 1637—which we consider as the original version—and another from 1888—which we consider as the modern version (using 19th century Dutch as if it were *modern Dutch*).

El Quijote [7]: the well-known 17th century Spanish novel by Miguel de Cervantes, and its correspondent 21st century version.

OE-ME [35]: contains the original 11th century English text *The Homilies of the Anglo-Saxon Church* and a 19th century version—which we consider as *modern English*.

Additionally, to enrich the neural models we made use of the following *modern documents*: the collection of Dutch books available at the *Digitale Bibliotheek voor de Nederlandse letteren*[3], for Dutch; and OpenSubtitles [21]—a collection of movie subtitles in different languages—for Spanish and English. Table 1 contains the corpora statistics.

Table 1. Corpora statistics. $|S|$ stands for number of sentences, $|T|$ for number of tokens and $|V|$ for size of the vocabulary. *Modern documents* refers to the monolingual data used to create the synthetic data. M denotes millions and K thousands.

		Dutch Bible		El Quijote		OE-ME	
		Original	Modernized	Original	Modernized	Original	Modernized
Train	$|S|$	35.2K		10K		2716	
	$|T|$	870.4K	862.4K	283.3K	283.2K	64.3K	69.6K
	$|V|$	53.8K	42.8K	31.7K	31.3K	13.3K	8.6K
Validation	$|S|$	2000		2000		500	
	$|T|$	56.4K	54.8K	53.2K	53.2K	12.2K	13.3K
	$|V|$	9.1K	7.8K	10.7K	10.6K	4.2K	3.2K
Test	$|S|$	5000		2000		500	
	$|T|$	145.8K	140.8K	41.8K	42.0K	11.9K	12.9K
	$|V|$	10.5K	9.0K	8.9K	9.0K	4.1K	3.2K
Modern documents	$|S|$	3.0M		2.0M		6.0M	
	$|T|$	76.1M	74.1M	22.3M	22.2M	67.5M	71.6M
	$|V|$	1.7M	1.7M	210.1K	211.7K	290.2K	287.4K

5.4 Metrics

In order to assess our proposal, we make use of the following well-known metrics:

Word Stroke Ratio (WSR) [42]: measures the number of words edited by the user, normalized by the number of words in the final translation.

Mouse Action Ratio (MAR) [3]: measures the number of mouse actions made by the user, normalized by the number of characters in the final translation.

---

[3] http://dbnl.nl/.

Additionally, to assess the initial quality of the modernization systems, we used the following well-known metrics:

BiLingual Evaluation Understudy (BLEU) [25]: computes the geometric average of the modified n-gram precision, multiplied by a brevity factor that penalizes short sentences. In order to ensure consistent BLEU scores, we used *sacreBLEU* [30] for computing this metric.

Translation Error Rate (TER) [37]: computes the number of word edit operations (insertion, substitution, deletion and swapping), normalized by the number of words in the final translation.

Finally, we applied approximate randomization testing (ART) [31]—with $10,000$ repetitions and using a p-value of 0.05—to determine whether two systems presented statistically significance.

6   Results

Table 2 presents the experimental results. It presents the initial quality of each modernization system and compares their performance on a prefix-based or a segment-based framework.

Table 2. Experimental results. The initial modernization quality is meant to be a starting point comparison of each system. All results are significantly different between all approaches except those denoted with †. Given the same approach, all results are significantly different between the different IMT protocols except those denoted with ‡. [↓] indicates that the lowest the value the highest the quality. [↑] indicates that the highest the value the highest the quality. Best results are denoted in **bold**.

Corpus	Approach	Modernization quality		Prefix-based		Segment-based	
		TER	BLEU	WSR	MAR	WSR	MAR
		[↓]	[↑]	[↓]	[↓]	[↓]	[↓]
Dutch Bible	SMT	11.5	77.5	14.3	4.4	**9.0**	**10.8**
	NMT$_{LSTM}$	50.7†	43.4	42.6‡	9.2	42.6‡	50.9
	NMT$_{Transformer}$	50.3†	35.8	49.2‡	10.4	49.2‡	48.3
El Quijote	SMT	30.7	58.3	38.8	10.9	**22.0**	**19.7**
	NMT$_{LSTM}$	42.9	50.4	68.9‡	11.8	68.9‡	47.8
	NMT$_{Transformer}$	47.3	46.1	73.2‡	13.4	73.2‡	50.5
OE-ME	SMT	39.6	39.6	58.2	15.5	**28.2**	**26.1**
	NMT$_{LSTM}$	56.4	30.3	72.1†	12.8‡	72.1‡	59.5
	NMT$_{Transformer}$	58.9	28.2	73.5‡	13.3†	73.5‡	49.5

In all cases, the SMT approach yielded the best results with a great difference. The prefix-based protocol successfully reduces the human effort of creating error-free modernizations. Moreover, the segment-based protocol reduced the typing effort even more, at the expenses of a small increase in the use of the mouse, which is believed to have a smaller impact in the human effort [10].

With respect to the NMT approaches, while all of them also successfully decreased the human effort, these diminishes are significantly smaller than with the SMT approach. Moreover, the segment-based protocol does not offer any benefit with respect to the prefix-based: both protocols have the same typing effort. However, the segment-based protocol has a significant increase in the mouse usage. Finally, is it worth noting how the initial quality of the systems is considerably lower than the SMT approach, which has an impact in the IMT performance.

Table 3. Example of a prefix-based IMT session. The session starts with the system proposing an initial modernization. The user, then, looks for the leftmost wrong word and corrects it (*foresees* instead of *foresceawað*). Inherently, they are validating the prefix *All thing he*. Immediately, the system reacts to this feedback by suggesting a new hypothesis. The process is repeated until the user finds the system's hypothesis satisfactory.

Source (x): Ealle ðing he foresceawað and wát, and ealra ðeoda gereord he cann		
Target translation (ŷ): All things he foresees and knows, and he understands the tongues of all nations		
IT-0	MT	All things he foresceawað and knows, and of all nations language he understands
IT-1	User	All things he **foresees** and knows, and of all nations language he understands.
	MT	All things he foresees and knows, and of all nations language he understands
IT-2	User	All things he foresees and knows, and **he** all nations language he understands
	MT	All things he foresees and knows, and he understands of all nations language
IT-3	User	All things he foresees and knows, and he understands **the** all nations language
	MT	All things he foresees and knows, and he understands the beginning of all nations language
IT-4	User	All things he foresees and knows, and he understands the **tongues** of all nations language
	MT	All things he foresees and knows, and he understands the tongues all
IT-5	User	All things he foresees and knows, and he understands the tongues **of**
	MT	All things he foresees and knows, and he understands the tongues of all
IT-6	User	All things he foresees and knows, and he understands the tongues of all **nations**
	MT	All things he foresees and knows, and he understands the tongues of all nations
END	User	All things he foresees and knows, and he understands the tongues of all nations

6.1 Quality Analysis

Finally, we show some examples which reflect the benefits of using an interactive framework. Table 3 showcases a prefix-based IMT session for modernazing the language of a sentence from an old English document. Modernizing the sentence from scratch has an associated cost of 14 word strokes and one mouse action, while correcting the automatic modernization has an associated cost of 7 word strokes and 7 mouse actions. However, with the prefix-based protocol this cost is reduced to 6 word strokes and 6 mouse actions. Furthermore, with the segment-based protocol (see Table 4) this cost is reduced to only 3 word strokes at the expenses of increasing the mouse actions—which have a smaller impact in the human effort—to 15.

Table 4. Example of a segment-based IMT session. The session starts with the system proposing an initial modernization. The user, then, reviews the hypothesis and selects all the word segments that considers to be correct (All things he , and knows , and and of all nations). Then they make a word correction (*foresees* instead of *foresceawað*). Immediately, the system reacts to this feedback by suggesting a new hypothesis. The process is repeated until the user finds the system's hypothesis satisfactory.

Source (x): Ealle ðing he foresceawað and wát, and ealra ðeoda gereord he cann		
Target translation (ŷ): All things he foresees and knows, and he understands the tongues of all nations		
IT-0	MT	All things he foresceawað and knows, and of all nations language he understands
IT-1	User	All things he **foresees** and knows, and of all nations language he understands
	MT	All things he foresceawað foresees and knows, and language he understand of all nations
IT-2	User	All things he foresees and knows, and he understand **the** of all nations
	MT	All things he foresees and knows, and he understand the language of all nations
IT-3	User	All things he foresees and knows, and he understand the **tongues** of all nations
	MT	All things he foresees and knows, and he understand the tongues of all nations
END	User	All things he foresees and knows, and he understands the tongues of all nations

7 Conclusions and Future Work

In this work we have deployed the interactive framework into language modernization in order to help scholar generate error-free modernizations. We deployed two different protocols to SMT and NMT modernization approaches.

Under simulated conditions, we observed that, while the IMT framework always succeeded in reducing the human effort, the SMT approach yielded the best results. Moreover, while the segment-based protocol performed significantly better than the prefix-based protocol for the SMT approach, there was no significantly difference for the NMT approaches.

Finally, in a future work we would like to conduct a human evaluation with the help of scholars to better assess the benefits of the interactive language modernization framework.

Acknowledgements. The research leading to these results has received funding from *Generalitat Valenciana* under project *PROMETEO/2019/121*. We gratefully acknowledge *Andrés Trapiello* and *Ediciones Destino* for granting us permission to use their book in our research.

References

1. Apostolico, A., Guerra, C.: The longest common subsequence problem revisited. Algorithmica **2**, 315–336 (1987)
2. Bahdanau, D., Cho, K., Bengio, Y.: Neural machine translation by jointly learning to align and translate (2015). arXiv:1409.0473
3. Barrachina, S., et al.: Statistical approaches to computer-assisted translation. Comput. Linguist. **35**, 3–28 (2009)

4. Biçici, E., Yuret, D.: Optimizing instance selection for statistical machine translation with feature decay algorithms. IEEE/ACM Trans. Audio Speech Lang. Process. **23**(2), 339–350 (2015)
5. Bollmann, M., Søgaard, A.: Improving historical spelling normalization with bi-directional LSTMs and multi-task learning. In: Proceedings of the International Conference on the Computational Linguistics, pp. 131–139 (2016)
6. Brown, P.F., Pietra, V.J.D., Pietra, S.A.D., Mercer, R.L.: The mathematics of statistical machine translation: parameter estimation. Comput. Linguist. **19**(2), 263–311 (1993)
7. Domingo, M., Casacuberta, F.: A machine translation approach for modernizing historical documents using back translation. In: Proceedings of the International Workshop on Spoken Language Translation, pp. 39–47 (2018)
8. Domingo, M., Casacuberta, F.: Modernizing historical documents: a user study. Patt. Recogn. Lett. **133**, 151–157 (2020). https://doi.org/10.1016/j.patrec.2020.02.027
9. Domingo, M., Chinea-Rios, M., Casacuberta, F.: Historical documents modernization. Prague Bull. Math. Linguist. **108**, 295–306 (2017)
10. Domingo, M., Peris, Á., Casacuberta, F.: Segment-based interactive-predictive machine translation. Mach. Transl. **31**, 1–23 (2017)
11. Foster, G., Isabelle, P., Plamondon, P.: Target-text mediated interactive machine translation. Mach. Transl. **12**, 175–194 (1997)
12. Gage, P.: A new algorithm for data compression. C Users J. **12**(2), 23–38 (1994)
13. Gehring, J., Auli, M., Grangier, D., Yarats, D., Dauphin, Y.N.: Convolutional sequence to sequence learning. In: Proceedings of the International Conference on Machine Learning, pp. 1243–1252 (2017)
14. Gers, F.A., Schmidhuber, J., Cummins, F.: Learning to forget: continual prediction with LSTM. Neural Comput. **12**(10), 2451–2471 (2000)
15. Given, M.D.: A discussion of bible translations and biblical scholarship. http://courses.missouristate.edu/markgiven/rel102/bt.htm (2015)
16. Hochreiter, S., Schmidhuber, J.: Long short-term memory. Neural Comput. **9**(8), 1735–1780 (1997)
17. Kingma, D.P., Ba, J.: Adam: a method for stochastic optimization. arXiv preprint arXiv:1412.6980 (2014)
18. Knowles, R., Koehn, P.: Neural interactive translation prediction. In: Proceedings of the Association for Machine Translation in the Americas, pp. 107–120 (2016)
19. Koehn, P., et al.: Moses: Open source toolkit for statistical machine translation. In: Proceedings of the Annual Meeting of the Association for Computational Linguistics, pp. 177–180 (2007)
20. Lam, T.K., Schamoni, S., Riezler, S.: Interactive-predictive neural machine translation through reinforcement and imitation. In: Proceedings of Machine Translation Summit, pp. 96–106 (2019)
21. Lison, P., Tiedemann, J.: Opensubtitles 2016: extracting large parallel corpora from movie and TV subtitles. In: Proceedings of the International Conference on Language Resources Association, pp. 923–929 (2016)
22. Marie, B., Max, A.: Touch-based pre-post-editing of machine translation output. In: Proceedings of the Conference on Empirical Methods in Natural Language Processing, pp. 1040–1045 (2015)
23. Och, F.J.: Minimum error rate training in statistical machine translation. In: Proceedings of the Annual Meeting of the Association for Computational Linguistics, pp. 160–167 (2003)
24. Och, F.J., Ney, H.: Discriminative training and maximum entropy models for statistical machine translation. In: Proceedings of the Annual Meeting of the Association for Computational Linguistics, pp. 295–302 (2002)

25. Papineni, K., Roukos, S., Ward, T., Zhu, W.J.: BLEU: a method for automatic evaluation of machine translation. In: Proceedings of the Annual Meeting of the Association for Computational Linguistics, pp. 311–318 (2002)
26. Peng, X., Zheng, Y., Lin, C., Siddharthan, A.: Summarising historical text in modern languages. In: Proceedings of the Conference of the European Chapter of the Association for Computational Linguistics, pp. 3123–3142 (2021)
27. Peris, A., Casacuberta, F.: NMT-Keras: a very flexible toolkit with a focus on interactive NMT and online learning. Prague Bull. Math. Linguist. **111**, 113–124 (2018)
28. Peris, Á., Casacuberta, F.: Online learning for effort reduction in interactive neural machine translation. Comput. Speech Lang. **58**, 98–126 (2019)
29. Peris, Á., Domingo, M., Casacuberta, F.: Interactive neural machine translation. Comput. Speech Lang. **45**, 201–220 (2017)
30. Post, M.: A call for clarity in reporting bleu scores. In: Proceedings of the Third Conference on Machine Translation, pp. 186–191 (2018)
31. Riezler, S., Maxwell, J.T.: On some pitfalls in automatic evaluation and significance testing for MT. In: Proceedings of the Workshop on Intrinsic and Extrinsic Evaluation Measures for Machine Translation and/or Summarization, pp. 57–64 (2005)
32. Robbins, H., Monro, S.: A stochastic approximation method. Ann. Math. Statist. 400–407 (1951)
33. Marcos, J.R.: Un 'Quijote' moderno (2015). https://elpais.com/cultura/2015/05/27/babelia/1432726379_211033.html
34. Sanchis-Trilles, G., Ortiz-Martínez, D., Civera, J., Casacuberta, F., Vidal, E., Hoang, H.: Improving interactive machine translation via mouse actions. In: Proceedings of the Conference on Empirical Methods in Natural Language Processing, pp. 485–494 (2008)
35. Sen, S., Hasanuzzaman, M., Ekbal, A., Bhattacharyya, P., Way, A.: Take help from elder brother: Old to modern English NMT with phrase pair feedback. In: Proceedings of the International Conference on Computational Linguistics and Intelligent Text Processing (2019). (in press)
36. Sennrich, R., Haddow, B., Birch, A.: Neural machine translation of rare words with subword units. In: Proceedings of the Annual Meeting of the Association for Computational Linguistics, pp. 1715–1725 (2016)
37. Snover, M., Dorr, B., Schwartz, R., Micciulla, L., Makhoul, J.: A study of translation edit rate with targeted human annotation. In: Proceedings of the Association for Machine Translation in the Americas, pp. 223–231 (2006)
38. Stolcke, A.: SRILM - an extensible language modeling toolkit. In: Proceedings of the International Conference on Spoken Language Processing, pp. 257–286 (2002)
39. Sutskever, I., Vinyals, O., Le, Q.V.: Sequence to sequence learning with neural networks. In: Proceedings of the Advances in Neural Information Processing Systems, vol. 27, pp. 3104–3112 (2014)
40. Szegedy, C., et al.: Going deeper with convolutions. In: Proceedings of the IEEE Conference on Computer Vision and Pattern Recognition, pp. 1–9 (2015)
41. Sang, E.T.K., et al.: The CLIN27 shared task: translating historical text to contemporary language for improving automatic linguistic annotation. Comput. Linguist. Netherlands J. **7**, 53–64 (2017)
42. Tomás, J., Casacuberta, F.: Statistical phrase-based models for interactive computer-assisted translation. In: Proceedings of the International Conference on Computational Linguistics/Association for Computational Linguistics, pp. 835–841 (2006)
43. Torregrosa, D., Forcada, M.L., Pérez-Ortiz, J.A.: An open-source web-based tool for resource-agnostic interactive translation prediction. Prague Bull. Math. Linguist. **102**, 69–80 (2014)

44. Vaswani, A., et al.: Attention is all you need. In: Advances in Neural Information Processing Systems, pp. 5998–6008 (2017)
45. Zens, R., Och, F.J., Ney, H.: Phrase-based statistical machine translation. In: Jarke, M., Lakemeyer, G., Koehler, J. (eds.) KI 2002. LNCS (LNAI), vol. 2479, pp. 18–32. Springer, Heidelberg (2002). https://doi.org/10.1007/3-540-45751-8_2

From Captions to Explanations: A Multimodal Transformer-based Architecture for Natural Language Explanation Generation

Isabel Rio-Torto[1,2]([⊠]), Jaime S. Cardoso[1,3], and Luís F. Teixeira[1,3]

[1] INESC TEC, Porto, Portugal
isabel.riotorto@inesctec.pt
[2] Faculdade de Ciências da Universidade do Porto, Porto, Portugal
[3] Faculdade de Engenharia da Universidade Porto, Porto, Portugal
{jaime.cardoso,luisft}@fe.up.pt

Abstract. The growing importance of the Explainable Artificial Intelligence (XAI) field has led to the proposal of several methods for producing visual heatmaps of the classification decisions of deep learning models. However, visual explanations are not sufficient because different end-users have different backgrounds and preferences. Natural language explanations (NLEs) are inherently understandable by humans and, thus, can complement visual explanations. Therefore, we introduce a novel architecture based on multimodal Transformers to enable the generation of NLEs for image classification tasks. Contrary to the current literature, which models NLE generation as a supervised image captioning problem, we propose to learn to generate these textual explanations without their direct supervision, by starting from image captions and evolving to classification-relevant text. Preliminary experiments on a novel dataset where there is a clear demarcation between captions and NLEs show the potential of the approach and shed light on how it can be improved.

Keywords: Natural language explanations · Explainable AI · Transformers · Natural language generation · Image captioning

1 Introduction

Much of the research in XAI has focused on post-model visual explanations, also known as saliency/heat maps, i.e., visual explanations that highlight the most relevant pixels for the final predictive output. The existence of several off-the-shelf open-source implementations [2,10] of these methods for different machine learning frameworks, coupled with their ease of use since they do not require retraining the models, has led to their widespread application, despite their known problems [1,7,18,19]. In-model visual explainability methods are scarcer, but proposals like [17] have demonstrated that these approaches are possible and

© Springer Nature Switzerland AG 2022
A. J. Pinho et al. (Eds.): IbPRIA 2022, LNCS 13256, pp. 54–65, 2022.
https://doi.org/10.1007/978-3-031-04881-4_5

even outperformed *post-hoc* methods, with the advantage of solving some of the aforementioned problems, mainly regarding model-independence given that, in this case, explanations are model-dependent by design.

However, visual explanations might not suffice: even an ideal visual explanation might not be as interpretable as an NLE. Visual explanations are more precise spatially but are less semantically meaningful than a textual description of the reasons behind a predictive outcome. Moreover, NLEs have the advantage of being inherently understandable by humans, whereas other types of explanations sometimes can be hard to understand [8]. At the end of the day, different audiences with distinct backgrounds have diverse preferences, so systems should be versatile enough to leverage the advantages and complementarity of various modalities of explanations.

In the literature, several works [9,14,15,23] approach the problem of generating NLEs as traditional supervised image captioning tasks, where the model learns to produce some human collected ground-truth explanations. We consider that as long as a model is directly trained to generate some known textual sequence instead of explaining how the main decision-making network reached its output, it does not constitute a true NLE generation framework but rather a traditional image captioning method; in other words, NLE generation can not be a directly supervised process. Furthermore, there should be a clear distinction between a caption and an explanation: the first should describe in detail all objects of an image, while the latter should only take into account parts of the image relevant for a given decision.

Taking these aspects into consideration, we propose a multimodal network based on Transformers [21], the state-of-the-art in natural language generation, to progress from captions to explanations. In a first stage, an encoder-decoder Transformer is trained for supervised image captioning to ensure that the decoder, i.e., the language model can generate coherent text. In a second stage, both visual and textual latent representations are concatenated to perform classification. By making the whole network learn to classify, the language model learns to modify the generated image captions to only produce text relevant for the decision-making process and, thus, learns to generate NLEs.

Therefore, our main contributions are twofold:

– A dataset where there is a clear distinction between captions and NLEs, facilitating objective assessment of performance and debugging of NLE generation methods without requiring expert knowledge.
– An in-model multimodal Transformer-based architecture for generating NLEs without direct supervision.

The rest of this manuscript is structured as follows: Sect. 2 addresses the related work and state-of-the-art both on image captioning and NLE generation, Sect. 3 introduces the proposed dataset and multimodal architecture, Sect. 4 describes the results obtained in both training phases, and Sect. 5 presents the main conclusions and next steps.

2   Related Work

2.1   Image Captioning

Image captioning consists of describing an image in natural language. Since this task involves connecting vision and language, existing architectures usually involve a visual understanding system and a language model. The first methods used CNNs as high-level feature extractors whose representations were used to condition LSTM-based language models [22]. Naturally, more recently Transformers have taken the image captioning field by storm with incredible performance improvements; the latest works include both traditional encoder-decoder [13] and BERT-like architectures [27]. In encoder-decoder approaches [13], the encoder is based on the Vision Transformer (ViT) [5], so the image is first divided into patches before being given to the model, and the decoder is an auto-regressive or causal Transformer, in which a given word only attends to previous words in the input sequence. In BERT-like architectures [27], both image patches and word tokens are given to a unified Transformer as a single sequence divided by a special separation token. This unified Transformer is usually trained with a bidirectional objective in which a given token can attend to both present and past tokens of the sequence. The current state-of-the-art in the COCO image captioning task [12] is a BERT-like architecture that, in a similar way to how image captioning evolved from using CNN representations of whole images to using CNN representations of regions of interest indicated by object detection networks, jointly models image region features, object tags and word tokens [11, 25].

2.2   Natural Language Explanation Generation

One of the first works and perhaps still one of the most relevant on NLEs for visual classification tasks is the seminal paper by Hendricks et al. [6]. The authors introduce an architecture composed of a compact bilinear classifier for image feature extraction and an LSTM conditioned on the predicted class label. This architecture distinguishes itself from image captioning scenarios because the recurrent explanation generator is conditioned on the class and due to a reinforcement learning-based discriminative loss that rewards explanations with class discriminative properties by employing a sentence classifier to predict the class label. This loss is accompanied by a traditional relevance loss, in which the model is trained to predict the next word in a ground-truth sentence, ensuring that the produced sentences are image-relevant. The results showed that the produced explanations are more class discriminative and image-relevant than simple image or class descriptions, respectively.

Recent works [9, 14, 15, 23] are more distant from our view of NLE generation, not only because they focused on visual question answering rather than traditional image classification, but mainly because they rely on human-annotated explanations for training. Therefore, we consider the ideas of Hendricks et al. [6]

to be the most related to ours, with several key differences: we replace convolutional classifiers and LSTMs with Transformer-based models, our equivalent of the relevance loss is achieved through image captioning pre-training, we directly use both image and text embeddings to perform classification and no extra sentence classifier nor reinforcement-based losses are needed.

3 Proposed Methodology

3.1 A Synthetic Dataset to Distinguish Captions from Explanations

The same way a visual explanation differs from a segmentation map [20], so does a textual explanation differ from a caption [6]. While the caption constitutes a detailed description of an image, ideally mentioning all the existing objects, an explanation should only refer to (part of) objects that are relevant for a certain classification outcome.

In traditional image captioning datasets, such as COCO [12], captions tend to overlap with explanations because images usually have a single object present and classification is based on the presence of that object; for example a caption could be "An image of a dog laying down." while the explanation could be "There is a dog in the image.". Furthermore, some efforts have been made to collect human-annotated explanations for existing datasets, but these are either for purely textual data [3], activity recognition [15], visual question answering [15], or visual entailment [24]. Therefore, there is a clear lack of image classification data with associated NLEs.

Given this shortage of datasets with NLEs for classification and where a clear distinction between the captioning and the classification tasks is present, we propose a dataset that allows faster prototyping, and easy debugging and evaluation of the developed solutions without needing specific expert knowledge. The dataset is composed of images containing triangles and/or squares of different colours. An image is labelled as positive if there is at least one triangle on the left of the leftmost square, and is considered negative otherwise (if no squares/triangles exist or the leftmost figure is not a triangle). As such, there is a clear contrast between a caption, which ideally should account for all the polygons and their relative positions, and an explanation for the classification label, which should only need to identify the presence of the leftmost square and focus on the objects on its left. Each image in the dataset is also accompanied by three possible captions and an explanation. We generated 2000 images for training, 600 for validation and 200 for testing (all sets are evenly balanced); each image has at least 2 and at most 4 polygons. Examples can be seen in Fig. 1.

3.2 An Encoder-Decoder Vision Transformer for Natural Language Explanation Generation

As previously mentioned, several works [9,14,15,23] approach the problem of generating NLEs for classification problems as traditional supervised image

Label: negative
Caption: There are a total of 3 shapes, of which 1 is a square and of which 2 are triangles. The square is purple. One triangle is red and the other is yellow. The polygon on the right is the yellow triangle and the polygon on the left is the purple square. The figure in the middle is the red triangle.
Explanation: The purple square is the leftmost figure.

(a) Negative dataset instance, with an example caption and explanation.

Label: positive
Caption: Four geometric figures, 2 squares and 2 triangles. One triangle is purple and another is green. One square is orange and another is blue. The rightmost polygon is the orange square and the leftmost polygon is the purple triangle. The green triangle and the blue square are in the middle of the orange square and the purple triangle. The green triangle is on the right of the blue square.
Explanation: The purple triangle is on the left of the blue square.

(b) Positive dataset instance, with an example caption and explanation.

Fig. 1. Examples from the proposed dataset, with the corresponding labels, captions and explanations.

captioning tasks, where the model learns to produce some ground-truth explanations.

However, we argue that for an explanation to reflect the decision process made by a model, the model cannot be directly trained to generate said explanation. By doing so, we would simply obtain a *captioning model*, when in fact what we want is to be able to produce textual explanations for the predictions of a *classification model*. Therefore, our main focus lies in generating such explanations without direct supervision, i.e., without ever learning from ground-truth explanations. Nevertheless, having some sort of ground-truth explanations is useful to objectively measure the explanations produced by such a system. That is exactly why the aforementioned dataset includes one possible explanation for each instance.

According to [6], an explanation should be simultaneously image- and class-relevant. We can think of it as being a trade-off between both sides of a spectrum, where on one side we have image captions/descriptions, which are image-relevant but not necessarily class-relevant, and on the other side lie class descriptions, which are class-relevant but not necessarily image-relevant. Keeping this in mind, we propose the architecture depicted in Fig. 2. Its main building blocks are an encoder-decoder Transformer using a ViT [5] as encoder and an autoregressive/causal language model as decoder. The hidden states from both modules are fed to linear projection layers, responsible for ensuring both representations have the same dimensions (in this case 768). Their outputs are concatenated and given to a Multilayer Perceptron (MLP) with 3 Linear+ReLU layers for the

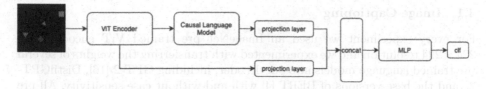

Fig. 2. Diagram representing the proposed Transformer-based architecture for self-explanatory classification with NLEs. Briefly, it is composed of an encoder-decoder Transformer structure with subsequent projection layers and a multimodal classification sub-network (MLP).

final classification. To obtain the ViT hidden states we use its CLS token, since it constitutes a latent representation of the whole image that can be used for classification. For the decoder using the CLS token is only possible if it follows a BERT-like architecture. Furthermore, Zhang et al. [26] concluded that using max-pooling achieved better results when compared to using the CLS representation, something that we also concluded in preliminary experiments. Thus, we also employed a max-pooling strategy over all the token representations of the last layer of the decoder.

The main rationale behind this architecture is to guarantee both image and class dependence by, respectively, generating text from the image and using that text to influence the classification outcome. Similarly to previous work developed in the scope of the in-model generation of visual explanations [17], if the explanations contribute directly to the classification, then the classification loss will alter the explanations accordingly, making them reflect what is important for the classification task.

Notwithstanding, the classification loss alone might not be a strong enough supervisory signal to ensure that coherent text is produced. Thus, we introduce a first training step in which the Transformer is trained for supervised image captioning and then transfer the resulting weights to the second training step, where the whole network is fine-tuned for classification.

Finally, to regularise the training, we randomly ignore one of the input modalities, i.e., sometimes both image and text embeddings are used, sometimes only the text counterpart is used, and sometimes only the image representation is used. When a single modality is employed, we repeat its projection layer's resulting vector to be able to keep the structure of the concatenation block.

4   Results and Discussion

Our proposed methodology is composed of two training stages: first a standard image captioning phase, followed by the classification and NLE generation phase. As such, this section is subdivided into two sections, one for each training phase. Regarding evaluation, we employ traditional image captioning metrics such as BLEU, METEOR, ROUGE, CIDER and SPICE to evaluate both generated captions and NLEs, comparing them to their corresponding ground-truths.

4.1 Image Captioning

For every experiment we used an ImageNet pre trained ViT encoder with 224×224 resolution and we experimented with transferring the weights of several pre trained language models into our decoder, including GPT-2 [16], DistilGPT-2[1] and the base versions of BERT [4] with and without case sensitivity. All pre trained models were obtained from https://huggingface.co and fine-tuned with cross entropy for 20 epochs with a batch size of 8 and an initial learning rate of 5×10^{-5}, linearly decayed. During training we consider sequences until 95 tokens and during inference sentences can have 100 tokens at maximum. Finally, we use beam search decoding with 4 beams.

From Table 1, where the results of this first image captioning stage are reproduced, we can conclude that using the weights from BERT for the decoder yields better results across all metrics. This can be explained by the fact that GPT-2 is originally trained for open-end text generation, so it becomes more difficult to learn when to stop producing tokens. In fact, we verified that with GPT-2-based models the end-of-sentence (EOS) token is never reached and the produced captions always present the maximum allowed number of tokens, while with BERT the produced sentences have different lengths; in both GPT-2 and BERT the first sentences are coherent and similar to the ground-truth captions, but after a certain number of tokens BERT stops the generation and GPT-2 starts producing incoherent text until its output is truncated. In the end, these unnecessary extra sentences hinder GPT-2-based models' performance. Naturally, the cased version of BERT performs slightly better than the uncased version simply because our ground-truth captions are themselves cased. As such, we opt for the model trained with the cased version of BERT as our initialisation strategy for the second training phase, whose results are detailed in the next section.

Table 1. Image captioning results obtained on the proposed dataset. For all experiments we used a ViT encoder pre trained on ImageNet and different weights from pre trained language models for the decoder.

Decoder	BLEU1	BLEU2	BLEU3	BLEU4	METEOR	ROUGE	CIDER	SPICE
GPT-2	0.591	0.550	0.495	0.434	0.422	0.566	0.034	0.517
DistilGPT-2	0.600	0.562	0.509	0.449	0.430	0.572	0.041	0.525
BERT (uncased)	0.901	0.821	0.733	0.648	0.434	0.707	**2.589**	0.556
BERT (cased)	**0.903**	**0.833**	**0.755**	**0.677**	**0.446**	**0.717**	2.477	**0.573**

4.2 Natural Language Explanations

For the second training phase, models were trained for 10 epochs with a batch size of 16 and an initial learning rate of 1×10^{-4}, linearly decayed. We experimented with randomly changing which embeddings were given to the MLP, either only

---

[1] https://huggingface.co/distilgpt2.

the image, or only the text or concatenating both (which we call Multimodal). We also experimented with combinations of the previous alternatives, for example switching between multimodal and text-only with 50% probability each or using all three with 33% probability each. In terms of classification accuracy all models perform well within a small number of epochs, achieving 99% accuracy, as can be seen in Fig. 3. This is not surprising, given the fact that the proposed dataset is balanced and the classification problem is not purposely difficult, since the main focus of this work is not the classification task.

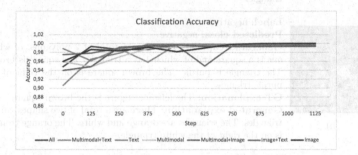

Fig. 3. Classification accuracy obtained during training on the validation set of the proposed data. We present the combinations between the different modalities used during training. When two modalities are used each has a 50% probability and when all three are used ("All") the probability for each is 33%.

Regarding the performance of the system in terms of the generated explanations, there is a slight improvement of the model trained with all three modalities with equal probability. Nevertheless, the results of all models can be greatly improved (see Table 2 and Fig. 4). The explanations differ from the captions obtained in the first training stage, which is expected considering that in this second phase they are directly influenced by the classification. However, the

Table 2. Objective evaluation of the generated NLEs. The "text", "image" and "multimodal" (concatenation of image and text embeddings) columns refer to the different modalities used during training and their respective probabilities. The model which was trained using all modalities with equal probability performs slightly better.

Text	Image	Multimodal	BLEU1	BLEU2	BLEU3	BLEU4	METEOR	ROUGE	CIDER	SPICE
1.0	0	0	0.113	0.077	0.047	0.025	0.177	0.200	0.000	0.206
0	1.0	0	0.147	0.105	0.077	0.059	0.208	0.241	0.000	0.271
0	0	1.0	0.143	0.100	0.065	0.041	0.193	0.244	**0.002**	0.192
0.5	0	0.5	0.143	0.105	0.068	0.043	0.202	0.236	0.000	0.226
0	0.5	0.5	0.130	0.087	0.055	0.031	0.200	0.199	0.000	0.213
0.5	0.5	0	0.129	0.079	0.046	0.022	0.177	0.209	0.000	0.193
0.33	0.33	0.33	**0.163**	**0.124**	**0.095**	**0.075**	**0.228**	**0.272**	0.000	**0.319**

(a)

Label: negative
Predicted class: negative
Generated caption: There are a total of three polygons, all of which are triangles. The triangles are blue, purple and red. The leftmost figure is the blue triangle and the rightmost is the purple triangle. The shape in the middle is the red triangle.
GT explanation: No squares present.
Generated explanation: There are a total all triangles. The triangles are red, orange, blue and yellow. The rightmost polygon is the yellow triangle and the leftmost p all is the red triangle. The polycate is in all of all are triangles only triangles all are only triangle.

(b)

Label: negative
Predicted class: negative
Generated caption: There are a total of 2 polygons, of which one is a square and of which 1 is a triangle. The square is brown. The triangle is purple. The brown square is on the right of the purple triangle.
GT explanation: The leftmost figure is the brown square.
Generated explanation: There are a total of two shapes, all are triangles. The triangles are orange and white. The orange triangle is on the right of the white triangle.

(c)

Label: positive
Predicted class: positive
Generated caption: There are a total of three polygons, of which 1 is a triangle and of which 2 are squares. One square is orange and the other is blue. The triangle is purple. The rightmost figure is the blue square and the leftmost figure on the right is the purple triangle. The orange square is in the middle.
GT explanation: The purple triangle is on the left of the orange square.
Generated explanation: There are a total of 2 polygons, of which one is a triangle and of which 1 is a square. The triangle is brown. The square is blue. The brown triangle is on the left of the square.

(d)

Label: positive
Predicted class: positive
Generated caption: There are a total of 4 polygons, of which 2 are triangles and 2 are squares. One triangle is green and the other one is brown. One square is white and another is blue. The figure on the left is the green triangle and on the right is the blue square. The brown triangle and the white square are in the middle of the green square and the yellow triangle. The white square is on the shape of the brown triangle.
GT explanation: The green triangle is on the left of the blue square.
Generated explanation: 2 polygons in total, of which one is a triangle and of which 1 is a square. The triangle is brown. The square is blue. The brown triangle is on the left of the square.

Fig. 4. NLE generation results. For each example we present the ground-truth (GT) and predicted classification labels, the caption generated in the first training stage, as well as a possible ground-truth and the generated explanations. The generated explanations refer to training with all three alternatives (image-only, text-only and multimodal) with equal probability. In green (red) are highlighted (mis)matches between the generated text and the corresponding image. (Color figure online)

explanations lose their image relevance, for example, the generated text describes more polygons than the ones actually present or switches their colour. It is also interesting to note that in the majority of images of the positive class the explanations only mention the existence of two polygons, which is consistent with the original rationale that to identify the positive class one only needs to pay attention to the leftmost square and to one triangle on its left. Furthermore, in every image described as having two polygons some variation of the sentence "There are a total of 2 shapes, of which one is a triangle and of which 1 is a square." occurs. There is also a recurrence of the sentences "The triangle is brown." and "The brown triangle is on the left of the square." in about 20% of the positive instances. This suggests that the network might be learning that is it sufficient to say that one triangle is on the left of the square in order for the image to be classified as positive. As such, it seems that the current challenge is to teach the network to specify which relationship between polygons is the one responsible for the correct classification, guaranteeing that relationship is different from image to image. In more general terms, it is necessary to ensure that image relevance is not lost when updating the network parameters learned by the image captioning task in the first training phase.

5   Conclusion and Future Work

We introduced a novel multimodal synthetic dataset where there is a clear distinction between an image caption and an NLE for a classification outcome. Besides allowing faster prototyping and easy usage without requiring expert knowledge, it encourages a more objective and automatic assessment of NLEs. Alongside this contribution, we also proposed a novel multimodal Transformer-based architecture to produce NLEs for classification decisions without direct supervision of those explanations. By using both image and text to perform the classification we expect the captions generated in the first training phase to be modified to include more class discriminative information, thus progressing from captions towards NLEs. Preliminary experiments show that indeed the generated text is more class dicriminative than the original captions but at the cost of losing image-relevance. Therefore, work in the near future involves correcting this loss of image-relevance and possible alternatives include introducing a (cyclic) reconstruction step to force the generated text to contain enough image-relevant information to be able to reconstruct to some extent the original input image or its classification relevant regions.

Acknowledgements. This work was funded by the Portuguese Foundation for Science and Technology (FCT) under the PhD grant "2020.07034.BD".

References

1. Adebayo, J., Gilmer, J., Muelly, M., Goodfellow, I., Hardt, M., Kim, B.: Sanity checks for saliency maps. In: Proceedings of the 32nd International Conference on Neural Information Processing Systems. NIPS 2018, Red Hook, NY, USA, pp. 9525–9536. Curran Associates Inc. (2018)
2. Alber, M., et al.: innvestigate neural networks! J. Mach. Learn. Res. **20**(93), 1–8 (2019). http://jmlr.org/papers/v20/18-540.html
3. Bowman, S.R., Angeli, G., Potts, C., Manning, C.D.: A large annotated corpus for learning natural language inference. In: Proceedings of the 2015 Conference on Empirical Methods in Natural Language Processing, Lisbon, Portugal, pp. 632–642. Association for Computational Linguistics, September 2015. https://doi.org/10.18653/v1/D15-1075, https://aclanthology.org/D15-1075
4. Devlin, J., Chang, M.W., Lee, K., Toutanova, K.: BERT: pre-training of deep bidirectional transformers for language understanding. In: NAACL HLT 2019–2019 Conference of the North American Chapter of the Association for Computational Linguistics: Human Language Technologies - Proceedings of the Conference, vol. 1, pp. 4171–4186 (2019)
5. Dosovitskiy, A., et al.: An image is worth 16×16 words: transformers for image recognition at scale. In: International Conference on Learning Representations (2021). https://openreview.net/forum?id=YicbFdNTTy
6. Hendricks, L.A., Akata, Z., Rohrbach, M., Donahue, J., Schiele, B., Darrell, T.: Generating visual explanations. In: Leibe, B., Matas, J., Sebe, N., Welling, M. (eds.) ECCV 2016. LNCS, vol. 9908, pp. 3–19. Springer, Cham (2016). https://doi.org/10.1007/978-3-319-46493-0_1
7. Hooker, S., Erhan, D., Kindermans, P.J., Kim, B.: A benchmark for interpretability methods in deep neural networks. In: Advances in Neural Information Processing Systems 32(NeurIPS) (2019)
8. Kaur, H., Nori, H., Jenkins, S., Caruana, R., Wallach, H., Wortman Vaughan, J.: Interpreting interpretability: understanding data scientists' use of interpretability tools for machine learning. In: Proceedings of the 2020 CHI Conference on Human Factors in Computing Systems, pp. 1–14 (2020)
9. Kayser, M., et al.: e-ViL: a dataset and benchmark for natural language explanations in vision-language tasks (2021). http://arxiv.org/abs/2105.03761
10. Kokhlikyan, N., et al.: Captum: a unified and generic model interpretability library for pytorch. arXiv preprint arXiv:2009.07896 (2020)
11. Li, X., et al.: OSCAR: object-semantics aligned pre-training for vision-language tasks. In: Vedaldi, A., Bischof, H., Brox, T., Frahm, J.-M. (eds.) ECCV 2020. LNCS, vol. 12375, pp. 121–137. Springer, Cham (2020). https://doi.org/10.1007/978-3-030-58577-8_8
12. Lin, T.-Y., et al.: Microsoft COCO: common objects in context. In: Fleet, D., Pajdla, T., Schiele, B., Tuytelaars, T. (eds.) ECCV 2014. LNCS, vol. 8693, pp. 740–755. Springer, Cham (2014). https://doi.org/10.1007/978-3-319-10602-1_48
13. Liu, W., Chen, S., Guo, L., Zhu, X., Liu, J.: CPTR: full transformer network for image captioning. arXiv preprint arXiv:2101.10804 (2021)
14. Marasović, A., Bhagavatula, C., Park, J.S., Le Bras, R., Smith, N.A., Choi, Y.: Natural language rationales with full-stack visual reasoning: from pixels to semantic frames to commonsense graphs. In: Findings of the Association for Computational Linguistics: EMNLP 2020, pp. 2810–2829. Association for Computational Linguistics, November 2020. https://doi.org/10.18653/v1/2020.findings-emnlp.253, https://aclanthology.org/2020.findings-emnlp.253

15. Park, D.H., et al.: Multimodal explanations: justifying decisions and pointing to the evidence. In: 2018 IEEE/CVF Conference on Computer Vision and Pattern Recognition. pp. 8779–8788. IEEE, June 2018. https://doi.org/10.1109/CVPR. 2018.00915, https://ieeexplore.ieee.org/document/8579013/

16. Radford, A., Wu, J., Child, R., Luan, D., Amodei, D., Sutskever, I.: Language Models are Unsupervised Multitask Learners, July 2019

17. Rio-Torto, I., Fernandes, K., Teixeira, L.F.: Understanding the decisions of CNNs: an in-model approach. Pattern Recogn. Lett. **133**(C), 373–380 (2020). https://doi.org/10.1016/j.patrec.2020.04.004, http://www.sciencedirect.com/science/article/pii/S0167865520301240

18. Rudin, C.: Stop explaining black box machine learning models for high stakes decisions and use interpretable models instead. Nat. Mach. Intell. **1**(5), 206–215 (2019). https://doi.org/10.1038/s42256-019-0048-x, http://www.nature.com/articles/s42256-019-0048-x

19. Rudin, C., Chen, C., Chen, Z., Huang, H., Semenova, L., Zhong, C.: Interpretable machine learning: fundamental principles and 10 grand challenges, pp. 1–80, March 2021. http://arxiv.org/abs/2103.11251

20. Samek, W., Binder, A., Montavon, G., Lapuschkin, S., Müller, K.R.: Evaluating the visualization of what a deep neural network has learned. IEEE Trans. Neural Networks Learn. Syst. **28**(11), 2660–2673 (2017). https://doi.org/10.1109/TNNLS. 2016.2599820

21. Vaswani, A., et al.: Attention is all you need. In: Advances in Neural Information Processing Systems, pp. 5999–6009 (2017)

22. Vinyals, O., Toshev, A., Bengio, S., Erhan, D.: Show and tell: a neural image caption generator. In: 2015 IEEE Conference on Computer Vision and Pattern Recognition (CVPR), pp. 3156–3164. IEEE, June 2015. https://doi.org/10.1109/CVPR.2015.7298935, http://ieeexplore.ieee.org/document/7298935/

23. Wu, J., Mooney, R.: Faithful multimodal explanation for visual question answering. In: Proceedings of the 2019 ACL Workshop BlackboxNLP: Analyzing and Interpreting Neural Networks for NLP, pp. 103–112 (2019)

24. Xie, N., Lai, F., Doran, D., Kadav, A.: Visual entailment: a novel task for fine-grained image understanding. arXiv preprint arXiv:1901.06706 (2019)

25. Zhang, P., et al.: Vinvl: making visual representations matter in vision-language models. In: CVPR 2021 (2021)

26. Zhang, Y., Jiang, H., Miura, Y., Manning, C.D., Langlotz, C.P.: Contrastive learning of medical visual representations from paired images and text. arXiv preprint arXiv:2010.00747 (2020)

27. Zhou, L., Palangi, H., Zhang, L., Hu, H., Corso, J., Gao, J.: Unified vision-language pre-training for image captioning and VQA. In: Proceedings of the AAAI Conference on Artificial Intelligence, vol. 34, pp. 13041–13049 (2020)

Medical Image Processing

Diagnosis of Skin Cancer Using Hierarchical Neural Networks and Metadata

Beatriz Alves[✉], Catarina Barata, and Jorge S. Marques

Institute for Systems and Robotics, Instituto Superior Técnico, Lisbon, Portugal
beatriz.c.alves@tecnico.ulisboa.pt

Abstract. Skin cancer cases have been increasing over the years, making it one of the most common cancers. To reduce the high mortality rate, an early and correct diagnosis is necessary. Doctors divide skin lesions into different hierarchical levels: first melanocytic and non-melanocytic and then as malignant or benign. Each lesion is also assessed taking into consideration additional patient information (*e.g.*, age and anatomic location of the lesion). However, few automatic systems explore such complementary medical information. This work aims to explore the hierarchical structure and the patient metadata to determine if the combination of these two types of information improves the diagnostic performance of an automatic system. To approach this problem, we implemented a hierarchical model, which resorts to intermediary decisions and simultaneously processes dermoscopy images and metadata. We also investigated the fusion of a flat and hierarchical model to see if their advantages could be brought together. Our results showed that the inclusion of metadata has a positive impact in the performance of the system. Despite hierarchical models performing slightly worse than flat models, they improved certain lesion classes, and can narrow down the lesion to a sub type, as opposed to the flat model.

Keywords: Skin lesion diagnosis · Hierarchical model · Metadata · Deep neural networks

1 Introduction

Skin cancer impact has been increasing due to the alarming growth in the number of cases over the years. According to the World Health Organization, 2/3 million cases of non-melanoma and 132,000 cases of melanoma occur worldwide every year. These numbers reinforce that skin cancer is one of the most common cancers [1]. Melanoma in particular has a high mortality rate when detected in the latest stages. Therefore, an early and accurate diagnosis must be achieved.

This work was supported by the FCT project and multi-year funding [CEECIND/00326/2017] and LARSyS - FCT Plurianual funding 2020–2023; and by a Google Research Award'21.

© Springer Nature Switzerland AG 2022
A. J. Pinho et al. (Eds.): IbPRIA 2022, LNCS 13256, pp. 69–80, 2022.
https://doi.org/10.1007/978-3-031-04881-4_6

The creation of the ISIC Challenges [2] promoted the development of the skin cancer classification models using deep learning techniques. While a variety of deep learning methods were proposed to diagnose dermoscopy images, they were still found to lack in one or more aspects. Deep learning techniques do not understand the information they are dealing with, they simply try to detect patterns or correlations among the different skin lesions. When dealing with only images, important traits can be neglected, such as the anatomical site, which can be a decisive diagnostic feature in some lesions. This conveys that a significant portion of the knowledge acquired by doctors during their medical training is not being put to use in the existing models. This is also true for lesion taxonomy, *i.e.*, the hierarchical organization of lesion types defined by dermatologists which could be used to provide a better understanding of the diagnosis made by the system. While some works on recent literature try to innovate how hierarchical classifiers are used [3,4], others study which method between flat and hierarchical models is better [5], or which hierarchical model prevails among all the different hierarchical structures [5,6]. However, none of these works has explored the incorporation of patient metadata.

Most skin lesion diagnosis systems are based on the analysis of dermoscopic images using flat classifiers. Due to the lack of two previously mentioned topics in the current literature, this work has two main goals. First, it aims to complement dermoscopic images with clinical information/metadata (age, gender, location of the lesion) and to evaluate differences in performance. The second objective is to use medical information about the taxonomic structure of the lesions shown in Fig. 1, using hierarchical classifiers. It is also intended to combine these two types of information and assess whether they allow to improve the performance of the system. The hierarchical structure that this work used can be seen in Fig. 1. This figure displays the eight skin lesions found in ISIC 2019 training dataset [7–9], which are Nevus (NV), Melanoma (MEL), Dermatofibroma (DF), Basal cell carcinoma (BCC), Squamous cell carcinoma (SCC), Actinic keratosis (AK), Benign keratosis (BKL) and Vascular (VASC).

This document is organized as follows. Section 2 explains all the methodologies used to classify skin lesions with and without metadata. Section 3 describes the experimental setup and presents the results with their associated analysis. Lastly, Sect. 4 discusses the conclusion and future works topics.

2   Methodology

In this work we compared 3 approaches: (i) a flat model, that performs a single decision, (ii) a hierarchical model, which resorts to intermediary decision before predicting the final diagnosis, and (iii) a mixed model which combines the two previous models. Additionally, we compare different ways to combine images and metadata for each of the previous approaches. Note that, in all models, images are processed using a convolutional neural network (CNN), while metadata are processed using fully connected layers (FCLs).

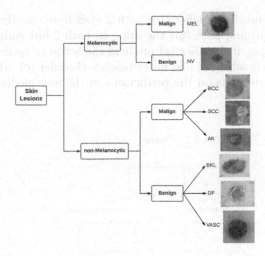

Fig. 1. Hierarchical organization of skin lesions in ISIC 2019 dataset. Dermoscopy images taken from [7–9].

2.1 Flat Classifier

Flat classifiers are simple, straight-forward models. They only need a single classifier to predict all of the categories as it does not take into account the inherent hierarchy among them. Therefore, this model is only required to, given an image and the corresponding metadata, predict one of the classes out of the eight-possible.

In the training phase, the flat classifier is trained with all the images and/or metadata of the eight classes. To train the model we use the training dataset and to choose hyperparameter values we use the validation dataset. The training is done with categorical cross-entropy loss function and Adam optimizer [10].

2.2 Hierarchical Classifier

Unlike the flat model, this one takes into account the taxonomy of several lesions by making intermediary decisions before reaching the final decision. With Fig. 1 in mind, we consider 3 levels of hierarchy: (i) melanocytic vs non-melanocytic, (ii) benign vs malignant, and (iii) the final diagnosis. We can see the visual representation of the proposed hierarchical model in Fig. 2. The model is an aggregation of 5 different classifiers (a, b, c, d and e) with (a) and (c) as intermediary decision classifiers and (b), (d) and (e) as final decision classifiers. The inside architecture of the individual classifiers is the same as its flat counterpart, with the only difference being in the *softmax* block. Since classifiers (a), (b) and (c) distinguish between 2 types of lesion, their *softmax* has 2 neurons and classifiers (d) and (e) have three, because they separate three lesions.

To predict a lesion, it is not mandatory that the data passes through all classifiers. There are 3 possible paths that it can go through. Path 1 starts in classifier (a) and ends in classifier (b), path 2 goes from classifier (a) to (c) and ends in classifier (d) and path 3 is the same as path 2 but ends in classifier (e) instead. For example, if classifier (a) predicts the lesion to be melanocytic, then the lesion goes to classifier (b) and never reaches classifier (c), (d) and (e). Thus, the adopted path depends on the predictions of the intermediate classifiers (a) and (c).

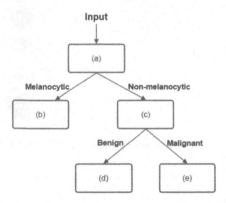

Fig. 2. Generic structure of the classifier block diagram in the hierarchical model. Classifier (a) distinguishes melanocytic lesions from non-melanocytic, (b) NV from MEL, (c) benign non-melanocytic lesions from malignant non-melanocytic lesions, (d) distinguishes between BKL, DF and VASC lesions and classifier (e) sets apart AK, BCC and SCC.

The training phase of the hierarchical models has the same loss function and optimizer as the flat classifier. However, the hierarchical model's classifiers (a)-(e) are only trained with the corresponding subset of the lesions it diagnoses. For example, classifier (d) is trying to diagnose BKL, DF and VASC so it is only trained with data of these three lesions. It is important to note that each classifier is independently trained.

2.3   Methods to Combine Images and Metadata

To combine the metadata with images, we opted to use only early fusion methods, where the combination is done at the feature level. We developed 3 different methods.

The first method is the **concatenation** between both features. First, the features are extracted from the images and from the metadata. Then they are concatenated into a single vector and sent to the classification block, which returns the diagnosis.

The second method is the **multiplication of features** as suggested in [11]. It tries to replicate an attention module by making the model learn which feature maps are less relevant and assigning lower values or even zero to their respective positions. Here, the metadata goes through a FCL of d neurons, where d represents the dimension of the last convolutional layer of the CNN. After the feature extraction, they are multiplied element-wise and the corresponding vector is sent to the classification block, which returns a diagnosis.

The third method was inspired on [12] and it consists on **reducing the number of image features**. It fixates the number of metadata features, m, and changes the number of image features according to hyperparameter r. Its relation is shown in Eq. 1, where n is the number of the reduced image features, m is the number of metadata features and r is the ratio (0–1) of image features present in the combined feature vector.

$$n = \frac{m}{1-r} - m \tag{1}$$

We also set the number of metadata features, m, as an additional hyperparameter. To reach the number of reduced image features, n, the output of the CNN goes through an extra FCL layer with n neurons and ReLU activation. Simultaneously, the metadata goes through a similar layer but with m neurons instead. Both results are concatenated to form the combined feature vector and sent to the classification block to diagnose the lesion. Figure 3 illustrates this method.

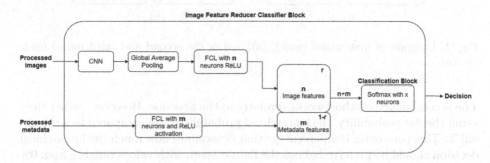

Fig. 3. Classifier block diagram of third method, image feature reducer.

2.4   Selection Between Flat and Hierarchical Models

Hierarchical and flat classifiers have different strengths. By considering the diagnosis from both classifiers, we tested whether both models' advantages can be brought together and if their individual weaknesses can be eliminated. For this, we created three new models which are based on the confidence of the diagnosis of both models.

The first mixed method is a direct competition between the hierarchical and the flat classifiers and it returns the decision of the model with the higher confidence. In this case, each classifier returns the class with the highest probability. Note that the probability is given by the *softmax* layer, present in both models. The higher the probability, the higher is the confidence of the model in its decision. For each data, the flat probability is compared to the correspondent hierarchical individual probabilities. If one of these probabilities is lower than the flat, the flat classifier makes the final decision. Otherwise, the decision falls upon the hierarchical classifier. In this and the following mixed models, the flat model has only one probability, while the hierarchical model has either 2 or 3, depending on which path the data goes through. If the data follows path 1, then it will have 2 probabilities, correspondent to classifiers (a) and (b). Otherwise, it will have 3 individual probabilities, which are from classifier (a), (c) and (d) or (e).

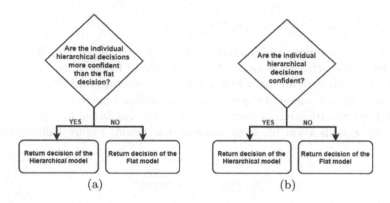

Fig. 4. Diagram of first mixed model, (a), and of the second and third mixed models, (b).

The second mixed method works similarly to the first one. However, rather than using the flat probability, the hierarchical probabilities are compared to a threshold T. This parameter is a percentage that determines how much the hierarchical decision should be preferred above the flat decision, with values ranging from 0% to 100%. If all of the equivalent hierarchical individual probabilities are higher than T, then the hierarchical classifier makes the final decision. Otherwise, the flat model makes it.

In the third mixed model, the confidence of the decision relies in the difference between the two highest *softmax* probabilities outputted by each classifier. The greater the difference, the higher is the likelihood of the decision being correct. In order to find what is the value of the optimal difference, we created a parameter P. Same as mixed 2, it represents a percentage used to figure out to what extent the hierarchical decision should be prioritized over the flat decision. Similar to mixed 2, the decision is made by the hierarchical classifier unless one of its probabilities is below P, in which case the decision passes to the flat classifier. Figure 4 presents the diagrams of the mixed models.

3   Results

This chapter starts by introducing the dataset, followed by the metrics used to evaluate all the results. Then, it presents the experimental results and its discussion of all the methods proposed. In all tests, 5 CNN architectures were considered ResNet50 [13], ResNet101 [13], DenseNet121 [14], EfficientNet-B0 [15] and EfficientNet-B2 [15] and were evaluated using the validation set unless otherwise specified.[1]

3.1   Dataset

This work used the database of ISIC 2019 challenge [7–9] to train and evaluate the proposed models. The training data has 25,331 dermoscopic images and 8 different classes: NV, MEL, DF, VASC, BCC, SCC, AK, BKL. To perform all of the experiments we divided the original training dataset into 2 subsets, the training set with 80% and the validation set with 20%. The best model obtained after training will be tested using the held out test set that is also provided by the challenge organizers. Table 1 presents the total number of images and how they are divided among all skin lesion classes in the reduced training set, the validation and test sets. No labels are provided for the test set, so there is no information about the number of cases in each class.

Table 1. The total number of samples in training, validation and test sets. The number of samples per class in the training and validation set.

Dataset	MEL	NV	BCC	AK	BKL	VASC	DF	SCC	Total
Train	3,617	10,300	2,658	693	2,099	202	191	502	20,262
Validation	905	2,575	665	174	525	51	48	126	5,069
Test	-	-	-	-	-	-	-	-	8,238

3.2   Performance Metrics

In this work we will use two metrics to evaluate the results. Sensibility (SE), also known as recall or True positive (TP) rate, measures the ratio of all the positive samples that were correctly classified as positive for each class.

To give equal importance to all classes, we used Balanced Accuracy ($BACC$) instead of the weighted accuracy. $BACC$ is the average of the SE obtained for each class and it is given by Eq. 2, where N represents the number of classes. In this work, N is set to eight, as there are eight classes.

$$BACC = \frac{\sum_{i=0}^{N-1} SE_i}{N} \tag{2}$$

---

[1] Source code to reproduce all experiments is available at https://github.com/Bia55//Skin_cancer_AI.

3.3   Impact of Metadata on Hierarchical and Flat Models

We tested the five previously mentioned CNN as well as different combinations among them in the hierarchical classifiers for all the methods. These combinations proved to be better and their results can be seen in table 2 alongside the results of the models that only use metadata or images. Table 3 reports the CNNs used in each configuration.[2]

Table 2. *BACC* scores of all the models best performance. Cells highlighted in green represent the best score of each classifier. The highlights of the individual classifiers, (a)-(e), also represent the classifiers of the Combined 4 model. Note that the flat score is from the method the Combined model belongs to.

Model\Classifier	(a)	(b)	(c)	(d)	(e)	Final score	Flat
Only Metadata	74.55	71.85	59.75	59.73	49.93	27.83	35.30
Only Images	90.69	86.95	86.66	91.75	69.30	66.48	73.05
Concatenation	92.21	88.10	89.40	94.64	77.76	71.98	79.05
Multiplication	91.94	87.75	90.54	95.05	80.57	73.99	79.02
Image feature reducer	91.84	88.00	90.12	96.12	78.96	72.92	79.08
Combined 4	92.21	88.10	90.54	96.12	80.57	73.82	-

Table 3. *CNN* configurations of the best hierarchical and flat models shown in table 2. Only configurations of the models that incorporate images are presented.

Model\Classifier	(a)	(b)	(c)	(d)	(e)	Flat
Only Images	EffNetB2				ResNet101	EffNetB2
Concatenation	DenseNet	EffNetB2	ResNet50	EffNetB0		ResNet101
Multiplication				ResNet101		
Image feature reducer			ResNet101	ResNet50		
Combined 4			ResNet50			-

There were 3 CNN that stood through. When dealing with only images, the EfficientNet-B2 distinguished itself from the other networks with the best results in every classifier except classifier (e) and the flat one. However, when we add metadata into the mixture, the two ResNet networks take over as the best networks. These two together lead to 5 out of 7 best performances. While ResNet-101 has the best results in the individual classifiers (c) and (d), ResNet-50 has the best results in the flat, the overall hierarchical classifier and classifier (e). Classifier (a) best score belongs to DenseNet121 and EfficientNet-B2 holds the best score for classifier (b).

The inclusion of the metadata proved to be beneficial regardless of the combination method or the CNN in use. It performed particularly well in classifier

---

[2] The results for each CNN model can be found in our supplementary material https://github.com/Bia55/Skin_cancer_AI/blob/main/Supplementary_Results.pdf.

(e) and the flat classifier. Despite this, the flat models consistently outperformed their hierarchical model counterpart. While the hierarchical model has a higher accuracy in diagnosing melanocytic lesions (MEL and NV), the flat classifier performs better in non-melanocytic lesions, particularly malignant ones.

Additionally, we created the hierarchical model Combined 4, which besides combining different CNN, it also combines different image-metadata fusion approaches. Despite having a slightly lower $BACC$ score than the multiplication method, it misdiagnosed MEL as a benign lesion less frequently so we considered it the best hierarchical model.

3.4 Comparison of the Mixed Models

Using the mixed models, we tested if it was possible to improve the results by combining the flat and hierarchical models (recall Sect. 2.4). We used the ResNet101 flat classifier from the image feature reducer model with $r = 0.8$ and $m = 200$, and Combined 4 model as the hierarchical classifier. Table 4 presents the results.

Table 4. Comparison of the mixed models with the best hierarchical and flat models. Flat transfers represent the number of cases, in percentage, that the hierarchical model passed to the flat model and the last column represents the $BACC$ of the flat model in the transferred lesions. The cells highlighted in green represent the best result for each column.

Model	BACC (%)	Flat transfers (%)	Flat BACC (%)
Mixed 1	80.76	59.12	84.58
Mixed 2	80.62	49.60	70.40
Mixed 3	80.57	49.10	71.40
Flat	79.08	-	-
Hierarchical	73.53	-	-

The mixed models performed very well with mixed 1 being the best one. They had a 7% improvement from the hierarchical model and outperformed the flat model by 1% to 2%. The model with more transfers from the hierarchical to the flat model was mixed 1. However, this model has a 15% higher chance of a lesion being correctly diagnosed when sent to the flat classifier. The results of mixed 2 and 3 were nearly identical. Thus, the third mixed method becomes redundant.

We also observed that the mixed 1 model diagnosed malignant lesions as malignant more often than benign. Hence, even if the diagnosis is incorrect, the model determines that it is a detrimental lesion. On the other hand, the false positives of the malignant lesions are not so concerning, as they could be further analyzed by a pathologist and end up being correctly diagnosed by them.

3.5   Elimination of Classifiers (d) and (e)

Throughout the hierarchical model experiments we observed that classifier (c) confuses malignant with benign lesions very often. We decided to investigate what would happen if all the non-melanocytic lesions were to be diagnosed in classifier (c), thereby eliminating classifiers (d) and (e). Three different architectures were tested for the new classifier (c). They were chosen due to each one being the best model for the previous individual classifiers (c), (d) and (e). The final score uses the same classifiers (a) and (b) as the model Combined 4.

The new classifier (c) with the best performance belongs to ResNet50 using the multiplication method with a ReLU activation, which is the same model as the best previous classifier (c). Its final score is extremely similar to Combined 4, less than 1% of difference. This modified hierarchy only altered the performance of the AK class with a 6% improvement and BKL with a decrease of the same order.

We also tested the modified hierarchy with the previous best model, mixed 1. It reached a performance of 80.45%, just 0.3% less than the original hierarchy. Overall, the number of malignant lesions diagnosed as malignant increased and significant changes can be seen in MEL, where this lesion was 4% worse and greatly augmented the chances of being diagnosed as a benign lesion, especially BKL and NV. These results show that the division of malignant and benign non-melanocytic lesions does not have the impact previously thought and may not be needed. Although more testing would be necessary to know for certain, these findings cause one to reconsider the previously defined hierarchy.

3.6   Evaluation on Held-out Test Set

As mentioned previously, ISIC Challenge provides a test dataset with 8,238 images with no ground truth. The evaluation of the best models established previously was performed in the Challenge online platform [2]. Tables 5 and 6 present the results using the test and validation datasets, respectively.

Overall, the models performed in a similar way in both datasets regarding their order from best to worst, that is, the best model is still mixed 1, followed by the flat and then the hierarchical models. However, in the test set they suffered a decrease of roughly 25% in their final $BACC$ scores. The individual performance of the lesions decreased sharply in every lesion except NV, MEL and BCC for all the different models.

NV and BCC were the lesions with the best performance in the test set as opposed to the validation set. MEL closely followed them having the third best perfomance. The other lesions suffered very drastic downgrades with performances below 50%. While SCC remained one of the hardest lesions to diagnose, VASC went from being the best in the validation set to one of the worst in the test set.

In the test set, the original hierarchy was better at diagnosing malignant lesions and the modified hierarchy was better at diagnosing benign lesions. While in the validation set there is no clear best mixed 1 model, in the test set it is

Table 5. SE scores of each lesion for the best hierarchical and flat image and metadata models using the test dataset. Cells highlighted in green represent the best result in each column.

	MEL	NV	BCC	AK	BKL	DF	VASC	SCC	BACC
Mixed 1	65.60	79.20	76.70	47.10	45.40	48.90	49.50	38.90	56.41
Mixed 1 with new Hier	66.10	79.30	75.60	47.60	43.80	54.40	50.50	40.10	57.18
Flat	59.80	77.10	74.90	46.80	42.20	51.10	47.50	32.50	53.99
Hier	65.60	75.30	68.10	38.00	39.70	36.70	31.70	28.70	47.98
New Hier	62.60	78.20	63.20	38.20	37.20	42.20	39.60	26.80	48.50

Table 6. SE scores of each lesion for the best hierarchical and flat image and metadata models using the validation dataset. Cells highlighted in green represent the best result in each column.

	MEL	NV	BCC	AK	BKL	DF	VASC	SCC	BACC
Mixed	78.67	87.77	85.86	70.11	77.52	83.33	92.16	70.63	80.76
Mixed 1 with new Hier	74.25	89.20	85.86	72.99	75.62	83.33	94.12	68.25	80.45
Flat	73.59	84.82	86.17	71.84	74.10	83.33	92.16	66.67	79.08
Hier	77.90	86.83	81.65	59.20	70.10	66.67	82.35	65.87	73.82
New Hier	77.90	86.83	78.95	65.52	65.33	66.67	82.35	64.23	73.53

a different story. Here, the majority of the lesions do better with the modified hierarchy in the mixed 1 model.

4   Conclusion

This work addressed the shortage of the current literature on hierarchical models and metadata. To achieve this, we used two types of models, hierarchical and flat. Each model was tested using only metadata, only image, and image with metadata. Furthermore, we developed three models that combine hierarchical and flat models and tested their performances using images and metadata.

The experimental setup consisted in the comparison of five different CNN architectures in the hierarchical and flat models of only images and image and metadata. Our results show that the inclusion of metadata has a deep positive impact either in flat or hierarchical models as it always improves their performances. However, it is not sufficient on its own. Despite hierarchical models performing slightly worse than flat models, they provide a rationale for the decision.

The first mixed model ended up being the model with the best performance, which makes the models compete directly with each other. This shows that the combination of flat and hierarchical models can increase both the individual performance of the lesions and the overall score of the model.

Additionally, we studied the prospect of reducing the original three-level hierarchy to a two-level hierarchy, eliminating the intermediary decision of non-melanocytic lesions between malignant and benign. The modified hierarchy showed that the pruning the original tree can be beneficial to some lesions as it ended up performing better in the test set.

We believe that the findings of this work point towards several research directions, in particular: i) using the hierarchical model to see if it helps with the diagnosis of the unknown category/lesion; and ii) explore a different hierarchical structure, where the first level is benign and malignant lesions and the second melanocytic and non-melanocytic.

References

1. Skin cancer statistics from skin cancer foundation. https://www.skincancer.org/skin-cancer-information/skin-cancer-facts/. Accessed November 2020
2. ISIC Challenge. https://challenge.isic-archive.com/. Accessed November 2020
3. Barata, C., Marques, J. S., Emre Celebi, M.: Deep attention model for the hierarchical diagnosis of skin lesions. In: Proceedings of the IEEE/CVF Conference on Computer Vision and Pattern Recognition (CVPR) Workshops, June 2019
4. Esteva, A., et al.: Dermatologist level classification of skin cancer with deep neural networks. Nature **542**(7639), 115–118 (2017)
5. Barata, C., Marques, J. S.: Deep learning for skin cancer diagnosis with hierarchical architectures. In: 2019 IEEE 16th International Symposium on Biomedical Imaging (ISBI 2019), pp. 841–845 (2019)
6. Kulhalli, R., Savadikar, C., Garware, B.: A hierarchical approach to skin lesion classificationn. ser. CoDS-COMAD 2019, New York, NY, USA: Association for Computing Machinery (2019). https://doi.org/10.1145/3297001.3297033
7. Tschandl, P., Rosendahl, C., Kittler, H.: The HAM10000 dataset, a large collection of multi-source dermatoscopic images of common pigmented skin lesions. Sci. Data **5**, 180161 (2018). https://doi.org/10.1038/sdata.2018.161
8. Codella, N.C.F, et al.: Skin Lesion Analysis Toward Melanoma Detection: A Challenge at the 2017 International Symposium on Biomedical Imaging (ISBI), Hosted by the International Skin Imaging Collaboration (ISIC) (2017). arXiv:1710.05006
9. Combalia, Marc, et al.: BCN20000: Dermoscopic Lesions in the Wild (2019). arXiv:1908.02288
10. Kingma, D. P., Ba, J.: Adam: A method for stochastic optimization. arXiv preprint arXiv:1412.6980 (2014)
11. Li, W., Zhuang, J., Wang, R., Zhang, J., Zheng W. S.: Fusing metadata and dermoscopy images for skin disease diagnosis. In: 17th IEEE International Symposium on Biomedical Imaging (ISBI), pp. 1996–2000. (2020). https://doi.org/10.1109/ISBI45749.2020.9098645
12. Pacheco, A.G., Krohling, R.A.: The impact of patient clinical information on automated skin cancer detection. Comput. Biol. Med. **116**, 103545 (2020)
13. He, K., Zhang, X., Ren, S., Sun, J.: Deep residual learning for image recognition. In: Proceedings of the IEEE Conference on Computer Vision and Pattern Recognition (CVPR) (2016)
14. Huang, G., Liu, Z., van der Maaten, L., Weinberger, K. Q.: Densely connected convolutional networks. In: Proceedings of the IEEE Conference on Computer Vision and Pattern Recognition (CVPR) (2017)
15. Tan, M., Le, Q.: EfficientNet: rethinking model scaling for convolutional neural networks. In: Chaudhuri, K., Salakhutdinov, R. (eds.) Proceedings of the 36th International Conference on Machine Learning, ser. Proceedings of Machine Learning Research. PMLR, vol. 97, 09–15, pp. 6105–6114 (2019)

Lesion-Based Chest Radiography Image Retrieval for Explainability in Pathology Detection

João Pedrosa[1,2]✉ ⓘ, Pedro Sousa[3], Joana Silva[4,5], Ana Maria Mendonça[1,2] ⓘ, and Aurélio Campilho[1,2] ⓘ

[1] Institute for Systems and Computer Engineering, Technology and Science (INESC TEC), Porto, Portugal
joao.m.pedrosa@inesctec.pt
[2] Faculty of Engineering of the University of Porto (FEUP), Porto, Portugal
[3] Centro Hospitalar de Vila Nova de Gaia/Espinho, Vila Nova de Gaia, Portugal
[4] Instituto Português de Oncologia do Porto Francisco Gentil, Porto, Portugal
[5] Administração Regional De Saúde Do Norte (ARSN), Porto, Portugal

Abstract. Chest radiography is one of the most common medical imaging modalites. However, chest radiography interpretation is a complex task that requires significant expertise. As such, the development of automatic systems for pathology detection has been proposed in literature, particularly using deep learning. However, these techniques suffer from a lack of explainability, which hinders their adoption in clinical scenarios. One technique commonly used by radiologists to support and explain decisions is to search for cases with similar findings for direct comparison. However, this process is extremely time-consuming and can be prone to confirmation bias. Automatic image retrieval methods have been proposed in literature but typically extract features from the whole image, failing to focus on the lesion in which the radiologist is interested. In order to overcome these issues, a novel framework LXIR for lesion-based image retrieval is proposed in this study, based on a state of the art object detection framework (YOLOv5) for the detection of relevant lesions as well as feature representation of those lesions. It is shown that the proposed method can successfully identify lesions and extract features which accurately describe high-order characteristics of each lesion, allowing to retrieve lesions of the same pathological class. Furthermore, it is show that in comparison to SSIM-based retrieval, a classical perceptual metric, and random retrieval of lesions, the proposed method retrieves the most relevant lesions 81% of times, according to the evaluation of two independent radiologists, in comparison to 42% of times by SSIM.

This work was funded by the ERDF - European Regional Development Fund, through the Programa Operacional Regional do Norte (NORTE 2020) and by National Funds through the FCT - Portuguese Foundation for Science and Technology, I.P. within the scope of the CMU Portugal Program (NORTE-01-0247-FEDER-045905) and UIDB/50014/2020.

Keywords: Chest radiography · Image retrieval · Deep learning · Explainability

1   Introduction

Chest X-ray (CXR), or chest radiography, is one of the most ubiquitous medical imaging modalities globally playing an essential role in the screening, diagnosis and management of disease [1]. However, CXR interpretation is a complex task, requiring years of training. The complexity of this task, associated to the high number of exams that require reporting, makes CXR analysis an extremely time- and resource-consuming task for radiologists [4]. As such, computer-aided diagnosis (CAD) systems for CXR analysis have been proposed. The advent of deep learning has fostered the development of multi-disease approaches through the use of convolutional neural networks (CNNs) trained in multi-label classification scenarios, namely DenseNet [5,7,9], ResNet [10] among others. In spite of the promising results obtained, the lack of explainability of these models is a major hurdle towards their adoption in clinical practice.

While a degree of explainability can be obtained through saliency methods such as GradCAM [16], these tools can only provide information regarding the region of the image that led to the decision of the model. Nevertheless, it might not always be self-evident why a particular location was considered to be pathological. One strategy commonly employed by radiologists to support and explain decisions when confronted with difficult or dubious cases is to use evidence from previously analysed CXRs, comparing the current CXR to similar previous patients. However this process implies manually searching for similar cases on public or internal databases. Such a process is naturally time-consuming and inefficient as radiologists have to manually search for similar CXRs and rely on their memory of previous cases. Furthermore, such a process can be prone to confirmation bias, as the search for information is defined by the radiologist and cases that confirm the initial suspicion are more likely to be searched for.

The development of content-based image retrieval tools that automatically search existing databases for images that are similar to the one being analysed have thus been explored in literature. These tools are usually composed of two parts, feature representation (FR) - where relevant features are extracted from the image - and feature indexing - where the distance between the query image and all other images in the feature space is computed to obtain a ranking of the most similar images [11]. Early approaches relied fully on handcrafted features, such as Scale-Invariant Feature Transform (SIFT) [13], among others. Perceptually motivated distance metrics, that aim to measure how similar two images are in a way that matches human judgement, have also received significant attention, with the most widely used metric being Structural Similarity Index (SSIM) [21]. More recently, learned features using deep learning have been proposed, allowing to use large datasets without a need for domain expert knowledge. Although these can achieve excellent performance, a large amount of data annotated by experts is required, which is prohibitively expensive. As such, autoenconder structures [17] or auxilliary supervised tasks, such as classification to the presence of a pathology [6,22], are often used to train deep learning networks, thus automatically generating features. After FR, feature ranking can be approached as a nearest-neighbor search among the generated feature library by pairwise comparison according to a predetermined distance metric such as L^2-Norm [23].

In spite of the promising results shown in literature, current approaches are based on features extracted from the full CXR, thus failing to take into account the regions that are determinant for a particular pathology. While the extraction of features used in image classification for pathology detection will naturally represent the pathologies of interest, they are also bound to represent general image features, which are irrelevant for the retrieval of similar findings. The use of saliency maps has been proposed as an attention mechanism for the FR, focussing the generated features on the pathological characteristics [18]. While this improves FR and leads to the retrieval of more relevant CXRs, it does not allow radiologists to query cases based on specific lesions. In fact, the extraction of features from the whole CXR disregards the possibility of multiple lesions, e.g. if a CXR presents multiple nodules, current approaches do not allow to retrieve examples similar to one of those nodules in specific, since the extracted FR represents all occurrences of nodules that led to the model's decision.

In order to overcome these issues, the aim of this work was to design and validate a framework for CXR image retrieval based on unique lesions, rather than on full image FRs. In this work, an object detection framework is thus used to simultaneously identify abnormal regions - that thus require FR - and perform the FR itself, by extracting the features of each predicted object at the corresponding grid location. This in turn allows to efficiently retrieve CXRs based on a specific lesion query, making it possible to retrieve CXRs that present lesions similar to the query lesion, independent of other lesions/aspects of the image, thus automatically retrieving more relevant examples for radiologists.

2 Methods

2.1 Dataset

The CXRs used are from the VinDr-CXR dataset [14], which consists of 18.000 postero-anterior CXRs, of which 15.000 were manually annotated by three radiologists. Annotations include the location of each finding through bounding boxes, where each bounding box is assigned a pathology class from 14 possible labels: Aortic Enlargement (AoE), Atelectasis (Atl), Calcification (Clc), Cardiomegalym (Cmg), Consolidation (Cns), Infiltration (Inf), Interstitial Lung Disease (ILD), Lung Opacity (LOp), Nodule/Mass (Nod), Other Lesion (OtL), Pleural Effusion (PlE), Pleural Thickening (PlT), Pneumothorax (Pnm) and Pulmonary Fibrosis (PuF).

2.2   CXR Pathology Object Detection

Detection of pathologies in CXR was performed using a YOLOv5x object detection architecture [8], an extension of previous YOLO architectures (Fig. 1). The YOLOv5 is composed of a CSPDarkNet backbone, which incorporates cross stage partial network (CSPNet) [19] in DarkNet [15], followed by a path aggregation network (PANet) [20]. At the output level, YOLOv5 integrates multi-scale predictions at three different sizes.

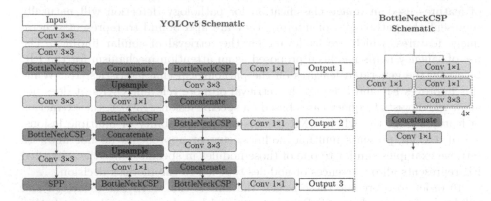

Fig. 1. Overview of the YOLOv5x architecture and the BottleNeckCSP module. *Conv* 1×1 and *Conv* 3×3 indicate convolutional layers with 1×1 and 3×3 kernels respectively; *BottleNeckCSP* indicates a cross-stage partial network bottleneck [19]; *SPP* indicates a spatial pyramid pooling layer [8]; *Upsample* indicates a spatial upsampling operation with a factor of 2; and *Concatenate* indicates a concatenation of inputs across channels. The dashed box indicates that the operation is repeated 4 times.

While 14 pathology classes are available in VinDr-CXR, some pathologies have low representation and present significant variability between radiologists in the identification of the nature of the pathology, i.e. regions that are identified by multiple radiologists as abnormal but receive different pathology labels. As such, some pathologies were grouped in terms of their anatomical region and/or appearance, resulting in 6 pathology group classes: **AoE**: AoE; **Cmg**: Cmg; **OtL**: OtL; **Parenchymal Lesion (PaL)**: Atl, Clc, Cns, ILD, Inf, LOp, Nod, PuF; **Pleural Lesion (PlL)**: PlE, PlT; **Pnm**: Pnm.

2.3   Lesion-Based CXR Image Retrieval

Figure 2 shows the workflow of the proposed lesion-based CXR image retrieval (LXIR) framework. In brief, a YOLOv5x CXR pathology object detection network is first trained on annotated CXRs. The trained YOLOv5x is then used to predict lesions in a library set of CXRs and from each predicted lesion, a FR is extracted, generating a lesion library of FRs. When a new test CXR (absent from the library set) is processed by YOLOv5x, any of the predicted lesions can be selected. An FR is then extracted for the selected lesion and the distance between the query FR and each of the FRs on the library is computed. This distance can then be used to rank the FR library and present to the user the CXR(s) and corresponding lesion(s) most similar to the query lesion.

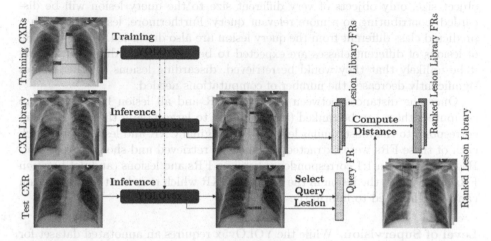

Fig. 2. Overview of the proposed lesion-based CXR image retrieval workflow.

Feature Representation. One of the basic concepts of YOLO is that an input image is divided into an $M \times M$ grid and bounding boxes are predicted at each of these grid positions. This means that a unique FR will be generated for each grid position in the last convolutional layers, which can be seen as a semantic description of the characteristics of the predicted lesion. As such, after training the YOLOv5x network, any CXR image can be processed to predict lesions and an FR for each of these lesions can be extracted at the corresponding grid positions. The FR for each bounding box was extracted from the features before and after the last 1×1 convolutional layer of the BottleNeckCSP's of each of the output branches. The last convolutional layers were chosen as these are known to have higher-order semantic features, enabling an efficient FR. Note that a unique FR will be extracted from each lesion at the corresponding grid position, meaning that multiple FRs can be extracted from a single CXR, each of them describing an independent lesion. Given the architecture of YOLOv5x, the number of features extracted for a given bounding box will depend on the output branch that has generated it. Specifically, 640, 1280 and 2560 features will be obtained from output branches 1, 2 and 3 respectively. For each output branch, FRs were normalized to zero mean and standard deviation 1.

Feature Indexing. Once a lesion library of FRs has been established and a predicted lesion is selected from a test CXR, the feature indexing is done by computing the L^2-Norm between the query FR and each of the lesion library FRs. Note that due to the different FR sizes of the YOLOv5x output branches, distances can only be computed between FRs of the same output branch. However, because the YOLOv5x output branch is closely correlated to the predicted object size, only objects of very different size to the query lesion will be discarded, contributing to a more relevant query. Furthermore, lesion FRs with a predicted class different from the query lesion are also discarded. While the FRs of lesions of different classes are expected to be significantly different, making it be unlikely that they would be retrieved, discarding lesions of other classes significantly decreases the number of computations needed.

Once the distances between the query FR and all lesion library FRs are computed, these can be ranked (from smallest to largest) and the first FRs will correspond to the most similar lesions. The original CXR and lesion from which each of these FRs were extracted can then be retrieved and shown to the user. Because there is a 1:1 correspondence between FRs and lesions (and not between FRs and CXRs), the lesion in the retrieved CXR which matches the query lesion can be highlighted to inform the user.

Level of Supervision. While the YOLOv5x requires an annotated dataset for training, this is not the case for the retrieval process itself. Because YOLOv5x automatically identifies lesions and extracts the corresponding FRs, the construction of the lesion FR library and retrieval can be performed on an unannotated dataset in an unsupervised manner. Additional information such as medical reports could then be consulted to infer on the dubious lesion under study.

In spite of the excellent performance of YOLOv5x, under such an unsupervised retrieval, there is no guarantee that the retrieved lesion is a true lesion. To prevent the retrieval of false positive lesions, a supervised retrieval can also be considered. Starting from an annotated CXR dataset such as VinDR-CXR, the predicted lesions are compared to the lesions annotated by the radiologists and only predicted lesions that match annotated lesions are included in the lesion library. The matching criterion considered was an intersection over union (IoU) ≥ 0.4, which is a criterion and value commonly used in CXR lesion detection [14]. Because only annotated lesions can thus be retrieved, a supervised retrieval provides additional security when studying lesions, which can be important for example in an independent learning scenario. However, there are significant advantages to an unsupervised retrieval. Most importantly, the number of annotated CXRs currently available is extremely small when compared to the available public datasets, or even to any hospital picture archiving system, which can be extremely important when querying rare lesions.

2.4 Structural Similarity Lesion-Based CXR Image Retrieval

For the purpose of comparison to LXIR, a lesion-based image retrieval based on SSIM was also used [21]. In order to avoid the computational expense of retrieving a high number of image patches from all CXRs, the bounding boxes annotated by the radiologists on VinDr-CXR were used to extract lesion image patches. Image retrieval was then performed by computing the SSIM between the query lesion image patch and each of the library lesion image patches. The library lesions can then be ranked (from highest to lowest SSIM) to obtain the most similar lesions to the query lesion. Because this method is based on the radiologist annotations, only annotated CXRs can be used.

3   Experiments

3.1   CXR Pathology Object Detection

For training of the YOLOv5x, the VinDr-CXR data was divided into train, validation and test sets in a 60-20-20 ratio. The division was performed through random selection of images, where the ratio of cases of each pathology was maintained as much as possible. Because patient identification is not available in VinDr-CXR, that information could not be taken into account during this division. The YOLOv5x network was initialized using weights pretrained on the COCO dataset [12] and training was performed using a stochastic gradient descent optimizer with an initial learning rate of 0.01 for a duration of 150 epochs.

3.2   Lesion-Based CXR Image Retrieval

In order to construct the lesion FR library, the YOLOv5x was used to predict bounding boxes on the train and validation sets. In order to maximize the diversity in the lesion FR library, all lesions with a confidence score above 0.1 were considered and no post-processing was applied to the predicted bounding boxes.

For all experiments, queries were performed with the test set. To simulate queries on real-world inference situations, the optimal operation point for each class as determined through the validation set was obtained and YOLOv5x predictions with a confidence score below this level were discarded. Non-maximum suppression (NMS) was then applied for each CXR. In brief, predicted boxes were compared and, for pairs of boxes with an IoU ≥ 0.6, the predicted box with lower confidence score was removed. In this way, overlapping predictions are removed so that only lesions of high confidence were considered. The IoU value used during NMS was chosen empirically by comparing the output predictions of random CXRs. The number of lesions considered to construct the library and test set are described in Table 1.

Table 1. Number of lesions annotated by radiologists on the whole dataset (VinDr-CXR) compared to the lesions predicted by YOLOv5x on the train and validation (Library) and the test set (Test).

Group	AoE	Cmg	OtL	PaL								PlL		Pnm
Class				Atl	Clc	Cns	ILD	Inf	LOp	Nod	PuF	PlE	PlT	
VinDr-CXR	7,162	5,427	2,203	\multicolumn{8}{c}{11,525}								7,318		226
				279	960	556	1,000	1,247	2,483	2,580	4,655	2,476	4,842	
Library	107,842	39,666	24,147	155,150								91,377		3,013
Test	886	495	252	2,476								1,695		41

3.3 Quantitative Evaluation

In order to quantitatively evaluate the performance of the proposed LXIR framework, two experiments were performed. In the first experiment, using unsupervised retrieval, the top-5 most similar lesions for each predicted lesion in the test set were retrieved using LXIR. The retrieved lesions were then compared with the ground truth annotations to compute the ratio of top-5 retrieved lesions that match an annotated lesion of the same pathology class group. This ratio is hereinafter referred to as retrieval precision (RP_5).

In the second experiment, using supervised retrieval, the top-5 most similar lesions for each predicted lesion in the test set were retrieved. Because supervised retrieval was used, all retrieved lesions match annotated bounding boxes for which the pathology classes is known. As such, the pathology class of the query lesion is compared to the pathology class of the retrieved lesions to compute the ratio of top-5 retrieved lesions that match at least one annotated lesion of the same pathology class. This measure is hereinafter referred to as class consistency (CC_5). Note that pathology class refers to the 14 pathology classes and not to the 6 pathology groups. In this way, it is tested if for example, when a Nod lesion is queried within the pool of PaL's, how often a Nod is retrieved by LXIR. In order to establish a baseline for comparison, this experiment was also performed using the SSIM retrieval (Sect. 2.4) as well as a random retrieval of lesions. Note that CC_5 can only be computed for PaL and PlL pathology groups as other groups are comprised of a single pathology class.

For all experiments, a match between two lesions was considered for IoU $\geq$ 0.4.

3.4 Qualitative Evaluation

In order to qualitatively evaluate the relevance of the retrieved examples for specific query lesions, two radiologists were asked to independently rate the relevance of the retrieved examples for 50 random PaL's predicted by the YOLOv5x from the test set. For each query lesion, the top lesion retrieved according to LXIR, SSIM and random retrieval were obtained. Given that an absolute classification of each retrieved lesion into relevant/not relevant can be subjective, a comparison-based approach was used. Each expert was shown the query lesion and the retrieved lesions and asked to rank the retrieved lesions from the most similar to the least similar. Experts were blinded to the method used for the retrieval of each lesion. A standard competition ranking was used so that equally similar/dissimilar lesions received the same rank and a gap was left in the rank following it (e.g. two retrieved lesions equally similar to the query lesion followed by a less similar lesion were ranked as 1-1-3).

4   Results

Figure 3 shows the results obtained in the quantitative evaluation. It can be seen that a high RP_5 is obtained for all pathology classes. In terms of CC_5, random retrieval generates, as expected, the worst results while LXIR outperforms SSIM in nearly every pathology class. Table 2 shows the results of the qualitative evaluation. It can be seen that, for the majority of query lesions (81%), LXIR was ranked as the best retrieval method, significantly outperforming SSIM (42%). Figure 4 shows examples of query and retrieved PaL's for each of the methods.

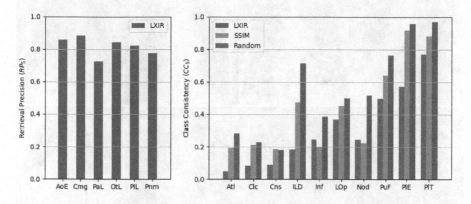

Fig. 3. RP_5 and CC_5 for each pathology class/group.

Table 2. Number of lesions ranked as best (1st), second best (2nd) or third best (3rd) for each method and average position in the ranking. Number of lesions shown as $R_1/R_2(P)$ where R_1 and R_2 is the number of lesions by each radiologist and P is the average percentage of lesions.

Position	LXIR	SSIM	Random
1st	43/38 (81%)	22/20 (42%)	15/10 (25%)
2nd	5/9 (14%)	17/22 (39%)	16/20 (36%)
3rd	2/3 (5%)	11/8 (19%)	19/20 (39%)
Average position	1.24	1.77	2.14

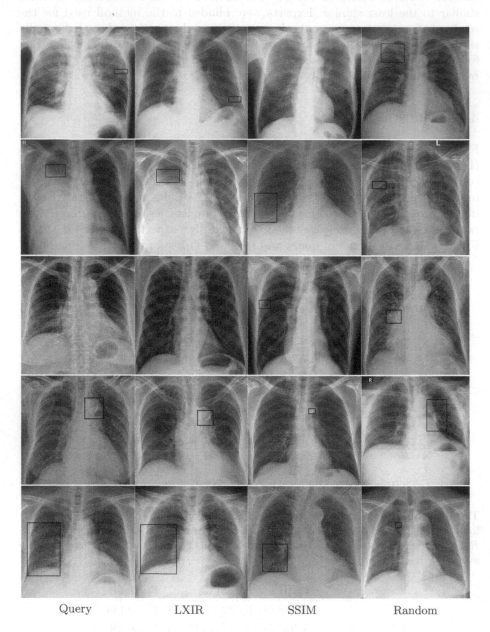

Query   LXIR   SSIM   Random

Fig. 4. Examples of query and retrieved PaL's for each of the evaluated methods. The rankings given by radiologists to the LXIR, SSIM and Random methods were respectively, 1-2-2 for rows 1–3, 1-1-3 for row 4 and 2-1-3 for row 5.

5   Discussion

In this work, a framework for lesion-based CXR image retrieval is proposed, allowing to efficiently and in an unsupervised manner extract relevant patches from CXR images and provide an adequate FR for each independent lesion, thus allowing to retrieve similar examples to an individual query lesion independent of other CXR features or findings.

Regarding the quantitative evaluation, a high RP_5 (above 0.7) was obtained for all pathology groups, meaning that a high percentage of all unsupervised retrieved lesions match an annotated object and could thus be a relevant lesion. The lowest RP_5 was obtained for the PaL group, which is probably due to the fact that it the most diverse group covering multiple findings and also because it includes smaller findings which may have been missed or underannotated in VinDr-CXR. Regarding CC_5, the superior performance of LXIR in retrieving relevant examples is clear, outperforming SSIM in almost all pathology classes. It can be seen that CC_5 within a pathology group is closely related to the prevalence of each pathology class in the dataset, such as for Atl, Clc and Cns. It is important to note that while CC_5 can give a comparative measure of performance, little meaning can be derived from the absolute values of CC_5 as they are a reflection of many factors. Importantly, class variability in the annotations can play a large role as lesions often have multiple labels. Nod findings, for example, are much more prevalent than Clc findings and, while some calcified Nod findings receive both Nod and Clc findings, this is not the case for the majority of calcified Nod findings. As such, a query for a calcified nodule might easily retrieve lesions which do not have an associated Clc label, lowering the CC_5, when in fact the retrieved lesion might be deemed relevant upon inspection.

Regarding the qualitative evaluation, Table 2 clearly shows the superiority of LXIR. The examples shown in rows 1–4 of Fig. 4 highlight the performance of LXIR in retrieving examples of similar lesions, capturing lesion nature and appearance, as well as size and independently from the general appearance and other features of the query and retrieved CXRs. It should be noted that the features extracted from YOLOv5x and used for the FR were trained to predict not only lesion class and confidence score but also lesion dimensions. As such, each lesion FR captures information on lesion appearance but also on its relative size, which leads to more relevant retrievals: a similar lesion of similar size is more relevant than a similar lesion of a different size. Furthermore, while the FR is trained to capture information on the lesion nature and size, it will also extract information regarding other image features at that grid position, such as anatomical cues (presence of ribs, clavicles, proximity to the edge of the lung, etc.), which can improve the relevance of retrieved lesions. For example, in rows 3 of Fig. 4 not only is the retrieved lesion similar in appearance and size, but the relative anatomical position - near the edge of the lung field - is also similar. For SSIM on the other hand, a lower percentage of lesions were ranked as the most relevant and that percentage decreases even further for random retrieval. While some of the SSIM retrieved lesions are relevant, the fact that SSIM is a purely perceptual metric means that lesion-based features are not explicitly extracted, which is detrimental to performance. Anatomical cues unrelated to the lesion itself are often used instead of lesion-specific features, particularly for small or tenuous lesions as in rows 1 and 3 of Fig. 4.

In spite of the promising results obtained in this study, there are limitations which should be taken into account in the interpretation of these results and addressed in future work. In particular, the limited data available might have an impact on the results obtained. For one, on the training of YOLOv5x, which would benefit from additional data during training so that more general and relevant features could be obtained, particularly for less represented pathology classes within PaL. Secondly, and most importantly, on the lesion retrieval process itself. By being limited to the train/validation sets of VinDr-CXR for retrieval, it can be hypothesized that a query for a particular rare lesion might not lead to any relevant retrievals, merely due to the fact that a similar lesion is not found in this population. While this problem is attenuated by performing a comparative evaluation (which method retrieves the most similar example) rather than an absolute evaluation (is the retrieved lesion relevant), expanding the dataset to have a larger library might prove beneficial for lesion retrieval. Finally, while the entanglement of lesion appearance, size and anatomical cues was beneficial for LXIR performance, it would be a major contribution to enable the disentanglement of lesion FRs so that a guided retrieval, based on criteria defined by the radiologist, becomes possible. When querying a lesion, a radiologist might deem more relevant, for example, the lesion appearance rather than size itself, and separating these features within the LXIR framework is currently not possible. Furthermore, a disentanglement of lesion appearance features into specific radiologist-defined interpretable features would allow for a fully guided lesion retrieval, which would be extremely relevant for radiologists.

6   Conclusion

In conclusion, a novel framework for lesion-based CXR image retrieval (LXIR) based on a state of the art object detection is proposed, allowing to efficiently detect lesion regions, extract relevant features and use these to retrieve similar examples from a library of CXRs. It is shown through quantitative and qualitative validation with radiologists that the proposed framework outperforms classical perceptual methods in the retrieval of relevant lesions. As such, it could play a role in improving the explainability of 2nd opinion automatic systems as well as teaching and case discussion scenarios.

References

1. Bansal, G.: Digital radiography. A comparison with modern conventional imaging. Postgrad. Med. J. **82**(969), 425–428 (2006)
2. Bay, H., Ess, A., Tuytelaars, T., Van Gool, L.: Speeded-up robust features (SURF). Comput. Vis. Image Underst. **110**(3), 346–359 (2008)

3. Chen, B., Li, J., Guo, X., Lu, G.: DualCheXNet: dual asymmetric feature learning for thoracic disease classification in chest X-rays. Biomed. Signal Process. Control **53**, 101554 (2019)
4. Cowan, I.A., MacDonald, S.L., Floyd, R.A.: Measuring and managing radiologist workload: measuring radiologist reporting times using data from a Radiology Information System. J. Med. Imaging Radiat. Oncol. **57**(5), 558–566 (2013)
5. Gündel, S., Grbic, S., Georgescu, B., Liu, S., Maier, A., Comaniciu, D.: Learning to recognize abnormalities in chest X-rays with location-aware dense networks. In: Vera-Rodriguez, R., Fierrez, J., Morales, A. (eds.) CIARP 2018. LNCS, vol. 11401, pp. 757–765. Springer, Cham (2019). https://doi.org/10.1007/978-3-030-13469-3_88
6. Hofmanninger, J., Langs, G.: Mapping visual features to semantic profiles for retrieval in medical imaging. In: Proceedings of the IEEE Conference on Computer Vision and Pattern Recognition, pp. 457–465 (2015)
7. Irvin, J., et al.: CheXpert: a large chest radiograph dataset with uncertainty labels and expert comparison. In: Proceedings of the AAAI Conference on Artificial Intelligence, vol. 33, pp. 590–597 (2019)
8. Jocher, G., et al.: Ultralytics/YOLOv5 (2021). https://github.com/ultralytics/yolov5
9. Kumar, P., Grewal, M., Srivastava, M.M.: Boosted cascaded convnets for multilabel classification of thoracic diseases in chest radiographs. In: Campilho, A., Karray, F., ter Haar Romeny, B. (eds.) ICIAR 2018. LNCS, vol. 10882, pp. 546–552. Springer, Cham (2018). https://doi.org/10.1007/978-3-319-93000-8_62
10. Li, Z., et al.: Thoracic disease identification and localization with limited supervision. In: Proceedings of the IEEE Conference on Computer Vision and Pattern Recognition, pp. 8290–8299 (2018)
11. Li, Z., Zhang, X., Müller, H., Zhang, S.: Large-scale retrieval for medical image analytics: a comprehensive review. Med. Image Anal. **43**, 66–84 (2018)
12. Lin, T.-Y., et al.: Microsoft COCO: common objects in context. In: Fleet, D., Pajdla, T., Schiele, B., Tuytelaars, T. (eds.) ECCV 2014. LNCS, vol. 8693, pp. 740–755. Springer, Cham (2014). https://doi.org/10.1007/978-3-319-10602-1_48
13. Lowe, D.G.: Distinctive image features from scale-invariant keypoints. Int. J. Comput. Vision **60**(2), 91–110 (2004)
14. Nguyen, H.Q., et al.: VinDr-CXR: An open dataset of chest X-rays with radiologist's annotations. arXiv preprint arXiv:2012.15029 (2020)
15. Redmon, J., Farhadi, A.: Yolov3: an incremental improvement. arXiv preprint arXiv:1804.02767 (2018)
16. Selvaraju, R.R., Cogswell, M., Das, A., Vedantam, R., Parikh, D., Batra, D.: Grad-CAM: visual explanations from deep networks via gradient-based localization. In: Proceedings of the IEEE International Conference on Computer Vision, pp. 618–626 (2017)
17. Shin, H.C., Orton, M.R., Collins, D.J., Doran, S.J., Leach, M.O.: Stacked autoencoders for unsupervised feature learning and multiple organ detection in a pilot study using 4d patient data. IEEE Trans. Pattern Anal. Mach. Intell. **35**(8), 1930–1943 (2012)
18. Silva, W., Poellinger, A., Cardoso, J.S., Reyes, M.: Interpretability-guided content-based medical image retrieval. In: Martel, A.L., et al. (eds.) MICCAI 2020. LNCS, vol. 12261, pp. 305–314. Springer, Cham (2020). https://doi.org/10.1007/978-3-030-59710-8_30

19. Wang, C.Y., Liao, H.Y.M., Wu, Y.H., Chen, P.Y., Hsieh, J.W., Yeh, I.H.: CSPNet: a new backbone that can enhance learning capability of CNN. In: Proceedings of the IEEE/CVF Conference on Computer Vision and Pattern Recognition Workshops, pp. 390–391 (2020)

20. Wang, K., Liew, J.H., Zou, Y., Zhou, D., Feng, J.: PANet: few-shot image semantic segmentation with prototype alignment. In: Proceedings of the IEEE/CVF International Conference on Computer Vision, pp. 9197–9206 (2019)

21. Wang, Z., Bovik, A.C., Sheikh, H.R., Simoncelli, E.P.: Image quality assessment: from error visibility to structural similarity. IEEE Trans. Image Process. **13**(4), 600–612 (2004)

22. Wolterink, J.M., Leiner, T., Viergever, M.A., Išgum, I.: Automatic coronary calcium scoring in cardiac CT angiography using convolutional neural networks. In: Navab, N., Hornegger, J., Wells, W.M., Frangi, A.F. (eds.) MICCAI 2015. LNCS, vol. 9349, pp. 589–596. Springer, Cham (2015). https://doi.org/10.1007/978-3-319-24553-9_72

23. Zhang, R., Isola, P., Efros, A.A., Shechtman, E., Wang, O.: The unreasonable effectiveness of deep features as a perceptual metric. In: Proceedings of the IEEE Conference on Computer Vision and Pattern Recognition, pp. 586–595 (2018)

Deep Learning for Diagnosis of Alzheimer's Disease with FDG-PET Neuroimaging

José Bastos, Filipe Silva(iD), and Petia Georgieva(✉)(iD)

University of Aveiro, Aveiro, Portugal
{bastosjose,fmsilva,petia}@ua.pt
http://www.ua.pt

Abstract. Alzheimer's Disease (AD) imposes a heavy burden on health services both due to the large number of people affected as well as the high costs of medical care. Recent research efforts have been dedicated to the development of computational tools to support medical doctors in the early diagnosis of AD. This paper is focused into studying the capacity of Deep Learning (DL) techniques to automatically identify AD based on PET neuroimaging. PET images of the cerebral metabolism of glucose with fluorodeoxyglucose (^{18}F-FGD) were obtained from the Alzheimer's Disease Neuroimaging Initiative (ADNI) database. Two DL approaches are compared: a 2D Inception V3 pre-trained model and a custom end-to-end trained 3D CNN to take advantage of the spatial patterns of the full FDG-PET volumes. The results achieved demonstrate that the PET imaging modality is suitable indeed to detect early symptoms of AD. Further to that, the carefully tuned custom 3D CNN model brings computational advantages, while keeping the same discrimination capacity as the exhaustively pre-trained 2D Inception V3 model.

Keywords: Alzheimer's disease · FDG-PET neuroimaging · Convolutional Neural Networks · ADNI dataset

1 Introduction

Neurodegenerative diseases are a spectrum of brain disorders that cause a progressive loss of neurological function and structure, such as Alzheimer's disease (AD), Parkinson's disease (PD), Huntington's disease (HD) and amyotrophic lateral sclerosis (ALS). Amongst them, AD is documented as the most common cause of dementia worldwide (responsible for 60 to 80% of cases), affecting roughly 30% of people over the age of 85 [1]. Dementia refers to a set of symptoms marked by decline in memory, reasoning or other cognitive functions. Nowadays, there is a broad consensus that AD appears decades before its first manifestation.

This research work is funded by National Funds through the FCT - Foundation for Science and Technology, in the context of the project UIDB/00127/2020.

© Springer Nature Switzerland AG 2022
A. J. Pinho et al. (Eds.): IbPRIA 2022, LNCS 13256, pp. 95–107, 2022.
https://doi.org/10.1007/978-3-031-04881-4_8

Apart from the search for a cure, the most recent efforts are aimed at developing computational tools to support the medical decision. Over the last few years, deep learning (DL) - based methods have made important contributions in medical imaging. They proved to be a valuable technology to assist the preventive healthcare with computerized diagnosis. In this context, Magnetic Resonance Imaging (MRI) and Positron Emission Tomography (PET) are the most common neuroimaging modalities useful for the AD diagnosis.

Several studies have highlighted the importance of DL-based diagnostic systems using either MRI or PET scans [10]. Others address the integration of multimodal information, such as PET and T1-weighted MRI images [12]. Recently, ^{18}F-FDG PET revealed to have a potential to assess the risk of AD at a very early stage [9]. PET images of the cerebral metabolism of glucose with ^{18}F-FDG provide representations of neuronal activity closely linked to the initial manifestations of AD [20].

These recent findings motivated the present work aiming to explore the potential of DL techniques in the diagnosis of AD with ^{18}F-FDG PET images. The study focuses on how to leverage convolutional neural networks (CNNs) for classifying healthy versus AD patients, with a limited dataset collected from the ADNI. For this purpose, two CNN models are compared in terms of predictive performance. The first CNN model explores transfer learning as a promising solution to the data challenge using a pre-trained model. The second model involves a custom developed 3D-CNN to take advantage of spatial patterns on the full PET volumes by using 3D filters and 3D pooling layers.

The rest of the paper is organised as follows. Section 2 reviews related works. Section 3 explains the proposed CNN framework. The results using the ADNI dataset are presented and discussed in Sect. 4. Section 5 summarises the work.

2 Related Work

The current diagnosis of AD relies on neuropsychological tests and neuroimaging biomarkers. The diagnostics can be performed in an early stage, even in the prodromal stage of the disease also referred to as mild cognitive impairment (MCI). The biomarkers for early AD diagnosis that are currently in use reflect the deposition of amyloid (CSF $A\beta$1-42 or PET with amyloid ligands), formation of neurofibrillary tangles (CSF P-tau), neuronal degeneration (CSF T-tau), changes in brain metabolism (FDG-PET), as well as volumetric changes in brain structures that cause the disease's symptoms, such as the hippocampus.

PET is an imaging modality that involves the application of a radioactive substance, called a radioactive tracer, into the body and the posterior observation of the emitted radiation in the organ or tissue being examined. Fluorine-18 is radioactive tracer commonly attached to compounds like glucose, as is the case with ^{18}F-FDG, for the measurement of brain metabolism.

Decreased brain glucose consumption, known as hypometabolism, is seen as one of the earliest signs of neural degeneration, being associated with AD progression. FDG-PET represents a valuable and unique tool able to estimate local cerebral rate of glucose consumption. Thereby, PET may point out biochemical changes that underlie the onset of a disease before anatomical changes can be detected by other modalities such as CT or MRI.

Table 1. CNN applications in brain medical imaging.

Task	Modality	Reference
Tissue necrosis after CVA prediction	MRI	Stier et al. [26]
PD identification	SPECT	Choi et al. [6]
Brain tumor segmentation	MRI	Havaei et al. [11]
Brain lesion segmentation	MRI	Kamnitsas et al. [13]
Brain age prediction	MRI	Cole et al. [8]

Note: CVA = Cerebrovascular Accident

MRI = Magnetic Resonance Imaging, PD = Parkinson's Disease

SPECT = Single-Photon Emission Computed Tomography

CNNs are an important tool in medical imaging [5] and disease diagnostics, as shortly summarised in Table 1. A systematic review of deep learning techniques for the automatic detection of AD can be found in [10]. Authors emphasize important aspects to understand the whole scenario of AD diagnosis. First, approximately 73% of neuroimaging studies have been performed with single-modality data, around 83% of the studies are based on MRI, 9% refer to fMRI and only 8% to PET scans.

Second, a significant part of the studies, summarised in Table 2, address the binary classification problem, i.e., consider normal cognitive (NC) state against AD. A more challenging task is to discriminate between early and late stages of mild cognitive impairment (MCI). MCI is sometimes subdivided into sMCI (Stable Mild Cognitive Impairment) and pMCI (Progressive Mild Cognitive Impairment) which will eventually develop into AD.

Third, a common approach is to convert the volumetric data into a 2D image to be applied at the input of a 2D-CNN. Most of the studies transfer the weights from pre-trained networks on the ImageNet database to the target medical task. This process, known as transfer learning, speeds up training and reduces costs by leveraging previous knowledge.

Although the application of DL techniques in AD diagnosis is still in their initial stage, recent works [9,14] demonstrate that deep neural networks can outperform radiologist abilities. The coming years may determine the feasibility of these models as a support tool to help clinicians reach an appropriate decision in real clinical environments.

3 Methodology

As illustrated in Table 2, MRI is the most frequently used imaging modality as well as a 2D CNN as the discrimination model. In contrast, in this work we aim to leverage the full information by exploring 3D CNN for learning representations from the less explored FDG-PET data. For that purpose, a comparative study will be carried out centred on two CNN models: fine tuning of a pre-trained 2D-CNN model against an end-to-end trained from scratch custom 3D-CNN.

Table 2. CNN application for diagnosis of Alzheimer's Disease (AD).

Modality	Classes	Score	Reference
MRI	MCI vs CN	83%[a]	Qiu et al. [22]
MRI	Multi-class[b]	57%	Valliani et al. [27]
MRI	AD vs CN	91%	Liu et al. [19]
MRI	MCI vs CN	74%[a]	Li et al. [15]
MRI	sMCI vs pMCI	74%	Liu et al. [18]
MRI	sMCI vs pMCI	80%[a]	Lian et al. [16]
MRI	AD vs CN	91%	Aderghal et al. [2]
MRI	AD vs MCI	70%	Aderghal et al. [2]
MRI	MCI vs CN	66%	Aderghal et al. [2]
MRI	sMCI vs pMCI	73%	Lin et al. [17]
MRI	AD vs CN	90%	Bäckström et al. [3]
MRI	AD vs MCI	76%	Senanayake et al. [23]
MRI	MCI vs CN	75%	Senanayake et al. [23]
MRI	sMCI vs pMCI	62%	Shmulev et al. [24]
AV-45 PET	AD vs CN	85%	Punjabi et al. [21]
AV-45 PET + MRI	AD vs CN	92%	Punjabi et al. [21]
AV-45 + FDG PET	sMCI vs pMCI	84%	Choi et al. [7]

Note: CN = Cognitively Normal, AD = Alzheimer's Disease
MCI = Mild Cognitive Impairment, sMCI = Stable MCI, pMCI = Progressive MCI
[a] = Severely Imbalanced Dataset, [b] = AD vs MCI vs CN

3.1 2D Slice-Level CNN Model

The first approach is implemented by the Google 2D Inception V3 model, pre-trained on the ImageNet dataset and fine-tuned with the ADNI dataset. This approach requires a pre-processing step in which the 3D PET volume is converted into a 2D image which is the input to the pre-trained model. Inspired by the work of Ding et al. [9], a 2D collage of a grid of 4 by 4 FDG-PET scan slices were used as the inputs to a deep model, as shown in Fig. 1.

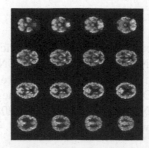

Fig. 1. Collage of 2D slices extracted from volumetric PET scans.

The advantage of this approach is that pre-trained models exist and they can be quickly updated to fit new target data [4]. Further to that, the training data is increased since larger number of 2D slices can be obtained from a single 3D sample.

3.2 3D Subject-Level CNN Model

The second approach is a custom 3D-CNN to take advantage of the spatial patterns of the full PET volumes for each subject. It is referred here as the 3D subject-level CNN model (see Fig. 2). In contrast to the 2D Slice -level model, where the raw PET scans were transformed into 2D patches, here the data is first pre-processed into 3D tensors and then loaded into the network.

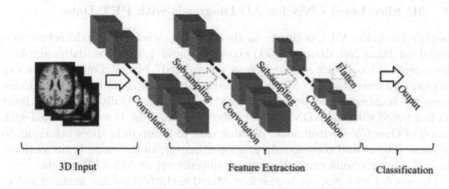

3D Input   Feature Extraction   Classification

Fig. 2. 3D subject-level approach to dealing with volumetric input data [25].

4   Experiments and Results

This section presents the experiments for automatic diagnosis of Alzheimer's Disease (AD) using the ADNI database. The two CNN-based classifiers, introduced in the previous section, attempt to discriminate between Cognitively Normal (CN) and Alzheimer's Disease (AD) classes. The models were trained on a remote server supported by NVIDIA GeForce RTX 2080 Ti graphics cards using the Keras and the Tensorflow environments.

4.1   Dataset

The ADNI dataset consists of FDG-PET scans saved in the NII file format (or NIfTI), typically used for neuroimaging data. NIfTI stands for Neuroimaging Informatics Technology Initiative. The scans have been collected from different machines, with different resolutions, ranging from $128 \times 128 \times 35$ up to $400 \times 400 \times 144$ voxels, with the average resolution around $150 \times 150 \times 70$ voxels. The pixel intensity was normalized into the 0–255 interval and the images were cropped to a certain consistent resolution.

Data consists of 1355 total samples from witch 866 CN samples (63.91%), 489 AD samples (36.09%), 796 male patients (58.75%) and 559 female ones (41.25%). Data was divided into 1250 training samples and 105 testing samples (66 from the CN class and 39 from the AD class). The 1250 training samples were splitted into 10 folds for K-fold Cross Validation, 80 CN samples and 45 AD samples for each fold.

4.2   2D Slice-Level CNN for AD Diagnosis with PET Data

Google's Inception V3 was chosen as the 2D Slice -level CNN architecture pre-trained on ImageNet dataset (1000 classes, around 1.3 million data samples). The inception blocks are also known as the "mixed" blocks. Four variations of Inception V3 were trained - Mixed7, Mixed8, Mixed9, Mixed10, where the index means the number of the inception blocks. Only the last (fully connected) layer was fine tuned with the ADNI dataset. The 2D slices (Fig. 1) were obtained with the aid of OpenCV environment. We took care to group only slices belonging to the same subject and corresponding, approximately, to the same brain sections. The collages have uniformed dimensions, initially set at 512×512 pixels.

The results with respect to the four Mixed architectures are summarized in Fig. 3. The models were trained with SGD (Stochastic Gradient Descent) and the trainable parameters (in the last fully connected layer) were initialized with the pre-trained ImageNet weights. Although all models suffer from overfitting, the Mixed8 model outperforms the other architectures and, therefore, it was selected for further tuning.

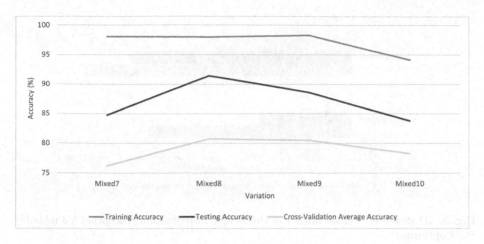

Fig. 3. 2D slice-level models: impact of the Inception V3 architecture (SGD optimizer; initialization with pre-trained ImageNet weights).

The importance of the optimization method and the impact of the parameter initialization were validated for all architectures. Figure 4 and Fig. 5 depict the results only for Mixed8 model. SGD was the most favourable optimizer and set up for the next experiments. Starting from the optimal parameters obtained at the pre-trained stage with the ImageNet reveals to be advantageous compared to random weights initialization.

Though the Mixed8 model reached a promising testing accuracy of 91.43% it still suffers overfitting. This problem was tackled by adding a dropout layer. Figure 6 depicts the classifier performance for a range of dropout rates. Note that 50% dropout rate appears to be a reasonable compromise between the overfitting and the fast convergence of the loss function as shown in Fig. 7. Smaller the dropout rate, faster the convergence, however more prone to overfitting.

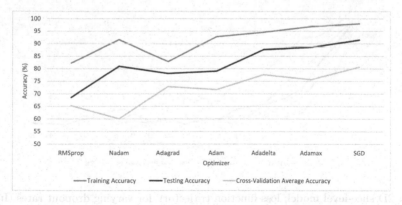

Fig. 4. 2D slice-level model: impact of the optimizer (Inception V3 mixed8; initialization with pre-trained ImageNet weights).

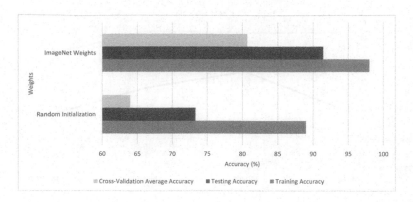

Fig. 5. 2D slice-level model: impact of the weight initialization (Inception V3 mixed8; SGD optimizer).

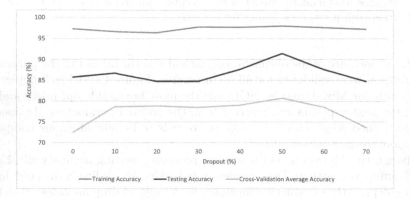

Fig. 6. 2D slice-level model: impact of dropout rates (Inception V3 mixed8; initialization with pre-trained ImageNet weights; SGD optimizer).

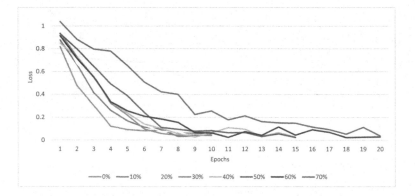

Fig. 7. 2D slice-level model: loss function trajectory for varying dropout rates (Inception V3 mixed8; initialization with pre-trained ImageNet weights; SGD optimizer).

4.3 3D Subject-level CNN for AD Diagnosis with PET Data

The CNN model used for this experiment was end-to-end designed and optimized. Exhaustive search for the optimal architecture is computationally infeasible. Instead, we selected a similar topology to the one proposed in [25]. The base structure consists of two groups of two 3D Conv layers and a 3D max-pooling layer, followed by a batch normalization and a flatten layer. The implementation code for the base architecture is shown in Fig. 8. Four variations of the base architecture (see Table 3) were trained with the binary cross-entropy loss function, SGD optimizer and PET images with dimension of $75 \times 75 \times 30$ voxels. The variations, basically, consist in changing the number of the Conv filters.

```
print("##################### NETWORK ARCHITECTURE #####################")

input_format = Input((75, 75, 30, 1))

conv1 = Conv3D(filters=16, kernel_size=(3, 3, 3), activation='relu')(input_format)
conv2 = Conv3D(filters=8, kernel_size=(3, 3, 3), activation='relu')(conv1)

max_pool1 = MaxPool3D(pool_size=(2, 2, 2))(conv2)

conv3 = Conv3D(filters=16, kernel_size=(3, 3, 3), activation='relu')(max_pool1)
conv4 = Conv3D(filters=8, kernel_size=(3, 3, 3), activation='relu')(conv3)

max_pool2 = MaxPool3D(pool_size=(2, 2, 2))(conv4)

norm1 = BatchNormalization()(max_pool2)
flat1 = Flatten()(norm1)

fc1 = Dense(units=512, activation='relu')(flat1)

drop1 = Dropout(0.4)(fc1)

fc2 = Dense(units=128, activation='relu')(drop1)

drop2 = Dropout(0.4)(fc2)

fc3_out = Dense(units=1, activation='sigmoid')(drop2)
```

Fig. 8. 3D binary classification - base custom 3D-CNN architecture.

Table 3. 3D CNN - custom variations.

Model	Description
Custom 1	conv1(16 filters); conv 2(8 filters); conv3 (16 filters); conv4(8 filters)
Custom 2	conv1(8 filters); conv 2(16 filters); conv3 (8 filters); conv4(16 filters)
Custom 3	conv1(16 filters); conv 2(16 filters); conv3 (8 filters); conv4(8 filters)
Custom 4	conv1(8 filters); conv 2(8 filters); conv3 (16 filters); conv4(16 filters)

The results in terms of training, cross-validation and testing accuracy are depicted in Fig. 9. Custom4 model outperforms the other models and it is used in the next experiments. The impact of the 3D PET image resolution and the batch size were analysed as shown in Fig. 10 and Fig. 11. Based on these results, the Custom4 model trained with the $75 \times 75 \times 30$ voxels input image resolution and batch size 2 was considered as the optimal training configuration. Similarly to the 2D approach, the overfitting issue was tackled through the variation of the

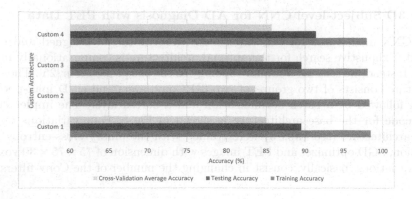

Fig. 9. 3D subject-level model: impact of the custom architecture ($75 \times 75 \times 30$ voxels; SGD optimizer; 0.001 learning rate; 50% dropout rate; batch size = 2).

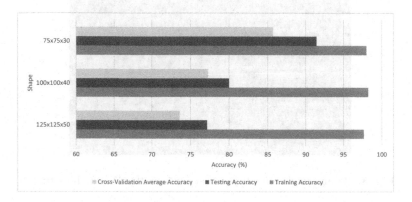

Fig. 10. 3D subject-level model - impact of the 3D PET image resolution (SGD optimizer; 0.001 learning rate; Custom4 model; 50% dropout rate; batch size = 2).

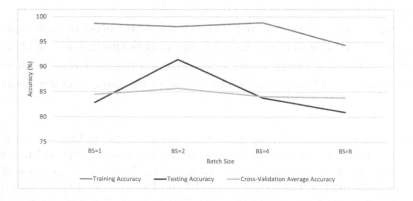

Fig. 11. 3D subject-level model - impact of the batch size ($75 \times 75 \times 30$ voxels; SGD optimizer; 0.001 learning rate; Custom4 model; 50% dropout rate).

dropout rate (see Fig. 12). The model struggles to converge for higher dropout rates, achieving a remarkable 91.43% testing accuracy for a 50% dropout rate (Fig. 13).

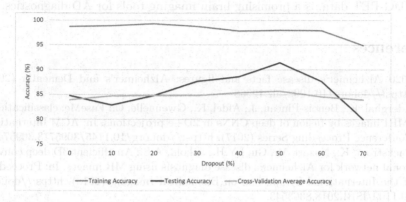

Fig. 12. 3D subject-level model - impact of the dropout rate ($75 \times 75 \times 30$ voxels; SGD optimizer; 0.001 learning rate; Custom4 model; batch size = 2).

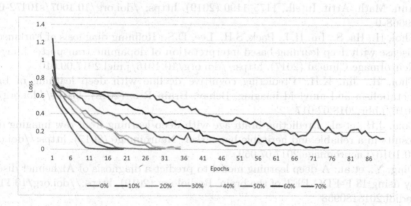

Fig. 13. 3D subject-level model: - loss function trajectory for varying dropout rate ($75 \times 75 \times 30$ voxels; SGD optimizer; 0.001 learning rate; Custom4 model; batch size = 2).

5   Conclusions

The primary objective of this paper was to study the potential of ^{18}F-FDG PET neuroimaging as a AD biomarker for classifying healthy versus AD patients. The first CNN model explores transfer learning with a pre-trained 2D Inception V3 model, as a typical solution in medical imaging. The second solution involves a custom 3D-CNN designed and trained from scratch. Both models achieved competitive performance (testing accuracy above 91%), with scores above most of the referred works in Table 2. The custom 3D-CNN is computationally more attractive because it has less conv layers and, therefore, less number of parameters. Further to that, this study demonstrates that the 3D-CNN, provided with the FDG-PET data, is a promising brain imaging tools for AD diagnostics.

References

1. 2020 Alzheimer's disease facts and figures: Alzheimer's and Dementia (2020). https://doi.org/10.1002/alz.12068
2. Aderghal, K., Benois-Pineau, J., Afdel, K., Gwenaëlle, C.: FuseMe: classification of sMRI images by fusion of deep CNNs in 2D+e projections. In: ACM International Conference Proceeding Series (2017). https://doi.org/10.1145/3095713.3095749
3. Backstrom, K., Nazari, M., Gu, I.Y.H., Jakola, A.S.: An efficient 3D deep convolutional network for Alzheimer's disease diagnosis using MR images. In: Proceedings of the International Symposium on Biomedical Imaging (2018). https://doi.org/10.1109/ISBI.2018.8363543
4. Bozhkov, L., Georgieva, P.: Overview of deep learning architectures for EEG-based brain imaging. In: 2018 International Joint Conference on Neural Networks (IJCNN). IEEE (2018)
5. Bozhkov, L., Georgieva, P.: Deep learning models for brain machine interfaces. Ann. Math. Artif. Intell., 1175–1190 (2019). https://doi.org/10.1007/s10472-019-09668-0
6. Choi, H., Ha, S., Im, H.J., Paek, S.H., Lee, D.S.: Refining diagnosis of Parkinson's disease with deep learning-based interpretation of dopamine transporter imaging. NeuroImage Clinical (2017). https://doi.org/10.1016/j.nicl.2017.09.010
7. Choi, H., Jin, K.H.: Predicting cognitive decline with deep learning of brain metabolism and amyloid imaging. Behav. Brain Res. (2018). https://doi.org/10.1016/j.bbr.2018.02.017
8. Cole, J.H., et al.: Predicting brain age with deep learning from raw imaging data results in a reliable and heritable biomarker. Neuroimage (2017). https://doi.org/10.1016/j.neuroimage.2017.07.059
9. Ding, Y., et al.: A deep learning model to predict a diagnosis of Alzheimer disease by using 18 F-FDG PET of the brain. Radiology (2019). https://doi.org/10.1148/radiol.2018180958
10. Ebrahimighahnavieh, M.A., Luo, S., Chiong, R.: Deep learning to detect Alzheimer's disease from neuroimaging: a systematic literature review. Comput. Methods Programs Biomed. (2020). https://doi.org/10.1016/j.cmpb.2019.105242
11. Havaei, M., et al.: Brain tumor segmentation with Deep Neural Networks. Med. Image Anal. (2017). https://doi.org/10.1016/j.media.2016.05.004
12. Huang, Y., Xu, J., Zhou, Y., Tong, T., Zhuang, X.: Diagnosis of Alzheimer's disease via multi-modality 3D convolutional neural network. Front. Neurosci. (2019). https://doi.org/10.3389/fnins.2019.00509
13. Kamnitsas, K., et al.: Efficient multi-scale 3D CNN with fully connected CRF for accurate brain lesion segmentation. Med. Image Anal. (2017). https://doi.org/10.1016/j.media.2016.10.004
14. Klöppel, S., et al.: Automatic classification of MR scans in Alzheimer's disease. Brain (2008). https://doi.org/10.1093/brain/awm319

15. Li, F., Liu, M.: Alzheimer's disease diagnosis based on multiple cluster dense convolutional networks. Comput. Med. Imaging Graph. (2018). https://doi.org/10.1016/j.compmedimag.2018.09.009
16. Lian, C., Liu, M., Zhang, J., Shen, D.: Hierarchical fully convolutional network for joint atrophy localization and Alzheimer's disease diagnosis using structural MRI. IEEE Trans. Pattern Anal. Mach. Intell. (2020). https://doi.org/10.1109/TPAMI.2018.2889096
17. Lin, W., et al.: Convolutional neural networks-based MRI image analysis for the Alzheimer's disease prediction from mild cognitive impairment. Front. Neurosci. (2018). https://doi.org/10.3389/fnins.2018.00777
18. Liu, M., Cheng, D., Wang, K., Wang, Y.: Multi-modality cascaded convolutional neural networks for Alzheimer's disease diagnosis. Neuroinformatics, 295–308 (2018). https://doi.org/10.1007/s12021-018-9370-4
19. Liu, M., Zhang, J., Adeli, E., Shen, D.: Landmark-based deep multi-instance learning for brain disease diagnosis. Med. Image Anal. (2018). https://doi.org/10.1016/j.media.2017.10.005
20. Marcus, C., Mena, E., Subramaniam, R.M.: Brain PET in the diagnosis of Alzheimer's disease (2014). https://doi.org/10.1097/RLU.0000000000000547
21. Punjabi, A., Martersteck, A., Wang, Y., Parrish, T.B., Katsaggelos, A.K.: Neuroimaging modality fusion in Alzheimer's classification using convolutional neural networks. PLoS ONE (2019). https://doi.org/10.1371/journal.pone.0225759
22. Qiu, S., Chang, G.H., Panagia, M., Gopal, D.M., Au, R., Kolachalama, V.B.: Fusion of deep learning models of MRI scans, Mini-Mental State Examination, and logical memory test enhances diagnosis of mild cognitive impairment. Alzheimer's Dement. Diagn. Assess. Dis. Monit. (2018). https://doi.org/10.1016/j.dadm.2018.08.013
23. Senanayake, U., Sowmya, A., Dawes, L.: Deep fusion pipeline for mild cognitive impairment diagnosis. In: Proceedings of the International Symposium on Biomedical Imaging (2018). https://doi.org/10.1109/ISBI.2018.8363832
24. Shmulev, Y., Belyaev, M.: Predicting conversion of mild cognitive impairments to Alzheimer's disease and exploring impact of neuroimaging. In: Stoyanov, D., et al. (eds.) GRAIL/Beyond MIC 2018. LNCS, vol. 11044, pp. 83–91. Springer, Cham (2018). https://doi.org/10.1007/978-3-030-00689-1_9
25. Singh, S.P., Wang, L., Gupta, S., Goli, H., Padmanabhan, P., Gulyás, B.: 3d deep learning on medical images: a review (2020). https://doi.org/10.3390/s20185097
26. Stier, N., Vincent, N., Liebeskind, D., Scalzo, F.: Deep learning of tissue fate features in acute ischemic stroke. In: Proceedings - 2015 IEEE International Conference on Bioinformatics and Biomedicine, BIBM 2015 (2015). https://doi.org/10.1109/BIBM.2015.7359869
27. Valliani, A., Soni, A.: Deep residual nets for improved Alzheimer's diagnosis. In: ACM-BCB 2017 - Proceedings of the 8th ACM International Conference on Bioinformatics, Computational Biology, and Health Informatics (2017). https://doi.org/10.1145/3107411.3108224

Deep Aesthetic Assessment and Retrieval of Breast Cancer Treatment Outcomes

Wilson Silva[1,2]([✉]), Maria Carvalho[1], Carlos Mavioso[2,3], Maria J. Cardoso[2,3,4], and Jaime S. Cardoso[1,2]

[1] Faculty of Engineering, University of Porto, Porto, Portugal
[2] INESC TEC, Porto, Portugal
wilson.j.silva@inesctec.pt
[3] Breast Unit, Champalimaud Foundation, Lisbon, Portugal
[4] Nova Medical School, Lisbon, Portugal

Abstract. Treatments for breast cancer have continued to evolve and improve in recent years, resulting in a substantial increase in survival rates, with approximately 80% of patients having a 10-year survival period. Given the serious that impact breast cancer treatments can have on a patient's body image, consequently affecting her self-confidence and sexual and intimate relationships, it is paramount to ensure that women receive the treatment that optimizes both survival and aesthetic outcomes. Currently, there is no gold standard for evaluating the aesthetic outcome of breast cancer treatment. In addition, there is no standard way to show patients the potential outcome of surgery. The presentation of similar cases from the past would be extremely important to manage women's expectations of the possible outcome. In this work, we propose a deep neural network to perform the aesthetic evaluation. As a proof-of-concept, we focus on a binary aesthetic evaluation. Besides its use for classification, this deep neural network can also be used to find the most similar past cases by searching for nearest neighbours in the high-semantic space before classification. We performed the experiments on a dataset consisting of 143 photos of women after conservative treatment for breast cancer. The results for accuracy and balanced accuracy showed the superior performance of our proposed model compared to the state of the art in aesthetic evaluation of breast cancer treatments. In addition, the model showed a good ability to retrieve similar previous cases, with the retrieved cases having the same or adjacent class (in the 4-class setting) and having similar types of asymmetry. Finally, a qualitative interpretability assessment was also performed to analyse the robustness and trustworthiness of the model.

Keywords: Aesthetic evaluation · Breast cancer · Deep learning · Image retrieval · Interpretability

1 Introduction

Breast cancer is an increasingly treatable disease, with 10-year survival rates now exceeding 80% [2]. This high survival rate has led to increased interest in

© Springer Nature Switzerland AG 2022
A. J. Pinho et al. (Eds.): IbPRIA 2022, LNCS 13256, pp. 108–118, 2022.
https://doi.org/10.1007/978-3-031-04881-4_9

quality of life after treatment, particularly with regard to aesthetic outcome. In addition, with the development of new surgical options and radiation therapies, it has become even more important to have means to compare cosmetic outcomes. In order to refine current and emerging techniques, identify factors that have a significant impact on aesthetic outcome [3], and compare breast units, an objective and reproducible method for evaluating aesthetic outcome is essential. Currently, there is no accepted gold standard method for evaluating the aesthetic outcome of a treatment. Nevertheless, we find in the literature two groups of methods used to evaluate the aesthetic outcome of breast cancer treatments: subjective and objective methods.

The first methods to emerge were the subjective methods. In these methods, the assessment is made by the patient or by one or more observers. Currently, subjective evaluation by one or more observers is the most commonly used form of assessment of the aesthetic outcome. This evaluation can be done by direct observation of the patient or by photographic representations. However, since professionals involved in the treatment are often present in the group of observers, impartiality is not guaranteed. In addition, the direct observation of the patient can be very uncomfortable. It should also be pointed out that even professionals tend to disagree on the outcome of the assessment, making the reproducibility of the method doubtful.

To overcome the problem of reproducibility of the previous methods, objective methods for the assessment of the breast cancer conservative treatment (BCCT) were introduced. Originally, measurements were taken directly on patients or on patients' photographs to compare the changes induced by the treatments. The Breast Retraction Assessment - BRA and Upward Nipple Retraction - UNR are two examples of the measurements used. More than two decades later, the 2D systems developed by Fitzal *et al.* [4] and by Cardoso and Cardoso [2] and the 3D system developed by Oliveira *et al.* [10] are the most relevant works. In [4], a software, Breast Analysing Tool - BAT© (see Fig. 2 (a)), which measures the differences between left and right breast sizes from a patient's digital image, was presented. In this work, the aesthetic outcome of BCCT is evaluated using a Breast Symmetry Index (BSI). With this software, it is possible to effectively distinguish between good and fair classes. However, excellent and good classes and fair and poor classes are not differentiated. Around the same time, Cardoso and Cardoso [2] introduced the BCCT.core software, being able to differentiate between the four classes (see Fig. 2 (b)). In addition to breast symmetry, color differences and scar visibility are also considered as relevant features for the pattern classifier. Here, the influence of scars is quantified by local differences in color.

Although these works represented a breakthrough, they were still not considered the most appropriate convention for the aesthetic evaluation of breast cancer treatments because they require manual identification of fiducial points for aesthetic evaluation, are applicable only to classic conservative treatments, and have limited performance [12]. Therefore, there is a need to develop new

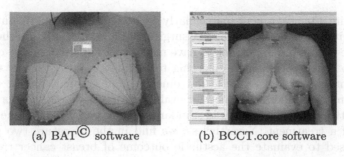

(a) BAT© software   (b) BCCT.core software

Fig. 1. Aesthetic evaluation softwares. Images from [9]

methods that automatically identify relevant features in images and perform the aesthetic assessment.

Apart from the fact that there is no gold standard for the evaluation of aesthetic outcomes, patients' expectations are not properly managed which, combined with some objective deficiencies, results in nearly 30% of patients undergoing breast cancer treatment being dissatisfied with the results obtained [11]. Therefore, it is very important that patients are aware of realistic outcomes and feel engaged with those results. To fulfill this goal, the presentation of photographs of breast cancer treatment outcomes from patients with similar characteristics is of utmost importance. In this context, to automatically find the most similar images, we need to develop content-based image retrieval systems [14], and, ideally, adapt them, to retrieve a new, generated image, that retains the realistic aesthetic outcome but shares the biometric characteristics of the patient requiring treatment. However, this will only be possible if we have a model for the aesthetic assessment of breast cancer treatments that can be integrated into this ideal end-to-end model for generating and retrieving realistic probable outcomes.

In this work, we propose a deep neural network that performs the aesthetic evaluation automatically and therefore does not require any manual annotation. Moreover, the network also retrieves the most similar past cases from the dataset, by searching in a high semantic space previous to classification. We also analyse the interpretability saliency maps generated by Layer-wise Relevance Propagation (LRP) [1] to find out whether the model is robust and trustworthy.

2   Materials and Methods

2.1   Data

For the experiments, we used 143 photographs of women who had undergone breast cancer conservative treatment. These 143 images came from two previously acquired datasets (PORTO and TSIO) [12]. These images were then graded by a highly experienced breast surgeon in terms of aesthetic outcome into one of four classes: Excellent, Good, Fair, and Poor. In this work, due to the small dimension of the dataset, we only used the binary labels ({Excellent, Good} vs. {Fair, Poor}) to evaluate classification performance; this work is the first to explore the use of a deep neural network to solve this problem. Nevertheless, the original four classes were used to evaluate the quality of the retrieval. Of the original 143 images, we used 80% for training and model/hyperparameter selection, and 20% for test. To reduce the complexity of the data, the images were resized to (384 × 256), while retaining the original three-channel RGB nature.

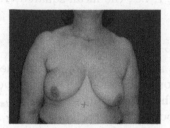 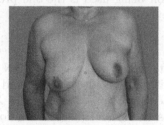

(a) Photography from PORTO dataset   (b) Photography from TSIO dataset

Fig. 2. Example of images used in this work.

2.2 Method

The state-of-the-art in the aesthetic evaluation of breast cancer treatments is the method proposed by Cardoso and Cardoso [2], which uses an SVM as the machine learning method to perform the classification. However, the SVM requires a first step involving a semi-automatic annotation of keypoints (such as breast contour and nipple positions) and a computation of asymmetry features (based on dimension and colour differences). Therefore, there is a need for human intervention. Moreover, it also can't be integrated into an end-to-end image generation model for the retrieval of biometrically-morphed probable outcomes.

Our proposed method (Fig. 3), which is inspired by the ideas described in [3], uses a highly regularized deep neural network to assess the aesthetic outcome automatically. It is highly regularized because the dataset dimension is very small, and the aesthetic result is very subjective, thus prone to variation, term. After having performed experiments with standard CNN networks, such as DenseNet-121 [6] and ResNet50 [5], we concluded that we had to considerably reduce the number of parameters to learn to overcome the intense overfitting. Thus, we designed a much simpler deep neural network, following the traditional "conv-conv-pooling" scheme, totalizing 262,908 learnable parameters (already including the fully-connected layers). However, reducing the number of parameters was not enough to prevent overfitting, leading us to the introduction of intermediate supervision in regard to the detection of important keypoints, and to the integration of pre-defined functions to translate those keypoints to asymmetry measures, namely, LBC (difference between the lower breast contours), BCE (difference between inframammary fold distances), UNR (difference between nipple levels), and BRA (breast retraction assessment). All these functions were integrated into the network by the use of "Lambda" layers (FTS Computation Functions in Fig. 3).

The training process was divided into two steps. First, we train the CNN model to learn to detect the keypoints (8 coordinates describing the positions of the nipples, levels of inferior breast contour, and sternal notch). The loss function being used is the one present in Eq. 1, i.e., the mean-squared error of the keypoint coordinates.

$$\mathcal{L}_{model} = \mathcal{L}_{keypoints} \tag{1}$$

Afterwards, we train the CNN model in a multitask fashion, simultaneously optimizing keypoint detection and classification performance, which can be represented by Eq. 2, where λ_k and λ_c weight the different losses, $\mathcal{L}_{keypoints}$ represents the mean-squared error loss for keypoint detection, and $\mathcal{L}_{classification}$ represents the binary cross-entropy loss for classification.

$$\mathcal{L}_{model} = \lambda_k \mathcal{L}_{keypoints} + \lambda_c \mathcal{L}_{classification} \tag{2}$$

All images were pre-processed using the same procedure as for the ImageNet data. The model was first trained for 350 epochs (with the last fully-connected layers frozen), following the loss function presented in Eq. 1. Afterwards, the model was trained for 250 epochs (with the first convolutional layers frozen), following the loss function presented in Eq. 2. In both steps, we used the Adadelta optimizer [15], and a batch size of 16. The model at the end of the first step was the one that led to the lowest mean-squared error in the validation data. The final model was selected based on the binary classification performance in the validation data.

2.3   Evaluation

Baseline: As baselines we considered four SVM models, such as the one used in Cardoso and Cardoso [2]. These four SVM models resulted from variations to the inputs and kernels being used. We performed the experiments using both a linear and an RBF kernel, and giving as input to the SVM either the entire set of symmetry features, or only the four features being implicitly used by our deep neural network (i.e., LBC, BCE, UNR, and BRA).

Performance Assessment: We considered two types of evaluation, one for classification performance using accuracy and balanced accuracy (due to the imbalanced nature of the data), and another for retrieval, where we checked whether the top-3 retrieved images belonged to the same class or to a neighbouring class (here, considering the original four classes). In addition, we generated saliency maps for the test images to understand the origin of the deep model's decisions.

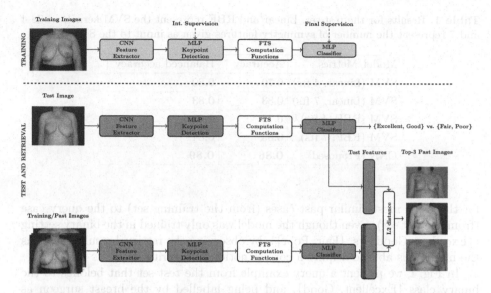

Fig. 3. Overview of the proposed approach. Blocks in light gray mean deep neural networks are being trained (i.e., weights are being updated), whereas blocks in dark gray represent trained deep neural networks (i.e., weights are fixed). The block in white means there are no weights being learnt. The "L2 distance" is computed based on the features from the previous to last layer of the network, i.e., exactly before the classification decision. (Color figure online)

3   Results

In Table 1, we present the results in terms of accuracy and balanced accuracy for all models considered, i.e., SVM baselines and our proposed model. For all SVM models, the parameter C, which weights the trade-off between misclassifying the data and the achieved margin, was optimized using 5-fold cross-validation, following the same search space as originally used by Cardoso and Cardoso [2] (i.e., exponentially growing sequences of C: $C = 1.25^{-1}, 1.25^{0}, ..., 1.25^{30}$). The *gamma* value for the SVM with *RBF* kernel was set to 3, also as done in Cardoso and Cardoso [2]. For all SVM models, class weights were set to "balanced", meaning that the misclassifications were weighted by the inverse of the class frequency. For our proposed CNN model, we used data augmentation (horizontal flips and translations), and also weighted the misclassifications by the inverse of the class frequency.

As can be seen in Table 1, our proposed model outperforms all SVM models both in terms of accuracy and balanced accuracy. Only regarding the SVM models, the ones that used all asymmetry features (7 fts) were able to achieve a higher performance. When comparing SVMs with different kernels, there was only a slight improvement with the RBF kernel in terms of balanced accuracy.

Besides comparing our model with the state-of-the-art, we were also interested in exploring the retrieval quality of the model. To measure that, we looked

Table 1. Results for the test set. Linear and RBF represent the SVM kernel, while 4 and 7 represent the number of symmetry features given as input to the SVM.

Model/Metrics	Accuracy ↑	Balanced accuracy ↑
SVM (Linear, 4 fts)	0.79	0.80
SVM (Linear, 7 fts)	0.83	0.83
SVM (RBF, 4 fts)	0.79	0.82
SVM (RBF, 7 fts)	0.83	0.84
CNN (Proposed)	**0.86**	**0.89**

for the top-3 most similar past cases (from the training set) to the query case (from the test set). Even though the model was only trained in the binary setting ({Excellent, Good} vs. {Fair, Poor}), by observing the retrieval results, it seems the model was able to acquire a correct notion of severity.

In Fig. 4, we present a query example from the test set that belongs to the binary class {Excellent, Good}, and being labelled by the breast surgeon as having an Excellent aesthetic outcome. The LRP saliency map demonstrates that the algorithm is paying attention to a region of interest (breast contour and nipple), which increases trust in the model. All the top-3 images retrieved were labelled as either Excellent or Good (i.e., the same class or a neighbouring class).

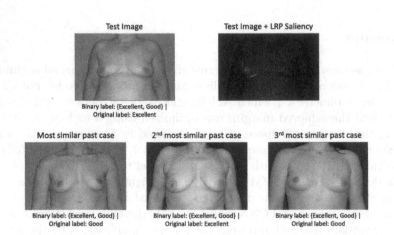

Fig. 4. Example of query and retrieved images for an Excellent aesthetic outcome. Binary label means class belongs to set {Excellent, Good}. Original label is the ordinal label previous to binarization (Excellent, Good, Fair, or Poor). LRP saliency map is also shown for the test image.

In Fig. 5, we also present a query example from the test set belonging to the binary class {Excellent, Good}, but this time having being labelled by the breast surgeon as having a Good aesthetic outcome. The LRP saliency map presented also demonstrates the algorithm is paying attention to a region of interest. All the top-3 images retrieved were labelled with the same class of the query (i.e., Good).

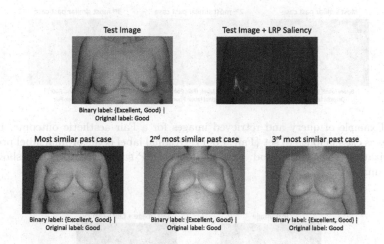

Fig. 5. Binary label means class belongs to set {Excellent, Good}. Original label is the ordinal label previous to binarization (Excellent, Good, Fair, or Poor). LRP saliency map is also shown for the test image.

A query example from the test set belonging to the binary class {Fair, Poor}, and having been labelled as Fair is presented in Fig. 6. This time, the LRP saliency map highlights the breast contour of both breasts, which makes sense as the difference between the two breasts is what is impacting more the lack of aesthetic quality in this particular case. The top-3 images retrieved were labelled as either Fair or Poor (i.e., same class or neighbouring class).

The last query example we present in this work belongs to the binary class {Fair, Poor}, and was labelled as Poor, meaning the worst aesthetic outcome. Quite interestingly, the LRP saliency map points to the breast retraction more than to the breast contour, which also makes clinical sense, as it is the most determinant factor for the poor aesthetic outcome. The top-3 images retrieved were labelled as either Poor or Fair (i.e., same class or neighbouring class).

Even though we only presented four examples of query images and their respective top-3 retrieved most similar past cases, the results were similar for all the other query/test images, in the sense that all saliency maps were focused on clinically relevant regions (breast, breast contour, nipples), and that all the identified most similar past cases belonged to one of the neighbouring original classes.

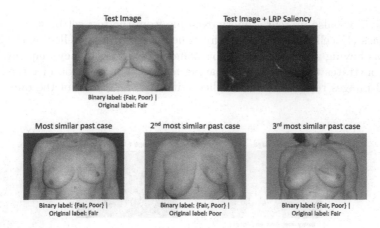

Fig. 6. Example of query and retrieved images for a Fair aesthetic outcome. Binary label means class belongs to set {Fair, Poor}. Original label is the ordinal label previous to binarization (Excellent, Good, Fair, or Poor). LRP saliency map is also shown for the test image.

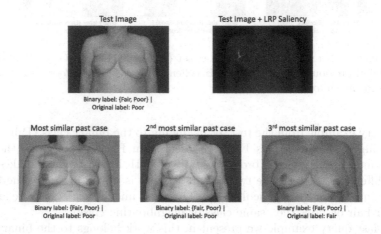

Fig. 7. Example of query and retrieved images for a Poor aesthetic outcome. Binary label means class belongs to set {Fair, Poor}. Original label is the ordinal label previous to binarization (Excellent, Good, Fair, or Poor). LRP saliency map is also shown for the test image.

4   Discussion and Conclusions

We have proposed to improve the automatic assessment of the aesthetic outcome of breast cancer treatments using deep neural networks. In addition to improving performance, the use of a deep neural network allows a natural semantic search for similar cases and an easy integration into future image generation models. As presented in Table 1, our proposed model outperforms state-of-the-art methods for aesthetic evaluation and does not require manual or semi-automatic pre-processing during inference. Furthermore, as illustrated by Figs. 4, 5, 6, and 7, it has the capacity to find meaningful past cases, what can be extremely useful for teaching purposes (for instance, new breast surgeons, or nurses) and for management of expectations (for the patients).

This was the first work using deep neural networks to assess the aesthetic outcome of breast cancer treatments. In future work, we plan to extend this model to the original ordinal scenario in order to completely replace the SVM models currently being used, and integrate it in a web application that will be openly accessible to any breast unit in the world. Moreover, we want to deepen the explainability of the model, exploring the inherent interpretability generated by the intermediate supervision and representation, in order to provide multimodal explanations (by complementing the retrieval with the importance given by the high-level concepts learnt in the semantic space, similarly to what is done in Silva et al. [13]). Finally, we will explore image generation techniques currently used for privacy-preserving case-based explanations [7,8] to adapt the retrieved past cases to the biometric characteristics of the query image in order to maximize patient engagement and acceptance.

Acknowledgement. This work was partially funded by the Project TAMI - Transparent Artificial Medical Intelligence (NORTE-01-0247-FEDER-045905) financed by ERDF - European Regional Fund through the North Portugal Regional Operational Program - NORTE 2020 and by the Portuguese Foundation for Science and Technology - FCT under the CMU - Portugal International Partnership, and also by the Portuguese Foundation for Science and Technology - FCT within PhD grant number SFRH/BD/139468/2018.

References

1. Bach, S., Binder, A., Montavon, G., Klauschen, F., Müller, K.R., Samek, W.: On pixel-wise explanations for non-linear classifier decisions by layer-wise relevance propagation. PLoS ONE **10**(7), e0130140 (2015)
2. Cardoso, J.S., Cardoso, M.J.: Towards an intelligent medical system for the aesthetic evaluation of breast cancer conservative treatment. Artif. Intell. Med. **40**(2), 115–126 (2007)
3. Cardoso, J.S., Silva, W., Cardoso, M.J.: Evolution, current challenges, and future possibilities in the objective assessment of aesthetic outcome of breast cancer locoregional treatment. Breast **49**, 123–130 (2020)
4. Fitzal, F., et al.: The use of a breast symmetry index for objective evaluation of breast cosmesis. Breast **16**(4), 429–435 (2007)
5. He, K., Zhang, X., Ren, S., Sun, J.: Deep residual learning for image recognition. In: Proceedings of the IEEE Conference on Computer Vision and Pattern Recognition, pp. 770–778 (2016)
6. Huang, G., Liu, Z., Van Der Maaten, L., Weinberger, K.Q.: Densely connected convolutional networks. In: Proceedings of the IEEE Conference on Computer Vision and Pattern Recognition, pp. 4700–4708 (2017)

7. Montenegro, H., Silva, W., Cardoso, J.S.: Privacy-preserving generative adversarial network for case-based explainability in medical image analysis. IEEE Access **9**, 148037–148047 (2021)
8. Montenegro, H., Silva, W., Cardoso, J.S.: Towards privacy-preserving explanations in medical image analysis. In: Interpretable Machine Learning in Healthcare Workshop at ICML 2021 (2021)
9. Oliveira, H.P., Cardoso, J.S., Magalhães, A., Cardoso, M.J.: Methods for the aesthetic evaluation of breast cancer conservation treatment: a technological review. Curr. Med. Imaging Rev. **9**(1), 32–46 (2013)
10. Oliveira, H.P., Patete, P., Baroni, G., Cardoso, J.S.: Development of a BCCT quantitative 3D evaluation system through low-cost solutions. In: Proceedings of the 2nd International Conference on 3D Body Scanning Technologies, pp. 16–27 (2011)
11. Ovadia, D.: AI will help Cinderella to see herself in the mirror. https://cancerworld. net/ai-will-help-cinderella-to-see-herself-in-the-mirror/
12. Silva, W., Castro, E., Cardoso, M.J., Fitzal, F., Cardoso, J.S.: Deep keypoint detection for the aesthetic evaluation of breast cancer surgery outcomes. In: 2019 IEEE 16th International Symposium on Biomedical Imaging (ISBI 2019), pp. 1082–1086. IEEE (2019)
13. Silva, W., Fernandes, K., Cardoso, M.J., Cardoso, J.S.: Towards complementary explanations using deep neural networks. In: Stoyanov, D., et al. (eds.) MLCN/DLF/IMIMIC -2018. LNCS, vol. 11038, pp. 133–140. Springer, Cham (2018). https://doi.org/10.1007/978-3-030-02628-8_15
14. Silva, W., Poellinger, A., Cardoso, J.S., Reyes, M.: Interpretability-guided content-based medical image retrieval. In: Martel, A.L., et al. (eds.) MICCAI 2020. LNCS, vol. 12261, pp. 305–314. Springer, Cham (2020). https://doi.org/10.1007/978-3-030-59710-8_30
15. Zeiler, M.D.: ADADELTA: an adaptive learning rate method. arXiv preprint arXiv:1212.5701 (2012)

Increased Robustness in Chest X-Ray Classification Through Clinical Report-Driven Regularization

Diogo Mata[1]([✉]), Wilson Silva[1,2], and Jaime S. Cardoso[1,2]

[1] Faculdade de Engenharia, Universidade do Porto, Porto, Portugal
diogo.20mata@gmail.com
[2] INESC TEC, Porto, Portugal

Abstract. In highly regulated areas such as healthcare there is a demand for explainable and trustworthy systems that are capable of providing some sort of foundation or logical reasoning to their functionality. Therefore, deep learning applications associated with such industry are increasingly required by this sense of accountability regarding their production value. Additionally, it is of utter importance to take advantage of all possible data resources, in order to achieve a greater amount of efficiency respecting such intelligent frameworks, while maintaining a realistic medical scenario. As a way to explore this issue, we propose two models trained with information retained in chest radiographs and regularized by the associated medical reports. We argue that the knowledge extracted from the free-radiology text, in a multimodal training context, promotes more coherence, leading to better decisions and interpretability saliency maps. Our proposed approach demonstrated to be more robust than their baseline counterparts, showing better classification performances, and also ensuring more concise, consistent and less dispersed saliency maps. Our proof-of-concept experiments were done using the publicly available multimodal radiology dataset MIMIC-CXR that contains a myriad of chest X-rays and its correspondent free-text reports.

Keywords: Chest X-ray · Multimodal deep learning · Interpretability

1 Introduction

The increasing availability of documentation related to support and diagnostic services opens up a plethora of clinical decision-making options. The development of a wider network of medical examples, as well as greater transversality concerning the information that a clinician has at his disposal translates into an opportunity not only to maximise the extraction of knowledge from data, but also to optimise its use. Additionally, there is a demand for a clear and understandable study inherent to an area as regulated and controlled as healthcare. Thus, the interest in capable auxiliary automated systems, that are able to support their decisions and explain their mechanisms, is naturally growing.

© Springer Nature Switzerland AG 2022
A. J. Pinho et al. (Eds.): IbPRIA 2022, LNCS 13256, pp. 119–128, 2022.
https://doi.org/10.1007/978-3-031-04881-4_10

Works such as [9,11,12] explore image data in order to achieve in-depth explanations that would, feasibly, fulfill stakeholder's expectations. Furthermore, [11] examines the adoption of explainability techniques as a means of directly improving such system's functionality. Despite the undeniable contribution of these approaches, there is a need to take advantage of resources from other modalities, if possible. This account is based on the idea that the integration of different source representations can favour proposed tasks. Moreover, beholding a medical context, where diagnostic images are often accompanied by medical reports, that usually address more detailed information, it is intuitive to assume that textual representations can likely carry certain knowledge to a multimodal approach, that the individual visual embedding cannot define. Even though the contained information in text data can also be somewhat expressed image-wise, in this framework, the textual component is more processed and "treated", thus composing a more semantic configuration. Such premise is proved to have potential in [13], as the advanced multimodal methodology for Content Based Image Retrieval (CBIR), on radiology images, shows performance improvements over traditional unimodal CBIR solutions.

The key behind this work is also established in connection to real scenarios. When analysing a new case, a professional usually relies on image-based diagnostic evidence. However, the possible transfer of knowledge, impregnated in textual documentation, to a system that purely manages image inputs can reveal to be crucial and decisive in terms of performance and robustness. However, we aim to explore if the increased robustness given by the clinical report-driven regularization can be observed in the interpretability saliency maps generated. Our approach follows a coordinated representation setting [1], where the textual information is used to regularize the embedding learnt by the visual encoder, meaning that for inference only one modality is needed (i.e., the Chest X-ray images).

2   Materials and Methods

2.1   Data

For the experiments, we used the MIMIC (*MIMIC-CXR-JPG*) database [3,5], which is a publicly accessible collection of JPG-format chest radiographs annexed with structured labels obtained from corresponding free-text radiology reports. The set includes 377,110 JPG files and the associated 227,827 free-text radiology reports. A common reference for data splits is supplied and the dataset is de-identified. The large size, the multimodality setup and the convenience associated with the prevailing image format, were determining factors respecting the preference for our use of *MIMIC-CXR-JPG*.

In this study, we are interested in analysing the interpretability saliency maps generated for the different models. In order to have a clear and consistent comparison, only images acquired in an anterior-posterior (AP) orientation and the respective clinical reports were selected.

Regarding the images (Fig. 1(a)), we resized them to a (224×224) proportion, and kept the original 3 channels (RGB). Afterwards, images were normalized following the same procedure as done for the ImageNet pre-training. To prevent overfitting, we used data augmentation, namely, rotations within a $10^{\underline{o}}$ to $20^{\underline{o}}$ range, and random horizontal translations up to a maximum of 0.5 of the image width. For the clinical reports (Fig. 1(b)), we first removed all non-alphanumeric elements. Afterwards, we applied the tokenization, following the same procedure used by Toutanova *et al.* [2]. Special tokens were created for the beginning and the end of each sentence, and every word was mapped to an *id* (*token id*), and an attention mask, which signals the model where to focus. For our task, the token type *ids*, that are formally used in the tokenization process, remain irrelevant since, each input sequence is not paired.

We will focus on a binary classification setting, namely in detecting the presence of pleural effusion, which is a condition broadly categorized as an excess of fluid within the pleural cavity or within the different layers of the pleura [8].

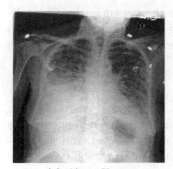

FINAL REPORT CHEST RADIOGRAPH PERFORMED ON ___ Comparison is made with a CT chest from ___ and a chest radiograph from ___ CLINICAL HISTORY Cough metastatic non small cell lung cancer assess for cause of new cough FINDINGS AP upright portable chest radiograph is obtained Overall there is no significant change from the recent CT performed ___ with innumerable metastatic nodularity involving both lungs and large consolidation occupying the right lower lung with a small to moderate right pleural effusion There is no new area of atelectasis or new area of confluent opacity to suggest a superimposed pneumonia though given the extensive underlying lung disease a subtle acute process would be impossible to exclude Heart size cannot be assessed Mediastinal contour is stable No pneumothorax is seen Bony structures appear stable Known metastatic lesions involving the inferior scapulae are not clearly visualized as well as the recently diagnosed nondisplaced fracture involving the right posterior eighth rib IMPRESSION Overall stable exam with extensive metastatic disease to the lungs with right pleural effusion and right basal consolidation

(a) Chest X-ray   (b) Clinical Report

Fig. 1. Test example - Chest X-ray and associated clinical report. Pleural effusion case.

2.2   Method

Our proposed methodology (Fig. 2) makes use of the available clinical reports to regularize the semantic space previous to classification. Moreover, it is designed in a way that does not require the availability of the clinical reports in inference.

Training: In the training process, our network receives two inputs: Chest X-ray images, and clinical reports. For each modality, a specific branch is built. For the image branch, we use a Convolutional Neural Network to extract the relevant embedding features (I_{emb}). On the other hand, for the clinical report branch, we use a Natural Language Processing classification network to extract the relevant embedding features (T_{emb}). Through an embedding loss function (Eq. 1), we promote a similar semantic representation for both branches. We assume that it is easier to learn a good embedding representation for the natural language branch than for the image branch, since the clinical reports already result from a human description of the image information. Thus, we aim to improve image classification by incorporating this knowledge into the network (via the embedding loss).

$$\mathcal{L}_{emb} = MSE(I_{emb}, T_{emb}) \tag{1}$$

Besides this embedding loss function, there is the main classification loss, which is the binary cross-entropy loss (we are dealing with the classification task Pleural Effusion vs. Non-Pleural Effusion). Thus, the final loss function being used to optimize the architecture's parameters is the one described in Eq. 2, where $\mathcal{L}_{clf}(y_{true}, y_{pred})$ represents the binary cross-entropy loss term and $\mathcal{L}_{emb}(I_{emb}, T_{emb})$ the embedding loss term, with λ_{clf} and λ_{emb} weighting the importance of each loss term.

$$\mathcal{L}_{final} = \lambda_{clf}\mathcal{L}_{clf}(y_{true}, y_{pred}) + \lambda_{emb}\mathcal{L}_{emb}(I_{emb}, T_{emb}) \tag{2}$$

Test: For test, only one type of input is given to the network, the Chest X-ray images. Thus, the inference process occurring is similar to the one of a conventional CNN approach. This is important since at test time, clinical reports are not available.

2.3 Evaluation

The evaluation process we conducted is divided into two main components: the study of classification task performances (Accuracy and F1 score), and the analysis of the impact of the regularization process in the interpretability saliency maps generated.

We considered two different CNN architectures and compared the performance of both, with and without the clinical reports' regularization (working as baselines/ablation).

3   Results

For the experiments, we considered two different architectures for our CNN model, the well-known DenseNet-121 [4] (pre-trained on ImageNet data) and a simpler CNN model (with three blocks of "Conv-MaxPooling", followed by three fully-connected layers). As the NLP model, we used the standard state-of-the-art BERT model [2]. Since CNN and NLP models do not generate embeddings of the same size, we added a layer in order to have equal size embeddings (for image and text). All models considered in these experiments were trained during 20 epochs, using a batch size of 4, and the Adam optimizer [6].

Since we were interested in observing the impact of the clinical report regularization in the robustness of the models, we also computed the interpretability saliency maps. For that, we used the DeepLift [10] implementation available in Pytorch's Captum library [7].

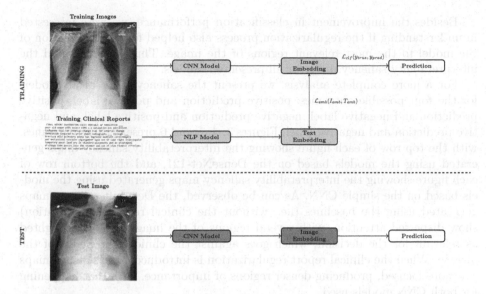

Fig. 2. Overview of the proposed approach. Blocks in light gray mean deep neural networks are being trained (i.e., weights are being updated), whereas blocks in dark gray represent trained deep neural networks (i.e., weights are fixed). The block in white means there are no weights being learnt. $\mathcal{L}_{emb}(I_{emb}, T_{emb})$ represents the embedding loss, which depends on the Image Embedding (I_{emb}) and Text Embedding (T_{emb}). $\mathcal{L}_{clf}(y_{true}, y_{pred})$ represents the classification loss, which depends on the ground-truth labels (y_{true}) and predicted classes (y_{pred}).

In this work, we used four models: our baselines, DenseNet-121 and the simple CNN; and the same models with the regularization component given by the textual embeddings computed by the BERT model. After the pre-processing described in the previous section, we ended up with 27697 training samples, 238 validation samples, and 966 test samples. It is important to note that in test, only images are used, even for the proposed regularized versions.

The classification task results are presented in Table 1 according to standard Accuracy and F1 score metrics. As can be observed, in both CNN architectures, the regularization promoted by the textual embeddings led to a significant improvement in both Accuracy and F1 scores.

Table 1. Accuracy and F1 test scores.

	Accuracy score	F1 score
DenseNet121	0.849	0.909
DenseNet121 + BERT Embedding Representation	**0.867**	**0.920**
Simple CNN	0.804	0.862
Simple CNN + BERT Embedding Representation	0.817	0.898

Besides the improvement in classification performance, we were interested in understanding if the regularization process also helped focus the attention of the model to the most relevant regions of the image. Thus, we computed the interpretability saliency maps for all presented models.

For a more complete analysis, we present the saliency maps of all models for the four possible outcomes: positive prediction and positive label, positive prediction and negative label, negative prediction and positive label, and negative prediction and negative label. Figures 3, 4, 5 and 6 present those outcomes, with the top row of each figure showing the interpretability saliency maps generated using the models based on the DenseNet-121, and the bottom row of each figure showing the interpretability saliency maps generated using the models based on the simple CNN. As can be observed, the DeepLift saliency maps generated using the baselines (i.e., without the clinical report regularization) show dispersed attention, with several regions of the images being highlighted as relevant for the decision, which goes against the clinical knowledge of the disease. When the clinical report regularization is introduced, the saliency maps get more focused, producing denser regions of importance, with that happening for both CNN models used.

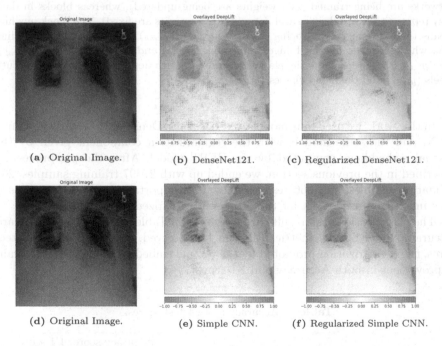

(a) Original Image.   (b) DenseNet121.   (c) Regularized DenseNet121.

(d) Original Image.   (e) Simple CNN.   (f) Regularized Simple CNN.

Fig. 3. DeepLift saliency maps for positive prediction and positive label. Green and red pixels stand for positive and negative prediction contributions, respectively. (Color figure online)

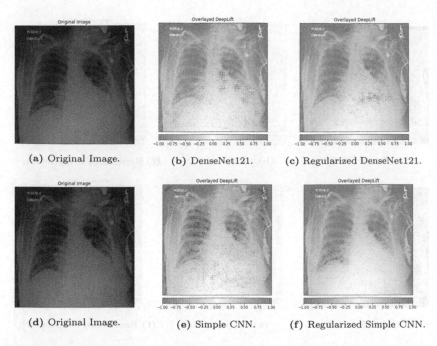

(a) Original Image.   (b) DenseNet121.   (c) Regularized DenseNet121.

(d) Original Image.   (e) Simple CNN.   (f) Regularized Simple CNN.

Fig. 4. DeepLift saliency maps for positive prediction and negative label. Green and red pixels stand for positive and negative prediction contributions, respectively. (Color figure online)

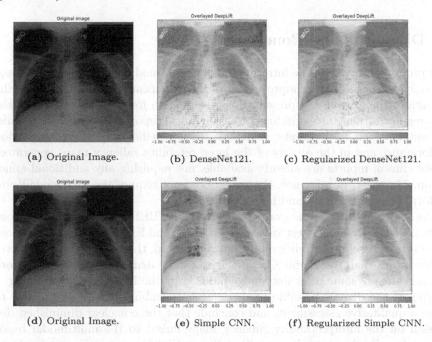

(a) Original Image.   (b) DenseNet121.   (c) Regularized DenseNet121.

(d) Original Image.   (e) Simple CNN.   (f) Regularized Simple CNN.

Fig. 5. DeepLift saliency maps for negative prediction and positive label. Green and red pixels stand for positive and negative prediction contributions, respectively. (Color figure online)

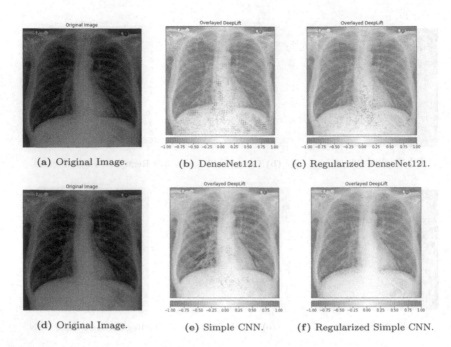

(a) Original Image.   (b) DenseNet121.   (c) Regularized DenseNet121.

(d) Original Image.   (e) Simple CNN.   (f) Regularized Simple CNN.

Fig. 6. DeepLift saliency maps for negative prediction and negative label. Green and red pixels stand for positive and negative prediction contributions, respectively. (Color figure online)

4   Discussion and Conclusions

Our proposal centred on the introduction of a multimodal learning setup aiming, not only, the performance improvement of the classification task, but also in the learning of more robust representations, leading to more trustworthy models. The regularization process induced by the clinical reports' data reveals promising outcomes on both counts, as it led to improvements in the classification performance and in the quality of the interpretability saliency maps generated. These clinical reports are already available, not requiring any additional annotation. Moreover, in inference, the model does not require the text modality to perform the classification and help the diagnosis.

As shown in Table 1, the examples trained with BERT based representations show transversely higher test values for accuracy and F1 scores, when compared to their peer baselines. Additionally and as expected, the DenseNet-121 achieves higher scores than the simple CNN, as it is a more dense and complex network and has already some prior knowledge induced by the ImageNet pre-training.

Across the diverse possible prediction-label combinations, it is possible to make a qualitative assessment, and conclude that there is clear diminished dispersion on the interpretability saliency maps linked to the multimodal based approaches. Furthermore, it seems that the spotlight is more circumscribed and usually present on the lower portion of the lungs, where a manifestation of the addressed condition should prevail, indicating an excess of fluid. Overall, we can conclude that the proposed approach came to fruition, leading to better classification performance and robustness (verified via interpretability saliency maps).

Concerning future advances, we will mainly focus on two different topics. Firstly we would like to apply our approach in a Content Based Image Retrieval (CBIR) context, expanding the work of [11], and secondly, we intend to explore the multimodality problematic even further by automatically generating the clinical reports/textual explanations for the test images.

Acknowledgement. This work was partially funded by the Project TAMI - Transparent Artificial Medical Intelligence (NORTE-01-0247-FEDER-045905) financed by ERDF - European Regional Fund through the North Portugal Regional Operational Program - NORTE 2020 and by the Portuguese Foundation for Science and Technology - FCT under the CMU - Portugal International Partnership, and also by the Portuguese Foundation for Science and Technology - FCT within PhD grant number SFRH/BD/139468/2018.

References

1. Baltrušaitis, T., Ahuja, C., Morency, L.P.: Multimodal machine learning: a survey and taxonomy. IEEE Trans. Pattern Anal. Mach. Intell. **41**(2), 423–443 (2018)
2. Devlin, J., Chang, M.W., Lee, K., Toutanova, K.: BERT: pre-training of deep bidirectional transformers for language understanding (2019)
3. Goldberger, A., et al.: Components of a new research resource for complex physiologic signals. PhysioNet **101** (2000)
4. Huang, G., Liu, Z., van der Maaten, L., Weinberger, K.Q.: Densely connected convolutional networks (2018)
5. Johnson, A.E.W., et al.: MIMIC-CXR-JPG: a large publicly available database of labeled chest radiographs. CoRR abs/1901.07042 (2019). http://arxiv.org/abs/1901.07042
6. Kingma, D.P., Ba, J.: Adam: a method for stochastic optimization. arXiv preprint arXiv:1412.6980 (2014)
7. Kokhlikyan, N., et al.: Captum: A unified and generic model interpretability library for PyTorch (2020)
8. Li, Y., Tian, S., Huang, Y., Dong, W.: Driverless artificial intelligence framework for the identification of malignant pleural effusion. Transl. Oncol. **14**(1), 100896 (2021). https://doi.org/10.1016/j.tranon.2020.100896. https://www.sciencedirect.com/science/article/pii/S1936523320303880
9. Lucieri, A., Dengel, A., Ahmed, S.: Deep learning based decision support for medicine - a case study on skin cancer diagnosis (2021)
10. Shrikumar, A., Greenside, P., Kundaje, A.: Learning important features through propagating activation differences. In: International Conference on Machine Learning, pp. 3145–3153. PMLR (2017)
11. Silva, W., Poellinger, A., Cardoso, J.S., Reyes, M.: Interpretability-guided content-based medical image retrieval. In: Martel, A.L., et al. (eds.) MICCAI 2020. LNCS, vol. 12261, pp. 305–314. Springer, Cham (2020). https://doi.org/10.1007/978-3-030-59710-8_30

12. Tjoa, E., Guan, C.: A survey on explainable artificial intelligence (XAI): toward medical XAI. IEEE Trans. Neural Netw. Learn. Syst. **32**(11), 4793–4813 (2021)
13. Yu, Y., Hu, P., Lin, J., Krishnaswamy, P.: Multimodal multitask deep learning for X-ray image retrieval. In: de Bruijne, M., et al. (eds.) MICCAI 2021. LNCS, vol. 12905, pp. 603–613. Springer, Cham (2021). https://doi.org/10.1007/978-3-030-87240-3_58

Medical Applications

Deep Detection Models for Measuring Epidermal Bladder Cells

Angela Casado-García[1](✉) (iD), Aitor Agirresarobe[2](iD), Jon Miranda-Apodaca[2],
Jónathan Heras[1](iD), and Usue Pérez-López[2](iD)

[1] Department of Mathematics and Computer Science, University of La Rioja,
Ed. CCT, C. Madre de Dios, 53, 26004 Logroño, La Rioja, Spain
{angela.casado,jonathan.heras}@unirioja.es
[2] Department of Plant Biology and Ecology, Faculty of Science and Technology,
University of the Basque Country (UPV/EHU), 48080 Bilbao, Bizkaia, Spain
{aitor.aguirresarobe,usue.perez}@ehu.es

Abstract. Epidermal bladder cells (EBC) are specialized structures of halophyte plants that accumulate salt and other metabolites, and are thought to be involved in salinity tolerance, as well as in UV-B protection, drought and other stresses tolerance. However, the role of the EBC size, density or volume remains to be confirmed since few studies have addressed the relevance of these traits. This is due to the fact that those measurements are mostly carried out manually. In this work, we have tackled this problem by conducting a statistical analysis of several deep learning algorithms to detect EBC on images. From such a study, we have obtained a model, trained with the YOLOv4 algorithm, that achieves a F1-score of 91.80%. Moreover, we have proved the reliability of this model to other varieties and organs using different datasets than the ones used for training the model. In order to facilitate the use of the YOLO model, we have developed LabelGlandula, an open-source and simple-to-use graphical user interface that employs the YOLO model. The tool presented in this work will help to understand the functioning of EBC and the molecular mechanisms of EBC formation and salt accumulation, and to transfer this knowledge to crops one day.

Keywords: Epidermal bladder cells · Object detection · Deep learning · Software

1 Introduction

Currently, salinity presents one of the major problems for agricultural production and tremendous efforts are made to provide salt tolerance to crops [14,24]. One

This work was partially supported by Ministerio de Economía y Competitividad [MTM2017-88804-P], Ministerio de Ciencia e Innovación [PID2020-115225RB-I00] and GRUPO Gobierno Vasco-IT1022-16. Ángela Casado-García has a FPI grant from Community of La Rioja 2020, and Aitor Agirresarobe is the recipient of a predoctoral fellowship from the Gobierno Vasco (Spain).

© Springer Nature Switzerland AG 2022
A. J. Pinho et al. (Eds.): IbPRIA 2022, LNCS 13256, pp. 131–142, 2022.
https://doi.org/10.1007/978-3-031-04881-4_11

of the mechanisms involved in the response of halophyte plants to high saline soils are the epidermal bladder cells (EBC), which are specialized structures that accumulate salt and other metabolites [3]. EBC are present only in Aizoaceae and Amaranthaceae [10] and are thought to be involved in salinity tolerance [1,15], as well as in UV-B protection, drought and other stresses tolerance [13,23].

Recent interest has grown to understand the functioning of EBC and the molecular mechanisms of EBC formation and salt accumulation. However, the role of the EBC size, density or volume remains to be confirmed [16], as few studies have addressed the relevance of these traits, since these measurements are mostly manual using computer programs like ImageJ that have little to no automatisation [13,16,18,23]. Those measurements are repeated over multiple images from different plant species, organs and growth conditions. This is a tedious, error-prone, time-consuming and, in some cases, subjective task due to the large number of EBC in each image. In this paper, we tackle this problem by employing computer vision techniques based on deep learning models.

Counting and measuring EBC can be solved by means of object detection algorithms, that locate the position of multiple objects in a given image. Currently, object detection tasks are mainly approached by means of deep learning techniques [4,28]. This is also the case in the agricultural context where this kind of algorithms has been employed to detect stomata [5], fruits [6,29], crops [19], or plant diseases [9]; and, it is also the approach followed in this work.

Namely, we have conducted a statistical study of 7 deep learning algorithms to elucidate which model produces the most accurate detection of EBC, see Sect. 2. From that analysis, we have concluded that the best model for this task is the YOLOv4 algorithm [4], that achieves a mean F1-score of 91.80%, see Sect. 3. In addition, we have developed LabelGlandula, an open-source and simple to use graphical tool for measuring EBC density using the developed YOLO model, see Sect. 4. LabelGlandula is freely available at https://github.com/ancasag/labelGlandula. Finally, using LabelGlandula, and its underlying YOLO model, we have shown that this tool generalises to images from varieties and conditions that were not seeing during the training process, see Sect. 5.

The tool presented in this work is the first, up to the best of our knowledge, tool for EBC detection, and it will help to understand the functioning of EBC and the molecular mechanisms of EBC formation and salt accumulation, and to transfer this knowledge to crops one day [10,13].

2   Materials and Methods

Images of leaf EBC from one quinoa variety (Marisma) grown under different salinity, elevated CO_2 and elevated temperature conditions were used. Besides, the density of EBC decreases as leaves become older, therefore, leaves of different age were used.

EBC can be easily removed when the plants are manipulated so the leaves were cut carefully from the plant and were hold with a tweezer. In order to obtain EBC images from both adaxial and abaxial surfaces, a cross cut was made, obtaining two leaf pieces, one for the analysis of each leaf side. Carefully and maintaining the EBC intact, from the central part of each leaf side, two leaf discs of 0.6 cm in diameter were cut in order to have a flat surface and, therefore, an image focused of the EBC.

Leaf EBC were observed using a Nikon SM645 Stereo Microscope $0.8X - 5X$ (Nikon Corporation, Japan) with a zoom range of $2X$. Images were captured with an Optikam ®Microscopy Digital USB Camera 4083.B3 (OPTIKA Microscopes, Italy) and Optika View 7.1 program (OPTIKA Microscopes, Italy). A total of 178 images of size 2048×1536 were acquired using this procedure and manually annotated using the LabelImg software [31], see Fig. 1 for some samples of the captured images.

Fig. 1. Images of leaf EBC

2.1   Computational Methods

Using the dataset presented previously, we have trained several models using different deep learning architectures for object detection. Namely, we have used the two-phase algorithms Faster R-CNN [22] and EfficientDet [28], and the one-phase algorithms FCOS [30], CSResnet [4], FSAF [32] and YOLO version 3 [21] and version 4 [4]. All the models were trained using the by-default parameters with a GPU GeForce RTX 2080 Ti. Moreover, due to the limited number of images, we applied both transfer learning [20] and data augmentation [27] to train our models.

During the training process, we have faced a common challenge that arises when working with microscopic images. Deep learning object detection algorithms are usually trained with images of size close to 600×600; however, the images of our datasets are considerably bigger, and resizing them is not a feasible option since the details of the EBC would be lost if the images are scaled. In order to deal with this problem, the images of the dataset were split into patches of size 600×600, and used for training. It is worth mentioning that the size problem only occurs during the training stage, since models can be employed to detect objects in images of bigger size without splitting them into patches.

2.2 Experimental Study

In order to validate the aforementioned detection models, a 5-fold cross validation approach has been employed. To evaluate the performance of the detection models, we have measured their precision, recall, F1-score and mAP. In our statistical study the results are taken as the mean and standard deviation of the different metrics for the 5 folds.

In order to determine whether the results obtained are statistically significant, several null hypothesis tests are performed using the methodology presented in [11,26]. In order to choose between a parametric or a non-parametric test to compare the models, we check three conditions: independence, normality and heteroscedasticity—the use of a parametric test is only appropriate when the three conditions are satisfied [11].

The independence condition is fulfilled in our study since we perform 5 different runs with independent folds. We use the Shapiro-Wilk test [25] to check normality—with the null hypothesis being that the data follow a normal distribution—and, a Levene test [17] to check heteroscedasticity—with the null hypothesis being that the results are heteroscedastic.

Since we compare more than two models, we will employ an ANOVA test [26] if the parametric conditions are fulfilled, and a Friedman test [26] otherwise. In both cases, the null hypothesis will be that all the models have the same performance. Once the test for checking whether a model is statistically better than the others is conducted, a post-hoc procedure is employed to address the multiple hypothesis testing among the different models. A Holm post-hoc procedure [12], in the non-parametric case, or a Bonferroni-Dunn post-hoc procedure [26], in the parametric case, is used for detecting significance of the multiple comparisons [11,26] and the p values should be corrected and adjusted. We have performed our experimental analysis with a level of confidence equal to 0.05. In addition, the size effect has been measured using Cohen's d [7] and Eta Squared [8].

3   Results

In this section, we present the results of our study of object detection algorithms for EBC. In Table 1, we have summarised the mean and standard deviation of the 5 folds for each detection algorithm. The model trained with the YOLO v4 algorithm achieves the best results in terms of recall, F1-score and mAP. The YOLO v3 and FasterRCNN algorithms produce models with a better precision than YOLO v4, but their performance in the other metrics is considerably worse than YOLO v4.

Table 1. Mean (and standard deviation) of detection models for EBC

	Precision	Recall	F1-score	mAP
YOLOv3	94.20 (1.4)	75.80 (10.2)	83.40 (6.8)	84.12 (1.5)
YOLOv4	92.80 (0.4)	**90.40 (0.4)**	**91.80 (0.4)**	**89.92 (0.1)**
FCOS	86.80 (1.1)	80.60 (2.4)	83.40 (1.3)	70.93 (3.3)
CSResnet	92.20 (0.4)	83.60 (2.0)	87.60 (1.0)	84.74 (0.8)
FSAF	86.80 (5.0)	79.20 (1.9)	83.60 (0.8)	70.74 (3.1)
EfficientDet	92.00 (1.2)	78.80 (2.4)	84.80 (1.1)	72.70 (3.2)
FasterRCNN	**94.80 (0.7)**	81.40 (1.4)	87.60 (0.8)	78.17 (0.4)

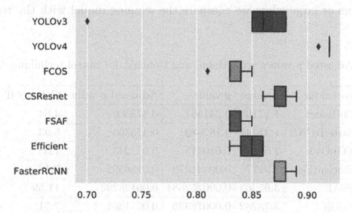

Fig. 2. Results from 5 independent runs in F1-score for selected object detection algorithms

We focus on the F1-score metric, that provides a trade-off between precision and recall, to perform the statistical analysis of the object detection models for EBC—the F1-score of the models is summarised in Fig. 2. In order to compare the trained detectors, the non-parametric Friedman's test is employed since the normality condition is not fulfilled (Shapiro-Wilk's test $W = 0.855043$;

$p = 0.000301$). The Friedman's test performs a ranking of the models compared (see Table 2), assuming as null hypothesis that all the models have the same performance. We obtain significant differences ($F = 14.85; p < 4.89 \times 10^{-7}$), with a large size effect eta squared $= 0.53$.

Table 2. Friedman's test for the F1-score of the object detection models

Technique	F1-score	Friedman's test average ranking
YOLOv4	91.80 (0.40)	7
CSResnet	87.60 (1.01)	5.4
FasterRCNN	87.60 (0.80)	5.2
YOLOv3	83.40 (6.8)	3.5
EfficientDet	84.80 (1.16)	3
FSAF	83.60 (0.80)	2
FCOS	83.40 (1.35)	1.9

The Holm algorithm was employed to compare the control model (winner) with all the other models adjusting the p value, results are shown in Table 3. As it can be observed in Table 3, there are three object detection algorithms with no significant differences as we failed to reject the null hypothesis. The size effect is also taken into account using Cohen's d, and as it is shown in Table 3, it is medium or large when we compare the winning model with the rest of the models.

Table 3. Adjusted p-values with Holm, and Cohen's d. Control technique: YOLOv4

Technique	Z value	p value	Adjusted p value	Cohen's d
CSResnet	1.17108	0.241567	0.375366	4.84
FasterRCNN	1.31747	0.187683	0.375366	5.93
YOLOv3	2.56174	0.010415	0.031245	1.55
EfficientDet	2.9277	0.00341479	0.0136592	7.18
FSAF	3.65963	0.000252584	0.00126292	11.59
FCOS	3.73282	0.00018935	0.0011361	7.51

As it can be seen in the results presented in this section, the algorithm that produces the best model is obtained with YOLO v4. However, using this model might be challenging for non-expert users since it requires experience working with deep learning libraries. We have addressed this issue by developing a user-friendly application, called LabelGlandula, that facilitates the dissemination and use of our methods.

4   LabelGlandula

LabelGlandula is a graphical tool implemented in Python, and developed using LabelImg [31], a widely used image annotation application, as a basis. Label-Glandula aims to facilitate the use of the EBC detection model developed in the previous section.

The first set of features included in LabelGlandula, see Fig. 3, is dedicated to measure the amount of EBC in a set of images. Given a folder with images, LabelGlandula detects the EBC in each image using the YOLO model. In addition, LabelGlandula provides the necessary features to modify the detections by adding and/or removing EBC from those detected by the model. Moreover, the box associated with each EBC can be adjusted. Finally, LabelGlandula not only provides the detections obtained for each image, but can also generate a summary of the results in the form of an Excel file. Such an excel file includes the number of EBC in each image and other statistics such as average area of EBC, folial area or covered folial area.

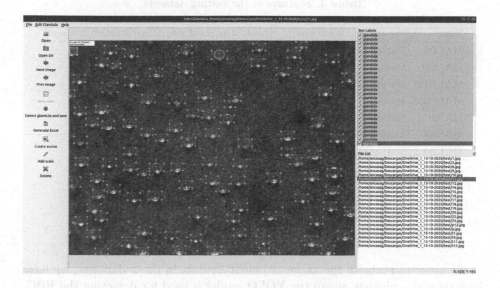

Fig. 3. LabelGlandula interface

In order to illustrate the feasibility of using LabelGlandula for counting and measuring EBC, a case study is presented in the following section.

5   Case Study

In this case study, we will analyse how LabelGlandula behaves with images from three experiments that are grown under different conditions and whose images were not used for training the original model. To this aim, we will use 11 datasets that are summarised in Table 4 and some samples from those datasets are provided in Fig. 4. From the same experiment used to train the original model, we selected images not used for training, or evaluating the precision of the model. Besides, in order to validate the model for conditions not explored in the training set, we used images of the same variety; but taken from the stems. For stem analysis, in the main stem section where the petiole of the youngest fully expanded leaf joins the stem, one thin slice of the epidermis containing EBC was cut longitudinally with a scalpel. The slice was carefully put on a slide avoiding the removal of the EBC. The stem slices were observed in the same way as the leaf pieces. In addition, we used images from another quinoa variety, concretely, Pasancalla. Pasancalla was characterized by having white and purple coloured EBC while Marisma showed only white coloured EBC. Pasancalla images were taken in the same way as explained in Sect. 2. All these factors affect EBC features, so the images between datasets are different among them.

Table 4. Features of the testing datasets

Name	Variety	Conditions	Leaf/Stem	Leaf side	Concentration	♯ images
MALB0	Marisma	Actual	Leaf	Abaxial	0 mM	8
MALD0	Marisma	Actual	Leaf	Adaxial	0 mM	8
MALB125	Marisma	Actual	Leaf	Abaxial	125 mM	8
MALD125	Marisma	Actual	Leaf	Adaxial	125 mM	8
MAS0	Marisma	Actual	Stem	–	0 mM	8
MAS500	Marisma	Actual	Stem	–	500 mM	8
MCLB0	Marisma	Climate change	Leaf	Abaxial	0 mM	8
MCLD0	Marisma	Climate change	Leaf	Adaxial	0 mM	8
MCLD500	Marisma	Climate change	Leaf	Adaxial	500 mM	8
PALB0	Pasancalla	Actual	Leaf	Abaxial	0 mM	8
PALD0	Pasancalla	Actual	Leaf	Adaxial	0 mM	6

In order to evaluate the usage of LabelGlandula in those datasets, we have investigated a pipeline where the YOLO model is used for detecting the EBC, and a user edits, by means of LabelGlandula, those detections to add missing EBC and remove those that are not correct. For our experiments, we have evaluated, see Table 5, the YOLO model using the same metrics presented in Sect. 3, and we have also included the number of True Positive (TP), False Positive (FP), and False Negative (FN) EBC.

Fig. 4. Samples from the testing datasets and predictions produced by LabelGlandula. *Top-Left*. Image from the MAS500 dataset. *Top-Right*. Image from the PALB0 dataset. *Bottom-Left*. Image from the MALD0 dataset. *Bottom-Right*. Image from the MCLD500 dataset

From those results, we can draw two main conclusions. First of all, the precision of the model is over 97% for all datasets, this indicates that most detections produced by the model are correct, and the user only needs to remove a few of them (in the worst case, a 2.64% of the detected EBC). The second conclusion is related to the number of EBC that must be added by the users; that is the recall. For this metric, the model achieves a value over 90% for all the datasets but 3 of them. The value for those datasets is over 80%; however, this means that almost a 20% of the EBC must be manually added. This happens because images taken with the conditions of those 3 datasets where not included in the training dataset. This issue is known as domain shift, and it is an open problem for deep learning models [2]. In our context, the problem is mitigated thanks to LabelGlandula that allows the user to easily add EBC; however, further research is needed in this context.

Table 5. Results for the case study

Dataset	Precision	Recall	F1-score	mAP	TP	FP	FN
MALB0	99.15	97.51	98.32	90.66	942	8	24
MALD0	98.52	93.48	95.94	90.07	603	9	42
MALB125	99.53	92.04	95.64	90.75	3656	17	316
MALD125	99.72	96.16	97.90	90.80	1803	5	72
MAS0	99.23	86.85	92.63	81.52	780	6	118
MAS500	97.37	80.87	88.36	79.99	2301	62	544
MCLB0	99.81	96.80	98.28	90.83	2178	4	72
MCLD0	99.74	93.02	96.26	90.71	1187	3	89
MCLD500	99.57	95.87	97.69	90.77	2327	10	100
PALB0	99.70	90.95	95.12	90.76	4676	14	465
PALD0	97.35	80.90	88.37	80.20	2793	76	659

6   Conclusions and Further Work

In this paper, we have presented LabelGlandula, an open-source application for automatically measuring the number of EBC and their size. The underlying model employed by LabelGlandula for automatically detecting EBC is based on the YOLO algorithm version 4 and has achieved a F1-score of 91.80. This algorithm has proven to be significantly better than other deep learning models for the same task. Moreover, we have proven the reliability of the model using images from datasets containing varieties and organs that were not seeing during the training process. Thanks to the development of this tool, the analysis of EBC will be more reliable and will allow plant biologists to advance their understanding of the role of the EBC size, density and volume in salinity, drought and other stress tolerance.

The main task that remains as further work is the development of an approach to deal with the domain shift problem. The performance of the YOLO model included in LabelGlandula slightly decays when tested with images from varieties and organs that were not seeing during the training process. Currently, this problem is tackled by providing the LabelGlandula interface that allows the user to add missing EBC. A different approach consists in fine-tuning the YOLO model with images from new varieties and organs, but this requires re-training the model for each of them; hence, further research remains to deal with this issue.

References

1. Agarie, S., et al.: Salt tolerance, salt accumulation, and ionic homeostasis in an epidermal bladder-cell-less mutant of the common ice plant Mesembryanthemum crystallinum. J. Exp. Bot. **58**, 1957–1967 (2007)
2. Arvidsson, I., et al.: Generalization of prostate cancer classification for multiple sites using deep learning. In: IEEE 15th International Symposium on Biomedical Imaging (ISBI 2018), pp. 191–194. IEEE (2018). https://doi.org/10.1109/ISBI. 2018.8363552
3. Barkla, B.J., Vera-Estrella, R.: Single cell-type comparative metabolomics of epidermal bladder cells from the halophyte Mesembryanthemum crystallinum. Front. Plant Sci. **6**, 435 (2015)
4. Bochkovskiy, A., et al.: YOLOv4: optimal speed and accuracy of object detection. CoRR abs/2004.10934 (2020)
5. Casado-García, A., et al.: LabelStoma: a tool for stomata detection based on the YOLO algorithm. Comput. Electron. Agric. **178**, 105751 (2020). https://doi.org/ 10.1016/j.compag.2020.105751
6. Chu, P., Li, Z., Lammers, K., Lu, R., Liu, X.: Deep learning-based apple detection using a suppression mask R-CNN. Pattern Recogn. Lett. **147**, 206–211 (2021). https://doi.org/10.1016/j.patrec.2021.04.022. https://www.sciencedirect. com/science/article/pii/S0167865521001616
7. Cohen, J.: Statistical Power Analysis for the Behavioral Sciences. Academic Press, Cambridge (1969)
8. Cohen, J.: Eta-squared and partial eta-squared in fixed factor ANOVA designs. Educ. Psychol. Measur. **33**, 107–112 (1973)
9. Cynthia, S.T., et al.: Automated detection of plant diseases using image processing and faster R-CNN algorithm. In: Proceedings of 2019 International Conference on Sustainable Technologies for Industry 4.0. STI 2019 (2019). https://doi.org/10. 1109/STI47673.2019.9068092
10. Dassanayake, M., Larkin, J.C.: Making plants break a sweat: the structure, function, and evolution of plant salt glands. Front. Plant Sci. **8**, 406 (2017)
11. Garcia, S., et al.: Advanced nonparametric tests for multiple comparisons in the design of experiments in computational intelligence and data mining: experimental analysis of power. Inf. Sci. **180**, 2044–2064 (2010)
12. Holm, O.S.: A simple sequentially rejective multiple test procedure. Scand. J. Stat. **6**, 65–70 (1979)
13. Imamura, T., et al.: A novel WD40-repeat protein involved in formation of epidermal bladder cells in the halophyte quinoa. Commun. Biol. **3**, 513 (2020)
14. Isayenkov, S.V.: Genetic sources for the development of salt tolerance in crops. Plant Growth Regul. **89**(1), 1–17 (2019). https://doi.org/10.1007/s10725-019-00519-w
15. Kiani-Pouya, A., et al.: Epidermal bladder cells confer salinity stress tolerance in the halophyte quinoa and Atriplex species. Plant Cell Environ. **40**, 1900–1915 (2017)
16. Kiani-Pouya, A., et al.: A large-scale screening of quinoa accessions reveals an important role of epidermal bladder cells and stomatal patterning in salinity tolerance. Environ. Exp. Bot. **168**, 103885 (2019)
17. Levene, H.: chap. Robust tests for equality of variances. In: Contributions to Probability and Statistics: Essays in Honor of Harold Hotelling, pp. 278–292. Stanford University Press, USA (1960)

18. Orsini, F., et al.: Beyond the ionic and osmotic response to salinity in Chenopodium quinoa: functional elements of successful halophytism. Funct. Plant Biol. **38**, 818–831 (2011)
19. Pratama, M.T., et al.: Deep learning-based object detection for crop monitoring in soybean fields. In: Proceedings of 2020 International Joint Conference on Neural Networks. IJCNN 2020 (2020). https://doi.org/10.1109/IJCNN48605.2020.9207400
20. Razavian, A.S., Azizpour, H., Sullivan, J., et al.: CNN features off-the-shelf: an astounding baseline for recognition. In: CVPRW 2014, pp. 512–519 (2014)
21. Redmon, J., Farhadi, A.: YOLOv3: an incremental improvement. CoRR abs/1804.02767 (2018)
22. Ren, S., He, K., Girshick, R., Sun, J.: Faster R-CNN: towards real-time object detection with region proposal networks. Adv. Neural. Inf. Process. Syst. **28**, 91–99 (2015)
23. Shabala, L., et al.: Oxidative stress protection and stomatal patterning as components of salinity tolerance mechanism in quinoa (Chenopodium quinoa). Physiol. Plant. **146**, 26–38 (2012)
24. Shabala, S.: Learning from halophytes: physiological basis and strategies to improve abiotic stress tolerance in crops. Ann. Bot. **112**, 1209–1221 (2013)
25. Shapiron, S.S., Wilk, M.B.: An analysis for variance test for normality (complete samples). Inf. Sci. **180**, 2044–2064 (1965)
26. Sheskin, D.: Handbook of Parametric and Nonparametric Statistical Procedures. CRC Press, London (2011)
27. Simard, P., Steinkraus, D., Platt, J.C.: Best practices for convolutional neural networks applied to visual document analysis. In: Proceedings of the International Conference on Document Analysis and Recognition. ICDAR 2003, vol. 2, pp. 958–964 (2003)
28. Tan, M., Pang, R., Le, Q.V.: EfficientDet: scalable and efficient object detection. In: Proceedings of 2020 IEEE/CVF Conference on Computer Vision and Pattern Recognition. CVPR 2020 (2020). https://doi.org/10.1109/CVPR42600.2020.01079
29. Tian, Y., et al.: Apple detection during different growth stages in orchards using the improved YOLO-V3 model. Comput. Electron. Agric. **157**, 417–426 (2019). https://doi.org/10.1016/j.compag.2019.01.012
30. Tian, Z., et al.: FCOS: fully convolutional one-stage object detection. CoRR abs/1904.01355 (2019)
31. Tzutalin, D.: LabelImg (2015). https://github.com/tzutalin/labelImg
32. Zhu, C., He, Y., Savvides, M.: Feature selective anchor-free module for single-shot object detection. CoRR abs/1903.00621 (2019)

On the Performance of Deep Learning Models for Respiratory Sound Classification Trained on Unbalanced Data

Carlos Castorena[✉][ID], Francesc J. Ferri[ID], and Maximo Cobos[ID]

Departament d'Informàtica, Universitat de València, 46100 Burjassot, Spain
{carlos.castorena,francesc.ferri,maximo.cobos}@uv.es

Abstract. The detection of abnormal breath sounds with a stethoscope is important for diagnosing respiratory diseases and providing first aid. However, accurate interpretation of breath sounds requires a great deal of experience on the part of the clinician. In the past few years, a number of deep learning models have been proposed to automate lung classification tasks in physical examination. Unfortunately, acquiring accurately annotated data for this problem is not straightforward and important issues arise, as the available examples of abnormal and normal sounds usually differ substantially. This work provides a comprehensive analysis of deep learning models making use of different class balancing methods during training, considering multiple network architectures and audio input features. The results show that good performance is achievable when applying random oversampling and a convolutional neural network operating over Mel-frequency cepstral coefficient (MFCC) representations.

Keywords: Deep learning · Class imbalance · Respiratory diseases

1 Introduction

The timely detection of respiratory diseases such as COPD (Chronic Obstructive Pulmonary Disease) and pneumonia has become especially important in the context of the COVID-19 pandemic, as patients suffering from respiratory issues may experience greater complications when infected by the SARS-CoV-2 [2, 19, 23]. COPD is a group of diseases that includes emphysema and bronchitis, characterized by the obstruction of the flow of air in the respiratory system, while pneumonia is caused by an infection that inflames the lungs. Sound datasets such as the Respiratory Sound Database of ICBHI (International Conference on Biomedical Health Informatics), which has more than 5.5 h of respiratory cycle records with different diagnoses [24], provide a framework to develop automatic prediction tools for these diseases.

Different works have been carried out that address the classification of respiratory audios with different architectures, mostly using multilayer perceptrons

© Springer Nature Switzerland AG 2022
A. J. Pinho et al. (Eds.): IbPRIA 2022, LNCS 13256, pp. 143–155, 2022.
https://doi.org/10.1007/978-3-031-04881-4_12

(MLPs) [16], convolutional neural networks (CNNs) [3,6,20,25] and recurrent neural networks (RNN) [9,18,21]. These deep learning methods have been shown to provide better results than other traditional classification schemes such as support vector machines (SVMs) or quadratic discriminators (QD) [17].

This work aims at comparing the above deep learning architectures (MLP, CNN and RNN) for the classification of respiratory sounds into three classes: *Healthy, COPD* and *Pneumonia*, a problem that is addressed by Naqvi et al. in 2020 [17] and where the best result is obtained by using a quadratic discriminator (QD) using 116-dimensional feature vectors (FV) computed from signal descriptors including time-averaged Mel frequency cepstral coefficients (MFCCs) and gammmatone frequency cepstral coefficients (GFCCs). Moreover, empirical mode decomposition (EMD) and discrete wavelet transform (DWT)-based techniques are used to denoise and segment the pulmonic signals [3,9,17].

Unlike in [17], architectures such as MLP, CNN and RNN can process higher numbers of features or even the entire time-dependent MFCC and GFCC spectra as input. These approaches have more information available, which in turn helps the classifier to identify relevant patterns more easily. In [16], an MLP is proposed for the classification of healthy patients and patients with respiratory symptoms by using 330 features extracted from the audios. Perna [21] uses an RNN with an LSTM layer to classify the database into two classes: chronic (COPD, asthma) and non-chronic (respiratory tract infection, pneumonia and bronchiolitis). In Perna's work, the standardized MFCCs are used as input, reaching the conclusion that the architectures that exploit the time series information show an improvement in the classification task. In [22], a similar architecture is used to detect four classes (Crackle, Wheeze, Normal and Crackle-Wheeze). Srivastava [25] uses a CNN to classify the database into three classes: COPD, asthma and respiratory tract infections, obtaining its best result when characterizing the audios with MFCCs and applying data augmentation techniques (Loudness, Mask, Shift and Speed).

One common problem within the classification frameworks described above is that the training data used is significantly unbalanced. The difference in the number of available samples per class clearly affects the performance of the classifier. In this context, sampling techniques, e.g. random oversampling (ROS), synthetic minority oversampling technique (SMOTE) and variations, are used in [6,16,17,20] to generate synthetic samples and reduce negative effects during training. Table 1 shows a compilation of related work and the different techniques used to classify respiratory audios.

The present work tries to shed light on the interaction between class imbalance and some state of the art deep neural models in the particular context of detecting respiratory diseases from audio signals.

2   Data Description

The ICBHI Scientific Challenge is a database of 920 lung audio files from 128 patients with different diagnosis and a total duration of 5.5 h [24]. Each sound

Table 1. Related Works on lung sounds organized by classifier and input type (rows) and sampling/augmentation techniques used (columns). Works using additional EMD/DWT-based signal denoising are marked with *.

Classifier	Input	Original	Aug	AugBal	ROS	SMOTE	ADASYN
QD	FV						[17]*
MLP	MFCC				[5]		
	GFCC	[18]					
	FV				[16]		
CNN	MFCC	[10, 25]		[3, 25]*		[6, 20]	[6]
	GFCC	[18]					
RNN	MFCC	[9, 21]*					
	GFCC	[9, 22]*					

file is accompanied by information regarding to its acquisition, the corresponding diagnosis of the patient and information on the time distribution of the cycles of breathing. In this work, a three-class subset of the whole dataset is processed by segmenting the audio files into individual respiratory cycles according to the provided temporal information, generating a total of 6353 sound segments of which 5746 belong to the COPD class, 332 correspond to the healthy class and 285 to the pneumonia class.

3 Experimental Setup

3.1 Noise Filtering

Naqvi [17] points out the use of EMD and the DWT to decompose the original signal and erase the noise components. EMD is a method that divides the audio signal into components represented by a set of so-called intrinsic mode functions (IMFs). According to [17], $IMF2$ to $IMF4$ must be selected to generate a new signal without noise components. On the other hand, DWT decomposes a signal into several time series of coefficients that describe the time evolution of the signal in the corresponding frequency band. Reference work [17] uses the mother wavelet Coiflets5 due to its morphological resemblance and reconstructs the signal from the low-frequency DWT coefficients only.

3.2 Input Audio Representations

The selection of an appropriate input representation is one of the most important stages when implementing automatic learning algorithms for audio data. MFCCs have been traditionally used for a wide range of applications and are a common choice. Basically, they are obtained by analyzing the signal over relatively small time frames, grouping the energy according to a number of linearly spaced sub-bands in the Mel frequency scale and subsequently taking the discrete cosine transform [15]. MFCCs are known for their capability to emphasize relevant signal components and reduce noise. GFCCs, unlike MFCCs, use Gammatone filters instead of the triangular Mel scale filters, and it is also one of the most commonly used time-dependent representations for audio classification [11].

The FV representation used in this work consist of the mean of the MFCCs and GFCCs on the temporal axis. In this case, 64 frequency bands are considered both for MFCC and GFCC representations, leading to a FV with shape 1×128. Publicly available python packages Librosa[1] and Spafe[2] have been used to compute MFCCs and GFCCs, respectively.

3.3 Sampling Methods

The term imbalance refers to a significant difference in the number of samples from each of the classes. It can be measured using the so-called imbalance ratio, $I_r = N_M/N_n$, where N_M is the number of samples of the majority class and N_m is the number of samples of the minority class. The imbalance affects the performance of most classifiers [14], and different proposals have been suggested to reduce these effects. Random over sampling (ROS) [14] and synthetic minority over-sampling technique (SMOTE) [4] generate new samples by duplicating and interpolating the original ones, respectively. These two methods constitute straightforward preprocessing alternatives that have been extensively used in practice. SMOTE with edition (SMOTE-ENN) [1], SMOTE borderline (SMOTE-BL) [8] and adaptive synthetic sampling (ADASYN) [7] are further extensions of the original SMOTE. SMOTE-ENN seeks to reduce noise using the edited nearest neighbour (ENN) rule [27] to remove noisy samples in overlapping regions. SMOTE-BL generates samples close to the decision boundary. ADASYN calculates the local distribution to generate the corresponding number of samples per class. All the sampling methods considered are taken from Imblearn[3] and have been used with the default parameter setting.

3.4 Data Augmentation

Data augmentation is a technique that increases the number of samples available by applying variations that help to reduce overfitting effects. In the case of audio signals, data augmentation can be done in many different ways. In this work, two techniques are used: Gaussian noise addition and time stretching. Time stretching is a process by which the duration of the signal is changed without affecting its amplitude and frequency [12]. Gaussian noise addition generates normally distributed random noise samples with variable standard deviation. Srivastava [25] used the time stretch method by changing the duration of the audios between 0.5 and 2.0 times the original duration in a random way. We keep this setting as well for this work. The Gaussian noise standard deviation changes

---

[1] https://librosa.org/doc/latest/index.html.
[2] https://spafe.readthedocs.io.
[3] https://imbalanced-learn.org/stable/index.html.

randomly between 0.001 and 0.015. We apply data augmentation in two different ways: by maintaining class frequencies (Aug) and balancing them (AugBal).

3.5 Classifiers

We consider 4 different classifier structures in our experiments: QD, MLP, CNN and RNN. While QD only accepts FV as input, and both CNN and RNN are meant to be fed with time-dependent descriptors as e.g. the whole MFCC/GFCC spectrogram, the MLP may accept an FV or a flattened spectrogram as input. In particular, the configurations of the classifiers/inputs used in the present work are as follows:

- Quadratic discriminant (QD): This algorithm generates a quadratic decision frontier. This classifier has shown to provide good results in previous works [17] and can be considered as a baseline method not requiring the adjustment of hyperparameters.
- Multilayer perceptron (MLP): It is a type of neural network where each layer is fully connected to the previous one and is used to solve problems that are not linearly separable. In this work, 5 layers are used with 20, 10, 6, 5, and 4 neurons respectively for each layer. ReLU has been used as activation function and the first 3 layers use a dropout rate of 0.2 as a regularizer.
- Convolutional neural network (CNN): The basis of this type of networks are the convolutional layers that work as an adaptive feature extractor. The particular architecture used is the one proposed by [25] that has 4 convolutional layers with 16, 32, 64 and 128 filters, respectively, with kernel size of 2×2 and ReLU activation, followed by a max pooling layer with a kernel size of 2×2 and a 0.02 rate dropout layer. The output feature maps of the last convolutional block are globally averaged.
- Recurrent neural network (RNN): The architecture used in this work has 4 1D convolutional layers operating on the temporal axis, with kernel size 4×1 using ReLU activation, and dropout of rate 0.1, 0.15, 0.2, 0.25 for each layer, adding as well batch normalization and average pooling [22]. The time-dependent output of this convolutional block is connected to a bidirectional GRU layer with 128 units whose output is finally connected to two dense layers of 1024 neurons using ReLU activation.

A final three-neuron dense layer with SoftMax activation is a added to the MLP, CNN and RNN architectures. Also, in these three cases the training considers categorical cross-entropy as loss function and the adam optimizer along 50 epochs, all implemented in Tensorflow[4]. All the code for the classifiers and sampling methods implementation can be found in https://github.com/ccastore/Respiratory-Sound.

---

[4] https://www.tensorflow.org/.

3.6   Validation and Metrics

For this work, 5 different metrics have been considered to compare the performance of the above classifiers: accuracy (ACC), which represents the percentage of samples classified correctly; (mean) Recall, which is the average of the accuracy for each class; G-Mean, that uses geometric instead of arithmetic mean; Kappa coefficient, which shows the correlation between two observers, and the receiver operating characteristic (ROC) curves, which illustrate graphically the behavior of the classifiers with different discrimination thresholds [26].

Using the confusion matrix (C) the multiclass Accuracy (ACC) is defined as [13]:

$$ACC = \frac{\sum_{k=1}^{K} C_{kk}}{c} \tag{1}$$

where C_{kk} is the true positive count for class k and c is the total number of samples ($c = \sum_{i=1}^{K} \sum_{j=1}^{K} C_{ij}$). The Recall and the G-mean measures are given by [13]:

$$\text{Recall} = \frac{\sum_{k=1}^{K} \frac{C_{kk}}{t_k}}{K}, \quad \text{G-mean} = \sqrt[\kappa]{\prod_{k=1}^{K} \frac{C_{kk}}{t_k}} \tag{2}$$

where t_k is the number of samples from class k ($t_k = \sum_{i=1}^{K} C_{ik}$). The Kappa coefficient is the correlation between the real class and predicted labels [13]:

$$Kappa = \frac{s \times c - \sum_{k=1}^{K} p_k \times t_k}{c^2 - \sum_{k=1}^{K} p_k \times t_k} \tag{3}$$

where p_k is the number of samples predicted as class k ($p_k = \sum_{i=1}^{K} C_{ki}$), and s is the total number of correctly predicted elements ($s = \sum_{k=1}^{K} C_{kk}$).

4   Results and Discussion

Table 2 shows the results of the experiments measured in G-mean and the standard deviation calculated when averaging for the 5 repetitions. The G-mean measure has been put forward first because it is not so susceptible to imbalance and penalizes low classification values for any of the classes. The corresponding standard deviation indicates how susceptible the classifier is to the randomness introduced in the different training/testing partitions, to different classifier initializations, and to the samples selected or generated by each sampling or data augmentation method. The combination of data augmentation using Gaussian noise addition and EMD/DWT-based denoising has not been considered in the experiments as these constitute antagonistic processes with conflicting goals.

In general terms, the results when applying EMD and DWT are below their pairs without the application of these filters, except in specific cases such as *CNN/MFCC/orig* and *RNN/MFCC/orig* that went from having a G-mean of 0.8006 and 0.6967 to 0.8037 and 0.7725 respectively, which only occurs when

Table 2. Averaged G-mean and standard deviation corresponding to 5 different holdout partitions.

Classifier	Input	Orig	Aug	AugBal	ROS	SMOTE	SMOTE-BL	SMOTE-ENN	ADASYN
				Without EMD and DWT					
QD	FV	0.7980	0.7016	0.7102	0.7951	0.7897	0.7839	0.8026	0.7944
MLP	MFCC	0.4736	0.0000	0.9183	0.8675	0.8695	0.8602	0.9028	0.8464
	GFCC	0.0000	0.0000	0.7136	0.5825	0.5154	0.5254	0.1219	0.5082
	FV	0.6480	0.1803	0.9086	0.9684	0.9625	0.9624	0.9660	0.9564
CNN	MFCC	0.8006	0.8193	0.9501	0.9636	0.9225	0.9329	0.9333	0.9270
	GFCC	0.7491	0.7084	0.8645	0.8640	0.8518	0.8607	0.6729	0.8438
RNN	MFCC	0.6967	0.2636	0.9617	0.8002	0.8034	0.8327	0.8841	0.8651
	GFCC	0.3802	0.2532	0.7547	0.4179	0.4086	0.3381	0.6027	0.3728
				EMD DWT					
QD	FV	0.5166			0.5168	0.5957	0.5786	0.6310	0.5545
MLP	MFCC	0.0449			0.8045	0.8008	0.7625	0.8653	0.8149
	GFCC	0.0000			0.5628	0.5938	0.6305	0.0000	0.5812
	FV	0.0000			0.8133	0.8566	0.8334	0.8308	0.8331
CNN	MFCC	0.8037			0.9363	0.9116	0.9021	0.9073	0.9042
	GFCC	0.3368			0.7737	0.7608	0.7876	0.7094	0.7612
RNN	MFCC	0.7725			0.7507	0.8209	0.7381	0.8330	0.7713
	GFCC	0.3176			0.4201	0.4876	0.4986	0.6532	0.4942

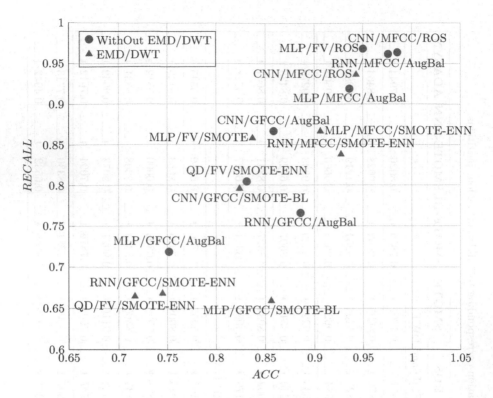

Fig. 1. Averaged ACC and (mean) Recall over 5 holdout partitions.

the classes are unbalanced. On the other hand, the standard deviation is correlated with the improvement of some results. For example, *MLP/MFCC/orig*, *MLP/FV/orig* and *RNN/MFCC/orig* have less dispersion so the results will tend to be similar regardless of the different variability factors involved. The best result with EMD and DWT is obtained using a *CNN* as classifier, *MFCC* as audio representation, and balancing the classes with *ROS* which leads to a G-mean of 0.9363 with a standard deviation of 0.0095. Not using the denoising filters raises this same result to 0.9684 with a standard deviation of 0.0100.

Table 2 shows that several *orig* and *Aug* results have high standard deviations, which indicates that algorithms like *MLP/MFCC*, *MLP/FV*, *CNN/MFCC*, *RNN/MFCC* are sensitive to the different randomness factors involved. On the other hand, results using different balancing options lead to generally lower standard deviations.

When using balanced data augmentation the best results are *MLP/MFCC*, *MLP/GFCC*, *CNN/GFCC*, *RNN/GFCC* and *RNN/MFCC*. The latter two have a significant advantage over using sampling techniques such as *ROS* or *SMOTE*, reaching G-mean of 0.7547 and 0.9617 respectively. This indicates that there is an advantage when using data augmentation for a *RNN* classifier, taking into account that the imbalance, like *MLP* and *CNN*, directly affects its performance.

It is worth noting that the *ROS* sampling method leads to the best overall results in terms of G-mean even though it is below data augmentation in most other configurations. In particular, *MLP/FV/ROS* and *CNN/MFCC/ROS* G-mean values are 0.9684 and 0.9636, respectively.

The remaining sampling methods *SMOTE*, *SMOTE-BL*, *SMOTE-ENN* and *ADASYN* lead to relatively similar results, below *ROS* and *Data Augmentation* but above using the original data.

Finally, it can be observed that filters like EMD and DWT harm the G-mean results, especially when using *GFCC* or *FV* as inputs since decomposition removes some of the information in the input signal.

From the whole group of results, it can be concluded that treating class imbalance has a significant impact on the different classifiers considered. But this also has a considerable impact on the computational burden in training.

The best results for each classifier and audio representation are presented in Fig. 1 in terms of ACC and mean Recall. The ACC, being a measure susceptible to imbalance, will tend to benefit the classification of the majority class, while the Recall will tend to benefit the two minority classes. Nevertheless, the three best methods in terms of G-mean are also the top ones here: *CNN/MFCC/ROS*, *RNN/MFCC/AugBal* and *MLP/FV/ROS*. In contrast, *RNN/GFCC/SMOTE-ENN* and *QD/FV/SMOTE-ENN*, both using EMD and DWT, are the worst performing ones.

For providing further insight, the average confusion matrices and false positive rates for all classes across the different partitions is shown in Fig. 2, expressed in percentages, for the previous best performing methods. Similarly, Table 3 presents as well the average values for all the considered performance metrics. Note that all three methods provide very competitive results. In fact, ACC and Kappa favor *CNN/MFCC/ROS* from a global perspective, while class-averaged metrics like G-mean and mean Recall, suggest *MLP/FV/ROS* as the best. However, one of the most relevant differences corresponds as well to a key aspect from a clinical perspective, which is the number of pathological examples erroneously classified as healthy. In fact, the *false healthy rate* is very low (0.002) for the CNN and more than eight times greater (0.017) in the case of MLP. This behavior is even more evident if we look at the two corresponding ROC curves shown in Fig. 3 where the false healthy rate of the plain classifiers is also plot along their standard deviations. The ROC curve corresponding to MLP degrades very quickly for lower values on the X axis. On the other hand, the CNN curve allows for far lower values of the false healthy rate while still keeping a convenient tradeoff. The behavior of the RNN lies somewhere between the previous two but is clearly below them according to all other performance measures shown in Table 3. Therefore, *CNN/MFCC/ROS* seems to be a more meaningful choice for this particular application.

Table 3. Average metrics for the best performing systems.

	$CNN/MFCC/ROS$	$RNN/MFCC/AugBal$	$MLP/VF/ROS$
ACC	**0.9855**	0.9759	0.9503
Recall	0.9639	0.9618	**0.9686**
G-mean	0.9636	0.9617	**0.9684**
Kappa	**0.9241**	0.8914	0.7775

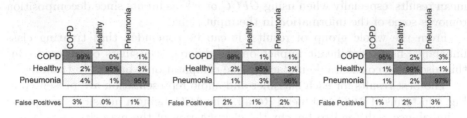

(a) CNN/MFCC/ROS   (b) RNN/MFCC/AugBal   (c) MLP/FV/ROS

Fig. 2. Average confusion matrices for the best performing systems, including false positive percentages for each class.

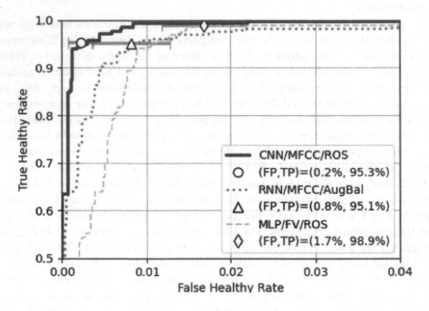

Fig. 3. ROC curves corresponding to the best systems considering *healthy* as the positive class.

5   Conclusions

This work has evaluated the use of several deep learning architectures (MLP, CNN and RNN) for addressing the problem of respiratory sound classification to detect pneumonia and COPD. Several strategies for tackling the class-imbalance problems of the available dataset were analyzed, as well as different pre-processing and input feature choices. The results show that, from the evaluated alternatives and specific configurations, deep learning architectures combined with the full MFCC description and class-balance algorithms lead to very competitive results. In particular, the CNN-based architecture using MFCC features and random oversampling performs best taking into account both global performance and the specific application at hand.

References

1. Batista, G., Prati, R., Monard, M.C.: A study of the behavior of several methods for balancing machine learning training data. SIGKDD Explor. **6**, 20–29 (2004). https://doi.org/10.1145/1007730.1007735
2. Bouazza, B., Hadj-Said, D., Pescatore, K.A., Chahed, R.: Are patients with asthma and chronic obstructive pulmonary disease preferred targets of COVID-19? Tuberc. Respir. Dis. **84**(1), 22–34 (2021). https://doi.org/10.4046/trd.2020.0101
3. Chanane, H., Bahoura, M.: Convolutional neural network-based model for lung sounds classification. In: 2021 IEEE International Midwest Symposium on Circuits and Systems (MWSCAS), pp. 555–558 (2021). https://doi.org/10.1109/MWSCAS47672.2021.9531887
4. Chawla, N.V., Bowyer, K.W., Hall, L.O., Kegelmeyer, W.P.: SMOTE: synthetic minority over-sampling technique. J. Artif. Intell. Res. **16**, 321–357 (2002). https://doi.org/10.1613/jair.953
5. Do, Q.T., Lipatov, K., Wang, H.Y., Pickering, B.W., Herasevich, V.: Classification of respiratory conditions using auscultation sound. In: 2021 43rd Annual International Conference of the IEEE Engineering in Medicine Biology Society (EMBC), pp. 1942–1945 (2021). https://doi.org/10.1109/EMBC46164.2021.9630294
6. García-Ordás, M.T., Benítez-Andrades, J.A., García-Rodríguez, I., Benavides, C., Alaiz-Moretón, H.: Detecting respiratory pathologies using convolutional neural networks and variational autoencoders for unbalancing data. Sensors **20**(4) (2020). https://doi.org/10.3390/s20041214
7. Haibo, H., Yang, B., Garcia, E.A., Shutao, L.: ADASYN: adaptive synthetic sampling approach for imbalanced learning. In: 2008 IEEE International Joint Conference on Neural Networks (IEEE World Congress on Computational Intelligence), pp. 1322–1328 (2008). https://doi.org/10.1109/IJCNN.2008.4633969
8. Han, H., Wang, W.-Y., Mao, B.-H.: Borderline-SMOTE: a new over-sampling method in imbalanced data sets learning. In: Huang, D.-S., Zhang, X.-P., Huang, G.-B. (eds.) ICIC 2005. LNCS, vol. 3644, pp. 878–887. Springer, Heidelberg (2005). https://doi.org/10.1007/11538059_91
9. Jayalakshmy, S., Sudha, G.: GTCC-based BiLSTM deep-learning framework for respiratory sound classification using empirical mode decomposition. Neural Comput. Appl. **33**, 17029–17040 (2021). https://doi.org/10.1007/s00521-021-06295-x

10. Kim, Y., Hyon, Y., Jung, S.: Respiratory sound classification for crackles, wheezes, and rhonchi in the clinical field using deep learning. Sci. Rep. **11** (2021). https://doi.org/10.1038/s41598-021-96724-7
11. Liu, G.K.: Evaluating gammatone frequency cepstral coefficients with neural networks for emotion recognition from speech. CoRR abs/1806.09010 (2018). http://arxiv.org/abs/1806.09010
12. Mahjoubfar, A., Churkin, D., Barland, S.: Time stretch and its applications. Nat. Photon. **11** (2017). https://doi.org/10.1038/nphoton.2017.76
13. Margherita, G., Enrico, B., Giorgio, V.: Metrics for multi-class classification: an overview (2020). https://arxiv.org/abs/2008.05756
14. Menardi, G., Torelli, N.: Training and assessing classification rules with imbalanced data. Data Min. Knowl. Disc. **28**(1), 92–122 (2012). https://doi.org/10.1007/s10618-012-0295-5
15. Molau, S., Pitz, M., Schluter, R., Ney, H.: Computing Mel-frequency cepstral coefficients on the power spectrum. In: 2001 IEEE International Conference on Acoustics, Speech, and Signal Processing. Proceedings (Cat. No.01CH37221), vol. 1, pp. 73–76 (2001). https://doi.org/10.1109/ICASSP.2001.940770
16. Monaco, A., Amoroso, N., Bellantuono, L., Pantaleo, E., Tangaro, S., Bellotti, R.: Multi-time-scale features for accurate respiratory sound classification. Appl. Sci. **10**(23) (2020). https://doi.org/10.3390/app10238606
17. Naqvi, S.Z.H., Choudhry, M.A.: An automated system for classification of chronic obstructive pulmonary disease and pneumonia patients using lung sound analysis. Sensors **20**(22) (2020). https://doi.org/10.3390/s20226512
18. Ngo, D., Pham, L., Nguyen, A., Phan, B., Tran, K., Nguyen, T.: Deep learning framework applied for predicting anomaly of respiratory sounds. In: 2021 International Symposium on Electrical and Electronics Engineering (ISEE), pp. 42–47 (2021). https://doi.org/10.1109/ISEE51682.2021.9418742
19. Olloquequi, J.: COVID-19 susceptibility in chronic obstructive pulmonary disease. Eur. J. Clin. Invest. **50**(10), e13382 (2020). https://doi.org/10.1111/eci.13382
20. Perna, D.: Convolutional neural networks learning from respiratory data. In: 2018 IEEE International Conference on Bioinformatics and Biomedicine (BIBM), pp. 2109–2113 (2018). https://doi.org/10.1109/BIBM.2018.8621273
21. Perna, D., Tagarelli, A.: Deep auscultation: predicting respiratory anomalies and diseases via recurrent neural networks. In: IEEE CBMS International Symposium on Computer-Based Medical Systems (2019). https://doi.org/10.1109/CBMS.2019.00020
22. Pham, L.D., McLoughlin, I.V., Phan, H., Tran, M., Nguyen, T., Palaniappan, R.: Robust deep learning framework for predicting respiratory anomalies and diseases. CoRR abs/2002.03894 (2020). https://arxiv.org/abs/2002.03894
23. Poggiali, E., Vercelli, A., Iannicelli, T., Tinelli, V., Celoni, L., Magnacavallo, A.: Covid-19, chronic obstructive pulmonary disease and pneumothorax: a frightening triad. Eur. J. Case Rep. Intern. Med. **7**(7) (2020). https://doi.org/10.12890/2020_001742
24. Rocha, B.M., Filos, D., Mendes, L.: A respiratory sound database for the development of automated classification. In: Precision Medicine Powered by pHealth and Connected Health, pp. 33–37 (2018). https://doi.org/10.1007/978-981-10-7419-6_6
25. Srivastava, A., Jain, S., Miranda, R., Patil, S., Pandya, S., Kotecha, K.: Deep learning based respiratory sound analysis for detection of chronic obstructive pulmonary disease. PeerJ. **7** (2021). https://doi.org/10.107717/peerj-cs.369

26. Vining, D.J., Gladish, G.W.: Receiver operating characteristic curves: a basic understanding. Radiographics **12**(6), 1147–1154 (1992). https://doi.org/10.1148/radiographics.12.6.1439017
27. Wilson, D.L.: Asymptotic properties of nearest neighbor rules using edited data. IEEE Trans. Syst. Man Cybern. **SMC-2**(3), 408–421 (1972). https://doi.org/10.1109/TSMC.1972.4309137

Automated Adequacy Assessment of Cervical Cytology Samples Using Deep Learning

Vladyslav Mosiichuk[1,2] , Paula Viana[2,3(✉)] , Tiago Oliveira[4], and Luís Rosado[1]

[1] Fraunhofer Portugal AICOS, 4200-135 Porto, Portugal
vladyslav.mosiichuk@aicos.fraunhofer.pt, luis.rosado@fraunhofer.pt
[2] School of Engineering, Polytechnic of Porto, 4249-015 Porto, Portugal
pmv@isep.ipp.pt
[3] INESC TEC, 4200-465 Porto, Portugal
[4] First Solutions - Sistemas de Informação S.A, 4450-102 Matosinhos, Portugal

Abstract. Cervical cancer has been among the most common causes of cancer death in women. Screening tests such as liquid-based cytology (LBC) were responsible for a substantial decrease in mortality rates. Still, visual examination of cervical cells on microscopic slides is a time-consuming, ambiguous and challenging task, aggravated by inadequate sample quality (e.g. low cellularity or the presence of obscuring factors like blood or inflammation). While most works in the literature are focused on the automated detection of cervical lesions to support diagnosis, to the best of our knowledge, none of them address the automated assessment of sample adequacy, as established by The Bethesda System (TBS) guidelines. This work proposes a new methodology for automated adequacy assessment of cervical cytology samples. Since the most common reason for rejecting samples is the low count of the squamous nucleus, our approach relies on a deep learning object detection model for the detection and counting of different types of nuclei present in LBC samples. A dataset of 41 samples with a total of 42387 nuclei manually annotated by experienced specialists was used, and the best solution proposed achieved promising results for the automated detection of squamous nuclei (AP of 82.4%, Accuracy of 79.8%, Recall of 73.8% and F1 score of 81.5%). Additionally, by merging the developed automated cell counting approach with the adequacy criteria stated by the TBS guidelines, we validated our approach by correctly classifying an entire subset of 12 samples as adequate or inadequate.

Keywords: Cervical cancer · Cervical cytology · Machine learning · Deep learning · Adequacy assessment · Nuclei detection

1 Introduction

Cervical cancer is the fourth most frequently diagnosed cancer and the fourth leading cause of cancer death in women. Nearly 605 000 new cases were

© Springer Nature Switzerland AG 2022
A. J. Pinho et al. (Eds.): IbPRIA 2022, LNCS 13256, pp. 156–170, 2022.
https://doi.org/10.1007/978-3-031-04881-4_13

registered in 2020, and about 342 000 died from cervical cancer worldwide [21,23]. Screening tests, such as cytology, have been responsible for a strong decrease in cervical cancer deaths over the past years. Screening reduced the incidence of cervical cancer by 60%–90%, and the death rate by 90% [13]. The increasing interest in the development of computer-aided diagnosis (CADx) systems for cervical screening is related to difficulties experienced by health facilities due to a shortage of specialized staff and equipment. CADx systems often use machine learning and deep learning techniques to reduce the dependence on manual input and increase their autonomy.

In a recent review article [2], the authors analyzed and discussed focus and adequacy assessment, segmentation and computational classification approaches used for the analysis of microscopic images acquired from cervical cytology smears. Among the general conclusion about the technical state of computer vision methods, the authors outlined smear adequacy assessment as a topic scarcely addressed in the literature. Most of the works ignore it, while others implement some techniques to detect and remove unwanted objects such as inflammatory cells, dirt, blood, or other artifacts. According to The Bethesda System (TBS) guidelines, the sample should present at least 5000, or 3.8 per microscopic field at 40× magnification, well-preserved squamous nuclei to be considered adequate.

This paper presents a new automated deep learning approach for the assessment of cytological specimen adequacy. Major focus was given to cellularity assessment, being low squamous cellularity the most common cause for the unsatisfactory specimens. In particular, this work aims to automatically count the number of squamous nuclei in a liquid-based cytology (LBC) sample, and consequently classify it as adequate or inadequate.

The rest of this paper is structured as follows: Sect. 2 highlights some of the works described in the literature that addressed similar problems or approaches. Section 3 presents the methodology, including the system overview, dataset description and details about the experimental setup. Section 4 describes the achieved results. And, finally, the conclusions and future work are drawn in Sect. 5.

2 Related Work

General cell detection, segmentation, and counting are rather well addressed in the literature, and most recent proposals mostly rely on machine learning and deep learning approaches. Particularly, [4,6,10] describe and propose using deep learning models and architectures such as U-net and FPN networks to produce cells detection and segmentation. In [8] the author describes using Single Shot Detector (SSD) in pair with a Convolutional Neural Network (CNN) to localize blood cells and then count them separately. On the other hand, [24] proposes a microscopy cell counting based on density estimation employing fully convolutional regression networks. While all of the approaches output the number of cells, the segmentation also outputs the mask from the detected objects. In decreasing order of output complexity, the detection task normally results only in the localization of the object and a bounding box around it. Lastly, density estimation only gives the final number of objects. In all approaches mentioned above, authors outline results with performance comparable with a human specialist.

A less explored field in the literature is to devise methods that enable implementing efficient smear quality control approaches that could contribute to increase the performance of the detection. In [25], the authors describe an AI assistive diagnostic solution to improve cervical liquid-based thin-layer cell smear diagnosis according to clinical TBS criteria. The developed system consists of five AI models which are employed to detect and classify the lesions. The quality assessment is done on the entire sample and comprises technical image characteristics, such as focus and contrast, and quantitative cell evaluation. In contrast to the detection and classification tasks, the cell counting task relied on simpler methods. The number of the cells was estimated by, firstly, engaging the Otsu thresholding method to separate the cells from the background and then by calculation of the cell to overall area ratio to obtain a rough number of present cells in the sample. The authors report a 99.11% average accuracy on the task of samples classification as satisfactory and unsatisfactory on the validation sets. However, it must be noticed that in their approach, the total number of cells was estimated, while the TBS guidelines specify a minimum threshold specifically for squamous nuclei.

Besides the quality of the data and the performance of the detection methods, some effort has also been put into developing low-cost, portable microscopes that enable supporting microscopy-based diagnosis in areas with limited access or without enough financial resources. As an alternative to benchtop microscopes, and due to the impressive evolution in the quality of the cameras, processing power and memory, smartphone-based solutions are good candidates for implementing a cost-effective platform for microscopic inspection of samples. The work in [5] describes a portable smartphone based brightfield microscope for screening blood smears. A wide range of applications have also been used to test the feasibility of affordable approaches based on smartphones. Examples include the detection of viruses [22], the quantification of immunoassays, the automated classification of parasites [19], etc. [7,17].

Regarding the cervical cytology use case, a device called μSmartScope [18] was recently adapted for the digitalization of cervical cytology samples [1], being then used for the automated detection of cervical lesions [20] (Fig. 1). This device is a fully automated 3D-printed smartphone microscope tailored to support microscopy-based diagnosis in areas with limited access. Being fully powered and controlled by a smartphone, in addition to the motorized stage, the device aims to decrease the burden of manual microscopy examination. Following this stream, we aim at developing a methodology for the automated adequacy assessment that can be deployed in a solution based on the μSmartScope device.

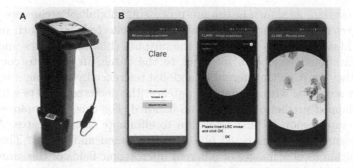

Fig. 1. Mobile-based framework for the automated detection of cervical lesions: (**A**) μSmartScope with smartphone attached and LBC sample inserted; (**B**) smartphone application screenshots (from [20]).

3   Methodology

3.1   System Overview

To cope with the low-cost requirements for the system under development, our solution shall be supported by the portable microscope μSmartScope [18], with all the software running on the smartphone. Since the device uses an objective magnification of 40×, several images of the different microscopic fields must be acquired for each slide, in order to cover a representative sample area. The most common approaches for cellularity assessment on LBC slides are based on the comparison with reference images or counting well-preserved squamous cells in a defined number of fields at high (40×) or low power (10×) magnification [3]. Since LBC slides provide relatively clean images where a human eye can distinguish most of the cells individually, as can be observed in Fig. 2, our approach will rely on counting those target objects separately. Figure 2 also provides illustrative microscopic fields of adequate (2a and 2b) and inadequate (2c and 2d) samples.

Figure 3 depicts the general pipeline of the proposed solution. The first operation slices the images into patches of fixed dimensions, with the optional pos-

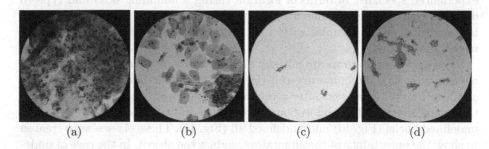

(a)   (b)   (c)   (d)

Fig. 2. Illustrative images acquired with the μSmartScope of adequate (a and b) and inadequate (c and d) LBC samples.

sibility of overlap, as shown in the pre-processing module. Patches are then fed to the object detection model, and outputs are collected. Since each patch will pass individually through the model, the outputs (bounding boxes and respective classifications) will be in reference to the patch. In order to count them properly, they must be transferred to a global reference by merging into a single image. For scenarios with overlapped patches, the detected objects with two or more overlapped predictions need to be handled. For this, we employed a non-maximum suppression (NMS) algorithm to eliminate the duplicates. After the post-processing, all the detected objects are counted and stored. The process will start again as new images of different microscopic fields of the same sample become available for analysis.

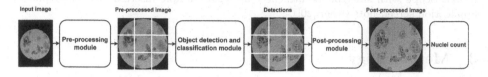

Fig. 3. General pipeline of the proposed solution.

3.2 Dataset

Even though there are some publicly available datasets with cervical cells annotations, such as Herlev [9], SIPaKMeD Database [15], Cervix93 [14], and ISBI Challenges [11,12], they are not adequate for our purpose. In the case of Herlev and SIPaKMeD, they only contain isolated images of cell nuclei, with annotations by abnormality of the cell. While ISBI Challenges, Cervix93, and the more recent CRIC [16] dataset offer images with the surrounding area, they are not annotated in terms of the type of nuclei.

Given that none of the available datasets satisfies our requirements, the µSmartScope was used to create a new dataset comprised of 139 samples with approximately 100 images in each, taken along the sampling area with amplification of 40×. At the image level, each nucleus was manually annotated by an experienced specialist in terms of location (using a bounding box) and type (6 different classes were considered). Currently, the annotated part of the dataset is constituted by 41 samples with a total of 765 annotated images and 42387 nuclei annotations.

The nuclei annotations are initially divided into four main classes: squamous nucleus (Fig. 4a), inflammatory cell (Fig. 4b), glandular nucleus (Fig. 4c), and artifact (Fig. 4f). However, after receiving feedback from the specialists during the initial stages of the annotation process, two additional classes were added: undefined nuclei (Fig. 4d) and undefined all (Fig. 4e). These classes were created to show the uncertainty of the annotators against the object. In the case of undefined nuclei, the annotator meant that it could be either squamous or glandular.

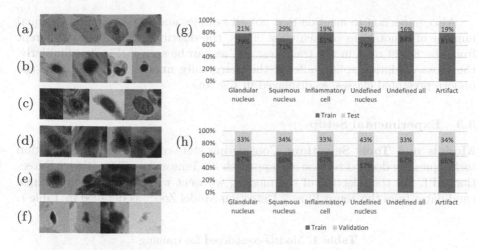

Fig. 4. Examples of present classes in the dataset: Squamous nucleus (a); Inflammatory cell (b); Glandular nucleus (c); Undefined nuclei (d); Undefined all (e); Artifact (f);. And distribution of annotations per class per train/test sets split (g) and train/validation subsets (h).

In the case of undefined all, it means that it can be squamous, inflammatory, or glandular. As per artifact, they can be defined as structures that are generally not present in the living tissue and should not be mistaken with nuclei.

At the sample level, the entire dataset of 139 samples is annotated in terms of abnormal changes and sample adequacy. The latter is particularly interesting for our case, since the annotation between adequate and inadequate samples enables an adequacy evaluation of the proposed approach at the sample level. Unfortunately, our dataset is significantly imbalanced regarding the adequacy annotation. Only 5 samples out of the 139 were labelled as inadequate (only 3 of them with nuclei annotations), an aspect that slightly limits our analysis at the sample level.

Notwithstanding, the part of the dataset that was annotated at the image level is still relatively big, totalling 42387 annotations. Predominant types such as squamous nuclei and inflammatory cells account for 21973 and 17914 annotations, respectively, whether the distribution of the other four classes is the following: Glandular nucleus: 684; Undefined all: 997; Undefined nucleus: 667; Artifact: 152.

For the training and evaluation purposes, the dataset was split at the sample level, with an 80% to 20% ratio. The split was done manually in order to ensure the presence of inadequate samples in the test set and to ensure the more or less homogeneous distribution of different classes. For the hyperparameter tuning, cross-validation was used with three folds with random split at the image level.

Also, important to mention that not all samples and images contain an equal number of annotations of each class. Thus, the distribution is slightly different from the 80/20 ratio in the train/test set, as can be seen in Fig. 4g.; Similarly, the cross-validation splits suffered the marginally uneven distribution in the underrepresented classes, as displayed in Fig. 4h.

3.3 Experimental Setup

Models and Tools Selection: Considering state-of-the-art object detection techniques, we decided to base our approach on TensorFlow (TF) Object Detection API. For the selection of the models, we went with the selection of three candidates from the TensorFlow 2 Detection Model Zoo[1], as detailed in Table 1.

Table 1. Models considered for training.

Model name	Input size	Speed (ms)	COCO mAP
SSD MobileNet V2 FPNLite	320 × 320	22	22.2
SSD ResNet50 V1 FPN	640 × 640	46	34.3
EfficientDet D0	512 × 512	39	33.6

The selection of these models was based mainly on the speed of inference and the performance on the COCO mAP metric[2]. Accounting for mobile-based platforms and the prioritization of speed also affected the selected input size of the models. To be able to perform any experiments with different batch sizes, we chose the models with 320 × 320 pixels of input size. It is important to note that the TensorFlow Model Zoo does not provide pre-trained ResNet50 and Efficient-Det models with the input size specified above. However, the configuration file allows changing the input size. Also, the raw size of images is much bigger compared to the input of the models. Therefore a slicing into 320 × 320 patches of raw images from the dataset was performed, taking into consideration the best preservation of annotations.

Hyperparameter Tuning and Training: The further experimental setup consisted of training the selected models and fine-tuning the hyperparameters. Among all training parameters, the batch size and learning rate (LR) seemed to be more promising to result in a performance increase. Through experimentation, we determined that the maximum batch size that allowed successful training for ResNet50 and MobileNet was 16, while for EfficientDet was 1. Thus, we established that for ResNet50 and MobileNet, we would try three different values

---

[1] https://github.com/tensorflow/models/blob/master/research/object_detection/ g3doc/tf2_detection_zoo.md.
[2] https://cocodataset.org/#detection-eval.

of batch size: 16, 8, and 4; and only batch size of 1 for EfficientDet. For the learning rate, we randomly generated five values on the logarithmic scale between values 1 and 1×10^{-6} and ensured an even distribution of the values, obtaining the following LR values: 3.06×10^{-1}; 2.50×10^{-2}; 5.29×10^{-4}; 1.81×10^{-5} and 1.24×10^{-6};

With a defined set of parameters to evaluate, there were still two variables that needed to be defined prior to the conduction of any experiments: stop criteria and the learning rate scheduler function. Several preliminary model trainings were conducted with different numbers of epochs to determine the optimal stop criteria. The test resulted in 50 epochs being the optimum value, both for letting the model achieve the plateau and not to extend the training time unnecessarily. As per learning rate schedulers, we tested schedulers like exponential decay, linear decay, and constant LR value, with and without the warm-up. The tests showed that there is a difference between constant value and scheduler with decay. In the case of the constant value, the model performed worse than with the decay, and while the first epochs showed almost no difference, at the last ones, the model with constant LR value was still unstable, whereas the model with decay was converging. The tests with warm-up and without didn't result in substantial differences, aside from being able to smooth a learning curve with higher LR values. However, since these variables were not so promising to improve the model quality, no further tests were conducted in regard to the scheduler or a maximum number of steps. As a result, the exponential decay scheduler was selected as the default scheduler for the rest of the experiments with the following parameters: 500 decay steps; decay factor of 0.97.

For other parameters such as optimizer and loss functions, we went with the default selection on the configuration file of pre-trained models: Momentum (with momentum value of 0.9) for optimizer; weighted smooth l1 for classification and weighted sigmoid focal for localization loss functions. We took into consideration other optimizers, such as Adam, for the fact of being widely recommended among the literature. However, as we discovered later, with some tests where the only variable was the optimizer, it seemed only to affect, as expected, the learning curve, i.e., how early the model will converge. In spite of these findings, for this set of tests, optimizer and other parameters were considered less important for the performance of the final model, or, at least, less promising to help achieve better results.

Then, the test ensemble for hyperparameter tuning consisted of testing three different batch sizes and five different values of the learning rate for each of the selected models on each cross-validation fold. The training process was evaluated continuously in order to ensure model convergence. Otherwise, the training process was aborted.

To cope with the clearly unbalanced dataset, we decided to test the model training with a reduced number of classes. The first approach consisted in eliminating the underrepresented classes and seeing if their presence does make any difference in the model performance. The second approach relied on merging underrepresented and similar classes. These two new datasets, then, consisted of (squamous nucleus and inflammatory cell) and (squamous nucleus, inflammatory cell and others) classes.

All experiments were conducted on a server with two AMD EPYC 7302 and several Nvidia Tesla T4 GPU. However, the configuration of the particular Virtual Machine (VM) consisted of only four virtualized CPU cores, 16 GB of RAM, and one shared Nvidia Tesla T4 with a little less than 8 GB of VRAM available.

3.4   Model Evaluation

The evaluation of the models was performed in two stages: during the model training and final system evaluation. While the first was carried out with the objective of hyperparameter tuning at the patch level on the validation set, the latter was required to assess final system performance on the test set at image and sample levels. During the model training, we used two main metrics to evaluate the performance: average precision (AP) per class and average recall (AR). However, during the final system evaluation, we included other metrics to deeply evaluate the system. These metrics consisted of Accuracy, Specificity, F1 score, Youden's index, and more basic approaches such as false negative (FN), false positive (FP) and true positive (TP), as well as critical analysis of the confusion matrix. Also, the final system offers three main variables to tune: percentage of overlap of sliced patches, non-maximum suppression (NMS) threshold and the minimum intersection over union (IoU) value for a true positive prediction. Another batch of tests was performed in order to encounter the best set of parameters for the final system.

Finally, a study was carried out to assess the adequacy classification at the sample level. The test subset consisted of 12 samples, five inadequate and seven adequate, with around 100 images in each. These samples lack nuclei annotations, thus, our approach relied on counting all detected nuclei in each sample and averaging the number of squamous nuclei per image. Applying the adequacy criteria stated by the TBS for an adequate sample, which specify at least 3.8 well preserved squamous nuclei per image, each sample was then classified as adequate or inadequate. Afterwards, each classification was compared to the ground truth, and the number of correct classifications was counted.

Given this, a total of 419 experiments were performed, being 135 for hyperparameters tuning, 278 for tuning the overlap, NMS, and IoU thresholds on the final system, and 6 for soft class imbalance handling. For the hyperparameters tuning and class imbalance, each experiment consisted of one training process and respective evaluation on the validation set. Regarding tuning final system parameters, each experiment consisted of detecting all present nuclei in the images of the tests set and comparing it to ground truth annotations.

4 Results and Discussion

Hyperparameter Tuning: The first difficulty encountered was regarding the model EfficientDet. Unfortunately, the change of the input size in the configuration only led to an early failing training process. Accounting for the fact that we only could run the model with the batch size of 1 and, having obtained mediocre results from one successful test, we decided to run the rest of the test only on Mobilenet V2 and ResNet50.

The general observation of the results from the hyperparameters test showed that the bigger batch size, namely 16, resulted in better model performance, as can be seen in Fig. 5a. Observation of the Fig. 5b showcase the best performing LR value: 5.29×10^{-4}. Also, from the observation of average precision per class, Fig. 5c, the poor performance of underrepresented classes is noticeable. We performed an additional set of tests with all classes merged except squamous and inflammatory. Also, tests with only one (squamous nucleus) and two (squamous nucleus and inflammatory cell) classes were conducted. In both cases, new tests resulted in similar or worse performance than in the test with all classes included.

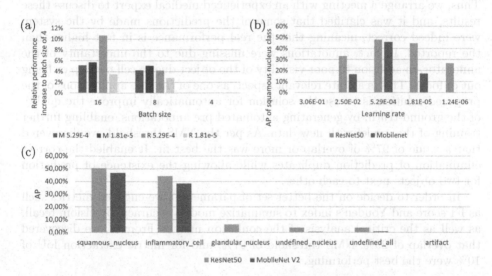

Fig. 5. Results of hyperparameters tests performed on cross-validation folds. Results are averaged across all folds. Image (a) outlines relative performance increase of bigger batch size values when compared to batch size of 4. Image (b) depicts AP of squamous nucleus class across different values of LR with batch size of 16. Image (c) shows AP across different classes with LR = 5.29×10^{-4} and batch size = 16.

With the best model (ResNet50) and parameters obtained in hyperparameter tuning tests, we trained the model on the entire training set and evaluated on the test set, obtaining at this stage the following performance metrics: squamous nucleus AP@50: 47.3%; inflammatory cell AP@50: 37.8%; AR@100: 43%.

Given the demarked imbalance in the dataset regarding the number of annotations of squamous and inflammatory versus the remaining classes, the poor results regarding the latter were already expected. Since the model appears only to be capable to properly generalize to these two highly represented classes, and being the squamous nuclei the key class for adequacy assessment, most of the following results and respective discussion are focused on this class.

Final System Parametrization: First tests and visual inspection of final model predictions revealed that some correct predictions were reported as false positives due to low intersection area and an IoU threshold set to 50%. Through testing, we discovered that a threshold of 10% enables correct reporting of true positives without compromising performance and increasing misclassifications.

We also tested the overlap of sliced patches and a threshold for the non-maximum suppression algorithm. It became apparent that an increasing overlap produced more predictions. By visually analyzing these results, we realized that a significant number of false positives detected by our model had very similar image characteristics to several annotated objects being detected as true positives. Thus, we arrange a meeting with an experienced medical expert to discuss these results, and it was clarified that some of the predictions made by the system were indeed correct, meaning that the real performance is in fact higher than the reported. In fact, annotations were missing due to the uncertainty of the annotator, in addition to poor visibility of the object due to cell overlap or being out of focus. This is a quite relevant aspect, as one of the side applications of the proposed solution is to use the solution for automatically improve the quality of the ground-truth, by generating automated pre-annotations, enabling further training of the model with new data. As per the NMS threshold, we discovered that a value of 97% of overlap or more was the best fit. It enabled the correct elimination of prediction duplicates while allowing the existence of prediction for two objects next to each other.

In order to decide on the better set of parameters, we engaged metrics such as F1 score and Youden's index to summarize models accuracy, precision, recall, as well as the critical analysis of the confusion matrix. From it, we discovered that overlap of 29%, NMS threshold of 97%, and for metric extraction IoU of 10% were the best performing.

Image Level Results: At the image level, the system reported an AP of 82.4%, Accuracy of 79.8%, and F1 score of 81.5% for the class of squamous nucleus. In terms of raw detection, the system made a total of 5216 predictions for 5483 existing annotations with a true positive rate of 74% and 473 false positives, which can be explored with more details in Table 2 and confusion matrix present in Fig. 6(c). Given the visual results shown in Fig. 6(b) and the previous conclusions concerning unannotated objects, these values are however expected to be higher, as some of them could be considered true positives if the dataset was correctly annotated.

It is worth mentioning that better results in terms of the recall could be achieved with a higher overlap value. However, aside from increasing processing requirements or time for assessment, the false positives rate will also increase. Thus, to correctly evaluate if overlap increasing brings performance benefits, we consider that the required dataset ground-truth quality improvement previously discussed should be handled first, to avoid tuning this parameter based on incorrectly unannotated nuclei.

Table 2. Performance metrics of the final system.

Class	Squamous	Inflammatory	Glandular	Und. n.	Und. all	Artifact
AP	0.8236	0.7072	0.2257	0.0	0.0	0.0
Recall	0.7380	0.6274	0.0347	0.0	0.0	0.0
Accuracy	0.7984	0.8310	0.9864	0.9837	0.9847	0.9973
Specificity	0.8893	0.9232	0.9993	0.9999	1.0	1.0
F1	0.8147	0.6982	0.0641	0.0	0.0	0.0
Youden index	0.6273	0.5505	0.0341	-0.0001	0.0	0.0
TP	4743	2091	5	0	0	0
FP	473	566	7	1	0	0
FN	1684	1242	139	173	164	29

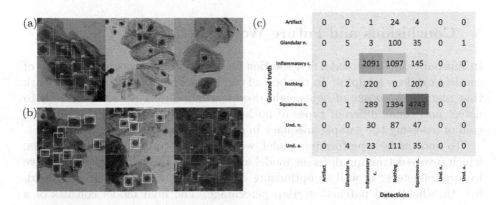

Fig. 6. Detections and classifications made by the final system (a) and (b). Image (a) depicts mostly correct detection and classifications (green and blue bounding boxes means correct classifications of two different classes), while on the image (b) yellow and red bounding boxes correspond to misclassification and false positives detections respectively. Image (c) shows the confusion matrix of the test set. "Nothing" corresponds to false positives and false negatives. (Color figure online)

Sample Level Results: After the evaluation of the proposed approach at the image level, a study to assess sample adequacy at the sample level was conducted. In particular, a set of 12 samples (5 inadequate and 7 adequate) was used for that purpose, with each sample consisting of around 100 images of different microscopic fields. The final model was then used to count the number of squamous nuclei on each image, and the average number of squamous nuclei per sample was computed. The results are depicted in Fig. 7a, where it is visible that we were able to classify all samples correctly as adequate or inadequate, by using the threshold of 3.8 squamous nuclei per image stated by the TBS guidelines.

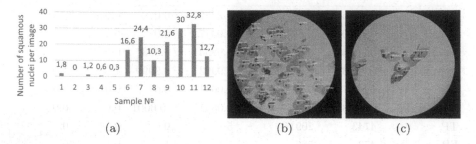

(a)   (b)   (c)

Fig. 7. Results for the proposed solution: (a) average number of detected squamous nuclei per sample (sample 1 to 5 are inadequate and 6 to 12 adequate); (b) and (c) illustrative images of automated nuclei detection at the image level.

5   Conclusions and Future Work

In this paper, we proposed a new solution for automated adequacy assessment of cervical cytology samples. To the best of our knowledge, this was the first work that merged the usage of lightweight deep learning object detection models to detect and count different types of nuclei on LBC samples, with the ultimate goal of assessing the sample adequacy in compliance to the TBS guidelines.

To find the best performing model, we executed a total of 419 experiments, which involved training different model architectures and tuning the respective hyperparameters, as well as optimizing other parameters like NMS threshold, IoU threshold and patches' overlap percentage. The final model consists of a SSD detector with a ResNet50 backbone, which yielded promising results for sample adequacy assessment, both at the image and sample levels. At the image level, interesting performance metrics were obtained for detecting and counting squamous nuclei, a crucial cellular structure to assess LBC samples' cellularity. At the sample level, our system was able to correctly classify 12 out of 12 LBC samples as adequate or inadequate.

Despite the apparent success of the proposed approach, several shortcomings of the current version of the dataset were identified and should be addressed in future works. Improving annotations quality, increasing dataset size, or balancing underrepresented classes will undoubtedly foster performance gains, both on automatic nuclei detection and adequacy assessment.

As future work, it would be interesting to explore the usage of the developed model to improve dataset quality in a semi-supervised manner, either by improving the already annotated data, or to automatically pre-annotate new data. Furthermore, additional model-centric techniques such as fine-tuning online data augmentation or anchor box optimization could also be explored. However, the importance of quality data to this task cannot be overstated, so we believe that innovative data-centric approaches will certainly hold higher potential for improvements.

Regarding the future deployment and usage of the proposed solution in a clinical context, further improvements and integration efforts are obviously required. Nevertheless, it becomes clear that this work corresponds to a significant step forward in the development of innovative solutions that bring real efficiency gains to current cervical cancer screening procedures.

Acknowledgements. This work was financially supported by the project Transparent Artificial Medical Intelligence (TAMI), co-funded by Portugal 2020 framed under the Operational Programme for Competitiveness and Internationalization (COMPETE 2020), Fundação para a Ciência and Technology (FCT), Carnegie Mellon University, and European Regional Development Fund under Grant 45905. The authors would like to thank the Anatomical Pathology Service of the Portuguese Oncology Institute - Porto (IPO-Porto).

References

1. Brandão, P., Silva, P.T., Parente, M., Rosado, L.: μsmartscope: towards a low-cost microscopic medical device for cervical cancer screening using additive manufacturing and optimization. Proc. Inst. Mech. Eng. Part L J. Mater. Des. Appl. (2021)
2. da Conceição, T., Braga, C., Rosado, L., Vasconcelos, M.J.M.: A review of computational methods for cervical cells segmentation and abnormality classification. Int. J. Mol. Sci. **20**, 5114 (2019)
3. Eurocytology: Criteria for adequacy of a cervical cytology sample. https://www.eurocytology.eu/en/course/1142. Accessed 02 Nov 2021
4. Falk, T., et al.: U-Net: deep learning for cell counting, detection, and morphometry. Nat. Methods **16** (2019). https://doi.org/10.1038/s41592-018-0261-2
5. de Haan, K., et al.: Automated screening of sickle cells using a smartphone-based microscope and deep learning. npj Digit. Med. **3**(1) (2020). https://doi.org/10.1038/s41746-020-0282-y
6. Hernández, C.X., Sultan, M.M., Pande, V.S.: Using deep learning for segmentation and counting within microscopy data (2018)
7. Holmström, O., et al.: Point-of-care mobile digital microscopy and deep learning for the detection of soil-transmitted helminths and schistosoma haematobium. Global Health Action **10**, 1337325 (2017). https://doi.org/10.1080/16549716.2017.1337325
8. Huh, I.: Blood cell detection using singleshot multibox detector (2018)
9. Jantzen, J., Norup, J., Dounias, G., Bjerregaard, B.: Pap-smear benchmark data for pattern classification. In: Nature Inspired Smart Information Systems (NiSIS), January 2005

10. Ke, J., Jiang, Z., Liu, C., Bednarz, T., Sowmya, A., Liang, X.: Selective detection and segmentation of cervical cells. In: ICBBT 2019: Proceedings of the 2019 11th International Conference on Bioinformatics and Biomedical Technology, pp. 55–61, May 2019. https://doi.org/10.1145/3340074.3340081

11. Lu, Z., et al.: Evaluation of three algorithms for the segmentation of overlapping cervical cells. IEEE J. Biomed. Health Inform. **21**, 1 (2016). https://doi.org/10.1109/JBHI.2016.2519686

12. Lu, Z., Carneiro, G., Bradley, A.: An improved joint optimization of multiple level set functions for the segmentation of overlapping cervical cells. IEEE Trans. Image Process. **24** (2015). https://doi.org/10.1109/TIP.2015.2389619. IEEE Signal Processing Society

13. Marth, C., Landoni, F., Mahner, S., McCormack, M., Gonzalez-Martin, A., Colombo, N.: Cervical cancer: ESMO clinical practice guidelines for diagnosis, treatment and follow-up. Ann. Oncol. **29** (2018). https://doi.org/10.1093/annonc/mdy160

14. Phoulady, H.A., Mouton, P.R.: A new cervical cytology dataset for nucleus detection and image classification (Cervix93) and methods for cervical nucleus detection. CoRR abs/1811.09651 (2018). http://arxiv.org/abs/1811.09651

15. Plissiti, M., Dimitrakopoulos, P., Sfikas, G., Nikou, C., Krikoni, O., Charchanti, A.: SIPAKMED: a new dataset for feature and image based classification of normal and pathological cervical cells in pap smear images. In: 25th IEEE International Conference on Image Processing (ICIP), pp. 3144–3148, January 2018. https://doi.org/10.1109/ICIP.2018.8451588

16. Rezende, M.T., et al.: Cric cervix cell classification (2020)

17. Rivenson, Y., et al.: Deep learning enhanced mobile-phone microscopy. ACS Photonics **5** (2017). https://doi.org/10.1021/acsphotonics.8b00146

18. Rosado, L., et al.: μSmartScope: towards a fully automated 3D-printed smartphone microscope with motorized stage. In: Peixoto, N., Silveira, M., Ali, H.H., Maciel, C., van den Broek, E.L. (eds.) BIOSTEC 2017. CCIS, vol. 881, pp. 19–44. Springer, Cham (2018). https://doi.org/10.1007/978-3-319-94806-5_2

19. Rosado, L., Correia da Costa, J.M., Elias, D., Cardoso, J.: Automated detection of malaria parasites on thick blood smears via mobile devices. Procedia Comput. Sci. **90**, 138–144 (2016). https://doi.org/10.1016/j.procs.2016.07.024

20. Sampaio, A.F., Rosado, L., Vasconcelos, M.J.M.: Towards the mobile detection of cervical lesions: a region-based approach for the analysis of microscopic images. IEEE Access **9**, 152188–152205 (2021)

21. Sung, H., et al.: Global cancer statistics 2020: GLOBOCAN estimates of incidence and mortality worldwide for 36 cancers in 185 countries. CA Cancer J. Clin. **71** (2021). https://doi.org/10.3322/caac.21660

22. Wei, Q., et al.: Fluorescent imaging of single nanoparticles and viruses on a smart phone. ACS Nano **7** (2013). https://doi.org/10.1021/nn4037706

23. WHO: Cancer today (2021). https://gco.iarc.fr/today/fact-sheets-cancers. Accessed 11 Feb 2021

24. Xie, W., Noble, J., Zisserman, A.: Microscopy cell counting and detection with fully convolutional regression networks. Comput. Methods Biomech. Biomed. Eng. Imaging Visual. 1–10 (2016). https://doi.org/10.1080/21681163.2016.1149104

25. Zhu, X., et al.: Hybrid AI-assistive diagnostic model permits rapid TBS classification of cervical liquid-based thin-layer cell smears. Nat. Commun. **12**, 3541 (2021). https://doi.org/10.1038/s41467-021-23913-3

Exploring Alterations in Electrocardiogram During the Postoperative Pain

Daniela Pais[✉][iD], Susana Brás[iD], and Raquel Sebastião[iD]

Institute of Electronics and Informatics Engineering of Aveiro (IEETA) Department of Electronics, Telecommunications and Informatics (DETI), University of Aveiro, 3810-193 Aveiro, Portugal
{danielapais,susana.bras,raquel.sebastiao}@ua.pt

Abstract. The assessment of postoperative pain after a surgical procedure is a critical step to guarantee a suitable analgesic control of pain, and currently, it is based on the self-reports of the patients. However, these assessment methods are subjective, discontinuous, and inadequate for evaluating the pain of patients unable or with limited ability to communicate verbally. Developing an objective and continuous tool for assessing and monitoring postoperative pain, which does not require patient reports could assist pain management during the patient stay in the post-surgery care unit and, ultimately, promote better recovery. In the last years, the evaluation of pain through physiological indicators has been investigated. In the present work, electrocardiogram (ECG) signals collected from 19 patients during the postoperative period were studied in order to find relationships between physiological alterations and pain and identify which ECG-feature or combination of ECG-features better describe postoperative pain.

Considering a multivariate approach, analysing the performance of sets of two or more ECG-features using clustering algorithms proved to be promising, allowing the identification of different pain characteristics based on the extracted features from the ECG signals.

Keywords: Clustering · Electrocardiogram · Data acquisition · Feature extraction · Feature selection · Physiological pain assessment · Postoperative pain

1 Introduction

After surgery, patients commonly suffer from postoperative pain, which combines nociceptive and inflammatory mechanisms. This type of pain, with a long duration, arises from recent surgical intervention due to tissue injury and trauma and due to the release of inflammatory chemical factors [17–19].

The inadequate management of pain can lead to psychological and physiological adverse effects. Undertreatment may contribute to tachycardia and

A. J. Pinho et al. (Eds.): IbPRIA 2022, LNCS 13256, pp. 171–181, 2022.
https://doi.org/10.1007/978-3-031-04881-4_14

increased cardiac metabolic demand, whereas overtreatment may lead to respiratory complications, delirium, oversedation, urinary retention, ileus, and postoperative nausea and vomiting [13]. Inadequate treatment is also related to chronic pain, deep vein thrombosis, pulmonary embolism, coronary ischemia, myocardial infarction, pneumonia, poor wound healing, insomnia, demoralization, postoperative stress, and opioid addiction [1,17,19]. It is also believed that children and neonates exposed to untreated pain can experience long-term consequences, such as altered pain sensitivity and maladaptive pain [4,16]. Therefore, postoperative pain should be alleviated rapidly and managed effectively to decrease patient discomfort, promote better recovery, and reduce postoperative morbidity and mortality [1,13,17].

A suitable choice of analgesic intervention for pain control is dependent on an accurate assessment [15]. Nonetheless, the use of self-reports from the patient to assess pain, which consists of the current standard method, may be hazarded by the age, cognitive condition, and verbal communication capabilities of the patient [14,20]. Moreover, when evaluating the pain of children, elderly, or cognitive-impaired patients, for example, pain needs to be assessed by an observer who can recognize behaviors, such as facial expressions or body movements, exhibited by the patient when in distress, and detect indicators in physiological measures, including heart rate (HR), blood pressure, and respiratory rate [16,20]. However, the assessment tools used in such cases produce a third-person subjective report, and may be, therefore, inaccurate due to observational bias. The assessment can be a challenging task for the healthcare provider due to individual differences in pain expression [6,11]. Besides, when the assessment scales describe several parameters, as most tools for this group of patients, the task can be complex and time-consuming [16,20].

The main goal of this study consisted of the investigation of physiological alterations of the electrocardiogram (ECG) signal in response to postoperative pain while exploring if the use of a higher number of ECG-features improves the description of postoperative pain. Our aims are to investigate the influence of pain on the features of the ECG signal and identify combinations of relevant features associated with pain.

The paper is organized as follows: the next section presents the related work concerning the objective measurement of postoperative pain through physiological signals. Section 3 presents the proposed methodology. Sections 4 and 5 describe and discuss the obtained results. Conclusions and future work are presented in the last section.

2   Related Literature

In recent years, objective measurement of pain has been a topic under investigation as it could provide an important complement to self-reports or provide a measure for those who cannot communicate. The use of physiological signals as objective markers of nociception and pain is one of the strategies proposed in the literature as they could allow monitoring changes in the nervous system induced by pain [7].

Regarding the ECG, HR and heart rate variability (HRV) have been explored. HR has shown variable correlation with numerical rating scale (NRS) in the postoperative period [12,14]. In addition, it did not change after administration of analgesia and had low performance in discriminating intense pain [12]. However, HRV measures, such as the Analgesia Nociception Index (ANI), correlated with the pain scale implemented in the studies [2,3,5,9].

Furthermore, the number of fluctuations of skin conductance per second (NFSC), which is extracted from electrodermal activity (EDA), was investigated in adults and children, showing a significant correlation with the pain scales [10,14,15]. Additionally, NFSC was significantly different between no, mild, moderate, and severe pain reports ($p < 0.0001$) [14,15], and was significantly lower after administration of analgesic for pain management ($p < 0.013$) [15].

Lastly, surface electromyography (sEMG) signals and accelerometry activity were studied in infants to assess muscle activity and body movement, respectively. Accelerometry-based overall extremity activity and EMG-based wrist flexor activity showed a significant correlation with the pain scale. However, the accelerometery-EMG based indicator showed better results in discriminating pain than the individual signals and, therefore, authors conclude that the combined use of more than one signal could be valuable in the evaluation of pain [18].

Several strategies produced important developments and findings in the field of objective indicators of postoperative pain, but most of them are based on the analysis of only one feature and one physiological signal.

3 Methods

This section describes the methodology used to accomplish the proposed goals. The experiments were performed in Python, mainly using Neurokit2[1] (which provides biosignal processing routines), SciPy[2] (which provides algorithms for statistics), and scikit-learn[3] (which supports supervised and unsupervised learning).

3.1 Data Collection

After approval by the Ethics Committee for Health of Centro Hospitalar Tondela-Viseu (CHTV) and by the Ethics and Deontology Council of the University of Aveiro, and after obtaining written informed consent from participants, 19 patients (with ages ranging from 23 to 83 years) undergoing elective abdominal surgery at CHTV were included.

At the entrance in the recovery room, after being stabilized by the clinical team, electrodes were placed on the body of the patient, and the positioning was maintained through the recording. The ECG signal was collected (sampling

---

[1] https://neurokit2.readthedocs.io/en/latest/.

[2] https://scipy.org/.

[3] https://scikit-learn.org/stable/.

frequency of 500 Hz), with minimally invasive equipment VitalJacket® [8], during the standard clinical practices of analgesia and without compromising the patient's wellbeing.

After arrival in the recovery room, and once able to communicate, patients were asked to rate their pain at different time points on a verbal scale and using descriptive phrases and adjectives. The clinical interventions, including pain assessment and management with analgesic therapeutic, were registered through a trigger manually annotated.

As the clinical team considered that the patient was stable for discharge, the data collection ended, and the electrodes were removed.

3.2 Preprocessing

The preprocessing of the ECG signals included removal of the last samples of the ECG signals that consisted of periods with excessive noise, filtering, and normalization in relation to the patient baseline (which, in this study, is considered to correspond to the last 10 min of useful ECG) to reduce inter-individual differences. The noise from ECG signals was filtered with a bandpass Butterworth filter, of order 4, with cut-off frequencies of 0.5 Hz and 40 Hz.

3.3 Feature Extraction and Transformation

Feature extraction was done using time intervals of 5 and 10 min associated with the assessment of pain. A total of 44 features were calculated, which are enumerated in Table 1.

Previous to the multivariate analysis, all features were normalized into a range of [0, 1] in order to transform the extracted features into the same scale to ensure a fair comparison, avoid misleading results, and improve the performance of the machine learning models.

3.4 Univariate Approach

All features were tested for normal distribution using the Shapiro-Wilk test. Spearman correlation coefficient (ρ) was used to describe the correlation between features. In addition, the Mann-Whitney U test (with statistical significance level set at 5%) was used to test differences of the data within tolerable pain, reported as no or mild pain, and intense pain, reported as moderate or severe pain.

The features were also studied through boxplots, using notched boxes, to visualize the distribution of the features and assess differences according to pain reported.

3.5 Unsupervised Multivariate Approach

An unsupervised multivariate approach was implemented to inspect the combination of features information and how the different subset of features perform in binary clustering tasks.

Table 1. Extracted features from collected ECG signals (using the Neurokit2 package).

ECG	Extracted Features
	Mean of RR intervals (MeanNN); median of the absolute values of the successive differences between RR intervals (MedianNN); standard deviation of RR intervals (SDNN); standard deviation of successive differences of RR intervals (SDSD); root mean square of successive differences of RR intervals (RMSSD); SDNN/MeanNN (CVNN); RMSSD/MeanNN (CVSD); proportion of RR intervals greater than 50ms, out of the total number of RR intervals (pNN50); proportion of RR intervals greater than 20ms, out of the total number of RR intervals (pNN20); baseline width of the RR intervals distribution obtained by triangular interpolation (TINN); total number of RR intervals divided by the height of the RR intervals histogram (HTI); spectral power density of the low-frequency band (LF); spectral power density of the high-frequency band (HF); LF/HF; natural logarithm of HF (LnHF); normalized low frequency (LFn); normalized high frequency (HFn); sample entropy (SampEn); approximate entropy (ApEn); Poincaré plot standard deviation perpendicular the line of identity (SD1); Poincaré plot standard deviation along the line of identity (SD2); SD1/SD2; mean, median, maximum, minimum, standard deviation and the difference between the highest and lowest HR; mean and median of successive differences of RR, SS, and TT intervals; mean and median of R, S, and T waves peaks amplitude; mean and median area of all cardiac cycles; sample and approximate entropy of ECG

Although the data is labelled, our goal in performing unsupervised learning was to uncover similarities between samples, independently of the labels, and perceive if the algorithms recognize associations accordingly with the labels or if they discover other relations that could be connected to the subjectivity of pain associated with inter-individual differences. In short, the aim is to extract useful information concerning postoperative pain and expose groups that are similar.

This multivariate analysis allows exploring if the features present similar behavior and are grouped accordingly to the reported pain intensity. Therefore, relying on the group assigned by the clustering models, we compare with respect to the pain level, evaluating the performance using f1-score, sensitivity, and specificity.

Three clustering algorithms were tested: k-means, agglomerative and spectral techniques. The k-means algorithm was randomly initialized and were performed 10 runs with different centroid seeds, with 300 maximum iterations for each initialization of the algorithm. The agglomerative clustering used the euclidean distance as the distance metric between the data points and the ward linkage to calculate the proximity of clusters, which minimizes the variance of the clusters being merged. Lastly, for the spectral algorithm, the affinity matrix was constructed using radial basis function (RBF) kernel and the labels in the embedding space were assigned by k-means, with the number of run times defined as 10.

Before clustering, feature selection was accomplished to remove redundant features and reduce the complexity and run time. Firstly, the correlation between features was considered, and from a pair of strongly correlated features, the one with higher variance was selected. Features were considered strongly correlated if the correlation coefficient was superior to 0.9. Afterwards, backward feature selection was used to select sets composed of 2 to 10 features. Backward elimination starts with the set of all available features, and one feature at a time is removed. This algorithm performs k-fold cross-validation when removing the feature at each iteration to find the best one to remove from the set based on a greedy procedure. Support vector machine (SVM) was used as an estimator.

When groups with more than two features achieved the best results, principal component analysis (PCA) was applied, in order to reduce dimensionality by projecting the data to a 2D dimensional space, for data visualization purposes.

4 Results

4.1 Dataset

The dataset is composed of ECG signals and data regarding the patient and clinical interventions, namely patient's age and gender, type of surgical intervention, and the procedures performed during the patient stay in the recovery room, including self-reports of pain, pain relief therapeutics, and patient repositioning.

A variable number of pain assessment points was obtained for each patient during ECG recording, which were grouped into four categories: "no pain", "mild", "moderate", and "severe". A total of 32 pain readings were obtained, with 12 assessments in patients with no pain, 4 in those with mild pain, 11 with moderate pain, and 5 in those with severe pain.

4.2 Univariate Approach

Initially, all the computed features were studied individually according to the level of pain reported. Among the extracted features, significant differences between tolerable and intense pain reports were found only in entropy measures.

Concerning the features from the 10-min intervals, the Mann-Whitney U test did not report significant differences between the groups (with the exception of the ECG_SampEn).

Regarding the features extracted from 5-min intervals, the returned p-values ($p < 0.05$) of the Mann-Whitney U test indicate the rejection of the null hypothesis for the two nonlinear HRV metrics, HRV_SampEn and HRV_ApEn, and for the ECG_SampEn. Thus, it allows concluding that the median values of these features from both levels of pain are significantly different.

Figure 1 represents the distribution of these three features. It can be noticed that ECG_SampEn presents smaller values for tolerable pain level, while the median of HRV_SampEn and HRV_ApEn are smaller for intense pain level, for both time intervals of 5 and 10 min. It can also be observed that HRV_ApEn and ECG_SampEn presented higher variance in the intense pain group, oppositely to HRV_SampEn, which tolerable pain group had superior variance.

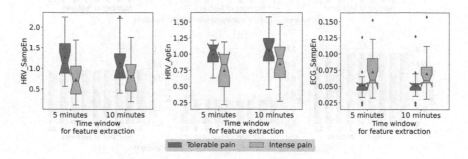

Fig. 1. Boxplots of ECG features representing the distribution among tolerable pain and intense pain. The red triangle represents the mean. (Color figure online)

4.3   Unsupervised Multivariate Approach

Using a multivariate approach, an exploration of how the clustering performs to separate tolerable pain (16 samples) from intense pain (16 samples) was conducted. Figure 2 presents the assessment metrics when matching the clustering group results with the 2 levels of pain, varying the number of selected features from 2 to 10, for the 5-min and 10-min data intervals.

It can be observed that the features selected from the 10-min analysis provided inferior results (which was already disclosed through the univariate analysis). Concerning the 5-min features, the best overall result was obtained using the agglomerative algorithm with eight and nine features (f1-score = 73.3%, sensitivity = 68.8%, and specificity = 81.2%). However, pain could be described with similar performance with only two features (HRV_SampEn and MaxMin_HR) and using any of the three clustering algorithms (f1-score = 71.4%, sensitivity = 62.5%, and specificity = 87.5%). Regarding the 10-min analysis, the best performance is respective to the k-means clustering with four features, including Mean_Heartbeats_Area, Median_Amplitude_T_waves, pNN20, and LnHF (f1-score = 73.2%, sensitivity=93.8%, and specificity = 37.5%).

Figure 3 presents scatter plots, showing the relationship between the different sets of features that achieved the best results, and representing the data from the tolerable pain and intense pain with blue and red, respectively.

The data corresponding to the two 5-min features is distributed in the scatterplot in Fig. 3a. This plot is divided into the two regions of each cluster of the k-means. In Fig. 3, middle and right, it is also represented the projection into the two principal components of the sets with eight and with four selected features, respectively. For visualization purposes, PCA was applied to reduce the dimension of the set composed by the eight 5-min features

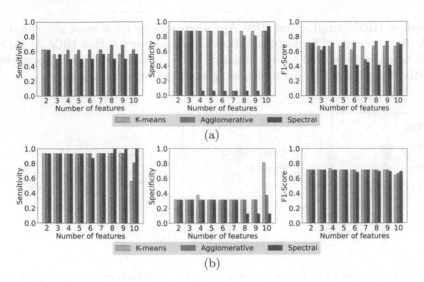

Fig. 2. Sensitivity, specificity, and f1-score obtained for the study of tolerable and intense pain using features extracted from (a) 5-min and (b) 10-min intervals.

(HRV_SampEn, MaxMin_HR, pNN20, HTI, LF, MinHR, MaxHR, and StdevHR) and of the set composed by the four 10-min features (Mean_Heartbeats_Area, Median_Amplitude_T_waves, pNN20, and LnHF). When observing the distribution of the two components obtained with PCA regarding the four 10-min features, in Fig. 3c, although it is possible to identify two possible groups, the scatter plot does not show a distinctive boundary between the two pain classes.

5 Discussion

This work investigated the potential of using the physiological signal ECG as an assessment tool of nociception and postoperative pain. The concept is that a painful stimulus will induce autonomic responses that can be assessed by physiological markers of postoperative pain.

In the postoperative period, the 5-min entropy measures (HRV_SampEn, HRV_ApEn, and ECG_SampEn) exhibited differences among tolerable and intense pain ($p < 0.05$ returned from the Mann-Whitney U test). Thus, relying on the statistical test and the respective distribution, the obtained results lead to believe that the physiological characteristics exposed by these features are feasible in discriminating among pain levels.

With respect to the multivariate analysis, the best results were obtained with the 5-min features, allowing a pain description through the combination of only two features, similar to the one achieved by a greater number of features. Although the eight features provided a superior f1-score, their two-dimensional representation in Fig. 3b, through the projection into the 2 principal components, does not enable to distinguish two groups of pain among the data points.

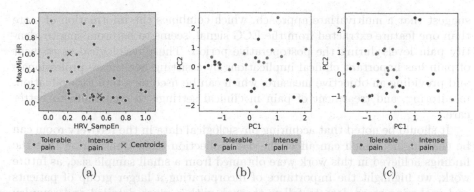

Fig. 3. Scatter plots representing the sets of features which achieved the best results for each time analysis regarding tolerable and intense pain. (a) Distribution of the set of two features extracted from 5-min intervals. (b) Two dimensional representation of the set of eight features extracted from 5-min intervals. (c) Two dimensional representation of the set of four features extracted from 10-min intervals. (Color figure online)

Nevertheless, the scatter plot of the set of two features in Fig. 3a reveals more noticeable differences between tolerable and intense pain. There are several observations from both classes, six "intense pain" reports and fourteen "tolerable pain" reports, concentrated in lower MaxMin_HR values, that were placed in the "tolerable pain" group. The other data points in the opposite cluster have higher MaxMin_HR values. The majority of them (10) correspond to "intense pain" reports, which indicates that a wider range between extremes in the HR could be characteristic of more severe pain. Even though, a sensitivity of 62.5% is low, as several intense pain samples fall incorrectly in the "tolerable pain" group (FN = 6), which would correspond to undertreatment.

Different features were selected when comparing both time analyses, and, despite not being possible to perfectly separate the two groups, several feature groups provided satisfactory and feasible results. Although the multivariate approach did not allow to separate the two pain groups, the univariate analysis showed that intense pain expose physiological behaviors distinct to the ones found on patients who reported tolerable pain. Nonetheless, it is essential to acknowledge the main limitations of this study and interpret the findings carefully. One substantial setback in the study is the limited number of participants and, consequently, the number of samples is also smaller, which limited the research scope.

6   Conclusions and Future Work

Patients commonly manifest postoperative pain after surgery. It consists of an unpleasant experience that needs to be treated effectively. Considering that the choice of analgesic technique for pain treatment depends on assessing the pain intensity, an adequate assessment is critical. This study analysed the physiological response of ECG signals during postoperative pain. Overall the results suggest that a multivariate approach, which combines the information of more than one feature extracted from the ECG signal, seems to be promising to identify pain levels during the postoperative period. The physiological assessment of pain has important clinical implications by assisting healthcare professionals and providing an objective measure, which can be used as a tool for evaluation, monitoring, and treatment of pain in clinical settings to contribute to better care.

It should be noted that acquiring physiological data in the recovery room can be challenging, which can affect the data collection task. Considering that the findings achieved in this work were obtained from a small sample size, as future work, we highlight the importance of incorporating a larger group of patients and gathering more data to follow through with a more detailed and complex analysis.

Pain is a subjective experience due to inter-individual variability, as pain perception is affected by several factors, such as genetics, age, emotional state, personality, and cognitive components. Thus, more patients should be considered to enable deeper research. A larger group of individuals would permit additional sex and aged based studies to find group-based differences and the consideration of the impact of the surgical procedure and the anesthesia protocol in the postoperative pain experience.

Considering the complexity and multidimensionality of postoperative pain in which physiological reaction manifests from autonomic, psychological, and behavioral responses, combining different physiological signals as a multisignal approach may be valuable and a more reliable method by representing different information of the body's reaction to pain as the human body mechanisms make up a complex system. In addition to the ECG signal, EDA and EMG signals, which have shown promising results on the topic, can also be collected.

Acknowledgments. This work was funded by national funds through FCT - Fundação para a Ciência e a Tecnologia, I.P., under the Scientific Employment Stimulus - Individual Call - CEECIND/03986/2018, and is also supported by the FCT through national funds, within IEETA/UA R&D unit (UIDB/00127/2020). This work is also funded by national funds, European Regional Development Fund, FSE through COMPETE2020, through FCT, in the scope of the framework contract foreseen in the numbers 4, 5, and 6 of the article 23, of the Decree-Law 57/2016, of August 29, changed by Law 57/2017, of July 19.

Particular thanks are due to the clinical team for allowing and supporting the researchers of this work during the procedure of data collection. The authors also acknowledge all volunteers that participated in this study.

References

1. Apfelbaum, J.L., Chen, C., Mehta, S.S., Gan, T.J.: Postoperative pain experience: results from a national survey suggest postoperative pain continues to be undermanaged. Anesth. Analg. **97**(2), 534–540 (2003)

2. Boselli, E., et al.: Prediction of immediate postoperative pain using the analgesia/nociception index: a prospective observational study. Br. J. Anaesth. **112**(4), 715–721 (2014)
3. Boselli, E., et al.: Prospective observational study of the non-invasive assessment of immediate postoperative pain using the analgesia/nociception index (ani). Br. J. Anaesth. **111**(3), 453–459 (2013)
4. Brand, K., Al-Rais, A.: Pain assessment in children. Intensive Care Med. **20**(6), 314–317 (2019)
5. Chang, L.H., Ma, T.C., Tsay, S.L., Jong, G.P.: Relationships between pain intensity and heart rate variability in patients after abdominal surgery: a pilot study. Chin. Med. J. **125**(11), 1964–1969 (2012)
6. Coghill, R.C.: Individual differences in the subjective experience of pain: New insights into mechanisms and models. Headache **50**(9), 1531–1535 (2010)
7. Cowen, R., Stasiowska, M.K., Laycock, H., Bantel, C.: Assessing pain objectively: the use of physiological markers. Anaesthesia **70**(7), 828–847 (2015)
8. Cunha, J.P., Cunha, B., Pereira, A., Xavier, W., Ferreira, N., Meireles, L.: Vitaljacket®: A wearable wireless vital signs monitor for patients' mobility in cardiology and sports. In: 4th International ICST Conference on Pervasive Computing Technologies for Healthcare (PervasiveHealth), pp. 1–2 (2010)
9. Gall, O., et al.: Postoperative pain assessment in children: a pilot study of the usefulness of the analgesia nociception index. Br. J. Anaesth. **115**(6), 890–895 (2015)
10. Hullet, B., et al.: Monitoring electrical skin conductance: a tool for the assessment of postoperative pain in children? Anesthesiology **111**(3), 513–517 (2009)
11. Hummel, P., van Dijk, M.: Pain assessment: current status and challenges. Semin. Fetal Neonatal. Med. **11**(4), 237–245 (2006)
12. Kantor, E., Montravers, P., Longrois, D., Guglielminotti, J.: Pain assessment in the postanaesthesia care unit using pupillometry. Eur. J. Anaesthesiol. **31**(2), 91–97 (2014)
13. Kwon, A.H., Flood, P.: Genetics and gender in acute pain and perioperative opioid analgesia. Anesthesiol. Clin. **38**(2), 341–355 (2020)
14. Ledowski, T., Bromilow, J., Paech, M.J., Storm, H., Hacking, R., Schug, S.A.: Monitoring of skin conductance to assess postoperative pain intensity. Br. J. Anaesth. **97**(6), 862–865 (2006)
15. Ledowski, T., Bromilow, J., Wu, J., Paech, M.J., Storm, H., Schug, S.A.: The assessment of postoperative pain by monitoring skin conductance: results of a prospective study. Anaesthesia **62**(10), 989–993 (2007)
16. Maxwell, L.G., Fraga, M.V., Malavolta, C.P.: Assessment of pain in the newborn: An update. Clin. Perinatol. **46**(4), 693–707 (2019)
17. Pogatzki-Zahn, E.M., Segelcke, D., Schugb, S.A.: Postoperative pain-from mechanisms to treatment. Pain Reports 2 (2017)
18. Schasfoort, F.C., Formanoy, M., Bussmann, J., Peters, J., Tibboel, D., Stam, H.J.: Objective and continuous measurement of peripheral motor indicators of pain in hospitalized infants: a feasibility study. Pain **137**(2), 323–331 (2008)
19. Segelcke, D., Pradier, B., Pogatzki-Zahn, E.: Advances in assessment of pain behaviors and mechanisms of post-operative pain models. Curr. Opin. Physio. **11**, 85–92 (2019)
20. Storm, H.: Changes in skin conductance as a tool to monitor nociceptive stimulation and pain. Current Opin. Anesthesiology **21**(6), 796–804 (2008)

Differential Gene Expression Analysis of the Most Relevant Genes for Lung Cancer Prediction and Sub-type Classification

Bernardo Ramos[1], Tania Pereira[1] (ID), Francisco Silva[1,2], José Luis Costa[3,4,5] (ID),
and Hélder P. Oliveira[1,2(✉)] (ID)

[1] INESC TEC - Institute for Systems and Computer Engineering,
Technology and Science, Porto, Portugal
{tania.pereira,helder.f.oliveira}@inesctec.pt
[2] FCUP - Faculty of Science, University of Porto, Porto, Portugal
[3] i3S - Instituto de Investigação e Inovação em Saúde, University of Porto,
Porto, Portugal
[4] IPATIMUP - Institute of Molecular Pathology and Immunology of the University
of Porto, Porto, Portugal
[5] FMUP - Faculty of Medicine, University of Porto, Porto, Portugal

Abstract. An early diagnosis of cancer is essential for a good prognosis, and the identification of differentially expressed genes can enable a better personalization of the treatment plan that can target those genes in therapy. This work proposes a pipeline that predicts the presence of lung cancer and the subtype allowing the identification of differentially expressed genes for lung cancer adenocarcinoma and squamous cell carcinoma subtypes. A gradient boosted tree model is used for the classification tasks based on RNA-seq data. The analysis of gene expressions that better differentiate cancerous from normal tissue, and features that distinguish between lung subtypes is the main focus of the present work. Differential expressed genes are analyzed by performing hierarchical clustering in order to identify gene signatures that are commonly regulated and biological signatures associated with a specific subtype. This analysis highlighted patterns of commonly regulated genes already known in the literature as cancer or subtype-specific genes, and others that are not yet documented in the literature.

Keywords: AI-based prediction models · Feature importance · Gene signatures · Lung cancer

1 Introduction

Lung cancer is the deadliest cancer type, responsible for approximately 20% of the cancer-related deaths in 2007 [2]. The most frequent histological subtypes are small cell lung cancer (SCLC) and non-small cell lung cancer (NSCLC), being the

© Springer Nature Switzerland AG 2022
A. J. Pinho et al. (Eds.): IbPRIA 2022, LNCS 13256, pp. 182–191, 2022.
https://doi.org/10.1007/978-3-031-04881-4_15

later divided into adenocarcinoma (LUAD), squamous cell carcinoma (LUSC), and large cell carcinoma (LCC) cases. The complexity and heterogeneity of the physiological phenomena and genetic profiles found even within the same histological subtype have been comprehensively studied [8]. The early diagnosis of cancer is a key factor for higher survival chances, and an accurate histological subtype diagnosis plays an important role in the treatment plan definition, which is based on the specific biomarkers to be targeted in therapy. Ribonucleic acid-sequencing (RNA-seq) is a technique that examines the quantity and sequences of ribonucleic acid (RNA) in a sample using next-generation sequencing (NGS). It analyzes the transcriptome of gene expression patterns encoded within RNA [20]. Gene signatures have been studied to complement pathological factors in diagnosis, prognostic information, patient stratification, prediction of therapeutic response, and better surveillance strategies [3,13]. The gene signatures of cancer will allow to address clinical questions and optimize the use of therapeutic resources. The powerful AI-based tools could push the multi-omics analysis to improve the understanding of cancer biology and clinical outcomes.

Previous works have done a huge effort to improve the predictive performance of type and histological subtype of lung cancer. Yuan et al. [11] studied the association between somatic point mutations and cancer types/subtypes. A deep learning (DL) approach achieved an accuracy of 0.601 on the classification and allowed to extract the high level features between combinatorial somatic point mutations and cancer types. Danaee et al. [5] proposed learning models for breast cancer detection, with the identification of genes critical for the diagnosis. A stacked denoising autoencoder was used for dimensionality reduction and to generate a meaningful representation, which was followed by machine learning algorithms to observe how the new compacted features were effective for the classification task. The accuracy value obtained with the neural network was 0.917 and 0.947 for the SVM. Ye et al. [19] used unsupervised learning techniques to identify gene signatures for accurate NSCLC subtype classification. A set of 17 genes was isolated, and multiple classifiers were employed for prediction, with decision trees achieving best performance with an accuracy value of 0.922. Beyond the type and subtype prediction, there is a need to identify novel biomarkers based on the most relevant genes for cancer development. Ahn et al. [1] used a feed-forward network for cancer prediction using gene expression data for 24 different cancer types. An accuracy value of 0.979 was achieved, and the importance of an individual gene was computed to provide some explainability insights of the classifier. Based on feature selection techniques, the weight of gene contribution was computed by observing changes in the classifier outcome caused by manipulations in the gene of interest. Differentially expressed genes analysis have been used to identify genes that can be over or underexpressed, which reveals there importance and as consequence, they can be potential therapeutic targets and prognosis markers [4,15,18].

The proposed work aims to provide a method for cancer prediction and subtype classification, while providing the identification of the cancer driver genes, which is essential for a personalized treatment plan and prognosis. A differential expressed genes (DEG) analysis was performed using hierarchical clustering in order to identify commonly regulated cancer and subtype-specific gene signatures. The present work represents an extension of a preliminary work (presented at "43rd Annual International Conference of the IEEE Engineering in Medicine and Biology Society - 2021") [14] dedicated to the classification of lung cancer and subtype. The current study adds the DEG analysis and a deep discussion of the results considering the literature dedicated to the assessment of gene signatures.

2   Materials and Methods

This section presents the dataset used in the current study, the classification method, and the feature analysis performed to interpret the model decision. An overview of the pipeline proposed in this work is represented in Fig. 1.

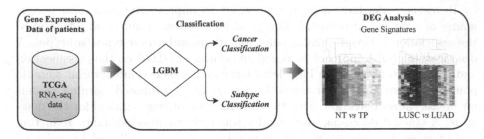

Fig. 1. Lung cancer (Normal Type (NT) vs Tumor Primary (TP)) and subtype (lung squamous cell carcinoma (LUSC) and lung adenocarcinoma (LUAD)) classification using Light Gradient Boosting Machine (LGBM) followed by Differential Expressed Genes (DEG) analysis.

2.1   Dataset and Preprocessing

The data used in this work belongs to The Cancer Genome Atlas (TCGA) project. The data was retrieved from the Genomic Data Commons (GDC) legacy database. The tissue types, primary solid tumour (TP) and solid tissue normal (NT), were retrieved for the LUAD and LUSC TCGA projects. Using TCGABiolinks function *GDCprepare*, the clinical information for the patients was added, and duplicated patient records were removed. A total of 575 samples were retrived for LUAD dataset, out of which 60 (10.43%) are normal tissue (NT) samples. The LUSC dataset has 502 primary tumour samples (TP) and 52 (9.39%) NT samples (see Table 1).

The expression analysis on Table 1 highlights differences between histologic subtypes and between tissue types as it would be expected. The minimum number of expression for both subtypes and tissue types is 0, which translates to no

Table 1. Analysis of minimum, maximum, average and standard deviation of LUAD and LUSC gene expression values by tissue type.

Subtype	LUAD		LUSC	
Tissue type	TP	NT	TP	NT
Min	0	0	0	0
Max	1432694.16	1324252.01	1737511.59	1036891.47
Average	969.57	1024.19	984.87	1049.01
S.d.	695.85	353.83	678.58	375.58

measurable expression for that gene. An average of 4896 genes present expression values close to 0 for NT type, and an average of 7052 genes present close to null expression for LUAD and LUSC subtypes. The maximum expression value belongs to the LUSC subtype, relating to the ADAM6 gene and on LUAD a value of 1432694.16 associated with the SFTPB gene. Both LUAD and LUSC primary tumour samples present higher maximum values than their normal tissue counterparts. Regarding the aggregated expression values, normal tissue types show higher average expression values and lower standard deviations than LUAD and LUSC primary tissue samples, which tells that the distributions of expression are more disperse for the primary tumour types. The average expression value is relatively close between histologic subtypes, showing disperse distributions with similar standard deviations.

2.2 Cancer and Subtype Classification

The classifier used in this work is the light gradient boosting machine (LGBM) [14]. For cancer prediction, a split ratio of {70,15,15}% was used for train, validation and test sets, respectively. Stratified sampling was used to maintain the 9:1 ratio of positive samples in the validation and test sets. To overcome the imbalance on the dataset, oversampling of the negative class was done for each fold of the optimisation process. For the subtype classification, a split ratio of {80,10,10}% was used. For this task there was a balanced ratio of labels, so there was no need to perform data augmentation. The larger size of the independent test size for cancer prediction is due to the imbalance data; therefore we need to guarantee a minimum amount of negative samples for support in the test and validation sets. Hyper-parameters were tuned with a Bayesian optimizer using 5-fold cross-validation. The train and test splits were executed 100 times, randomizing the split seed to reduce variability and overcome skewness caused by the short sample size and class imbalance. The binary cross-entropy (logloss) and area under the ROC Curve (AUC) were used as evaluation metrics to assess training performance and control overfitting by early stopping.

2.3 Differential Expressed Genes Analysis

Differential gene expression analysis refers to the study and interpretation of differences in the abundance of gene transcripts within a transcriptome. The genes selected as the most relevant for the cancer prediction and subtype classification task were used to analyze the changes in the gene expression across the entire samples. Hierarchical clustering of the genes was used to select subgroups of genes whose distribution was most similar in each population. This type of clustering mechanism allows for identifying common-regulated genes, which can serve as a mechanism to identify groups of gene signatures associated with specific diseases.

3 Results and Discussion

3.1 Classification Results

The cancer classification model showed an AUC of 0.983 ± 0.017 [14]. The average number of independent test of the samples was 155 for the positive class and 18 for the negative class. The cancer subtype classification model showed an AUC of 0.971 ± 0.018 [14]. The average for the positive class was 52 and 50 for the negative.

3.2 Differential Expressed Genes Analysis

Figure 2 and 3 present the heatmaps showing the expression of the top 20 genes for cancer prediction and subtype classification, respectively. This type of analyses allow to identify statistically significant gene expression changes across the entire samples. log_2 normalised values were used for counts of the expression of genes for differences in sequencing depth and composition bias between the samples to generate the heatmaps. They are organised as follows: the data is displayed in a grid where each row represents a sample, and each column represents a gene signature. In the heatmaps (Figs. 2 and 3), the intensity colour represents gene expression changes across samples, being the colour red overexpression and the colour blue under-expression, while white represents no visible changes in expression. Furthermore, there are two hierarchical clusters rowwise and column-wise that serve to group similar samples or genes together, respectively. This type of clustering mechanism based on the similarity of gene expression patterns can be useful for identifying genes that are commonly regulated, or biological signatures associated with a particular condition, such as a disease [7].

Cancer Prediction. Regarding the most important genes for cancer prediction, on Fig. 2(a) it is presented the analysis of a total of 110 normal tissue samples and on Fig. 2(b) is presented the analysis of 1016 cancerous samples.

The normal tissue samples (Fig. 2(a)) present an almost homogeneous structure with no visible row-wise clusters, which is expected for NT samples. The columns are sorted by cluster proximity, and on the NT samples it is possible to observe 4 distinct macro clusters from left to right: ECSCR to STX1A comprising mainly under-expression genes; SFTPC up to TGFBR2 containing the highest expression values; C13orf15 up to FHL1 containing more moderate values of over-expression, and TNNC1 till S1PR1 containing less visible changes in gene expression. These clusters are further divided into sub-clusters according to their inner cluster proximity: for example for the under-expression cluster, STX1A and EFNA3 are the bottom-most cluster and therefore are the closest in terms of expression levels, following C16orf59 that is the closest presenting the most visible under-expression and finally and ECSCR, being the furthest away presenting very moderate under-expression.

On Fig. 2(b), it is possible to observe two well defined row-wise clusters, which refer to the LUAD samples on top and the LUSC samples on the bottom-cluster. The cancer tissue samples provide a more heterogeneous composition with five macro column-clusters and less defined subclusters. Furthermore, the distinction between specific genes of the subtype is evident, e.g. SFTPC is more expressed in LUAD samples, and the STX1A, EFNA3 and C16orf59 are on the middle cluster that is more under-expressed in LUAD samples.

The previous works identified that SFTPC encodes the pulmonary-associated Surfactant Protein C (SPC) and its deficiency is associated with interstitial lung disease [12]. SFTPC downregulation might be involved in the progression of lung cancer [9]. Considering the C16orf59 or TEDC2, the high expression of these gene is related with an unfavourable prognostic in lung cancer [17]. PDLIM2 is a putative tumour suppressor protein, and its decreased expression is associated with several malignancies including breast cancer and adult T-cell leukaemia [12].

Subtype Prediction. For the subtype prediction task, on Fig. 3(a) there are a total of 515 LUAD samples represented, and on Fig. 3(b) 501 LUSC samples. Both subtypes present three well-defined parent clusters, ordered from left to right by under-expressed genes first, then the ones over-expressed, and finally genes that present no visible over or under expression.

For LUAD, the most relatively expressed genes are MACC1, PVRL1, SIAH2, SLC44A4, DDAH1 and ERBB3. PVRL1 and SIAH2 are genes that present high relative expression in-between subtypes, and by repeating the same analysis on the LUAD and LUSC populations together, relative values of these genes were higher amongst LUSC samples. This type of analysis presents validation on the selected features of the model since there is the same type of mutual exclusion between the high and low-expressed features in the histologic subtypes.

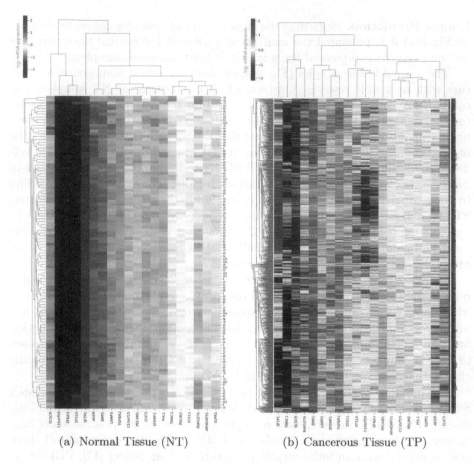

(a) Normal Tissue (NT)   (b) Cancerous Tissue (TP)

Fig. 2. Analysis of log_2 mRNA expression values for the top 20 most important genes for the LGBM cancer classifier. The rows represent samples and the columns genes, and each individual square represents changes in gene expresion from under-expression (blue) to over (red). (Color figure online)

Comparing the obtained results and the literature, it is possible to verify that alterations in genes with known roles in LUSC were found, including over-expression and amplification of TP63 [16]. Elevated MACC1 expression has been implicated in the progression of LUAD [10], and ERBB3 or HER3 is an important paralog of gene ERBB2, whose aberrations have been found to drive LUAD tumours [6].

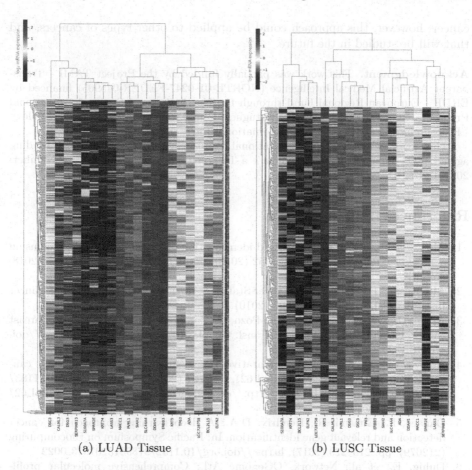

(a) LUAD Tissue   (b) LUSC Tissue

Fig. 3. Analysis of log_2 mRNA expression values for the top 20 most important genes for the LGBM subtype classifier. The rows represent samples and the columns genes, and each individual square represents changes in gene expression from under-expression (blue) to over (red). (Color figure online)

4 Conclusions

This work proposes a methodology for lung cancer prediction and subtype classification based on gradient boosted trees. Two feature sets were extracted using model interpretability that should provide biological insight on differences in gene expression between cancerous and healthy tissue, and LUAD and LUSC subtypes. Hierarchical clustering identified common-regulated genes, which allowed to find some patterns of over and under-regulation for genes specific to each histologic subtype which were validated according to relevant literature. In future work, further validation of these present results will be implemented by extending the learners to other cancer types and performing validation to datasets outside of the TCGA scope. This work was implemented for lung cancer; however, this approach could be applied to other types of cancers, and that will be studied in the future.

Acknowledgment. This work was partially funded by the Project TAMI - Transparent Artificial Medical Intelligence (NORTE-01-0247-FEDER-045905) financed by ERDF - European Regional Fund through the North Portugal Regional Operational Program - NORTE 2020 and by the Portuguese Foundation for Science and Technology - FCT under the CMU - Portugal International Partnership.

This work is also financed by National Funds through the Portuguese funding agency, FCT-Fundação para a Ciência e a Tecnologia, within a PhD Grant Number: 2021.05767.BD.

References

1. Ahn, T., et al.: Deep learning-based identification of cancer or normal tissue using gene expression data, pp. 1748–1752 (2018). https://doi.org/10.1109/BIBM.2018. 8621108
2. Altekruse, S.F., et al.: SEER Cancer Statistics Review 1975–2007 National Cancer Institute. Cancer, pp. 1975–2007 (2010)
3. Arranz, E.E., Vara, J.Á.F., Gámez-Pozo, A., Zamora, P.: Gene signatures in breast cancer: current and future uses. Transl. Oncol. **5**(6), 398–403 (2012). https://doi. org/10.1593/tlo.12244
4. Arroyo Varela, M., et al.: Comparative gene expression analysis in lung cancer. Europ. Respiratory J. **52**(suppl 62), PA2797 (2018). https://doi.org/10.1183/ 13993003.congress-2018.PA2797. http://erj.ersjournals.com/content/52/suppl_62/ PA2797.abstract
5. Danaee, P., Ghaeini, R., Hendrix, D.A.: A deep learning approach for cancer detection and relevant gene identification. In: Pacific Symposium on Biocomputing (212679), pp. 219–229 (2017). https://doi.org/10.1142/9789813207813_0022
6. Duhig, E., et al.: Network, CGenome Atl,: Comprehensive molecular profiling of lung adenocarcinoma: the cancer genome atlas research network. Nature **511**(7511), 543–550 (2014). https://doi.org/10.1038/nature13385
7. Grant, G.R., Manduchi, E., Stoeckert, C.J.: Analysis and management of microarray gene expression data. Current protocols in molecular biology Chapter 19, Unit 19.6, January 2007. https://doi.org/10.1002/0471142727.mb1906s77
8. Inamura, K.: Lung cancer: understanding its molecular pathology and the 2015 wHO classification. Front. Oncol. **7**, 1–7 (2017). https://doi.org/10.3389/fonc.2017. 00193
9. Li, B., et al.: Mir-629-3p-induced downregulation of SFTPC promotes cell proliferation and predicts poor survival in lung adenocarcinoma. Artif. Cells Nanomed. Biotechnol. **47**(1), 3286–3296 (2019). https://doi.org/10.1080/21691401.2019. 1648283. pMID: 31379200
10. Li, Z., et al.: MACC1 overexpression in carcinoma-associated fibroblasts induces the invasion of lung adenocarcinoma cells via paracrine signaling. Int. J. Oncol. **54**(4), 1367–1375 (2019). https://doi.org/10.3892/ijo.2019.4702
11. Liang, M., Li, Z., Chen, T., Zeng, J.: Integrative data analysis of multi-platform cancer data with a multimodal deep learning approach. IEEE/ACM Trans. Comput. Biol. Bioinf. **12**(4), 928–937 (2015). https://doi.org/10.1109/TCBB.2014. 2377729

12. O'Leary, N.A., et al.: Reference sequence (RefSeq) database at NCBI: current status, taxonomic expansion, and functional annotation. Nucleic Acids Res. **44**(D1), D733–D745 (2015). https://doi.org/10.1093/nar/gkv1189
13. Qian, Y., et al.: Prognostic cancer gene expression signatures: current status and challenges. Cells **10**(3), 648 (2021). https://doi.org/10.3390/cells10030648
14. Ramos, B., Pereira, T., Moranguinho, J., Morgado, J., Costa, J.L., Oliveira, H.P.: An interpretable approach for lung cancer prediction and subtype classification using gene expression. In: 2021 43rd Annual International Conference of the IEEE Engineering in Medicine Biology Society (EMBC), pp. 1707–1710 (2021). https://doi.org/10.1109/EMBC46164.2021.9630775
15. Shriwash, N., Singh, P., Arora, S., Ali, S.M., Ali, S., Dohare, R.: Identification of differentially expressed genes in small and non-small cell lung cancer based on meta-analysis of MRNA. Heliyon **5**(6), e01707 (2019). https://doi.org/10.1016/j.heliyon.2019.e01707
16. The Cancer Genome Atlas Research Network: Comprehensive genomic characterization of squamous cell lung cancers. Nature **489**(7417), 519–525 (2012). https://doi.org/10.1038/nature11404
17. Uhlén, M., et al.: Tissue-based map of the human proteome. Science **347**(6220), 1260419 (2015). https://doi.org/10.1126/science.1260419. https://www.science.org/doi/abs/10.1126/science.1260419
18. Yang, R., Zhou, Y., Du, C., Wu, Y.: Bioinformatics analysis of differentially expressed genes in tumor and paracancerous tissues of patients with lung adenocarcinoma. J. Thoracic Disease **12**(12) (2020). https://jtd.amegroups.com/article/view/47626
19. Ye, X., Zhang, W., Sakurai, T.: Adaptive unsupervised feature learning for gene signature identification in non-small-cell lung cancer. IEEE Access **8**, 154354–154362, e01707 (2020). https://doi.org/10.1109/ACCESS.2020.3018480
20. Zhong Wang, M.G., Snyder, M.: RNA-Seq: a revolutionary tool for transcriptomics. Nat. Rev. Genet. **10**(1), 57–63 (2009). https://doi.org/10.1038/nrg2484

Detection of Epilepsy in EEGs Using Deep Sequence Models – A Comparative Study

Miguel Marques[1](✉) ⓘ, Catarina da Silva Lourenço[2] ⓘ, and Luís F. Teixeira[1] ⓘ

[1] INESC TEC, Faculdade de Engenharia da Universidade do Porto,
R. Dr. Roberto Frias, 4200-605 Porto, Portugal
up201603957@edu.fe.up.pt

[2] Clinical Neurophysiology, University of Twente Drienerlolaan 5, 7522 NB Enschede,
The Netherlands

Abstract. The automation of interictal epileptiform discharges through deep learning models can increase assertiveness and reduce the time spent on epilepsy diagnosis, making the process faster and more reliable. It was demonstrated that deep sequence networks can be a useful type of algorithm to effectively detect IEDs. Several different deep networks were tested, of which the best three architectures reached average AUC values of 0.96, 0.95 and 0.94, with convergence of test specificity and sensitivity values around 90%, which indicates a good ability to detect IED samples in EEG records.

Keywords: Deep learning · Interictal epileptiform discharges · Recurrent neural networks · Electroencephalogram

1 Introduction

Epilepsy is a neurologic disorder characterized by an enduring predisposition to generate epileptic seizures, that generates neurobiologic, cognitive, psychological, and social consequences, affecting the patient's lifestyle. An epileptic seizure is a "transient occurrence of signs and/or symptoms due to abnormal excessive or synchronous neuronal activity in the brain." [10]. This event can be expressed by various body reactions, depending on which part of the brain gets involved (such as loss of awareness with body shaking, confusion, and difficulty responding; visual or other sensory symptoms; isolated posturing of a single limb) [14].

Nowadays, the diagnosis of epilepsy is done by visual assessment of Electroencephalography (EEG) recordings [17]. Regarding the analysis of EEG recording, interictal epileptiform discharge (IED) detection is one of the most prominent indicatives for the presence of epilepsy disorder. Between seizures, the brain of some patients with epilepsy generates pathological patterns of activity that are clearly distinguished from the activity observed during the seizure itself and normal brain activity. The detection of these patterns is a prolonged and demanding

© Springer Nature Switzerland AG 2022
A. J. Pinho et al. (Eds.): IbPRIA 2022, LNCS 13256, pp. 192–203, 2022.
https://doi.org/10.1007/978-3-031-04881-4_16

process. In addition to this, subjectivity and intra/inter-variability are also factors that affect the diagnosis [15]. All these reasons often lead to difficulties in a precise diagnosis, making, unfortunately, misdiagnosis common, with 20%–30% of epileptic patients misdiagnosed [2, 3].

To improve the detection of interictal epileptiform discharges in EEGs, diminishing time and increasing the accuracy of diagnosis, deep learning algorithms might be the solution needed. These algorithms have been employed in various fields and, their application in the health field is becoming more usual and is gaining importance. Within all the deep learning models, sequential models were the ones chosen to be explored in this work. These deep frameworks are specialized in deep sequence processing, being able to scale to much longer sequences [6].

The implementation of deep sequence models for IED detection was studied several times. Firstly, [12] implemented an Elman Neural Network to test a three-class classification (normal vs interictal vs ictal events), reaching sensitivity for IED segments equals 96.9%. For a pre-ictal vs interictal classification, [20] implemented three LSTM networks concatenated with Fully Connected and a final Dense layers. For an event-based evaluation, sensitivity and false positive rate per hour were 100.0% for all window segments. In 2018, [19] developed a CNN and hybrid CNN+LSTM architecture, with a different number of layers, to identify interictal epileptiform discharges. The 2D CNN and 2D CNN+LSTM architectures, which reached the best results, obtained an average AUC of 0.94, with the corresponding value of sensitivity for the IED samples, for all architectures, around 47.4% (range 18.8–78.4%). Regarding the work developed by [8], a Dense Convolutional Neural Network concatenated with a Bidirectional LSTM layer and a Deep Convolutional Stacked Autoencoder (DCAE) concatenated with a Bidirectional LSTM network were implemented, achieved a sensitivity of 99.7% and false positive rate per hour of 0.004 h^{-1} for both methods, under patient-specific detection. Lastly, [4] implemented a Recurrent Neural Network composed of LSTM layers for the classification between ictal and interictal segments, with an overall accuracy of 97.4%, 19.7% for sensitivity and specificity of 99.7%.

A common issue with EEG databases is the lack of IED samples present, which can affect the correct performance of the deep models. Deep learning algorithms, such as the models mentioned above, require a great amount of data to be trained correctly and to obtain a good predictive performance. So, to provide a larger number of IED samples to the deep networks, several data augmentation techniques can be used. Data overlap with a sliding window or the application of Synthetic Minority Over-sampling Technique (SMOTE) [5] are examples of commonly used methods. Another alternative is the attribution of different weights to the classes present in the database, so the network can give more importance to the minority class and decrease the predictive errors related to the lack of examples of that class.

This study aimed to design and test different deep sequence architectures to automatize the detection of IEDs in EEG recordings, improving the efficiency of the EEG patterns classification. This work precedes the one developed by [16] and [7], in which convolutional neural networks were tested within the same issue.

2   Methods

2.1   EEG Data and Pre-processing

The database is comprised of two-second EEG epochs with EEG patterns from two different classes. Normal EGGs, from 67 healthy subjects without any epileptic episodes and interictal epileptiform discharges from focal (50 patients) and generalized (49 patients) epilepsy. As pre-processing steps, the data were filtered in the 0.5–35 Hz range to reduce artifacts and downsampled 125 Hz to reduce input size (and consequently computational load). The acquisition of the EEG signals was performed under the longitudinal bipolar electrode montage and, afterwards, the signals were converted to the common average and source derivation montages. After that, each recording was split into non-overlapping epochs, yielding an 18×250 (channels $\times$ time) matrix for each epoch.

2.2   Database Structuring

For the development of this work, four different datasets were created. First, Set A is constituted by all the normal EEG samples as the negative class and focal and generalized IED samples as the positive class. Set B was created with the same signals presented in Set A, but with the records from the three different montages rather than just from one, as in Set A. After that, the other two sets were created, with the intent to artificially increase the number of positive class samples. Set C was created with the IED samples increased by the implementation of SMOTE [5] technique and Set D with the application of a data overlap around 70% with a sliding window. In the Table 1 a description of all datasets can be seen.

The data were randomly split 80% for training and validation and 20% for testing. To decrease the difference between positive and negative samples, the negative samples present in the set for train and validation were divided into 5 different parts. All the positive samples were saved in the 5 parts and, on the other hand, negative class samples were distributed similarly between the five parts, to undersample the negative class. Five-cross validation was implemented and the evaluation results were calculated as an average of the five parts.

2.3   Deep Learning Models

All the deep models were implemented at python 3.7, using Keras 2.4.3 and a CUDA-enabled NVIDIA GPU. For the deep sequence models (Recurrent and hybrid Recurrent Convolutional Neural Networks), Stochastic Optimization was calculated with Adam optimizer, with a learning rate of 10^{-3}, β_1 of 0.9, β_2 of 0,99

Table 1. Databases description (number of epochs for train, validation and test)

Dataset	Train & Validation			Test		
	Epochs		Duration (h)	Epochs		Duration (h)
	Negative	Positive		Negative	Positive	
Set A	43589		24.2	15716		8.7
	41369	2220		15414	302	
Set B	145709		80.9	21285		11.8
	139373	6336		19900	1385	
Set C	88320		49.1	15716		8.7
	44160	44160		15414	302	
Set D	50249		27.9	15716		8.7
	41369	8880		15414	302	

and ϵ equal to 10^{-7}. Five different neural network architectures were explored, three deep sequence models and two attention-based networks. Regarding the deep sequence models, the first network designed was a five-layer Long Short-term Memory neural network [11], named LSTM1, with a total of 256 units per layer and a Dropout layer with 0.35 of dropout rate associated with each LSTM layer.

The other two models are hybrid Recurrent Convolutional Neural Networks, with a similar design, inspired by [1]. The first one, called CNN+LSTM1, was constituted by four 1D convolutional layers, alternating with three 1D Max-pooling layers. In turn, a bi-directional LSTM layer was connected as the back-end of the deep neural network. For this network, Batch Normalization and Dropout with a dropout rate of 0.1 layers were applied after each layer. For the last model, called CNN+LSTM2, the main difference in regards to CNN+LSTM1 was the application of 2D Convolutional and Max-pooling layers instead of 1-dimensional functions.

Concerning the attention-based networks also called Transformer, it is a recent type of deep network, first published in 2017, that works with different mechanisms used by other deep models since it has an attention-based process. It achieved remarkable results in several areas like natural language processing, image processing, among others. On the other hand, there are few studies made with the application of these deep structures dealing with physiological signal processing and, consequently, IED detection [9,13,18,22]. Regarding the architectures designed to test these attention-based networks, these designs were motivated by [21]. The architecture designed included six Attention Modules, working as an encoder and, after that, a 1D Convolutional layer, a Flatten layer and Multi Layer Perceptron was used to perform two-class classification. The modifications on the architecture focused on the number of units in the Dense layers, the dimensions of key and value matrices and the number of heads in the Multi Head function, under a correlation rule used in [21], where the dimensions of key and value matrices are equal to the number of units of Dense layers divided by the number of heads used in the Multi Head Function. So, maintaining the same number of four heads, the units of Feed Forward layers and dimensions of the Multi Head function changed between 64 and 16 and between 16 and 4, respectively.

A batch size of 512 was used for all experiments, and the networks were trained for 60 epochs.

2.4 Performance Assessment

Regarding the binary classification problem, several evaluation metrics were calculated to assess the performance obtained by the deep networks. Test Sensitivity, Specificity, AUC, the True Positive and False Positive rate per hour were calculated, using two different decision thresholds: a threshold to acquire a specificity of 99% and an intersection between sensitivity and specificity. In addition to this, an average ROC+AUC curve was obtained for several five-fold cross-validation loops.

3 Results

Regarding the results obtained for the LSTM1 model, several experiments were made, under the implementation of Set A, in order to design the more efficient architecture to distinguish properly IED samples from normal EEG epochs. This came up with a LSTM network containing five layers, as mentioned in Sect. 2.3. For Set A, the test sensitivity value reached 11.4% for 99.0% specificity, and intersection between sensitivity and specificity of 80.5%. After the experiments made with Set A, LSTM1 was tested with the three remaining datasets, to study if the increase of total number of positive class samples could have a positive effect in the network's performance. Regarding Set B results, the sensitivity value reached 12.1% for 99.0% specificity and the convergence value for sensitivity and specificity reached 88.7%. With the implementation of Set B, the average AUC value improved from 0.89 to 0.94 In Table 2 , all the results are shown, concerning Sets A,B,C and D.

Table 2. Final Results for the LSTM1 model, with the implementation of Sets A, B, C and D

		Specificity = 99.0%				Specificity = Sensitivity				
Class weight	Dataset	Sensitivity	Specificity	TP/hour	FP/hour	Sensitivity	Specificity	TP/hour	FP/hour	AUC
1:30	Set A	11.4%	99.0%	3.96	17.64	80.5%	80.5%	27.95	347.92	0.89
1:10	Set B	12.1%	99.0%	14.11	16.74	88.7%	88.7%	104.23	195.39	0.94
1:10	Set C	1.9%	99.0%	0.64	17.64	78.0%	78.0%	27.06	393.89	0.84
1:20	Set D	3.4%	99.0%	1.19	17.64	73.4%	73.4%	25.42	474.45	0.76

Secondly, similar experiments were performed over CNN+LSTM1 and CNN +LSTM2, reaching two different designs. The 1D CNN+LSTM network reached its best performance with 32 units on the Convolutional and Max Pooling layers. On the other hand, CNN+LSTM2 reached the best performance with 128 units per layer. For the trials performed with the CNN+LSTM1 model, the performance, under the implementation of Set A, reached 29.6% of sensitivity for 99.0% specificity and intersection of sensitivity and specificity of 85.2%. Secondly, the best performance was achieved with the application of Set B, with AUC of 0.95, 21.2% of sensitivity for 99.0% specificity and the intersection of 90.4% between sensitivity and specificity. For the implementation of Set C, the model reached, for 99.0% specificity, sensitivity of 9.2% and equal value between sensitivity and specificity of 82.9%, for an average AUC of 0.90. Ultimately, for Set D, the best results reached an average AUC of 0.75, with a convergence of sensitivity and specificity of 71.0% and 2.9% test sensitivity for 99.0% specificity. In Table 3, all the results can be seen.

Table 3. Results for best performances of CNN+LSTM1 model, with the implementation of Sets A, B, C and D

| Class weight | Dataset | Specificity = 99.0% | | | | Specificity = Sensitivity | | | | |
		Sensitivity	Specificity	TP/hour	FP/hour	Sensitivity	Specificity	TP/hour	FP/hour	AUC
1:20	Set A	29.6%	99.0%	10.25	17.64	85.2%	85.2%	29.63	269.03	0.92
1:20	Set B	21.2%	99.0%	24.80	16.74	90.4%	90.4%	106.50	165.10	0.95
1:1	Set C	9.2%	99.0%	3.18	17.64	82.9%	82.9%	28.77	305.61	0.90
1:10	Set D	2.9%	99.0%	1.01	17.64	71.0%	71.0%	24.60	513.59	0.75

At last, for the implementation of Set A on CNN+LSMT2 model, at a 99.0% specificity criteria, sensitivity reached 27.9%, with a convergence of sensitivity and specificity of 87.9%, corresponding to an average AUC value of 0.94. Looking at the implementation of the other three datasets, the best performance was obtained with the implementation of Set B, reaching an overall AUC of 0.96, sensitivity of 26.1% to 99.0% specificity and convergence of specificity and sensitivity around 90.6%. Regarding the application of Set C, the performance with best results achieved an average AUC of 0.90, reaching 14.7% of sensitivity to 99.0% specificity and 82.6% for specificity equal to sensitivity. At last, the results for Set D's application are 6.0% for 99.0% specificity, 74.2% for sensitivity and specificity equality criteria and AUC of 0.76. In Table 4, all the results for CNN+LSTM2 model can be seen.

Table 4. Results for best performances of CNN+LSTM2 model , with the implementation of Sets A, B, C and D

| Class weight | Dataset | Specificity = 99.0% | | | | Specificity = Sensitivity | | | | |
		Sensitivity	Specificity	TP/hour	FP/hour	Sensitivity	Specificity	TP/hour	FP/hour	AUC
1:4	Set A	27.3%	99.0%	9.42	17.64	87.9%	87.9%	30.49	218.31	0.94
1:10	Set B	26.4%	99.0%	30.95	16.74	90.6%	90.6%	106.52	157.85	0.96
1:10	Set C	14.7%	99.0%	5.07	17.64	82.6%	82.6%	28.67	308.56	0.90
1:20	Set D	6.0%	99.0%	2.08	17.64	74.2%	74.2%	25.71	459.04	0.76

4   Discussion

The main objective of this work was to explore deep sequence architectures capable of detect effectively IED samples in epileptic EEG records. A main obstacle was the imbalance of the first dataset, Set A. The total presence of positive class samples corresponded to 4.2% of total EEG samples. Some decisions were made to decrease the difference of class samples, to achieve a better performance. The first decision was relative to the organization of the datasets in six different parts, one for test and the rest for train and validation. The particularity about the latter five parts was that, while the negative samples were similarly distributed throughout these parts, the positive samples were all repeated at each one of them. This distribution was made to undersample the negative class samples at each training phase. Besides this decision, other techniques were implemented to overcome the imbalance issue. One example of this was the attribution of different weights to the class samples and the creation of other datasets with a increase of IED samples total number (in this case, Sets B, C and D).

Different class weights were attributed at every experiment performed, in order to survey the influence of these weights in the ability of the models to learn the principal features of the EEG samples. The class weight pair that matched with the best performance was different across the experiments, between 1:10, 1:20 and 1:30. In this way, no standardized or common behavior between the weight of the positive class and the algorithm's ability to detect IEDs was confirmed, i.e., not always the highest weight attributed to the positive class leads to a better algorithm performance.

Regarding the comparison between the three deep sequence models, CNN+LSTM2 achieved the best performances for most of the datasets, with the highest AUC values reached for Set A, B and D around 0.94, 0.96 and 0.76 respectively. The two-dimensional characteristic of this architecture was better suited for the structure of the data and leveraged the available spatial context, which led to this model having better performance in IED detection. Even though CNN+LSTM2 model achieved the best results, other models also obtained a remarkable performance.

Looking at the utilization of the datasets, the implementation of Set B was the only dataset that improved the performance of the deep models, when compared with Set A's results. The presence of data samples from the three montages allowed the networks to learn more effectively specific patterns from the IEDs and normal EEG samples, which improved the ability of the models to distinguish these two distinct physiological patterns. On the other hand, the application of Sets C and D did not develop the same success as the application of Set B. These two sets were generated with the intent of increase the total number of positive class samples through an artificial technique (SMOTE and data overlap with sliding window). For both datasets, the performance achieved was worse than the one reached with the implementation of Sets A and B. Neither the creation of artificial positive samples from the original epochs nor the creation of new segments through the copy of parts of the original positive samples enabled the neural network to attain a better predictive ability. Concerning the influence of these two datasets, Set C reached superior results compared to Set D, showing that SMOTE technique is a more efficient method to produce artificially more positive class samples rather than overlap the information already existent in the original IED samples.

As regards to the TP/hour and FP/hour results, these values are of high importance, as it helps to better understand the behavior of the models at the clinical level, making the results obtained more relatable to clinical reality and less technical. Looking at Tables 6 and 7, the TP/hour rates obtained for the best performances are higher and, on the other hand, FP/hour rates are lower, which indicates that, in a clinical environment, the physician does not have to look at many samples that are not IEDs. These rates are crucial for clinical environment, because they guarantee that, for a high TP/hour rate, the probability of misdiagnosis decreases, as the correct assessment of IED samples' presence occurs repeatedly and, for a low FP/hour rate, the amount of time needed for the diagnosis decreases, because most of the detections are going to be right, so the detection of IED pattern in EEG samples will most certain indicate the presence of epilepsy in that patient.

Taking this into consideration, all this work allowed the establishment of a map with the various architectures that can deal effectively with the EEG datasets. These models have distinct features, with more or less parameters, time training and classification performance, which can be significant to decide if a model can fit better the needs of a certain situation. For example, even though the baseline VGG model and CNN+LSTM2 reached better performances, the design CNN+LSTM1 also achieved satisfactory results, with a simpler architecture. The baseline model has 1000 times more parameters and CNN+LSTM2 presents a training time 5 times longer, when compared to the lighter model CNN+LSTM1. This comparison shows that for situations where computational power and resources are limited, it is possible to find lighter solutions with comparable performance. For a better comprehension of this diversity, Table 5 retains the number of parameters and training time of the best models and Figs. 1 and 2 show a plot with the different ROC curves derived from the neural networks that achieved the best performances under implementation of Sets A and B, respectively. This work was developed in continuity with the studies made by [16] and [7], which performed similar experiments on a VGG model using the same Sets A and B. Regarding Set A implementation, the baseline model achieved better performance than the deep sequence models, with a sensitivity of 79.0% for 99.0% specificity and a convergence between sensitivity and specificity of 93.0%, corresponding to an average AUC value of 0.96, the highest in comparison with the other AUC values obtained for Set A, as can be seen in Table 6. Similar results were obtained for Set B implementation on the baseline model. This network reached 74.2% of sensitivity for 99% specificity. For the intersection between sensitivity and specificity, that convergence did not occur. Instead, close values of sensitivity and specificity were obtained, which were 89.5% and 93.5%, respectively.

Table 5. Training time, for Set A and B, and total number of parameters for the deep models with the best performances

Model	Training time per epoch		Total number of parameters
	Set A	Set B	
Baseline	4 s	13 s	45570306
LSTM1	9 s	28 s	2383105
CNN+LSTM1	1 s	4 s	27777
CNN+LSTM2	6 s	19 s	1389411

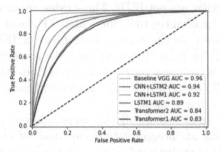

Fig. 1. Plot of the ROC curve correspondent to the three models with best performance with Set A

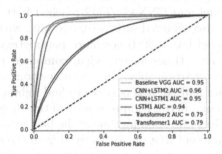

Fig. 2. Plot of the ROC curve correspondent to the three models with best performance with Set B

Table 6. Best performance results obtained for Set A, including baseline model from [16]

Model	Set	99.0% Specificity			Specificity = Sensitivity				
		Sensitivity	Specificity	TP/hour	Sensitivity	Specificity	TP/hour	FP/hour	AUC
Baseline	A	79.0%	99.0%	40.27	93.0%	93.%	47.72	122.41	0.96
LSTM1	A	11.4%	99.0%	3.96	80.5%	80.5%	27.95	347.92	0.89
CNN+LSTM1	A	29.6%	99.0%	10.25	85.2%	85.2%	29.63	269.03	0.92
CNN+LSTM2	A	27.3%	99.0%	9.42	87.9%	87.9%	30.49	218.31	0.94

Table 7. Best performance results obtained for Set B, including baseline model from [7]

Model	Set	99.0% Specificity			Specificity = Sensitivity				
		Sensitivity	Specificity	TP/hour	Sensitivity	Specificity	TP/hour	FP/hour	AUC
Baseline	B	74.2%	99.0%	38.05	89.5%	93.5%	45.90	114.00	0.95
LSTM1	B	12.1%	99.0%	14.11	88.7%	88.7%	104.23	195.39	0.94
CNN+LSTM1	B	21.2%	99.0%	24.80	90.4%	90.4%	106.50	165.10	0.95
CNN+LSTM2	B	26.4%	99.0%	30.95	90.6%	90.6%	106.52	157.85	0.96

5 Conclusion

A main conclusion that can be drawn is that it is feasible to explore and implement deep sequence models for the classification of interictal epileptiform discharges between normal EEG patterns, in order to assess the presence or absence of epilepsy. Several distinct models that were designed achieved a good predictive ability, independently the degree of complexity associated to them (either in relation to the architecture or to the total number of parameters).

As said in Sect. 4, the work was developed in continuity with the exploration of pure CNN models, with highlight to a VGG network, which achieved remarkable results. The objective of this comparison was to infer if it was possible to address the problem exploring alternative methods (in this case, sequential and attention-based networks), achieving a similar performance. Looking at the results obtained, the aforementioned inference was confirmed.

Furthermore,the work allowed the investigation of a common issue related to EEG signals databases, which is the imbalance of the various EEG patterns present in the database, and to explore several techniques to overcome this issue and potentiate the performance of the networks, in order to achieves a reliable predictive ability. From under-sampling the majority class examples, to artificially increasing the total number of positive class examples, to inflating the weight given to the minority class examples, a wide variety of techniques were explored, and we conclude that using different montages and changing class weights can have a positive impact on models performance.

References

1. Abdelhameed, A.M., Daoud, H.G., Bayoumi, M.: Deep convolutional bidirectional LSTM recurrent neural network for epileptic seizure detection. In: 2018 16th IEEE International New Circuits and Systems Conference, NEWCAS 2018, pp. 139–143 (2018). https://doi.org/10.1109/NEWCAS.2018.8585542
2. Beach, R., Reading, R.: The importance of acknowledging clinical uncertainty in the diagnosis of epilepsy and non-epileptic events. Arch. Dis. Childhood **90**(12), 1219–1222 (2005). https://doi.org/10.1136/adc.2004.065441
3. Benbadis, S.: The differential diagnosis of epilepsy: a critical review. Epilepsy Behav. **15**(1), 15–21 (2009). https://doi.org/10.1016/j.yebeh.2009.02.024, http://dx.DOI.org/10.1016/j.yebeh.2009.02.024

4. Bongiorni, L., Balbinot, A.: Evaluation of recurrent neural networks as epileptic seizure predictor. Array **8**, 100038 (2020). https://doi.org/10.1016/j.array.2020.100038

5. Chawla, N.V., Bowyer, K.W., Hall, L.O., Kegelmeyer, W.P.: SMOTE: synthetic minority over-sampling technique. J. Artif. Intell. Res. **16**(February 2017), 321–357 (2002). https://doi.org/10.1613/jair.953

6. Chelgani, S.C., Shahbazi, B., Hadavandi, E.: Learning representations by back-propagating errors. Meas. J. Int. Meas. Confederation **114**(2), 102–108 (2018). https://doi.org/10.1016/j.measurement.2017.09.025

7. da Silva Lourenço, C., Tjepkema-Cloostermans, M.C., van Putten, M.J.: Efficient use of clinical EEG data for deep learning in epilepsy. Clin. Neurophysiol. **132**(6), 1234–1240 (2021). https://doi.org/10.1016/j.clinph.2021.01.035

8. Daoud, H., Bayoumi, M.A.: Efficient Epileptic Seizure Prediction Based on Deep Learning. IEEE Trans. Biomed. Circ. Syst. **13**(5), 804–813 (2019). https://doi.org/10.1109/TBCAS.2019.2929053

9. Dosovitskiy, A., et al.: An Image is Worth 16×16 Words: Transformers for Image Recognition at Scale. CoRR, pp. 1–21 (2020). http://arxiv.org/abs/2010.11929

10. Fisher, R.S., et al.: Epileptic seizures and epilepsy: definitions proposed by the International League Against Epilepsy (ILAE) and the International Bureau for Epilepsy (IBE). Epilepsia **46**(10), 1701–1702 (2005). https://doi.org/10.1111/j.1528-1167.2005.00273_4.x

11. Goodfellow, I., Bengio, Y., Courville, A.: Deep learning, vol. 29. MIT Press (2016). http://www.deeplearningbook.org

12. Guler, N.F., Ubeyli, E.D., Guler, I.: Recurrent neural networks employing Lyapunov exponents for EEG signals classification. Expert Syst. Appl. **29**(3), 506–514 (2005). https://doi.org/10.1016/j.eswa.2005.04.011

13. Jiang, K., Liang, S., Meng, L., Zhang, Y., Wang, P., Wang, W.: A two-level attention-based sequence-to-sequence model for accurate inter-patient arrhythmia detection. In: Proceedings - 2020 IEEE International Conference on Bioinformatics and Biomedicine, BIBM 2020, pp. 1029–1033 (2020). https://doi.org/10.1109/BIBM49941.2020.9313453

14. Johnson, E.L.: Seizures and Epilepsy. Med. Clin. North Am. **103**(2), 309–324 (2019). https://doi.org/10.1016/j.mcna.2018.10.002

15. Lodder, S., Putten, V., Antonius, M.J.: Automated EEG analysis: characterizing the posterior dominant rhythm. J. Neurosci. Meth. **200**(1), 86–93 (2011). https://doi.org/10.1016/j.jneumeth.2011.06.008

16. Lourenço, C., Tjepkema-Cloostermans, M.C., Teixeira, L.F., van Putten, M.J.: Deep learning for interictal epileptiform discharge detection from Scalp EEG recordings. IFMBE Proc. **76**, 1984–1997 (2020). https://doi.org/10.1007/978-3-030-31635-8_237

17. Rosenow, F., Klein, K.M., Hamer, H.M.: Non-invasive EEG evaluation in epilepsy diagnosis. Expert Rev. Neurother. **15**(4), 425–444 (2015). https://doi.org/10.1586/14737175.2015.1025382

18. Song, H., Rajan, D., Thiagarajan, J.J., Spanias, A.: Attend and diagnose: clinical time series analysis using attention models. In: 32nd AAAI Conference on Artificial Intelligence, AAAI 2018, pp. 4091–4098 (2018)

19. Tjepkema-Cloostermans, M.C., de Carvalho, R.C., van Putten, M.J.: Deep learning for detection of focal epileptiform discharges from scalp EEG recordings. Clin. Neurophysiol. **129**(10), 2191–2196 (2018). https://doi.org/10.1016/j.clinph.2018.06.024

20. Tsiouris, K.M., Pezoulas, V.C., Zervakis, M., Konitsiotis, S., Koutsouris, D.D., Fotiadis, D.I.: A long short-term memory deep learning network for the prediction of epileptic seizures using EEG signals. Comput. Bio. Med. **99**, 24–37 (2018). https://doi.org/10.1016/j.compbiomed.2018.05.019

21. Vaswani, A., et al.: Attention is all you need. Adv. Neural Inf. Process. Syst. 2017-December(Nips), 5999–6009 (2017)

22. Yan, G., Liang, S., Zhang, Y., Liu, F.: Fusing transformer model with temporal features for ECG heartbeat classification. In: Proceedings - 2019 IEEE International Conference on Bioinformatics and Biomedicine, BIBM 2019, pp. 898–905 (2019). https://doi.org/10.1109/BIBM47256.2019.8983326

Biometrics

Facial Emotion Recognition for Sentiment Analysis of Social Media Data

Diandre de Paula$^{(\boxtimes)}$ [iD] and Luís A. Alexandre [iD]

Departamento de Informática and NOVA LINCS, Universidade da Beira Interior,
Covilhã, Portugal
diandre.paula.cavalini@ubi.pt

Abstract. Despite the diversity of work done in the area of image sentiment analysis, it is still a challenging task. Several factors contribute to the difficulty, like socio-cultural issues, the difficulty in finding reliable and properly labeled data to be used, as well as problems faced during classification (e.g. the presence of irony) that affect the accuracy of the developed models. In order to overcome these problems, a multitasking model was developed, which considers the entire image information, information from the salient areas in the images, and the facial expressions of faces contained in the images, together with textual information, so that each component complements the others during classification. The experiments showed that the use of the proposed model can improve the image sentiment classification, surpassing the results of several recent social media emotion recognition methods.

Keywords: Image sentiment analysis · Multimodal · Facial expression recognition · Salient areas · Text sentiment analysis

1 Introduction

We are increasingly witnessing the growth of the online community, where users seek ways to express themselves beyond the use of words, often using images to reach their goal. Thus, social media has posts with both text and images, that convey different (positive and negative) feelings. There are many factors to take into account when we analyze the sentiment transmitted by an image, for instance, the socio-cultural issues. Several other features can help us to identify the sentiment of an image, for example, the prevailing colors in the image, the type of objects in the image, and the metadata (e.g. image's caption) that are associated with it. This work aims to develop a multimodal approach that classifies the image sentiment to identify posts that may represent negative and strongly negative situations, since we are interested in predicting when possible

This work was supported by NOVA LINCS (UIDB/04516/2020) with the financial support of FCT – Fundação para a Ciência e a Tecnologia, through national funds and by the project MOVES – Monitoring Virtual Crowds in Smart Cities (PTDC/EEI-AUT/28918/2017), also financed by FCT.

© Springer Nature Switzerland AG 2022
A. J. Pinho et al. (Eds.): IbPRIA 2022, LNCS 13256, pp. 207–217, 2022.
https://doi.org/10.1007/978-3-031-04881-4_17

strongly negative events are going to take place, through the analysis of social media posts. This prediction will be obtained not just with the image information from the social media posts, but also with textual information. Several previous works have been done in this area [1–3,9,10], and we have identified a place where current models can be improved: the inclusion of a Face Emotion Recognition (FER) model can be used to clarify situations where the emotion conveyed by the text is not in agreement with the emotion in the image and also when the overall image has a positive sentiment if the face emotions are not taken into consideration. Besides this, we also experiment with the use of an image classifier for salient regions of the image, as to complement the information provided by a global image classier. Finally, we explore different ways to fuse the decisions from the proposed classifiers and present experiments on a large social media data set that show the strengths of our proposal.

2   Related Work

The work [10] approaches the image sentiment analysis (without considering text), using a Plutchik's wheel of emotions approach. It also addresses challenging issues like implementing supervised learning with weakly labeled training data, in other words, data that was labeled through a model (and not labeled by a human), and handles the image sentiment classification generalizability.

To predict the image sentiments, the authors of [9] proposed a model that combines global and local information. The work proposes a framework to leverage local regions and global information to estimate the sentiment conveyed by images, where the same pre-trained Convolutional Neural Network (CNN) model is used, but it is fine-tuned using different training sets: a first one addressing the entire images, and another addressing the sub-images; in the end, both predictions are fused to obtain the final sentiment prediction.

The work [2] aims to reduce the image classification's dependence on the text content. The proposed model was divided into three parts (in each one there is a specific task) and in the end, all parts are fused using a weighted sum, which is capable of predicting the polarity of a sentiment level (positive, neutral, and negative).

The authors in [1] propose a method based on a multi-task framework to combine multi-modal information whenever it's available. The proposed model contains one classifier for each task: i) text classification, ii) image classification, iii) prediction based on the fusion of both modalities. The authors evaluated the advantages of their multi-task approach on the generalization of each three tasks: text, image, and multi-modal classification. The authors concluded that their model is robust to a missing modality.

The work in [3] proposed a novel multi-modal approach, which uses both textual data and images from social media to perform the classification into three classes: positive, neutral, and negative. The approach consists of the classification of the textual and image components, followed by the fusion of both classifications into a final one using an Automated Machine Learning (AutoML) approach, which performs a random search to determine the best model to perform the final classification.

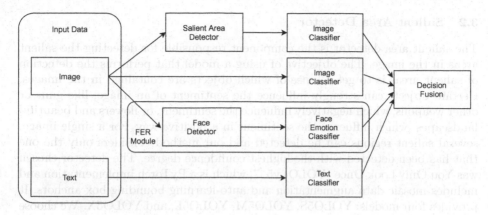

Fig. 1. Overview of the proposed method's architecture and its components.

We notice that, several works employed models with image and text classification. However, none of them employed an approach that could handle images, salient regions, textual data, and facial expressions. Regarding the final output, the majority of the work employed polarity as the final classification, using either two classes (positive and negative), or three classes (positive, neutral, and negative).

3   Proposed Method

Figure 1 shows an overview of the method that was developed. It's composed of an image classifier, an image salient area detector, a text classifier, and a facial expression module that contains both a face detector and a face emotion classifier. The outputs of the classifiers are fused to produce the final decision. The details are described in the next sub-sections.

3.1   Image Classifier

The image classifier model is responsible for analysing the sentiment of the original image. This is a mandatory model, that is, the information returned by this model will be always considered for the final fused decision. For the image classifier, the architecture proposed in [2] was used, since the proposed configurations obtained good results, better than the previous state-of-the-art [8]. A pre-trained Residual Network (ResNet) 152 was used, with the last layer being fully-connected, accompanied by a softmax layer, with 3 outputs, which represent the probability of each class (negative, neutral, and positive), in the range [0,1], where 1 represents that the image belongs to that respective class and 0 that it does not. The used model was trained with the B-T4SA data set (described in the Experiments section below). The model receives an image, and it predicts a class and its respective probability, which can be seen as the degree of certainty with which that class was predicted.

3.2   Salient Area Detector

The salient area detector is the component responsible for detecting the salient areas in the image. The objective of using a model that performs the detection of salient areas is to get a sense of which objects are contained in the images. Certain objects can strongly influence the sentiment of an image, like guns or other weapons, which negatively influence the sentiment, or flowers and beautiful landscapes, which influence the sentiment in a positive way. For a single image, several salient regions can be detected and our method considers only the one that has been detected with the highest confidence degree. The detector chosen was You Only Look Once (YOLO) v5 [7] which is a PyTorch implementation and includes mosaic data augmentation and auto-learning bounding box anchors. It provides four models: YOLO5S, YOLO5M, YOLO5L, and YOLO5X. We choose to use the largest model, YOLO5X, to aim for the best detection rates.

We used the VOC data set to train the detector. It contains 21,503 images with annotations. The data was split into 16,551 (77%) images for train and 4,952 (23%) for validation. The model was trained with a batch size of 64. The mean Average Precision (mAP)0.5 was 83.1%, the mAP@0.95 was 62.7% and the time it took to train the model was 27 min.

3.3   Text Classification

Since another important part of a social media post is the text, it also should be evaluated w.r.t. its sentiment.

We use the Regionbased Convolutional Neural Networks (R-CNN) text model proposed in [3], which uses an embedding layer, and a bi-directional Long Short Term Memory (LSTM) layer with input size equal to the dimension of the embedding, hidden size of 256 and a dropout of 0.8. The final embedding vector is the concatenation of its embedding and left and right contextual embeddings, which in this case is the hidden vector of the LSTM. This concatenated vector is then passed to a linear layer which maps the input vector back to a vector with a size equal to the hidden size of the LSTM, 256. This is passed through a 1D max-pooling layer, and finally, the output from this layer is sent to a linear layer that maps the input to a classification vector.

Tweets involve a lot of noise, such as emojis/emoticons, links, numbers, etc. Therefore, before going through the text evaluation method, the tweet must be cleaned, in order to remove this noise. First, the tweet will be processed by Beautiful Soup, which is a Python library for obtaining data from HyperText Markup Language (HTML) and eXtensible Markup Language (XML) files. It is used to decode HTML encoding that has not been converted to text, and ended up in the text field, such as '&amp' or '&quot'. The second part of the preparation is dealing with @mention. Even though @mention carries some information (which user the tweet mentioned), this information does not add value to build a sentiment analysis model. The third part is dealing with Uniform Resource Locators (URL) links, that although they can carry some information for sentiment analysis purposes, they will be ignored. There is also the possibility of

Unicode Transformation Format (UTF)-8 BOM character issues. The UTF-8 BOM is an array of bytes (EF BB BF) that allows the reader to recognize a file as being encoded in UTF-8. To avoid unfamiliar characters, we used a text decoder that replaces them by the symbol "?". Sometimes the text used with hashtags can give useful information about the tweet. So it was decided to leave the text intact and just remove the symbol (#) cleaning all the non-letter characters (including numbers). Then the text is transformed to lower case. During the letters-only process, unnecessary white space is created, so redundant white space is removed.

3.4 Facial Expression Recognition Module

Global features can give hints regarding the image sentiment, but some works faced difficulties with the models getting an erroneous classification due to global features [5,6,12]. Therefore, the objective of using a model that performs the classification of facial expressions would be to address these issues. Since in a single image we can have several faces, the information to be considered will be the one that has been obtained with the highest confidence degree. The Facial Expression Recognition Module is responsible for two tasks: i) detecting faces in the images; ii) classifying the detected faces' expressions. To make the detection, the Multi-task Cascaded Convolutional Networks (MTCNN) is used. This model has three convolutional networks (Proposal Network (P-Net), Refinement Network (R-Net), and Output Network (O-Net)). Upon receiving an image, the model will create an image pyramid, in order to detect faces of different sizes. Then, it is possible to split the MTCNN operation into three stages [11]:

- Stage 1: A fully convolutional network (P-Net) was used to obtain the candidate facial windows and their bounding box regression vectors. Candidates are calibrated based on the estimated bounding box regression vectors. Then, non-maximum suppression (NMS) is employed to merge highly overlapped candidates;
- Stage 2: All candidates are fed to another CNN (R-Net), which further rejects a large number of false candidates, performs calibration with bounding box regression, and conducts NMS;
- Stage 3: This stage is similar to the previous one, but it is aimed at identifying face regions with more supervision. In particular, the O-Net will output five facial landmarks' positions.

The FER model receives an image, and using the MTCNN, produces the bounding boxes together with the confidence degrees of each detected face in the image. Then, if faces were detected in the image, each identified bounding box is cropped. The resolution of the cropped image will be checked in order to maintain a certain quality level of the image crops and prevent images of very low quality (and which would not add any utility to the model) from being kept. We used the rule of only keeping images with a resolution greater than or equal to 30×30 pixels.

After the face detector, the emotion recognition is performed. For this task, a CNN was created with a ResNet9, which increases (gradually) the number of channels of facial data and decreases the dimension, followed by a fully connected layer responsible for returning an array with the values describing the probability of belonging to each class. The learning rate scheduler, 1Cycle, was used so that the learning rate was not manually set. It starts with a very low learning rate, increases, and decreases it again.

3.5   Decision Fusion

To make the fusion of the information of each model, we propose two different methods: i) considering the average of all models and ii) using a voting system. To obtain the average of all models, each class obtained by the model and its respective accuracy will be multiplied, and this value will be divided by the number of models that were evoked, that is, to consider only the information of the models that were actually evoked:

$$av = \frac{1}{n} \sum_{i=1}^{n} X_i p_i \tag{1}$$

where n is the number of used models, X_i is the polarity and p_i is the accuracy values obtained by the global image, salient areas, text, and FER models, respectively in the validation set.

After obtaining the average result, the class is defined by:

$$class = \begin{cases} 0, & \text{if } av \leq 0.34 \\ 1, & \text{if } 0.34 < av \leq 0.67 \\ 2, & \text{if } av > 0.67 \end{cases} \tag{2}$$

the values 0.34 and 0.67 were chosen to divide the [0,1] interval into three equal intervals.

For the voting system, the votes for each class are counted, and to avoid any tie, the accuracy value of each model will be considered, when necessary. Therefore, a tuple is created, which stores the vote count of each class and the sum of the accuracy values of each model that voted in this same class:

$$(v_i, s_i), \quad i = 0, 1, 2, \tag{3}$$

where v_i is the vote count for class i and s_i is the sum of the accuracy values for that class. The selected class is given by:

$$class = \arg\max_i v_i \tag{4}$$

when there is no draw between the votes, and in the case of a draw, the class is given by:

$$class = \arg\max_i s_i, \tag{5}$$

where the index i runs through the drawn classes only. Then, to obtain the winner class, we will consider the index of the tuple with the highest value. If there is no tie, the tuple with the highest number of votes is chosen otherwise, the tuple (among those that are tied) with the highest sum is chosen.

4 Experiments

4.1 Training the Facial Expression Recognition Model

For training the FER model, we created a data set from three different sources.

First, the FER2013 data set was used. This data set consists of 48 × 48 pixel grayscale images of faces. The faces have been automatically registered so that the face is approximately centered and occupies the same amount of space in each image. The labels are divided into 7 types: 0 = Angry, 1 = Disgust, 2 = Fear, 3 = Happy, 4 = Sad, 5 = Surprise, 6 = Neutral. The training set consists of 28,709 examples and the validation and test sets consist of 3,589 examples each. The model achieved approximately 68.82% accuracy in the test set.

Another data set was prepared with social media images, namely from Twitter, in order to assess the model's behavior when exposed to social media images, which may or may not have larger resolutions. The accuracy obtained with the Twitter image test data set was approximately 18.22%.

The final data set used for training the FER model included all the images from the two previous data sets (FER2013 and Twitter) plus the Japanese Female Facial Expression (JAFFE) data set [4]. It contains 50,783 images, which were divided into: 40,627 samples for training (80%), 5,078 samples for testing (10%), and 5,080 samples for validation (10%). Figure 2 presents some of the images that compose the final data set used. The model was trained for 55 epochs, and the accuracy on the test set was 72.75%.

4.2 Data Set for the Full Model Evaluation

For the experiments, a variation of the B-T4SA validation set was used. In [8], the authors trained a model for visual sentiment classification starting from a large set of user-generated and unlabeled contents. They collected more than 3 million tweets containing both text and images. The authors used Twitter's Sample Application Programming Interface (API) to access a random 1% sample of the stream of all globally produced tweets, discarding tweets not containing any static image or other media, tweets not written in the English language,

Fig. 2. Sample images of the data set used for training the FER model. Sentiment from left to right (label, class, polarity): (surprise, 5, 1), (sad, 4, 0), (happy, 3, 2), (angry, 0, 0), and (neutral, 6, 1).

Table 1. Accuracy on the validation set (top 4 rows) and test set (last row) of several combinations of the available modules, with the training and evaluation times. We show the results for the fusion with the mean and voting approaches. Experiments A, B, C and D used the training and validation data, and the final experiment, E, was done using training and test data with the configuration that yielded the best results with the validation data.

Exp.	Text clf.	Image clf.	Salient areas clf.	FER	Fusion Acc. [%] Mean	Vot.	Time [hours] Mean	Vot.
A	Y	Y	N	N	60.22	–	0.30	–
B	Y	Y	Y	Y	59.31	73.19	9.15	9.15
C	Y	Y	Y	N	62.25	72.74	8.97	8.97
D	Y	Y	N	Y	62.63	**82.90**	0.51	0.51
E	Y	Y	N	Y	–	**80.86**	–	5.18

whose text was less than 5 words long, and retweets. At the end of the data collection process, the total number of tweets in the T4SA data set was about 3.4 million. Each tweet (text and associated images) was labeled according to the sentiment polarity of the text (negative = 0, neutral = 1, positive = 2) predicted by their tandem LSTM-Support Vector Machines (SVM) architecture. The corrupted and near-duplicate images were removed, and they selected a balanced subset of images, named B-T4SA, that was used to train their visual classifiers.

The original B-T4SA validation set contains 51,000 samples. However, due to the hardware limitations, this set was randomly decreased to approximately 17% of its original size, resulting on a new validation set containing 9,064 samples.

Table 2. Comparison between the results obtained from the models that used the B-T4SA data set, all evaluated in the same test set.

Work	Accuracy [%]
VGG-T4SA FT-A [8]	51.30
VGG-T4SA FT-F [8]	50.60
Hybrid-T4SA FT-A [8]	49.10
Hybrid-T4SA FT-F [8]	49.90
Random classifier [8]	33.30
Multimodal approach [2]	52.34
Multimodal approach [3]	**95.19**
Ours	80.86

4.3 Results

The experiments were run in a computer with an AMD Ryzen 7 2700 (Octacore, 16 Threads) 3.2 GHz CPU, 16 GB Random Access Memory (RAM), NVIDIA 1080ti, a 3TB HDD and a 256 GB SSD.

We ran the first batch of experiments (corresponding to the first three rows of Table 1, experiments A, B, C and D) using only the training and validation data sets to study which was the best configuration in terms of model components to use with the final test data set, presented in the last row of Table 1, experiment E. The test data set contains 51,000 samples.

Test A was made in order to evaluate the model's accuracy without using the proposed methods. Since it only uses 2 models, the voting system was not used as the decision will be the same as if only the most accurate model was used.

From the validation set experiments, we found that the best results were achieved when not considering the salient region classifier, hence, in the final experiment E, with the test data, the configuration of our approach included only the text, global image and face emotion recognition modules. Regarding the two evaluated fusion approaches, the voting approach presented consistently over 10% better results than the fusion using the mean, as was the one evaluated in experiment E. Regarding the time that it took to train and evaluate the models, the table also shows that the salient area classifier was very costly and its removal significantly increase the speed of the system. The time increase from experiment D to E is due to the larger size of the test set when compared to the validation set.

We compared the results obtained in the test set with the results from other approaches in the literature that used the B-T4SA data set. The values are in Table 2 and show that our approach is able to improve on most of the previous results by a large margin, from around 50% to 80%, with the exception of the proposal in [3]. This points to the possibility of the module decision fusion used in that work, an AutoML approach, to be responsible for that large boost in accuracy, and to be a good alternative to the approaches explored in this paper.

5 Conclusions

Social media sentiment classification is a very demanding task. An approach that has been increasingly used is the analysis of both text and image (multi-modal) information to achieve improved results. In this paper, we also propose a multi-modal approach that uses text and image data from tweets to evaluate the post sentiment. We propose a system that explores the image data in several ways to try to overcome the ambiguity that can appear in the image sentiment evaluation. First, we use a global image classifier that is reused to process also salient image regions. From the experiments, we concluded that the salient regions' contribution to the final decision was not improving the overall classification results. We also proposed the use of a facial expression recognition (FER) module in the model. This approach has not been employed yet, or wasn't used with the three other models. The idea is to evaluate the emotion of the persons that might be present in the image and use this as a complement to the overall sentiment evaluation. The faces are detected and their emotions are obtained and we consider only the one with the highest confidence. To train the FER module, a data set was created, which contained images of faces in controlled environments and in the wild, in order to present a variety of possible situations, image quality and poses during the model's training. This module produced a good contribution to the final test set accuracy of the B-T4SA data set, of over 80%. The overall result largely improved previously obtained results with one exception only, that used AutoML to create a decision fusion from several models. In future work, we will study ways of improving our proposal that can compete with this approach.

References

1. Fortin, M., Chaib-Draa, B.: Multimodal sentiment analysis: a multitask learning approach. In: Proceedings of the 8th International Conference on Pattern Recognition Applications and Methods - ICPRAM, pp. 368–376. INSTICC, SciTePress (2019). https://doi.org/10.5220/0007313503680376
2. Gaspar, A., Alexandre, L.A.: A multimodal approach to image sentiment analysis. In: Yin, H., Camacho, D., Tino, P., Tallón-Ballesteros, A.J., Menezes, R., Allmendinger, R. (eds.) IDEAL 2019. LNCS, vol. 11871, pp. 302–309. Springer, Cham (2019). https://doi.org/10.1007/978-3-030-33607-3_33
3. Lopes, V., Gaspar, A., Alexandre, L.A., Cordeiro, J.: An AutoML-based approach to multimodal image sentiment analysis. In: 2021 International Joint Conference on Neural Networks (IJCNN), pp. 1–9 (2021). https://doi.org/10.1109/IJCNN52387.2021.9533552
4. Lyons, M., Akamatsu, S., Kamachi, M., Gyoba, J.: Coding facial expressions with Gabor wavelets. In: Proceedings of the Third IEEE International Conference on Automatic Face and Gesture Recognition, April 1998, 200–205, May 1998. https://doi.org/10.1109/AFGR.1998.670949
5. Machajdik, J., Hanbury, A.: Affective image classification using features inspired by psychology and art theory. In: Proceedings of the 18th ACM International Conference on Multimedia, pp. 83–92. MM 2010. Association for Computing Machinery, New York, NY, USA (2010). https://doi.org/10.1145/1873951.1873965

6. Ortis, A., Farinella, G.M., Battiato, S.: Survey on visual sentiment analysis. IET Image Process. **14**(8), 1440–1456 (2020). https://doi.org/10.1049/iet-ipr.2019. 1270, http://dx.doi.org/10.1049/iet-ipr.2019.1270

7. Ultralytics: Yolov5. https://github.com/ultralytics/yolov5

8. Vadicamo, L., et al.: Cross-media learning for image sentiment analysis in the wild. In: 2017 IEEE International Conference on Computer Vision Workshops (ICCVW), pp. 308–317 (2017)

9. Wu, L., Qi, M., Jian, M., Zhang, H.: Visual sentiment analysis by combining global and local information. Neural Process. Lett. **51**(3), 2063–2075 (2019). https://doi. org/10.1007/s11063-019-10027-7

10. You, Q., Luo, J., Jin, H., Yang, J.: Robust image sentiment analysis using progressively trained and domain transferred deep networks. In: Proceedings of the Twenty-Ninth AAAI Conference on Artificial Intelligence, pp. 381–388. AAAI 2015. AAAI Press (2015)

11. Zhang, K., Zhang, Z., Li, Z., Qiao, Y.: Joint face detection and alignment using multitask cascaded convolutional networks. IEEE Sign. Process. Lett. **23**(10), 1499–1503 (2016)

12. Zhang, K., Zhu, Y., Zhang, W., Zhu, Y.: Cross-modal image sentiment analysis via deep correlation of textual semantic. Knowl.-Based Syst. **216**, 106803 (2021). https://doi.org/10.1016/j.knosys.2021.106803, https://www.sciencedirect. com/science/article/pii/S0950705121000666

Heartbeat Selection Based on Outlier Removal

Miguel Carvalho[1,3] and Susana Brás[2,3]([✉]) [iD]

[1] Department of Physics, University of Aveiro, Aveiro, Portugal
[2] Department of Electronics, Telecommunication and Informatics (DETI), University of Aveiro, Aveiro, Portugal
susana.bras@ua.pt
[3] Institute of Electronics and Informatics Engineering (IEETA), University of Aveiro, Aveiro, Portugal

Abstract. The relevance of the Electrocardiogram (ECG) in clinical settings as well as in biometric identification systems prompted the necessity to develop automatic methodologies to remove irregular and noisy heartbeats. In addition, and regarding the field of biometrics, it is equally important to ensure a low degree of intra-subject variability in the extracted heartbeats in order to enhance the ability of classifiers to differentiate individuals. Therefore, the present work focuses on the development of a new algorithm aimed at extracting outliers and, simultaneously, maximizing the similarity between heartbeats. To this end, the developed method entails two distinct phases, the first one foreseeing the elimination of irregular heartbeats according to the variance of pre-defined features while the second phase selects heartbeats through the application of Normalized Cross-Correlation. An analysis of the results based on the average variance and average range of the obtained heartbeats demonstrates the capacity of the algorithm to reject noisy segments and guarantee the extraction of heartbeats with highly homogeneous characteristics. As such, the developed methodology holds considerable promise to be integrated in ECG-based biometric systems or medical applications.

Keywords: ECG · Outlier removal · Heartbeat selection

1 Introduction

According to the World Health Organization, cardiovascular diseases are the dominant cause of mortality globally, accounting for 32% of deaths worldwide in 2019 [1]. As such, electrocardiogram (ECG) analysis stands out as a powerful non-invasive methodology for diagnosing a multitude of cardiovascular pathologies, including early contractions of the atria (PAC) or ventricles (PVC), atrial fibrillation (AF), myocardial infarction (MI), arrhythmias, among others [2]. Simultaneously, it is fundamental to take into account the growing development of medical devices, especially in the wearable class, which allow constant ECG acquisition, culminating in a significant volume of medical data. Thus, one can realize that the conventional analysis of the ECG by health professionals becomes unfeasible in this context, thereby triggering the need to develop systems to detect ECG segments indicative of cardiac abnormalities [2, 3]. On the other hand, ECG has been

© Springer Nature Switzerland AG 2022
A. J. Pinho et al. (Eds.): IbPRIA 2022, LNCS 13256, pp. 218–229, 2022.
https://doi.org/10.1007/978-3-031-04881-4_18

mobilizing efforts on the part of the scientific community due to its potential in the area of biometric identification, in view of the inter-variability of its characteristics among distinct individuals [4]. Within this context, there is also the need to ensure the existence of a low degree of variability in the input provided to the classifier which enables the materialization of this system, thus enhancing the distinction between individuals [5]. Yet again, we find the necessity to automatically distinguish valid from invalid segments for further classification. This recurrent theme of implementing automatic ECG analysis tools has led to the development of multiple algorithms, such as DMEAN and DBSCAN [6], or other methodologies based on machine learning [2, 7]. Generally, outlier detection algorithms can be subdivided into two main categories, supervised and unsupervised. The first assumes the characterisation of data considered "normal" based on labelled data, where deviations from the learned patterns suggest the presence of irregular segments. Conversely, there are outlier detection methodologies which do not require prior knowledge about the data, typically implying the use of clustering and/or thresholding techniques on the basis of pre-defined criteria [6]. In light of the relevance associated with this topic, this work aims to present a new algorithm, based on pre-existing methods, for outlier extraction which intends to respond to some limitations of pre-existing approaches. The developed algorithm was designed and applied in a biometric identification context.

2 Proposed Approach

The proposed pre-processing methodology is composed of two complementary phases, the first one being the application of an outlier removal algorithm which operates on the basis of the features' variance, designated Modified DMEAN (mDMEAN). Afterwards, the highest values of Normalized Cross-Correlation are utilized to extract the desired number of heartbeats. The described phases aim to promote the removal of outliers and, simultaneously, to guarantee a high degree of similarity between the extracted heartbeats so as to guarantee reduced intra-subject variability.

2.1 Modified DMEAN: Formal Description

Firstly, we assume $X = \{x_1, \ldots, x_n\}$ as a set of individual heartbeats, which can be further described by a set of features, i.e., $x_i = (x_i(1), \ldots, x_i(m)) \in \mathbb{R}^m$. The present approach consists in applying a function $f : x_i \rightarrow \{0, 1\}$, where the representation of the heartbeat in the feature space is classified as *Normal* (0) or *Outlier* (1). As proposed by Lourenço et al. [6], the present methodology is based on the assumption that the irregular heartbeats are found in sparse regions in the feature space. Additionally, it is assumed that the points closer to the mean value, being this proximity evaluated as a function of the variance of the features, will correspond to heartbeats with a higher degree of similarity. Note that if we consider a two-dimensional feature space, where feature A has a smaller variance than feature B, then it is reasonable to claim that a threshold designed for feature B is less restrictive along feature A. This reasoning motivates the delineation of an adaptive threshold to account for the variance of the features. As such, this algorithm is not limited exclusively to outlier extraction, but also to the selection of ECG segments that minimize the intra-subject variability.

Therefore, the proposed method addresses the limitation of DMEAN in dealing with features with significant discrepancies in their variance, due to its selection of valid heartbeats within a hypersphere. Thus, we propose the application of Principal Component Analysis in order to represent the feature space on a basis in which its axes correspond to the direction of greatest variance of the data and, simultaneously, that the origin corresponds to the midpoint. Secondly, an n-dimensional ellipsoid is defined, whose axis coincide with the defined basis and have a length proportional to the variance along each dimension. Finally, the heartbeats are considered normal if they lie within the previously defined n-dimensional ellipsoid.

The proposed technique assumes the following steps:

1. Definition of the features which are going to represent the heartbeat. We highlight the relevance of defining features with low correlation degree in order to minimize the introduction of redundancy in the system.
2. Elimination of values above the 97.5 percentile and below the 2.5 percentile for all features. This procedure arises from the sensitivity of PCA to outliers in the definition of its principal components. To achieve this goal, while simultaneously reducing the percentage of data extracted, a general mask is defined which accounts for all the points eliminated, according to the aforementioned methodology, in all feature dimensions. Finally, all points which do not comply with the restriction imposed (>2.5 and <97.5 percentile) in at least one of their features, will be eliminated. The described strategy assumes that the irregular beats likely present irregular values in the different features characterizing it, that is, there will be points which are in the percentile to be eliminated in their different features. By doing so, the removal of 5% of the data for each feature is avoided.
3. Application of the PCA and transformation of the data to the new base.
4. Selection of valid beats according to the following expression:

$$
f(x_i) = \begin{cases} 0, & \text{if } \sum_{k}^{m} \dfrac{x_i(k)^2}{(Var_{k\_component} \times Threshold)^2} \leq 1 \\ 1, & \text{if } \sum_{k}^{m} \dfrac{x_i(k)^2}{(Var_{k\_component} \times Threshold)^2} > 1 \end{cases}
$$

The mathematical formulation presented above corresponds to the selection of the points inside the n-dimentional ellipsoid stated above. Note that the presented threshold is bound to define the degree of restrictiveness in the selection of heartbeats, as well as, the volume of extracted data. In terms of graphical analysis, this threshold establishes the dimensions of the axes of the n-dimentional ellipsoid, where higher values imply a greater number of points inside it.

2.2 Algorithm Tunning: Feature and Threshold Selection

As previously discussed, the practical implementation of this algorithm is subject to the definition of the number and type of features used to represent the heartbeats in the feature space, together with a suitable threshold. The selection of the aforementioned parameters must be governed by two criteria, whose behavior is contradictory and complementary, which are the quality and volume of extracted data. In this context, the quality of the extracted data is defined according to the primary objective of pre-processing in ECG biometrics, that is, the degree of similarity between heartbeats stemming from the same individual [4].

Regarding the type and number of features, we must notice that the number of features is directly correlated to the number of heartbeats considered valid, as long as the threshold remains unchanged. This statement is based on the principle that the greater the number of features used to describe a heartbeat, the less likely it is for the heartbeats to exhibit similar values in all dimensions, leading to an increasingly more disperse feature space. Therefore, approaches that rely on a more exhaustive representation of the heartbeats in the feature space should be complemented with more inclusive thresholds and vice-versa. As for the type of features, we intend to select metrics that are extremely susceptible to the influence of artefacts in order to endow the algorithm with sensitivity to distinguish normal heartbeats from outliers.

This kind of adaptability makes the algorithm a prime candidate for applications outside the biometrics field, as for instance in medical use, as long as its characteristics are adjusted in accordance with its purpose. Thus, the calibration of this methodology in any given setting should be guided by the following steps: (**1**) selection of relevant features; (**2**) selection of the number of features; (**3**) elimination of correlated features; (**4**) definition of an adequate cost function; (**5**) testing various thresholds and combinations of features in order to identify the optimum value with respect to the predefined cost function.

2.3 Normalized Cross-Correlation

Following the application of Modified DMEAN to extract outliers, a complementary pre-processing phase is implemented to maximize the similarity between the extracted heartbeats. Therefore, and on the basis of similar approaches reported in the literature [8], 200 heartbeats are selected based on the highest Cross-correlation values relative to a template heartbeat. This default heartbeat is obtained by computing the average of the heartbeats considered valid after applying Modified DMEAN.

3 Methodology

3.1 Pre-processing

The signal acquisition was performed with the 4-channel biosignalsplux, from the Biosignalsplux Explorer Research Kit. Afterwards, in an attempt to determine the best signal denoising approach, a set of criteria was defined to assess the quality of this procedure, in particular Higher Cross-Correlation, Lowest Root Mean Squared Error and Highest Signal-to-Noise Ratio [9]. In this manner, several methodologies commonly reported in the literature in the context of on-the-person acquisition were tested, such as band-pass filtering with a passband of (1–40), (2–40), (1–30), (2–30) [3] as well as discrete wavelet transform (DWT) [10] with different mother wavelets, decompositions levels and thresholding methods. Ultimately, in our approach, it became clear that DWT (coif5, level 4 and soft thresholding) was superior in maximizing the quality of the signal. Regarding R-peak detection, the Engelse-Zeelenberg algorithm [11] was chosen since methodologies based on signal differentiation have a higher accuracy rate in the correct identification of R-peaks [5].

3.2 Modified DMEAN in ECG Biometrics

Taking into account the described guidelines and the scope of the present work, the adopted pre-processing protocol consisted, in a first instance, of restricting the number of features to four in order to allow a compromise between selectivity and volume of extracted data, as shown in Table 1. Then, 7 features were selected as candidates to represent the heartbeats in the feature space. The reasoning behind the feature selection is the reduction of the need to introduce the detection of points beyond the already determined R-peak, thus mitigating the inevitable introduction of errors. Therefore, we chose metrics reported in the literature in the field of outlier extraction [6], such as the R-peak and the S-peak, as well as metrics between R-peaks, for instance, the standard deviation or the mean.

Table 1. Analysis of data quality and volume based upon the number of selected features. In order to guarantee comparability of results, the threshold was kept constant. It should also be noted that the smaller the percentage of extracted data, the closer the selected heartbeats are to the average point and, consequently, the probability of outliers being selected is inferior.

#Features	Feature set	Extracted data (%)
2	(0, 1)	57,47
3	(0, 1, 2)	50,47
4	**(0, 1, 2, 3)**	**11,03**
5	(0, 1, 2, 3, 4)	3,22

- **(1)** Peak R amplitude;
- **(2)** Maximum value between Peak R;
- **(3)** Peak S amplitude;
- **(4)** Entropy between R-peaks;
- **(5)** Mean value between R-peaks;
- **(6)** Standard Deviation between R-Peaks;
- **(7)** Square root of the number of inflection points between R-peaks;

Subsequently, a high degree of correlation was detected between features **(4)** and **(6)**, leading to the exclusion of feature **(4)** for being the most correlated with the remaining features. In a following phase, and with the objective of minimizing intra-subject variability, a cost function was defined to guide the choice of optimal parameters based on the lowest variance between valid heartbeats, as well as, the minimum range value in each time instance, as described in expression (1). By adding sensitivity to extreme values in the heartbeats, the cost function seeks to mitigate the permeability to outliers. To enable the computation of the cost function, Modified DMEAN was firstly applied with pre-defined parameters (threshold and feature set) to obtain the normal heartbeats. Then, the heartbeats were segmented by defining a window of 800 ms centered around the respective R peak, allowing the computation of the variance and range for each time instance spanned in the defined window.

$$f_{cost}\left(thr, set_{features}\right)$$
$$= argmin\{ \frac{\sum_{j=0}^{j}(\sum_{i=0}^{i} \frac{(x_{(i,j)}-\mu_{(:,j)})^2}{N_{valid_{heartbeat}}}) \cdot (max(x_{(:,j)})-min(x_{(:,j)}))^2}{N_{windowsamples}} \} \tag{1}$$

where $x_{(i,j)}$ corresponds to the valid heartbeats' matrix with dimensions (nº valid heartbeats X window samples).

As such, all possible combinations of 4 features out of 6 (15 total combinations) were generated and, for each combination, the cost function was evaluated at 95 distinct and unitarily spaced thresholds with values ranging from 5 to 100. By doing so, it was possible to identify the combination that guaranteed the best results and, simultaneously, the extraction of a certain percentage of data (10%, 20%, etc.), enabling the analysis of the algorithm's behavior in different scenarios, as described in Table 2.

Table 2. Optimal features combinations and respective thresholds for each data percentage of participant ID 2.

Data percentage (>%)	Combination	Threshold	Cost function
10	11	12	0,005
20	4	16	0,007
30	4	20	0,008
40	4	24	0,010
50	4	28	0,011
60	4	33	0,012
70	4	39	0,015

Additionally, with the purpose of ensuring a rigorous and representative assessment of the best feature combinations, the whole described procedure was performed on ten randomly selected participants, being the results schematized in Table 3.

Initially, the evaluation of the results showed the superiority of features 0 and 1 in minimizing the presence of outliers, given that 74.3% of the combinations considered optimal contained these two features. Subsequently, the combination (0, 1, 5, 6) was found to have the highest incidence rate as the optimal combination, being thus selected for the present pre-processing phase. Finally, it is important to note that there was no combination that stood out significantly from the others and that, concurrently, 11 of the 15 combinations tested were at least once considered ideal. Therefore, it is plausible to assume that the number of features employed holds a more important role in the quality of the pre-processing than the type of features selected.

Table 3. Assessment of the best feature combination. Note that the combinations that were never considered optimal are not described in the table, namely (0, 1, 2, 5), (0, 2, 4, 6), (1, 2, 4, 5) and (2, 4, 5, 6).

Combination	Incidence (%)
(0, 1, 2, 4)	18,6
(0, 1, 2, 6)	10,0
(0, 1, 4, 5)	4,3
(0, 1, 4, 6)	20,0
(0, 1, 5, 6)	21,4
(0, 2, 4, 5)	1,4
(0, 2, 5, 6)	1,4
(0, 4, 5, 6)	1,4
(1, 2, 4, 6)	2,9
(1, 2, 5, 6)	1,4
(1, 4, 5, 6)	17,1

After establishing the most suitable combination, it was necessary to select the threshold that best fitted the biometric system requirements. Due to the significant volume of data in the database, arising from ECG acquisitions with durations of approximately 40 min, a threshold of lower values was chosen, thus ensuring a greater degree of similarity between heartbeats and, simultaneously, a significant number of valid heartbeats. This fact is corroborated by Table 2, where one can observe that lower threshold values guarantee a higher degree of resemblance between heartbeats. Thus, the threshold which ensured an average extraction of 40% of the heartbeats, based on the study reference above, was selected, with a value of 28.66.

4 Performance Analysis

4.1 Performance Analysis Between Modified DMEAN and DMEAN

With the aim of gauging the performance of the developed algorithm, a comparative analysis was conducted with the method underpinning this work, DMEAN. To this end, and to guarantee the comparability of results, a new condition was introduced to DMEAN that implies the selection of the n-points closest to the average value, in order to ensure the extraction of the same volume of data for both algorithms. Having said that, it is evident that the fulfilment of the remaining restrictions described by Lourenço et al.

[6] was simultaneously guaranteed. Therefore, the two algorithms were performed for different percentages of removed data and their behavior was evaluated, as described in Table 4. As for the performance evaluation criterion, the mean range and mean variance were chosen due to their ability to monitor the similarity between heartbeats and the presence of outliers.

Table 4. Performance analysis of outlier removal with DMEAN and modified DMEAN.

Threshold	Mean variance DMEAN	Mean variance mDMEAN	Mean range DMEAN	Mean range mDMEAN
323.9 ± 258.2 Max: 932 Min: 54	0.020 ± 0.008 Max: 0.040 Min: 0.011	0.023 ± 0.010 Max: 0.046 Min: 0.011	0.785 ± 0.397 Max: 1.902 Min: 0.429	0.832 ± 0.509 Max: 2.148 Min: 0.471
530.3 ± 335.7 Max: 1257 Min: 110	0.022 ± 0.008 Max: 0.042 Min: 0.012	0.023 ± 0.009 Max: 0.042 Min: 0.013	0.871 ± 0.400 Max: 1.969 Min: 0.462	0.882 ± 0.471 Max: 2.081 Min: 0.513
720.3 ± 406.2 Max: 1668 Min: 301	0.023 ± 0.009 Max: 0.046 Min: 0.013	0.024 ± 0.010 Max: 0.048 Min: 0.014	0.958 ± 0.462 Max: 2.260 Min: 0.504	0.967 ± 0.513 Max: 2.333 Min: 0.545
1027.1 ± 480.4 Max: 2065 Min: 450	0.025 ± 0.010 Max: 0.051 Min: 0.014	0.026 ± 0.011 Max: 0.053 Min: 0.015	1.041 ± 0.496 Max: 2.447 Min: 0.566	1.038 ± 0.528 Max: 2.441 Min: 0.615
1458.6 ± 386.5 Max: 1981 Min: 877	0.026 ± 0.008 Max: 0.046 Min: 0.015	0.026 ± 0.010 Max: 0.047 Min: 0.015	1.194 ± 0.506 Max: 2.260 Min: 0.616	1.134 ± 0.474 Max: 2.202 Min: 0.632
1684.7 ± 407.3 Max: 2302 Min: 1154	0.025 ± 0.006 Max: 0.037 Min: 0.016	0.025 ± 0.007 Max: 0.037 Min: 0.015	1.181 ± 0.445 Max: 2.212 Min: 0.640	1.062 ± 0.347 Max: 1.853 Min: 0.664
1924.4 ± 395.1 Max: 2574 Min: 1414	0.027 ± 0.007 Max: 0.041 Min: 0.017	0.026 ± 0.007 Max: 0.040 Min: 0.016	1.248 ± 0.496 Max: 2.433 Min: 0.688	1.087 ± 0.367 Max: 1.948 Min: 0.678

Overall, a maintenance of the average variance (0,03%) and a decrease of the average range in approximately 5,00% were observed. However, and taking into consideration the theoretical foundation of the method at hand, it is worthwhile examining the performance variation in both low (<50%) and high percentages (>50%) of extracted data. As such, a reduction in the average variance of 4.36% and a reduction in the average range of 11.74% was verified for higher volumes of extracted data, whereas for lower values, there was an increase of 2.30% in the average variance and a reduction of 2.30% in the average range. Considering this, we can clearly observe a pattern, where the performance difference between the two methods becomes more pronounced as the data volume increases. This fact is congruent with the theoretical prediction when the method was designed, given that for lower values of data all the extracted heartbeats are relatively close to the average point, mitigating the need to consider asymmetries in the variance.

In a subsequent phase, and aiming to have a trustworthy understanding of the importance of heartbeat extraction according to variance, a new comparison between the two methods was carried out, with the addition of the removal of the percentile in the case of DMEAN. In this manner, not only is the extraction of the same volume of data guaranteed, but also the feature space is equal in both situations, enabling the comparative analysis of the fundamental concepts that govern each method. The results are schematized in Table 5.

Table 5. Performance analysis of outlier removal with DMEAN (with percentile extraction) and modified DMEAN.

#Heartbeats	Mean variance DMEAN	Mean variance mDMEAN	Mean range DMEAN	Mean range mDMEAN
323.9 ± 258.2 Max: 932 Min: 54	0.020 ± 0.008 Max: 0.040 Min: 0.011	0.023 ± 0.010 Max: 0.046 Min: 0.011	0.785 ± 0.397 Max: 1.903 Min: 0.427	0.832 ± 0.509 Max: 2.148 Min: 0.471
530.3 ± 335.7 Max: 1257 Min: 110	0.022 ± 0.008 Max: 0.042 Min: 0.012	0.023 ± 0.009 Max: 0.042 Min: 0.013	0.866 ± 0.404 Max: 1.970 Min: 0.462	0.882 ± 0.471 Max: 2.081 Min: 0.513
720.3 ± 406.2 Max: 1668 Min: 301	0.023 ± 0.009 Max: 0.046 Min: 0.012	0.024 ± 0.010 Max: 0.048 Min: 0.014	0.948 ± 0.446 Max: 2.200 Min: 0.504	0.967 ± 0.513 Max: 2.333 Min: 0.545
1027.1 ± 480.4 Max: 2065 Min: 450	0.024 ± 0.009 Max: 0.050 Min: 0.014	0.026 ± 0.011 Max: 0.053 Min: 0.015	1.023 ± 0.486 Max: 2.394 Min: 0.565	1.038 ± 0.528 Max: 2.441 Min: 0.615
1458.6 ± 386.5 Max: 1981 Min: 877	0.026 ± 0.008 Max: 0.046 Min: 0.015	0.026 ± 0.010 Max: 0.047 Min: 0.015	1.115 ± 0.429 Max: 2.200 Min: 0.605	1.134 ± 0.474 Max: 2.202 Min: 0.632
1684.7 ± 407.3 Max: 2302 Min: 1154	0.025 ± 0.006 Max: 0.037 Min: 0.015	0.025 ± 0.007 Max: 0.037 Min: 0.015	1.074 ± 0.342 Max: 1.900 Min: 0.628	1.062 ± 0.347 Max: 1.853 Min: 0.664
1924.4 ± 395.1 Max: 2574 Min: 1414	0.026 ± 0.007 Max: 0.041 Min: 0.016	0.026 ± 0.007 Max: 0.040 Min: 0.016	1.120 ± 0.367 Max: 1.978 Min: 0.647	1.087 ± 0.367 Max: 1.948 Min: 0.678

By analyzing the results, it is possible to identify the same tendency outlined above, however, the magnitude of the performance variation is reduced. As such, we are able to conclude that the percentile reduction is in fact a crucial part of the developed algorithm, promoting a significant reduction of outlier presence. Simultaneously, the usefulness of considering the variance in selecting valid heartbeats is still evidenced, since a decrease in the mean variance (0.47%) and mean range (2.66%) continues to be verified for high extracted data percentages. Additionally, it is possible to have an insight on the relative importance of the different stages of the algorithm, namely the extraction of points based on the percentile and the point selection method, as compared to DMEAN. With this, it was possible ascertain that the extraction of the percentile has a higher percentual weight than the consideration of the variance on the algorithm's performance.

On the other hand, we intend to extend the present discussion with the premise that the greater the asymmetries between the variance of the different features, the greater the utility and performance variation of mDMEAN with respect to DMEAN. In this way, the same tests presented above were conducted, with the exception of using the combination $(0, 1, 2, 4)$, which corresponds to the third combination with the highest incidence rate as shown in Table 3. The selection of this combination lies on the fact that the variance of feature 2 is significantly higher than the others, rendering it a perfect candidate to attest the mDMEAN behavior. Results show a decrease of the average variance in 21.56% and a diminishment of the average range in 32.47%, while for the comparison with DMEAN coupled with the percentile extraction, a decrease of the average variance in 10.23% and a reduction of the average range in 8.825% were found. Once again, the applicability of the developed methodology in the removal of outliers is confirmed. Note that all the reported percentages are computed as the relative variation to the mDMEAN output.

Lastly, it is worth mentioning some limitations associated with the developed technique. The first point to note is that the degree of selectivity of this algorithm is contingent on the product between a pre-defined threshold as well as the variance of the principal components of the data, and the latter may fluctuate considerably between individuals. Thus, and as seen in the standard deviations of Tables 4 and 5, the number of extracted heartbeats can vary significantly for similar threshold values. Likewise, and considering that distinct physiological data acquisition schemes influence the quality of data and, consequently, the distribution and variance of data in feature space, it is always recommended that the calibration of the algorithm is performed in order to define the thresholds required for the extraction of the desired data percentage.

4.2 Performance Analysis of Modified DMEAN and NCC

Although the performance differences between Modified DMEAN and DMEAN are not very significant in the percentage range of data extracted, mDMEAN appears to guarantee a smaller range of values, thus ensuring that the computation of the average heartbeat is less prone to be skewed by the presence of remaining outliers. Moreover, the introduction of this new segment allows to counteract the variability found in the heartbeats extracted by mDMEAN, hence guaranteeing the obtainment of a constant number of heartbeats for all individuals. In order to ascertain the impact of the introduction of this phase, a graphical analysis was conducted in the different stages of the pre-processing procedure, that is, (1) before the application of modified DMEAN, (2) after the application of modified DMEAN and (3) after the application of modified DMEAN and NCC (Fig. 1).

As shown in the illustration, the introduction of NCC mitigates the incidence of heartbeats with irregular segments, therefore ensuring the obtainment of 200 near-identical heartbeats.

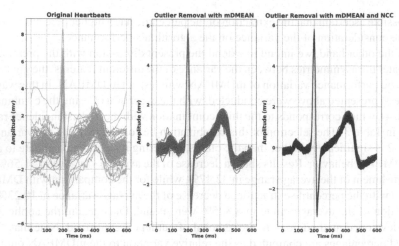

Fig. 1. Representation of the different stages of the proposed pre-processing method. In the first image, there are represented 3112 heartbeats, whereas in the second image only 1267 heartbeats are displayed. At last, and as stated in the previous section, only 200 heartbeats are illustrated in the third image.

5 Conclusions

To summarize, the first section of the developed algorithm (mDMEAN) holds a superior capacity to mitigate the selection of outliers specially on high values of extracted data when compared to DMEAN, being this fact illustrated by the decrease of mean variance and mean ranges. On the other hand, the second section of the algorithm ensures the selection of a fixed number of heartbeats with a high degree of similarity, thus maximizing the quality of the obtained data. Therefore, the relevance of the development and application of this methodology is justified, since the reduction of the variability inherent to each individual (lower mean variance and mean range) endows the classifier of a biometrical system with a greater ability to clearly distinguish them.

In terms of future perspectives, it is worth considering investigating the suitability of the developed method in ECG signals obtained from off-the-person systems with the aim of analyzing its behavior in environments with elevated noise levels. Simultaneously, it would be crucial to compare the obtained results with the proposed algorithm and other outlier removal methodologies described in the literature.

Acknowledgments. This work is funded by national funds (OE), the European Regional Development Fund, FSE through COMPETE2020, through FCT (Fundação para a Ciência e a Tecnologia, I.P.), in the scope of the framework contract foreseen in the numbers 4, 5 and 6 of the article 23, of the Decree- Law 57/2016, of August 29, changed by Law 57/2017, of July 19; and in the scope of the projects UIDB/00127/2020 (IEETA/UA).

References

1. World Health Organization, Monitoring Health for the SDGs Sustainable Development Goals, vol. 7, no. 1 (2021)
2. Hong, S., Zhou, Y., Shang, J., Xiao, C., Sun, J.: Opportunities and challenges of deep learning methods for electrocardiogram data: a systematic review. Comput. Biol. Med. **122**(April), 103801 (2020). https://doi.org/10.1016/j.compbiomed.2020.103801
3. Berkaya, S.K., Uysal, A.K., Gunal, E.S., Ergin, S., Gunal, S., Gulmezoglu, M.B.: A survey on ECG analysis. Biomed. Sig. Process. Control **43**, 216–235 (2018). https://doi.org/10.1016/j.bspc.2018.03.003
4. Uwaechia, A.N., Ramli, D.A.: A Comprehensive survey on ECG signals as new biometric modality for human authentication: recent advances and future challenges. IEEE Access **9**, 97760–97802 (2021). https://doi.org/10.1109/ACCESS.2021.3095248
5. Pinto, J.R., Member, S.: Evolution, current challenges, and future possibilities in ECG biometrics. **6** (2020). https://doi.org/10.1109/ACCESS.2018.2849870
6. Silva, H., Carreiras, C.: Outlier Detection in Non-intrusive, pp. 43–52 (2013)
7. Wu, B., Yang, G., Yang, L., Yin, Y.: Robust ECG biometrics using two-stage model. Proc. Int. Conf. Pattern Recognit. **2018**(August), 1062–1067 (2018). https://doi.org/10.1109/ICPR.2018.8545285
8. Pinto, J.R., Cardoso, J.S., Lourenço, A., Carreiras, C.: Towards a continuous biometric system based on ECG signals acquired on the steering wheel. Sensors (Switzerland) **17**(10), 1–14 (2017). https://doi.org/10.3390/s17102228
9. Chun, S.Y.: Single pulse ECG-based small scale user authentication using guided filtering. In: 2016 Interntional Conference on Biometrics, ICB 2016 (2016). https://doi.org/10.1109/ICB.2016.7550065
10. Bassiouni, M.M., El-Dahshan, E.-S., Khalefa, W., Salem, A.M.: Intelligent hybrid approaches for human ECG signals identification. Sig. Image Video Process **12**(5), 941–949 (2018). https://doi.org/10.1007/s11760-018-1237-5
11. Engelse, W.A.H., Zeelenberg, C.: A single scan algorithm for QRS-detection and feature extraction. Comput. Cardiol. **6**(1979), 37–42 (1979)

Characterization of Emotions Through Facial Electromyogram Signals

Lara Pereira[1]([✉]), Susana Brás[2,3][ID], and Raquel Sebastião[2,3][ID]

[1] Department of Physics (DFis), University of Aveiro, 3810-193 Aveiro, Portugal
`laraifp@ua.pt`
[2] Institute of Electronics and Informatics Engineering of Aveiro (IEETA),
Aveiro, Portugal
[3] Department of Electronics, Telecommunications and Informatics (DETI),
University of Aveiro, 3810-193 Aveiro, Portugal
`{susana.bras,raquel.sebastiao}@ua.pt`

Abstract. Emotions are a high interesting subject for the development of areas such as health and education. As a result, methods that allow their understanding, characterization, and classification have been under the attention in recent years. The main objective of this work is to investigate the feasibility of characterizing emotions from facial electromyogram (EMG) signals. For that, we rely on the EMG signals, from the frontal and zygomatic muscles, collected on 37 participants while emotional conditions were induced by visual content, namely fear, joy, or neutral. Using only the entropy of the EMG signals, from the frontal and zygomatic muscles, we can distinguish, respectively, neutral and joy conditions for 70% and 84% of the participants, fear and joy conditions for 81% and 92% of the participants and neutral, and fear conditions for 65% and 70% of the participants. These results show that opposite emotional conditions are easier to distinguish through the information of EMG signals. Moreover, we can also conclude that the information from the zygomatic muscle allowed to characterized more participants with respect to the 3 emotional conditions induced. The characterization of emotions through EMG signals opens the possibility for a classification system for emotion classification relying only on EMG information. This has the advantages of micro-expressions detection, signal constant collection, and no need to acquire face images. This work is a first step towards the automatic classification of emotions based solely on facial EMG.

Keywords: EMG · Emotion characterization · Entropy

1 Introduction

The meaning of the word "emotion" has been built along the times through many controversies, remaining under the spotlight nowadays. Emotions can be defined as a complex set of chemical and neural reactions and, although their expression is affected by cultural and personal aspects, it is still a process based

© Springer Nature Switzerland AG 2022
A. J. Pinho et al. (Eds.): IbPRIA 2022, LNCS 13256, pp. 230–241, 2022.
https://doi.org/10.1007/978-3-031-04881-4_19

on pre-established cerebral mechanisms [5]. Emotions play an important role in our lives, influencing our actions and decisions and having a great impact on health, in the way people learn, work, or interact. These are the main reasons that support the importance of studying and understanding them, to use it in our favor [8].

One of the emotion theories is that exists seven universal emotions: anger, fear, sadness, joy, surprise, disgust, and contempt. Universal emotions are distinct from the others because, independent of the language, ethnicity, culture, and place of birth, they are expressed, in general, in the same way in every person [4]. The expression of any kind of emotion is related to three parameters and it may present some variations according to personal experience and physiological and behavioral responses. Relatively to the first parameter, depending on the way each person was raised and their cultural environment, the same emotion can be expressed in different ways, confirming their subjective nature [13]. Considering the physiological response, experiencing different emotions can cause various reactions in the autonomic nervous system (ANS), which controls involuntary actions such as heart rate, blood flow, and sweating. Finally, the behavioral response is related to changes in facial expression, tone of voice, or body posture according to the emotion the person is feeling [10].

The last two parameters - physiological and behavioral responses - form the basis of emotion characterization. Relatively to behavioral response, the facial expression is the one that gets more attention from the scientific community due to its high emotional significance and for being one of the most relevant factors in communication between human beings [1]. The emotion recognition based on electromyogram (EMG) signal, collected from the face muscles, which is a physiological response, is intimately connected to emotion recognition using facial expression, which is a behavioral response. The facial expressions triggered by different emotions lead to the activation of different facial muscles. In this case associated with joy and fear, which will be studied along this project, together with neutral expression, where, theoretically, all the facial muscles are relaxed and consequently not activated. For fear, we will focus on frontal muscle and for joy on the zygomatic major muscle, which suffers great activation in the presence of these emotions. The EMG signals are collected with surface electrodes, which are non-invasive and allow constant data acquisition. Data acquisition for emotion characterization through facial EMG, nevertheless more intrusive, presents several advantages over recognition from facial expression, namely impartial regarding light conditions, detachment of frontal perspectives from the camera, provides continuous signal collection, and allows to capture the micro-expressions, being capable of measuring even weak emotional states. Also, according to other studies, facial EMG is capable of registering a response even in the absence of emotional expressions that are noticeable under normal viewing conditions or when participants are asked to inhibit their emotional expression, for example, in emotion regulation studies [3,9,16].

Considering this information, the main objective of this work is to find out the feasibility of emotion characterization based on EMG signals collected from facial muscles. Thus, Sect. 2 exposes the dataset used, along with the proposed methodology to accomplish emotion characterization. Thereafter, Sect. 3 presents and clarifies the obtained results, while in Sect. 4 conclusions of this study are drawn.

2   Dataset and Methodology

2.1   Dataset

The data from 37 participants used in this study was previously collected. The age of the participants was between 18 and 28 years (21.17 ± 2.65) and 19 of them were male.

The dataset used for this study was previously collected and contains information from a total number of 53 participants. It is composed of ECG, EMG, EDA, and heartbeat signals, but in this study, only the data of the EMG signals were used. The healthy participants that volunteered for acquisition were subjected to the visualization of three different videos, on three separate days, at least one week apart, which ensures the absence of emotional contamination between sessions. The videos watched aimed to induce three different emotional states: fear, joy, and neutral. Finally, although all the 53 participants agreed to participate in the collection, only 47 attended all sessions and only 37 of the data sets acquired had the necessary characteristics to be used [2].

The data acquisition session started with a short interview to obtain the participants' informed consent, as well as to provide some information about the procedure. Besides, the data collection instruments were connected, and for the EMG, electrodes were placed on the frontal and zygomatic muscles, having in mind that these are associated with expressions of fear and joy, respectively. After that, videos started, as well as the signal acquisition.

Along with the videos visualization, the EMG was acquired, with a sampling frequency of 1000 HzHz, with electrodes placed on the previously mentioned muscles of the face. The emotion-inducing stimuli consisted, each one of them, on a series of 8 to 12 film cuts which took approximately 30 min in total. Each one of the series was intended to induce fear, joy, or neutral, and were based, respectively, on horror, comedy, and documentary films. In addition, 5 min of neutral video were presented at the beginning of the data collection to obtain a baseline for each individual in each session. Finally, each participant was asked to answer a questionnaire, before and after data collection, to record their state of mind on both occasions, enabling to assess whether the induction was effective [2].

2.2   Methodology

The signals under study was subjected to a series of pre-processing and processing methodologies, applied with Matlab R2021a (MATLAB R2021a & Simulink R2021a) [12], to verify the viability of using facial EMG for characterizing emotions.

At first, the EMG signals were split into intervals corresponding to the baseline and the emotion-inducing video. The extraction of this information is important as it allows to compare the "normal" state of the participants with the "emotional induced" state.

Afterwards, the signal was filtered. Taking into account the EMG physiological characteristics, it is usually applied a high-pass filter and a notch filter to eliminate the noise. The application of the high-pass filter aims to discard the noise coming from the interaction of the skin surface with the electrodes, which, generally, does not exceed 10 Hz [14]. EMG also suffers from the presence of interference in 50 Hz range, originating from the electrical appliances used in data collection. Thus, to ensure a reliable analysis of useful EMG information, different types of IIR (Infinite Impulse Response) high-pass filters (Butterworth and Chebyshev type I), with various orders and cut-off frequency 10 Hz, were implemented and their stability was assessed. These were combined with a notch filter 50 Hz, as it is the most suitable to exclude the powerline interference without affecting the neighborhood [9]. Besides analyzing the filters' stability, an evaluation using the absolute error, relative error, and coefficient of variation [6,7] was also conducted.

The physiological response of an individual to a particular emotion should be identical each time it is induced, although it may vary depending on the person and context [13]. In the same way, it can be assumed that while watching a video, inducing a particular emotion, the EMG signal collected has a characteristic response triggered by this emotion, which should consist of the existence of activation values that stand out from the others. Accordingly, we proposed to study the characterization of emotions from the EMG signal, based on the activation values of facial muscles triggered by emotional induction. For that, we rely on the entropy of the signal, which should allow verifying if the activation points provide enough information to characterize emotions.

In this work, a time-dependent measurement of the entropy was performed using sliding windows of 5 s to compute the entropy. This resulted in a time series of entropy values obtained from the evaluation of the information contained in the EMG time series [11], which allows the detection of energy changes along with the signal. The computation of time-dependent entropy is based on the division of the signal amplitude interval into 1000 horizontal partitions. For that, we used fixed and adaptive partitions. In the fixed partitions, the amplitude of each interval (partition) is defined by equally dividing the signal amplitude according to the number of partitions. Accordingly to the EMG signal variation values of all participants, the amplitude of the fixed partition was set to 0.16, varying between −0.08 and 0.08. Regarding the adaptive partition, the amplitude of each interval (partition) is set individually for each time window, based on the maximum and minimum value recorded in that time slot, having the advantage of detecting transient events [15].

According to the number of subdivisions created, the probability distribution is determined, by time window, through the quotient between the number of samples per partition, N_P, and the total number of samples, N_T (Eq. 1) [15].

Next, the entropy of the signal is computed from the Eq. 2.

$$P = \frac{N_P}{N_T} \tag{1}$$

$$E_S = -\sum_{i=1}^{m} P log_2(P) \tag{2}$$

At first, a temporal analysis was performed to understand if the amplitude of the entropy stood out when the emotional state-inducing videos were watched, relatively to the baseline data.

This temporal analysis allows understanding that we need to perform normalization. Considering that each individual differently express his/her emotions according to the subjective nature of emotions, to reduce the inter-participant dependency, the entropy from the EMG signals collected during the "emotional induced" states were normalized using the entropy of the EMG signals, respectively, from the baseline state.

It is worthwhile to notice that the entropy of the baseline provide information about the initial state of the participant, working as a relevant term of comparison in relation to the data collected in the induction stage, where the participant is subjected to a video that intends to induce an emotional state. Therefore, if the muscle activation values in the induction stage do not stand out concerning the values from the first stage (baseline), it becomes impossible to characterize emotions. In this sense, the normalization of entropy using the baseline highlight the activation values. Thus, the normalization of the entropy of induction stages was performed subtracting the mean entropy of the baseline and dividing it by the standard deviation.

The Eq. 3 demonstrates the calculation performed, where S_N is the normalized entropy vector, S_0 is the initial entropy vector and μ and σ are, respectively, the mean and the standard deviation of the entropy vector referring to the session under study baseline.

$$S_N = \frac{S_0 - \mu}{\sigma} \tag{3}$$

Therefore, to analyze the differences between the 3 emotional conditions induced, we applied the Friedman Test, a non-parametric test, that allows deciding if the samples from the 3 emotional induced are originated from the same distribution, by comparing the mean ranks of the 3 groups of emotional induction. The analysis was performed in a paired manner, taking into account the muscle under analysis. Moreover, besides performing this analysis using the data from all the participants, it was also performed individually. This would allow evaluating which participants have different physiological responses that deviated from the overall group.

This test aims to detect differences in the median in, at least, one pair under evaluation, i.e., the null hypothesis that the three conditions have no significant differences between them is tested against an alternative hypothesis that at least two of these conditions are distinct from each other.

In the case of differences between the 3 emotional induced conditions (rejection of the null hypothesis), we performed a post hoc analysis with the Wilcoxon signed-rank test to evaluate which conditions were significantly different from each other. By performing multiple comparisons, the Bonferroni correction was used. Moreover, we further analyze those cases through multiple comparisons between the emotional conditions. For that, we use the multcompare function from MATLAB. Besides returning the pairwise comparison results based on the statistics outputted from the Friedman test, it assesses under which conditions there are significant differences between the medians of the respective entropy, making it possible to confirm the differentiation in-between the three conditions. Furthermore, it also allows interactive graphical multiple comparisons of the groups, displaying the mean rank estimates for each group and comparison intervals. Thus, through a graphic evaluation it is possible to verify that whenever intervals overlap between conditions, it is not possible to reject the equal median hypothesis. On the contrary, if there is no overlap, the values show significant differences.

3   Results

From the several filters implemented and through the metrics analyzed, the choice relied on Butterworth filter. Thus, for both EMG signals, the filtering process combined a high-pass filter, 10 Hz, followed by a notch filter 50 Hz.

The entropy from the filtered signals was computed, using fixed and adaptive partitions, and performed a temporal analysis of the entropy. Through the temporal response, it is clear that there are activation values that stand out in relation to the others.

However, this difference does not allow to take conclusions, since it varies a lot and without apparent logic. Also, the temporal analysis of entropy does not allow us to identify changes in entropy values according to the emotion induced by the respective video. Therefore, to confirm these results, the distribution of the non-normalized entropy, for each participant, was analyzed through box plots. Figure 1 shows the distribution of entropy, for participants 302 and 337, respectively.

From this figure it is clear that each individual has an intrinsic physiological response to similar emotional induction videos, showing that emotions are expressed in different ways from person to person [13]. Taking into account that the box plots correspond to two different participants, there are fair differences in the intensity of the same emotion from one individual to the other. For both facial EMGs, participant 302 shows lower values of entropy during the fear video induction, while participant 337 presents higher values in the similar video emotional induction. In the left figure (participant 302) it can be seen that fear is the emotion with the lowest entropy, and the emotion that stands out relative to the others is the neutral emotional state, while joy is close to fear. On the other hand, at the right (participant 337), fear is the emotion with the highest entropy and the values for neutral and joy are quite close to each other. This observation may be justified by the fact that the individual is more or less used to feel a

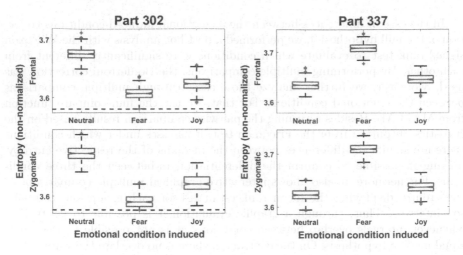

Fig. 1. Box plots of the non normalized EMG entropy, with adaptive partitioning, for the frontal and zygomatic muscles, relative to the videos inducing the different emotions. The left plots refer to participant 302 and right plots refer to participant 337.

certain type of emotion or the level of stress or tiredness, for example, may cause the activation of muscles even without any type of induction. It should be noted that this behavior in the entropy of EMG is true for the two muscles studied, for both participants.

This stands out the importance of normalizing the entropy of the EMG signals for each participant accordingly to the respective baseline content. Therefore, for each participant, the entropy values were normalized using the baseline to vanish the inter-participant dependency due to the subjective nature of emotions. This would allow a group comparison on the EMG response during the "emotional induced", enabling the characterization of emotions through facial EMG signals.

An analysis of the distribution of the normalized entropy, for each participant and both muscles, indicated that the individual influence on the expression of emotions was removed, or attenuated.

The box plots in Fig. 2 visually present the summary statistics of the normalized entropy of all the participants for the 3 emotional induced conditions (due to the presence of several outliers, bottom figures present a closer view of the box plots). It can be noticed that the neutral emotional state presents values slightly higher than the remaining two emotional conditions and that the entropy distribution of fear and joy are similar. In addition, all conditions show similar variance and the median values, for the 3 conditions, are close to zero due to normalization.

The conclusions drawn above were based on a visual analysis of the box plots, so there is a need of making a more reliable evaluation of these same values. In this sense, to validate the hypothesis of differences in the median entropy between the different conditions, the Friedman test was applied to the entropy values (computed with fixed and adaptive partitioning) of the 3 emotional induced

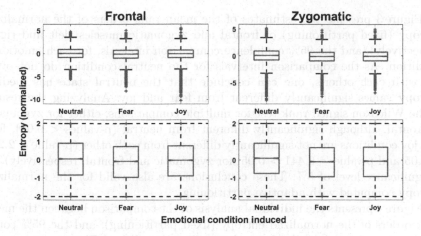

Fig. 2. Box plots for normalized entropy (adaptive partitioning) for all the participants, per emotional condition.

conditions. This analysis was performed individually for each participant and between all the participants, by comparing the mean ranks of the 3 groups of emotional induction.

With respect to the Friedman test for all participants, for normalized entropy (either computed with fixed or adaptive partitioning) of the EMG from zygomatic and frontal muscles, the returned p-values (<0.05) indicate that, at a significance level of 5%, the null hypothesis that the entropy from the three emotional conditions that come from the same distribution is rejected. As the Friedman tests allowed us to conclude that the median values of the entropy from the 3 conditions are significantly different, we performed multiple comparisons, through the Wilcoxon signed-rank test, to reveal which from the 3 conditions are significantly different from the others.

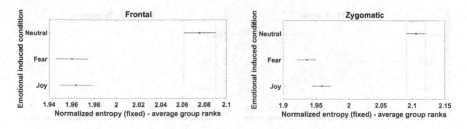

Fig. 3. Multi-comparison graphics for the mean ranks of the normalized entropy (fixed partitioning) of frontal and zygomatic muscles (left and right, respectively), for each emotional condition.

Figure 3 presents the estimates of the mean rank orders of the normalized entropy (fixed partitioning) of frontal and zygomatic muscles (left and right, respectively), and the 95% confidence comparison intervals, for each emotional condition. As the comparison intervals for the neutral condition do not overlap with each other's, one can conclude that the neutral state has median entropy values significantly different from fear and joy. Analyzing the results of the Wilcoxon signed-rank test for multiple comparisons, either for zygomatic or frontal, although significantly different from neutral (p-values < 0.05), fear and joy conditions are not significantly different from each other (p-value = 2.209 > 0.05 and p-value = 0.441 > 0.05, for zygomatic and frontal, respectively), at a significance level of 5%. These conclusions are also valid for the normalized entropy computed with adaptive partitioning.

Figure 4 presents the individual analysis of the comparison between the mean rank orders of the normalized entropy (fixed partitioning), and the 95% confidence comparison intervals, for each emotional condition. The top figure corresponds to participant 302 and the bottom to participant 317 (frontal and zygomatic muscles at the left and right, respectively). This figure shows that, for participant 317, the entropy of the EGM from zygomatic muscle allows to distinguish the 3 emotional conditions from each other (p-values < 0.05), while the information from the frontal muscle does not allow to differentiate joy from neutral (p-value = 0.981). As for participant 302, the entropy from both muscles, allow distinguishing the 3 emotional induced conditions.

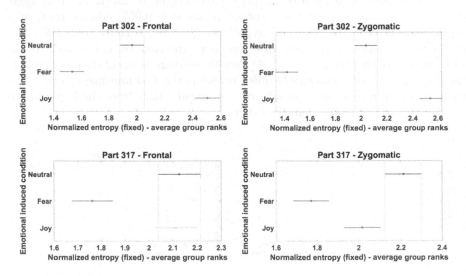

Fig. 4. Multi-comparison graphics for the mean ranks of the normalized entropy (fixed partitioning) of frontal and zygomatic muscles (left and right, respectively), for each emotional condition. The top plots refer to participant 302 and bottom plots refer to participant 317.

Concerning the individual results of the Wilcoxon signed-rank tests, and with regard to the entropy calculated with fixed partitions, in the frontal muscle, it was found, with a significance level of 5%, that there are no significant differences:

- Between the neutral emotion and fear induction conditions for 13 participants (35%);
- Between the neutral emotion and joy induction conditions for 11 participants (29%);
- Between the conditions of induction of fear and joy emotions for 7 participants (19%);

For the zygomatic muscle, using the entropy calculated with fixed partitions, it was found, with a significance level of 5%, that there are no significant differences:

- Between the neutral emotion and fear induction conditions for 11 participants (29%);
- Between the neutral emotion and joy induction conditions for 6 participants (16%);
- Between the conditions of induction of fear and joy emotions for 3 participants (8%);

Therefore, we may conclude that although the analysis of the entropy from facial EMG did not allow to distinguish between fear and joy conditions, an individual analysis, with frontal and zygomatic muscles, allowed to differentiate both for 30 and for 34 participants, respectively (when computing entropy with fixed partitions). Moreover, when using the entropy computed with adaptive partitions, we distinguish less emotional states. It can also be concluded that the information from the zygomatic muscle allowed to characterized more participants with respect to the 3 emotional conditions induced.

Comparing the group and the individual analysis it is straightforward that, for both muscles, fear and joy are the easier conditions to characterize. When evaluating all the participants, we do not have statistical evidence to reject that these conditions are significantly different from each other.

Therefore, the discrepancy between the group and individual analysis, specially regarding the characterization of fear and joy, supports that the emotions variation induction did not always occur in the same direction.

4 Conclusion

In accordance with what has been reported throughout this article, it is concluded that the emotional characterization based on the facial EMG signal is feasible. It should be pointed out, that the physiological characterization of emotions through the entropy of EMG signals prevents the extraction of features, and subsequent feature selection. The analysis of the obtained results allows concluding that, the entropy computed with fixed partitions performed better in distinguishing among emotions.

According to the results, one can conclude that the participants' percentage where the distinction of emotions is possible is significant, and reveals the possibility of performing the characterization based only on zygomatic muscle. Following this, there is no need to combine information from the zygomatic and frontal muscles, since the increase in performance should be small.

These preliminary results, and limited to this dataset, point out the feasibility of an emotion characterization based only on zygomatic EMG. Therefore, these results support a future collection with more participants to allow a deep analysis with the data, and opens the possibility of emotion classification relying only on zygomatic EMG. Such a classification system presents several advantages over classification based on facial images and expressions and over classification approaches based on several physiological signals. This kind of approach would benefit from the ability of detecting micro-expressions and signal constant collection. Moreover, being a less intrusive approach, this would allow the application of this method to several problems, remembering the advantages associated with it. Among them, the issue of data protection, mainly regarding children and minors, registering responses even in emotion inhibition studies, and also the collection of a signal that is constant over time, without interruptions, which does not occur in methods that are based on behavioral responses, namely from images of the face.

Acknowledgments. This work was funded by national funds through FCT - Fundação para a Ciência e a Tecnologia, I.P., under the Scientific Employment Stimulus - Individual Call - CEECIND/03986/2018, and is also supported by the FCT through national funds, within IEETA/UA R&D unit (UIDB/00127/2020). This work is also funded by national funds, European Regional Development Fund, FSE through COMPETE2020, through FCT, in the scope of the framework contract foreseen in the numbers 4, 5, and 6 of the article 23, of the Decree-Law 57/2016, of August 29, changed by Law 57/2017, of July 19.

References

1. Alvarez, V.M., Velazquez, R., Gutierrez, S., Enriquez-Zarate, J.: A method for facial emotion recognition based on interest points. In: Proceedings of the 2018 3rd IEEE International Conference on Research in Intelligent and Computing in Engineering, RICE 2018 (2018)
2. Brás, S., Carvalho, J.M., Barros, F., Figueiredo, C., Soares, S.C., Pinho, A.J.: An information-theoretical method for emotion classification. In: Henriques, J., Neves, N., de Carvalho, P. (eds.) MEDICON 2019. IP, vol. 76, pp. 253–261. Springer, Cham (2020). https://doi.org/10.1007/978-3-030-31635-8_30
3. Cacioppo, J.T., Petty, R.E., Losch, M.E., Kim, H.S.: Electromyographic activity over facial muscle regions can differentiate the valence and intensity of affective reactions. J. Pers. Soc. Psychol. **50**(2), 260–268 (1986). https://doi.org/10.1037/0022-3514.50.2.260
4. Cowen, A.S., Keltner, D.: Self-report captures 27 distinct categories of emotion bridged by continuous gradients. Proc. Natl. Acad. Sci. U.S.A. **114**(38), E7900–E7909 (2017)

5. Damásio, A.: O mistério da consciência. Companhia das Letras, 2nd edn. (1999)
6. Departamento de Física da FCTUC: Introdução ao cálculo de erros nas medidas de grandezas físicas, pp. 1–14 (2003)
7. Education U.I.f.D.R. FAQ - What is the coefficient of variation. https://stats.idre.ucla.edu/other/mult-pkg/faq/general/faq-what-is-the-coefficient-of-variation/.. Accessed 3 July 2020
8. Ekman, P.: Universal Emotions. https://www.paulekman.com/universal-emotions/.. Accessed 1 July 2020
9. Gruebler, A., Suzuki, K.: Measurement of distal EMG signals using a wearable device for reading facial expressions. In: 2010 Annual International Conference of the IEEE Engineering in Medicine and Biology Society, EMBC 2010, pp. 4594–4597 (2010)
10. Ko, B.C.: A brief review of facial emotion recognition based on visual information. Sensors (Switzerland) **18**(2), 401 (2018)
11. Ligrone, R.: Biological Innovations that Built the World. Springer, Cham (2019). https://doi.org/10.1007/978-3-030-16057-9
12. MATLAB version 9.10.0.1684407 (R2021a). The Mathworks, Inc., Natick, Massachusetts: (2021)
13. Purves, D., et al.: Neuroscience. Sinauer Associates, 3rd edn. (2004)
14. Tan, J.W., et al.: Facial electromyography (fEMG) activities in response to affective visual stimulation. In: IEEE SSCI 2011 - Symposium Series on Computational Intelligence - WACI 2011: 2011 Workshop on Affective Computational Intelligence (2011)
15. Thakor, N.V., Tong, S.: Advances in quantitative electroencephalogram analysis methods. Annu. Rev. Biomed. Eng. **6**(1), 453–495 (2004)
16. Wilson, J.: What Is Facial EMG and How Does It Work? (2018). https://imotions.com/blog/facial-electromyography/. Accessed 8 July 2020

Multimodal Feature Evaluation and Fusion for Emotional Well-Being Monitorization

Irune Zubiaga[✉] and Raquel Justo[iD]

University of the Basque Country, Sarriena no number, 48940 Leioa, Spain
irune.zubiaga@ehu.eus
https://www.ehu.eus/es/

Abstract. Mental health is a global issue that plays an important roll in the overall well-being of a person. Because of this, it is important to preserve it, and conversational systems have proven to be helpful in this task. This research is framed in the MENHIR project, which aims at developing a conversational system for emotional well-being monitorization. As a first step for achieving this purpose, the goal of this paper is to select the features that can be helpful for training a model that aims to detect if a patient suffers from a mental illness. For that, we will use transcriptions extracted from conversational information gathered from people with different mental health conditions to create a data set. After the feature selection, the constructed data set will be fed to supervised learning algorithms and their performance will be evaluated. Concretely we will work with random forests, neural networks and BERT.

Keywords: Mental health · Feature selection · Machine learning

1 Introduction

Mental health is a global issue that plays an important roll in the overall well-being of a person. Depression and anxiety are very common afflictions nowadays, having become specially frequent during the last years due to the COVID-19 pandemic [10]. To maintain good health, it is not enough with fixing mental health problems, it is also necessary to preserve mental well-being. Conversational systems like chatbots have demonstrated to be helpful for this task [16, 18, 19], along with professional help.

This research is framed in the MENHIR project [2] whose end goal is to create a chatbot technology that provides symptom and mood management, personalized support and motivation, coping strategies, mental health education, signposting to online resources and complementing the support received by local services [11].

As a first step to develop this technology, in this project we will focus on feature selection to create a model whose aim is to evaluate if a person suffers

© Springer Nature Switzerland AG 2022
A. J. Pinho et al. (Eds.): IbPRIA 2022, LNCS 13256, pp. 242–254, 2022.
https://doi.org/10.1007/978-3-031-04881-4_20

from a mental illness or not. For this, we will use transcriptions extracted from conversational information gathered from people with different mental health conditions to create a data set. These transcriptions contain both semantic and paralinguistic information that will be analyzed for feature selection. Finally, the constructed data set will be fed to supervised learning algorithms and their performance will be evaluated. Concretely, we will work with random forests, neural networks and BERT.

Various studies have been made about feature selection for mental state evaluation. In said studies, the feature selection techniques and the used machine learning methods are diverse [7,31], just like the data sources; both visual and auditory information have been analyzed [17,25,30].

The main contributions of this work are the analysis of semantic and paralinguistic information extracted from transcriptions associated to spontaneous conversations (corpus further explained in Sect. 2) by means of clustering algorithms and the examination of the efficiency of these features for monitoring mental health by means of latest state-of-the-art machine learning approaches.

2   Task and Corpus

This work deals with real conversations among a counsellor and a potential patient. Specifically, clients from the Action Mental Health [5] foundation (AMH), diagnosed with anxiety and/or mild depression are involved. The corpus is built up of transcriptions of 60 interviews made to two different groups; the control group and the AMH group. The control group consists of 28 members that have never been diagnosed with any kind of mental illness whilst the AMH group consists of 32 members that have been diagnosed with different mental illnesses, depression and anxiety being the predominant ones. The transcriptions include the dialogues written in a literal way and paralinguistic information like *noises (music, footsteps, inhaling, etc.), interjections, yes/no sounds, pauses* and *laughs*. The corpus was gathered within the MENHIR project framework and the recording procedure was conducted according to the guidelines of the Declaration of Helsinki and approved by the Ethics Committee of the Ulster University.

The interviews consist of three main sections; in the first section the counsellor asks the patient 5 questions that lead to non-emotional conversation. The second part consists of fourteen affirmations from the Warwick-Edinburgh Mental Well-Being Scale [28]. The participants have 5 possible answers that go from *none of the time* to *all of the time* to indicate how often they feel the way that these affirmations express. In this part, in addition to the answer to the affirmations, sometimes the interviewees added some dialogue of their own to further explain the answer they gave. In the third and last part, the participants were asked to read a text passage of the popular tale *The Boy and the Wolf*.

To create the data set, only the interviewee's conversational information was considered. The reading phase was removed, because there is no distinguishing semantic information associated with it. Then, all the text associated with the remaining dialogues was gathered as a non-labeled corpus to start working with.

Table 1. Interviewer's annotations that give information about the surroundings or the file's state.

Token	Meaning
\<redacted\>, \</redacted\>	A piece of information was removed to preserve confidentiality
\<not_understood\>, \</not_understood\>	What the interviewee said was not understood
\</file_problem\>	There was a problem with file
{phone_ringing}	A phone started ringing
{footsteps}	The sound of footsteps was heard
{noise}	A noise was heard
{music}	Music started playing
{door}	The sound of a door opening/closing was heard
{other}	Some other kind of sound was heard

All in all, the corpus is composed of 6741 sentences; 4743 sentences from the AMH group and 1998 from the control group.

Some annotations regarding the paralinguistic information were removed because they only provided information about the surroundings or the state of the acoustic file. The removed information is summarized in Table 1.

Moreover, the words that held the same meaning were represented by a common token (see Table 2). The interviewee's answers that only consisted of a yes/no and the answers to the Warwick-Edinburgh Mental Well-Being Scale were removed, because these answers are completely dependent on the counsellor's questions and have no value on their own. The Python contractions library was used for expanding common English contractions and the typos in the transcriptions were corrected as far as possible.

At this point the corpus is composed of 3955 sentences; 2810 sentences from the AMH group and 1145 from the control group.

Table 2. Words with similar meaning and the tokens that represent them.

Words	Token
uh-huh, um-hum, mm-hm, yep, yeah	Yes
huh-uh, hum-um, uh-uh	No
em, ew, jeeze, oh, okay, uh-oh, um, whoa, whew, uh, mm, eh	{INTERJECTIONS}
a-ha	{SURPRISE}
huh	{PUZZLEMENT}
hum	{STALLING}

3   Feature Analysis

The correct selection of the features that will feed a machine learning algorithm, can make the difference when it comes to designing a predictive system [12, 23]. In this work, we look for the best representation of the text to distinguish among healthy subjects and subjects that suffer from a mental illness. The idea is to explore whether semantic information associated to the text extracted from conversational speech is relevant and can help to distinguish between the two groups.

3.1   Semantic and Paralinguistic Information

Firstly, an analysis of information from different sources was carried out (mainly semantic information but also paralinguistic one). The idea is to conclude whether they can complement each other or if one should be predominant when regarding the detection of mental illness.

The semantic information is gathered as word embeddings. FastText [8] word embeddings were used to represent each word. Specifically, 1 million pre-trained word vectors learned with subword information on Wikipedia 2017, UMBC webbase corpus and statmt.org news data set (16B tokens) were considered (https://fastText.cc/docs/en/english-vectors.html). The word vectors were achieved using a skip-gram model as described in [8]. This model predicts the surrounding words based on the given input words which are within the given distance. It aims to predict the context from the given word. Words occurring in similar contexts tend to have similar meaning [26]. Therefore, it can capture the semantic relationship between the words. Finally, the average of all the word vectors in a sentence was carried out to get a 300 dimensional vector representing each sentence. This way we obtained a vector representation for each sentence. FastText model, as explained in [8], outperforms methods that don't take subword information into account as well as methods relying in morphological analysis [24, 29].

In a second stage, the paralinguistic information that was added in the transcription procedure, such as *interjections, laugh, pause* etc. was taken into account in conjunction with the semantic information. The specific events that were considered are shown in Table 3. All of them are related to the flow of the discourse and we aim to explore whether they can provide any cue about the mental state of the speaker. These new features are added to the fastText representation as new dimensions. The new 15 dimensions are real values associated to the frequency of appearance of a specific event in a sentence. The only exception is the feature *pause at start*, which has a binary value (1 or 0) instead of a frequency. It will take 1 when a sentence starts with a pause and a 0 when it does not. This token was created because it is well known that depression can impair information processing and decision-making skills. Because of this, it is reasonable to think that people who suffer from it might take longer to answer to questions.

Table 3. Paralinguistic features and their meanings.

Paralinguistic features	Frequency in AMH	Frequency in control	Meaning
{INTERJECTIONS}	0,193	0,258	Uses an interjection
{CUT}	1,72e−2	2,58e−2	Stops speaking mid-word
{inhaling}	1,33e−2	1,97e−2	Inhales
{laugh}	1,3e−2	1,01e−2	Laughs
<pause>WORDS</pause>	1,22e−2	3,77e−2	Makes a pause, says something, then makes a pause again
{STALLING}	5,57e−3	10e−3	Makes a sound that denotes they are thinking about an answer
<laugh>WORDS</laugh>	3,68e−3	3,96e−3	Says something while laughing
{breathing}	3,59e−3	4,48e−3	Takes a deep breath
{tsk}	2,88e−3	0	Makes a flicking sound with their mouth
{STUTTER}	6,43e−4	11e−4	Stutters when saying a word
{PUZZLEMENT}	5,81e−4	6,33e−4	Makes a sound that denotes puzzlement
{cough}	2,94e−4	8,32e−4	Coughs
<breathing>WORDS</breathing>	1.05e−4	0	Says something while taking a deep breath
<pause>empty</pause>	5.37e−5	0	Makes a long pause
Pause at start	0,13897	0,15594	Gives information about the sentence starting with a pause

3.2 Clustering

In order to assess the contribution of the different features in the mental state detection task, unsupervised learning was used. The idea is to see which set of features provides a clustering of samples that better matches the discrimination among healthy and non-healthy participants. Since the label of each data sample was known, it was possible to use a confusion matrix and the corresponding F-score to evaluate each set's outcome. Another purpose of using unsupervised learning algorithms is to asses their performance for future works in which the data might be unlabeled. Two different clustering algorithms were used: k-means [22] and DBSCAN [15]. The k-means algorithm is an interesting approach because it is fast and the number of clusters must be predefined (in our task the number of clusters is known to be two). DBSCAN assumes that clusters are areas of high density separated by areas of low density. Due to this rather generic view, clusters found by DBSCAN can be any shape, as opposed to k-means which assumes that clusters are convex shaped. As a consequence, while k-means clustering does not work well with outliers and noisy data sets, DBSCAN handles them efficiently.

Several tests where ran with both algorithms and the best outcomes are summarized in Table 4.

Table 4. Clustering results.

fastText	Other features	F-score	
		k-means	DBSCAN
100d fastText		0.667	0.667
	{INTERJECTIONS}	0.674	0.669
	Pause at start	0.669	0.669
	All paralinguistic information	0.669	0.670
300d fastText		0.667	0.667
	{INTERJECTIONS}	0.667	0.667
	Pause at start	0.669	0.670
	All paralinguistic information	0.667	0.667
0d fastText	Paralinguistic information without pause at start	0.673	0.671
	All paralinguistic information	0.669	0.671

As seen in Table 4 there are no significant differences between any of the tested hypotheses and more studies are needed to conclude which one of them presents better performance. This being said, the results do seem to have a tendency toward improvement in some scenarios and thus it would be interesting to further analyze them. The best scenario is the one considering semantic information associated to 100 dimensional fastText and interjection related information. When working with 300 dimensional fastText, the best outcome is achieved when working with pause related information. When considering only fastText the results are very similar, what means that this representation can provide information about the mental state. However, it is also quite clear that paralinguistic information can be of help in this task. This can be seen by the fact that when working with only paralinguistic information we get very similar results of those obtained when working with fastText. It is worth mentioning that this happens even though there are a lot less features regarding paralinguistic information than regarding fastText (15 features instead of 300 or 100).

As Table 4 shows, both when working with 300 dimensional and with 100 dimensional fastText word embeddings, the F-score has a value of 0.667. A significant difference between 100 dimension and 300 dimension embedding is that in the second case the paralinguistic information looses its influence in the model's outcome, and both when taking paralinguistic information into account and when not the F-score value is the same.

Another set of experiments was carried out using DBSCAN. In this case the number of formed clusters cannot be previously defined. We know beforehand that there are only two possible outcomes; mentally ill and not mentally ill. Thus, only the cases in which 2 clusters were achieved were analyzed. The obtained results were similar to the ones obtained with k-means.

3.3   Novel Techniques for Semantic Information Extraction

Unlike recent language representation models, BERT [13], which stands for Bidirectional Encoder Representations from Transformers, is designed to pre-train deep bidirectional representations from unlabeled text by jointly conditioning on both left and right context in all layers. As a result, the pre-trained BERT model can be fine-tuned with just one additional output layer to create state-of-the-art models for a wide range of tasks, such as question answering and language inference, without substantial task-specific architecture modifications [6]. Thus, lastly we used a BERT to represent our sentences. Will be further explained in Sect. 4.3.

4   Supervised Learning for Classification Experiments

Once relevant features are selected the data set is fed to different supervised learning algorithms. The aim is to analyze which is the best approach to attain a good classification result.

4.1   Random Forest

Random forests [9] construct many individual decision trees at training and pool the predictions from all those trees to make a final prediction.

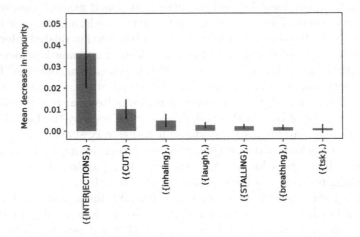

Fig. 1. Paralinguistic information with impact on the model's outcome.

Mean Decrease in Impurity (MDI) is one of the most widely used measures of feature importance. It calculates each feature's importance as the sum over the number of splits that include the feature, proportionally to the number of samples it splits. This is done across all trees. It is used to measure how much the model's accuracy decreases when a certain feature is excluded.

Random forest has shown to be able to handle the feature selection issue even with a high number of variables [12].

The random forest classifier had the best results when having no ramification limit. The 5608 samples were divided as follows; the train size was set to 90% and test size to 10%. There are 5608 samples instead of 5620 because some sentences consisted of typos and fastText could not handle them. Because of this, those sentences were removed.

A test was run to assess feature importance by using MDI. As in previous tests, the paralinguistic feature with the biggest impact on the model's outcome is the one with interjection information (see Fig. 1).

Feature relevance was once again evaluated using SHAP values [21]. Lundberg et al. [21] demonstrated that for MDI the importance of a feature does not always increase as the outcome becomes more dependent on that feature. To remedy this issue, they propose the tree SHAP feature importance. This focuses on giving consistent feature attributions to each sample. Once individual feature importance is obtained, overall feature importance is calculated by averaging the individual feature importance across samples. The most relevant features according to SHAP values are shown in Fig. 2. The features with numbers as names correspond to fastText dimensions. As in all previous tests, interjections are shown to have a considerable impact on the model's outcome.

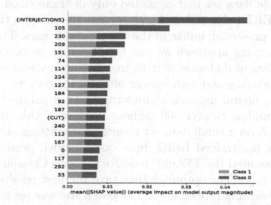

Fig. 2. Features with the highest SHAP values.

4.2   ANN

Artificial neural networks (ANNs) are biologically inspired computational networks. Multilayer perceptrons (MLPs) are fully connected feed-forward ANNs that comprise at least three layers of nodes: input, hidden, and output. Except for the input nodes, each node is a neuron that uses a nonlinear activation function. Because of having multiple layers and non-linear activation, MLPs have the ability to distinguish data that is not linearly separable.

The larger the data set, the better the ANNs performance. In fact, ANNs are shown to outperform traditional models when working with large data sets; traditional algorithms get to a saturation point and will not improve their outcome while ANNs are still capable of learning.

Several neural network architectures were tried out, but the one that gave the best result was a feed-forward neural network with an input layer, two hidden layers with the ReLu activation function and an output layer with the Sigmoid activation function. The first hidden layer had 500 neurons whilst the second layer had 300. The dropout method was used, setting its value in 0.6, to prevent the neural network from over-fitting. The 5608 samples were divided as follows; the training set size was set to 75%, the test set size to 15% and the validation set size to 10%.

To find the model with the optimal validation loss, we used the Keras *ModelCheckpoint* [3].

4.3   BERT

While the standard neural network algorithm is known to have the risk of exhibiting bias towards the majority class when working with imbalanced data, BERT is shown to be able to handle imbalanced classes with no data augmentation [27]. Because of this, oversampling was not performed and the network was trained with a 3955 sample data set that consisted only of transcribed sentences. As a consequence of BERT being able to handle typos, the samples that were formed by them could be preserved, unlike in the other experiments. This makes BERT a even more interesting approach for our task because, as described in Sect. 2, the data set consists of dialogues written in a literal way and as a consequence typos, written stuttering and such appear often in the corpus.

BERT is a big neural network architecture with a parameter number that can go from 100 million to over 300 million. Because of this, training a BERT model from scratch on a small data set results in overfitting. To overcome this, we fine-tuned the pre-trained BERT base cased model, presented in the [13] paper. For that, we used the TFAutoModelForSequenceClassification auto class from Hugging Face [1]. The optimizer that gave the best results was the Adam optimizer with a learning rate of $4e-5$. The batch size was set to 8 and the used loss function was Sparse Categorical Crossentropy.

The sentences that compose the data set were tokenized using Hugging Face's AutoTokenizer and then fed to the model. The training set size was once again set to 75%, the test set size to 15% and the validation set size to 10%.

The model was evaluated after every epoch and the one whose output had the optimal validation loss was chosen.

4.4   Results

As seen in Table 5, the model with the best performance is the random forest with a F-score of 0.93. We must note that this might be related to our data set being too small, since neural networks tend to have a better performance when working with big data sets and in this case it was limited to 5608 samples. We must contemplate the chance that, when working with bigger data sets, this might change.

We also notice that even if we thought that BERT would outperform the neural network the outcomes we got with both of them are very similar. This might be because the BERT model architecture that we used was not specifically designed for our task; while we ran a lot of test with the neural network to get the best possible outcome, in the case of BERT we chose the BERT AutoModel that seemed to better suit our task.

It has been noted the importance of the paralinguistic information. Both when working with unsupervised learning and with random forests we were able to see a strong relation between the presence of these features and the improvement of the model's outcome. The interjections feature was the most relevant one. In our corpus *em*, *um*, *uh* and *mm* are labeled as interjections, where *mm* and *em* are used in the same way as *um*. It is argued in [4] that *uh* and *um* are often used to initiate a delay in speaking that implies that the speaker is searching for a word or deciding what to say next. The use of these interjections could be a result of the brain fog that depression and anxiety often lead to. Several previous works on feature extraction for depression detection marked the influence of pauses to be the strongest [14,20]. The impact of pauses was mainly related to their length rather than their frequency. In our corpus the pause time was not considered, which might be an explanation of why this feature was not shown as relevant.

Also, in the case of the neural network, taking paralinguistic information into account made training go from the initial 70 epochs that the neural network took for training to 50 epochs. In terms of time this does not make a difference since each epoch took only a few seconds to finish, but it could be useful for future applications.

Table 5. Classification results.

Algorithm	Data	Loss	Accuracy	F-score	Precision	Recall
Random forest	fastText		0.92	0.92	0.92	0.92
	fastText with paralinguistic information		0.93	0.93	0.93	0.93
Neural network	fastText	0.42	0.83	0.82	0.86	0.79
	fastText with paralinguistic information	0.43	0.83	0.82	0.83	0.85
BERT	Sentences without paralinguistic information	0.42	0.79	0.82	0.85	0.79

5   Conclusions and Future Work

The main observation was the importance the paralinguistic information holds in this classification task, the presence of interjections being the most relevant feature. We did not try to add more weight to paralanguage, but it would be an interesting thing to try in future work. A more complex architecture of BERT might also be convenient.

This research focused on written data. Because of this, there is a lot of information in the audio files of the interviews that has not been used and that would improve the prediction accuracy remarkably as seen in audio centred researches. Also, it would be of interest to measure the duration of the pauses.

Acknowledgements. This work was partially funded by the European Commission, grant number 823907 and the Spanish Ministry of Science under grant TIN2017-85854-C4-3-R.

References

1. Auto   classes.   https://huggingface.co/docs/transformers/model_doc/auto. Accessed 20 Jan 2022
2. Mental health monitoring through interactive conversations. https://cordis.europa. eu/project/id/823907. Accessed 15 Jan 2022
3. Modelcheckpoint. https://keras.io/api/callbacks/model_checkpoint/. Accessed 20 Jan 2022
4. Using uh and um in spontaneous speaking: Cognition **84**(1), 73–111 (2002). https://doi.org/10.1016/S0010-0277(02)00017-3
5. Home: Action mental health (Dec 2021). https://www.amh.org.uk/. Accessed 15 Jan 2022
6. Alaparthi, S., Mishra, M.: BERT: a sentiment analysis odyssey. J. Mark. Anal. **9**(2), 118–126 (2021). https://doi.org/10.1057/s41270-021-00109-8
7. Alghowinem, S., Gedeon, T., Goecke, R., Cohn, J., Parker, G.: Interpretation of depression detection models via feature selection methods. IEEE Trans. Affect. Comput. 1 (2020). https://doi.org/10.1109/TAFFC.2020.3035535
8. Bojanowski, P., Grave, E., Joulin, A., Mikolov, T.: Enriching word vectors with subword information. Trans. Assoc. Comput. Linguist. **5**, 135–146 (2017). https:// doi.org/10.1162/tacl_a_00051
9. Breiman, L.: Random forests. Mach. Learn. **45**(1), 5–32 (2001). https://doi.org/ 10.1023/A:1010933404324
10. Bueno-Notivol, J., Gracia-García, P., Olaya, B., Lasheras, I., López-Antón, R., Santabárbara, J.: Prevalence of depression during the COVID-19 outbreak: a meta-analysis of community-based studies. Int. J. Clin. Health Psychol. **21**(1), 100196 (2021). https://doi.org/10.1016/j.ijchp.2020.07.007
11. Callejas, Z., et al.: Towards conversational technology to promote, monitor and protect mental health. In: IberSPEECH 2021, 24–25 March 2021, Valladolid, Spain, Proceedings. ISCA (2021)
12. Chen, R.-C., Dewi, C., Huang, S.-W., Caraka, R.E.: Selecting critical features for data classification based on machine learning methods. J. Big Data **7**(1), 1–26, 100196 (2020). https://doi.org/10.1186/s40537-020-00327-4

13. Devlin, J., Chang, M.W., Lee, K., Toutanova, K.: BERT: pre-training of deep bidirectional transformers for language understanding. In: Proceedings of the 2019 Conference of the North American Chapter of the Association for Computational Linguistics: Human Language Technologies, vol. 1 (Long and Short Papers), pp. 4171–4186. Association for Computational Linguistics, June 2019. https://doi.org/10.18653/v1/N19-1423

14. Esposito, A., Esposito, A., Likforman-Sulem, L., Maldonato, N., Vinciarelli, A.: On the Significance of Speech Pauses in Depressive Disorders: Results on Read and Spontaneous Narratives, pp. 73–82, 01 2016. https://doi.org/10.1007/978-3-319-28109-4_8

15. Ester, M., Kriegel, H.P., Sander, J., Xu, X.: A density-based algorithm for discovering clusters in large spatial databases with noise. In: Proceedings of the Second International Conference on Knowledge Discovery and Data Mining, pp. 226–231. KDD 1996. AAAI Press (1996)

16. Fitzpatrick, K.K., Darcy, A., Vierhile, M.: Delivering cognitive behavior therapy to young adults with symptoms of depression and anxiety using a fully automated conversational agent (Woebot): a randomized controlled trial. JMIR Ment. Health 4(2), e19, June 2017. https://doi.org/10.2196/mental.7785

17. He, L., et al.: Deep learning for depression recognition with audiovisual cues: a review. Inf. Fus. 80, 56–86, 100196 (2022). https://doi.org/10.1016/j.inffus.2021.10.012

18. Inkster, B., Sarda, S., Subramanian, V.: A real-world mixed methods data evaluation of an empathy-driven, conversational artificial intelligence agent for digital mental wellbeing. JMIR mHealth and uHealth 6 (2018). https://doi.org/10.2196/12106

19. Joerin, A., Rauws, M., Ackerman, M.: Psychological artificial intelligence service, tess: delivering on-demand support to patients and their caregivers: Technical report. Cureus 11, 01 2019. https://doi.org/10.7759/cureus.3972

20. Liu, Z., Kang, H., Feng, L., Zhang, L.: Speech pause time: a potential biomarker for depression detection, pp. 2020–2025, 11 2017. https://doi.org/10.1109/BIBM.2017.8217971

21. Lundberg, S.M., Lee, S.I.: A unified approach to interpreting model predictions. In: Advances in Neural Information Processing Systems, vol. 30. Curran Associates, Inc. (2017)

22. MacQueen, J.B.: Some methods for classification and analysis of multivariate observations. In: Proceedings of the Fifth Berkeley Symposium on Mathematical Statistics and Probability, vol. 1, pp. 281–297 (1967)

23. Miao, J., Niu, L.: A survey on feature selection. Proc. Comput. Sci. 91, 919–926 (2016). https://doi.org/10.1016/j.procs.2016.07.111

24. Mikolov, T., Grave, E., Bojanowski, P., Puhrsch, C., Joulin, A.: Advances in pre-training distributed word representations. In: Proceedings of the Eleventh International Conference on Language Resources and Evaluation (LREC 2018). European Language Resources Association (ELRA), May 2018

25. Pampouchidou, A., et al.: Automatic assessment of depression based on visual cues: a systematic review. IEEE Trans. Affect. Comput. 10(4), 445–470, 100196 (2019). https://doi.org/10.1109/TAFFC.2017.2724035

26. Rubenstein, H., Goodenough, J.: Contextual correlates of synonymy. Commun. ACM 8, 627–633 (1965). https://doi.org/10.1145/365628.365657

27. Madabushi, H.T., Kochkina, E., Castelle, M.: Cost-sensitive BERT for generalisable sentence classification on imbalanced data. In: Proceedings of the Second Workshop on Natural Language Processing for Internet Freedom: Censorship, Disinformation, and Propaganda, pp. 125–134. Association for Computational Linguistics, November 2019. https://doi.org/10.18653/v1/D19-5018
28. Tennant, R., et al.: The Warwick-Edinburgh mental well-being scale (WEMWBS): development and UK validation. Health Qual. Life Outcomes 5(1) (2007). https://doi.org/10.1186/1477-7525-5-63
29. Toshevska, M., Stojanovska, F., Kalajdjieski, J.: Comparative analysis of word embeddings for capturing word similarities. In: 6th International Conference on Natural Language Processing (NATP 2020), April 2020. https://doi.org/10.5121/csit.2020.100402
30. Zhu, Y., Shang, Y., Shao, Z., Guo, G.: Automated depression diagnosis based on deep networks to encode facial appearance and dynamics. IEEE Trans. Affect. Comput. 9(4), 578–584 (2018). https://doi.org/10.1109/TAFFC.2017.2650899
31. Zulfiker, M.S., Kabir, N., Biswas, A.A., Nazneen, T., Uddin, M.S.: An in-depth analysis of machine learning approaches to predict depression. Current Res. Behav. Sci. 2, e100044–e100044, 100196 (2021). https://doi.org/10.1016/j.crbeha.2021.100044

Temporal Convolutional Networks for Robust Face Liveness Detection

Ruslan Padnevych, David Carmo, David Semedo[iD], and João Magalhães[✉][iD]

NOVA LINCS, School of Science and Technology, Universidade NOVA de Lisboa,
Caparica, Portugal
{r.padnevych,dm.carmo}@campus.fct.unl.pt,
{df.semedo,jm.magalhaes}@fct.unl.pt

Abstract. Face authentication and biometrics are becoming a commodity in many situations of our society. As its application becomes widespread, vulnerability to attacks becomes a challenge that needs to be tackled. In this paper, we propose a non-intrusive on the fly liveness detection system, based on 1D convolutional neural networks, that given pulse signals estimated through skin color variation from face videos, classify each signal as genuine or as an attack. We assess how fundamentally different approaches – sequence and non-sequence modelling – perform in detecting presentation attacks through liveness detection. For this, we leverage on the Temporal Convolutional Network (TCN) architecture, and exploit distinct and TCN grounded types of convolution and architectural design schemes. Experiments show that our TCN model provides the best balance in terms of usability and attack detection performance, achieving up to 90% AUC. We further verify that while our 1D-CNN with a residual block variant performs on par with the TCN model in detecting fake pulses, it underperforms in detecting genuine ones, leading to the conclusion that the TCN model is the most adequate for a production environment. The dataset will be made publicly available to foster research on the topic. (https://github.com/novasearch/Mobile-1-D-Face-Liveness-Detection)

Keywords: Temporal convolutional networks · Liveness detection ·
Mobile presentation attacks · 1D convolutional neural networks

1 Introduction

Due to the rise in popularity of face unlocking mechanisms on smartphones (and subsequently on other mediums), the risk of face spoofing - by using a copy of someone's face to bypass security measures - becomes higher [13]. One possible reason for this risk is the fact that face spoofing is a very low-cost attack, since basically anyone has access to a camera or a printer, and a picture of the person they want to attack [18]. Hence, mechanisms like liveness biometrics can help mitigating this risk, since the presence of a real person in time and space implies that it is not a presentation attack (PA) [12].

© Springer Nature Switzerland AG 2022
A. J. Pinho et al. (Eds.): IbPRIA 2022, LNCS 13256, pp. 255–267, 2022.
https://doi.org/10.1007/978-3-031-04881-4_21

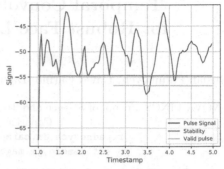

(a) **Genuine Scenario**: User face video registration and corresponding pulse.

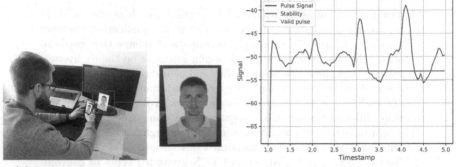

(b) **Attack Scenario**: Photo attack video registration and corresponding pulse.

Fig. 1. Comparison between genuine and fake (photo attack) scenarios. At the right, we show facial pulse signals, extracted in a non-intrusive manner from a mobile RGB camera using an EVM approach.

Face authentication and biometric systems have been widely adopted, making face liveness detection an actively researched field. The use of highly specialised hardware such as thermal cameras [2], hand geometry, depth cameras and fingerprinting scanners [6,15], among others, is not feasible in many scenarios due to the added cost. Alternatively, face liveness detection with RGB cameras, is a low-cost solution, where liveness can be assessed with the Eulerian Video Magnification method (EVM) [20]. This method is a non-intrusive manner, that estimates a person's cardio-vascular activity by analysing skin colour variations in face videos [1,11,13,16].

Among the existing approaches, convolutional neural networks (CNNs) have been widely used [9,21], due to their differential filter-based characteristics, and capacity to process spatial information (from 1D to 3D). These are usually applied directly to RGB video frames, leading to 2-D and 3-D CNNs, that look at blood volume changes, face texture and depth information to assess liveness.

In this paper we propose a face liveness detection system using 1D-CNNs, based on the Temporal Convolutional Network [3] (TCN), that consider pulse signal information obtained by magnifying blood volume change and retaining the green colour channel. Two fundamentally different approaches are considered, sequence and non-sequence based, with distinct convolution types, regular and dilated causal convolution, respectively. Experimental evaluation shows that a TCN-based approach provides the better trade-off between security and usability (i.e. detecting genuine pulses), ensuring security while retaining an high capacity of detecting genuine pulses, thus permitting users to securely and swiftly enrol through a smartphone their athentication photo.

In the next section we discuss previous work, in Sect. 3 we describes the pre-processing steps before the biometrics method is applied. Section 4 details the new TCN face biometrics method. Experiments are presented in Sect. 5.

2 Related Work

2.1 Non-intrusive Pulse Extraction

As observed from pulse oximetry [16], blood tends to absorb a higher percentage of light, when compared to the surrounding tissue. This happens due to the presence of a protein – haemoglobin – which is responsible for oxygen transport in blood, with an high absortion of green light [13], allowing for a semi-accurate blood circulation estimation of heart-rate through video processing [1,11,13,16]. In [11], Independent Component Analysis (ICA) was used as a means to recover the pulse wave from a RGB camera. The RGB sensors pick up a mixture of the pulse wave signal, along with some noise caused by motion or ambient light variations. Eulerian Video Magnification [1,19] (EVM), overcomes the ICA assumption regarding the multiple signal sources being combined as a linear mixture. Instead finds and amplifies small variations in RGB videos, leading to an effective technique for pulse estimation. Other non-intrusive techniques could be used such as facial temperature [7], using thermal infrared cameras. However, such specialised hardware is not available in most smartphones. Recently, CNNs have also been used to estimate pulse signals [4,5], however, in our setting these approaches are not feasible, since they require considerably large training datasets to generalise, while signal based blood flow estimation approaches such as EVM do not [1].

2.2 Face Liveness Detection

Face liveness detection is an highly relevant problem, that has been actively researched. Based on blood flow variation, in [8] pulse signal is extracted from face videos, and then used as features to train a classifier. In [14], the authors proposed a data-driven approach for face liveness, by analysing face movement (e.g. eyes blinking, lips movement, and other facial dynamics) and face texture. In [17], texture difference between red and green channel, is used along with local face regions' color distribution, to discriminate between live and spoofing data. Also focusing on local face regions but adding depth, in [21] the authors use a two-stream CNN approach to extract both local face features and depth maps. [9] also proposes to extract pulse signal, through multi-scale long-term statistical spectral features, but also considers local face texture information, using a contextual patch-based CNN.

3   Robust Liveness Detection on the Fly

When users register their face for future authentication, liveness detection is performed to overcome presentation attacks [10]. A user face video stream of 5 s is received, decoded and the pulse signal is extracted in real-time. For real-time face detection and alignment, the Multitask cascaded convolutional networks (MTCNN) method [22] is used, based on cascaded convolutional neural networks and trained under a multitask learning setting. Thus for each decoded frame, we obtain a face bounding box, and crop the original frame to keep only the user face. To estimate the heartbeat from the individual's face, frequency bands representing plausible human pulse are amplified, revealing the variation in redness as blood flows [1,19]. This results in 1D pulse signals, as the ones shown in Fig. 1. By obtaining the pulse rate it cannot be stated with complete certainty that it originated from a real person, as it is possible to simulate a heartbeat using presentation attacks more specifically *face-spoofs*. In the next section, we design robust 1D pulse signals classifiers based on convolutional neural networks, for liveness detection.

4   Temporal Convolutional Neural Networks for Liveness Detection

For liveness detection from 1D pulse signals (as the ones depicted in Fig. 1), two distinct approaches are considered: sequence-based and non-sequence models. In general, the sequential modelling task consists of given a sequence $x_0, \ldots, x_T$ as input, the goal is to predict the corresponding output values $y_0, \ldots, y_T$, such that to predict y_t, for a given instant of time t, the model is only allowed to use the previously observed input values $x_0, \ldots, x_t$ and not any of the future inputs values $x_{t+1}, \ldots, x_T$. In contrast, non-sequence models take as input the full sequence and generate a single prediction output. The two approaches can be formally defined as:

$$\underbrace{\hat{y} = f(x_0, \ldots, x_T)}_{\text{Non-Sequence}} \quad (1) \qquad \underbrace{\hat{y}_0, \ldots, \hat{y}_T = f(x_0, \ldots, x_T)}_{\text{Sequence-based}} \quad (2)$$

While RNNs are a natural choice for sequence modelling, recent results indicate that CNNs, specifically Temporal Convolutional Networks, can outperform RNNs on several sequence modelling tasks [3]. Departing from this hypothesis, we adopt the TCN as the target architectural design pattern. It combines simplicity, auto-regressive prediction and long context memory, i.e. the ability to

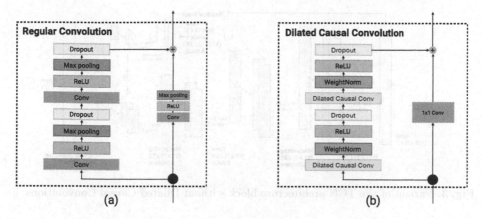

Fig. 2. Comparison between the CNN residual block with regular and dilated convolutions.

look far into the past to make predictions. Convolutions in the architecture are causal, which means that there is no leakage of information from the future to the past. Simple causal convolutions consider a history whose size is linear with the depth of the network, making it difficult to apply them in sequential tasks. Instead, in [3], dilated convolutions are used, which allow for an exponentially increase in the receptive field, which in turn, allows to cover a larger amount of the signal history. This can be seen as introducing a fixed space between every two adjacent elements of the filter.

We consider the TCN architectural block as a building block to obtain both a sequence-based and a non-sequence model for liveness detection. Figure 2 illustrates the two blocks. In terms of components, they have very similar structure, containing one main path with two convolutions and one residual convolution. These blocks may then be stacked to obtain deeper networks. A clear distinction between the two is that in the former (a), the output size of each convolution is different from the input, due to filters greater than 1×1 followed by max-pooling. In the later (b), dilated causal convolutions are used in the main block, and 1×1 filters in the residual, resulting in an output of the same size as the input at every layer.

4.1   Regular Convolution TCN Block

This TCN inspired block variant, Fig. 3(a), models the problem as non-sequential prediction, as described in Eq. 1. The model takes as input the whole (128-D) signal and outputs a prediction regarding being a genuine pulse or an attack. In this architecture, standard convolutions are used, with filters of size k, followed by a max-pooling operation. The convolution filters on the residual path were defined such that the output matches the output of the second convolution of the main path. Figure 3 depicts the full architecture in detail. The dimension of

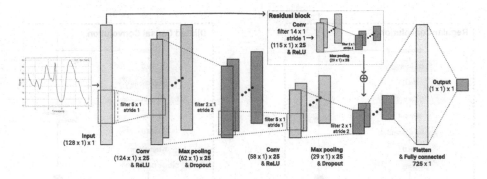

Fig. 3. Variant of the TCN architecture block without Dilated Causal Convolutions.

the filters directly impact the type of patterns captured, since they define the number of time instants of the signal considered.

4.2 Dilated Convolution TCN Block

Since pulse signals have their own set of characteristics that vary from individual to individual, this implies that liveness patterns can span over different sub-sequence lengths and occur at any time. In this case, dilated convolutions, Fig. 3(b) allow stacking multiple layers while always giving access to the whole input, which in turn can lead to a higher probability of capturing these distinct patterns. Therefore, this variant models the problem as a sequential prediction, as described in Eq. 2, leveraging on dilated convolutions learn filters that operate on parts with different sizes (including the whole signal). Namely, given a signal of size T, the output of the last sequence element, i.e. $\hat{y}_T$, is used as the final liveness prediction, which naturally considers the full signal.

By definition of the TCN architecture, the receptive field of its filters increase exponentially with the layers depth. Therefore, we determine the depth d (number of blocks stacked) of the model as $d = \log_2(T)$, so that the last prediction really depends on all the previous elements. As in the previous model, filter sizes are set to k.

Compared to the regular convolutions variant, this approach is capable of considering much larger receptive fields, potentially reducing the chances of missing an important discriminative pattern of pulse signals. Achieving this characteristic, requires stacking multiple blocks to obtain larger context windows. To assess this, in the experimental evaluation we will pay special attention to the impact of the model depth d.

5   Evaluation

5.1   Dataset

To evaluate the proposed models on our setting, i.e. face liveness detection on the fly, we collected a representative dataset, using a group of volunteers, simulating the described scenario. For each volunteer, we decided to gather 9 videos filming the face of a real individual, under different stress conditions, and 12 videos filming a portrait photograph, in different formats, of the same person (illustrated in Fig. 1b). This is a type of attack that can be easily performed by potentially everyone, using any given photo (e.g. obtained from internet). All the videos were filmed in different environments, with different illumination conditions (good and very low illumination), and frontal light, imitating the heart beat pattern, to simulate the real use case. Videos were captured under different situations like rest, after walking upstairs and after brisk walking, with each person participating in all activities. These are all activities that affect the heart rate, with some being more susceptible to create more notable pulse variations. For the presentation attack videos, we decided to use different sizes of photographs, for example, photo of a neutral face, moving the photo, face sized photo and mask, as these can be just some of the basic attacks that we will face in the future. In total, we obtained around 21 videos for each volunteer, of 2 min each, all of them with distinct characteristics. Subsequently, the recorded videos were cut into 5 s videos, meeting the requirements of our mobile user identification and authentication scenario. To perform these cuts, a slide window approach of 1 s was used, which gradually progresses until it slides through the full video. As a result, 2436 videos were obtained.

5.2 Protocol and Implementation Details

For evaluation, we take as the positive class an attack (respectively, class 1), and consider Recall and F1-score, evaluated on each individual class, and the Area Under the Curve (AUC). Given the nature of the problem, in terms of security it is important to minimise the false negatives, hence our focus on recall. The F1-score provides a metric that balances both recall and precision. For training and testing, we split the dataset in 15% for testing, and the remaining for development. We further split the development set in 15% for validation and the remaining for training, resulting in 1760 videos for training, 311 for validation, and 365 for testing.

We extract pulse signals using an EVM approach, resulting in 128-D signals that are then min-max normalised to the $[-1, 1]$ range. Both models were trained as a binary classification problem, using binary cross entropy. For hyperparameter tuning, we used a grid search approach. For dropout, we considered $\{0.0, 0.05, 0.5\}$, for learning rate $\{0.0001, 0.0005, 0.001, 0.005, 0.01, 0.1, 0.5\}$, for kernel size $k = \{3, 5, 7\}$, for hidden layer size $\{25, 50\}$, batch size $\{32, 64, 128\}$ and for the TCN depth level $d = \{1, 3, 5, 7\}$. All models were trained for 300 iterations, and we keep the model with the lowest validation loss.

Table 1. Liveness detection performance, with different filter sizes k, for both 1D-CNN variants and TCN. Both Recall-0/Recall-1, F1-score-0/1 (F1-0/F1-1) and AUC metrics are shown.

	Filter size k	Recall-0	F1-0	Recall-1	F1-1	AUC
FFNN	–	23.18	32.61	90.82	79.69	67.73
1D-CNN	3	55.59	61.99	88.53	84.32	86.03
	5	58.10	63.90	88.52	84.81	86.12
	7	60.34	64.38	86.91	84.35	85.66
1D-CNN + Residual block	3	<u>66.48</u>	70.21	88.93	86.71	89.88
	5	65.36	<u>71.02</u>	90.96	<u>87.59</u>	**90.21**
	7	57.82	65.92	**91.50**	86.37	89.58
TCN ($d = 5$)	3	**70.67**	70.57	85.70	85.75	89.00
	5	69.83	**72.89**	<u>89.47</u>	**87.70**	<u>90.17</u>
	7	64.80	68.74	88.53	86.15	89.42

5.3   Results

In this section we evaluate the two proposed approaches, that by definition, model the problem in a fundamentally different manner: sequence modelling (TCN) vs. non-sequence modelling (1D-CNN). Namely, we evaluate both the a) standard convolution variant and a residual block (1D-CNN+Res), with depth 1, and b) the dilated causal convolution TCN variant, with depth $d = 5$. As baseline, we also consider a 1D-CNN variant in which the residual path is removed (1D-CNN), and a simple classifier, implemented with a fully connected feedforward neural network (FFNN), with a *sigmoid* activation function.

Table 1 contains the results of the four approaches. For the CNN and TCN variants, we report the results for different filter sizes k. The results reported correspond to the models trained with the best hyper-parameters, which apart from the TCN's depth level, the set of hyper-parameters that allowed all models to obtain the best performance was the same: learning rate of 0.001, kernel size of 5, hidden layers with 50 neurons, batch size of 64 and dropout of 0.5.

Analysing the results, we observe that in terms of performance w.r.t. attack detection (Recall-1 and F1-1), the three models obtain fairly high performance, with 1D-CNN+Res block and TCN clearly demonstrating superior performance. 1D-CNN+Res block achieves a higher Recall-1, with the best configuration (i.e. $k = 7$) achieving 91.50%, but showing lower Precision-1 as indicated by the lower F1-1. TCN, on the other hand, achieved the highest F1-1, of 87.70%, also with $k = 5$. These results show that with both approaches, we are able to effectively detect presentation attacks from face videos recorded on the fly, however the TCN provides better balance between Precision-1 and Recall-1.

In our setting, detecting all the attacks is without no doubts the priority. However, in practice, it is not only important to have a high Recall-1, but also be capable of distinguishing between a fake and a genuine pulse. This can be seen

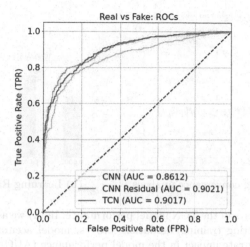

Fig. 4. Comparison of 1D-CNN, 1D-CNN+Residual and TCN presentation attack detectors ROC curves.

as a compromise between security (detecting attacks) and usability (correctly detecting genuine pulses to deliver swift genuine user registrations and minimise frustration). Therefore, we also report in Table 1 the performance in terms of the negative class (R-0, F1-0), i.e. a genuine pulse, and AUC. In terms of AUC, 1D-CNN+Res block and TCN demonstrate identical performance. However, in terms of F1-0, the TCN model achieved 72.89%, with filter size $k = 5$, outperforming all the other models. This means that in *terms of having the best trade-off between security and usability, the TCN model with a filter size of $k = 5$ should be the one considered.*

To further complement the results, in Fig. 4, we show the ROC curves for the three models, where it can be observed that the two top-performing models, 1D-CNN+Res and TCN, demonstrate a very good performance, as they are able to keep the Recall-1 quite high (91.5% and 89.47% respectively), while having solid overall performance (AUC) in both classes (90.21% and 90.17% respectively).

5.4 TCN Model Training Analysis

In this section, we analyse the training behaviour of the best performing model, the TCN. Namely, in Fig. 5a, we show the training and validation loss and accuracy, for the best model. It is noticeable that as the training progresses, there is a steady improvement in presentation attacks detection performance. However, we clearly see that due to the TCN depth (5 TCN blocks stacked), the validation loss stops improving around epoch 150, and the network starts to overfit.

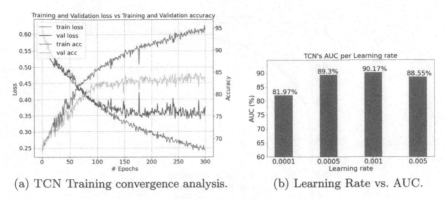

(a) TCN Training convergence analysis.   (b) Learning Rate vs. AUC.

Fig. 5. Further analisys of the TCN model performance. In (a) we assess TCN training convergence, by plotting training/validation loss vs. model accuracy, and in (b) we evaluate the learning rate impact in the model performance (AUC).

To further understand training convergence, we fixed all parameters (based on the top-performing configuration), and trained the TCN model with different learning rates. Figure 5b reports the AUC results, for each learning rate value. By observing these results, we observe that the best learning rate is *0.001*. Additionally, we observe that in general, the TCN delivers consistent performance with different learning rate values, unless the value is too low (e.g. 0.0001), for which the performance starts to deteriorate.

In Fig. 6, we evaluate how the TCN network depth d influences the results. We can observe that as the depth (number blocks) increases, its performance improves, despite the fact that with $d = 7$, the model achieved slightly worse AUC. This result is expected given the way dilated convolutions are exploited in the TCN. Namely, as depth increases, at each instant t, the model has access to more information of previous sequence neighbours $t - i$, by having wider history window, what allows the extraction of richer *low* and *high-level* features.

Another relevant aspect is that AUC of the 1D-CNN+Res (Table 1) is better w.r.t. the TCN with depth $d = 1$ (Fig. 6), even though the architectures in this scenario are quite similar (depth, number of convolutions, and residual block). We believe that this difference is due to the fact that the TCN with only 1 level has a very small receptive field. Therefore, with $d = 1$, the TCN is not being applied correctly, since the number of levels is not enough to take benefit of its main advantage, which is the ability to take into account a very large history.

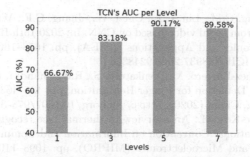

Fig. 6. Analysis of the impact of TCN network depth d. We evaluate the model using $d = 1, 3, 5, 7$ and report the obtained AUC.

6   Conclusions

In this paper we proposed a face liveness detection approach, based on a temporal convolutional neural network (TCN), for the specific scenario of on the fly liveness detection, through an user mobile smartphone. In this setting, pulse signals are estimated in a non-intrusive manner using an RGB camera, and directly used as 1D sequence features. We concluded that both regular and dilated casual convolutions were capable of delivering high performance. However, *the TCN model provides the best trade-off between performance and usability, making it the most suited solution for a production environment*. While the proposed regular convolution variant of the TCN achieved the highest attack detection performance, TCN yielded the best trade-off in terms of performance and usability.

We posit that the TCN may benefit from more data (quantity and diversity). Thus, in future work, we plan to exploit Generative Adversarial Networks to perform data augmentation and further improve the performance.

References

1. Abbas, G., Khan, M.J., Qureshi, R., Khurshid, K.: Scope of video magnification in human pulse rate estimation. In: 2017 International Conference on Machine Vision and Information Technology (CMVIT), pp. 69–75 (2017). https://doi.org/10.1109/CMVIT.2017.28
2. Agarwal, A., Yadav, D., Kohli, N., Singh, R., Vatsa, M., Noore, A.: Face presentation attack with latex masks in multispectral videos. In: 2017 IEEE Conference on Computer Vision and Pattern Recognition Workshops (CVPRW), pp. 275–283 (2017). https://doi.org/10.1109/CVPRW.2017.40
3. Bai, S., Kolter, J.Z., Koltun, V.: An empirical evaluation of generic convolutional and recurrent networks for sequence modeling. CoRR abs/1803.01271 (2018). http://arxiv.org/abs/1803.01271
4. Bousefsaf, F., Pruski, A., Maaoui, C.: 3D convolutional neural networks for remote pulse rate measurement and mapping from facial video. Appl. Sci. **9**, 4364 (2019). https://doi.org/10.3390/app9204364

5. Huang, B., Chang, C.M., Lin, C.L., Chen, W., Juang, C.F., Wu, X.: Visual heart rate estimation from facial video based on CNN. In: 2020 15th IEEE Conference on Industrial Electronics and Applications (ICIEA), pp. 1658–1662 (2020). https://doi.org/10.1109/ICIEA48937.2020.9248356
6. Kolberg, J., Gomez-Barrero, M., Venkatesh, S., Ramachandra, R., Busch, C.: Presentation Attack Detection for Finger Recognition, pp. 435–463. Springer International Publishing, Cham (2020). https://doi.org/10.1007/978-3-030-27731-4_14
7. Krišto, M., Ivasic-Kos, M.: An overview of thermal face recognition methods. In: 2018 41st International Convention on Information and Communication Technology, Electronics and Microelectronics (MIPRO), pp. 1098–1103 (2018). https://doi.org/10.23919/MIPRO.2018.8400200
8. Li, X., Komulainen, J., Zhao, G., Yuen, P., Pietikäinen, M.: Generalized face anti-spoofing by detecting pulse from face videos. In: 2016 23rd International Conference on Pattern Recognition (ICPR), pp. 4244–4249 (2016)
9. Lin, B., Li, X., Yu, Z., Zhao, G.: Face liveness detection by rPPG features and contextual patch-based CNN. In: Proceedings of the 2019 3rd International Conference on Biometric Engineering and Applications, pp. 61–68. ICBEA 2019. Association for Computing Machinery, New York, NY, USA (2019). https://doi.org/10.1145/3345336.3345345
10. Padnevych, R.: Smartyflow - biometria facial robusta para identificação virtual (2021)
11. Poh, M.Z., McDuff, D.J., Picard, R.W.: Non-contact, automated cardiac pulse measurements using video imaging and blind source separation. Opt. Express 18(10), 10762–10774 (2010)
12. Ramachandra, R., Busch, C.: Presentation attack detection methods for face recognition systems: a comprehensive survey. ACM Comput. Surv. (CSUR) 50(1), 1–37 (2017)
13. Tarassenko, L., Villarroel, M., Guazzi, A., Jorge, J., Clifton, D., Pugh, C.: Non-contact video-based vital sign monitoring using ambient light and auto-regressive models. Physiol. Meas. 35(5), 807 (2014)
14. Tirunagari, S., Poh, N., Windridge, D., Iorliam, A., Suki, N., Ho, A.: Detection of face spoofing using visual dynamics. IEEE Trans. Inf. Forensics Secur. 10, 762–777 (2015). https://doi.org/10.1109/TIFS.2015.2406533
15. Toosi, A., Bottino, A., Cumani, S., Negri, P., Sottile, P.L.: Feature fusion for fingerprint liveness detection: a comparative study. IEEE Access 5, 23695–23709 (2017). https://doi.org/10.1109/ACCESS.2017.2763419
16. Verkruysse, W., Svaasand, L.O., Nelson, J.S.: Remote plethysmographic imaging using ambient light. Opt. Express 16(26), 21434–21445 (2008)
17. Wang, S.Y., Yang, S.H., Chen, Y.P., Huang, J.W.: Face liveness detection based on skin blood flow analysis. Symmetry 9(12) (2017). https://doi.org/10.3390/sym9120305, https://www.mdpi.com/2073-8994/9/12/305
18. Wen, D., Han, H., Jain, A.K.: Face spoof detection with image distortion analysis. IEEE Trans. Inf. Forensics Secur. 10(4), 746–761 (2015)
19. Wu, H.Y., Rubinstein, M., Shih, E., Guttag, J., Durand, F., Freeman, W.: Eulerian video magnification for revealing subtle changes in the world. ACM Trans. Graph. 31(4) (2012). https://doi.org/10.1145/2185520.2185561
20. Wu, H.Y., Rubinstein, M., Shih, E., Guttag, J., Durand, F., Freeman, W.T.: Eulerian video magnification for revealing subtle changes in the world. ACM Trans. Graph. (Proc. SIGGRAPH 2012) 31(4), 1–8 (2012)

21. Yousef, A., Yaojie, L., Amin, J., Xiaoming, L.: Face anti-spoofing using patch and depth-based CNNs. In: 2017 IEEE International Joint Conference on Biometrics (IJCB), pp. 319–328 (2017)
22. Zhang, K., Zhang, Z., Li, Z., Qiao, Y.: Joint face detection and alignment using multitask cascaded convolutional networks. IEEE Signal Process. Lett. **23**(10), 1499–1503 (2016)

Pattern Recognition and Machine Learning

MaxDropoutV2: An Improved Method to Drop Out Neurons in Convolutional Neural Networks

Claudio Filipi Goncalves dos Santos[1] , Mateus Roder[2] ,
Leandro Aparecido Passos[3(✉)] , and João Paulo Papa[2]

[1] Federal University of São Carlos, São Carlos, Brazil
cfsantos@ufscar.br
[2] São Paulo State University, Bauru, Brazil
{mateus.roder,joao.papa}@unesp.br
[3] University of Wolverhampton, Wolverhampton, UK
l.passosjunior@wlv.ac.uk

Abstract. In the last decade, exponential data growth supplied the machine learning-based algorithms' capacity and enabled their usage in daily life activities. Additionally, such an improvement is partially explained due to the advent of deep learning techniques, i.e., stacks of simple architectures that end up in more complex models. Although both factors produce outstanding results, they also pose drawbacks regarding the learning process since training complex models denotes an expensive task and results are prone to overfit the training data. A supervised regularization technique called MaxDropout was recently proposed to tackle the latter, providing several improvements concerning traditional regularization approaches. In this paper, we present its improved version called MaxDropoutV2. Results considering two public datasets show that the model performs faster than the standard version and, in most cases, provides more accurate results.

1 Introduction

The last decades witnessed a true revolution in people's daily life habits. Computer-based approaches assume the central role in this process, exerting fundamental influence in basic human tasks, such as communication and interaction, entertainment, working, studying, driving, and so on. Among such approaches, machine learning techniques, especially a subfield usually called deep learning, occupy one of the top positions of importance in this context since they empower computers with the ability to act reasonably in an autonomous fashion.

The authors are grateful to FAPESP grants #2013/07375-0, #2014/12236-1, #2019/07665-4, Petrobras grant #2017/00285-6, CNPq grants #307066/2017-7, and #427968/2018-6, as well as the Engineering and Physical Sciences Research Council (EPSRC) grant EP/T021063/1.

A. J. Pinho et al. (Eds.): IbPRIA 2022, LNCS 13256, pp. 271–282, 2022.
https://doi.org/10.1007/978-3-031-04881-4_22

Deep learning regards a family of machine learning approaches that stacks an assortment of simpler models. The bottommost model's output feeds the next layer, and so on consecutively, with a set of possible intermediate operations among layers. The paradigm experienced exponential growth and magnificent popularity in the last years due to remarkable results in virtually any field of application, ranging from medicine [2,11,18] and biology [12] to speech recognition [10] and computer vision [14].

Despite the success mentioned above, deep learning approaches still suffer from a drawback very commonly observed in real-world applications, i.e., the lack of sufficient data for training the model. Such a constraint affects the learning procedure in two main aspects: (i) poor classification rates or (ii) overfitting to training data. The former is usually addressed by changing to a more robust model, which generally leads to the second problem, i.e., overfitting. Regarding the latter, many works tackled the problem using regularization approaches, such as the well-known batch normalization [5], which normalizes the data traveling from one layer to the other, and dropout [17], which randomly turns-off some neurons and forces the layer to generate sparse outputs.

Even though dropout presents itself as an elegant solution to solve overfitting issues, Santos et al. [15] claim that deactivating neurons at random may impact negatively in the learning process, slowing down the convergence. To alleviate this impact, the authors proposed the so-called MaxDropout, an alternative that considers deactivating only the most active neurons, forcing less active neurons to prosecute more intensively in the learning procedure and produce more informative features.

MaxDropout obtained significant results considering image classification's task, however, at the cost of considerable computational cost. This paper addresses such an issue by proposing MaxDropoutV2, an improved and optimized version of MaxDropout capable of obtaining similar results with higher performance and substantial reduction of the computational burden.

Therefore, the main contributions of this work are presented as follows:

– to propose a novel regularization approach called MaxDropoutV2, which stands for an improved and optimized version of MaxDropout;
– to evaluate MaxDropoutV2 overall accuracy and training time performance, comparing with the original MaxDropout and other regularization approaches; and
– to foster the literature regarding regularization algorithms and deep learning in general.

The remainder of this paper is organized as follows. Section 2 introduces the main works regarding Dropout and its variation, while Sect. 3 presents the proposed approach. Further, Sects. 4 and 5 describe the methodology adopted in this work and the experimental results, respectively. Finally, Sect. 6 states conclusions and future work.

2   Related Works

The employment of regularization methods for training deep neural networks (DNNs) architectures is a well-known practice, and its use is almost always considered by default. The focus of such approaches is helping DNNs to avoid or prevent overfitting problems, which reduce their generalization capability. Besides, regularization methods also allow DNNs to achieve better results considering the testing phase since the model becomes more robust to unseen data.

Batch Normalization (BN) is a well-known regularization method that employs the concept of normalizing the output of a given layer at every iteration in the training process. In its seminal work, Ioffe and Szegedy [5] demonstrated that the technique is capable of speeding up the convergence regarding the task of classification. Further, several other works [16,20,22] highlighted its importance, including the current state-of-the-art on image classification [19].

Among the most commonly employed techniques for DNN regularization is the Dropout, which is usually applied to train such networks in most of the frameworks used for the task. Developed by Srivastava et al. [17], Dropout shows significant improvements in a wide variety of applications of neural networks, like image classification, speech recognition, and more. The standard approach has a simple and efficient work procedure, in which a mask that directly multiplies the weight connections is created at training time for each batch. Such a mask follows a Bernoulli distribution, i.e., it assign values 0 with a probability p and 1 with a probability $1 - p$. The authors showed that the best value for p in hidden layers is 0.5. During training, the random mask varies, which means that some neurons will be deactivated while others will work normally.

Following the initial development of the Dropout method, Wang and Manning [21] focused on exploring different sampling strategies, considering that each batch corresponds to a new subnetwork taken into account since different units are dropped out. In this manner, the authors highlighted that the Dropout represents an approximation of a Markov chain executed several times during training time. Also, the Bernoulli distribution tends to a Normal distribution in a high dimensional space, such that Dropout performs best without sampling.

Similarly, Kingma et al. [6] proposed the Variational Dropout, which is a generalization of the Gaussian Dropout with the particularity of learning the dropout rate instead of randomly select one value. The authors aimed to reduce the variance of the stochastic gradients considering the variational Bayesian inference of a posterior over the model parameters, retaining the parallelization by investigating the reparametrization approach.

Further, Gal et al. [3] proposed a new Dropout variant to reinforcement learning models. Such a method aims to improve the performance and calibrate the uncertainties once it is an intrinsic property of the Dropout. The proposed approach allows the agent to adapt its uncertainty dynamically as more data is provided. Molchanov et al. [9] explored the Variational Dropout proposed by Kingma et al. [6]. The authors generalized the method to situations where the dropout rates are unbounded, giving very sparse solutions in fully-connected and convolutional layers. Moreover, they achieved a reduction in the number of parameters up to 280 times on LeNet architectures and up to 68 times on VGG-like networks with a small decrease in accuracy rates. Such a fact highlights the importance of sparsity for robustness and parameter reduction, while the overall performance for "simpler" models can be improved.

Another class of regularization methods emerged in parallel, i.e., techniques that change the neural network's input. Among such methods, one can refer to the Cutout [1], which works by cutting off/removing a region of the input image and setting such pixels at zero values. Such a simple approach provided relevant results in several datasets. In a similar fashion emerged the RandomErasing [23], which works by changing the pixel values at random for a given region in the input, instead of setting these values for zero.

Roder et al. [13] proposed the Energy-based Dropout, a method that makes conscious decisions whether a neuron should be dropped or not based on the energy analysis. The authors designed such a regularization method by correlating neurons and the model's energy as an index of importance level for further applying it to energy-based models, as Restricted Boltzmann Machines.

3   MaxDropoutV2 as an Improved Version of MaxDropout

This section provides an in-depth introduction to MaxDropout-based learning.

3.1   MaxDropout

MaxDropout [15] is a Dropout-inspired [17] regularization task designed to avoid overfitting on deep learning training methods. The main difference between both techniques is that, while Dropout randomly selects a set of neurons to be cut off according to a Bernoulli distribution, MaxDropout establishes a threshold value, in which only neurons whose activation values higher than this threshold are considered in the process. Results provided in [15] show that excluding neurons using their values instead of the likelihood from a stochastic distribution while training convolutional neuron networks produces more accurate classification rates.

Algorithm 1.1 implements the MaxDropout approach. Line 2 generates a normalized representation of the input tensor. Line 3 attributes the normalized value to a vector to be returned. Further, Lines 4 and 5 set this value to 0 where the normalized tensor is bigger than the threshold. This process is only performed during training. Concerning the inference, the original values of the tensor are used.

Algorithm 1.1. Original MaxDropout code

```
1   def MaxDropout(tensor, threshold):
2       norm_tensor = normalize(tensor)
3       return_tensor = norm_tensor
4       if norm_tensor > threshold:
```

```
5            return_tensor = 0 where
6        return return_tensor
```

Even though MaxDropout obtained satisfactory results for the task, it was not tailored-designed for Convolutional Neural Networks (CNNs), thus presenting two main drawbacks:

- it does not consider the feature map spacial distribution produced from a CNN layer output since it relies on individual neurons, independently of their location on a tensor; and
- it evaluates every single neuron from a tensor, which is computationally expensive.

Such drawbacks motivated the development of an improved version of the model, namely MaxDropoutV2, which addresses the issues mentioned above and provides a faster and more effective approach. The following section describes the technique.

3.2 MaxDropoutV2

The main difference between MaxDropout and MaxDropoutV2 is that the latter relies on a more representative feature space. While MaxDropout compares the values from each neuron directly, MaxDropoutV2 sums up these feature maps considering the depth axis, thus providing a bidimensional representation. In a nutshell, consider a CNN layer output tensor with dimensions $32 \times 32 \times 64$. The original MaxDropout performs $32 \times 32 \times 64$, i.e., $65,536$ comparisons. The proposed method sums up the values of the tensor over axis one (which would be the depth of the tensor) for each 32×32 kernel, thus performing only $1,024$ comparisons.

Algorithm 1.2 provides the implementation of the proposed approach. Line 2 performs the sum in the depth axis. Similar to Algorithm 1.1, Line 3 generates a normalized representation of the sum in depth of the input tensor. Line 4 creates the mask that defines what positions of the original tensor should be dropped, i.e., set to 0. Notice that the process is performed faster in MaxDropoutV2 due to the reduced dimensionality of the tensor. Further, in Lines 5 and 6, the tensor is unsqueezed and repeated so the mask can be used along all the tensor dimensions. Finally, the mask is applied to the tensor in Line 8 and returned in Line 9. These operations are only performed during training, similar to the original.

Algorithm 1.2. MaxDropoutV2 code

```
1  def MaxDropout_V2(tensor, threshold):
2      sum_axis = sum tensor along axis 1
3      sum_axis = normalized(sum_axis)
4      mask = 0 where sum_axis > threshold
5      mask_tensor = tensor.shape[0]
6      repetitions of mask
```

```
7
8      return_tensor = tensor * mask_tensor
9      return return_tensor
```

Figure 1 depicts an example of application, presenting an original image in Fig. 1a and a simulation of output colors considering Dropout, MaxDropout, and MaxDropoutV2, for Figs. 1b, 1c, and 1d, respectively.

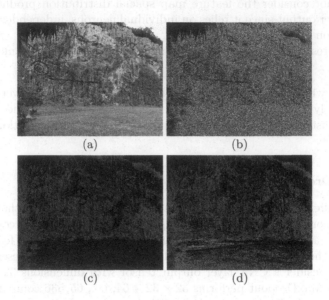

(a)   (b)

(c)   (d)

Fig. 1. Simulation using colored images. The original colored image is presented in (a), and its outcomes are bestowed after (b) Dropout, (c) MaxDropout, and (d) Max-DropoutV2 transformations using a dropout rate of 50%.

4   Methodology

This section provides a brief description of the datasets employed in this work, i.e., CIFAR-10 and CIFAR-100, as well all the setup considered during the experiments.

4.1   Dataset

In this work, we consider the public datasets CIFAR-10 and CIFAR-100 [7] to evaluate the performance of MaxDropoutV2 since both datasets are widely employed in similar regularization contexts [1,8,15,23]. Both datasets comprise 60,000 color images of animals, automobiles, and ships, to cite a few, with a size of 32×32 pixels. Such images are divided such that 50,000 instances are employed for training, and 10,000 samples are considered for evaluation purposes. The main difference between CIFAR-10 and CIFAR-100 regards the number of classes, i.e., CIFAR-10 comprises 10 classes while CIFAR-100 is composed of 100 classes.

4.2   Experimental Setup

To provide a fair comparison, we adopted the same protocol employed in several works in literature [1,15,23], which evaluate the proposed techniques over the ResNet-18 [4] neural network. Regarding the pre-processing steps, each image sample is resized to 32×32 pixels for further extracting random crops of size 28×28 pixels, with the addition of horizontal flip. The network hyperparameter setup employs the Stochastic Gradient Descent (SGD) with Nesterov momentum of 0.9 and a weight decay of 5×10^{-4}. The initial learning rate is initially set to 0.1 and updated on epochs 60, 120, and 160 by multiplying its value by 0.2. Finally, the training is performed during a total of 200 epochs and repeated during five rounds over each dataset to extract statistical measures. It is important to highlight that this protocol is used in several other works related to regularization on Deep Learning models [1,15,24]. In this work, we compare our proposed method against other regularizers that explicitly target to improve the results of CNNs.

Regarding the hardware setup, experiments were conducted using an Intel 2x Xeon®E5-2620 @ 2.20GHz with 40 cores, a GTX 1080 Ti GPU, and 128 GB of RAM.[1]

5   Experimental Results

This section provides an extensive set of experiments where MaxDropoutV2 is compared against several baselines considering both classification error rate and time efficiency. Additionally, it also evaluates combining MaxDropoutV2 with other regularization techniques.

5.1   Classification Error

Table 1 shows the average error rate for all models and architectures regarding the task of image classification. Highlighted values denote the best results, which were obtained over five independent repetitions.

From the results presented in Table 1, one can observe that Cutout obtained the lowest error rate over the CIFAR-10 dataset. Meanwhile, MaxDropoutV2 achieved the most accurate results considering the CIFAR-100 dataset, showing itself capable of outperforming its first version, i.e., MaxDropout, over more challenging tasks composed of a higher number of classes.

Additionally, Fig. 2 depicts the convergence evolution of MaxDropout and MaxDropoutV2 over the training and validation splits, in which the training partition comprises 50,000 samples, and the validation contains 10,000 samples. In Fig. 2, V1 stands for the MaxDropout method, and V2 stands for the proposed approach. One can notice that MaxDropoutV2 does not overpass the MaxDropout validation accuracy, mainly on the CIFAR-10 dataset. However, when dealing with more classes and the same number of training samples, both performances was almost the same, indicating the robustness of MaxDropoutV2.

---

[1] The code will be available in case of the paper acceptance.

Table 1. Average classification error rate (%) over CIFAR-10 and CIFAR-100 datasets. Notice ResNet18 results are provided as the baseline.

	CIFAR-10	CIFAR-100
ResNet-18 [4]	4.72 ± 0.21	22.46 ± 0.31
Cutout [1]	**3.99 ± 0.13**	21.96 ± 0.24
RandomErasing [24]	4.31 ± 0.07	24.03 ± 0.19
LocalDrop [8]	4.3	22.2
MaxDropout [15]	4.66 ± 0.13	21.93 ± 0.07
MaxDropoutV2 (ours)	4.63 ± 0.04	**21.92 ± 0.23**

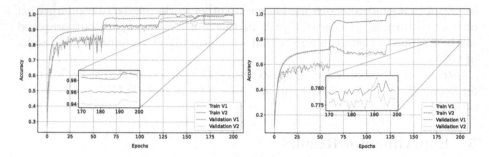

Fig. 2. Convergence analysis regarding the CIFAR-10 and CIFAR-100 datasets.

5.2   Combining Regularization Techiniques

A critical point about regularization concerns avoiding overfitting and improving the results of a given neural network architecture in any case. For instance, if some regularization approach is already applied, including another regularization should still improve the outcomes. In this context, MaxDropoutV2 performs this task with success, as shown in Table 2.

5.3   Performance Evaluation

The main advantage of MaxDropoutV2 over MaxDropout regards its computational time. In this context, Tables 3 and 4 provides the average time demanded to train both models considering each epoch and the total consumed time. Such results confirm the hypothesis stated in Sect. 3.2 since MaxDropoutV2 performed around 10% faster than the standard version.

5.4   Evaluating Distinct Drop Rate Scenarios

This section provides an in-depth analysis of MaxDropoutV2 and MaxDropout [15] results considering a proper selection of the drop rate parameter. Tables 5 and 6 present the models' results while varying the drop rate from 5% to 50% considering CIFAR-10 and CIFAR-100 datasets, respectively.

Table 2. Average classification error rate (%) over CIFAR-10 and CIFAR-100 datasets combining MaxDropout and MaxDropoutV2 with Cutout.

	CIFAR-10	CIFAR-100
ResNet-18 [4]	4.72 ± 0.21	22.46 ± 0.31
Cutout [1]	3.99 ± 0.13	21.96 ± 0.24
MaxDropout [15]	4.66 ± 0.13	21.93 ± 0.07
MaxDropoutV2 (ours)	4.63 ± 0.04	21.92 ± 0.23
MaxDropout [15] + Cutout [1]	$\mathbf{3.76 \pm 0.08}$	21.82 ± 0.13
MaxDropoutV2 (ours) + Cutout [1]	3.95 ± 0.13	$\mathbf{21.82 \pm 0.12}$

Table 3. Time consumed in seconds for training ResNet-18 in CIFAR-10 dataset.

	Seconds per Epoch	Total time
MaxDropout [15]	32.8	$6,563$
MaxDropoutV2 (ours)	**29.8**	**5,960**

Table 4. Time consumed in seconds for training ResNet-18 in CIFAR-100 dataset.

	Seconds per Epoch	Total time
MaxDropout [15]	33.1	$6,621$
MaxDropoutV2 (ours)	**30.2**	**6,038**

Table 5. Mean error (%) concerning CIFAR-10 dataset.

Drop rate	MaxDropoutV2	MaxDropout
5	$\mathbf{4.63 \pm 0.03}$	4.76 ± 0.09
10	4.67 ± 0.13	4.71 ± 0.09
15	4.76 ± 0.12	$\mathbf{4.63 \pm 0.11}$
20	4.66 ± 0.13	4.70 ± 0.08
25	4.75 ± 0.11	4.70 ± 0.06
30	4.63 ± 0.16	4.67 ± 0.12
35	4.70 ± 0.18	4.71 ± 0.16
40	4.74 ± 0.13	4.79 ± 0.20
45	4.65 ± 0.16	4.71 ± 0.11
50	4.71 ± 0.04	4.75 ± 0.10

Even though MaxDropoutV2 did not achieve the best results in Table 1, the results presented in Table 5 show the technique is capable of yielding satisfactory outcomes considering small drop rate values, i.e., 5%, while the standard model obtained its best results considering a drop rate of 15%. Additionally, one can notice that MaxDropoutV2 outperformed MaxDropout in eight-out-of-ten scenarios, demonstrating the advantage of the model over distinct circumstances.

Table 6. Mean error (%) concerning CIFAR-100 dataset.

Drop rate	MaxDropoutV2	MaxDropout
5	22.26 ± 0.31	22.05 ± 0.17
10	22.19 ± 0.13	22.06 ± 0.32
15	22.25 ± 0.23	22.16 ± 0.20
20	22.26 ± 0.30	21.98 ± 0.21
25	22.02 ± 0.13	**21.93 ± 0.23**
30	**21.92 ± 0.23**	22.07 ± 0.24
35	22.00 ± 0.07	22.10 ± 0.29
40	22.09 ± 0.16	22.16 ± 0.34
45	21.95 ± 0.15	22.31 ± 0.29
50	22.13 ± 0.19	22.33 ± 0.23

In a similar fashion, Table 6 provides the mean classification error considering distinct drop rate scenarios over CIFAR-100 dataset. In this context, both techniques required larger drop rates to obtain the best results, i.e., 25 and 30 for MaxDropout and MaxDropoutV2, respectively. Moreover, MaxDropoutV2 outperformed MaxDropout in all cases when the drop rates are greater or equal to 30, showing more complex problems demand higher drop rates.

5.5   Discussion

According to the provided results, the proposed method accomplishes at least equivalent outcomes to the original MaxDropout, outperforming it in terms of classification error in most cases. Moreover, MaxDropoutV2 presented itself as a more efficient alternative, performing around 10% faster than the previous version for the task of CNN training.

The main drawback regarding MaxDropoutV2 is that the model is cemented to the network architecture, while MaxDropout applicability is available to any network's architecture. In a nutshell, MaxDropoutV2 relies on a matrix or high dimensional tensors designed to accommodate CNNs' layers outputs, while the standard MaxDropout works well for any neural network structure, such as Multilayer Perceptrons and Transformers, for instance.

6 Conclusion and Future Works

This paper presented an improved version of the regularization method Max-Dropout, namely MaxDropoutV2, which stands for a tailored made regularization technique for convolutional neural networks. In short, the technique relies on a more representative feature space to accommodate the convolutional layer outputs.

Experimental results showed the method significantly reduced the time demanded to train the network, performing around 10% faster than the standard MaxDropout with similar or more accurate results. Moreover, it demonstrated that MaxDropoutV2 is more robust to the selection of the drop rate parameter. Regarding future work, we will evaluate MaxDropoutV2 in distinct contexts and applications, such as object detection and image denoising.

References

1. DeVries, T., Taylor, G.W.: Improved regularization of convolutional neural networks with cutout. arXiv preprint arXiv:1708.04552 (2017)
2. dos Santos, C.F.G., Passos, L.A., de Santana, M.C., Papa, J.P.: Normalizing images is good to improve computer-assisted COVID-19 diagnosis. In: Kose, U., Gupta, D., de Albuquerque, V.H.C., Khanna, A. (eds.) Data Science for COVID-19, pp. 51–62. Academic Press (2021). https://doi.org/10.1016/B978-0-12-824536-1.00033-2, https://www.sciencedirect.com/science/article/pii/B9780128245361000332
3. Gal, Y., Hron, J., Kendall, A.: Concrete dropout. In: Advances in Neural Information Processing Systems, pp. 3581–3590 (2017)
4. He, K., Zhang, X., Ren, S., Sun, J.: Deep residual learning for image recognition. In: Proceedings of the IEEE Conference on Computer Vision and Pattern Recognition, pp. 770–778 (2016)
5. Ioffe, S., Szegedy, C.: Batch normalization: accelerating deep network training by reducing internal covariate shift. arXiv preprint arXiv:1502.03167 (2015)
6. Kingma, D.P., Salimans, T., Welling, M.: Variational dropout and the local reparameterization trick. In: Advances in Neural Information Processing Systems, pp. 2575–2583 (2015)
7. Krizhevsky, A., Nair, V., Hinton, G.: CIFAR-10 and CIFAR-100 datasets. https://www.cs.toronto.edu/kriz/cifar.html 6, 1 (2009)
8. Lu, Z., Xu, C., Du, B., Ishida, T., Zhang, L., Sugiyama, M.: LocalDrop: a hybrid regularization for deep neural networks. IEEE Trans. Patt. Anal. Mach. Intell. (2021)
9. Molchanov, D., Ashukha, A., Vetrov, D.: Variational dropout sparsifies deep neural networks. In: Proceedings of the 34th International Conference on Machine Learning, vol. 70, pp. 2498–2507. JMLR. org (2017)
10. Noda, K., Yamaguchi, Y., Nakadai, K., Okuno, H.G., Ogata, T.: Audio-visual speech recognition using deep learning. Appl. Intell. 42(4), 722–737 (2014). https://doi.org/10.1007/s10489-014-0629-7
11. Passos, L.A., Santos, C., Pereira, C.R., Afonso, L.C.S., Papa, J.P.: A hybrid approach for breast mass categorization. In: ECCOMAS Thematic Conference on Computational Vision and Medical Image Processing, pp. 159–168. Springer (2019). https://doi.org/10.1007/978-3-030-32040-9_17

12. Roder, M., Passos, L.A., Ribeiro, L.C.F., Benato, B.C., Falcão, A.X., Papa, J.P.: Intestinal parasites classification using deep belief networks. In: Rutkowski, L., Scherer, R., Korytkowski, M., Pedrycz, W., Tadeusiewicz, R., Zurada, J.M. (eds.) ICAISC 2020. LNCS (LNAI), vol. 12415, pp. 242–251. Springer, Cham (2020). https://doi.org/10.1007/978-3-030-61401-0_23

13. Roder, M., de Rosa, G.H., de Albuquerque, V.H.C., Rossi, A.E.L.D., Papa, J.A.P.: Energy-based dropout in restricted Boltzmann machines: why not go random. IEEE Trans. Emerg. Topics Comput. Intell. 1–11 (2020). https://doi.org/10.1109/TETCI.2020.3043764

14. Santana, M.C., Passos, L.A., Moreira, T.P., Colombo, D., de Albuquerque, V.H.C., Papa, J.P.: A novel Siamese-based approach for scene change detection with applications to obstructed routes in hazardous environments. IEEE Intell. Syst. 35(1), 44–53 (2019)

15. Santos, C.F.G.D., Colombo, D., Roder, M., Papa, J.P.: MaxDropout: deep neural network regularization based on maximum output values. In: Proceedings of 25th International Conference on Pattern Recognition, ICPR 2020, 10–15 January 2021, Milan, Italy, pp. 2671–2676. IEEE Computer Society (2020)

16. Simon, M., Rodner, E., Denzler, J.: ImageNet pre-trained models with batch normalization. arXiv preprint arXiv:1612.01452 (2016)

17. Srivastava, N., Hinton, G., Krizhevsky, A., Sutskever, I., Salakhutdinov, R.: Dropout: a simple way to prevent neural networks from overfitting. J. Mach. Learn. Res. 15(1), 1929–1958 (2014)

18. Sun, Z., He, S.: Idiopathic interstitial pneumonias medical image detection using deep learning techniques: a survey. In: Proceedings of the 2019 ACM Southeast Conference, pp. 10–15 (2019)

19. Tan, M., Le, Q.V.: EfficientNet: rethinking model scaling for convolutional neural networks. arXiv preprint arXiv:1905.11946 (2019)

20. Wang, J., Hu, X.: Gated recurrent convolution neural network for OCR. In: Advances in Neural Information Processing Systems, pp. 335–344 (2017)

21. Wang, S., Manning, C.: Fast dropout training. In: International Conference on Machine Learning, pp. 118–126 (2013)

22. Zhang, K., Zuo, W., Chen, Y., Meng, D., Zhang, L.: Beyond a gaussian denoiser: residual learning of deep CNN for image denoising. IEEE Trans. Image Process. 26(7), 3142–3155 (2017)

23. Zhong, Z., Zheng, L., Kang, G., Li, S., Yang, Y.: Random erasing data augmentation. In: Proceedings of the AAAI Conference on Artificial Intelligence (AAAI) (2020)

24. Zhong, Z., Zheng, L., Kang, G., Li, S., Yang, Y.: Random erasing data augmentation. In: AAAI, pp. 13001–13008 (2020)

Transparent Management of Adjacencies in the Cubic Grid

Paola Magillo[1] and Lidija Čomić[2(✉)]

[1] DIBRIS, University of Genova, Genova, Italy
magillo@dibris.unige.it
[2] Faculty of Technical Sciences, University of Novi Sad, Novi Sad, Serbia
comic@uns.ac.rs

Abstract. We propose an integrated data structure which represents, at the same time, an image in the cubic grid and three well-composed images, homotopy equivalent to it with face-, edge- and vertex-adjacency. After providing an algorithm to build the structure, we present examples showing how, thanks to such data structure, image processing algorithms can be written in a transparent way w.r.t. the adjacency type. Applications include rapid prototyping and teaching.

Keywords: Well-composed images · Cubic grid · BCC grid · FCC grid · Data structure

1 Introduction

Given a polyhedral grid (e.g., a cubic grid), we denote as a 3D image a set of black voxels, while other voxels (background) are white. A 3D image is well-composed if its boundary surface is a 2-manifold. Intuitively, this means that, at each point, the surface is locally topologically equivalent to a disc.

In the cubic grid, not all 3D images are well composed, as two black cubes may share just an edge or just a vertex, and no face. Therefore, it is necessary to take into account three possible types of adjacency relations (vertex-, edge- or face- adjacency) for the definition of homotopy-related properties, such as connectedness, cavities or tunnels, and for the design of algorithms.

The process of transforming a 3D image in another one which is well-composed is called repairing. Recently [3,4], we have proposed approaches to repair a cubic image by transferring it into another polyhedral grid, ensuring that the resulting image is homotopy equivalent to the original one, with respect to the chosen type of adjacency relation, among the three possible ones. The resulting images are defined in polyhedral grids derived from the cubic one, namely the Body Centered Cubic (BCC) grid and the Face Centered Cubic (FCC) grid.

These grids have been effectively used as viable alternatives to the cubic grid in computer graphics, rendering and illumination [2,9,21], ray tracing and ray casting [12–14], discrete geometry [5,6], voxelization [10], reconstruction [7,18],

© Springer Nature Switzerland AG 2022
A. J. Pinho et al. (Eds.): IbPRIA 2022, LNCS 13256, pp. 283–294, 2022.
https://doi.org/10.1007/978-3-031-04881-4_23

simulation [20] distance transform [23], fast Fourier transform [24], and a software system for processing and viewing 3D data [8].

We propose a unified data structure which represents simultaneously the given 3D image in the cubic grid and its repaired versions in the BCC and FCC grids. Algorithms operating on any of the repaired images can be written in a simple and transparent way.

This paper is organized as follows. In Sect. 2, we introduce 3D grids and well-composed images. In Sect. 3, we describe the proposed unified data structure and in Sect. 4 a construction algorithm unifying the ones in [3,4]. In Sect. 5 we show how our unified data structure allows for writing image processing algorithms that may work with any of the three adjacency types in a transparent manner. Finally, in Sect. 6, we draw concluding remarks.

2   3D Grids and Well-composed Images

In general, we can define a 3D grid as any partition of the 3D space into convex polyhedral cells, called voxels. The voxels of a grid naturally define their faces (polygons where two voxels meet), edges (segments where three or more faces meet), and vertices (points where three or more edges meet). Voxels, faces, edges and vertices constitute a cell complex associated with the grid. Adjacency and incidence relations in the grid are then defined.

Within a grid, an image I is a finite set of voxels. Conventionally, we call black the voxels belonging to I and white the voxels belonging to its complement I^c (background). An image I is (continuously) well-composed [17] if the boundary surface of I, i.e., the surface made up of faces that are incident with exactly one black voxel, is a 2-manifold. Recall that a 2-manifold is a topological space in which each point has a neighborhood homeomorphic to the open unit disc.

The cubic grid, made up of unit cubes, can be obtained by placing a seed at each 3D point with integer coordinates, and considering the Voronoi diagram of such seeds. The resulting Voronoi cells are cubes. Two non-disjoint cubes may share exactly one face (and all its edges and vertices), exactly one edge (and its two vertices) and no face, or exactly one vertex and no edge or face. This leads to three types of adjacency relations: face-adjacency, edge-adjacency, and vertex-adjacency, respectively. As a cube has six face-adjacent cubes, eighteen edge-adjacent cubes, and 26 vertex-adjacent cubes, the three adjacency types are also known as 6-adjacency, 18-adjacency, and 26-adjacency, respectively.

An image I in the cubic grid is (digitally) well-composed [17] if and only if there is no occurrence of either critical edges or critical vertices (see Fig. 1). A critical vertex v is a vertex having two (six) incident black cubes and six (two) incident white cubes, where the two black (white) cubes share just the vertex v. A critical edge e is an edge having two incident black cubes and two incident white cubes, where the two black (white) cubes share just the edge e. The two characterizations of well-composed images (continuous and digital) in the cubic grid are equivalent [1].

At critical edges and vertices, the image has different topology if we consider it with 6-, 18-, or 26–adjacency. Therefore, algorithms on general cubic

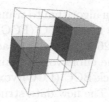

Fig. 1. Two black cubes sharing a critical vertex and a critical edge. The configurations with exchanged cube colors are also critical.

images need to take all the possible critical configurations into account, with a different treatment depending on the considered adjacency type. Moreover, for consistency reasons, an image I and its complement I^c need to be considered with different types of adjacency relations, for example 6-adjacency for I and 26-adjacency for I^c.

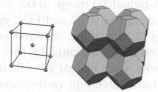

Fig. 2. Placement of seeds and growing process of seeds to create the BCC grid (left) and the FCC grid (right).

The Body Centered Cubic (BCC) grid is obtained from the cubic grid by placing a seed at the center of each cubic voxel, and a seed at each cube vertex. In the Voronoi diagram of such seeds, the Voronoi cells are truncated octahedra (see Fig. 2), and these are the voxels of the BCC grid. The BCC grid has just one type of adjacency, i.e., face-adjacency, and any 3D image in such grid is well-composed. In fact, if two truncated octahedra share a vertex, they must share a face as well.

The Face Centered Cubic (FCC) grid is obtained from the cubic grid by placing a seed at the center of each cubic voxel, and a seed at the midpoint of each edge. In the Voronoi diagram of such seeds, the Voronoi cells are rhombic dodecahedra (see Fig. 2), and these are the voxels of the FCC grid. Not all images in the FCC grid are well-composed, as pairs of rhombic dodecahedra, sharing just a vertex and no face or edge, exist. On the other hand, two rhombic dodecahedra cannot share an edge, without sharing a face as well.

Note that a subset of voxels in both the BCC and the FCC grid corresponds to the voxels of the cubic grid. Other voxels of the BCC grid correspond to the vertices of the cubic voxels, while other voxels of the FCC grid correspond to the edges of the cubic voxels.

In [3] (respectively, [4]), we have proposed an algorithm to transform an image I in the cubic grid into a well-composed image in the BCC (FCC) grid, which is homotopy equivalent to I according to 6- or 26-adjacency (18-adjacency). The BCC voxels (FCC voxels) corresponding to the cubic voxels maintain the same color they had in the cubic grid. All other black BCC (FCC) voxels correspond to some of the vertices (edges) of black cubes. The new data structure proposed here will store the original cubic image and all its repaired versions in the BCC and FCC grids, in a compact way.

3   Data Structure

We propose a unified data structure which can mark as black or white the cubic voxels (equivalently, the voxels of the BCC and FCC grid corresponding to them), the FCC voxels corresponding to cube edges, and the BCC voxels corresponding to cube vertices (with two marks, as they may get a different color in the repaired BCC image, equivalent to the cubic one with 26- or 6-adjacency). In this way, we can represent the original 3D image in the cubic grid and, simultaneously, its well-composed versions in the BCC and FCC grids.

Assuming that the cubic grid is made up of unit cubes centered at points with integer Cartesian coordinates, we multiply all coordinates by two, so that we can identify all cubes, faces, edges, vertices, with integer coordinates [16]: cubes have three even coordinates; vertices have three odd coordinates; edges have one even and two odd coordinates (the even coordinate corresponds to the axis the edge is parallel to), and faces have two even and one odd coordinate (the odd coordinate corresponds to the axis the face is orthogonal to).

Each cube has eight vertices, each shared by eight cubes. So, we can associate each cube with one of its vertices, by convention the one with maximum Cartesian coordinates. If (x, y, z) denotes a cube, the associated vertex is $(x+1, y+1, z+1)$. Each cube has twelve edges, each shared by four cubes. So, we can associate each cube with three of its edges, by convention the ones incident with the vertex with maximum Cartesian coordinates. If (x, y, z) denotes a cube, the associated edges are $(x, y+1, z+1)$, $(x+1, y, z+1)$, and $(x+1, y+1, z)$.

In a black and white (binary) image, the color of a cell (cube, edge or vertex) can be stored in just one bit: 1 for black and 0 for white. For representing the color of a cube, the color of the three associated edges, and the two colors of the associated vertex, we need six bits, which fit in one byte. Given a triplet (i, j, k), with $i, j, k \in \mathbb{Z}^3$, the bits of the associated byte contain:

- bit 0 is the color of the cube $b = (2i, 2j, 2k)$, i.e., the color of the BCC and of the FCC voxels corresponding to b;
- bits 1 and 2 are not used;
- bit 3 is the color of the edge $e_x = (2i, 2j+1, 2k+1)$ of b, i.e., the color of the FCC voxel corresponding to e_x, in the equivalent well-composed FCC image with 18-adjacency;

- bit 4 is the color of the edge $e_y = (2i + 1, 2j, 2k + 1)$ of b, i.e., the color of the FCC voxel corresponding to e_y, in the equivalent well-composed FCC image with 18-adjacency;
- bit 5 is the color of the edge $e_z = (2i + 1, 2j + 1, 2k)$ of b, i.e., the color of the FCC voxel corresponding to e_z, in the equivalent well-composed FCC image with 18-adjacency;
- bit 6 is the color of the vertex $v = (2i + 1, 2j + 1, 2k + 1)$ of b, i.e., the color of the BCC voxel corresponding to v, in the equivalent well-composed BCC image with 26-adjacency;
- bit 7 is the color of the same vertex (i.e., the BCC voxel) v in the equivalent well-composed BCC image with 6-adjacency.

We have developed our implementation in Python. Python is a high-level language providing the dictionary as a built-in type. The implementation uses a dictionary, where the keys are tuples of three integers, and the values are bytes. In another programming language, we would use a 3D matrix. For such purpose, we would select a subset of $\mathbb{Z}^3$, for example we can consider the bounding box of black voxels in the cubic grid, plus an extra layer of white voxels in all six directions. In addition, we would shift coordinates to ensure that they are non-negative, and can thus be used as matrix indexes.

Given a cell $c = (x, y, z)$, which may be a cube, an edge, or a vertex (or, equivalently, given a voxel in the cubic, FCC, or BCC grids), the dictionary key is obtained by simply dividing the coordinates of c by two with integer division. Figure 3 shows the Python code of the function returning the bit index, inside the byte, storing the color of a cell. The two Python functions for reading and setting the color of a cell (x, y, z) are shown in Fig. 4. If not present in the dictionary, a triplet (i, j, k) is considered as having byte 00000000 as value. Therefore elements of the dictionary having zero value do not need to be stored. That is why, in Fig. 4, a new element is added only if the value to be set is 1.

From the point of view of the spatial complexity, just one byte is stored for each cube belonging to the bounding box of the object (plus a layer of white cubes around it), and the single bits inside it are used to code the color of the other cells (BCC and FCC voxels corresponding to cube vertices and edges). As one byte is the smallest storage unit, it is not possible to use less memory than this, even for representing just the color of the cubes. Therefore, the storage cost of our unified data structure can be considered as optimal.

4 Integrated Image Repairing Algorithm

The data structure, described in Sect. 3, can be filled by running the algorithms proposed in [3] and [4], which specify the vertices and edges, respectively, to be set as black in order to have a well-composed image, equivalent to the given one with 6-, 26- and 18–adjacency, depending on the case. In the following, we describe a new integrated algorithm.

We consider a vertex v and two collinear incident edges e_1 and e_2 extending in negative and in positive axis direction from v, respectively (see Fig. 5). Referring

```
def getBitIndex(x,y,z, face_adj=False):
    rx, ry, rz = x%2, y%2, z%2
    if face_adj: return rx + 2*ry + 3*rz + (rx*ry*rz)
    else: return rx + 2*ry + 3*rz
```

Fig. 3. Python code of the function returning the bit index for a cell (x, y, z), which may be a cube, an edge or a vertex. The returned value is an integer from 0 to 7. The value of `face_adj` is only relevant if (x, y, z) denotes a vertex, i.e., if all three coordinates are odd. True means 6-adjacency, False means 26-adjacency. The operator % denotes the remainder of integer division.

```
def getCellColor(x,y,z, adj=0):
    ind1 = getMainIndex(x,y,z)
    if not ind1 in colorMatrix.keys(): return 0 # default is white
    ind2 = getBitIndex(x,y,z, adj==6)
    return getBitFrom(ind2, colorMatrix[ind1])


def setCellColor(value, x,y,z, adj=0):
    ind1 = getMainIndex(x,y,z)
    if not ind1 in colorMatrix.keys():
        if value==0: return 0
        else: colorMatrix[ind1] = 0x0
    ind2 = getBitIndex(x,y,z, adj==6)
    colorMatrix[ind1] = setBitInto(value, ind2, colorMatrix[ind1])
    return value
```

Fig. 4. Python functions `getCellColor` to return and `setCellColor` to set the color of a cell (x, y, z), which can be a cube, an edge, or a vertex. The color is 1 for black and 0 for white. Parameter `adj` is relevant only if (x, y, z) is a vertex, as a vertex may have a different color with 6-adjacency (`adj==6`) or 26-adjacency (any other value). Variable `colorMatrix` is a dictionary associating triplets with bytes. The function call `getMainIndex(x,y,z)` returns the dictionary key for cell (x, y, z). The function call `getBitFrom(triplet,byte)` returns the bit of the given triplet within the given byte. The function call `setBitInto(value,triplet,byte)` sets to value the bit of given triplet within the given byte, and returns the modified byte.

to such a configuration, we rewrite the rules used in [3] and in [4] to decide whether (the BCC voxel corresponding to) v and (the FCC voxel corresponding to) e_1 must become black. In [3], v becomes black if:

1. edge e_1 has four incident black cubes,
2. edge e_1 is critical;
3. edge e_2 is critical, and edge e_1 does not have four incident white cubes;
4. v is a critical vertex;

Rule 1 also applies for producing an equivalent BCC image to the given one with 6-adjacency. These rules must be applied for all three axes. Rule 4 is applied only when edges e_1 and e_2 are parallel to the x-axis, otherwise we would check it three times. The algorithm in [4] performs in two stages. In the first stage:

- if e_1 is incident with three or four black cubes, then e_1 becomes black;
- if e_1 is critical, then e_1 becomes black;
- in configurations where e_1 and e_2 are both critical, with two pairs of 6-adjacent black cubes, then two more edges e_3 and e_4, incident with v, become black. Edge e_3 is chosen conventionally, depending on the supporting axis of e_1, e_2, and e_4 is chosen in such a way that e_3 and e_4 belong to the same two white cubes (see [4] for details).

As before, these rules must be repeated for all three axes. The second stage examines all faces f shared by two black cubes. If no edge of f is black, or exactly two opposite edges of f are black, then one (more) edge of f becomes black. Such edge is chosen conventionally, depending on the normal axis of f, and on the supporting axis of its two black edges (see [4] for details).

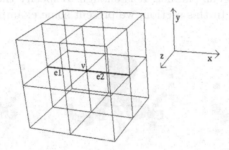

Fig. 5. A vertex v and its incident edges e_1, e_2 in negative and positive axis directions, respectively (here for the x-axis).

Our integrated algorithm factorizes the configurations checked by both algorithms. The first stage considers a vertex v and, in turn, the three Cartesian directions. We describe the procedure for the x-parallel direction. Let e_1 and e_2 be the edges incident with v in negative and positive direction with respect to v (as in Fig. 5). Based on the configuration at e_1, e_2, we change the color of v, e_1 to black according to the following rules:

1. If e_1 is incident with four white cubes, we skip.
2. If e_1 is incident with four black cubes, then e_1 becomes black and v becomes black (with both 26- and 6-adjacency).
3. If e_1 is incident with three black cubes, then e_1 becomes black.
4. If e_2 is critical (for Rule 1, here e_1 cannot have four incident white cubes), then v becomes black with 26-adjacency (e_2 will become black when processing its second endpoint).
5. If e_1 is critical, then e_1 becomes black, and v becomes black with 26-adjacency.

6. If e_1 and e_2 are both critical with the same configuration, then two more edges e_3 and e_4 become black, as specified in [4].
7. If v is critical, then v becomes black with 26-adjacency. This last condition is only checked for e_1, e_2 parallel to the x-axis (otherwise we would check it three times).

The main algorithm iterates the above described procedure three times, once for each axis, for all vertices (x, y, z) within or on the boundary of the bounding box of the image. Then, it iterates once on the cubes $b = (x, y, z)$, and, only if the cube b is black, executes the second stage of [4] for the three faces of b in the positive axis directions, i.e., the faces $(x + 1, y, z), (x, y + 1, z), (x, y, z + 1)$.

5 Applications

Thanks to the unified data structure, many image processing tasks, which need to consider a 3D image with one of 26-, 18- or 6-adjacency, can be performed in a simple and transparent manner. It is enough to specify the desired adjacency type as a parameter. In this section, we present some examples.

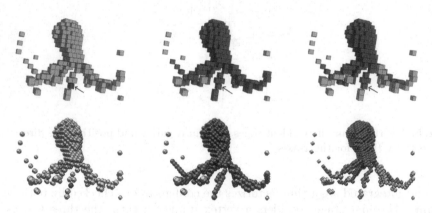

Fig. 6. A raw cubic discretization of an octopus, with many critical edges and vertices, and the connected component containing the cube pointed at by the arrow (which has a critical edge and a critical vertex) on the cubic image (upper row) and on its repaired version (lower row), with 6-, 26- and 18-adjacency, from left to right.

The first example is the extraction of the connected components. In a general cubic image, this operation will give a different result, depending on the considered adjacency type. After the three repaired images have been computed (see Sect. 4) and stored in our unified data structure (see Sect. 3), it is sufficient to choose which one we want to deal with. Figure 7 shows the Python function for this task. The different treatment of the three cases is hidden inside functions `adjacents` and `getCellColor`. Given a BBC voxel (a FCC voxel) e, `adjacents`(e) provides a list of BBC voxels (FCC voxels) if the adjacency type

```
def connectedComponents(cube_list, adj_type):
    components = dict() # empty dictionary
    comp_list = [] # empty list
    ind = -1 # index of last found connected component
    for b in cube_list:
        if not b in components.keys():
            # Start a new connected component from b:
            ind += 1
            comp_mark[b] = ind
            comp_list.append([b]) #init comp_list[ind] as [b]
            # Depth first search to find all cells in the connected component:
            stack = [b]
            while len(stack)>0:
                e = stack.pop()
                for d in adjacents(e, adj_type):
                    if getCellColor(d, adj_type)==1 and not d in comp_mark.keys():
                        comp_mark[d] = ind
                        stack.append(d)
                        if isCube(d): comp_list[ind].append(d)
    return (ind+1, comp_list, comp_mark)
```

Fig. 7. Python code of the function finding the connected components. cube_list is a list of black cubes, adj_type is 6, 18 or 26. Dictionary comp_marks will mark each black cell with the index of the connected component containing it. comp_list is a list of lists of cubes, each list comp_list[i] will contain the cubes belonging to the i-th connected component. In addition, the function returns the number of connected components. The process applies a standard depth first search using a stack, here a Python list with operations append and pop.

is 6 or 26 (18). Remember that a tuple representing a cube also represents a FCC and a BCC voxel. Function getCellColor returns the black or white color (1 or 0) of a cell, also depending on the adjacency type if the cell is a vertex.

A second example is extracting the boundary surface of a 3D image. In a well composed image, this is a 2-manifold surface consisting of one or more closed surfaces. The Python function shown in Fig. 8 collects all polygonal faces composing the boundary. Again, this is done in a transparent way with respect to the adjacency type.

Because the image is well-composed, the boundary surface can also be used to compute the Euler characteristic of the repaired image in the BCC or FCC grid (i.e., that of the original image in the cubic grid, with the corresponding adjacency type). In fact, given a 3-manifold object O, its boundary surface ∂O is a 2-manifold, and the Euler characteristics χ of the two sets are related by property $2\chi(O) = \chi(\partial O)$. The Euler characteristic of a polygonal surface can be computed as $\chi(\partial O) = n_0 - n_1 + n_2$, where n_0, n_1, n_2 denote the number of vertices, edges, and polygonal faces, respectively. A common representation of a polygonal surface is the so-called indexed representation, storing an array of vertices (without duplicates), and each face as a list of indexes within the vertex array. Numbers n_0 and n_2 are stored, and n_1 is easily retrieved by exploiting the

```
def findBoundary(cell_list, adj_type):
    boundary = Surface()
    for c in cell_list:
      for d in adjacents(c, adj_type):
          # If the adjacent voxel is white, the common face is on the boundary
          if getCellColor(d, adj_type)==0:
              boundary.addFace(commonPolygon(c,d))
    return boundary
```

Fig. 8. Python code of the function finding the boundary surface of the repaired image according to a given adjacency type. cell_list is a list of black cells, adj_type is 6, 18 or 26.

fact that, as the surface is a 2-manifold without boundary, an edge is shared by exactly two faces. Therefore $n_1 = \frac{1}{2} \sum_f \text{len}(f)$, where $\text{len}(f)$ denotes the number of edges of a face f.

Another example is the computation of digital distances in an image. Digital distances are used, for example, for computing the medial axis. Given a black cube b, the Euclidean distance of b from the white background is digitally approximated by taking the minimum length of a path of adjacent black cubes connecting b to a white cube. Of course, the digital distance depends on the considered adjacency type. Again, this can be done with our unified data structure in a transparent way. Many other image processing operations [11,15,16,19,22] might benefit from our unified data structure.

6   Concluding Remarks

We have proposed a new compact data structure which allows storing, at the same time, a cubic image and its repaired versions according to vertex-, face- and edge- adjacency. The specific contribution of this work is in the possibility of writing image processing algorithms in a uniform way, without the need for a different treatment of critical configurations in the three adjacency types. With almost no extra space, this simplifies the implementation of algorithms, and can be especially useful at a prototype level, when execution time is not yet an issue. Among possible application fields, we mention the design of new mathematical definitions and tools, and teaching. In the first case, it will be possible to get preliminary results and select the best adjacency type for a more efficient implementation. In teaching, students will be able to write code comparing the impact of the adjacency type on the known image processing operations.

Acknowledgment. This research has been partially supported by the Ministry of Education, Science and Technological Development through project no. 451-03-68/2022-14/ 200156 "Innovative scientific and artistic research from the FTS (activity) domain".

References

1. Boutry, N., Géraud, T., Najman, L.: A tutorial on well-composedness. J. Math. Imaging Vis. **60**(3), 443–478 (2017). https://doi.org/10.1007/s10851-017-0769-6
2. Brimkov, V.E., Barneva, R.P.: Analytical honeycomb geometry for raster and volume graphics. Comput. J. **48**, 180–199 (2005)
3. Čomić, L., Magillo, P.: Repairing 3D binary images using the BCC grid with a 4-valued combinatorial coordinate system. Inf. Sci. **499**, 47–61 (2019)
4. Čomić, L., Magillo, P.: Repairing 3D binary images using the FCC grid. J. Math. Imaging Vis. **61**, 1301–1321 (2019)
5. Čomić, L., Magillo, P.: On Hamiltonian cycles in the FCC grid. Comput. Graph. **89**, 88–93 (2020)
6. Čomić, L., Zrour, R., Largeteau-Skapin, G., Biswas, R., Andres, E.: Body centered cubic grid - coordinate system and discrete analytical plane definition. In: Lindblad, J., Malmberg, F., Sladoje, N. (eds.) DGMM 2021. LNCS, vol. 12708, pp. 152–163. Springer, Cham (2021). https://doi.org/10.1007/978-3-030-76657-3_10
7. Csébfalvi, B.: An evaluation of prefiltered B-spline reconstruction for quasi- interpolation on the body-centered cubic lattice. IEEE Trans. Visual. Comput. Graph. **16**(3), 499–512 (2010)
8. Linnér, E.S., Morén, M., Smed, K.-O., Nysjö, J., Strand, R.: LatticeLibrary and BccFccRaycaster: software for processing and viewing 3D data on optimal sampling lattices. SoftwareX **5**, 16–24 (2016)
9. Finkbeiner, B., Entezari, A., Van De Ville, D., Möller, T.: Efficient volume rendering on the body centered cubic lattice using box splines. Comput. Graph. **34**(4), 409–423 (2010)
10. He, L., Liu, Y., Wang, D., Yun, J.: A voxelization algorithm for 3D body-centered cubic line based on adjunct parallelepiped space. In: Information Computing and Applications - Third International Conference, ICICA, Part I, pp. 352–359 (2012)
11. Herman, G.T.: Geometry of Digital Spaces. Birkhauser, Boston (1998)
12. Ibáñez, L., Hamitouche, C., Roux, C.: Ray-tracing and 3D objects representation in the BCC and FCC grids. In: 7th International Workshop on Discrete Geometry for Computer Imagery, (DGCI), pp. 235–242 (1997)
13. Ibáñez, L., Hamitouche, C., Roux, C.: Ray casting in the BCC grid applied to 3D medical image visualization. In: Proceedings of the 20th Annual International Conference of the IEEE Engineering in Medicine and Biology Society, vol. 20, Biomedical Engineering Towards the Year 2000 and Beyond (Cat. No. 98CH36286), vol. 2, pp. 548–551 (1998)
14. Kim, M.: GPU isosurface raycasting of FCC datasets. Graph. Models **75**(2), 90–101 (2013)
15. Klette, R., Rosenfeld, A.: Digital geometry. Geometric methods for digital picture analysis, Chapter 2. Morgan Kaufmann Publishers, San Francisco, Amsterdam (2004)
16. Kovalevsky, V.A.: Geometry of Locally Finite Spaces (Computer Agreeable Topology and Algorithms for Computer Imagery), Chapter 3. Editing House Dr. Bärbel Kovalevski, Berlin (2008)
17. Latecki, L.J.: 3D Well-Composed Pictures. CVGIP: Graphical Model and Image Processing **59**(3), 164–172 (1997)
18. Meng, T., et al.: On visual quality of optimal 3D sampling and reconstruction. In: Graphics Interface 2007, pp. 265–272. ACM, New York, NY, USA (2007)

19. Pavlidis, T.: Algorithms for Graphics and Image Processing. Computer Science Press (1982)
20. Petkov, K., Qiu, F., Fan, Z., Kaufman, A.E., Mueller, K.: Efficient LBM visual simulation on face-centered cubic lattices. IEEE Trans. Vis. Comput. Graph. **15**(5), 802–814 (2009)
21. Qiu, F., Xu, F., Fan, Z., Neophytou, N., Kaufman, A.E., Mueller, K.: Lattice-based volumetric global illumination. IEEE Trans. Vis. Comput. Graph. **13**(6), 1576–1583 (2007)
22. Rosenfeld, A., Kak, A.C.: Digital Picture Processing. Academic Press, London (1982)
23. Strand, R.: The Euclidean distance transform applied to the FCC and BCC grids. In: Marques, J.S., Pérez de la Blanca, N., Pina, P. (eds.) IbPRIA 2005. LNCS, vol. 3522, pp. 243–250. Springer, Heidelberg (2005). https://doi.org/10.1007/11492429_30
24. Zheng, X., Gu, F.: Fast Fourier transform on FCC and BCC lattices with outputs on FCC and BCC lattices respectively. J. Math. Imaging Vis. **49**(3), 530–550 (2014). https://doi.org/10.1007/s10851-013-0485-9

Abbreviating Labelling Cost for Sentinel-2 Image Scene Classification Through Active Learning

Kashyap Raiyani$^{(\boxtimes)}$ ⓘ, Teresa Gonçalves ⓘ, and Luís Rato ⓘ

Department of Informatics, School of Science and Technology,
University of Évora, 7000-671 Évora, Portugal
{kshyp,tcg,lmr}@uevora.pt

Abstract. Over the years, due to the enrichment of paired-label datasets, supervised machine learning has become an important part of any problem-solving process. Active Learning gains importance when, given a large amount of freely available data, there's a lack of expert's manual labels. This paper proposes an active learning algorithm for selective choice of training samples in remote sensing image scene classification. Here, the classifier ranks the unlabeled pixels based on predefined heuristics and automatically selects those that are considered the most valuable for improvement; the expert then manually labels the selected pixels and the process is repeated. The system builds the optimal set of samples from a small and non-optimal training set, achieving a predefined classification accuracy. The experimental findings demonstrate that by adopting the proposed methodology, 0.02% of total training samples are required for Sentinel-2 Image Scene Classification while still reaching the same level of accuracy reached by complete training data sets. The advantages of the proposed method is highlighted by a comparison with the state-of-the-art active learning method named entropy sampling.

Keywords: Active learning · Supervised classification · Sentinel-2 image scene classification · Pattern recognition · Training data reduction

1 Introduction

Traditional supervised learning, such as binary or multi-class classification, is used to create high-predictive accuracy models from labeled training data. Labeled data, on the other hand, isn't cheap in terms of labeling costs, time spent, or the number of instants consumed [46]. As a result, the objective of active learning is to maximize the effective use of labeled data by allowing the learning algorithm to pick the instances that are most informative on their own [12]. In comparison to random sampling, the goal is to get better results with the same amount of training data or get the same results with fewer data [39].

© Springer Nature Switzerland AG 2022
A. J. Pinho et al. (Eds.): IbPRIA 2022, LNCS 13256, pp. 295–308, 2022.
https://doi.org/10.1007/978-3-031-04881-4_24

Active Learning is an iterative process that cycles over selecting new examples and retraining models. In each iteration, the value of candidate instances is calculated in terms of a usefulness score, and the ones with the highest scores are queried once [42] and its corresponding label is retrieved. The instance's value usually refers to the reduction of uncertainty in the context of "To what extent does knowing the label of a particular instance aids the learner reducing the ambiguity over instances that are similar?" [11]. In uncertainty sampling, one of the most common methods measures the instance's value in terms of predictive uncertainty [33], which leads the active learner to choose the cases where its current prediction is the most ambiguous. Almost all predictions are probabilistic, as are the measurements used to quantify the level of uncertainty, such as entropy [20].

For many real-world learning challenges where there is a limited collection of labeled data and a large amount of unlabeled data, pool-based active learning can be used. Here, samples are chosen greedily from a closed (i.e., static or non-changing) pool using an information measure such as entropy [21]. Many real-world machine learning areas have been examined using the pool-based active learning; these include (but are not limited to) text classification [19,37], information extraction [34], image classification and retrieval [44], video classification and retrieval [18], speech recognition [38], and cancer detection [23].

On the other hand, uncertainty sampling in active learning is perhaps the most basic and often used query framework. In this paradigm, an active learner inquires about situations for which there is the least amount of certainty of how to classify them [40]. If the underlying data distribution can be completely categorized by some hypothesis, then drawing $O(1/\varepsilon)$ random labeled examples, where ε is the maximum desirable error rate, is enough, according to the presumably approximately accurate (PAC) learning model [10]. According to the research [3,33], considering a pool-based active learning scenario where we can get some number of unlabeled examples for free (or very cheaply) from distribution; The (unknown) labels of these locations on the real line are a sequence of zeros followed by ones, and objective is to find the place (decision boundary) where the transition happens while paying as little as possible for labels. Because all additional labels can be inferred, a classifier with an error less than ε can be attained with just $O(log1/\varepsilon)$ queries, leading to an exponential decrease in the number of classified cases. Of course, this is a basic binary toy learning challenge that is one-dimensional and noiseless.

Scene classifying, or the classification of areas of high-resolution optical satellite images into morphological categories (e.g., land, ocean, cloud, etc.) has lately become a focus of many researchers [25] in the field of remote sensing. Scene classification is important in urban and regional planning [17], environmental vulnerability and impact assessment [26], and natural catastrophes and hazard monitoring [7]. Given the importance of scene classification, the study described here provides an active learning strategy for reducing Sentinel-2 image scene classification training data.

1.1 Problem Statement

This paper is an extension of previous work where the authors conjecture that, given a train-test split and different classifiers built over the training set, it is possible to find an Evidence Function for the prediction error using the relation between the training and the test sets [29]. The goal of the current paper is to elaborate more broadly on the usefulness of Evidence Functions for uncertainty sampling, comparing their performance in active learning.

Consider the following: having N labeled training points from a set of classes (C) described by a set of attributes (A) and T testing points, "Is it possible to use fewer labeled samples $(S \ll N)$ during the training phase and achieve the same accuracy over the test set?". (Note that A and C can have any number of attributes and classes, respectively).

The idea is to start by using $X \subset N$ points for training, and predict the rest of the unlabeled points $U = \{u_1, u_2, ...\}$ (where, $U = N \backslash X$) and T test points. Over the T points, the trained model will achieve some performance. Subsequently, among the U predicted samples, find the E points with highest uncertainty using an heuristic approach (Probability, Entropy) and/or incorporating the Evidence Function Model (EFM) (relation between train-set and test-set) approach. Here, E is chosen not at random but rather according to a problem-oriented heuristic that aims to maximize the classifiers' performance. Then, the E points identified as the most informative are added to X and the model is retrained and the prediction is repeated. This process will continue until the trained model reaches a performance higher (or equal) to the one obtained with the full training set resulting in optimal S samples.

The remainder of this paper is organised as follows: the experimental dataset and the general framework of uncertainty sampling are explored in the next section along with the introduction of several sampling methods and the experimental setup; Sect. 3 describes the results and provides an in-depth discussion; finally, the conclusions and future steps are exposed in Sect. 4.

2 Methods and Materials

This section focuses on (a) material acquisition, (b) development of evidence function model, and finally, (c) experimental setup.

2.1 Material

Raiyani et al. [30] provided an expanded collection of manually annotated Sentinel-2 spectra with various Surface Reflectance values (surface reflectance is defined as the proportion of incoming solar energy reflected from Earth's surface for a certain incident or observing condition). The dataset was created from images collected all across the world and contains 6.6 million points (precise 6,628,478 points) divided into six categories (Water, Shadow, Cirrus, Cloud, Snow, Other). Table 1 introduces the dataset.

Table 1. Structure of the extended manually labeled Sentinel-2 dataset [30].

Header	Attributes
Product ID	1 attribute (id; string)
Coordinates	2 attributes (latitude and longitude; real)
Bands/features	13 attributes (band 1 to 12 and 8A; real)
Label	1 attribute; values: Water, Shadow, Cirrus, Cloud, Snow, Other

The collection contains data from 60 items/images from five continents, including a wide range of surface types and geometry: Europe (22 products), Africa (14 products), America (12 products), Asia (6 products), and Oceania (6 products). Aside from that, three different classifier models based on K-Nearest Neighbors (KNN) [45], Extra Trees (ET) [13] and Convolutional Neural Networks (CNN) [22] were also published, along with the training set (50 products/images with $N = 5,716,330$ observations) and test set (10 products/images with $T = 912,148$ observations) were also published. The micro-F1 [27] performances assessed across the test set are provided in Table 2. These classifiers were fine-tuned using a RandomizedSearchCV [2] with 5 folds cross-validation.

Table 2. Micro-F1 over the test set for KNN, ET and CNN models.

Class	KNN	ET	CNN	Support
Other	0.68	0.83	0.82	174,369
Water	0.84	0.90	0.90	117,010
Shadow	0.68	0.81	0.83	155,715
Cirrus	0.61	0.79	0.76	175,988
Cloud	0.76	0.86	0.84	134,315
Snow	0.84	0.90	0.91	154,751
micro-F1	0.73	0.84	0.84	912,148

2.2 Method

This paper follows a pool-based sampling selection method. The most informative data samples are selected from the pool of unlabeled data based on some sampling strategy or informativeness measure which are discussed next. Irrespective of the sampling strategy, the following Generalized Sampling Algorithm 1 is used. (f1_score library[1].) (Here, micro-F1$_{ac}$ represents the micro-F1 score obtained using active learning methods.)

---

[1] https://scikit-learn.org/stable/modules/generated/sklearn.metrics.f1_score.html.

Algorithm 1. Generalized Sampling Algorithm

Input:

—Unlabeled dataset U.

—Initial train-set set X. Where, $X <<< U$

—Test-set T (with data-label pair).

Output:

—Optimal dataset S. Where, $X < S < U$

1: **procedure** SAMPLING_STRATEGIES(U, X, T)
2:     micro-F1$_{ac}$ = 0
3:     $S = X$
4:     **for** $S < U$ **do**
5:         Train classifier C over S.
6:         Using C calculate U_{pred} and T_{pred}.
7:         micro-F1$_{ac}$ = f1_score(T_{true}, T_{pred})   ▷ f1_score is a library from sklearn.metrics
8:         **if** micro-F1$_{ac}$ > $F1$ **then return** F ▷ $F1$ is a predefined micro-F1 score over test-set (84.0%)
9:         **else**
10:             select E samples from U using $M_i(U, U_{pred})$. Here, for each class in U, E samples are selected.   ▷ M_i is a sampling strategy discussed in the next subsections
11:         $S = S \cup E$ ▷ Updating S with the E chosen samples
12:         $U = U \setminus E$ ▷ Removing E chosen samples from U
13:     **end if**
14:     **end for**
15: **end procedure**

M_{en}: **Entropy-Based Method.** Entropy is an information-theoretic measure of a random variable uncertainty. The entropy measurement, which is computed using Eq. 1, is the most prominent measurement used in active learning pool-based sampling processes [35].

$$H(Y|x) = -\sum_{y \in Y} P(y|x) \log P(y|x) \tag{1}$$

where y is the class, with $y \in Y = y_1, y_2, ..., y_k$ and $P(y|x)$ is the a posteriori probability [32]. $H(x)$ is the uncertainty measurement function based on the classifier's posterior distribution entropy estimation [31]. This measure, also known as traditional uncertainty sampling, will be the baseline for the following experiments.

Entropy is computed for each prediction using Eq. 1 in step-10 of the generalized Algorithm 1, provided with U and U_{pred}. Following that, a total of E samples are selected for the highest entropy value.

M_{md}: **Mahalanobis Distance-Based Method.** Assume the training set $X = (x_1, x_2, ..., x_n)$ contains n observations, each $x_i = (t_i; g_i)$, where t_i is the predictor or feature and g_i is the response. In [29] a best-fit method based on X is proposed.

It outputs the likelihood of the classifier prediction (P) being wrong or, in other words, outputs the prediction uncertainty over unseen observations using the constructed classifiers $C(X)$. Knowing the prediction uncertainty can be used as an additional feature in the field of 'Disagreement-based Active Learning'. Here, idea is to query from dense regions where $\hat{P}(X)$ could disagree a lot with $P(X) \in P$ [8]. The reasoning is that measuring the uncertainty can aid in the generation of labeled datasets, with human input required only for data with a higher level of error [16].

Given the experimental dataset, the initial training set X is further split into training and test subsets and a classifier is built over the first; for this specific application, the training set X was split with a train subset of 49 products as one distribution D and the remaining product as the test subset. Then, for each point of the test subset, the Mahalanobis distance to distribution D is calculated (using μ_D and Σ_D), the Extra-Tree model is built over the training subset and tested over the test subset.

The Mahalanobis Distance Δ [9] between a point x_i and a distribution D with mean μ and covariance matrix Σ is given by Eq. (2).

$$\Delta = \sqrt{(x_i - \mu)\Sigma^{-1}(x_i - \mu)^\top} \tag{2}$$

The above process is repeated for all products in X as a test subset. As a result, for each data point in training set X, we have a set of Mahalanobis distances, one for each class distribution, and a prediction X_{pred}. As we know the true labels of training set X, a mismatch between the true label and the predicted label is referred to as 'classification prediction error' or 'uncertainty of the prediction'. The Mahalanobis distance is computed for each point in X to U using Eq. 2 in step-10 of the generalized Algorithm 1, provided with X, U and U_{pred}. Following that, a total of E samples are selected using their probabilistic measure based on the discovered prediction error points over U. In other words, either the top E samples with the low prediction uncertainty (i.e. high EFM confidence; here onward refereed as efm_max) or the high prediction uncertainty (i.e. low EFM confidence; here onward refereed as efm_min) are chosen.

2.3   Experimental Setup

The information about the experimental setup used to build the proposed solution is presented in Table 3. Table 4 details the Extra Trees algorithm parameters (Step-5 of the Generalized Sampling Algorithm 1). Note that, for each class, E manual labels are used in the initialization (i.e. during the initial training); following that, every iteration within Algorithm 1, adds $E * 6$ manual labels (by Oracle) from the U_{pred} class. These samples do not have to be exactly E samples for each class.

3   Results and Discussion

In this section, the use of the Evidence Function Model in active learning is evaluated against the Entropy-based sampling method.

Table 3. Experimental setup.

Attribute	Value
Language and library	Python and Scikit-learn [28]
System specification	Intel(R) Xeon(R) Silver 4110 CPU @ 2.10 GHz
Evidence function setup	Same setup as [29]
(Batch size, Max iteration) (E, N)	(10, 2000), (50, 400), (100, 200), (200, 100), and (500, 40)
Max. no. of samples labelled	120,000 (initially $E * 6$ upto $20,000 * 6$)

Table 4. Extra Trees (ET) algorithm parameters.

Parameter	T_{pred} [30]	U_{pred}
criterion	gini	gini
n_estimators	177	1000
min_samples_split	20	1
min_samples_leaf	1	1
max_features	sqrt	sqrt
max_depth	24	None
bootstrap	True	True

Table 5 shows the total number of labels required to attain micro-F1$_{ac}$ > micro-F1 for various batch sizes E (note that the total number of labels added would never exceed $20,000 * 6$, the original number of examples.) (Here, micro-F1$_{ac}$ represents the micro-F1 score obtained using active learning methods.) Consider the following example for a better understanding of the results: For method M_{en}, with batch size $E = 50$, the micro-F1$_{ac}$ reached 0.82% after 400 iterations (i.e. after adding 20,000 * 6 labels. The number 6 denotes the number of classes in the dataset); with batch size $E = 100$, the micro-F1$_{ac}$ was able to attain (the desired micro-F1) 0.84% after 81 iterations (i.e. after adding 8,100 * 6 labels). The table also includes the computational time required to achieved micro-F1$_{ac}$.

Table 5. Total number of labels added to reach micro-F1 for different batch sizes E.

Method	Batch size E	# Iterations N	# Labels added $E * N * 6$	micro-F1$_{ac}$	Training time
M_{en}	10	2,000	20,000 * 6	0.7218	108 (hrs)
	50	400	20,000 * 6	0.8275	21 (hrs)
	100	81	8,100 * 6	0.8403	4 (hrs)
	200	12	2,400 * 6	0.8401	36 (mins)
	500	0	0	0.8519	NA
M_{md}	10	2,000	20,000 * 6	0.8008	388 (hrs)
	50	10	500 * 6	0.8424	3 (hrs)
	100	2	200 * 6	0.8419	28 (mins)
	200	3	600 * 6	0.8465	41 (mins)
	500	0	0	0.8519	NA

Figure 1 illustrates the performance obtained over the test set on each iteration for M_{en} and M_{md} methods for different batch sizes E. (Y axes: micro-F1 over Test-set and X axes: No. of Training Samples.) After analysing Table 5, Algorithm 1 and Fig. 1, the following observations and discussions are made:

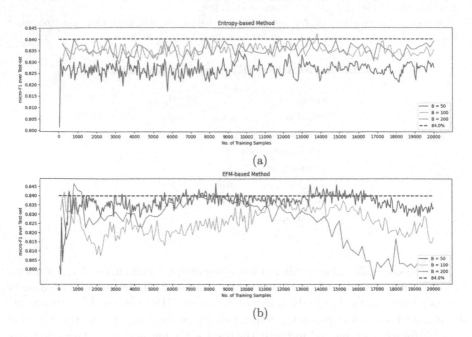

(a)

(b)

Fig. 1. A graphical comparison of methods Entropy (M_{en}) and Mahalanobis distance (M_{md}) based sampling strategy for various batch sizes E.

Training Label Cost. From Table 5 one can say that M_{md} (EFM) outperforms M_{en} by 79 and 9 iterations for batch size $E = 100$ and 200 (i.e. 47,400 and 10,800 labels), respectively. This means that for these batch sizes, by using Mahalanobis distance-based sampling the required number of training samples are reduced by 99.98% and 99.92% while achieving the same level of accuracy as using the complete training data; for the entropy-based sampling method the reduction was 99.15% and 99.73%, respectively. Moreover, M_{md} is able to achieve 84.0% micro-F1 in 10 iterations (i.e. 3000 labels) while M_{en} was unable to achieve it for a batch size $E = 50$. After 2000 iterations, no strategy was able to achieve the same level of accuracy as the full training data for batch sizes of $E = 10$; for a batch size of $E = 500$ as the initial training dataset, the same degree of accuracy as the whole training data set was obtained, hence no further labels were added.

Statistically [6], the 'active learning result curve' should rise as more informative labels are added, however, this is not the case with Fig. 1 (micro-F1 result curve). Here, compared to the M_{md}, the M_{en} approach has less variance between the beginning and end of the iteration. A fair explanation would be that for method M_{md}, added points have a greater heterogeneity to test-set, also there might be a presence of numerous clusters containing similar information among test-set data points. When M_{md} adds a new point to the train-set, which is a heterogeneous point to the test-set point, all the related points (within test-set) are misclassified as well, resulting in a greater variance between the beginning and end of the iteration (or dropping of the active learning result curve). (As part of future development,) within the train and test sets, similar informative points should be replaced with only those that are distinctive. As a result, when new points are added through active learning, regardless of the method, the learning model is always generalized, and the curve would follow an upward trajectory with reduced variance.

Initial Training Sample Selection. Active learning performance could be increased by carefully selecting the initial training samples. Using a fuzzy-c clustering approach [1], three initial training data selection processes were proposed [43]: center-based, border-based and hybrid. As initial training data, center-based selection chooses samples with a high degree of membership in each cluster; border-based selection selects samples from the clusters' edges; hybrid selection combines center-based and border-based selection. According to their findings, the hybrid selection is able to significantly improve the effectiveness of active learning when compared to randomly selected initial training samples.

In our experiments, the initial training samples are selected randomly using the train_test_split function[2] of sklearn; we kept shuffle = True and a fixed random seed for reproducibility. Also, to observe the effect of initial training sample selection, we kept Shuffle = False, meaning the selection is not random, and the grouped E samples from each image are considered for selection. Figure 2 shows the performance of both (randomly and grouped) selection methods for different numbers of initial samples. As can be seen, the random sample selection approach outperforms the grouped one. Furthermore, for random selection, using 250+ samples or more the performance obtained is always above 84.0%, implying that the initial training sample size does not need to exceed 250+ using random selection for this dataset.

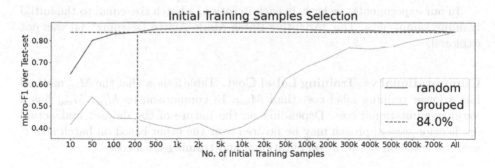

Fig. 2. Initial training samples selected (random vs grouped).

---

[2] https://scikit-learn.org/stable/modules/generated/sklearn.model_selection. train_test_split.html.

Myopic vs. Batch Mode Active Learning. In Myopic Active Learning (MAL) a single instance is queried at a time, whereas in Batch Mode Active Learning (BMAL) a batch of samples is picked and labeled concurrently. Single instance selection techniques require retraining the classifier for each classified instance, whereas BMAL offers the benefit [15] of not requiring the model to be retrained numerous times throughout the selection phase. On the other hand, BMAL faces various barriers [41]: choosing E samples from a pool of U instances might cause computing issues since the number of possible batches C_E^U can be rather high, depending on the values of U and E; additionally, designing an appropriate method to measure the overall information carried by a batch of samples can be quite challenging; finally for each iteration, one needs to ensure low information redundancy within a batch of chosen instances. Yingjie et al. [14] discussed sample selection with the highest density and least redundancy. For example, dense regions are supposed to be representative and informative, whereas the chosen instances could be not benefitial because of the redundancy among them, since various instances may include similar information.

In this study, we did not evaluate whether the picked points were information redundant or not; irrespective of the method (i.e. M_{en} or M_{md}), the generalized sampling Algorithm 1, selects the top E samples without considering their informative relation. The drawback of choosing the top E samples is that some of the selected samples $\hat{E} \subset E$ might provide enough information to the learner regarding remaining samples (i.e. $E \setminus \hat{E}$), leading to redundancy among the selected samples and generating extra labeling. Some of the recent research [5] uses a clustering step after selecting the top E to diversify and select only $\hat{E}$ samples.

Batch Size E. According to the Jingyu et al. [36], when more samples are chosen at the beginning of the training process, fewer samples may be used in later phases to exploit data recommendations. If more samples were allocated to later iterations, the model would have higher variation in the early iterations but a better chance of biasing samples for active learning in the later rounds. Lourentzou et al. [24], on the other hand, states that the optimal batch size is determined by the dataset and machine learning application to be addressed.

In our experiments, we kept, in each iteration, a batch size equal to the initial training sample size. The "Adaptive Batch Mode Active Learning" [4] was not explored.

Computational vs. Training Label Cost. Table 5 show that the M_{en} method has a lower training label cost than M_{md}. In comparison to M_{en}, M_{md} has a lower computational cost. Depending on the nature of the dataset and actual application, one approach may be favored over the other based on batch size E and the trade-off between computational and training label cost.

Reasoning for High Results. In remote sensing, image classification differs from traditional classification problems as the labelled dataset often consist of dense regions. This means that for any given class (say 'Water'), labelled images would have ample similar (high dense) data points representing 'Water' pixels. To verify our claim, Fig. 3 presents the class-wise surface reflectance value distribution of the training set (50 products/images with 5,716,330 observations) using the violin plot.

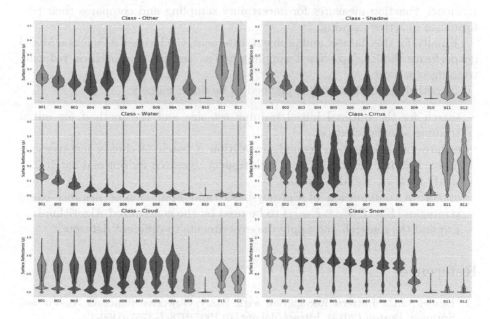

Fig. 3. 50 products/images Dataset: class-wise surface reflectance value distribution over 13 bands [30].

Here we can observe that for each class and feature, (nearly 5.7 m points), the data points are highly dense (i.e. the surface reflectance values range with 0.1 difference). Thus, by only selecting a few S samples from the dense region, the proposed learner was able to achieve the same level of performance as the complete training data sets. Although, depending upon different data-set and their distribution, this might not be always true.

4 Conclusion and Future Work

There are several factors to consider in active learning, including the initial training sample, the batch mode and size, the training label and computational costs, the problem statement and dataset. From this research, one can conclude that using active learning does reduce the overall training label cost, especially when the dataset comprises high-density regions (such as multiple pixels in satellite images). Furthermore, the results provide an in-depth comparison of two approaches on the above-mentioned factors, notably the initial training sample selection and the batch size usefulness.

This work demonstrates that it's possible to reduce the necessary training dataset for Sentinel-2 image scene classification using a state-of-the-art active learning approach along with Entropy Sampling or the suggested Evidence Function Model.

Moreover, this work expands previous work [29] and proves the utility of Evidence Function measures for uncertainty sampling and compares their performance in active learning.

Finally, for this dataset and problem statement, the Evidence Function Model outperforms the Entropy-based sampling active learning approach on reducing training label cost.

Nonetheless, further work is planned to be done, namely:

- search for the optimal initial training set size and the method for initial training sample selection;
- ensure low information redundancy within a batch of chosen instances on active learning iterations. Also, with respect to search for optimal batch size, study fixed vs. variable batches.
- explore a combination of Entropy-based and Mahalanobis distance-based strategies ($E = E_{M_{en}} \cap E_{M_{md}}$) to be used by the generalized Algorithm 1.
- Expand the research and apply the experiments to different datasets.

References

1. Bezdek, J.C.: Pattern Recognition with Fuzzy Objective Function Algorithms. Springer, Boston (2013). https://doi.org/10.1007/978-1-4757-0450-1
2. Buitinck, L., et al.: API design for machine learning software: experiences from the scikit-learn project. arXiv preprint arXiv:1309.0238 (2013)
3. Carbonneau, M.A., Granger, E., Gagnon, G.: Bag-level aggregation for multiple-instance active learning in instance classification problems. IEEE Trans. Neural Netw. Learn. Syst. **30**(5), 1441–1451 (2018)
4. Chakraborty, S., Balasubramanian, V., Panchanathan, S.: Adaptive batch mode active learning. IEEE Trans. Neural Netw. Learn. Syst. **26**(8), 1747–1760 (2015). https://doi.org/10.1109/TNNLS.2014.2356470
5. Citovsky, G., et al.: Batch active learning at scale. Adv. Neural Inf. Process. Syst. **34** (2021)
6. Cohn, D.A., Ghahramani, Z., Jordan, M.I.: Active learning with statistical models. CoRR cs.AI/9603104 (1996). https://arxiv.org/abs/cs/9603104
7. Dao, P.D., Liou, Y.A.: Object-based flood mapping and affected rice field estimation with Landsat 8 OLI and MODIS data. Remote Sens. **7**(5), 5077–5097 (2015)
8. Dasgupta, S., Langford, J.: Active learning tutorial, icml 2009." (2009)
9. De Maesschalck, R., Jouan-Rimbaud, D., Massart, D.L.: The mahalanobis distance. Chemom. Intell. Lab. Syst. **50**(1), 1–18 (2000)
10. Devonport, A., Saoud, A., Arcak, M.: Symbolic abstractions from data: a PAC learning approach. arXiv preprint arXiv:2104.13901 (2021)
11. Fu, Y., Li, B., Zhu, X., Zhang, C.: Active learning without knowing individual instance labels: a pairwise label homogeneity query approach. IEEE Trans. Knowl. Data Eng. **26**(4), 808–822 (2014). https://doi.org/10.1109/TKDE.2013.165

12. Fu, Y., Zhu, X., Li, B.: A survey on instance selection for active learning. Knowl. Inf. Syst. **35**(2), 249–283 (2013)
13. Geurts, P., Ernst, D., Wehenkel, L.: Extremely randomized trees. Mach. Learn. **63**(1), 3–42 (2006)
14. Gu, Y., Jin, Z., Chiu, S.C.: Active learning with maximum density and minimum redundancy. In: Loo, C.K., Yap, K.S., Wong, K.W., Teoh, A., Huang, K. (eds.) ICONIP 2014. LNCS, vol. 8834, pp. 103–110. Springer, Cham (2014). https://doi.org/10.1007/978-3-319-12637-1_13
15. Gui, X., Lu, X., Yu, G.: Cost-effective batch-mode multi-label active learning. Neurocomputing **463**, 355–367 (2021)
16. Hanneke, S.: Theory of disagreement-based active learning. Found. Trends® Mach. Learn. **7**(2–3), 131–309 (2014). https://doi.org/10.1561/2200000037
17. Hashem, N., Balakrishnan, P.: Change analysis of land use/land cover and modelling urban growth in greater Doha, Qatar. Ann. GIS **21**(3), 233–247 (2015)
18. Hauptmann, A.G., Lin, W.H., Yan, R., Yang, J., Chen, M.Y.: Extreme video retrieval: joint maximization of human and computer performance. In: Proceedings of the 14th ACM International Conference on Multimedia, pp. 385–394 (2006)
19. Hoi, S.C., Jin, R., Lyu, M.R.: Large-scale text categorization by batch mode active learning. In: Proceedings of the 15th international conference on World Wide Web, pp. 633–642 (2006)
20. Hüllermeier, E., Waegeman, W.: Aleatoric and epistemic uncertainty in machine learning: an introduction to concepts and methods. Mach. Learn. **110**(3), 457–506 (2021)
21. Lewis, D.D., Gale, W.A.: A sequential algorithm for training text classifiers. In: Croft, B.W., van Rijsbergen, C.J. (eds.) SIGIR 1994, pp. 3–12. Springer, London (1994). https://doi.org/10.1007/978-1-4471-2099-5_1
22. Liu, Y., Zhong, Y., Fei, F., Zhu, Q., Qin, Q.: Scene classification based on a deep random-scale stretched convolutional neural network. Remote Sens. **10**(3), 444 (2018)
23. Liu, Y.: Active learning with support vector machine applied to gene expression data for cancer classification. J. Chem. Inf. Comput. Sci. **44**(6), 1936–1941 (2004)
24. Lourentzou, I., Gruhl, D., Welch, S.: Exploring the efficiency of batch active learning for human-in-the-loop relation extraction. In: Companion Proceedings of the Web Conference 2018, pp. 1131–1138 (2018)
25. Mohajerani, S., Krammer, T.A., Saeedi, P.: A cloud detection algorithm for remote sensing images using fully convolutional neural networks. In: 2018 IEEE 20th International Workshop on Multimedia Signal Processing (MMSP), pp. 1–5 (2018). https://doi.org/10.1109/MMSP.2018.8547095
26. Nguyen, K.A., Liou, Y.A.: Mapping global eco-environment vulnerability due to human and nature disturbances. MethodsX **6**, 862–875 (2019)
27. Opitz, J., Burst, S.: Macro F1 and macro F1. CoRR abs/1911.03347 (2019). http://arxiv.org/abs/1911.03347
28. Pedregosa, F., et al.: Scikit-learn: machine learning in Python. J. Mach. Learn. Res. **12**, 2825–2830 (2011)
29. Raiyani, K., Gonçalves, T., Rato, L., Barão, M.: Mahalanobis distance based accuracy prediction models for sentinel-2 image scene classification. Int. J. Remote Sens. 1–26 (2022). https://doi.org/10.1080/01431161.2021.2013575
30. Raiyani, K., Gonçalves, T., Rato, L., Salgueiro, P., Marques da Silva, J.R.: Sentinel-2 image scene classification: a comparison between sen2cor and a machine learning approach. Remote Sens. **13**(2) (2021). https://doi.org/10.3390/rs13020300, https://www.mdpi.com/2072-4292/13/2/300

31. Roy, N., McCallum, A.: Toward optimal active learning through monte Carlo estimation of error reduction. ICML, Williamstown **2**, 441–448 (2001)
32. Seifert, C., Granitzer, M.: User-based active learning. In: 2010 IEEE International Conference on Data Mining Workshops, pp. 418–425. IEEE (2010)
33. Settles, B.: Active learning literature survey. Computer Sciences Technical report 1648, University of Wisconsin-Madison (2009). http://axon.cs.byu.edu/~martinez/classes/778/Papers/settles.activelearning.pdf
34. Settles, B., Craven, M.: An analysis of active learning strategies for sequence labeling tasks. In: Proceedings of the 2008 Conference on Empirical Methods in Natural Language Processing, pp. 1070–1079 (2008)
35. Shannon, C.E.: A mathematical theory of communication. Bell Syst. Tech. J. **27**(3), 379–423 (1948). https://doi.org/10.1002/j.1538-7305.1948.tb01338.x
36. Shao, J., Wang, Q., Liu, F.: Learning to sample: an active learning framework. CoRR abs/1909.03585 (2019). http://arxiv.org/abs/1909.03585
37. Tong, S., Koller, D.: Support vector machine active learning with applications to text classification. J. Mach. Learn. Res. **2**(Nov), 45–66 (2001)
38. Tur, G., Hakkani-Tür, D., Schapire, R.E.: Combining active and semi-supervised learning for spoken language understanding. Speech Commun. **45**(2), 171–186 (2005)
39. Vapnik, V.N.: An overview of statistical learning theory. IEEE Trans. Neural Netw. **10**(5), 988–999 (1999)
40. Wang, Z., Brenning, A.: Active-learning approaches for landslide mapping using support vector machines. Remote Sens. **13**(13), 2588 (2021)
41. Yang, Y., Yin, X., Zhao, Y., Lei, J., Li, W., Shu, Z.: Batch mode active learning based on multi-set clustering. IEEE Access **9**, 51452–51463 (2021). https://doi.org/10.1109/ACCESS.2021.3053003
42. Yu, G., et al.: CMAL: cost-effective multi-label active learning by querying subexamples. IEEE Trans. Knowl. Data Eng. 1 (2020). https://doi.org/10.1109/TKDE.2020.3003899
43. Yuan, W., Han, Y., Guan, D., Lee, S., Lee, Y.K.: Initial training data selection for active learning. In: Proceedings of the 5th International Conference on Ubiquitous Information Management and Communication, pp. 1–7 (2011)
44. Zhang, C., Chen, T.: An active learning framework for content-based information retrieval. IEEE Trans. Multimedia **4**(2), 260–268 (2002)
45. Zhang, M.L., Zhou, Z.H.: ML-KNN: a lazy learning approach to multi-label learning. Pattern Recogn. **40**(7), 2038–2048 (2007)
46. Zou, H., Hastie, T.: Regularization and variable selection via the elastic net. J. Roy. Stat. Soc. Ser. B (Stat. Methodol.) **67**(2), 301–320 (2005)

Feature-Based Classification of Archaeal Sequences Using Compression-Based Methods

Jorge Miguel Silva[1](✉) ⓘ, Diogo Pratas[1,2,3] ⓘ, Tânia Caetano[4] ⓘ,
and Sérgio Matos[1,2] ⓘ

[1] Institute of Electronics and Informatics Engineering of Aveiro,
University of Aveiro, Aveiro, Portugal
`jorge.miguel.ferreira.silva@ua.pt`

[2] Department of Electronics, Telecommunications and Informatics,
University of Aveiro, Aveiro, Portugal

[3] Department of Virology, University of Helsinki, Helsinki, Finland

[4] CESAM and Department of Biology, University of Aveiro, Aveiro, Portugal

Abstract. Archaea are single-celled organisms found in practically every habitat and serve essential functions in the ecosystem, such as carbon fixation and nitrogen cycling. The classification of these organisms is challenging because most have not been isolated in a laboratory and are only found in ambient samples by their gene sequences. This paper presents an automated classification approach for any taxonomic level based on an ensemble method using non-comparative features. This methodology overcomes the problems of reference-based classification since it classifies sequences without resorting directly to the reference genomes, using the features of the biological sequences instead. Overall we obtained high results for classification at different taxonomic levels. For example, the Phylum classification task achieved 96% accuracy, whereas 91% accuracy was achieved in the genus identification task of archaea in a pool of 55 different genera. These results show that the proposed methodology is a fast, highly-accurate solution for archaea identification and classification, being particularly interesting in the applied case due to the challenging classification of these organisms. The method and complete study are freely available, under the GPLv3 license, at https://github.com/jorgeMFS/Archaea2.

Keywords: Archaeal sequences · Feature-based classification · Taxonomic identification · Data compression · Feature selection

1 Introduction

In recent years, metagenomics analysis has been growing exponentially and becoming a prominent instrument in the medical, forensic, and exobiology fields [1–5]. The main driver for this expansion was next-generation sequencing (NGS), which can, among other applications, perform gene expression quantification, genotyping, genome reconstruction, detection of human vial communities, or

© Springer Nature Switzerland AG 2022
A. J. Pinho et al. (Eds.): IbPRIA 2022, LNCS 13256, pp. 309–320, 2022.
https://doi.org/10.1007/978-3-031-04881-4_25

unveil genome rearrangements [6–8]. Furthermore, NGS technologies differentiate themselves from other previously existing methods by being a more time-cost-effective option for rapid screening of larger genomes [9].

The identification of organisms from metagenomic samples requires several computing-laboratory steps, namely sequencing, trimming, filtering, assembly, and analysis. During this process, the cloning and enrichment of specific reconstructed regions to achieve better quality and completeness can lead to a computing-laboratory spiral [10]. Nonetheless, the results obtained can be inconclusive when employing referential comparison approaches after sequence reconstruction. Causes for this issue can be the divergence between the sequences of known organisms, the presence of irregularities introduced during the reconstruction process, or simply the fact that a new organism has been sequenced [11,12]. Despite these challenges, reference-based comparison approaches are used by the vast majority of classification pipelines [13–23], where the reconstructed sequence is compared to a collection of references stored in a database. However, despite its popularity, when confronted with the obstacles described previously, this method proves to be ineffective.

Some recent works have tried to leverage a different strategy to perform metagenomic identification of organisms. For instance, Karlicki et al. [24] created Tiara, a deep learning-based classification system for eukaryotic sequences. This program functions as a two-step classification process that enables the classification of nuclear and organellar eukaryotic fractions and separates organellar sequences into plastidial and mitochondrial [24].

More recently, Lourenço et al. [25] described a computational pipeline for the reconstruction and classification of unknown sequences in metagenomic samples. For classification of groups/domains (virus, bacteria, archaea, fungi, plant, protozoa, mitochondria, plastid), it employed feature-based classification, using compression-based measures of the genome and proteome, as well as sequence length and GC-content. This type of classification has shown state-of-the-art results and high flexibility since it does not resort directly to the reference genomes but instead to features that the biological sequences share. In this paper, we follow this line, improving this methodology to provide capability to perform analysis at any taxonomic level. Additionally, we added more features and classifiers. Moreover, we apply this methodology specifically to archaeal sequences, known to be one of the most challenging domains to perform taxonomic classification.

Archaea are single-celled organisms that do not have a nucleus. However, their cells contain characteristics that differentiate them from bacteria and eukarya. Despite being initially discovered in extreme environments, such as hot springs and salt lakes, they exist in practically every habitat. These organisms are also part of the human microbiome, where they play critical roles, especially in the intestine. In terms of size and form, archaea and bacteria are very similar. Nevertheless, despite their structural resemblance to bacteria, archaea have genes and metabolic pathways that are more closely related to eukaryotes, particularly for transcription and translation enzymes. In addition, archaeal biochemistry is distinct, as proven by the isoprene-composed and ether-bond phospholipids of their cell membranes.

The reference-based classification problem is even more pronounced in archaea since most of these organisms have not been isolated in a laboratory and have only been detected by their gene sequences in environmental samples. Other alternative classification methods might be more advantageous for this problem. As a result, in resemblance with metagenomic identification works [25], we leverage compressor-based features conjugated with simple metrics to create an accurate archaea taxonomic classification and identification at different depths. The complete pipeline can be fully replicated using the repository available at https://github.com/jorgeMFS/Archaea2.

2 Methods

The methodology described in this paper is based on the work developed by Lourenço *et al.* [25]. Herein, we improve this methodology by providing additional features, classifiers, and flexibility to classify any taxonomic level. Furthermore, this work is applied to a different problem: identification of archaea at different taxonomic levels.

In this section we describe the methods used to perform the feature-based archaea's taxonomic classification at different levels; first we describe the dataset and the pipeline used, followed by the features selected to perform classification.

2.1 Data Description

The dataset comprises all archaea genome samples (FASTA format) present in the NCBI database on 10 of January 2022. In total, 2,437 samples were retrieved using the dataset download tool (for replication see *download_dataset.sh* at the repository). The download included the genomic sequence of the sample and, when available, their respective proteomic sequence. The metadata header was removed from each genomic and proteomic sequence using the GTO toolkit [26]. Also, in the genomic sequences, any nucleotide outside the quaternary alphabet $\{A, C, G, T\}$ was replaced by a random nucleotide from the quaternary alphabet. Furthermore, the taxonomic description of each sample was retrieved using Entrez-direct [27]. After retrieving all the data from the NCBI database, the dataset was filtered, discarding files that did not provide proteomic sequences and sequences without any taxonomic information on the NCBI repository. As a result of the filtering process, a total of 667 (of the initial 2,437) samples remained for classification. All these samples contain proteomic and genomic sequences and taxonomic descriptions.

2.2 Pipeline

The methodology used to perform the classification is shown in Fig. 1. Initially, the samples are retrieved from the NCBI database. Then, after the samples metadata containing their taxonomic description is retrieved and the samples filtered, each sample genomic and proteomic files are stripped from their headers and turned into single line files with only genomic and proteomic alphabetic characters. Afterwards, relevant features are retrieved from each sample's genomic and proteomic files, then fed to different classifiers to obtain the classification.

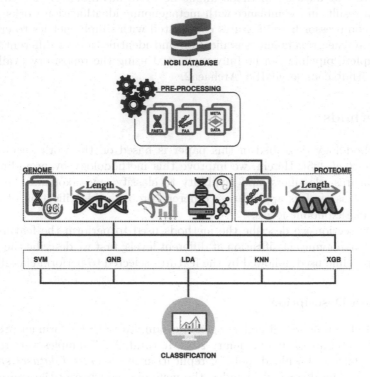

Fig. 1. Figure depicting the methodology for archaea's taxonomic classification.

2.3   Predictors

The feature-based classification uses multiple predictors to perform classification. The features obtained for each sample were: the Normalized Compression (NC) and the Sequence Length (SL) for the genomic and proteomic sequences, and the percentage of each nucleotide and GC-content in the genome.

Normalized Compression

An efficient compression of a string x $(C(x))$ provides an upper bound approximation for the Kolmogorov complexity of the same string, $K(x)$, where $K(x) < C(x) \leq |x| \log_2 (\theta)$, $|x|$ is the length of string x, and θ is the cardinality of the alphabet of the string x. Therefore, if the compressor is efficient, it can approximate the quantity of probabilistic-algorithmic information in the string. The normalized version, known as the Normalized Compression (NC), is defined by

$$\mathrm{NC}(x) = \frac{C(x)}{|x| \log_2 |\theta|}, \tag{1}$$

where $C(x)$ is the compressed size of x in bits. This normalization enables the comparison of the proportions of information contained in the strings independently from their sizes [28]. In this paper, to compute the NC, we use the genome compressor, GeCo3 [29], and the proteomic compressor AC [30] to compress individual genomic and proteomic sequences, respectively.

Nucleotide Percentage, GC-Content, and Sequence Length

The other measures used to perform the taxonomic classification of archaea are the percentage of each nucleotide (NP), GC-content (GC) of the genome and the length of the genome and the proteome sequence.

The percentage of each nucleotide is proportional to the number of adenine, cytosine, guanine, and thymine ($\{A, C, G, T\}$, respectively) in the sequence. GC-content (GC) is the proportion of guanine (G) and cytosine (C) in the sequence. GC-content is a relevant feature since it differs between organisms and is correlated with the organism's life-history traits, genome size [31], genome stability [32], and GC-biased gene conversion [33].

2.4 Classification

Classifiers

The features were fed into several machine learning algorithms to perform the taxonomic classification task. Specifically, the classifiers used were the Support Vector Machine (SVM) [34], Gaussian Naive Bayes (GNB) [35], Linear Discriminant Analysis (LDA) [36], K-Nearest Neighbors (KNN) [37], and XGBoost classifier (XGB)[38]. Support Vector machines are supervised learning models with associated learning algorithms that construct a hyperplane in a high-dimensional space using data, and then performs classification [34]. Gaussian Naive Bayes is a supervised machine learning classification algorithm based on the Bayes theorem following Gaussian normal distribution [35]. Linear Discriminant Analysis is a generalization of Fisher's linear discriminant, a method used in statistics to find a linear combination of features that separates classes of objects. The resulting combination can be used as a linear classifier [36]. K-Nearest Neighbors uses distance functions and performs classification predictions based on the majority vote of its neighbours [37]. Finally, XGBoost [38] is an efficient open-source implementation of the gradient boosted trees algorithm. Gradient boosting is a supervised learning algorithm that predicts a target variable by combining the estimates of a set of simpler models. Specifically, new models are created that predict the residuals or errors of prior models and then added together to make the final prediction. This task uses a gradient descent algorithm to minimize the loss when adding new models. XGBoost uses this approach in both regression and classification predictive modelling problems.

Classification Methodology

We performed two separate tests to determine the optimal settings for classification. One test determined the best classifier to use, and another determined the best set of features to feed the classifier. To determine the best classifier all features were feed to the classifiers and the evaluation metrics were retrieved. On the other hand, to determine the best set of features to feed the classifiers, different combinations of features were fed to the best classifier and for each the evaluation metrics were retrieved. The sets of features used were:

- The genome sequence Normalized Compression (NC_g);
- The genome and proteome Normalized Compression ($NC_g + NC_p$);
- The proteome Sequence Length and Normalized Compression ($SL_p + NC_p$);
- All the genome feature: the genome Sequence Length, GC-content, the percentage of each nucleotide, and Normalized Compression. Described as $SL_g + GC + NP_g + NC_g$ or All_g;
- All the features ($SL_g + GC + NP_g + NC_g + SL_p + NC_p$ or All).

We performed a random 80–20 train-test split on the dataset to perform classification. In addition, several actions were performed due to classes being imbalanced in the dataset. First, we did not consider classes with less than four samples. As such, depending on the classification task, the number of samples decreased from 667 to the values shown in Table 1 (samples column). Secondly, we performed the train-test split in a stratified way to ensure the representability of each label in the train and test sets. Finally, instead of performing k-fold cross-validation, we performed the random train-test split fifty times, and we retrieved the average of the evaluation metrics.

Evaluation Metrics

The accuracy and weighted F1-score were used to select and evaluate the classification performance of the measures. Accuracy is the proportion between correct and the total number of classifications, while the F1-score is computed using the precision and recall of the test. We utilized the weighted version of the F1-Score due to imbalanced classes of the dataset.

For comparison of the obtained results, we assessed the outcomes obtained using a random classifier. For that purpose, we determined the probability of a random sequence being correctly classified (p_{hit}) as

$$p_{hit} = \sum_{i=0}^{n} [p(c_i) * p_{correct}(c_i)], \tag{2}$$

where $p(c_i)$ is the probability of each class, determined as

$$p(c_i) = \frac{|samples_{class}|}{|samples_{total}|}.$$

On the other hand, $p_{correct}(c_i)$ is the probability of that class being correctly classified. In the case of a random classifier,

$$p_{correct}(c_i) = \frac{1}{|classes|}.$$

3   Results

Using the archaea dataset, we performed taxonomic classification of the archaea regarding their phylum, class, order, family, and genus.

For benchmarking our methodology, we did not compare the results with other state-of-the-art solutions for two reasons. First, reference-based classification solutions were not evaluated since they require each sequence to be compared with a collection of reference sequences stored in the database. In our work, we use exclusively the features of the biological sequences without comparative measures to perform this direct classification (reference-free). Secondly, methodologies such as the one used by Lourenço et al. [25] can not be directly compared with ours because of current restrictions in groups/domains (high-level classification: virus, bacteria, archaea, fungi, etc.). In contrast, our work is applied at any taxonomic level of archaea. Moreover, our classification methodology is an improvement of the one used by Lourenço et al. [25].

A total of five types of classifiers were tested: Linear Discriminant Analysis (LDA) [36], Gaussian Naive Bayes (GNB) [35], K-Nearest Neighbors (KNN) [37], Support Vector Machine (SVM) [34], and XGBoost classifier (XGB)[38]. We computed the Accuracy and the Weighted F1-score to select the best performing method. The results of feeding all genomic and proteomic features to the classifier is represented in Table 1.

For all classification tasks, the best performing classifier was the XGBoost classifier, being that the second-best results were obtained using LDA for the majority of the classification tasks.

Table 1. Accuracy (ACC) and F1-score (F1) results for taxonomic group classification using all features. The classifiers used were Linear Discriminant Analysis (LDA), Gaussian Naive Bayes (GNB), K-Nearest Neighbors (KNN), Support Vector Machine (SVM), and XGBoost classifier (XGB).

Group	Classes	Samples	SVM		GNB		KNN		LDA		XGB	
			Acc	F1-score	Acc	F1-score	Acc	F1-score	Acc	F1-score	Acc	F1-score
Phylum	5	660	44.36	0.4436	48.35	0.4835	68.45	0.6845	60.11	0.5600	**95.95**	**0.9591**
Class	10	615	29.10	0.2910	42.33	0.4233	52.41	0.5241	59.76	0.5674	**93.53**	**0.9328**
Order	20	634	23.64	0.2364	31.01	0.3101	36.53	0.3653	52.33	0.4804	**89.91**	**0.8948**
Family	29	623	22.84	0.2284	31.77	0.3177	33.92	0.3392	51.82	0.4703	**89.55**	**0.8882**
Genus	55	543	12.47	0.1247	27.69	0.2769	17.47	0.1747	60.67	0.5753	**91.58**	**0.9039**

On the other hand, to determine the best features to be used, different features were fed to XGBoost classifier. For each set of features, the accuracy and weighted F1-score were determined. These results are depicted in Table 2 and 3.

Overall, the best results were obtained for all feature groups in the phylum classification task. Additionally, the worst classification was present in the family classification task despite not having the highest number of classes. In contrast, remarkable results were obtained for the genus classification task, which possesses the highest number of classes.

Table 2. Accuracy results from taxonomic group classification using XGBoost classifier. The features used were the genome's Sequence Length (SL_g), the percentage of each nucleotide (NP_g), GC-content (GC) and Normalized Compression (NC_g), as well as the proteome's Sequence Length (SL_p), and Normalized Compression (NC_p). Accuracy is depicted as Acc, and the probability of a random sequence being correctly classified using a random classifier as p_{hit}.

Group	Classes	Samples	P_{hit}	$Acc_{(NC_g)}$	$Acc_{(NC_{g+p})}$	$Acc_{(SL_p+NC_p)}$	$Acc_{(All_g)}$	$Acc_{(All)}$
Phylum	5	660	20.00	65.76	76.45	75.62	**96.09**	95.95
Class	10	615	10.00	50.65	63.27	61.32	93.35	**93.53**
Order	20	634	5.00	41.75	56.83	50.60	89.67	**89.91**
Family	29	623	3.45	41.57	56.72	48.98	89.50	**89.55**
Genus	55	543	1.82	37.47	52.50	47.83	91.50	**91.58**

Table 3. F1-score results from taxonomic group classification using XGBoost classifier. The features used were the genome's Sequence Length (SL_g), the percentage of each nucleotide (NP_g), GC-content (GC), and Normalized Compression (NC_g), as well as the proteome's Sequence Length (SL_p) and Normalized Compression (NC_p).

Group	Classes	Samples	$F1_{(NC_g)}$	$F1_{(NC_{g+p})}$	$F1_{(SL_p+NC_p)}$	$F1_{(All_g)}$	$F1_{(All_{p+g})}$
Phylum	5	660	0.6423	0.7592	0.7513	**0.9606**	0.9591
Class	10	615	0.4933	0.6220	0.5951	0.9314	**0.9328**
Order	20	634	0.3963	0.5541	0.4850	0.8930	**0.8948**
Family	29	623	0.3872	0.5477	0.4665	**0.8889**	0.8882
Genus	55	543	0.3263	0.4801	0.4294	**0.9048**	0.9039

For comparison purposes, we assessed the outcomes obtained using a random classifier. Specifically, for each task, we determined the probability of a random sequence being correctly classified (p_{hit}). Overall, there is a substantial improvement relative to the random classifier, showing the features' importance in the classification process. For instance, the probability of a random classifier correctly identifying the genus of a given sequence was 1.82%, whereas, in our best classification, we obtained 91.58% accuracy. The results are particularly encouraging given the small sample size and the diversity of labels in the dataset.

In the genomic features, despite the NC being the most relevant, the ensemble of the GC-content, Sequence Length and the features related to the percentage of each nucleotide (All_g) improved the accuracy and F1-score in all tasks.

The results obtained from using only proteomic features indicate that although they are relevant features, genomic features are more relevant for the classification of archaea sequences since the accuracy and F1-score were higher on the latter. The results were mixed when we combined the genomic features with the proteomic features. Although the accuracy improved slightly for all classification tasks (phylum exception), the F1-score was better for the phylum, family and genera classification tasks using only genomic features. These results indicate that despite all features being relevant, mixing the proteomic features with the genomic features only improves classification in some instances.

4   Discussion

This manuscript describes and evaluates a methodology to perform feature-based classification and identification of archaea's taxonomic groups.

The features-based classification classifies sequences without resorting directly to the reference genomes, but rather it uses the biological sequences' features (reference-free classification). This process is computationally fast, and in the case of archaea, it seems to require few samples.

We used all archaea's samples in the NCBI database with proteomic and genomic sequences that contain a taxonomic description to perform this classification. Simple predictive genomic and proteomic features were extracted and fed to several classifiers from this data. Specifically, we considered the following genomic features: the length of each genome, the genome's Normalized Complexity, the genome's percentage of each nucleotide and GC-content. On the other hand, we considered the length of each proteomic sequence and the proteome's Normalized Complexity as proteomic features.

The best results were obtained using the XGBoost classifier. Regarding feature selection, genomic features showed to be highly efficient descriptors of the data, obtaining the best results for phylum identification.

In contrast to previous metagenomic works [25], where proteomic features were highly relevant for organism identification, in archaea's depth taxonomic classification, the results were mixed. Specifically, when proteomic features were combined with genomic features, in most classification tasks, accuracy increased, and the weighted F1-score decreased. As such, it seems that rich genomic features suffice to perform an accurate identification in the problem of archaea's depth taxonomic classification. Furthermore, the fact that we obtained high accuracy results for the genus identification task, characterized by having a high number of classes, demonstrates that these features are efficient descriptors of the sequences.

Furthermore, some possible inaccuracies in classification could be explained by errors in the assembly process of the samples, or eventual sub-sequence contamination of parts of the genomes [39,40], especially given that we worked directly on uncomplete samples (contig, scaffold and incomplete assembly sequences). Other inaccuracies could be due to several genomes being reconstructed using older methods that have been improved since then. Finally, this methodology is not restricted to the future addiction on more features, for example, signature features or nucleotide distances [41].

5 Conclusions

This manuscript describes and evaluates a methodology to perform feature-based classification and identification of archaea's taxonomic groups. The results show that the efficient approximation of the Kolmogorov complexities of archaea sequences as complexity measures impacts archaea taxonomic identification and classification. Regarding the features selected, surprisingly, rich genomic features suffice to perform a correct identification of archaea's taxon even with a large number of labels. We showed that the devised pipeline could automatically classify and identify archaea samples at different taxonomic levels with high accuracy. These results are significant given that this classification is challenging since a set of these archaea have not been isolated in a laboratory and were only detected by their gene sequences in environmental samples. We conclude that this methodology can serve as a fast tool to perform archaea identification in challenging scenarios.

Funding

This work was partially funded by National Funds through the FCT - Foundation for Science and Technology, in the context of the project UIDB/00127/2020. We acknowledge financial support to CESAM by FCT/MCTES (UIDP/50017/2020 + UIDB/50017/2020 + LA/P/0094/2020) through national funds. J.M.S. acknowledges the FCT grant SFRH/BD/141851/2018. D.P. is funded by national funds through FCT - Fundação para a Ciência e a Tecnologia, I.P., under the Scientific Employment Stimulus - Institutional Call - reference CEECINST/00026/2018. T.C is supported by POPH and national funds (OE), through FCT - Foundation for Science and Technology, I.P., under the scope Scientific Employment Stimulus (CEECIND/01463/2017).

References

1. Biesecker, L.G., Burke, W., Kohane, I., Plon, S.E., Zimmern, R.: Next-generation sequencing in the clinic: are we ready? Nat. Rev. Genet. **13**(11), 818–824 (2012)
2. Chiu, C.Y., Miller, S.A.: Clinical metagenomics. Nat. Rev. Genet. **20**(6), 341–355 (2019)
3. Hampton-Marcell, J.T., Lopez, J.V., Gilbert, J.A.: The human microbiome: an emerging tool in forensics. Microbial Biotechnol. **10**(2), 228–230 (2017)
4. Amorim, A., Pereira, F., Alves, C., García, O.: Species assignment in forensics and the challenge of hybrids. Forensic Sci. Int. Genet. **48**, 102333 (2020)
5. Eloe-Fadrosh, E.A., et al.: Global metagenomic survey reveals a new bacterial candidate phylum in geothermal springs. Nat. Commun. **7**(1), 1–10 (2016)
6. Del Fabbro, C., Scalabrin, S., Morgante, M., Giorgi, F.M.: An extensive evaluation of read trimming effects on illumina NGS data analysis. PLoS ONE **8**(12) (2013)
7. Toppinen, M., Sajantila, A., Pratas, D., Hedman, K., Perdomo, M.F.: The human bone marrow is host to the DNAs of several viruses. Front. Cell. Infect. Microbiol. **11**, 329 (2021)

8. Hosseini, M., Pratas, D., Morgenstern, B., Pinho, A.J.: Smash++: an alignment-free and memory-efficient tool to find genomic rearrangements. GigaScience **9**(5), giaa048 (2020)

9. Mardis, E.R.: DNA sequencing technologies: 2006–2016. Nat. Protoc. **12**(2), 213–218 (2017)

10. Thomas, T., Gilbert, J., Meyer, F.: Metagenomics - a guide from sampling to data analysis. Microb. Inf. Exp. **2**(1), 1–12 (2012)

11. Abnizova, I., et al.: Analysis of context-dependent errors for illumina sequencing. J. Bioinform. Comput. Biol. **10**(2) (2012)

12. Boekhorst, R.T., et al.: Computational problems of analysis of short next generation sequencing reads. Vavilov J. Genet. Breed. **20**(6), 746–755 (2016)

13. Breitwieser, F.P., Lu, J., J., Salzberg, J., A review of methods and databases for metagenomic classification and assembly. Brief. Bioinform. **20**(4), 1–15 (2017)

14. Chen, S., He, C., Li, Y., Li, Z., Charles III, E.M.: A computational toolset for rapid identification of SARS-CoV-2, other viruses, and microorganisms from sequencing data. Brief. Bioinform. **22**(2), 924–935 (2021)

15. Pickett, B.E., et al.: ViPR: an open bioinformatics database and analysis resource for virology research. Nucl. Acids Res. **40**(D1), D593–D598 (2012)

16. Khan, A., et al.: Detection of human papillomavirus in cases of head and neck squamous cell carcinoma by RNA-Seq and VirTect. Mol. Oncol. (13), 829–839 (2018)

17. Chen, X., et al.: A virome-wide clonal integration analysis platform for discovering cancer viral etiology. Genome Res. (2019)

18. Vilsker, M., et al.: Genome detective: an automated system for virus identification from high-throughput sequencing data. Bioinformatics **35**(5), 871–873 (2019)

19. Piro, V.C., Dadi, T.H., Seiler, E., Reinert, K., Renard, B.Y.: Ganon: precise metagenomics classification against large and up-to-date sets of reference sequences. Bioinformatics **36**, i12–i20 (2020)

20. Meyer, F., et al.: The metagenomics RAST server-a public resource for the automatic phylogenetic and functional analysis of metagenomes. BMC Bioinform. **9**(1), 1–8 (2008)

21. Huson, D.H., Auch, A.F., Qi, J., Schuster, S.C.: MEGAN analysis of metagenomic data. Genome Res. **17**(3), 377–386 (2007)

22. Brown, S.M., et al.: MGS-fast: metagenomic shotgun data fast annotation using microbial gene catalogs. GigaScience **8**(4), giz020 (2019)

23. Truong, D.T., et al.: MetaPhlAn2 for enhanced metagenomic taxonomic profiling. Nat. Methods **12**(10), 902–903 (2015)

24. Karlicki, M., Antonowicz, S., Karnkowska, A.: Tiara: deep learning-based classification system for eukaryotic sequences. Bioinformatics **38**(2), 344–350 (2022)

25. Lourenço, A.: Reconstruction and classification of unknown DNA sequences. Master dissertation (2021)

26. Almeida, J.R., Pinho, A.J., Oliveira, J.L., Fajarda, O., Pratas, D.: GTO: a toolkit to unify pipelines in genomic and proteomic research. SoftwareX **12**, 100535 (2020)

27. Kans, J.: Entrez direct: e-utilities on the UNIX command line. National Center for Biotechnology Information (US) (2020)

28. Pratas, D., Pinho, A.J.: On the approximation of the Kolmogorov complexity for DNA sequences. In: Alexandre, L.A., Salvador Sánchez, J., Rodrigues, J.M.F. (eds.) IbPRIA 2017. LNCS, vol. 10255, pp. 259–266. Springer, Cham (2017). https://doi.org/10.1007/978-3-319-58838-4_29

29. Silva, M., Pratas, D., Pinho, A.J.: Efficient DNA sequence compression with neural networks. GigaScience **9**(11), 11. giaa119 (2020)

30. Hosseini, M., Pratas, D., Pinho, A.J.: AC: a compression tool for amino acid sequences. Interdisc. Sci. Comput. Life Sci. **11**(1), 68–76 (2019)
31. Romiguier, J., Ranwez, V., Douzery, E.J.P., Galtier, N.: Contrasting GC-content dynamics across 33 mammalian genomes: relationship with life-history traits and chromosome sizes. Genome Res. **20**(8), 1001–1009 (2010)
32. Chen, H., Skylaris, C.-K.: Analysis of DNA interactions and GC content with energy decomposition in large-scale quantum mechanical calculations. Phys. Chem. Chem. Phys. **23**(14), 8891–8899, 102333 (2021)
33. Duret, L., Galtier, N.: Biased gene conversion and the evolution of mammalian genomic landscapes. Annu. Rev. Genomics Hum. Genet. **10**, 285–311 (2009)
34. Cristianini, N., Shawe-Taylor, J., et al.: An Introduction to Support Vector Machines and Other Kernel-Based Learning Methods. Cambridge University Press, Cambridge (2000)
35. Rish, I., et al.: An empirical study of the Naive Bayes classifier. In: IJCAI 2001 Workshop on Empirical Methods in Artificial Intelligence, vol. 3, pp. 41–46 (2001)
36. McLachlan, G.J.: Discriminant Analysis and Statistical Pattern Recognition, vol. 544. Wiley, New York (2004)
37. Guo, G., Wang, H., Bell, D., Bi, Y., Greer, K.: KNN model-based approach in classification. In: Meersman, R., Tari, Z., Schmidt, D.C. (eds.) OTM 2003. LNCS, vol. 2888, pp. 986–996. Springer, Heidelberg (2003). https://doi.org/10.1007/978-3-540-39964-3_62
38. Chen, T., Guestrin, C.: XGBoost: a scalable tree boosting system. In: Proceedings of the 22nd ACM SIGKDD International Conference on Knowledge Discovery and Data Mining, KDD 2016, pp. 785–794. ACM, New York (2016)
39. Lu, J., Salzberg, S.L.: Removing contaminants from databases of draft genomes. PLoS Comput. Biol. **14**(6), e1006277 (2018)
40. Cornet, L., Baurain, D.: Contamination detection in genomic data: more is not enough. Genome Biol. (2022)
41. Tavares, A.H.M.P., et al.: DNA word analysis based on the distribution of the distances between symmetric words. Sci. Rep. **7**(1), 1–11 (2017)

A First Approach to Image Transformation Sequence Retrieval

Enrique Mas-Candela, Antonio Ríos-Vila, and Jorge Calvo-Zaragoza[✉]

U.I. for Computer Research, University of Alicante, Alicante, Spain
emc89@alu.ua.es, {arios,jcalvo}@dlsi.ua.es

Abstract. Detecting the corresponding editions from just a pair of input-output images represents an interesting task for artificial intelligence. If the possible image transformations are known, the task can be easily solved by enumeration with brute force, yet this becomes an unfeasible solution for long sequences. There are several state-of-the-art approaches, mostly in the field of image forensics, which aim to detect those transformations; however, all related research is focused on detecting single transformations instead of a sequence of them. In this work, we present the Image Transformation Sequence Retrieval (ITSR) problem and describe a first attempt to solve it by considering existing technology. Our results demonstrate the huge difficulty of obtaining a good performance—being even worse than a random guess in some cases—and the necessity of developing specific solutions for ITSR.

Keywords: Computer vision · Image transformations · Deep learning

1 Introduction

Detecting image transformations between a pair of images is a task that can be useful in many applications. For example, to automate editing processes from a single example that has been done manually or the automatic detection of manipulations on images. However, such detection can be generally difficult, even for humans. Figure 1 depicts an example of sequential transformations to a given image.

The research field of image forensics is the one that has approached this challenge of detecting transformations on images. Most of the existing approaches in this regard use traditional computer vision algorithms for detecting specific transformations on images [6,10]. Bayar *et al.* [2] proposed a universal method for detecting different transformation on images. Similarly, Mazdumar *et al.* [11] presented a one-shot learning approach that is able to classify transformations from images not seen during the training stage. It works with just a reference

This paper is part of the project I+D+i PID2020-118447RA-I00, funded by MCIN/AEI/10.13039/501100011033. The second author is supported by grant ACIF/2021/356 from "Programa I+D+i de la Generalitat Valenciana".

A. J. Pinho et al. (Eds.): IbPRIA 2022, LNCS 13256, pp. 321–332, 2022.
https://doi.org/10.1007/978-3-031-04881-4_26

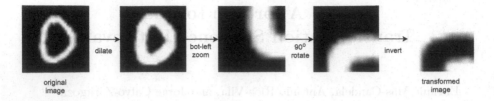

Fig. 1. Example of transformation sequence applied to an MNIST sample.

example for each possible transformation. Despite these previous efforts, all the presented approaches are focused on classification: they detect single transformations, or at most, a very constrained set of combinations. The Image Transformation Sequence Retrieval (ITSR) task, defined as decoding a multi-stage edition process represented by a sequence of transformations from input-output images, is yet to be explored.

As there is no previous research on this task, the main contribution of this work is to properly define the challenge of ITSR for the first time. This includes proposals of datasets and evaluation protocols. Then, we present an initial approach to serve as a baseline for future works. The approach adapts image-to-sequence technology, which outputs a text sequence from a given input image. Within the context of this work, it outputs a sequence of tokens that represents the ordered edition process from two (input and output) reference images.

We present results over several experimental scenarios regarding the transformations allowed, as well as the use of two datasets of heterogeneous characteristics. Our results demonstrate that the task is extremely complex for existing approaches. On some occasions, in fact, they do not outperform a random guess approach. However, the models do exhibit some ability to learn the challenge, especially when they must order the possible hypotheses according to their probability. In any case, the results point to a need for developing specific approaches to address ITSR with guarantees.

The rest of the paper is organized as follows: in Sect. 2 we formally define the problem of ITSR, and then describe a first approximation to address it. In Sect. 3, we describe the corpora, metrics, and details of the experiments designed to evaluate the proposed methodology. Results are reported in Sect. 4. Finally, in Sect. 5 we conclude the paper and present some ideas for future work.

2   Methodology

To the best of our knowledge, this is the first paper addressing the problem of ITSR. In this section, we formally define the challenge and then present a baseline approach to deal with it.

2.1   Problem Definition

Let $\mathcal{X}$ be an image space and $\mathcal{T}$ be a fixed set of possible image transformations in the form of $\mathcal{X} \to \mathcal{X}$. Let us denote by $\mathbf{t} = (t_1, t_2, \ldots, t_k)$ a sequence of k transformations, where $\mathbf{t} \in \mathcal{T}^k$. We define $\tau_\mathbf{t}(x)$ as a function that sequentially applies the sequence $\mathbf{t}$ to x; that is, $\tau_\mathbf{t}(x) = t_k(\cdots t_2(t_1(x)))$.

The problem of ITSR can be formally defined as follows: given a pair of images (x, x'), the goal is to retrieve the sequence $\mathbf{t}$ that satisfies $\tau_\mathbf{t}(x) = x'$. Note that the solution is not necessarily unique, as there might be different sequences that produce x' from x. Approaches to ITSR must estimate a function $M : \mathcal{X} \times \mathcal{X} \to \mathcal{T}^*$, that receives a pair (x, x') and maps them onto a sequence of transformations. Figure 2 graphically summarizes the formulation.

There might be many ways of approaching the estimation of M. Below, we describe our first attempts at this challenge employing deep neural networks.

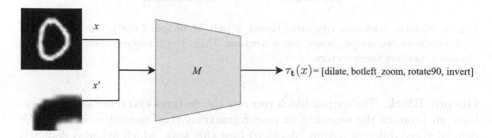

Fig. 2. A graphical overview of the problem of ITSR.

2.2   Baseline Approach

The proposed baselines to solve this formulation consist of neural approaches that map an image pair (x, x') into a sequence of transformations. Conceptually we separate the models in two parts (i) a *backbone*, which is in charge of extracting features of the input images, and (ii) an *output block* which classifies these features into the sequence of operations t that transforms x into x'.

Backbone. Our backbone is fed with an image pair and outputs a vector that represents the learned features of the transformations applied within this pair.

To implement the backbone, we consider a siamese Convolutional Neural Network (CNN) as it works well in similar image forensics tasks [9,11]. Siamese CNNs are a neural network architecture composed of two CNNs with shared weights. The output tensors of the CNNs are flattened and merged by subtraction, as it is an operation used in similar approaches [9] and we empirically found it performs better than other merge operations–like sum or concatenation. A graphical example of this implementation can be found in Fig. 3.

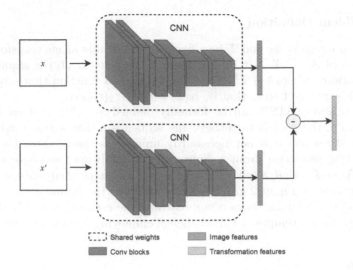

Fig. 3. Siamese backbone diagram. Given a pair of images (x, x'), our neural network produces two *image feature vector* vectors. Then, the *transformation features* are obtained merging these vectors.

Output Block. The output block receives the features extracted by the backbone to produce the sequence of transformations that convert x into x'. We defined three different output blocks to face this task, which are also depicted in Fig. 4:

- **Fully Connected** (FC) network with a classification output that predicts the next transformation in the sequence. For predicting a complete transformation sequence, on each step the predicted transformation is applied to the source image, and this new version is used to feed the model as source input in the next step. The sequence is completed when the transformed source image is equal to the target one (or a maximum number of transformations is reached). The block is conformed by two fully connected layers, each one with a ReLU activation function and finally followed by a classification layer.
- **Recurrent Neural Network** (RNN) which predicts the complete sequence in a single stage. Our RNN output block gets the transformation features as the initial hidden state and outputs a token representing the transformation in the sequence for each timestep, as it is done in the image captioning field [17]. We implement the RNN with a GRU layer followed by a classification block built with two fully connected layers. To avoid overfitting and predictions instability, we feed the layer with vectors with meaningless arbitrary values, instead of the previously predicted token.
- **RNN with attention.** To enhance the output block predictions, we included an attention mechanism to the RNN. The attention layer receives both the backbone output and the previous timesteps predictions t_{i-1} to generate an attention matrix that filters the information of the backbone output.

This filtered information with the initial state is then fed into the current timestep t_i GRU layer.

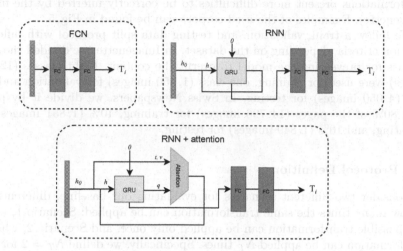

Fig. 4. Diagram of output blocks.

3 Experimental Setup

In this section, we describe the experimental setup designed to carry out the experimentation of this work. We explain the used corpora, the protocol followed during the experimentation process, the used metrics, and the implementation details of the proposed models to solve the problem.[1]

3.1 Corpus

We consider two different corpora in this work:

- **Swiss Newspapers** [1]. This corpus consists of images from three Swiss newspapers: the *Journal de Genève* (JDG, 1826-1994), the *Gazette de Lausanne* (GDL, 1804-1991), and *the Impartial* (IMP, 1881-2017). The dataset contains a total of 4,624 binarized newspaper pages.
- **Imagenette** [8], a subset of ImageNet [4]. This dataset consists of 13,394 images from the ten most easily classified categories from ImageNet (tench, English springer, cassette player, chain saw, church, French horn, garbage truck, gas pump, golf ball, and parachute).

These datasets allow us to evaluate our approach in two different graphic domains: (i) binarized images with homogeneous content where morphological

---

[1] For the sake of reproducibility, we release all the code related to these experiments in the following repository: https://github.com/emascandela/itsr.

operations should be easier to detect (Swiss Newspapers), and (ii) a more complex dataset with photographed images of heterogeneous content, where subtle transformations present more difficulties to be correctly inferred by the model (Imagenette). Examples of the used corpora can be found in Fig. 5.

We follow a train, validation, and testing data split protocol with different partition criteria, depending on the dataset. In Imagenette, we divided the data by class for preventing the model to overfit the content. Eight classes (12, 044 images) were used for training, one class (1, 350 images) for validation and one class (4, 050 images) for testing. In Swiss Newspapers, we divide it by pages using 80% of the pages (14, 796 images) for training, 10% (1, 851 images) for validating, and 10% (1, 849 images) for testing.

3.2   Protocol Definition

We consider two different scenarios for evaluating our baseline, differentiated by how many times the same transformation can be applied: Scenario 1, where each possible transformation can be applied only once, and Scenario 2, where a transformation can be applied N_T times. Specifically, we define $N_T = 2$ for Scenario 2. The transformations that are used in each scenario are listed in Table 1. Note that there are some transformations of Scenario 1 that are not present in Scenario 2. The reason is that the combination of some of these transformations can generate confusion with others of the set. For example, the result of two 90° rotations is exactly the same as a single 180° rotation. In both scenarios, the transformation sequences have a length in the range $[1, M_T]$ where M_T is the maximum number of transformations. The number of different transformations is 11 and 7 in Scenario 1 and 2, respectively.

As the possible transformations are known, the solution of the problem may seem trivial by enumeration (brute force), as we can simply apply all the possible sequence combinations to the original image and compare it with the target to check the correctness. However, the total number of possible solutions for each problem is $\sum_{i=1}^{M_T} i!$, which yields 6.7×10^9 in Scenario 1 and 4.0×10^6 in Scenario 2. These values demonstrate that the problem is unfeasible to be solved by brute force.

3.3   Metrics

As this problem can be only considered to be solved when the transformation sequence is predicted correctly in its whole, the performance of the algorithm should be measured with the accuracy of the sequence prediction. However, as the accuracy could be too low on long sequences and it may not provide enough information about the overall performance of the model, we consider some additional metrics: the *Translation Edit Rate* (TER) [14] and the *Top-K* accuracy.

Table 1. List of all possible transformations and the scenario where they are used. Note that in Scenario 1, each transformation is only applied once, while transformations can be repeated up to twice in Scenario 2. Transformations are the following: Gaussian Blur (GB), Erode (E), Dilate (D), Zoom Top-Left (ZTL), Zoom Bottom-Left (ZBL), Zoom Top-Right (ZTR), Zoom Bottom-Right (ZBR), Horizontal Flip (HF), Vertical Flip (VF), Rotate 90° (R90) and Rotate 180° (R180).

	GB	E	D	ZTL	ZBL	ZTR	ZBR	HF	VF	R90	R180
Scenario 1	x	x	x	x	x	x	x	x	x	x	x
Scenario 2	x	x	x	x	x	x	x				

On the one hand, the TER measures the ratio of editing operations that have to be performed to a predicted output so that it matches the ground truth. These operations include deletions, insertions, substitutions, and shifts—the movement of a part of the sequence that is correctly predicted but in a wrong timestep. Although this metric is mostly used in machine translation, it fits with the nature of our problem; as these sequence fixing operations can be applied to both cases.

On the other hand, the Top-K accuracy measures the ratio of correct predictions within the K most probable solutions. As all hypotheses from the models could be evaluated by checking whether the transformation sequence is correct, this metric estimates how many attempts are needed to find the solution in all cases.

3.4   Implementation Details

For our experiments, we used the ResNet34 [7] architecture without the classification layer for the feature extractor in the backbone block, as it has been demonstrated to work well in many computer vision tasks. We evaluate the performance of our approach with the output blocks defined in Sect. 2. We initialized the weights of the feature extractor with the ones of the ImageNet pre-trained model.

Models are trained through a Stochastic Gradient Descent (SGD) [12] optimizer with a momentum [13] of 0.9. Input sizes of the backbone have a 224×224 size with 3 RGB channels normalized to the range $[-1, 1]$. The images of the Imagenette dataset are directly resized to this size, while the ones from Swiss Newspapers are cropped, as the pages are too big to be directly used.

The images of the datasets described in Sect. 3.1 were used as the original images. The transformation sequences are randomly generated and then applied to the original image for generating the target one. For evaluation and validation, a single transformed image is generated for each original sample while, during training, different transformed images are generated for each epoch.

(a) Erode, zoom top-right, zoom bot-left, v.flip, 90° rotation, blur.

(b) Erode, blur, 180° rotation, dilate.

(c) Erode, zoom bot-left, 90° rotation, zoom bot-right, blur, dilate, 180° rotation.

(d) 90° rotation, zoom bot-right, h.flip, dilate, 180° rotation, zoom top-left, zoom top-right.

Fig. 5. Examples of image pairs (x, x') and its ground-truth labels **t** on both datasets: Swiss Newspapers on the left and Imagenette on the right.

4   Results

In this section, we present and discuss the results obtained in our experiments. As there is no baseline to compare, we established a Random Guess (RG) solution to assess the feasibility of solving the problem via the proposed approaches. This RG approach just generates a sequence of transformations by random choice (all possible sequences are equiprobable).

Table 2 shows the TER (%) obtained in these experiments. According to the results, the considered approaches can be considered unsatisfactory for ITSR. This is reflected in the fact that all TER figures are above 100%, which means that the models do not calibrate the number of operations well and end up suggesting more transformations than necessary (regardless of whether some of them are correct).

Concerning the specific models, the RNN without attention shows the most robust operation, being better than the RG in most cases (except for Imagenette on Scenario 2). Both the FC and the RNN with attention perform worse than the RG, in general, yet the latter one shows a fair performance on Scenario 2.

We observe that the results in Scenario 1 are better than in Scenario 2, although it seemed a harder challenge. From this result, we can claim that the models find it easier to predict the correct transformations when the set of possible ones is smaller, in spite of having longer sequence transformations.

Concerning the datasets, the error on Swiss Newspapers is slightly lower than the ones on Imagenette, possibly due to the simplicity of the content of the images.

Table 2. Results in terms of TER (%) from the different scenarios (Scenario 1 and Scenario 2) and datasets (SN: Swiss Newspapers, IMG: Imagenette). Note that the Random Guess (RG) is not conditioned to the specific dataset.

Approach	Scenario 1		Scenario 2	
	SN	IMG	SN	IMG
FC	204	223	210	203
RNN	162	166	195	214
RNN + Attention	215	213	129	196
RG	204		202	

We also evaluated the performance of the approaches on sequences of different lengths. Figure 6 provides the TER obtained in the evaluation of the RNN model—which we consider that gets the best general results on the previous experimentation—in both scenarios. The results from RG are also reported as a reference. We can clearly see that the RNN rapidly degrades as the number of transformations increases. For the shortest sequences, the result is much better than RG, not being the case for transformation sequences of 5 or 6 (depending on the dataset).

As previously mentioned, instead of asking the model for the correct solution directly, an alternative is to ask for a list of sequence transformation hypotheses ordered by probability. This is interesting because it would be possible to verify the correctness of each hypothesis until finding the solution. Table 3 provides the results in terms of TopK accuracy, for increasing values of K. Given the high cost of this experiment, we only provide the results of the RNN model, as it is the one that reported the most robust performance according to the TER. In this case, to compute the results of the RG, we analytically compute the probability of random sampling the correct sequence and multiply it by K.

From the results, we can clearly observe that the RNN does not achieve a fair performance from this perspective either. It is very unlikely to find the correct sequence in its first hypothesis (0% in all cases except for Imagenette on Scenario 1, with a 0.7%). However, the model does help to find the correct sequence more smartly, as the accuracy is generally much higher than the RG for this purpose, especially on Scenario 1. On Scenario 2, however, the RG moves closer to the results of the RNN for the highest values of K. This might be explained by the fact that Scenario 2 has a fewer number of solutions than Scenario 1, and so the probability of sampling the correct sequence is higher for the same K. As an example of the best possible performance, the accuracy of including the correct sequence in the 100,000 best hypotheses goes up to 15% for Imagenette on Scenario 1, which at least indicates that the RNN is able to partially learn the challenge under favorable conditions.

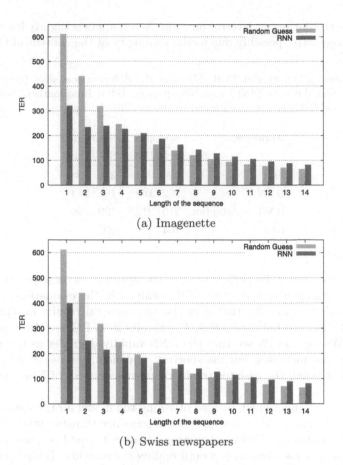

(a) Imagenette

(b) Swiss newspapers

Fig. 6. Results in terms of TER (%) with respect to the length of the actual transformation sequence.

Table 3. TopK accuracy of the RNN model for different values of K. For each value of K, the results of the model in both scenarios (Scenario 1 and Scenario 2) and both datasets (SN: Swiss Newspapers, IMG: Imagenette) are shown and compared with the Random Guess (RG) results.

K	Scenario 1			Scenario 2		
	SN	IMG	RG	SN	IMG	RG
1	0.0000	0.0079	0.0000	0.0000	0.0000	0.0000
10	0.0005	0.0338	0.0000	0.0000	0.0010	0.0000
100	0.0005	0.0622	0.0000	0.0016	0.0025	0.0000
1,000	0.0032	0.0874	0.0000	0.0059	0.0059	0.0002
10,000	0.0076	0.1200	0.0000	0.0124	0.0160	0.0025
100,000	0.0216	0.1568	0.0000	0.0222	0.0319	0.0248

5 Conclusions and Future Work

Detection of image transformations has been a center of attention for the scientific community through the years. Different approaches try to solve this problem; however, these usually focus on detecting a single transformation on an image. In this work, we take a step forward and reformulate the topic: we define a new problem—the Image Transformation Sequence Retrieval (ITSR)—which consists in the detection of sequences of transformations instead of a single one. In addition, we study a first approach for solving the challenge, which can be used as a baseline in future works.

The presented results from our experimentation showcase the difficulty of ITSR. The considered baselines are largely insufficient to deal with the challenge, even reporting a worse performance than a random strategy in some cases. For short sequences, the baselines manage to correctly retrieve some sequences, but they completely fail with longer ones. According to these results, we may conclude that the problem cannot be directly solved using standard architectures of other image-to-sequence problems, which opens the doors to novel research avenues for developing specific approaches to ITSR.

A possible way for improving the performance of the current approach could be using curriculum learning [3]: where the model starts learning short sequences and incrementally learns to predict longer ones, as they have been proved to be harder to retrieve correctly. Furthermore, using the attention mechanism more effectively, like applying it before the merging of the backbone outputs, might help to extract more specific features about the transformations on the images and become of better service to detect them. Other attention-based approaches like using a Transformer [16] decoder as an output block, or even using a Vision Transformer [5] as a backbone structure to extract image edition features.

It is also important to remark that, in this particular problem, we can easily check if a solution proposed by a model is correct or not, because we can apply the sequence of transformations retrieved to the input image x and compare it with the target x'. We are not using this information in the proposed baseline; however, it can be relevant at the time of designing a new approach, as it can be integrated into the system to help in some way the prediction of the sequence. A possible strategy to include this information is by using search techniques supported with machine learning; for example, using model-based reinforcement learning [15], such as the Monte Carlo Tree Search algorithm.

References

1. Barman, R., Ehrmann, M., Clematide, S., Oliveira, S.A., Kaplan, F.: Combining visual and textual features for semantic segmentation of historical newspapers. J. Data Min. Digit. Human. HistoInformatics (2021)
2. Bayar, B., Stamm, M.C.: Constrained convolutional neural networks: a new approach towards general purpose image manipulation detection. IEEE Trans. Inf. Forensics Secur. **13**(11), 2691–2706 (2018)

3. Bengio, Y., Louradour, J., Collobert, R., Weston, J.: Curriculum learning. In: Proceedings of the 26th Annual International Conference on Machine Learning, pp. 41–48. Association for Computing Machinery, New York (2009)
4. Deng, J., Dong, W., Socher, R., Li, L.J., Li, K., Fei-Fei, L.: ImageNet: a large-scale hierarchical image database. In: 2009 IEEE Conference on Computer Vision and Pattern Recognition, pp. 248–255. IEEE (2009)
5. Dosovitskiy, A., et al.: An image is worth 16x16 words: transformers for image recognition at scale. CoRR abs/2010.11929 (2020)
6. Feng, X., Cox, I.J., Doerr, G.: Normalized energy density-based forensic detection of resampled images. IEEE Trans. Multimedia 14(3), 536–545 (2012)
7. He, K., Zhang, X., Ren, S., Sun, J.: Deep residual learning for image recognition. In: 2016 IEEE Conference on Computer Vision and Pattern Recognition (CVPR), pp. 770–778 (2016)
8. Howard, J.: ImageNette. https://github.com/fastai/imagenette/
9. Hu, B., Zhou, N., Zhou, Q., Wang, X., Liu, W.: DiffNet: a learning to compare deep network for product recognition. IEEE Access 8, 19336–19344 (2020)
10. Kang, X., Stamm, M.C., Peng, A., Liu, K.J.R.: Robust median filtering forensics using an autoregressive model. IEEE Trans. Inf. Forensics Secur. 8(9), 1456–1468 (2013)
11. Mazumdar, A., Bora, P.K.: Siamese convolutional neural network-based approach towards universal image forensics. IET Image Process. 14(13), 3105–3116 (2020)
12. Robbins, H.E.: A stochastic approximation method. Ann. Math. Stat. 22, 400–407 (2007)
13. Rumelhart, D.E., Hinton, G.E., Williams, R.J.: Learning representations by back-propagating errors. Nature 323(6088), 533–536 (1986)
14. Snover, M., Dorr, B., Schwartz, R., Micciulla, L., Makhoul, J.: A study of translation edit rate with targeted human annotation. In: Proceedings of the 7th Conference of the Association for Machine Translation in the Americas: Technical Papers, pp. 223–231. Association for Machine Translation in the Americas, Cambridge, 8–12 August 2006
15. Sutton, R.S., Barto, A.G.: Reinforcement Learning: An Introduction. MIT Press, Cambridge (2018)
16. Vaswani, A., et al.: Attention is all you need. In: Advances in Neural Information Processing Systems. vol. 30, pp. 5998–6008. Curran Associates, Inc. (2017)
17. Xu, K., et al.: Show, attend and tell: neural image caption generation with visual attention. In: Bach, F., Blei, D. (eds.) Proceedings of the 32nd International Conference on Machine Learning. Proceedings of Machine Learning Research, vol. 37, pp. 2048–2057. PMLR, Lille, 07–09 July 2015

Discriminative Learning of Two-Dimensional Probabilistic Context-Free Grammars for Mathematical Expression Recognition and Retrieval

Ernesto Noya, José Miguel Benedí, Joan Andreu Sánchez(✉), and Dan Anitei

Pattern Recognition and Human Language Technologies Research Center,
Universitat Politècnica València, 46022 Valencia, Spain
{noya.ernesto,jmbenedi,jandreu,danitei}@prhlt.upv.es

Abstract. We present a discriminative learning algorithm for the estimation of two-dimensional Probabilistic Context-Free Grammars in the context of Mathematical Expressions Recognition and Retrieval. This algorithm is based on a generalization of the H-criterion, as the objective function, and the growth transformations as the optimization method. In addition, experimental results are reported on the *Im2Latex-100k* dataset, studying the performance of the estimated models depending on the length of the mathematical expressions and the number of admissible errors in the metric used.

Keywords: Discriminative learning · Probabilistic Context-Free Grammars · Mathematical expression retrieval · Probabilistic indexing

1 Introduction

Syntactic models have demonstrated to be a very important formalism for Pattern Recognition since they introduce effective restrictions in the solution search space. Thus, finite-state language models provide a prior probability in many current applications, like Automatic Speech Recognition (ASR), Machine Translation (MT) and Handwritten Text Recognition (HTR), that makes the decoding problem easier in real time. A noticeable characteristic of syntactic models is that they can provide a sorted set of alternative solutions for the same input. The computation of N-best solutions for stochastic finite-state models [14] and Probabilistic Context-Free Grammars (PCFG) has been studied in the past [11]. These N-best solutions can be represented as a word-graph [17] or a hyperforest [16] that is able to generalize and provide alternative hypotheses not previously included in the N-best solutions.

Word-graphs obtained from stochastic finite-state models are important representations that are used in ASR, MT and HTR since they can be used for

A. J. Pinho et al. (Eds.): IbPRIA 2022, LNCS 13256, pp. 333–347, 2022.
https://doi.org/10.1007/978-3-031-04881-4_27

obtaining confidence measures at word level or sentence level. Hyper-forest computed from PCFG can be used for the same purpose and they have been used in the past for interactive parsing [20] and Mathematical Expression Recognition [16]. The major problem related to these models is that they have to be trained from samples. Preparing training data is currently a bottle-neck for any kind of Machine Learning-based formalism. This training process is usually performed by optimizing a goal function and using some statistical optimization framework. It is of paramount importance to take profit as much as possible of data in the case of limited amount of training samples. Training PCFG has been researched in the past by using the maximum-likelihood criterion and optimizing this function by using growth transformations [5].

PCFG are a powerful formalism for parsing Mathematical Expressions (ME). These models are appropriate for capturing long-term dependencies between the different elements in a ME. In this paper, we consider a 2-dimensional extension of PCFG (2D-PCFG) [1,3] that will allow us to deal with the ambiguity associated with parsing ME. There are not many training datasets for training 2D-PCFG [8]. Recently, a new large dataset has been made publicly available that can be used for training a 2D-PCFG [2]. Even with these datasets, the amount of training data is not enough to represent the large variability that can exist in ME. Therefore, more efficient methods have to be devised that take more profit of the training data. This paper researches the use of discriminative techniques for the probabilistic estimation of 2D-SCFG [10,13]. In this approach the correct interpretation of the ME and the incorrect interpretation are combined in a discriminative way [22] in contrast to generative methods where incorrect interpretations are not explicitly used.

Current results on ME recognition are far from being perfect to searching purposes in large image datasets, and having accurate enough results may be impossible given the intrinsic ambiguity that are present in ME. Therefore more flexible approaches for searching ME have to be devised. Similar problems have been researched for HTR [21] and the adopted solution is based on getting an adaptive list of hypotheses for each word image that are obtained from word-graphs. This solution is based on Probabilistic Indexing (PrIx). We intend to follow a similar PrIx research approach for ME searching. In this sense, it is relevant to take into account the type of queries that users could make for searching ME. The concept of "word" is not defined for ME and we assume in this paper that users may be interested in searching not only full ME but also sub-expressions contained in large ME. Therefore, it is important to evaluate the recognition system not only at whole ME level but also at sub-expression level. The experiments and evaluation methods reported in Sec. 4 are stated with the PrIx approach in mind.

2   Problem Formulation and Notation

The input domain in the printed mathematical expression recognition and retrieval is the set of images or regions of an image that can contain a ME.

Given an input image, the first step is to determine a possible representation function that maps the image to another representation more suitable for solving the problem. Due to the nature of the input printed images, the selected representation function consists of the extraction of the connected components, $x = \{x_1, \ldots, x_{|x|}\}$, from the input image. Figure 1 shows an input ME and its representation in terms of connected components.

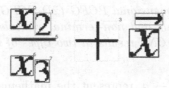

Fig. 1. Input image example for math expression $\frac{x_2}{x_3} + \vec{x}$ and its set of associated connected components.

Given the well-known structure of ME, many approaches are based on grammar models because they constitute a natural way to model this kind of problem. The fundamentals of ME recognition and retrieval based on probabilistic grammar models were proposed in [7,19] and further developed in [1], among other works. We pose the ME recognition and retrieval as a structural parsing problem, such that the main goal is to obtain the set of symbols and the structure that defines the relationships among them from an input image. Formally, let x be a set of connected components from an input ME to be recognized or as a search query. The aim is to obtain the most likely sequence of mathematical symbols $s \in S$ related among them according to the most likely syntactic parse $t \in T$ that accounts for x, where S is the set of all possible sequences of (pre-terminals) symbols and T represents the set of all possible syntactic parses, such that $s = \text{yield}(t)$. This can be approximated as follows:

$$(\hat{t}, \hat{s}) \approx \underset{\substack{t \in T, s \in S \\ s = \text{yield}(t)}}{\arg\max} \; p(t, s \mid x) \approx \underset{\substack{t \in T, s \in S \\ s = \text{yield}(t)}}{\arg\max} \; p(s \mid x) \cdot p(t \mid s) \qquad (1)$$

where $p(s \mid x)$ represents the observation (symbol) likelihood and $p(t \mid s)$ represents the structural probability. We consider Eq. (1) as a holistic search problem, where symbol segmentation, symbol recognition, and the structural analysis of the input expression are globally achieved [1].

In this paper, we will focus on the parsing problem associated with the computation of the structural probability $p(t \mid s)$ and especially on the matter of estimating the grammatical models used to tackle Eq. (1). We first introduce the notation used in this work.

Definition 1. *A Context-Free Grammar (CFG), $G = (\mathcal{N}, \Sigma, S, \mathcal{P})$, is a tuple where $\mathcal{N}$ is a finite set of non-terminal symbols, Σ is a finite set of terminal symbols ($\mathcal{N} \cap \Sigma = \emptyset$), $S \in \mathcal{N}$ is the start symbol of the grammar, and $\mathcal{P}$ is a finite set of rules: $A \to \alpha, \quad A \in \mathcal{N}$ and $\alpha \in (\mathcal{N} \cup \Sigma)^+$.*

A CFG in Chomsky Normal Form (CNF) is a CFG in which the rules are of the form $A \to BC$ or $A \to a$, where $A, B, C \in \mathcal{N}$ and $a \in \Sigma$.

Definition 2. *A Probabilistic CFG (PCFG) is defined as a pair (G, p), where G is a CFG and $p : P \to]0, 1]$ is a probability function of rule application such that $\forall A \in \mathcal{N} : \sum_{i=1}^{n_A} p(A \to \alpha_i) = 1$, where n_A is the number of rules associated with non-terminal symbol A.*

Definition 3. *A Two-Dimensional PCFG (2D-PCFG), $\mathbb{G}$, is a generalization of a PCFG, where terminal and non-terminal symbols describe bi-dimensional regions. This grammar in CNF results in two types of rules: terminal rules and binary rules.*

The terminal rules, $A \to a$, represent the mathematical symbols which are ultimately the terminal symbols of 2D-PCFG. Second, the binary rules, $A \xrightarrow{r} BC$, have an additional parameter, r, that represents a given spatial relationship and its interpretation is that A is a solution of the subproblems associated with B and C regions compatible with the spatial relationship r. In this work, we consider six spatial relationships: right, below, subscript, superscript, inside, and mroot [1].

Let $\mathbb{G}$ be a 2D-PCFG and let x a set of input connected components, we denote $\mathcal{T}_x$ as the set of all possible parse trees for x. The expression $N(A \to \alpha, t_x)$ represents the number of times that the rule $A \to \alpha$ has been used in the parse tree $t_x \in \mathcal{T}_x$, and $N(A, t_x)$ is the number of times that the non-terminal A has been used in t_x. Obviously, it is satisfied that $N(A, t_x) = \sum_{i=1}^{n_A} N(A \to \alpha_i, t_x)$. With all that, we define the following expressions:

– *Probability of the parse tree t_x of x,*

$$P(x, t_x) = \prod_{\forall (A \to \alpha) \in \mathcal{P}} p(A \to \alpha)^{N(A \to \alpha, t_x)}.$$

– *Probability of x,*

$$P(x) = \sum_{\forall t_x \in \mathcal{T}_x} P(x, t_x). \tag{2}$$

– *Probability of the best parse tree of x,*

$$\widehat{P}(x) = \max_{\forall t_x \in \mathcal{T}_x} P(x, t_x). \tag{3}$$

– *Best parse tree of x,*

$$\widehat{t}_x = \arg \max_{\forall t_x \in \mathcal{T}_x} P(x, tr_x).$$

The expressions (2) and (3) can be calculated respectively using modified versions of the well-known *Inside* [12] and *Viterbi* [15] algorithms para 2D-PCFG [16]. We can also calculate the n-best parse trees for 2D-PCFG [16]. Furthermore, given $\Delta_x \subseteq \mathcal{T}_x$, a finite subset of derivations of x, we can also define:

- Probability of x with respect to Δ_x,

$$P(x, \Delta_x) = \sum_{\forall t_x \in \Delta_x} P(x, t_x). \tag{4}$$

- Probability of the best parse tree of x with respect to Δ_x,

$$\widehat{P}(x, \Delta_x) = \max_{\forall t_x \in \Delta_x} P(x, t_x). \tag{5}$$

These expressions respectively coincide with expression (2) and (3) when $\Delta_x = \mathcal{T}_x$.

3 Discriminative Learning of 2D-PCFGs

Given a representative training sample Ω, and given a 2D-PCFG, $\mathbb{G}$, defined by its set of parameters θ, the problem of estimating the parameters θ of $\mathbb{G}$ can state as follow:

$$\widehat{\theta} = \arg\max_{\theta} f_\theta(\Omega),$$

where $f_\theta(.)$ is the *objective function* to be optimized. Two issues have to be considered: the optimization method and the selection of the objective function. In this paper, we consider an optimization method, based on the *growth transformation* (GT) framework [4,6], and an objective function derived from a generalization of the H-criterion [9,13].

3.1 H-criterion

The H-criterion-based learning framework was proposed by Gopalakrishnan et al. in [9] as a generalization of the estimators of maximum likelihood (ML), maximum mutual information (MMI), and conditional maximum likelihood (CML).

Let $\Omega = \{(x_i, y_i)\}_{i=1}^N$ be the training sample, where x_i are the input observations and y_i are the reference interpretations. And let θ be the parameters of the model to be estimated. An H-estimator, $\widehat{\theta}(a, b, c)$, can be obtained by minimizing the H-criterion.

$$H_{a,b,c}(\theta; \Omega) = -\frac{1}{n} \sum_{i=1}^{n} \log p_\theta^a(x_i, y_i) \, p_\theta^b(x_i) \, p_\theta^c(y_i), \tag{6}$$

where a, b, c are constants with $a > 0$. Therefore, the ML estimator can be represented by $\widehat{\theta}(1, 0, 0)$, the MMI estimator by $\widehat{\theta}(1, -1, -1)$, and the CML estimator by $\widehat{\theta}(1, 0, -1)$ [9].

3.2 Generalized H-criterion for 2D-PCFGs

Here we will propose a discriminative learning method to estimate the parameters (probabilities of the rules) of a 2D-PCFG through a generalization of the H-criterion [9,13]. Given a 2D-PCFG, $\mathbb{G}$ (Definition 3), a training sample Ω, and a set of parse trees, Δ_x ($\forall x \in \Omega$), the estimation of the parameters of $\mathbb{G}$ can be obtained through the generalized H-criterion minimizing the following estimator (see (6)),

$$H_{1,-h,0}(\mathcal{G},\Omega) = -\frac{1}{|\Omega|} \log \widetilde{F}_h(\mathbb{G},\Omega) = -\frac{1}{|\Omega|} \log \prod_{x \in \Omega} \frac{P^\eta(x,\Delta_x^r)}{P(x,\Delta_x^c)^h}, \quad (7)$$

where $0 < \eta$, $0 \leq h < 1$ and $\Delta_x^r \subset \Delta_x^c$. The set Δ_x^r must contain only correct parse trees of the sentence x, while the set Δ_x^c must contain any competing parse trees of the sentence x. If $h > 0$ the generalized H-criterion can be viewed as a discriminative learning method. The exponent h aims to establish the degree to which the competing parse trees discriminate against the correct parse trees. An optimization of generalized H-criterion attempts simultaneously to maximize the numerator term $P^\eta(x,\Delta_x^r)$ and to minimize denominator term $P(x,\Delta_x^c)^h$ for each observation $x \in \Omega$ of the training sample.

3.3 Growth Transformations for Generalized H-criterion

The objective function, obtained from the generalization of the H-criterion (7), can be optimized using growth transformations for rational functions. Since $\widetilde{F}_h$ is a rational function, the reduction of the case of rational functions to polynomial functions proposed in [10] can be applied.

$$Q_\pi(\mathbb{G},\Omega) = \prod_{x \in \Omega} P^\eta(x,\Delta_x^r) - \left(\widetilde{F}_h(\mathbb{G},\Omega)\right)_\pi \prod_{x \in \Omega} P(x,\Delta_x^c)^h. \quad (8)$$

$\left(\widetilde{F}_h(\mathbb{G},\Omega)\right)_\pi$ is the constant that results from evaluating $\widetilde{F}_h(\mathbb{G},\Omega)$ at the point π [10]. Where π is a point of the domain (in our case, π will be the probabilities of the rules of 2D-PCFG). The complete development can be found in [13], and the final expressions are as follows,

$$\bar{p}(A \rightarrow \alpha) = \frac{D_{A \rightarrow \alpha}(\Delta_x^r) - h\,D_{A \rightarrow \alpha}(\Delta_x^c) + p(A \rightarrow \alpha)\,\widetilde{C}}{D_A(\Delta_x^r) - h\,D_A(\Delta_x^c) + \widetilde{C}} \quad (9)$$

where,

$$D_{A \rightarrow \alpha}(\Delta_x) = \sum_{x \in \Omega} \frac{1}{P^\eta(x,\Delta_x)} \sum_{t_x \in \Delta_x} N(A \rightarrow \alpha, t_x)\,P^\eta(x,t_x) \quad (10)$$

$$D_A(\Delta_x = \sum_{x \in \Omega} \frac{1}{P^\eta(x,\Delta_x)} \sum_{t_x \in \Delta_x} N(A,t_x)\,P^\eta(x,t_x) \quad (11)$$

Following Gopalakrishnan et al. in [10], to obtain a fast convergence, the constant $\widetilde{C}$ should be calculated using the approximation,

$$\widetilde{C} = \max\left\{ \max_{p(A \rightarrow \alpha)} \left\{ -\frac{[D_{A \rightarrow \alpha}(\Delta_x^r) - h\,D_{A \rightarrow \alpha}(\Delta_x^c)]}{p(A \rightarrow \alpha)} \right\}_\pi, 0 \right\} + \epsilon$$

where ϵ is a small positive constant.

3.4 Discriminative Algorithms Based on Generalized H-criterion

From transformations (9), (10), and (11), a family of discriminative learning algorithms for 2D-PCFGs can be defined. This family of algorithms depends on how the respective sets of correct trees, Δ_x^r, and competing trees, Δ_x^c, are obtained and what are the values of the parameters η and h. For the algorithms studied here, we only analyze the effect of h over the optimization framework and fix $\eta = 1$.

The first issue to address is how the set of competing trees can be obtained. In this paper, Δ_x^c will be the set of n-best parse trees, calculated from an algorithm proposed in [11] and adapted to the 2D-PCDGs in [16]. The second issue to consider is how the set of correct parse trees, Δ_x^r, is obtained. In any case, it must be satisfied that $\Delta_x^r \subset \Delta_x^c$. In this paper, we propose to obtain the set of correct trees, Δ_x^r, from the set of the n-best parse trees by considering only the parse trees compatible with the ground truth. Below we will show the experiments carried out with the estimation algorithm, derived from the implementation of Eqs. (9), (10), and (11), of discriminative learning of 2D-PCFGs based on generalized H-criterion.

4 Experimentation

The empirical research was carried out to assess the effectiveness of the discriminative learning of 2D-PCFG and their possible application in PrIx tasks.

4.1 Datasets and Assessment Measures

All the experiments undertaken in this paper were performed with the *Im2Latex-100k* dataset [8]. This dataset consists of approximately 100 000 LATEX formulas, parsed from published articles aggregated in the KDD cup datasets.[1] In LATEX, the same rendered ME can have more than one different way of writing it. In order to reduce the structural ambiguity, the authors of the *im2latex* dataset suggested a normalization pre-processing of the data. In this pre-process, the order of sub/superscripts is fixed and matrices are converted to arrays.

Besides the normalization steps suggested in [8], we have developed two additional new normalization and filtering processes. In the first filtering process (STEP-1 in the Table 1), we implemented a parser that converts the LATEX markup to an abstract syntax tree (AST) for normalization purposes. This was done by expanding the code of the LuaTeX package *nodetree*,[2] that traverses and visualizes the structure of node lists as parsed by the LuaTeX engine.[3] From this resulting AST, we have performed some additional normalization steps. First, we disregarded the optional groupings defined by users to improve the readability of their LATEX code. Second, we normalized the font representations of the ME

---

[1] KDD cup: https://kdd.org/kdd-cup.

[2] Nodetree: https://ctan.org/pkg/nodetree?lang=en.

[3] LuaTeX: http://www.luatex.org/documentation.html.

symbols by mapping the glyphs to be rendered to LaTeX commands/symbols, in a many-to-one relation. Finally, the different user-defined horizontal spacing commands were mapped to the \hspace command.

The second filtering process (STEP-2 in the Table 1) is related to the fact that our original 2D-PCFG is not able to process all the expressions in the *Im2Latex-100k* dataset. We ignored all MEs from the training and validation sets that can not be parsed by the 2D-PCFG. This filter is not applied to the test set. Table 1 shows the number of samples of the training, validation, and test sets for the original partitions proposed in [8], and after processing STEP-1 and STEP-2. From hereinafter, all the experiments undertaken in this paper were performed with the extended normalization and filtering process we suggest.

Table 1. Number of samples in each partition. Numbers in bold face show the amount of data that was used in this paper.

	Train	Validation	Test
Original partitions [8]	83 883	9 319	10 354
STEP-1	71 099	7 908	**9 596**
STEP-2	**50 110**	**5 515**	–

Next, we analyze the average length of the MEs in the training set. Figure 2 shows the histogram indicating the number of MEs in the different length (number of connected components or symbols) intervals. As it can be seen in this histogram, there are very few expressions with more than 70 symbols in the training set. This same behavior was also observed in the validation set.

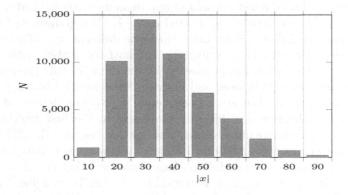

Fig. 2. Histogram representing the number of MEs (N) of the training set as a function of different number of LaTeX symbols $|x|$ of the MEs.

For the evaluation of the impact of the estimation algorithms on 2D-PCFG and the viability of our proposal, we consider three different metrics:

- **Exact Accuracy** (*ExAcc*): this metric measures the number of MEs in which the generated hypothesis (1-best or n-best) is an exact match with the ground truth (LaTeX expression) of the dataset. Note that this metric is very pessimistic given the average size of the EMs (see Fig. 2).
- **Bleu**: *Bleu* score [18] measures the difference between the best model prediction and the reference in terms of n-grams precision. This metric measures how far the predictions are from the references when the model does not generate an exact match. This measure may be very relevant in PrIx because it measures the precision at sub-expression level.[4]
- **Levenshtein Distance** (*LevD*): This distance measures the number of insertions, deletions, and substitutions required to match a hypothesis to the reference.

Note that *Bleu* and *LevD* are used in this paper because in the PrIx context it is very relevant to evaluate at sub-expression level as we mentioned in Sect. 1.

4.2 Estimation of 2D-PCFG and Parameter Setting

We start from a first model (2D-PCFG) obtained from the SESHAT system [1].[5] Considering the special features of our *Im2Latex-100k* dataset, we have extended this first model to include all symbols and relations appearing in the training set (see Sect. 4.1). The resulting model will be our initial baseline model ($\mathbb{G}_i$).

Next, we estimate our 2D-PCFG using the discriminative learning algorithm based on the generalized H-criterion. As discussed in Sect. 3, this estimation algorithm implements Eqs. (9), (10), and (11). To do this, the first step is to describe how to calculate the set of correct parse trees Δ_x^r, and the set of competing parse trees Δ_x^c. To obtain Δ_x^c, we use a new version of the N-best parsing algorithm of 2D-PCFG that is described in [16]. The experiments reported in this research have been carried out for $N = 50$. To get Δ_x^r, we develop a forced version of the parsing algorithm that uses the GT expression and the ME image to generate the most likely reference parse tree using our 2D-PCFG. The estimated model will be our final model ($\mathbb{G}_d$).

The first experiment analyzes the convergence of the estimation algorithm on the validation set. To do this, we compute the *Bleu* and *ExAcc* scores for each one of the iterations in the estimation algorithm's convergence process. Figure 3 plots the evolution of the *Bleu* and *ExAcc* scores for different models obtained in each iteration. In each of the plot of Fig. 3, the results of *Bleu* and *ExAcc* are respectively represented, considering for their calculation a set of N-best hypotheses ($\{1, 5, 10, 25,$ and $50\}$-best) generated by final estimated 2D-PCFG. We report the results for N-best hypotheses because this is relevant in the PrIx context: the 1-best solution could not be the correct solution but it could include

---

[4] It is usual in MT to compute the *Bleu* up to 4-grams when dealing with words. We consider an open problem to be researched in future to decide the appropriate n of the *Bleu* score for ME evaluation because of the misconception of "word" in ME recognition.

[5] https://github.com/falvaro/seshat.

in one of the following N-best hypotheses. As can be seen, the general results improve significantly from the 5-best hypotheses on. This means that obtaining at least the 5-best hypotheses would good enough to compose a hyper-forest for further use in PrIx.

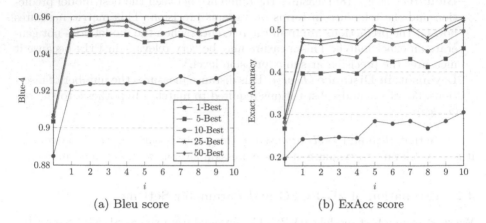

(a) Bleu score   (b) ExAcc score

Fig. 3. *Bleu* and *ExAcc* scores for each one of the iterations in the estimation process.

4.3   Experiments Depend on the Length of MEs

To further analyze the effect of expression length on the model outputs, we calculate *ExAcc* and *Bleu* scores for ME according to their size and Figs. 4 and 5 show the results. We also compare in both figures the performance of the discriminatively estimated model ($\mathbb{G}_d$) with our initial model ($\mathbb{G}_i$). In addition, these results have been obtained with the 1-best hypothesis and with the 50-best hypotheses. As can be seen, the results of the estimated model ($\mathbb{G}_d$) are consistently better than those of the initial model ($\mathbb{G}_i$), except for size 10 and considering the 50-best hypotheses. Also as expected, the results considering the 50-best hypotheses always improve those obtained with the 1-best hypothesis.

The performance of the models concerning the size of the expressions decreases dramatically with increasing sizes. This result is unsatisfactory for ME recognition systems. However, in PrIx problems, where the queries are ME or sub-expressions, it is reasonable to assume that these queries will have reduced sizes and therefore the results shown are encouraging.

4.4   Error-Dependent Precision

As mentioned above, precision is a very harsh metric as it requires an exact comparison symbol by symbol and at the same positions. In this section, we explore the Levenshtein distance as an evaluation metric based on the number of admissible errors. Figure 6 shows the precision of the discriminatively estimated model ($\mathbb{G}_d$) and the initial model ($\mathbb{G}_i$) varying the number of admissible errors (k). As in previous cases, these results have been obtained with the 1-best

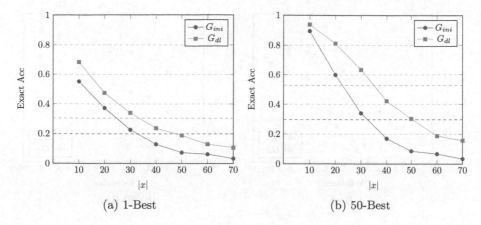

Fig. 4. Comparison of *ExAcc* score wit the $\mathbb{G}_i$ and $\mathbb{G}_d$ models for MEs of different length: (a) using only the 1-best hypothesis, and (b) using the 50-best. Dotted lines represent the global accuracy of each model for all lengths.

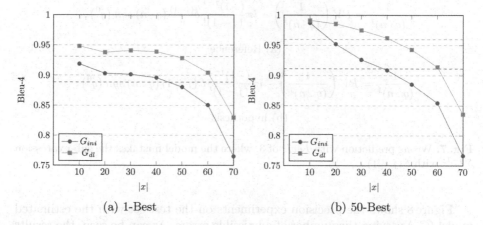

Fig. 5. Comparison of *Bleu* score with the $\mathbb{G}_i$ and $\mathbb{G}_d$ models for MEs of different length: (a) using only the 1-best hypothesis, and (b) using the 50-best. Dotted lines represent the global Bleu score of each model for all lengths.

hypothesis and with the 50-best hypotheses. The plot shows how most results, even for very large expressions, are very close from the reference, reinforcing the practical use for PrIx where this metric could be used as a way to optimize the precision-recall of a search engine.

Figure 7 provides one example of a large expression where the model hypothesis is incorrect but only by one relationship error. In PrIx, most of the queries (subexpressions of this expression) can still find the reference.

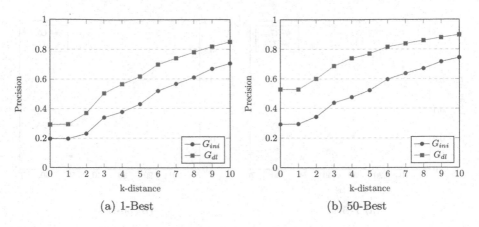

Fig. 6. Precision of the $\mathbb{G}_i$ and $\mathbb{G}_d$ models for different error values in the *LevD*: (a) using only the 1-best hypothesis, and (b) using the 50-best.

$$\frac{1}{(q \cdot n)^j} = PV \left(\frac{1}{(q \cdot n)^j} \right) - i\pi \frac{(-1)^{j-1}}{(j-1)!} \delta^{(j-1)} (q \cdot n) \, sgn \left(q^0 \right),$$

(a) Reference

$$\frac{1}{(q \cdot n)^j} = PV \left(\frac{1}{(q \cdot n)^j} \right) - i\pi \frac{(-1)^{j-1}}{(j-1)!} \delta (j-1) (q \cdot n) \, sgn \left(q^0 \right),$$

(b) hypothesis

Fig. 7. Wrong prediction with *LevD* of 3, where the model mistakes the sub-expression $\delta^{(j-1)}$ with $\delta (j-1)$.

Figure 8 shows the precision experiments on the test set with the estimated model ($\mathbb{G}_d$) varying the number of admissible errors. As can be seen, the results are reasonable, although somewhat worse than those reported on the validation set. However, it should be noted that the STEP-2 filter is not applied to the test set (see Table 1).

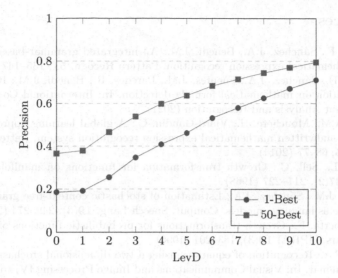

Fig. 8. Precision of the $\mathbb{G}_d$ model on the test set for different error values in the *LevD* using the 1-best and 50-best on the test set.

5 Conclusions

In this paper, we have presented a discriminative learning algorithm to estimate a 2D-PCFG through a generalization of the H-criterion, as the objective function, and the growth transformations as the optimization method. Several experiments have been reported on a well-known dataset. In the experiments that analyze the convergence of the estimation algorithm, in general, the results improve significantly from the 5-best hypotheses. These results would be enough to compose a hyper-forest for further use in PrIx.

In the experiments related to the length of the MEs, the model precision drops significantly beyond a ME length of |30|. However, using a structural model allows us to easily generate multiple hypotheses and decompose large expressions into all their correct sub-expressions. Thanks to this and given that queries in PrIx may be of short length (less than 20 LaTeX symbols) the mistakes produced by the model could not affect the majority of queries. Taking all these into account we think our model provides a good approximation to the PrIx problem in massive collections of digitized scientific documents. We expect to research on this PrIx framework as future work.

Acknowledgment. This work was partially supported by the Generalitat Valenciana under the grant PROMETEO/2019/121 (DeepPattern) and by a grant PID2020-116813RB-I00a funded by MCIN/AEI/ 10.13039/501100011033 (SimancasSearch). Furthermore, the work of the fourth author is financed by a grant FPI-Sub-1 from program PAID-01-21 of the Universitat Politècnica de València.

References

1. Álvaro, F., Sánchez, J.A., Benedí, J.M.: An integrated grammar-based approach for mathematical expression recognition. Pattern Recogn. **51**, 135–147 (2016)
2. Anitei, D., Sánchez, J.A., Fuentes, J.M., Paredes, R., Benedí, J.M.: ICDAR2021 competition on mathematical formula detection. In: International Conference on Document Analysis and Recognition (2021)
3. Awal, A.M., Mouchere, H., Viard-Gaudin, C.: A global learning approach for an online handwritten mathematical expression recognition system. Pattern Recogn. Lett. **35**, 68–77 (2014)
4. Baum, L., Sell, G.: Growth transformation for functions on manifolds. Pac. J. Math. **27**(2), 211–227 (1968)
5. Benedí, J.M., Sánchez, J.A.: Estimation of stochastic context-free grammars and their use as language models. Comput. Speech Lang. **19**(3), 249–274 (2005)
6. Casacuberta, F.: Growth transformations for probabilistic functions of stochastic grammars. IJPRAI **10**(3), 183–201 (1996)
7. Chou, P.A.: Recognition of equations using a two-dimensional stochastic context-free grammar. In: Visual Communications and Image Processing IV, vol. 1199, pp. 852–863 (1989)
8. Deng, Y., Kanervisto, A., Ling, J., Rush, A.M.: Image-to-markup generation with coarse-to-fine attention. In: Proceedings of the ICML-2017, pp. 980–989 (2017)
9. Gopalakrishnan, P., Kanevsky, D., Nadas, A., Nahamoo, D., Picheny, M.: Decoder selection based on cross-entropies. In: ICASSP-1988, pp. 20–23 (1988)
10. Gopalakrishnan, P., Kanevsky, D., Nadas, A., Nahamoo, D.: An inequality for rational functions with applications to some statistical estimation problems. IEEE Trans. Inf. Theory **37**(1), 107–113 (1991)
11. Jiménez, V.M., Marzal, A.: Computation of the N best parse trees for weighted and stochastic context-free grammars. In: Ferri, F.J., Iñesta, J.M., Amin, A., Pudil, P. (eds.) SSPR /SPR 2000. LNCS, vol. 1876, pp. 183–192. Springer, Heidelberg (2000). https://doi.org/10.1007/3-540-44522-6_19
12. Lari, K., Young, S.: Applications of stochatic context-free grammars using the inside-outside algorithm. Comput. Speech Lang. 237–257 (1991)
13. Maca, M., Benedí, J.M., Sánchez, J.A.: Discriminative learning for probabilistic context-free grammars based on generalized H-criterion. Preprint arXiv:2103.08656 (2021)
14. Marzal, A.: Cálculo de las K Mejores Soluciones a Problemas de Programación Dinámica. Ph.D. thesis, Universidad Politécnica de Valencia (1993)
15. Ney, H.: Stochastic grammars and pattern recognition. In: Laface, P., Mori, R.D. (eds.) Speech Recognition and Understanding. NATO ASI Series, vol. 75, pp. 319–344. Springer, Heidelberg (1992). https://doi.org/10.1007/978-3-642-76626-8_34
16. Noya, E., Sánchez, J.A., Benedí, J.M.: Generation of hypergraphs from the N-best parsing of 2D-probabilistic context-free grammars for mathematical expression recognition. In: ICPR, pp. 5696–5703 (2021)
17. Ortmanns, S., Ney, H., Aubert, X.: A word graph algorithm for large vocabulary continuous speech recognition. Comput. Speech Lang. **11**(1), 43–72 (1997)
18. Papineni, K., Roukos, S., Ward, T., Zhu, W.: BLEU: a method for automatic evaluation of machine translation. In: ACL, pp. 311–318 (2002)
19. Prǎša, D., Hlaváč, V.: Mathematical formulae recognition using 2d grammars. In: International Conference on Document Analysis and Recognition, vol. 2, pp. 849–853 (2007)

20. Sánchez-Sáez, R., Sánchez, J.A., Benedí, J.M.: Confidence measures for error discrimination in an interactive predictive parsing framework. In: Coling, pp. 1220–1228 (2010)
21. Toselli, A.H., Vidal, E., Puigcerver, J., Noya-García, E.: Probabilistic multi-word spotting in handwritten text images. Pattern Anal. Appl. **22**, 23–32 (2019)
22. Woodland, P., Povey, D.: Large scale discriminative training of hidden Markov models for speech recognition. Comput. Speech Lang. **16**(1), 25–47 (2002)

Computer Vision

Golf Swing Sequencing Using Computer Vision

Marc Marais[✉][iD] and Dane Brown[iD]

Department of Computer Science, Rhodes University,
Grahamstown 6140, South Africa
marcmarais07@outlook.com, d.brown@ru.ac.za

Abstract. Analysis of golf swing events is a valuable tool to aid all golfers in improving their swing. Image processing and machine learning enable an automated system to perform golf swing sequencing using images. The majority of swing sequencing systems implemented involve using expensive camera equipment or a motion capture suit. An image-based swing classification system is proposed and evaluated on the GolfDB dataset. The system implements an automated golfer detector combined with traditional machine learning algorithms and a CNN to classify swing events.

The best performing classifier, the LinearSVM, achieved a recall score of 88.3% on the entire GolfDB dataset when combined with the golfer detector. However, without golfer detection, the pruned VGGNet achieved a recall score of 87.9%, significantly better (>10.7%) than the traditional machine learning models. The results are promising as the proposed system outperformed a Bi-LSTM deep learning approach to achieve swing sequencing, which achieved a recall score of 76.1% on the same GolfDB dataset. Overall, the results were promising and worked towards a system that can assist all golfers in swing sequencing without expensive equipment.

Keywords: Golf swing sequencing · Histogram of oriented gradients · Linear Discriminate Analysis · Support Vector Machines · Gradient Boosting Machines · Deep learning

1 Introduction

Golf is one of the more complex and challenging sports today and is highly popular as more than 80 million people play the sport worldwide [16]. The fine-tuning and complexity behind the golf swing itself comes under much scrutiny. With little research on the effect of different swings, much of the debate regarding the correct body posture and swing technique stems from professional golfers and coaches' personal opinions [20]. Correct body posture and golf ball alignment at each swing event is one of the key elements to a good swing [10, p. 42].

This study was funded by the National Research Foundation of South Africa. This work was undertaken in the Distributed Multimedia CoE at Rhodes University.

A. J. Pinho et al. (Eds.): IbPRIA 2022, LNCS 13256, pp. 351–365, 2022.
https://doi.org/10.1007/978-3-031-04881-4_28

The golf swing action can be recognised using machine learning, deep learning, and neural network algorithms. This study will attempt to detect and classify golf swing events. Golf swing sequencing has recently been achieved using a deep neural network [16]. Their study, which uses a labelled video database to sequence golf swing events, enables head alignment to be checked for synchronisation with the swing sequence events.

Many studies around golf swing sequencing use sensors and optical cameras to detect the swing pattern and events [12,14,16]. Sensors and optical cameras with hardware processing are costly implementations to study swing events and sequencing. On the other hand, computer vision and machine learning can analyse golf swing events in a practical way that is accessible to all golfers.

Therefore, a cost-effective and easily accessible solution in the field of computer vision can help aid golfers in the analysis of their golf swing events.

This paper aims to provide an image-based system that implements golf swing event classification for the various phases during a golf swing, using the GolfDB dataset [16]. The events are as follows: address, toe-up, mid-backswing, top, mid-downswing, impact, mid-follow-through and finish. The proposed system incorporates different feature extraction methods, a variety of classification algorithms and a Histogram of Oriented Gradients (HOG) golfer detector. This led to contributions towards applying HOG-based person detection for the purpose of golfer detection.

The rest of the paper is structured as follows: Sect. 2 analyses related studies, Sect. 3 and 4 detail the methodology and implementation of the proposed system, including applying the HOG golfer detector. The analyses of the dimensionality reduction techniques and results of the proposed system are discussed in Sect. 5. Section 6 concludes the paper and discusses future work.

2   Related Studies

Golf swing analysis by McNally *et al.* [16] aimed to create a sequence of golf swing events through localising each event to a single frame. Events included the address, toe-up, mid-backswing, top, mid-downswing, impact, mid-follow-through and finish. These eight events were analysed and enabled consistent evaluation of the golf swing.

Using a benchmark database GolfDB, consisting of 1400 labelled videos and a combination of SwingNet, McNally *et al.* [16] implemented a lightweight deep learning neural network to analyse the performance of golf swings. SwingNet compromises a network architecture design, incorporating lightweight convolutional neural networks (CNNs) to enable effective mobile deployment.

The CNN, SwingNet, averaged a rate of 76.1% Percentage of Correct Events[1] (PCE) at detecting all eight events in the golf swing. A 91.8% PCE was achieved when detecting 6 out of 8 events, excluding the address and finish events. These

---

[1] PCE closely relates to the recall, the ratio of the number of true positives to the combined number of true positives and false negatives, metric used to measure machine learning models performance.

events were often misclassified with the start and end frames, respectively. Four random splits were generated for cross-validation to ensure the system generalised well.

Ko and Pan [14] also looked into swing sequencing but included body sway analysis during the swing. Using a single frontal facing camera and a motion capture suit to perform 3D analysis of the swing and body motion. The regression model is based on a Bidirectional Long Short Term Memory[2] (Bi-LSTM) neural network.

Swing events are captured and extracted as a sequence of images using the Sequence Feature Extraction and Classification Network (SFEC-net). SFEC-net is made up of three convolution layers, three pooling layers, and two fully connected layers. After extracting the swing event images, head-up analysis is conducted, creating three-dimensional and three-axis rotation angles of head movement.

Gehrig *et al.* [9] used single frames to robustly fit a golf club's location to a swing trajectory model.

Implementing traditional machine learning algorithms or simple CNNs combined with a histogram of oriented gradients golfer detector appears to be top candidates for yielding the best accuracy performance while minimising computational processing time for the proposed swing event classification system. Therefore, if CNNs were to be implemented, the frame foreground requires focusing purely on the golfer with minimal background variation. If there tends to be a lot of variation in the background and foreground, as is the case with the changing terrain on golf courses and the changing angle of viewing the golf swing, then using state-of-the-art models such as YOLO [17] may be warranted.

3 Methodology

3.1 Methodology Overview

The system is broken down into two stages to detect the eight golf swing events: an image processing stage and a machine learning stage. It was developed to accurately detect each golf swing event from the frames extracted from the GolfDB video dataset. The system will only analyse full golf swings based on the GolfDB dataset.

The main focus of image processing was to detect the golfer in the extracted frames and crop out the region of interest, which consisted of the golfer and the golf club swing region. The different camera views, down-the-line, face-on and other, were taken into account. To detect the golfer in an image, a 90×120 sliding window object detector was combined with the HOG feature descriptor for golfer detection, which is discussed in Sect. 3.4. If the golfer was located within the sliding window, illustrated in Fig. 1a, the image was cropped using the sliding

---

[2] Bi-LSTM models fall into the category of Bidirectional Recurrent Neural Networks [18].

window as a bounding box, illustrated in Fig. 1b. Determining the optimal sliding window for the golfer location is based on Non-Maxima Suppression (NMS) window scoring.

(a) Golfer localisation and bounding box sliding window.

(b) Cropped golfer image, 90 × 120.

Fig. 1. Automated golfer detector.

For the machine learning phase of the system, feature extraction was applied to the data to reduce dimensionality. Fitting and transforming the model using LDA reduces the dimensionality of the data and improves linear separability. The feature extraction stage was not applied to the pruned VGGNet deep learning model.

The proposed system aimed to compare the limits of traditional machine learning models when combined with image processing techniques to the pruned VGGNet as well as the related studies. State-of-the-art detectors for golfer detection were not explored due to the computational power and training time required by these systems. Machine learning algorithms considered were LinearSVM, CatBoost, Decision Tree Classifier, Random Forests and KNN.

3.2   Multi-class Classification

The eight golf swing events are made up as follows:

1. *Address (A)*. The frame before the initial backswing movement begins.
2. *Toe-up (TU)*. Backswing stage where the golf club shaft is parallel to the ground.
3. *Mid-backswing (MB)*. Arm is parallel to the ground during the backswing.
4. *Top (T)* The stage when the golf club changes directions and transitions from backswing to downswing.
5. *Mid-downswing (MD)*. Arm parallel to the ground during the downswing phase.

6. *Impact (I)*. The stage when the golf club head makes contact with the golf ball.
7. *Mid-follow-through (MFT)*. Golf club shaft is parallel to the ground during the follow-through phase of the swing.
8. *Finish (F)*. The frame before the golfer relaxes their final pose.

3.3 Frame Extraction from GolfDB Video Dataset

The GolfDB dataset was developed by McNally *et al.* [16] as a benchmark dataset. The dataset consists of 1400 labelled full swing golf video samples. The viewing angles of the videos vary between down-the-line, face-on and other views, illustrated in Fig. 2.

The video sample frames were looped over, and where the frame corresponded to an element of the events array[3], the frame was extracted and added to the corresponding event class folder creating the images of the different swing event classes; creating the image-based golf swing event dataset.

(a) Down-the-line View.

(b) Face-on View.

(c) Other View.

Fig. 2. The three different golf swing views.

This study's main focus was the different viewing angles. Consequently, an attempt was made to build a model that generalises well across the different viewing angles. An 80:20 train test split was implemented for validation and testing data for all test models.

3.4 HOG Golfer Detection

Histogram of Oriented Gradients (HOG) object detection involves machine learning and image processing techniques to detect semantic object instances within an image or video. Popular domains of object detection include face detection or pedestrian detection. HOG is a feature descriptor for images, used for object detection and recognition tasks in computer vision and machine learning [6].

---

[3] An array containing ten items each corresponding to the frame of an event, [SF, A, TU, MB, T, MD, I, MFT, F, EF].

A feature descriptor defines an image by analysing pixel intensities and the gradients of pixels. Through this, the vital information from an image is extracted.

A HOG-based golfer detector can perform person or non-person classification [6]. The HOG-based golfer detector uses a 64×128 sliding window containing the combined HOG feature vectors, fed into a LinearSVM for final classification. There are three elements to the human detector: winStride, padding, and scale. WinStride is a tuple (x, y) that defines the horizontal and vertical shifts of the window. Padding pads the sliding window with pixels and scale refers to the resize factor for each image layer of the resolution pyramid.

A smaller winStride and scale generally generate better results but are more computationally expensive. NMS is used to select the highest scoring window. NMS aims to retain only one window by eliminating low-score windows by thresholding those overlapping. NMS's limitation is that it is a greedy algorithm that does not always find the optimal solution.

3.5   Feature Extraction

Feature extraction is a form of dimensionality reduction where an image with a large number of pixels is efficiently represented so that the informative aspects of the image are represented effectively.

Principal Component Analysis(PCA) involves computing the Eigenvectors using linear algebra and implementing them to transform the basis on the data [22]. Typically the first few principal components are used, as the most significant explained variance is stored in the first components.

The total scatter matrix S_T is defined as [2]:

$$S_T = \sum_{k=1}^{N} (\mathbf{x}_k - \mu)(\mathbf{x}_k - \mu)^T \tag{1}$$

where N is the number of sample images x_k, and $\mu \in \mathbb{R}^n$ is the mean image obtained from the sample images.

Linear Discriminate Analysis (LDA) is a generalisation of Fisher's linear discriminant [21] and aims to learn a linear combination of features to consider the between-class and within-class scatter matrix differences.

The between-class scatter matrix S_B for C number of classes is defined as [2]:

$$S_B = \sum_{i=1}^{c} N_i (\mu_i - \mu)(\mu_i - \mu)^T \tag{2}$$

and the within-class scatter matrix S_W is defined as:

$$S_W = \sum_{i=1}^{c} \sum_{\mathbf{x}_k \in X_i} (\mathbf{x}_k - \mu_i)(\mathbf{x}_k - \mu_i)^T \tag{3}$$

where μ_i is the mean image and N_i refers to the number of samples of class X_i.

3.6   Classification

Based on validation data, classification is implemented to identify which category a new observation belongs to. Multi-class classification is applied to the system as there are more than two classes in the classification problem.

Support Vector Machine (SVM) is a supervised kernel-based machine learning model used for classification and regression, using discriminative classification. The algorithm separates validation data into their respective labelled classes by drawing a hyperplane.

Gradient Boosting Machines (GBMs) are based on a statistical framework to minimise the loss of a numerical optimisation problem through the addition of weak learners using a gradient descent procedure and boosting models [8]. Boosting machine models consists of iteratively learning weak classifiers to a specific distribution of data and adding them to a final robust classifier. Three commonly implemented GBMs are XGBoost, LightGBM and CatBoost [5, 7, 13].

Decision Trees (DTs) are implemented for either classification or regression and form part of non-parametric supervised learning algorithms. Through learning simple decision rules based on the data features, a model is generated to predict the value of a target variable. DTs implement recursive partitioning based on the attribute value learnt from the partition to construct the tree nodes.

Random Forests builds on decision trees through the application of a large number of individual decision trees implemented as an ensemble method [3]. An ensemble method uses multiple learning algorithms to improve predictive performance. Therefore, random forest implements a combination of tree predictors where the values of each tree are dependent on a random vector sampled independently with equal distribution across all trees in the forest.

K-Nearest Neighbors (KNN) classification model is a distance-based supervised learning algorithm where classification is determined through a simple majority vote of the nearest neighbours of each point [1].

Convolutional Neural Networks (CNN) form part of deep neural network algorithms utilising convolutional layers. CNNs are efficient at solving complex image pattern recognition tasks. The architecture of the VGG-16 [19] CNN consists of many convolutional layers containing 3×3 filters. Maximum pooling layers are applied in between the various convolutional layers. The architecture concludes with two fully connected layers followed by a softmax layer for output.

VGG-16 is predominately designed for problems relating to the much larger ImageNet [15] dataset. Therefore, this paper implements a pruned VGGNet [4].

The CNN architecture adopted is a pruned structure using the first 2/5 blocks of the VGG-16, one fully connected layer and a softmax layer. Each block consists of a convolution followed by batch normalisation and a dropout layer of 10%. Dropout regularisation randomly takes each hidden neuron and sets the output to zero based on a certain probability.

Images were augmented to aid the model's generalisation ability on unseen data. The data augmentation generates random transformations with a zoom of ±5% and horizontal and vertical shifts of ±10%. Data augmentation increases the amount of data through minor modifications of the existing data.

4   Evaluation Metrics

Accuracy is defined as the proportion of test samples for which the actual label matched the predicted label [11], the calculation is illustrated in Eq. 4.

$$Accuracy = \frac{TruePositives + TrueNegatives}{TruePositives + FalsePositives + TrueNegatives + FalseNegatives} \tag{4}$$

Precision defines the reliability of a classifier's predictions and is a good measure to determine, when the cost of false positives is high [11].

$$Precision = \frac{TruePositives}{TruePositives + FalsePositives} \tag{5}$$

Recall is a good measure when determining how comprehensive the classifier is in finding cases that are actually positive [11].

$$Recall = \frac{TruePositives}{TruePositives + FalseNegatives} \tag{6}$$

F-score is a combination of precision and recall, representing a harmonic mean between the two [11]. The F-score provides a much more balanced measure of a models performance.

$$F\text{-}score = 2 \times \frac{Precision \times Recall}{Precision + Recall} \tag{7}$$

Accuracy works well only if there is an even distribution of samples between the classes. The F-score metric is the preferred metric for measuring a model's performance when presented with an imbalanced dataset.

5   Results

5.1   Validation of Parameters

All classifier and dimensionality reduction parameters were validated using the entire GolfDB dataset, and all the views were included without applying the HOG golfer detector. The implementation was done in this manner to ensure better robustness and generalisation ability.

Dimensionality Reduction and Model Selection. Hyperparameter tuning is critical for technical comparison between PCA and LDA dimensionality reduction techniques. The critical parameter for both PCA and LDA is the number of components. Hence, a grid search is implemented to determine the optimal number of components.

PCA poorly separates the different swing event classes, as illustrated in Fig. 3a. In contrast, LDA groups the different classes in a linearly separable way, Fig. 3b. Therefore, LDA using seven components was chosen as the dimensionality reduction technique for the system. As the LDA transformation was applied to the entire dataset, data leakage[4] occurred in the traditional machine learning models.

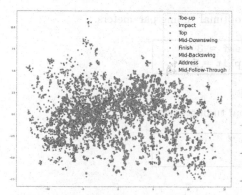 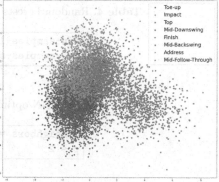

(a) Scatter plot of first two components for PCA on the entire GolfDB dataset using 500 components.

(b) Scatter plot of first two components for LDA on the entire GolfDB dataset using 7 components.

Fig. 3. PCA vs LDA dimensionality reduction, on all viewing angles of GolfDB validation dataset.

Classification Model Hyperparameter Tuning. All machine learning classifiers were tuned using a grid search to optimise the hyperparameters. Each grid search applied four-fold cross-validation. To reduce the bias of the classifiers and for comparability to McNally *et al.*'s study [16]. Tables 1, 2, 3, 4 and 5 describe the optimal parameters found for each machine learning classifier.

Table 1. LinearSVM optimal tuning parameters.

```
C = 10
kernel = linear
multi-class = crammer-singer
```

---

[4] Data from outside the training dataset is used to create the model, sharing information between the validation and training data sets [23, p. 93].

Table 2. CatBoost optimal tuning parameters.

```
learning-rate = 0.03
depth = 4
bagging-temperature = 1.5
l2-leaf-reg = 7
```

Table 3. Decision Tree Classifier optimal tuning parameters.

```
min-samples-leaf = 5
min-samples-split = 3
```

Table 4. Random Forests optimal tuning parameters.

```
min-samples-leaf = 3
min-samples-split = 3
```

Table 5. KNN optimal tuning parameters.

```
n-neighbors = 7
p = 2
```

5.2   Test Models

This Section evaluates the various experiments to gauge the overall performance of the golf swing classification system. The system evaluates the multi-class classification on the entire image without implementing the golfer detector. These results are compared to the classification results achieved with the golfer detector.

Experiment 1: Entire GolfDB Dataset Without Golfer Detection. This Section evaluates the systems classification strength on the entire GolfDB dataset without golfer detection and cropping of the golfer region of interest. Experiment 1 takes in uncropped images of size 160×160. The multi-class classification problem aims to address the system's generalisation ability by evaluating the system's identification accuracy of all the different viewing angles.

Table 6. Experiment 1 performance metrics.

Model	Accuracy	Recall	Precision	F-score
LinearSVM	77.1	77.1	77.3	77.2
CatBoost	77.0	77.0	77.3	77.1
Decision Tree	68.3	68.3	68.4	68.3
Random Forests	75.4	75.4	75.7	75.4
KNN	73.1	73.1	73.6	73.3
Pruned VGGNet	87.9	87.9	88.0	87.9

Using F-score, the pruned VGGNet significantly outperformed the LinearSVM machine learning algorithm by 10.7%. Table 6 shows that amongst the traditional machine learning models, the LinearSVM marginally outperformed CatBoost achieving an accuracy of 77.1% compared to the 77.0% of CatBoost. The KNN classifier yielded an accuracy of 73.1%. Random Forests and the decision tree classifier yielded accuracy scores of 75.4% and 68.3%, respectively.

The pruned VGGNet with data augmentation achieved an accuracy score of 87.9, which was 10% better than the pruned VGGNet with no data augmentation.

Experiment 2: Entire GolfDB Dataset with Golfer Detection. The golfer detection crops the region of interest containing the golfer and swing radius, using a sliding window of size 90 × 120. The golfer detector successfully predicted 7786 images containing a golfer, using a 30% confidence level. Some residual false detections were manually observed using visual inspection. These residual false detections were not manually removed and may negatively affect the subsequent classification results.

Table 7. Experiment 2 performance metrics.

Model	Accuracy	Recall	Precision	F-score
LinearSVM	88.3	88.3	88.3	88.3
CatBoost	87.9	87.9	88.0	87.9
Decision Tree	78.8	78.8	79.2	78.9
Random Forests	86.6	86.6	86.6	86.6
KNN	85.9	85.9	85.9	85.9
Pruned VGGNet	86.3	86.3	86.3	86.3

The best performing model was the LinearSVM, again narrowly outperforming the Catboost classifier with an F-score of 88.3%, compared to the 87.9% F-score achieved by CatBoost, as described in Table 7. Figure 4 illustrates the training and validation accuracy of the pruned VGGNet for experiments 1 and 2.

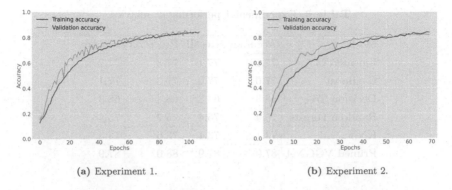

(a) Experiment 1.   (b) Experiment 2.

Fig. 4. Training and validation accuracy of pruned VGGNet.

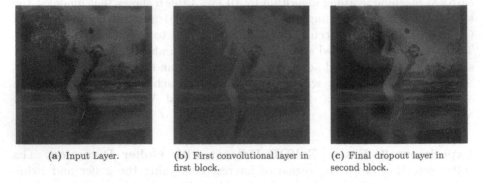

(a) Input Layer.   (b) First convolutional layer in   (c) Final dropout layer in
                          first block.   second block.

Fig. 5. Heat map activations of the pruned VGGNet showing improved localisation of
salient features with the convolutional and dropout layers.

Figure 5 illustrates the heat maps of the activations found through the progression in layers of the pruned VGGNet. Activations were visualised using the Keract[5] python library. The activation on the input layer, Fig. 5a, is an example of a poor activation, failing to focus the golfer. However, as the VGGNet progresses through the layers, the localisation of the golfer and swing region vastly improves, as seen in Fig. 5b and c, respectively.

5.3   Discussion

Overall, the multi-class golf swing classification system performed well. The initial hyperparameter tuning to optimise the proposed system's classification models was critical to improving the system's performance. LDA significantly outperformed PCA due to its ability to generate linearly separable classes. The golfer detector increased the accuracy and F-score of the system by 10%, even though falsely detected ROIs were included.

---

[5] https://github.com/philipperemy/keract.

The pruned VGGNet significantly outperformed the machine learning models in experiment 1. However, when implemented on the golfer detector images, the pruned VGGNet declined in performance, reinforcing the concept that CNNs are effective on busy backgrounds compared to other feature extraction and traditional machine learning approaches.

Table 8. Comparison of the different Swing Sequencing Systems using Recall. The best performing classification algorithm's recall scores were selected from each experiment.

System	A	TU	MB	T	MD	I	MFT	F	Recall
Experiment 1: Pruned VGGNet	93.3	86.5	84.9	86.3	81.8	86.2	89.5	93.7	87.9
Experiment 2: LinearSVM	90.2	83.8	80.2	89.6	88.2	90.1	89.8	92.6	88.3
McNally et al. [16]	31.7	84.2	88.7	83.9	98.1	98.4	97.6	26.5	76.1
Ko and Pan [14]	93.9	N/A	96.9	92.2	89.2	84.9	97.0	99.2	93.3

Table 8 compares the proposed system to McNally et al. and Ko and Pan's studies [14,16]. Experiment 1 and 2 significantly outperformed McNally et al.'s system with recall scores of 87.9% and 88.3%, respectively, compared to the 76.1% achieved by McNally et al. Both Ko and Pan and McNally et al. implemented a time series. However, McNally et al.'s system struggled with the address and finish classes due to difficulty precisely localising the two events temporally. Ko and Pan's system benefited from a motion-capture suit. However, the proposed system avoided it to allow for easier access to all golfers without incurring high costs.

The proposed system can generalise well across the various swing event classes. Exception for the consecutive toe-up and mid-backswing classes, classifying a video with time information may improve the classification between consecutive classes. The main area of concern within the system is the golfer detector, which does not handle images with multiple people in the background compared to neural networks. Neural networks are especially effective on busy backgrounds compared to other feature extraction and machine learning approaches. However, the proposed system showed that traditional machine learning algorithms combined with feature extraction can still achieve results on par with or better than deep learning techniques when carried out systematically.

6 Conclusion

Overall the proposed system performed well and, when implemented with no golfer detection, achieved an F-score of 87.9% using the pruned VGGNet. The application of the golfer detection to the GolfDB dataset improved the LinearSVMs F-score by 11.1%, achieving an F-score of 88.3%. The pruned VGGNet declined in performance when implemented with the golfer detector. Misclassifications were mainly present between the toe-up and mid-backswing for the LinearSVM classifier, which was expected due to similarity between the two swing events. However, the proposed system significantly outperformed McNally's *et al.* [16] deep learning study under the same conditions.

This study notably contributed to applying a HOG-based golfer detector for golfer detection to improve event classification, which has not been explored before. The proposed golf swing classification is thus promising as a system that can be made available to golfers and is a starting point towards golf swing form analysis. The system also shows that image processing combined with traditional machine learning models remain relevant and, when implemented effectively, can outperform lightweight deep learning models.

References

1. Altman, N.S.: An introduction to kernel and nearest-neighbor nonparametric regression. Am. Stat. **46**(3), 175–185 (1992)
2. Belhumeur, P.N., Hespanha, J.P., Kriegman, D.J.: Eigenfaces vs. fisherfaces: recognition using class specific linear projection. IEEE Trans. Pattern Anal. Mach. Intell. **19**(7), 711–720 (1997)
3. Breiman, L.: Random forests. Mach. Learn. **45**(1), 5–32 (2001)
4. Brown, D., Bradshaw, K.: Deep palmprint recognition with alignment and augmentation of limited training samples. SN Comput. Sci. **3**(1), 1–17 (2022)
5. Chen, T., Guestrin, C.: XGBoost: a scalable tree boosting system. In: Proceedings of the 22nd ACM SIGKDD International Conference on Knowledge Discovery and Data Mining, pp. 785–794 (2016)
6. Dalal, N., Triggs, B.: Histograms of oriented gradients for human detection. In: Proceedings - 2005 IEEE Computer Society Conference on Computer Vision and Pattern Recognition, CVPR 2005, vol. I, pp. 886–893 (2005). https://doi.org/10.1109/CVPR.2005.177
7. Dorogush, A.V., Ershov, V., Gulin, A.: CatBoost: gradient boosting with categorical features support. arXiv preprint arXiv:1810.11363 (2018)
8. Friedman, J.H.: Stochastic gradient boosting. Comput. Stat. Data Anal. **38**(4), 367–378 (2002). https://doi.org/10.1016/S0167-9473(01)00065-2
9. Gehrig, N., Lepetit, V., Fua, P.: Visual golf club tracking for enhanced swing analysis. In: British Machine Vision Conference (BMVC), pp. 1–10 (2003)
10. Glazier, P., Lamp, P.: Golf science: optimum performance from tee to green. In: Golf Science: Optimum Performance From Tee to Green, chap. The swing. University of Chicago Press (2013)
11. Hossin, M., Sulaiman, M.N.: A review on evaluation metrics for data classification evaluations. Int. J. Data Min. Knowl. Manag. Process **5**(2), 1 (2015)
12. Hsu, Y.L., Chen, Y.T., Chou, P.H., Kou, Y.C., Chen, Y.C., Su, H.Y.: Golf swing motion detection using an inertial-sensor-based portable instrument. In: 2016 IEEE International Conference on Consumer Electronics-Taiwan (ICCE-TW), pp. 1–2. IEEE (2016)
13. Ke, G., et al.: LightGBM: a highly efficient gradient boosting decision tree. Adv. Neural. Inf. Process. Syst. **30**, 3146–3154 (2017)
14. Ko, K.-R., Pan, S.B.: CNN and Bi-LSTM based 3D golf swing analysis by frontal swing sequence images. Multimedia Tools Appl. **80**(6), 8957–8972 (2020). https://doi.org/10.1007/s11042-020-10096-0
15. Krizhevsky, A., Sutskever, I., Hinton, G.E.: ImageNet classification with deep convolutional neural networks. Commun. ACM **60**(6), 84–90 (2017)

16. McNally, W., Vats, K., Pinto, T., Dulhanty, C., McPhee, J., Wong, A.: GolfDB: a video database for golf swing sequencing. In: IEEE Computer Society Conference on Computer Vision and Pattern Recognition Workshops, June 2019, pp. 2553–2562 (2019). https://doi.org/10.1109/CVPRW.2019.00311
17. Redmon, J., Divvala, S., Girshick, R., Farhadi, A.: You only look once: unified, real-time object detection. In: Proceedings of the IEEE Conference on Computer Vision and Pattern Recognition, pp. 779–788 (2016)
18. Schuster, M., Paliwal, K.: Bidirectional recurrent neural networks. IEEE Trans. Sig. Process. **45**, 2673–2681 (1997). https://doi.org/10.1109/78.650093
19. Simonyan, K., Zisserman, A.: Very deep convolutional networks for large-scale image recognition. arXiv preprint arXiv:1409.1556 (2014)
20. Smith, A., Roberts, J., Wallace, E., Forrester, S.: Professional golf coaches' perceptions of the key technical parameters in the golf swing. In: Procedia Engineering, vol. 34, pp. 224–229. Elsevier Ltd., Amsterdam (2012). https://doi.org/10.1016/j.proeng.2012.04.039
21. Tharwat, A., Gaber, T., Ibrahim, A., Hassanien, A.E.: Linear discriminant analysis: a detailed tutorial. AI Commun. **30**(2), 169–190 (2017)
22. Wold, S., Esbensen, K., Geladi, P.: Principal component analysis. Chemom. Intell. Lab. Syst. **2**(1–3), 37–52 (1987)
23. Zheng, A., Casari, A.: Feature Engineering for Machine Learning: Principles and Techniques for Data Scientists, 1st edn. O'Reilly Media Inc., Newton (2018)

Domain Adaptation in Robotics: A Study Case on Kitchen Utensil Recognition

Javier Sáez-Pérez, Antonio Javier Gallego(✉) ⓘ, Jose J. Valero-Mas ⓘ,
and Jorge Calvo Zaragoza ⓘ

Department of Software and Computing Systems, University of Alicante,
Carretera de San Vicente s/n, 03690 Alicante, Spain
{jgallego,jjvalero,jcalvo}@dlsi.ua.es

Abstract. Recognition methods based on Deep Learning (DL) currently represent the state of the art in a number of robot-related tasks as, for instance, computer vision for autonomous guidance or object manipulation. Nevertheless, the large requirements of annotated data constitute one of their main drawbacks, at least when considering supervised frameworks. In this context, the Domain Adaptation (DA) paradigm stands as a promising, yet largely unexplored, framework for tackling such cases avoiding the need for manually data annotation. This work presents a case of study of unsupervised DA algorithms for robot-based kitchenware manipulation. The influence of combining several source or target domains is also analyzed as a proposal to improve the adaptation process. For that, we evaluate three representative state-of-the-art techniques (DANN, PixelDA, and DeepCORAL) and assess them using four corpora of kitchen household images, being three of them specifically developed for this work. The results obtained show that the DeepCORAL strategy generally outperforms the rest of the cases. Moreover, the various scenarios posed suggest that the combination of source sets enhances the adaptation to novel target domains. Finally, the experimentation state the relevance of DA methods and their usefulness in robotic applications.

Keywords: Deep Learning · Domain Adaptation · Robotics · Computer vision

1 Introduction

The progressive increase of computing capacity and data storage as well as the emergence of new machine learning techniques have fostered the development of the so-called Deep Learning (DL) field [11]. While DL architectures are generally related to state-of-the art performances [12], they exhibit a clear drawback: the

This work was supported by the project I+D+i PID2020-118447RA-I00 (MultiScore), funded by MCIN/AEI/10.13039/501100011033, and the Generalitat Valenciana through project GV/2021/064. The third author is supported by grant APOSTD/2020/256 from "Programa I+D+i de la Generalitat Valenciana".

© Springer Nature Switzerland AG 2022
A. J. Pinho et al. (Eds.): IbPRIA 2022, LNCS 13256, pp. 366–377, 2022.
https://doi.org/10.1007/978-3-031-04881-4_29

need for a large amount of labeled data to be trained. Although the DL community argues that open source knowledge can make data collection easier, such statement is arguable when referring to labeled corpora since, for instance, most of the data available on the Internet is not labeled. Given that manual labeling tasks may be deemed as tedious and error-prone, there is great interest in developing methods that take advantage of the knowledge learned with data from a distribution for which labeling is available (which we will refer to as the *source domain*), to apply them to data from another distribution (*target domain*) for which no labeling is available. This type of techniques is generally called Transfer Learning (TL) and covers all cases where the information learned from a data domain is applied to others [7]. Within it, a case of particular interest is the Domain Adaptation (DA) one, which is when *source* and *target* domains deal with the same task but the distribution or appearance of the data differs (e.g., classifying the same type of labels but on a different dataset). This work studies the potential of DA in robotics when addressing kitchen utensil identification tasks.

There are different DA approaches depending on the amount of information available from the target domain, these are: 1) Unsupervised Domain Adaptation (UDA), in which the target labels are completely unknown, 2) Semi-supervised Domain Adaptation, in which some labels of the target domain are known, and 3) Supervised Domain Adaptation, in which all the labels of the target domain are known [10]. Out of these three approaches, UDA stands as the most adequate, yet challenging, case as it allows knowledge to be exploited without having to label the target domain. However, the existing UDA methods are not robust enough [8], hence remaining as an unsolved problem and open to new proposals.

Within UDA, there are three different ways of approaching the problem. One of the most used is based on the search for characteristics common to both source and target domains, so that they are invariant to domain changes. Some of the most relevant algorithms that apply this technique are: Domain-Adversarial Neural Networks (DANN) [8], which learns domain-invariant features by forcing the network to resort to those that do not allow domain differentiation; Visual-Adversarial Domain Adaptation (VADA) [13], which proposes a loss function to penalize differences between the internal representations learned by the method for these domains; Deep Reconstruction-Classification Network (DRCN), which forces data from both domains to be represented in the same way by reconstructing instances using a common architecture; or Domain Separation Network (DSN) [3], which is trained to classify input data into two subspaces— domain-specific and domain-independent—with the goal of improving the way domain-invariant features are learned.

The second commonly considered approach to perform UDA is based on the use Generative Adversarial Networks (GAN) [9]. Two of the most representative proposals within it are: Pixel Domain Adaptation (PixelDA) [2], which transforms the images from the source domain to resemble those from the target domain; and Deep Joint Distribution Optimal Transport (DJDOT) [4], which proposes a loss function based on optimal transport theory to learn an aligned representation between the source and target domains.

The third approach followed in the literature to perform UDA focuses on transforming and aligning the internal representations learned by the methods. Two of the most representative works based on this premise are: Subspace Alignment (SA) [5], which transforms features from one domain to another by using subspaces modeled through eigenvectors; and Deep CORelation ALignment (DeepCORAL) [15], which learns a nonlinear transformation to correlate the layers' activations of both domains.

Regardless of the strategy followed, it must be highlighted that all DA approaches are based on the use single source and target domains to carry out this adaptation process. However, as argued above, given that the large amount of information currently available, these methods could benefit from the combined use of several data sources for the adaptation process. In this way, more samples from different domains are available during the training process, which would improve the generalization capabilities of the model.

This paper assesses the use of DA methods for the classification of kitchen utensils. More precisely, we introduce three novel image corpora of such household items created for this work and, with an additional fourth reference data collection from the literature, we study the performance of three state-of-the-art UDA techniques in different source/target scenarios. The results obtained state the relevance of considering more than a single source domain for DA tasks as well as the influence of the UDA strategy in the success of the task.

Finally, note that the case study considered has potential applications in robotics. These mechatronic elements often use complex control algorithms for their operation, known as visual control when based on visual characteristics. In these, it is necessary to recognize the objects that appear in the images and forward this information to the robot controller. While the literature comprises a large number of proposals and methods for tackling such scenarios, they all show the need for a large set of labeled data to properly address the task. Furthermore, it should be noted that its application is limited to the domain of the data used for training. For this reason, the development of effective DA methods would be a great advance as it would allow the current techniques to be applied in more general and realistic scenarios.

2   Kitchen Utensil Corpora for Robotic-Oriented Tasks

This section introduces the kitchen utensil classification collections considered in this work, being three of them elaborated for this particular work. These latter corpora were created considering their application in the robotics case of study posed in this work based on the reference Edinburgh Kitchen Utensil Database (EKUD) set [6]. The three elaborated datasets can be downloaded from https://www.dlsi.ua.es/~jgallego/datasets/utensils.

Table 1 includes a summary of the characteristics of these datasets, including the total number of samples, and whether they have a uniform or a synthetic background. All corpora comprise color images of kitchen utensils classified into 12 possible classes and with a spatial resolution of 64 × 64 pixels. It is important to note that these collections were prepared to contain different characteristics, but with realistic domain shift variations that represent real cases in robotic tasks. Some examples of each dataset are shown in Fig. 1. Below, a comprehensive description of these four corpora is provided:

– Edinburgh Kitchen Utensil Database (EKUD): Dataset which comprises 897 color images of 12 kitchen utensils (about 75 images per class) created to train domestic assistance robots. These are real-world pictures of utensils with uniform backgrounds.
– EKUD-M Real Color (EKUD-M1): Corpus generated by the authors combining images of EKUD with patches from BSDS5000 [1], following an approach similar to that used to develop the MNIST-M collection [8]. In this case, only the background of the EKUD images was modified, keeping the original color of the objects.
– EKUD-M Not Real Color (EKUD-M2): Extension to EKUD-M1 in which the color of the objects was altered by being mixed with the color of the patches used for the background.
– Alicante Kitchenware Dataset for Robotic Manipulation (AKDRM): Collection developed by the authors by manually taking 1,480 photographs of the same 12 kitchen utensils contained in EKUD (about 123 images per class in this case). It contains real and varied backgrounds, different lighting levels, and considering different perspectives for the images.

Table 1. Summary of the characteristics of the four corpora considered for kitchenware classification in terms of the number of samples, background style, and color map of the utensil.

Corpus	Number of samples	Objects with real colors	Uniform background	Synthetic background
EKUD	897	✓	✓	
EKUD-M1	897	✓		✓
EKUD-M2	897			✓
AKDRM	1,480	✓		

3   Methodology

This section formally introduces the UDA framework considered for assessing the case of study previously posed. More precisely, we first present the gist of UDA as well as the proposal for combining several source and target domains for then introducing the three state-of-the-art methods contemplated in the work.

(a) EKUD   (b) EKUD-M1   (c) EKUD-M2   (d) AKDRM

Fig. 1. Examples of the different corpora introduced and considered in the work.

3.1 Problem Formalization

Let X denote the input data space and Y the output label space. Also, let $\mathcal{S} = \{(x_i, y_i) : x_i \in X, y_i \in Y\}_{i=1}^{|\mathcal{S}|}$ represent an annotated data collection data where each $x_i \in X$ is related to label $y_i \in Y$ by an underlying function $h : X \to Y$. The goal of a supervised DL algorithm is to learn the most accurate approximation $\hat{h}(\cdot)$ to that function using the data provided by the distribution $X \times Y$.

In a DA scenario, there exist two distributions over $X \times Y$: D_S and D_T, which are referred to as *source domain* and *target domain*, respectively. Moreover, when addressing UDA frameworks, we are only provided with a labeled source set $\mathcal{S} = \{(x_i, y_i)\} \sim D_S$ and a completely unlabeled target domain $\mathcal{T} = \{(x_i)\} \sim D_T$. In this context the goal of UDA is to learn the most accurate approximation $\hat{h}(\cdot)$ for D_T using the information provided in both $\mathcal{S}$ and $\mathcal{T}$.

In this paper, we study and assess the possibility of extending this definition by considering the combination of several source domains $\mathcal{S}_c = \bigcup_{i=1}^{n} \mathcal{S}_i$ or the combination of several target domains $\mathcal{T}_c = \bigcup_{i=1}^{n} \mathcal{T}_i$ when learning the approximation $\hat{h}(\cdot)$ to the function $h : X \to Y$.

3.2 Architectures

We now present the three UDA methods contemplated to study and benchmark the described corpora: DANN, PixelDA, and DeepCORAL. It must be noted that such proposals constitute representative approaches within the literature for UDA tasks.

DANN. Figure 2 shows a graphical representation of a DANN scheme. As it can be seen, it is divided in four main blocks: a feature extractor (G_f), a label classifier (G_y), a domain classifier (G_d), and a gradient reverse layer (GRL) [8].

The union of the feature extractor (G_f) and the label classifier (G_y) constitutes a standard feed-forward neural network that is trained to classify the labels $y \in Y$ from the input data $x \in X$. The last layer of G_y uses the Softmax activation function that assigns a probability that the input data x belongs to each of the possible labels in Y. The union of the feature extractor G_f, the GRL, and the domain classifier G_d constitutes another network that classifies the domain to which the input data belongs to ($d_i = 0$ if $x \sim D_S$ or $d_i = 1$ if $x \sim D_T$, being d_i a binary variable that indicates the domain of the input data).

The *GRL* layer forces the feature extractor (G_f) to learn features that are not related to the domain, i.e., domain-invariant features. This operation takes place during the backpropagation phase and consists of multiplying the gradient by a negative value ($-\lambda$). This process forces G_f to learn the domain-invariant features and, at the same time, G_y to classify both domains based on these features. In other words, these domain-invariant features allow G_y to also classify the target domain.

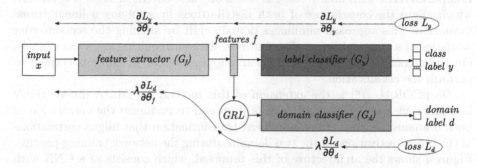

Fig. 2. Outline of the DANN architecture.

PixelDA. Figure 3 shows the outline of this architecture. Unlike other approaches, PixelDA [2] separates the DA step from the task for which the network is intended (i.e., classification, in our case). For this, a GAN network [9] is used to transform the images from the source domain (D_S) in a way that resembles the samples from the target domain (D_T). This network consist of a generator function $G(x, z; \theta_G) \to x^f$, parametrized by θ_G, that maps a source domain image $x \sim D_S$ and a noise vector $z \sim p_z$ to an adapted, or fake, image x^f. As usual in GAN, a discriminator network D is trained to differentiate the generated images $\mathcal{S}^f$ from the real ones (in this case using the target domain D_T). Therefore, G and D compete during the training phase, where the goal of G is to generate indistinguishable samples from the target domain.

Once the G network is trained, it is possible to create a new dataset $\mathcal{S}^f = \{G(x, z), y)\}$ of any size, where $\{x, y\} \subset D_S$ are labeled samples drawn from distribution D_S. Given this adapted dataset $\mathcal{S}^f$, a task-specific classifier (which learns the function $\hat{h}(\cdot)$ without the need for any further adaptation process) can be trained as if the training and test data were from the same distribution. It is important to note that, in this case, it is assumed that there are low-level differences in appearance between the domains (noise, resolution, illumination, color...) rather than high-level (object types, geometric variations...).

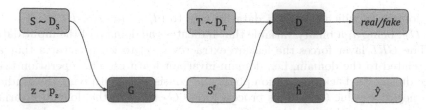

Fig. 3. Outline of the PixelDA architecture.

DeepCORAL. This architecture is based on the CORAL method [14] for DA, which aligns the covariances of both distributions by applying a linear transformation. This approach minimizes domain shift by aligning the second-order statistics of source and target distributions, without requiring any target labels. On these transformed features, a linear Support Vector Machine is trained to perform the classification.

DeepCORAL [15] is the extension of this method for DL. A differentiable loss function, namely *CORAL loss*, is proposed to minimize the correlation of both domains, i.e., it learns a nonlinear transformation that aligns correlations of the layers' activations of the two domains during the network training process. Figure 4 shows the architecture of this proposal, which consists of a CNN with two inputs. As it can be seen, to minimize the classification error of the source domain, a categorical loss function is combined with CORAL loss, while only CORAL loss is used for the target domain.

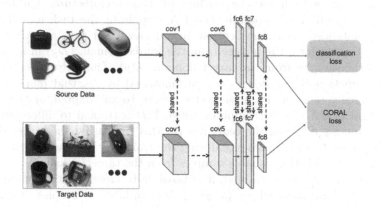

Fig. 4. Outline of DeepCORAL architecture [15].

4   Experimentation and Results

This section introduces the different experimental scenarios considered for assessing the capabilities of the introduced UDA methods with the four presented corpora, presents the results obtained for each one, and discusses the conclusions obtained.

In relation to the actual scenarios to study, we pose three different cases which differ in the number of source and target domains: first, we study that of using single source and target domains; then, we examine a scenario considering two source domains with a single target domain; finally, we explore a third case which focuses on a single source against two target domains. Note that, since the corpora contemplated for the study may be deemed as balanced[1], we considered the classification accuracy as the evaluation metric.

Regarding the parameters and training procedures of the UDA models, we resorted to those considered by the respective authors of the works. Specifically, DANN was trained for 300 epochs with a batch size of 64, PixelDA for 30,000 iterations with a batch size of 64, and DeepCORAL for 1,500 iterations and with a batch size of 32.

The rest of the section introduces and discusses the results obtained in each of the aforementioned experimental scenarios.

4.1 Single Source Domain

This first scenario addresses the case of single source and target domains. In this regard, Table 2 details the results obtained with the different DA methods studied for all possible combinations of the four considered corpora when used as both source and target domains. The case in which no adaptation strategy is applied—neural architecture used in DANN trained on the source domain and directly applied to the target one—is included for comparison.

The first point to remark is that the DeepCORAL UDA method achieves the best results in all source-target pairs except for the combination EKUD-M1→EKUD-M2 in which the PixelDA obtains the best result. The difference between DeepCORAL and the other two methods—DANN and PixelDA— is quite noticeable in most cases, especially when considering very different domains, as is the case of using AKDRM as the source or target domain due to it being the most challenging of the evaluated corpora. When the domains are similar (as in the case of EKUD, EKUD-M1, and EKUD-M2), good results are attained in all cases, although the DeepCORAL approach continues to stand out.

When comparing the UDA results against the cases in which no adaptation is performed, it can be observed that DA-based scenarios always report a superior performance. Nevertheless, note that the cases involving the AKDRM corpus in which the adaptation does not remarkably succeed (e.g., DANN method in AKDRM→EKUD or PixelDA in EKUD-M2→AKDRM), the difference among the classification figures is relatively low (around a 5% to 6% in absolute terms). This also suggests the inherent difficulty of the AKDRM corpus introduced in this work.

---

[1] Corpora are not exactly balanced, but the differences in terms of representation among the classes may be considered negligible.

Table 2. Results in terms of classification accuracy when considering single source and target domains for the different UDA strategies. For comparative purposes, **No adaptation** shows the cross-domain classification results when no UDA scheme is applied. Bold figures denote the best performing strategy for each pair of source and target domains.

Source domain	Target domain	No adaptation	DANN	PixelDA	DeepCORAL
EKUD	AKDRM	10.43	19.24	21.73	**55.31**
	EKUD-M1	26.43	67.48	62.36	**92.12**
	EKUD-M2	22.84	65.11	65.00	**95.78**
AKDRM	EKUD	11.81	16.47	17.55	**41.54**
	EKUD-M1	9.43	21.05	20.77	**39.45**
	EKUD-M2	10.43	19.47	20.45	**41.27**
EKUD-M1	AKDRM	9.42	15.44	20.45	**55.79**
	EKUD	28.18	78.12	77.41	**95.21**
	EKUD-M2	21.83	74.66	**76.28**	41.20
EKUD-M2	AKDRM	8.08	12.87	13.57	**59.31**
	EKUD	13.63	80.54	81.66	**97.98**
	EKUD-M1	22.98	79.32	84.42	**97.98**

4.2   Two Source Domains

The second scenario addresses the case of using two source domains for targeting a single one. Within this context, Table 3 details the results for the different UDA methods considered for all possible source and target combinations of the four studied corpora.

Attending to these results, the first point to highlight is that, in general, higher performance rates are observed with respect to the single source domain case (cf. Sect. 4.1). In particular, figures above 99% are obtained for EKUD, EKUD-M1, and EKUD-M2 datasets. Also, the case of the AKDRM corpus improves over a 67% in absolute terms compared to the single source domain case. Such results suggest that using more than a single source domain is beneficial for obtaining better adaptation rates in DA frameworks.

When comparing the different UDA strategies, it can be observed that Deep-CORAL stands out as the best method given that it achieves the highest classification rates for all source-target corpora combinations. This fact coupled to the previous case of using a single source domain suggests the robustness of this method in UDA cases.

Table 3. Results in terms of classification accuracy when considering two source domains against a single target one for the different UDA strategies. Bold figures denote the best performing strategy in each scenario.

Source domain	Target domain	DANN	PixelDA	DeepCORAL
AKDRM + EKUD	EKUD-M1	95.02	94.95	**98.99**
	EKUD-M2	94.78	95.12	**97.12**
AKDRM + EKUD-M1	EKUD	91.44	90.54	**99.56**
	EKUD-M2	91.77	90.00	**94.15**
AKDRM + EKUD-M2	EKUD	87.51	88.54	**95.65**
	EKUD-M1	90.02	91.85	**94.02**
EKUD + EKUD-M1	AKDRM	13.91	12.15	**67.35**
	EKUD-M2	91.02	92.05	**99.03**
EKUD + EKUD-M2	AKDRM	12.90	16.40	**67.15**
	EKUD-M1	92.02	93.41	**99.20**
EKUD-M1 + EKUD-M2	AKDRM	17.53	17.18	**65.40**
	EKUD	89.99	90.02	**98.52**

4.3 Overall Evaluation

The last scenario posed in the work considers the use of a single source domain against two target domains. In this regard, Table 4 shows the results obtained for such case and compares them against those of the two previous scenarios studied. Note that, for the sake of conciseness, we provide the average results obtained for all different combinations of source and target domains for the different scenarios considered.

As it may be observed, the scenario of a single source and two target domains depicts the least competitive results of the different cases considered. This is a somehow expected outcome given the remarkable difficulty of the configuration

Table 4. Results in terms of classification accuracy for the case of a single source domain against two target ones for the different DA strategies considered. Figures for each scenario represent the average of the individual results obtained for each combination of source and target domains. The results obtained in the two previous scenarios are included for comparative purposes, representing bold values the best performing DA strategy for each scenario posed.

Scenario	DANN	PixelDA	DeepCORAL
One source and two targets	33.74	40.05	**48.05**
One source and one target (Sect. 4.1)	45.81	45.10	**68.07**
Two sources and one target (Sect. 4.2)	81.56	82.42	**88.68**

at issue: a rather limited amount of source data against a challenging scenario comprising two target different domains.

Oppositely to that situation, the case of using two source domains achieves the best overall results, mainly due to having a greater amount of labeled data. Hence, the use of several source domains is of great help in DA approximations since it achieves almost a 20% improvement over the case of a single one.

Regarding the actual UDA methods considered, as in the previous cases, the DeepCORAL stands out as the best performing strategy, achieving PixelDA and DANN lower rates in most cases but being quite similar among them.

Finally, these experiments also prove the relevance of UDA in these types of schemes: on average, the performance achieved when not considering any DA scheme is 16.29%, whereas the lowest performance result obtained by any of the methods considered is 33.74% (DANN method with the single source and two target scenario), achieving a 88.68% in the best-case scenario (DeepCORAL with two source domains of data).

5   Conclusions

Deep Learning (DL) strategies are generally deemed to achieve state-of-the-art performances in recognition-related tasks. One important drawback, however, is their large requirements as regards the amount of annotated data to learn from. Among the different strategies in the scientific literature, the so-called Domain Adaptation (DA) paradigm stands as a promising framework for tackling this issue. Such strategies work on the basis that a model trained on a source domain could be adjusted to transfer its knowledge to a new and different target domain, as long as it shares the same set of categories.

This work evaluates the use of different unsupervised DA algorithms in the context of kitchen utensil recognition for robotics and studies the influence of contemplating several source and target domains. For that, we consider three state-of-the-art methods (DANN, PixelDA, and DeepCORAL) and assess them in different scenarios by combining four different corpora of kitchen utensil images (EKUD, EKUD-M1, EKUD-M2, and AKDRM). Note that three of these datasets have been developed by the authors, hence constituting a relevant contribution to both the DA and robotics communities which may use them for benchmarking purposes.

The results obtained provide a set of relevant conclusions for this case of study: a first one is that the DeepCORAL method proves to be the most suitable strategy for the task of kitchen utensil recognition; moreover, the use of several source data domains remarkably improves the base case of considering a single origin domain, thus stating its importance of such configuration in the overall success of the adaptation task; finally, it is also proved that, even in the worst-case scenario, DA always reports a remarkable improvement with respect to disregarding adaptation processes between source and target domains.

Future work considers extending this study to other state-of-the-art methods, such as VADA, DRCN, DSN, or DJDOT. Furthermore, we also aim to address the modification of the architectures so that they are capable of supporting several sources and target domains in their operation. As the last point, we also aim at extending these conclusions to other benchmarks corpora and cases of study.

References

1. Arbelaez, P., Maire, M., Fowlkes, C., Malik, J.: Contour detection and hierarchical image segmentation. IEEE Trans. Pattern Anal. Mach. Intell. **33**(5), 898–916 (2011). https://doi.org/10.1109/TPAMI.2010.161
2. Bousmalis, K., Silberman, N., Dohan, D., Erhan, D., Krishnan, D.: Unsupervised pixel-level domain adaptation with generative adversarial networks. In: 2017 IEEE Conference on Computer Vision and Pattern Recognition (CVPR), pp. 95–104 (2017). https://doi.org/10.1109/CVPR.2017.18
3. Bousmalis, K., Trigeorgis, G., Silberman, N., Krishnan, D., Erhan, D.: Domain separation networks. Adv. Neural. Inf. Process. Syst. **29**, 343–351 (2016)
4. Damodaran, B.B., Kellenberger, B., Flamary, R., Tuia, D., Courty, N.: DeepJDOT: deep joint distribution optimal transport for unsupervised domain adaptation. In: Proceedings of the European Conference on Computer Vision (ECCV), September 2018
5. Fernando, B., Habrard, A., Sebban, M., Tuytelaars, T.: Unsupervised visual domain adaptation using subspace alignment. In: 2013 IEEE International Conference on Computer Vision, pp. 2960–2967 (2013)
6. Fisher, R.: Edinburgh kitchen utensil database. https://homepages.inf.ed.ac.uk/rbf/UTENSILS
7. Gallego, A.J., Calvo-Zaragoza, J., Fisher, R.B.: Incremental unsupervised domain-adversarial training of neural networks. IEEE Trans. Neural Netw. Learn. Syst. **32**(11), 4864–4878 (2021). https://doi.org/10.1109/TNNLS.2020.3025954
8. Ganin, Y., et al.: Domain-adversarial training of neural networks. J. Mach. Learn. Res. **17**(1), 2096–2030 (2016)
9. Goodfellow, I., et al.: Generative adversarial networks. Adv. Neural Inf. Process. Syst. **3** (2014). https://doi.org/10.1145/3422622
10. Kouw, W.M., Loog, M.: A review of domain adaptation without target labels. IEEE Trans. Pattern Anal. Mach. Intell. **43**(3), 766–785 (2021). https://doi.org/10.1109/TPAMI.2019.2945942
11. OpenAI: AI and compute. OpenAI blog (2018)
12. O'Mahony, N., Campbell, S., Carvalho, A., Harapanahalli, S., Hernandez, G.V., Krpalkova, L., Riordan, D., Walsh, J.: Deep learning vs. traditional computer vision. In: Arai, K., Kapoor, S. (eds.) CVC 2019. AISC, vol. 943, pp. 128–144. Springer, Cham (2020). https://doi.org/10.1007/978-3-030-17795-9_10
13. Shu, R., Bui, H., Narui, H., Ermon, S.: A DIRT-t approach to unsupervised domain adaptation. In: International Conference on Learning Representations (2018)
14. Sun, B., Feng, J., Saenko, K.: Return of frustratingly easy domain adaptation. In: Proceedings of the AAAI Conference on Artificial Intelligence, vol. 30 (2016)
15. Sun, B., Saenko, K.: Deep CORAL: correlation alignment for deep domain adaptation. In: Hua, G., Jégou, H. (eds.) ECCV 2016. LNCS, vol. 9915, pp. 443–450. Springer, Cham (2016). https://doi.org/10.1007/978-3-319-49409-8_35

An Innovative Vision System for Floor-Cleaning Robots Based on YOLOv5

Daniel Canedo$^{(\boxtimes)}$, Pedro Fonseca , Petia Georgieva ,
and António J. R. Neves

IEETA/DETI, University of Aveiro, 3810-193 Aveiro, Portugal
{danielduartecanedo,pf,petia,an}ua.pt

Abstract. The implementation of a robust vision system in floor-cleaning robots enables them to optimize their navigation and analysing the surrounding floor, leading to a reduction on power, water and chemical products' consumption. In this paper, we propose a novel pipeline of a vision system to be integrated into floor-cleaning robots. This vision system was built upon the YOLOv5 framework, and its role is to detect dirty spots on the floor. The vision system is fed by two cameras: one on the front and the other on the back of the floor-cleaning robot. The goal of the front camera is to save energy and resources of the floor-cleaning robot, controlling its speed and how much water and detergent is spent according to the detected dirt. The goal of the back camera is to act as evaluation and aid the navigation node, since it helps the floor-cleaning robot to understand if the cleaning was effective and if it needs to go back later for a second sweep. A self-calibration algorithm was implemented on both cameras to stabilize image intensity and improve the robustness of the vision system. A YOLOv5 model was trained with carefully prepared training data. A new dataset was obtained in an automotive factory using the floor-cleaning robot. A hybrid training dataset was used, consisting on the Automation and Control Institute dataset (ACIN), the automotive factory dataset, and a synthetic dataset. Data augmentation was applied to increase the dataset and to balance the classes. Finally, our vision system attained a mean average precision (mAP) of 0.7 on the testing set.

Keywords: Computer vision · Object detection · Deep learning · Floor-cleaning robots

1 Introduction

Supporting floor-cleaning robots with a robust vision system is getting more popular thanks to the new functionalities it can provide. From camera based

Supported by the project "i-RoCS: Research and Development of an Intelligent Robotic Cleaning System" (Ref. POCI-01-0247-FEDER-039947), co-financed by COMPETE 2020 and Regional Operational Program Lisboa 2020, through Portugal 2020 and FEDER.

mapping [10] to detecting dirty spots on the floor [5], a robust vision system allows for a more efficient cleaning and navigation. Previous published works that managed to implement a vision system on their floor-cleaning robots, seem to have done it for several purposes: economize cleaning resources [8], distinguishing between different types of dirtiness [13], and distinguishing between dirty spots and useful objects [4]. In this work, we attempt to tackle all of these challenges. The vision system that we propose is able to distinguish between dirty spots and useful objects by carefully selecting training images where objects are present. We tackle three types of dirtiness: solid dirt, liquid dirt and stains. And, finally, we integrated this vision system into an autonomous cleaning robot prototype in order to economize the cleaning resources by controlling water, detergent and mechanical parts of the cleaning system based on the dirty information.

This document is an extension of our previous work [7], and presents the application of the proposed vision system in a real-world robot prototype. In the previous work, we explored the strengths of implementing a Deep Learning solution based on the YOLOv5 framework [9] to detect dirty spots. In that work, we tried to tackle the most relevant challenges pointed by the literature in this application. Those problems revolve around lack of data, complex floor patterns, extreme light intensities, blurred images caused by the robot movement, and dirt/clean discrimination. We tackled these problems by generating a synthetic dataset with complex floors that contain objects other than dirty spots, adding simulated light sources and shadows to the resulting artificial images. The main conclusion that we retrieved from that work was that generating synthetic data using complex floors that contain objects to train a YOLOv5 model is a viable solution to not only detect dirty spots, but also to distinguish between useful objects from dirt, as long as there is enough dirt variety. We also found that stains contributed to a considerable amount of false positives during the testing step, since this type of dirtiness is often overlooked in the literature and, therefore, not labelled.

In this document, the stains on the ACIN dataset [1] are annotated, complementing the annotations proposed in our previous work [7]. A real-world dataset was captured in an automotive factory using the floor-cleaning robot, and part of it was annotated. A synthetic dataset is generated using the tool [2] provided by [4], that we had to improve to use in this application. A data augmentation pipeline is proposed to balance the number of samples per class. Then, a YOLOv5 model is trained using a hybrid dataset consisting of the ACIN dataset, the automotive factory dataset and the synthetic dataset. The new annotations and the hybrid dataset are public, and the links provided at the end of the document. A self-calibration algorithm is adapted from [12] to stabilize the image intensity from the cameras installed on the floor-cleaning robot. Finally, the main results and conclusions are discussed.

This document is structured as follows: Sect. 2 presents the related work; Sect. 3 presents the methodology; Sect. 4 presents the results and discussion; Sect. 5 presents the conclusion.

2   Related Work

Building a vision system to aid floor-cleaning robots has been tackled in several different ways in the literature. Some works try to approach the problem with pre-processing and unsupervised techniques, which main advantage lies in avoiding a learning step. This approach, while not needing previous knowledge to be used, have some problems, such as detecting everything that is not within the floor pattern as dirt. Wires, objects, walls, doors, shoes, carpets, just to name a few examples, have a high chance of being detected as a dirty spot by this approach. More problems revolve around blurred images, uneven illumination, and floors with multiple patterns. Other works try to approach the problem with object detection techniques, mainly based on Deep Learning. The literature seems to indicate that this approach is more successful, however it also has some problems, specially if one does not know where the floor-cleaning robot will operate. It is quite difficult to organize a robust and varied training data that covers most real-word scenarios that the floor-cleaning robot might encounter. If the area and dirty spots that the floor-cleaning robot needs to cover is known, this problem can be overcome by capturing a dataset for the training step in that area. However, if it is unknown, the training dataset must be diverse both in floor patterns and dirt variety to enable the vision system to handle unknown circumstances with as much accuracy as possible.

Grünauer et al. [8] proposed an unsupervised approach based on Gaussian Mixture Models (GMMs) to detect dirty spots. Firstly, they do several pre-processing steps such as converting the captured images to the CIELAB color space, which main advantage lies in separating colour information from illumination. Then, the gradient is calculated, and the images are divided into blocks. The mean and standard deviation are calculated for each block, and those values are used by the GMMs to learn the floor pattern. If something in a given image breaks this pattern, it is considered as dirt. This approach suffers from some problems described above. However, it is a viable solution that does not require a learning step.

Ramalingam et al. [13] proposed a multi-stage approach to detect solid and liquid dirt based on a Single-Shot MultiBox Detector (SSD), a MobileNet, and a Support Vector Machine. The MobileNet extracts features, the SSD detects the dirty spots, and the SVM classifies liquid dirt based on size to identify spots that are harder to clean. Their strategy was to collect data that the robot might face during its cleaning operations, manually label it and use it to train the robot's vision system. This strategy allowed the robot to attain an accuracy higher than 96% in detecting solid and liquid dirt. The same first author proposed a three-layer filtering framework which includes a periodic pattern detection filter, edge detection, and noise filtering to detect dirty spots on complex floor patterns in another work [14]. The periodic pattern detection filter is able to identify the floor pattern and dirty spots, since floors generally have a defined pattern. The edge detection step is performed on the background subtracted images to sharpen the edges that may get blurred in the previous step. Finally, they filter the residual noise through a median filter. This work then proceeds to show promising results on some challenging images.

Bormann et al. [4] tackled the lack of data in this particular application by developing a tool to artificially generate data [2]. This tool is able to blend dirty spots into clean floors in random locations, add simulated light sources and shadows, and apply geometric transformations to both floors and dirty spots. Consequently, this tool enables the generation of large datasets from a small amount of images. Then, the authors proposed using the YOLOv3 framework to detect dirty spots. The YOLO family is basically an object detection algorithm supported by a CNN. This algorithm divides the image into a grid and outputs a bounding box and its probability of belonging to a certain class for each block of the grid. This approach allowed the YOLO framework to attain state-of-the-art results measured through the mAP in several benchmarks, such as COCO. And for the application of detecting dirty spots on the floor, this proposal also managed to obtain state-of-the-art results, demonstrating better performance than GMMs [8].

3   Methodology

This section is divided into subsections addressing the several steps carried out in this work, from data preparation to the experiments.

3.1   Vision System

The vision system of the robot was built based on the Robot Operating System (ROS). ROS provides a set of libraries and tools to build robot applications. We have built two nodes for our vision system: a node to access the cameras and a node to detect dirty spots. The node to access the cameras starts them up, calibrates the colormetric parameters of both cameras, accesses the captured images, and publishes them to the dirt detection node. Figure 1 illustrates the importance of calibrating the cameras.

Fig. 1. An example of an overexposed image on the left and a calibrated image on the right.

This is a result of a self-calibration algorithm adapted from [12] which is based on the image luminance histogram. With this histogram, it is possible to indicate if the image is underexposed or overexposed by dividing the histogram of the grayscale image into five regions to calculate its mean sample value (MSV). The histogram is uniform if MSV $\approx$ 2.5. Based on this value, we implemented a Proportional-Integral controller (PI) to regulate the gain and exposure of both cameras. This results in uniform images, which makes dirty spots more visible.

Afterwards, the dirt detection node receives those calibrated images, detects the dirty spots, calculates the dirty area present in the images, and publishes that information. Based on the dirty area calculated from the images captured by the front camera, the floor-cleaning robot will regulate its speed, water, and detergent. If significant dirty spots are detected on images captured by the back camera, it is an indication that the cleaning was not successful. All the regions that were unsuccessfully cleaned are mapped such that the floor-cleaning robot goes back for a second sweep at the end of its cleaning procedure. Figure 2 shows our floor-cleaning robot prototype.

Fig. 2. Floor-cleaning robot prototype where the vision system was implemented and tested.

3.2 Automotive Factory Dataset

A real-world dataset was captured using the floor-cleaning robot prototype under development. This dataset was captured in a challenging environment: an automotive factory. It was noted a huge variety in dirty spots, particularly in size. Since the overall dirty spots on the ACIN dataset are minuscule, part of the automotive factory dataset was used to train the YOLOv5 model. It is expected that by doing so, the model is capable of accurately detecting dirty spots independently of their size. 39 captured images were annotated using the LabelImg tool [3], resulting in 92 instances of solid dirt, 3 instances of liquid dirt, and 804 instances of stains. Figure 3 illustrates some samples of this dataset.

Fig. 3. Automotive factory dataset annotations using LabelImg.

3.3 Stain Annotations on the ACIN Dataset

As mentioned in Sect. 1, stains are also a type of dirtiness that is often overlooked during labeling. In this work, all stains on the ACIN dataset were annotated using the LabelImg tool. These new annotations complemented the ones proposed in our previous work [7], resulting in 1785 instances of solid dirt, 634 instances of liquid dirt, and 4162 instances of stains. The ACIN dataset consists of 968 images. Figure 4 illustrates some examples of stain annotations.

Fig. 4. Stain annotations using LabelImg.

3.4 Data Augmentation

Data augmentation is a well known technique used in Machine Learning to increase and enhance training data. This can be done by applying geometric transformations, color space augmentations, feature space augmentations, random erasing, and so on. This is particularly useful in tackling overfitting. Overfitting occurs whenever the network learns a function with very high variance to fit the training data, which makes it unreliable when facing new data. This phenomenon can happen when the network is too complex and/or the training data is too small.

Since the YOLOv5 network is complex, the ACIN dataset only has 968 images, and the automotive factory dataset only has 39 annotated images, it was expected to occur overfitting during the training step. Therefore, we performed data augmentation on both datasets. For this, we used a tool called

Albumentations [6]. This tool offers several image transformations. However, we had to make some modifications to the original code to fit our needs. Whenever we applied a perspective transformation to the image, sometimes dirty spots on the transformed image were out of bounds. The tool deals with these scenarios with a minimum visibility threshold, deleting any bounding box if the threshold is not met, however it was not working properly for our case. For this reason, changes were implemented in the tool such that no bounding box is deleted, and its position is returned even if it completely goes out of bounds. This change gave us the possibility to handle these situations better. Since some dirty spots are quite small, sometimes during a perspective transformation, dirty spots that were near the image borders hardly became visible. To avoid this partial visibility problem, we only generated images where dirty spots were fully visible or fully invisible.

We divided the ACIN dataset and the automotive factory dataset into training sets and validation sets. We only applied data augmentation on the training sets. From each ACIN training image, four were generated, and from each automotive factory training image, five were generated. This data augmentation was performed by applying the following transformations with a probability p:

- Flip either horizontally, vertically, or both horizontally and vertically. ($p = 0.75$).
- Randomly change hue, saturation and value, or randomly change gamma, or randomly change brightness and contrast ($p = 1$).
- Randomly shift values for each channel of the input RGB image ($p = 0.5$).
- Perform a random four point perspective transform ($p = 0.75$),

  Figure 5 illustrates an augmentation example.

 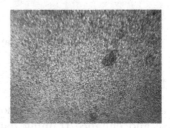

Fig. 5. Original image on the left, augmented image on the right.

3.5   Synthetic Dataset

In this work, we use an adapted version of the data generation tool proposed by [4]. As mentioned in Sect. 2, this tool is able to blend dirty spots into clean floors in random locations, add simulated light sources and shadows, and apply geometric transformations to both floors and dirty spots. We made some adaptations to this tool as mentioned in our previous work [7]. It is now able to generate liquid dirt by manipulating their transparency, it checks if there is a mask associated with the floor image such that dirty spots are only generated on the floor, and the output labels are converted to the YOLO format.

In order to generate these images, we needed floor images, solid dirt samples, and liquid dirt samples. Regarding floor images, 496 samples were obtained from Google Search. It was given priority to floor images that approximately simulated the distance from the floor (0.7 m) and angle (45 to 70 °C downwards) of the cameras placed on our floor-cleaning robot. Some images only represented a clean floor, while others had some objects in it such as shoes, wires, doors, carpets, walls, and so on. This helps the network to distinguish between a dirty spot and an object, reducing false positives. Regarding solid dirt, 141 samples were obtained from the Trashnet dataset [15] and Google Search. As for liquid dirt, 45 samples were obtained from Google Search. All of these samples were segmented using a tool proposed by Marcu et al. [11].

Before the data generation, we divided the floor images, solid dirt samples, and liquid dirt samples so the training set and validation set do not use the same images. It is possible to adjust the parameters of the synthetic data generation tool such as the number of dirty spots per image and the number of augmentations. These parameters were adjusted considering the amount of dirt instances on the ACIN dataset and the automotive factory dataset. This was done so that when we created the hybrid dataset by combining the three datasets, it should have a balanced number of samples per class. We have also increased the dirt size ceiling to compensate for the ACIN dataset small dirt size.

3.6 Training

For the experiment, we created a hybrid dataset. Specifically, the previously created training sets were combined, as well as the validation sets. Table 1 aims to provide a better insight on the training and validation sets.

Table 1. Training and validation sets.

Set	Images	Solid dirt	Liquid dirt	Stains
Training	9470	23670	23426	22367
Validation	388	506	584	599

There are several YOLOv5 models with different complexities. We chose the medium network (YOLOv5m6) since it was the best performing one for this type of data, as concluded in our previous work [7]. Transfer Learning was implemented by freezing the backbone, which helps to tackle overfitting. We used the Stochastic Gradient Descent (SGD) optimizer with a learning rate of 0.01 and a decay of 0.0005. The image size was set to 640×640, the batch size was set to 32, and the training was done over 50 epochs with early stopping, saving the best weights. This was performed with an Nvidia GeForce RTX 3080 GPU and an AMD Ryzen 5 5600X 6-Core 3.7GHz CPU.

3.7 Testing

We created a small testing set to test our network and the network from our previous proposal [7] for comparison. For this, we annotated a few images from the automotive factory dataset that are different from the ones used in the training, as well as a few images from the IPA dataset [1]. The following Table 2 aims to provide a better insight on the testing set.

Table 2. Testing set.

Set	Images	Solid dirt	Liquid dirt	Stains
Testing	24	80	2	212

The type of dirtiness on the automotive factory dataset mainly consists of stains and solid dirt, but only a few liquid dirt instances. That is why the classes on the testing set are not balanced. However, the testing step will provide an overall view of how the proposed vision system is capable of handling real-world settings. The testing was done for a binary classification (dirty or not dirty) since what is important in our application is the detection of dirty spots, and not so much the classification of those given spots.

4 Results and Discussion

Table 3 shows the mAP on the testing set of this work compared with our previous work. Figure 6 shows the respective precision-recall curves.

Table 3. Comparing the network of this work with our previous work.

	Previous YOLOv5m6 [7]	Current YOLOv5m6
mAP	0.29	**0.70**

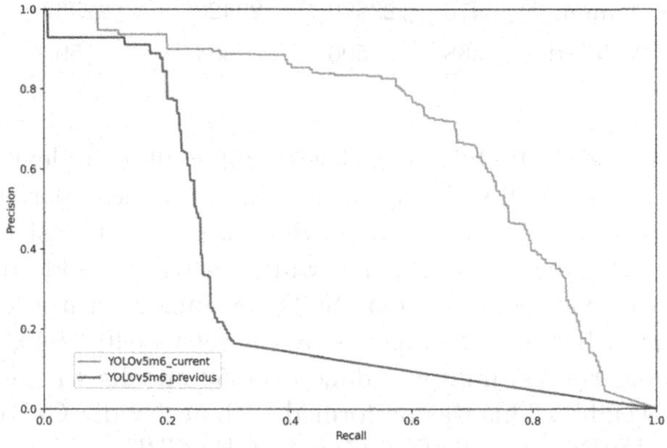

Fig. 6. Precision-recall curves on the testing set.

These results show how relevant stains are. This type of dirtiness is quite common because it is mostly caused by shoes. Although the previous network is able to detect some stains, it cannot outperform a network that was trained considering this type of dirtiness. Figure 7 illustrates the results of some images of the testing set for both networks, side by side.

It is possible to observe that the current network not only performs better in detecting stains, but also in detecting the other types of dirtiness. This was expected, since the previous network was only trained with a synthetic dataset with two classes: solid dirt and liquid dirt. The current network was trained on a hybrid dataset with one more class than the previous network: stains. Class balance and data augmentation was also considered in this work. Therefore, these results were expected and desired, since this is the network that will be used by our floor-cleaning robot.

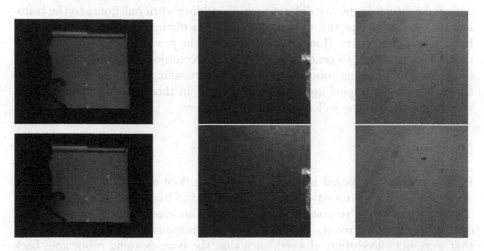

Fig. 7. Results on the testing set. Previous network on the top, current network on the bottom.

However, these results can still be improved by increasing the variety of floor images, solid dirt, liquid dirt, and stains of the training set. Dirty spots can come in different sizes, different colours, and complex shapes. Adding this to the limitless amount of floor patterns makes it sometimes difficult to detect dirty spots. During the testing step the network encountered some problems, mainly when the floor is worn out. Figure 8 illustrates an image where the network struggled to distinguish between holes and dirty spots.

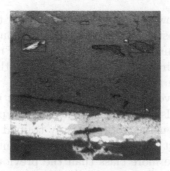

Fig. 8. Results on an image from the testing set.

As it is possible to observe, there are some holes in this image that are detected as dirty spots. We did not consider adding worn out floors to the training set, and therefore the network struggles to distinguish worn out features from actual dirty spots. However, this is one of the problems of trying to have a network that strives to generalize every single scenario. Generally, the application of a floor-cleaning robot is self-contained, meaning that the floor patterns and expected dirty spots are known in advance. In those cases, one can build a vision system that has a close to perfect efficiency.

5   Conclusion

In this work we proposed a vision system for a floor-cleaning robot. This is a novel approach that uses two cameras, one on the front and one on the back. The front camera is responsible for adjusting the speed, water, and detergent of the floor-cleaning robot. The back camera is responsible for mapping regions that were not successfully cleaned, such that the floor-cleaning robot goes back for a second sweep later on. The colormetric parameters of the cameras are autonomously calibrated to adapt the light conditions and floor type, a major contribution to spot dirt. A hybrid dataset was built using the ACIN dataset, the automotive factory dataset which was captured using our floor-cleaning robot, and a synthetic dataset. In this work, we paid special attention to a type of dirtiness that is often overlooked in the literature: stains. For this reason, we annotated all the stains on the ACIN dataset.

Finally, we trained a YOLOv5m6 network on the hybrid dataset. We then tested that network on a testing set which was built using a few images from the automotive factory dataset and the IPA dataset. We attained a mAP of 0.7, which was a considerable improvement over the result of our previous work: 0.29.

The ACIN annotations are available at https://tinyurl.com/3hjpxehw and the built Hybrid dataset is available at https://tinyurl.com/2p8ryr7s.

References

1. ACIN and IPA datasets. https://goo.gl/6UCBpR. Accessed 17 Jan 2022
2. IPA dirt detection. http://wiki.ros.org/ipa_dirt_detection. Accessed 12 Jan 2022
3. Tzutalin. labelimg. https://github.com/tzutalin/labelImg. Accessed 14 Jan 2022
4. Bormann, R., Wang, X., Xu, J., Schmidt, J.: DirtNet: visual dirt detection for autonomous cleaning robots. In: 2020 IEEE International Conference on Robotics and Automation (ICRA), pp. 1977–1983. IEEE (2020)
5. Bormann, R., Weisshardt, F., Arbeiter, G., Fischer, J.: Autonomous dirt detection for cleaning in office environments. In: 2013 IEEE International Conference on Robotics and Automation, pp. 1260–1267. IEEE (2013)
6. Buslaev, A., Iglovikov, V.I., Khvedchenya, E., Parinov, A., Druzhinin, M., Kalinin, A.A.: Albumentations: fast and flexible image augmentations. Information **11**(2), 125 (2020)
7. Canedo, D., Fonseca, P., Georgieva, P., Neves, A.J.: A deep learning-based dirt detection computer vision system for floor-cleaning robots with improved data collection. Technologies **9**(4), 94 (2021)
8. Grünauer, A., Halmetschlager-Funek, G., Prankl, J., Vincze, M.: The power of GMMs: unsupervised dirt spot detection for industrial floor cleaning robots. In: Gao, Y., Fallah, S., Jin, Y., Lekakou, C. (eds.) TAROS 2017. LNCS (LNAI), vol. 10454, pp. 436–449. Springer, Cham (2017). https://doi.org/10.1007/978-3-319-64107-2_34
9. Jocher, G., et al.: ultralytics/yolov5: v5.0 - YOLOv5-P6 1280 models, AWS, Supervise.ly and YouTube integrations, April 2021. https://doi.org/10.5281/zenodo.4679653
10. Kang, M.C., Kim, K.S., Noh, D.K., Han, J.W., Ko, S.J.: A robust obstacle detection method for robotic vacuum cleaners. IEEE Trans. Consum. Electron. **60**(4), 587–595 (2014)
11. Marcu, A., Licaret, V., Costea, D., Leordeanu, M.: Semantics through time: semi-supervised segmentation of aerial videos with iterative label propagation. In: Proceedings of the Asian Conference on Computer Vision (2020)
12. Neves, A.J.R., Trifan, A., Cunha, B.: Self-calibration of colormetric parameters in vision systems for autonomous soccer robots. In: Behnke, S., Veloso, M., Visser, A., Xiong, R. (eds.) RoboCup 2013. LNCS (LNAI), vol. 8371, pp. 183–194. Springer, Heidelberg (2014). https://doi.org/10.1007/978-3-662-44468-9_17
13. Ramalingam, B., Lakshmanan, A.K., Ilyas, M., Le, A.V., Elara, M.R.: Cascaded machine-learning technique for debris classification in floor-cleaning robot application. Appl. Sci. **8**(12), 2649 (2018)
14. Ramalingam, B., Veerajagadheswar, P., Ilyas, M., Elara, M.R., Manimuthu, A.: Vision-based dirt detection and adaptive tiling scheme for selective area coverage. J. Sens. **2018** (2018)
15. Yang, M., Thung, G.: Classification of trash for recyclability status. CS229 Project Report 2016 (2016)

LIDAR Signature Based Node Detection and Classification in Graph Topological Maps for Indoor Navigation

Sergio Lafuente-Arroyo[1]([✉]) [iD], Saturnino Maldonado-Bascón[1] [iD],
Roberto Javier López-Sastre[1] [iD], Alberto Jesús Molina-Cantero[2] [iD],
and Pilar Martín-Martín[1] [iD]

[1] GRAM, Departamento de Teoría de la Señal y Comunicaciones,
Universidad de Alcalá, Alcalá de Henares (Madrid), Spain
{sergio.lafuente,saturnino.maldonado,robertoj.lopez,p.martin}@uah.es
[2] Departamento de Tecnología Electrónica,
E.T.S.I. Informática. Universidad de Sevilla, Sevilla, Spain
almolina@us.es
https://gram.web.uah.es

Abstract. Topological map extraction is essential as an abstraction of the environment in many robotic navigation-related tasks. Although many algorithms have been proposed, an efficient and accurate modelling of a topological map is still challenging, especially in complex and symmetrical real environments. In order to detect relevant changes of trajectory or a robotic platform, we propose a feature extraction model based on LIDAR scans to classify the nodes of indoor structures. As a first approach we support the experiments in a preloaded metric map, which has been used as reference for locating the nodes. Experiments are conducted in a real scenario with a differential robot. Results demonstrate that our model is able to establish the graph automatically and with precision, and that it can be used as an efficient tool for patrolling.

Keywords: Assistive robot · Topological map · Node classification

1 Introduction

In many robotic navigation-related tasks, abstracting the real environment where mobile robots carry out some missions can be of a great benefit. In particular extracting a simple topological graph-like representation from a more complex detailed metric map is often required for path-planning and navigation. A topological graph, as defined by Simhon and Dudek [1], is a graph representation of an environment in which the important elements are defined along with the transitions among them. In complex real indoor environments, such as hospitals, residences and office buildings, the structure presents high level of symmetry and usually consists of many corridors in which rooms are distributed on both sides. The dynamic nature of these environments, which are generally frequented by

© Springer Nature Switzerland AG 2022
A. J. Pinho et al. (Eds.): IbPRIA 2022, LNCS 13256, pp. 390–401, 2022.
https://doi.org/10.1007/978-3-031-04881-4_31

many people, means that the topological map can undergo continuous changes. Thus, for example, the opening or closing of doors adaptively forces the topological map to create new paths and remove others, respectively. The dynamic and complex nature of these scenarios is a major challenge.

In this paper, we focus on the extraction of nodes in symmetrical indoor environments based on distributions of numerous corridors. In our approach, nodes represent relevant changes of direction, such as an end of aisle or a bifurcation, where several outlets are possible. As depth information for node classification, we rely on LIDAR scanning. Figure 1 shows a block diagram of the proposed approach. The complete process is performed automatically using the metric map as input and consists of three main modules: 1) pose estimation, which uses a particle filter to track the pose of a robot against a known map; 2) a classification of the scene to detect and classify the nodes (topological modelling); and 3) a local policy sub-system to determine the sequence of movements. The system can be in two operating modes: exploration or patrol. In the first one, the platform creates the topological map of the scenario. On the other hand, when the agent is on patrol performs a routine navigation based on the topological map already created. It is in patrol mode where the maintenance of the topological map is performed, since the robot is able to discover changes when detecting the presence of new nodes or the disappearance of previous ones.

Taking into account the pose estimations of the particle filter, the system positions the detected nodes in route on the pre-loaded metric map. The model has been evaluated by means of a case study using an area of a university building.

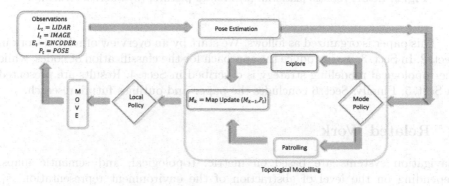

Fig. 1. Overview of the proposed system.

In summary, the main contributions of this work are as follows:

– We introduce a novel approach for topological mapping, adapted to symmetrical spaces based on corridor layouts, where nodes identify relevant changes of trajectory. Although similar approaches can be found in [2,3] based on classification of the positions of the robot (rooms, corridors, doorways, hall-ways, ...), we focus on the detection of corridor junctions through an efficient SVM-based approach.

- We use the semantic information from the objects found by means of a trained YOLO-v3 [4] based detector to automatically locate the nodes in the map.
- We present a thorough experimental evaluation embedding the proposed app-roach in our own low-cost assistive robotic platform (see Fig. 2). It is a dif-ferential wheeled robot, equipped with two motors and their corresponding encoders, which are all controlled with an open-source Arduino board. The sensing part is composed of an Intel RealSense D435 camera and a LIDAR. An on-board laptop with a Nvidia board Quadro RTX 5000 and/or a Nvidia Jetson TX2 board are provided for intensive computation.

Fig. 2. LOLA robotic platform. (a) Frontal picture. (b) Internal structure.

This paper is organized as follows. We start by an overview of related work in Sect. 2. In Sect. 3, we introduce the approach for the classification of nodes, while the topological modelling strategy is described in Sect. 4. Results are presented in Sect. 5. Finally, Sect. 6 concludes the paper and outlines future research.

2   Related Work

Navigation systems are based on metric, topological, and semantic maps, depending on the level of abstraction of the environment representation [5]. In the literature, a large variety of solutions to this problem is available. One intuitive way of formulating SLAM is to use a graph whose nodes correspond to the poses of the robot at different points in time and whose edges represent constraints between the poses.

Regarding topological navigation, since the first developments, the global conception of the system has attracted the interest of several authors. Surveys of models for indoor navigation are provided by [6–8] among others. Comparison of various graph-based models is provided in a very clear way in Kielar et al. in [9]. The main contribution of this work was the development of a low-cost process for building navigation graphs based only on geometry nodes. Topomap [10] is a framework which simplifies the navigation task by providing a map to the robot which is tailored for path planning use. Each vertex corresponds to a certain partially enclosed area within the environment (e.g. a room) which is connected to neighboring vertices. In our case, vertices represents points where a movement decision must be taken.

Enriched information can be provided by different sensors. A robot navigational method was presented in [11] based on an Extremum Seeking algorithm using Wireless Sensor Network topology maps. On the other hand, each node in the graph is associated with a panoramic image in [12]. The problem of image-goal navigation involves navigating in a novel previously unseen environment. For this task, a neural topological map was constructed. However, there is no additional information in our low-cost platform than the captured by the RGB-D camera or LIDAR.

A more similar to ours classification of space was given in [13] what they call types of corridors. They performed classification from different positions while the robot was moving and this allows to avoid wrong classifications. A Bayesian classifier was used to obtain the corridor types.

3   Classification of Nodes in the Topological Map

A topological map is a graph-based representation of the environment. Each node corresponds to a characteristic feature or zone of the environment. In our approach, we consider as points of interest those ones that imply a change of trajectory. Regarding to this criterion, common corridor structures can be classified into four node categories in our map:

- End node: there is no outlet at the front, neither from the left nor from the right of the corridor. The agent cannot follow the path.
- Node 'T': there are two outlets for the agent since the corridor presents two lateral bifurcations. This node type is also called Node 'Y' in other works as [13].
- Node 'L': the path presents an marked change of direction. We have considered in this category changes of direction involving angles greater than 40, regardless of whether the turn is from the left or from the right.
- Cross node: the corridor presents three or more outlets (two laterals: left and right, and another frontal one).

In addition, we introduce the category of a complementary type called No Node. It is used to incorporate additional context into the exploration, and refers to the zones of transition between the above mentioned categories, while the robot is moving along the corridors. Figure 3 represents the different categories considered in this work. Thus, for each type of node, we represent the location of the robot (X) and the possible trajectories depicted in green.

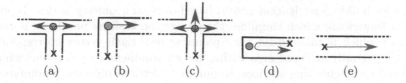

Fig. 3. Types of nodes considered in a building with dense distribution of corridors. (a) Node 'T'. (b) Node 'L'. (c) Cross node. (d) End node. (e) No node.

3.1 Extraction of Features

Different techniques can be utilized to obtain depth information. In this work we have used a laser range scanner, more specifically, the RPLIDAR A1 of the manufacturer Slamtec [14]. RPLIDAR is a low cost 2D LIDAR sensor suitable for indoor robotic applications. It provides 5.5 hz/10 hz rotating frequency within a 12-meter range distance. Each raw measurement is a tuple with the following format: quality, angle, distance. The quality indicates the reflected laser pulse strength, the measurement heading angle is given in degrees and object distances are related to the sensor's rotation center.

Even when the LIDAR sensor has one degree angular resolution, the angular resolution of RPLIDAR is not necessarily regular, that is, the spacing between two points is not necessarily the same. This is the reason why we need to fit a linear interpolation method to fill missing values. However, not all real depth curves are ideal. Due to sensor noises, irregular surfaces and obstacles, the capture contains fake information. Reflected sunlight from windows and directions of great depth provoke, in general, outliers of low strength. To overcome these disturbances, we propose a filtering process in which measurements of low reflected strength are discarded.

As an example, Fig. 4 shows an scan laser, where Fig. 4(a) represents the raw samples (black markers) and the quality is shown in Fig. 4(b). The signature (Fig. 4(c)) is obtained after having set a quality threshold of strength equal to 10. Note that we limit the angular range of the laser scans from $-90°$ to $90°$ because depending on the mobile platform structure the back side may intercept the beam.

The features play an important role in the classification algorithms to identify node types. We have established as parameter the number of bins N_b, which is related to the number of intervals into which we divide the angular range of measurements. Thus, the depth vector or LIDAR signature can be defined as $\mathbf{d} = (d_1, d_2, \ldots, d_{N_b})^T$. Two strategies have been considered to define the value of each feature d_i: 1) the normalized value of LIDAR signature at the center of the corresponding bin considering the maximum range distance of LIDAR (12 m); and 2) a binary feature that can be interpreted as a prediction of whether there is explorable area in the particular direction or not.

Figure 5 shows some examples of the proposed LIDAR signature for different nodes having fixed $N_b = 25$. For each node we can observe the panoramic image

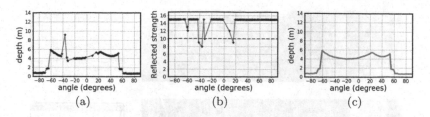

Fig. 4. Process of extraction of LIDAR signature. (a) Raw LIDAR scan. (b) Laser pulse strength. (c) Filtered signature.

in the range of interest, where the green arrows indicate the possible outlets, and the LIDAR signatures. Binary sequences have been obtained by setting a depth threshold $T_h = 4$ m, which is represented by a dotted line. The green and red boxes show explorable (free space) and non-explorable directions, respectively.

It is important to note that different patterns can be captured in the same type of node depending on the input access. Thus, the patterns are quite different in Figs. 5(b) and (c) even though they correspond to the same node, but with different robot poses. LIDAR signatures are highly dependent on the robot-corridor orientation. In this work, a robot controller ensures that during the reposition phase (such as the turning and obstacle avoiding phase), the classification process is suspended to prevent wrong observations. In this way, classification is only activated when the robot moves along the central line between the side walls.

The LIDAR signature feeds the input of our classifier whose output determines the node type. Because the efficiency of Support Vector Machines (SVMs) [15]-based approaches for classification has been widely tested in the scientific community, we have used this technique as a base of our classification module, also because of its generalization properties.

4 Topological Modelling

Taking into account the pose estimations of the particle filter when the robot is moving in the indoor environment, the system is able to positioning the detected nodes on a pre-loaded metric map. Figure 6(a) represents an example of detected nodes where our four categories, including No-Node, appear over an occupancy map. Here, white area denotes free locations, whereas black means occupied positions and grey means inaccessible. Each detected node is depicted by a different color depending on its category and transitions between relevant zones are classified as no-nodes.

When the mobile robot moves, it receives multiple observations of the same node and progressively obtains the classification. The nodes of the topological map (see Fig. 6(b)) are computed in real-time as the centroids of the resulting clusters using an agglomerative clustering algorithm with some restrictions. Thus, we set the maximum distance between nodes of the same type withing a

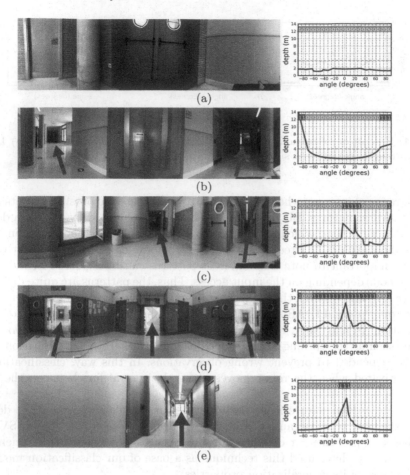

Fig. 5. Examples of node signatures. (a) End node. (b) Node T (example 1). (c) Node T (example 2). (d) Cross-node. (e) No node.

cluster to a value $d_{max} = 4$ m and set the minimum number of points per cluster as $N_{min} = 2$. The generated map is represented using a graph, which is denoted by G_t at time t. Two nodes n_i and n_j are connected by an edge $E_{i,j}$. In order to build the graph we take into account the fact that all accesses in each node must be connected to other nodes. The existence of an end-node implies the end of an exploring path. From these graphs, we can establish a patrolling system in which nodes constitute the list of way-points to cover.

Pose subsystem (see Fig. 1) estimates the location of nodes on the pre-loaded metric map. It is based on a particle filter that uses as input information in each iteration the previous estimated pose, the odometry and the image captured by the RGB-D camera. Multiple object detection allows for a visual interpretation of the scene in the task of estimating the robot pose. For this purpose, we have implemented an object detector based on the YOLO-v3 [4] network, with

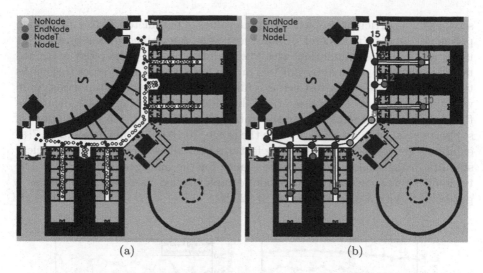

(a)   (b)

Fig. 6. Extraction of topological modelling. (a) Map of identification and localization of node clouds. (b) Topological map representation where each node represents a keypoint and the links connect adjacent nodes. (Color figure online)

categories of objects typically found in indoor environments. Specifically, we ran a fine-tuning process retraining YOLO to be adapted in the new domain of our own dataset, which includes the following ten categories: window, door, elevator, fire extinguisher, plant, bench, firehose, lightbox, column and toilet.

5   Results

In this work we collected our own dataset to evaluate the performance of the proposed node classification framework. Specifically, we have worked in the building of the Polytechnic School of the University of Alcalá, which is distributed in four floors of approximately 10,000 m². The map shown in Fig. 6 corresponds to one of the four similar areas of a floor. The dataset has been captured in this building and is composed of a training set and a test set, where the captures of both sets correspond to different areas. The signatures in this dataset cover a wide range of poses at each type of node. Table 1 presents the description about the dataset.

Table 1. Description of the indoor nodes dataset (number of samples)

	No node	End node	Node 'T'	Cross node	Node 'L'
Training set	137	138	133	44	107
Test set	68	69	67	22	54

One problem that faces the user of an SVM is how to choose a kernel and its specific parameters. Applications of an SVM require a search for the optimum settings for a particular problem. Optimal values of parameters C (regularization parameter) and γ (parameter of influence) were determined in our problem by using a 5-fold cross-validation with the training set. We generated a trained model using these optimal parameters on the full training set. Figure 7(a) shows the overall hit rate of the test set as a function of the number of intervals N_b in which the feature vector is discretized. In addition, as a significant indicator of the computational complexity and overfitting, Fig. 7(b) represents the number of support vectors as a function of N_b in the range [10,100]. In comparison, Fig. 7(a) exhibits a great superiority of analog features over binary ones and shows a clear stability of the analog pattern performance for the analyzed range of N_b. The trade-off between accuracy and model complexity led us to set $N_b = 50$ as a good choice with a global accuracy of 92.5%.

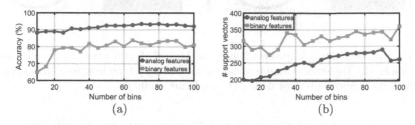

Fig. 7. Performance of the proposed method. (a) Overall accuracy of the node classification. (b) Number of support vectors

Table 2 details the node recognition results obtained on the test set with a total of 280 captures. Results are represented by means of the confusion matrix, using analog patterns with $N_b = 50$. In this case, optimal values of parameters were $\gamma = 1$ and $C = 5$. By inspection, we can observe that there is only seven cases of confusion between categories of specific nodes. Most misclassified examples are due to specific nodes assigned as non nodes. The reason is that the classification system assigns the signature to a specific node class a little before or a little after entering the zone of influence with respect to the labeling of the sample. Considering this circumstance, we can conclude that the performance of the classifier between specific classes, without considering No-Node class, achieves a good accuracy of 97.5%.

Table 2. Results of node classification (confusion matrix)

Category	No node	End node	Node 'T '	Cross node	Node 'L'
No node	61	1	1	0	5
End node	0	66	3	0	0
Node 'T'	2	0	63	1	1
Cross node'	2	0	2	18	0
Node 'L'	3	0	0	0	51

To validate the behaviour of our system, we have generated a demo video[1] that shows the construction of the topological map from the exploration in an area that includes two nodes-T and three End-nodes. In Fig. 8 some snapshots

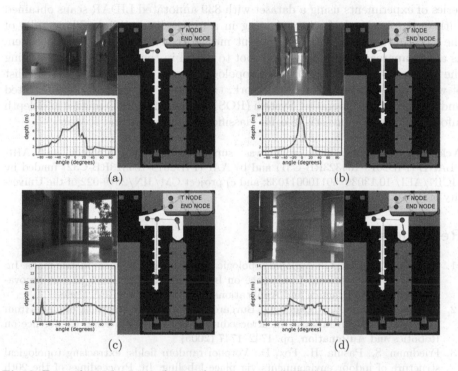

Fig. 8. Example exploration. Agent observations and LIDAR signatures are shown on the left (upper and bottom, respectively) and the topological map is shown on the right. The green arrow represents the current estimated pose of the robot. (a) $(t = 21)$ The model creates the first End-node. (b) $(t = 44)$ The model creates the second End-Node. (c) $(t = 72)$ The model creates the second node 'T'. (d) $(t = 77)$ The model creates the third End-node. (Color figure online)

---

[1] A video of our experiment is provided: https://universidaddealcala.sharepoint.com/ sites/Cadas/Documentos%20compartidos/General/video.mp4.

of the exploring at different iterations are depicted. The agent completes the exploration after an average of 74 iterations in a total of 4 trials and decides to take the stop action. Note that the algorithm does not represent a new node in the graph until it considers that the robot leaves its zone of influence.

6   Conclusions

An approach for topological modelling in symmetrical indoor buildings has been presented to address a main difficulty: detecting nodes adapted to large spaces based on corridor layouts, where nodes identify changes of trajectory. The presented multiclass node detection solutions is a real-time system that can classify LIDAR signatures into different categories as the robot moves during an exploration process and builds a topological structure of nodes and edge connections.

To evaluate the performance of the proposed method we have conducted a series of experiments using a dataset with 839 annotated LIDAR scans obtained with own robotic platform navigating in a university building. The results of the experiments show that the present method constitutes an efficient system as a first approach and allows the robot to patrol in the given scenario by using the metric map and the extracted topological map. Nodes constitute the list of way-points to cover. For future work, the system should be fully integrated under the Robotic Operating System (ROS). Fusion of RGB-D images and depth information will be investigated for classification of nodes and scenes.

Acknowledgements. This work was supported by the projects: (a) AIR-PLANE (PID2019-104323RB-C31) and b) AAI (PID2019-104323RB-C32) funded by MCIN/AEI/ 10.13039/501100011033; and c) project CM/JIN/2019-022 of the University of Alcalá.

References

1. Simhon, S., Dudek, G.: A global topological map formed by local metric maps. In: IEEE/RSJ International Conference on Intelligent Robots and Systems. Innovations in Theory, Practice and Applications, pp. 1708–1714. IEEE (1998)
2. Martínez-Mozos, O., Stachniss, C., Burgard, W.: Supervised learning of places from range data using AdaBoost. In: Proceedings - IEEE International Conference on Robotics and Automation, pp. 1742–1747 (2005)
3. Friedman, S., Pasula, H., Fox, D.: Voronoi random fields: extracting topological structure of indoor environments via place labeling. In: Proceedings of the 20th International Joint Conference on Artificial Intelligence (2007)
4. Redmon, J., Divvala, S., Girshick, R., Farhadi, A.: You only look once: unified, real-time object detection. In: Proceedings of the IEEE Conference on Computer Vision and Pattern Recognition (CVPR), pp. 779–788 (2016)
5. Levitt, T.S., Lawton, D.T.: Qualitative navigation for mobile robots. Artif. Intell. **44**, 305–360 (1990)
6. Barber, R., Crespo, J., Gómez, C., Hernández, A., Galli, M.: Mobile robot navigation in indoor environments: geometric, topological, and semantic navigation. In: Applications of Mobile Robots. IntechOpen, London (2018)

7. Fuqiang, G., et al.: Indoor localization improved by spatial context - a survey. ACM Comput. Surv. (CSUR) **52**(3), 1–35 (2019)
8. Morar, A., et al.: A comprehensive survey of indoor localization methods based on computer vision. Sensors **20**(9), 2641 (2020)
9. Kielar, P.M., Biedermann, D.H., Kneidl, A., Borrmann, A.: A unified pedestrian routing model combining multiple graph-based navigation methods. In: Knoop, V.L., Daamen, W. (eds.) Traffic and Granular Flow '15, pp. 241–248. Springer, Cham (2016). https://doi.org/10.1007/978-3-319-33482-0_31
10. Blochliger, F., Fehr, M., Dymczyk, M., Schneider, T., Siegwart, R.: Topomap: topological mapping and navigation based on visual slam maps. In: IEEE International Conference on Robotics and Automation (ICRA), pp. 3818–3825 (2018)
11. Gunathillake, A., Huang, H., Savkin, A.V.: Sensor-network-based navigation of a mobile robot for extremum seeking using a topology map. IEEE Trans. Industr. Inf. **15**(7), 3962–3972 (2019)
12. Devendra, S.C., Ruslan, S., Abhinav, G, Saurabh, G.: Neural topological SLAM for visual navigation. In: IEEE/CVF Conference on Computer Vision and Pattern Recognition (CVPR) (2020)
13. Cheng, H., Chen, H., Liu, Y.: Topological indoor localization and navigation for autonomous mobile robot. IEEE Trans. Autom. Sci. Eng. **12**(2), 729–738 (2014)
14. SlamTec RPLidar A1. http://www.slamtec.com/en/lidar/a1. Accessed 20 Dec 2021
15. Cortes, C., Vapnik, V.: Support vector machine. Mach. Learn. **20**(3), 273–297 (1995)

Visual Event-Based Egocentric Human Action Recognition

Francisco J. Moreno-Rodríguez[1], V. Javier Traver[2]([✉]) [ID],
Francisco Barranco[3] [ID], Mariella Dimiccoli[4] [ID], and Filiberto Pla[2] [ID]

[1] Universitat Jaume I, Castelló, Spain
[2] Institute of New Imaging Technologies, Universitat Jaume I, Castelló, Spain
{vtraver,pla}@uji.es
[3] Department of Computer Architecture and Technology, CITIC,
University of Granada, Granada, Spain
fbarranco@ugr.es
[4] Institut de Robótica i Informática Industrial (CSIC-UPC), Barcelona, Spain
mdimiccoli@iri.upc.edu

Abstract. This paper lies at the intersection of three research areas: human action recognition, egocentric vision, and visual event-based sensors. The main goal is the comparison of egocentric action recognition performance under either of two visual sources: conventional images, or event-based visual data. In this work, the events, as triggered by asynchronous event sensors or their simulation, are spatio-temporally aggregated into event frames (a grid-like representation). This allows to use exactly the same neural model for both visual sources, thus easing a fair comparison. Specifically, a hybrid neural architecture combining a convolutional neural network and a recurrent network is used. It is empirically found that this general architecture works for both, conventional gray-level frames, and event frames. This finding is relevant because it reveals that no modification or adaptation is strictly required to deal with event data for egocentric action classification. Interestingly, action recognition is found to perform better with event frames, suggesting that these data provide discriminative information that aids the neural model to learn good features.

Keywords: Egocentric view · Action recognition · Event vision

1 Introduction

In contrast to the more widespread third-person vision (3PV), first-person (egocentric) vision (1PV) provides unique insights of the scene as observed directly

Work supported by project UJI-B2018-44 from *Pla de promoció de la investigació de la Universitat Jaume I*, Spain, the research network RED2018-102511-T, from the Spanish *Ministerio de Ciencia, Innovación y Universidades*, and by National Grant PID2019-109434RA-I00/ SRA (State Research Agency /10.13039/501100011033).

A. J. Pinho et al. (Eds.): IbPRIA 2022, LNCS 13256, pp. 402–414, 2022.
https://doi.org/10.1007/978-3-031-04881-4_32

Fig. 1. Examples of RGB image (left) and simulated event frame (right) during a SitDown (up) and StandUp (down) sequences from the dataset used. The event frames have different size as a requirement of the video-to-event simulator used. The gray-level display of the event frames represents the aggregated polarity of the events as per Eq. (1).

from the privileged point of view of the camera wearer, as well as their activities and behaviour. Therefore, it is not surprising that gesture, action, or activity recognition, which have been widely explored in third-person contexts, have also been investigated over the last decade for the egocentric case. Although significant progress has been achieved, both 1PV and 3PV action recognition approaches have been dominated by the use of regular visual sensors. Recently, however, an alternative sensing paradigm, event-based neuromorphic visual sensors, or event cameras, have been receiving increasing attention. Unlike conventional cameras, these bio-inspired event-based sensors deliver a source of sparse and asynchronous flow of events corresponding to luminance changes detected with a very precise timing. Despite the potential benefits of event cameras in egocentric computational vision, action recognition with event-based visual data on egocentric visual streams has not been addressed. Actually, to the best of our knowledge, only one work exists (Sect. 2). The work reported in this paper combines these three research topics (1PV, action recognition, and event-based vision), and explores whether event-based representations can contribute to egocentric action recognition.

Please, note that the use of the term *event* throughout this paper refers to the low-level concept associated to brightness changes as delivered by event cameras. In particular, these events should not be confused to the higher-level events representing either temporally contiguous images identified as a unit, or to semantic concepts (such as "door open", "keys dropped") detected from a sequence of images.

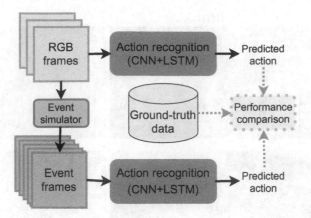

Fig. 2. Overview of the work. Note how event frames can be produced at a higher frame rate than the original RGB frames by using high-quality video interpolators. Exactly the same network (Fig. 3) is used in both visual sources (Fig. 1), but each is trained on their respective data.

Overview and Contributions. We propose the use of event frames [12,24] and a hybrid neural architecture (convolutional and recurrent neural networks) by leveraging authors' recent related work on egocentric gesture recognition using conventional (*not* event-based) eyewear cameras [17,18]. Specifically, the main contribution of this work is to compare conventional gray-level frames with event-based data (Fig. 1) for egocentric action recognition using exactly the same neural model (Fig. 2).

2   Related Work

Taking into account the problem and our particular approach, a brief overview of related work on the following areas is considered: motion estimation, visual events, egocentric vision, and action recognition.

(Ego)motion Estimation. Estimation of head motion [28] is generally relevant in the context of 1PV. More generally, homography estimation methods based on deep learning have been proposed recently [9,20,31] and have therefore the potential to properly characterise egocentric-related movements.

Event-Based Visual Data. Event cameras [12] offer a very distinct alternative to how conventional cameras operate, by delivering visual changes as a series of sparse and asynchronous events with a time resolution of a few microseconds. Each of such events typically encodes the sign of the brightness change and its spatial information. The main benefits of event cameras is a low-power consumption, high-temporal resolution, and low-latency event stream. Understandably, these cameras or their simulations have mostly been used for low-level

visual tasks such as tracking [3] or vehicle control [29], which most straightforwardly can exploit these properties. Generally speaking, tasks requiring ultra
fast visual processing are favoured by smart compression and high temporal
resolution of event cameras, for instance to overcome the undesirable motion
blur in traditional cameras. However, despite its potential advantages, the use
of these event-based visual data for other visual tasks has been limited so far.
Only recently, events have been studied for 3D pose estimation [25] and gesture recognition [2,30] of hands. Action recognition from 3PV datasets has also
been explored [14,15] with event data, which reinforces the rising interest of this
sensing paradigm.

Egocentric Vision and Action Recognition Egocentric visual streams can
be useful for a variety of tasks, including visual life-logging [4], hand analysis [6],
activity recognition [21,23], social interaction analysis [1,11], video summarisation [8,27], etc. To the best of our knowledge, only conventional cameras have
been used for action recognition in the context of egocentric vision, an exception
being a very recent work [22], which introduces N-EPIC-Kitchens dataset, an
event-based version of EPIC-Kitchens [7], a well-known dataset of egocentric
videos captured with conventional wearable cameras. Interestingly, the findings
of this parallel work align with ours (as discussed in Sect. 4.3) in that action
recognition performance with event-based visual data is on par or better than
with conventional imaging, even though different datasets and approaches have
been used in our respective works.

3   Methodology

How events are defined and event frames generated are described first (Sect. 3.1.
Then, the neural network architecture proposed for action recognition and training details are provided (Section 3.2).

3.1   Event Data Generation

Event cameras [12] trigger a flow of events $o_k, k \in \{1, 2, \ldots\}$, each represented by
a tuple $e_k = (\mathbf{x}_k, t_k, p_k)$, where $\mathbf{x}_k = (x_k, y_k)$ represents the spatial coordinates
where a brightness change higher than a threshold C has been detected at time
t_k since the last event at that same pixel location, and $p_k \in \{-1, +1\}$ is the
polarity (sign) of the change.

The dataset used in our work (Sect. 4.1) consists of conventional videos only.
Therefore, visual events were simulated from the regular video frames. Event
camera simulators are widely used in the literature [12,22,24] since they render
realistic events as compared to actual event cameras, and facilitate developing
and benchmarking algorithms under controlled conditions prior to the use of
specific event cameras. In our case, event simulation turns out to be very useful
for the comparison between conventional frame-based and event-based methods

for action recognition. Our choice was the simulator v2e [13], a Python tool for realistic event synthesis.

Since events are triggered asynchronously and correspond to spatio-temporal sparse data, there are several possibilities when it comes to representing and processing them. To process events one-by-one, spiking neural networks are possibly the straightforward choice [5], but these networks are more recent and far less known than conventional neural networks. Alternatively, we adopt a common approach [14,15,22] of aggregating these events into a grid representation. Events can essentially be aggregated by count (every n events) or by time (at regular time intervals). We aggregate events $\mathbf{e}_k = (\mathbf{x}_k, t_k, p_k)$ into 2D event frames $\mathbf{E}_t$ at regular time steps $t \in \{T, 2T, \ldots\}$ taking into account their polarity p_k [24],

$$\mathbf{E}_t(\mathbf{x}) = \sum_k p_k \cdot \frac{t_k - (t - T)}{T}, \quad \forall\, k \text{ such that } \mathbf{x}_k = \mathbf{x}, \text{and } t - T < t_k \leq t, \quad (1)$$

where $T = 5$ ms in our case, which is 8 times faster than the original frame rate (25 fps) due to video interpolation performed by the event simulator using high-end algorithms such as *Super SloMo* [16]. Examples of such event frames are given in Fig. 1.

Although these grid-like representations do not process events natively as they are triggered, and therefore a lag is introduced and some temporal resolution can be lost, grouping events helps exploit their spatio-temporal consistency, and allows existing and widely proven conventional neural models (such as convolutional neural networks) to be used straightforwardly. Also, for object classification and visual odometry tasks, architectures based on the grid representation have been reported to outperform specialised algorithms designed for event cameras [24]. For the purposes of our work this is also an ideal choice because it enables easier and fairer comparison between RGB and event-based visual data with a common learning model.

3.2 Action Recognition

A motion-based recognition approach is proposed that combines a convolutional neural network (CNN) and a recurrent network (Fig. 3), as in our recent work on egocentric gesture recognition with regular eyewear cameras [17,18]. The idea is to estimate head-induced motion with frame-to-frame homography estimation (i.e. an estimate of the global camera motion), and then use these estimates as input to a long short-term memory (LSTM) so that action-relevant visual dependencies more distant in time are captured by the overall network. For homography estimation, the chosen CNN model [31] features two favourable properties: it is unsupervised, and it is designed to be robust against independently moving objects not following the global camera motion.

Note that this exact model is used in both RGB (gray) frames and event frames (inputs I_t and I_{t-1} in Fig. 3). In both cases, input values are normalized to the range $[0, 1]$. We use the homography-estimation CNN as pretrained by their authors on sequences exhibiting motions different to those occurring in some of

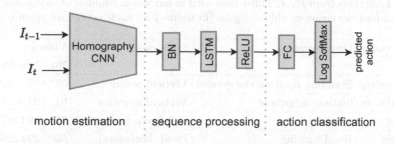

Fig. 3. Neural model for action recognition. The input is a pair of consecutive frames and the output is the predicted action. BN = Batch normalisation, FC = Fully Connected (dense) layer.

the actions in our experimental dataset. More importantly, only conventional images were used for training, *not* event frames. Although our purpose here is not obtaining accurate motion estimates, since only discriminative motion features are required, a natural question is whether using such CNN with event frames as input is a sensible choice. Therefore, as a sanity check, we briefly tested whether motion can be reasonably estimated with this off-the-shelf CNN even with event frames as input. It turned out that, compared to ground-truth motion, estimates were different with RGB and event frames, but both were reasonable. Therefore, the same pre-trained CNN was used for RGB and Events data, and only the LSTM and the fully connected (FC) layers are data-specific and supervisedly trained on the respective sequences. We leave as future work to train from scratch or fine-tune the homography CNN with data sequences for our specific task and visual formats.

As stated above, the event simulator produces event frames at higher temporal resolution than the available RGB videos. For the sake of fairer comparison, however, we temporally subsampled the event frames to match frame rates of the RGB and Events sequences.

Shorter and overlapped fragments (clips) from the training videos were used for training, thus having a larger training set. Classification is performed framewise, which is harder than video-wise classification because not the full video is observed, but this scenario lends itself to online and early action recognition. For video-level classification, majority voting is applied. The network was trained for 100 epochs, using a learning rate of 10^{-4}. The hidden size of the LSTM was set to 128.

4 Experimental Work

The dataset used is first introduced (Sect. 4.1). Next, the experiments and the results are described (Sect. 4.2) and discussed (Sect. 4.3).

Table 1. Actions from *First2Third-Pose* used in our work. Number of videos and mean and standard deviation of video lengths (in frames) for each action are given.

Action	Description	Main motion	Videos	
			No.	Length
StandUp	Standing up from the ground	Vertical positive	58	88 ± 22
SitDown	Sitting on ground	Vertical negative	61	94 ± 28
Rot	Rotating	Full-body, horizontal	71	194 ± 70
Turn	Head turning	Head, horizontal	59	294 ± 67

4.1 Dataset

A subset from *First2Third-Pose* dataset [10] has been selected for the purpose of this work. Although events could be generated from *Charades-Ego*, an egocentric action dataset [26], *First2Third-Pose*, was chosen as a part of a research collaboration[1], and the dataset N-EPIC-Kitchens [22], with available events simulated from egocentric videos of human activity, is very recent, and was not available by the time this work was developed [19], and it has not actually been released yet. *First2Third-Pose* features synchronised pairs of first- and third-view videos of 14 people performing 40 activities. Although mainly intended for 3D pose estimation, this dataset can also be useful for activity (action) recognition. Here we focus on the egocentric view and select four representative actions. The rationale for this choice was to have both dissimilar and similar actions, so that the proposed approach can be assessed in easier and harder scenarios. Therefore, two groups of two actions each were chosen (Table 1), with a total of 249 videos.

Videos of different actions tend to have significantly different lengths (last column in Table 1). In particular, StandUp and SitDown are shorter (about 1–6 s), which last about 9–20 s. These differences pose a practical challenge in terms of which video clip length to use for training. Long clips can be more informative, but fewer of them can be extracted from those original videos which are shorter. With shorter clips, more training instances can be sampled from short videos, but each clip characterises worse the action sequences, and therefore they will be harder to model. Overall, clips of size $S = 35$ frames with an overlap of $O = 25$ frames were found to be a good compromise. Clips with these (S, O) values will be referred to as 'short' clips. The dataset was split into 80%–20% training-validation sets in terms of individual clips. Additionally, since the same action is performed by different subjects and in different places (Fig. 4), specific subjects-based and location-based splits will be used in some tests, as detailed below.

---

[1] Red Española de Aprendizaje Automático y Visión Artificial para el Análisis de Personas y la Percepción Robótica (ReAViPeRo), https://www.init.uji.es/reavipero.

Indoor		Outdoor	
generic	specific (lab)	generic	specific (park)

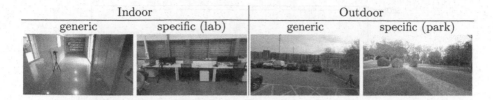

Fig. 4. Different places where actions in *First2Third-Pose* were recorded.

Table 2. Recognition results, accuracy (%) at video level.

Split	Test	RGB	Events
Clips (80%–20%)	1. Two-way, dissimilar, short (Fig. 5)	75.4	**86.7**
	2. Two-way, similar, short	**75.8**	65.48
	3. Two-way, similar, long	70.6	**81.8**
	4. Three-way, short	68.3	**75.8**
	5. Four-way, short (Fig. 6 (top))	52.7	**56.9**
	6. Four-way, long (Fig. 6 (below))	55.5	**64.5**
Places	7. Four-way, long	58.4	**67.4**
Subjects	8. Four-way, long	53.6	**74.6**

4.2   Results

To better understand the performance of the recognition system, different conditions are separately considered.

Two-Way Classifications. A simple binary classification between two dissimilar actions, Rot and SitDown, was first tested, and Events outperformed RGB (Table 2, Test 1). Arguably, Events deals better with this unbalanced class case. An example comparing RGB and Events at frame-level classification (Fig. 5) illustrates that with most frames correctly labelled, Events can correctly classify the sequence at video-level, whereas RGB cannot. For a harder scenario, two similar actions (Rot and Turn) are considered. Results (Table 2, Test 2) reveal that in this case RGB outperforms Events. One possible explanation is that these actions include faster motions, which might violate the small-baseline transformations assumption of the homography network. Under these circumstances, RGB might fare better due to the more similar nature to the images used for training the homography CNN. Interestingly, if the clip length and overlap are enlarged to $S = 84$ and $O = 80$, to account for the longer videos in these actions, performance (Table 2, Test 3) drops by about 5% points in RGB, and increases by 15 points in Events. This highlights the impact that the clip length has.

Three- and Four-Way Classification. When considering Rot, SitDown and Turn, results (Table 2, Test 4) are understandably worse for both RGB and Events.

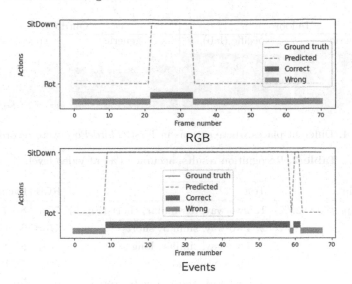

Fig. 5. Frame-level classification for a SitDown sequence (Fig. 1a) for RGB and Events. In RGB, most frames are mislabelled and therefore the action is misclassified at video level.

However, Events outperforms RGB. Finally, when considering the four actions, performance drops (Table 2, Test 5), but longer snippets help (Table 2, Test 6), particularly when using Events. The confusion matrices (Fig. 6) indicate that misclassifications mostly happen between the two similar-motion groups of actions (SitDown vs StandUp and Rot vs Turn). They also illustrate that, generally, better performance is obtained with long clips, both with RGB and Events. The overall trend of better recognition with Events over RGB can also be observed.

Generalisation Ability. Finally, we evaluate how the system generalises to different places and subjects. For places, the training split includes three cases ('outdoor', 'lab', 'park') and we evaluate on 'indoor'; this split represents roughly a 80%–20% split. While 'lab' is also an indoors place, its sequences contain more objects and texture that might facilitate homography estimates, whereas 'indoor' places are usually in texture-less corridor areas with many lighting artefact, or exterior parts of a building, with windows and light issues. Again, Events outperforms RGB (Table 2, Test 7), one possible reason being that light reflections might misguide RGB, whereas no (false) events are produced by these action-irrelevant data. Regarding the subject-split better results are also obtained with Events (Table 2, Test 8). As for why the performance of Events in this case (74.6%) is higher than in any other 4-way test considered, a tentative explanation is that by including all repetitions of the same subject in the training split, the model captures a useful variability that might possibly be not present with other train-val splits.

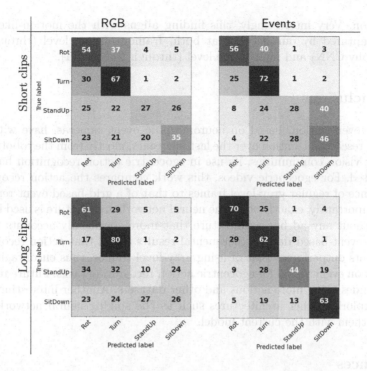

Fig. 6. Confusion matrices (values in %) for 4-way video-level classification with RGB (left) and Events (right) for short (top) and long (bottom) clips.

4.3 Interpretation and Discussion

As these results illustrate, better recognition performance is obtained with event visual data than with regular images. A likely explanation is that the event representation acts as feature selection and encodes motion-like information, thus helping the network to more directly focus on relevant and discriminative data. Related to this hypothesis, we plan to compare the performance of event-based data with edge-like or motion-like data precomputed on conventional frames. This will provide insights into what actually and intrinsically aids the recognition task.

The results are somehow surprising and particularly interesting because a generic learning model has been used, without any adaptation or specific design that takes into account the idiosyncrasies of event visual data. Furthermore, the CNN used as part of the architecture for homography estimation was completely "event agnostic", since it was trained on regular image sequences only.

The benefits in recognition performance using events observed in our work are in agreement with a recent study using different dataset and architectures [22]. Additionally, these authors explore the role of motion information from event frames by using existing neural architectures to extract temporal information. They conclude the significantly positive effect brought by this temporal information. Very interestingly, this finding aligns with the motion-like information captured by our network at both, frame-to-frame level (through the homography CNN) and inter-frame level (through the LSTM).

5   Conclusions

Although event vision, based on neuromorphic event cameras, have witnessed increased research attention over the last few years, mostly from the robotics and computer vision communities, its use in egocentric action recognition has been very limited. For egocentric videos, this work compares the action recognition performance of regular gray-level frames to that of a grid-based event representation. Importantly, exactly the same neural network architecture is used in both cases, without any ad hoc architecture that more specifically accounts for the nature of event-based data. Experimental results reveal that action recognition using events outperforms that of using gray-level frames, thus encouraging further work on event vision for egocentric action recognition tasks. Future research work includes using more actions and other datasets. Another interesting direction is exploring other architectures such as the spiking neural networks, and compare them with the current model.

References

1. Alletto, S., Serra, G., Calderara, S., Cucchiara, R.: Understanding social relationships in egocentric vision. Pattern Recogn. **48**(12), 4082–4096 (2015)
2. Amir, A., et al.: A low power, fully event-based gesture recognition system. In: IEEE Conference on Computer Vision and Pattern Recognition (CVPR), pp. 7388–7397 (2017)
3. Barranco, F., Fermuller, C., Ros, E.: Real-time clustering and multi-target tracking using event-based sensors. In: IEEE International Conference on Intelligent Robots and Systems (IROS), pp. 5764–5769 (2018)
4. Bolaños, M., Dimiccoli, M., Radeva, P.: Toward storytelling from visual lifelogging: an overview. IEEE Trans. Hum. Mach. Syst. **47**(1), 77–90 (2017)
5. Cordone, L., Miramond, B., Ferrante, S.: Learning from event cameras with sparse spiking convolutional neural networks. In: International Joint Conference on Neural Networks (IJCNN) (2021)
6. Cruz, S., Chan, A.: Is that my hand? An egocentric dataset for hand disambiguation **89**, 131–143 (2019)
7. Damen, D., et al.: The EPIC-KITCHENS dataset: collection, challenges and baselines. IEEE Trans. Pattern Anal. Mach. Intell. **43**(11), 4125–4141 (2021)
8. del Molino, A.G., Tan, C., Lim, J., Tan, A.: Summarization of egocentric videos: a comprehensive survey. IEEE Trans. Hum. Mach. Syst. **47**(1), 65–76 (2017)
9. DeTone, D., Malisiewicz, T., Rabinovich, A.: Deep image homography estimation. CoRR abs/1606.03798 (2016). https://arxiv.org/abs/1606.03798
10. Dhamanaskar, A., Dimiccoli, M., Corona, E., Pumarola, A., Moreno-Noguer, F.: Enhancing egocentric 3D pose estimation with third person views. CoRR abs/2201.02017 (2022). https://arxiv.org/abs/2201.02017

11. Felicioni, S., Dimiccoli, M.: Interaction-GCN: a graph convolutional network based framework for social interaction recognition in egocentric videos. In: IEEE International Conference on Image Processing (ICIP), pp. 2348–2352 (2021)
12. Gallego, G., et al.: Event-based vision: a survey. IEEE Trans. Pattern Anal. Mach. Intell. **44**, 154–180 (2022)
13. Hu, Y., Liu, S.C., Delbruck, T.: V2E: from video frames to realistic DVS events. In: IEEE Conference on Computer Vision and Pattern Recognition (CVPR), pp. 1312–1321 (2021)
14. Huang, C.: Event-based timestamp image encoding network for human action recognition and anticipation. In: International Joint Conference on Neural Networks (IJCNN) (2021)
15. Innocenti, S.U., Becattini, F., Pernici, F., Del Bimbo, A.: Temporal binary representation for event-based action recognition. In: International Conference on Pattern Recognition (ICPR) (2021)
16. Jiang, H., et al.: Super SloMo: high quality estimation of multiple intermediate frames for video interpolation. In: IEEE Conference on Computer Vision and Pattern Recognition (CVPR), pp. 9000–9008 (2018)
17. Marina-Miranda, J.: Head and eye egocentric gesture recognition for human-robot interaction using eyewear cameras. Master's thesis, Universitat Jaume I, Castellón, Spain, July 2021
18. Marina-Miranda, J., Traver, V.J.: Head and eye egocentric gesture recognition for human-robot interaction using eyewear cameras (submitted September 2021). Under review. Preprint at http://arxiv.org/abs/2201.11500
19. Moreno-Rodríguez, F.J.: Visual event-based egocentric human action recognition. Master's thesis, Universitat Jaume I, Castellón, Spain, July 2021
20. Nguyen, T., Chen, S.W., Shivakumar, S.S., Taylor, C.J., Kumar, V.: Unsupervised deep homography: a fast and robust homography estimation model. IEEE Robot. Autom. Lett. **3**(3), 2346–2353 (2018)
21. Núñcz-Marcos, A., Azkune, G., Arganda-Carreras, I.: Egocentric vision-based action recognition: a survey. Neurocomputing **472**, 175–197 (2022)
22. Plizzari, C., et al.: E^2(go)motion: motion augmented event stream for egocentric action recognition. CoRR abs/2112.03596 (2021). https://arxiv.org/abs/2112.03596
23. Possas, R., Caceres, S.P., Ramos, F.: Egocentric activity recognition on a budget. In: IEEE Conference on Computer Vision and Pattern Recognition (CVPR), pp. 5967–5976 (2018)
24. Rebecq, H., Ranftl, R., Koltun, V., Scaramuzza, D.: Events-to-video: bringing modern computer vision to event cameras. In: IEEE Conference on Computer Vision and Pattern Recognition (CVPR), pp. 3852–3861 (2019)
25. Rudnev, V., et al.: EventHands: real-time neural 3D hand pose estimation from an event stream. In: International Conference on Computer Vision (ICCV) (2021)
26. Sigurdsson, G.A., Gupta, A.K., Schmid, C., Farhadi, A., Karteek, A.: Actor and observer: joint modeling of first and third-person videos. IEEE Conference on Computer Vision and Pattern Recognition (CVPR), pp. 7396–7404 (2018)
27. Traver, V.J., Damen, D.: Egocentric video summarisation via purpose-oriented frame scoring and selection. Expert Syst. Appl. **189** (2022)
28. Tsutsui, S., Bambach, S., Crandall, D., Yu, C.: Estimating head motion from egocentric vision. In: ACM International Conference on Multimodal Interaction, pp. 342–346 (2018)

29. Vitale, A., Renner, A., Nauer, C., Scaramuzza, D., Sandamirskaya, Y.: Event-driven vision and control for UAVs on a neuromorphic chip. In: IEEE International Conference on Robotics and Automation (ICRA), pp. 103–109 (2021)
30. Xing, Y., Di Caterina, G., Soraghan, J.: A new spiking convolutional recurrent neural network (SCRNN) with applications to event-based hand gesture recognition. Front. Neurosci. **14** (2020)
31. Zhang, J., et al.: Content-aware unsupervised deep homography estimation. In: Vedaldi, A., Bischof, H., Brox, T., Frahm, J.-M. (eds.) ECCV 2020. LNCS, vol. 12346, pp. 653–669. Springer, Cham (2020). https://doi.org/10.1007/978-3-030-58452-8_38

An Edge-Based Computer Vision Approach for Determination of Sulfonamides in Water

Inês Rocha[1,2], Fábio Azevedo[1,3], Pedro H. Carvalho[1], Patrícia S. Peixoto[4],
Marcela A. Segundo[4], and Hélder P. Oliveira[1,2(✉)]

[1] INESC TEC - Institute for Systems and Computer Engineering Technology
and Science, Porto, Portugal
helder.f.oliveira@inesctec.pt
[2] Faculty of Sciences, University of Porto, Porto, Portugal
[3] Faculty of Engineering, University of Porto, Porto, Portugal
[4] LAQV - REQUIMTE, Faculty of Pharmacy, University of Porto, Porto, Portugal

Abstract. The consumption of antibiotics, such as sulfonamides, by
humans and animals has increased in recent decades, and with it their
presence in aquatic environments. This contribute to the increasing of
bacterial resistant genes, making the treatment of infectious diseases
more difficult. These antibiotics are usually detected by taking a water
sample to a laboratory and quantifying it using expensive methods.
Recently, digital colorimetry, has emerged as a new method for detecting
sulfonamides in water. When a reagent comes into contact water sample
containing sulfonamides, a color is produced from which we can infer the
concentration of sulfonamides. To ensure that the color is not affected
by the illumination when taking a photograph, a color reference target
is positioned next to the sample to correct the colors.

This method has already been implemented in smartphones to provide a faster and more practical tool that can be used immediately when
collecting water samples. Despite this improvement, the algorithms used
can still be outperformed by the use of machine learning. In this work,
we presented a machine learning approach and a mobile app to solve
the problem of sulfonamides quantification. The machine learning approach was designed to run locally in the mobile device, while the mobile
application is transversal to Android and iOS operation systems.

Keywords: Computer vision · Object Detection · Colorimetry ·
Mobile application · Sulfonamides

This work was partially funded by the Project TAMI - Transparent Artificial Med-
ical Intelligence (NORTE-01-0247-FEDER-045905) financed by ERDF - European
Regional Fund through the North Portugal Regional Operational Program - NORTE
2020 and by the Portuguese Foundation for Science and Technology - FCT under the
CMU - Portugal International Partnership.
I. Rocha and F. Azevedo—These authors contributed equally.

A. J. Pinho et al. (Eds.): IbPRIA 2022, LNCS 13256, pp. 415–429, 2022.
https://doi.org/10.1007/978-3-031-04881-4_33

1   Introduction

Thanks to scientific advances in the mid-1900s, there was an improvement in live-stock production that increased the consumption of animal protein per person by 50% from 1961 to 1998 [1]. This scientific knowledge brought new developments in breeding, nutrition, and animal health. An important development in animal health was the increased use of antibiotics to prevent the spread of disease. One of these antibiotics was sulfonamides. Sulfonamides comprise a class of antibiotics that have been frequently used with this aim [2]. Sulfonamides are antibiotics used to treat bacterial infections in humans and animals. Sulfonamides can natural reach the water by different pathways, such as human excretion, and hospital and industrial discharges [3]. When left in water streams, bacterial populations are exposed to repeated sublethal doses that cause them to become antibiotic resistant. As these new bacteria spread to humans, their infections will be more difficult to treat, cause higher medical costs, longer hospital stays, and increased mortality rates [4]. Thus, the detection and quantification of these sulfonamides is an important task. According to European Centre for Disease Prevention and Control, there has been a rapid and continuing rise in antibiotic-resistant infections [5] and according to the Centers for Disease Control and Prevention, at least 2.8 million people in the U.S. contract an antibiotic-resistant infection each year [6]. Considering that this antibiotic problem is a worldwide problem, it is therefore important to find suitable methods to detect antibiotics in the first place, thus preventing the emergence of resistant microorganims.

The conventional method of detecting and quantifying sulfonamides in water involves a researcher collecting a sample of water, transporting that sample to a laboratory, and then calculating the concentration using difficult and expensive equipment. This type of method makes it even more cumbersome to detect and control substances in a wild environment or in remote regions where the nearest laboratory is not as close to the water source [7–11]. In 2019 [12], we developed a new method for the detection of sulfonamides in water. This method is based on digital colorimetry and is simpler and less expensive than standard methods for detecting antibiotics. This method used a cost-efficient color correction approach to ensure color constancy between images [13], implemented as a mobile solution [14]. While this method was a great advance in detecting sulfonamides insitu, the algorithm used can be outperformed by using machine learning methods which can perform faster and obtain more accurate results. Additionally, the current mobile solution is attached to a singular OS (Android) which hinders other mobile users, such as iOS, to utilize the solution and does not include all the features expected by the user, necessary for use it in the daily routine. The objective of this work is to create a computer vision algorithm for the determination of sulfonamides in water that improves on the work of [12,13] by using machine learning models, and a cross-platform mobile application to support the colorometry algorithm.

2   A Computer Vision Approach for Determination of Sulfonamides in Water

2.1   Material and Methods

A dataset with 930 photographs of the *x-rite ColorChecker Passport* next to a sample of contaminated water was prepared in a laboratory setting, with the color chart used as the reference for the color correction and the target being the sample. Figure 1a is an example of a photograph from this dataset and Fig. 1b shows the colors the samples assume depending on the concentration of sulfonamides.

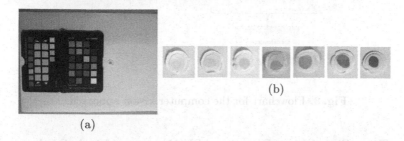

(a)

(b)

Fig. 1. a) Example photograph from dataset. b) Examples of samples of varying sulfonamide concentrations, from $0\,\mu g/L$ (left) to $150\,\mu g/L$ (right).

To create a realistic dataset we photograph the disks and color checkers with different smartphones under different lighting conditions, as shown in Fig. 2.

Fig. 2. Example of different lighting conditions.

The thought process to solve our problem started with detecting the color checker and its patches, detecting the disk and its color, and then color correction. This color correction model will be different that was done before [13], taken the advantage of the power of machine learning models. In the end, the color of the disk is transformed to a sulfonamides concentration value using the linear regression models presented in [12]. The solution thus developed is shown in Fig. 3.

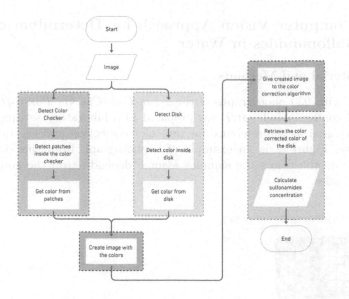

Fig. 3. Flowchart for the computer vision approach.

The TensorFlow Object Detection API [15] was used to build the machine learning models, which utilizes a specific input type called TFRecords. Even this framework has several pre-trained models, we trained our models from scratch using our own dataset. With the objective of running the algorithm into a mobile device, we converted our model into a TensorFlow Lite (TFLite) model. These models are specifically designed for use on mobile devices, as they are faster, smaller, and less computationally intensive.

2.2   Patches Detection

To train the model, we split the images and annotations into a training set (70%) and a test set (30%). The model used was the SSD Mobilenet V2, with the following configuration:

- Limit the number of classes to 24 (corresponding to the 24 color squares);
- Limit the model to only one detection per class;
- Batch size set to 32;
- Fine tuning checkpoint set to "detection".

The model outputs the result in following three components (see Fig. 4):

- Bounding boxes: the position of the detected patches;
- The class of the bounding box: class of the detected color;
- The score of the this prediction: confidence of the model.

We compared the methodology presented in this paper with our previous work [12], obtaining a detection error of 0.5% and 12% respectively.

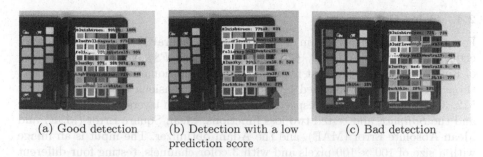

(a) Good detection   (b) Detection with a low   (c) Bad detection
                                prediction score

Fig. 4. Examples of patches detection results.

2.3 Disk and Color Sample Detection

The images were split into a training set (70%) and a test set (30%). For this task
we used a SSD MobileNet FPNLite model, taken into account that this model
uses a feature pyramid network, which increases accuracy for small objects [16].
This feature extractor accepts an image of arbitrary size as input and produces
appropriately sized feature maps. The model was trained with the following
parameters:

- Limit the number of classes to detect to 2 (disk + color sample);
- Limit the model to only one detection per class;
- Batch size set to 32;
- Fine tuning checkpoint set to "detection".

An example of the detection of the algorithm for disk and color sample detection can be seen in Fig. 5.

Fig. 5. Example of disk and color sample detection results.

We also compared the methodology presented in this paper with our previous
work [12]. For disk detection, we obtained a detection error of 0% in opposition
to 12.7% obtained with our previous work. For the sample detection a detection
error of 0.86% was obtained, while with the previous approach we had an error
of 15.5%.

2.4   Color Correction

For color correction algorithm we used a Deep Neural Network (DNN), with eight hidden layers (*C1, C2, C3, C4, C5, F6, F7, F8*), inspired in the work of Finlayson G. [17]. The first five layers are convolutional layers, while the last three layers are fully collected layers, and the dimensions used, were 64, 128, 512, 392, 392, 1024, 128, 3, respectively. For activation layers we tested the *Relu* and the *Softmax*, and tested two loss functions, Mean Square Error (MSE) and Mean Absolute Error (MAE), and the Adam optimizer. The input is an image with a size of 100 × 100 pixels and with 3 color channels, testing four different color spaces (*RGB, XYZ, HSV and CIELab*). Models were built in a sequential mode where each layer has only one input and one output. All convolutional layers are padded with zeros around the input so that the output has the same dimension as the input. To avoid overfitting, the training was stopped if they did not improve during 20 epochs. There were 16000 images used for training and the batch size was 32.

The input image, a 5 rows × 5 columns, corresponding to the 24 squares of the color checker, and the remaining it corresponds to the sample (see Fig. 6).

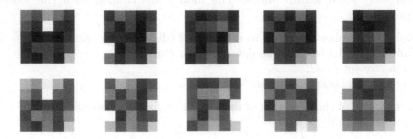

Fig. 6. Colors taken from the original photo (top), the true value the colors should had (bottom).

The results were obtained by calculating the standard deviation of the difference between the colors, of the 24 squares, obtained after running the algorithm with the ground truth values provided by the manufacturer. Taken into account the results obtained previously [12,13], the H and a^* values of the *HSV* and *CIELAB* color spaces, respectively, were used for comparison. For the best model, the eight hidden layers were represented by: *C1–C5 = Relu; C6–C8 = Softmax)*, with MAE as loss function. The standard deviation for H value was 0.046 with this new model, while in the previous work was 0.047. For a^* we obtained a standard deviation of 0.199, in comparison with 0.284 obtained with the previous model.

2.5   Summary

We developed 3 machine learning models suitable to be used in a mobile app: A modified SSD MobileNet V2 model to detect the color patches of the reference target; a modified SSD MobileNet FPNLite to detect the color of the sample and a DNN model to correct the colors of an image. These models improved the accuracy of the older algorithms [12,13].

3 A Cross-platform Mobile Application for Determination of Sulfonamides in Water

3.1 Material and Methods

Before starting implementing the mobile application, it was important to analyse the requirements for this specific tool, such as:

1. Take a photo or load an already taken from the gallery to be analysed;
2. Define the geographic location of the selected photo;
3. Save the selected photo to access it later and perform the analysis;
4. Visualize all the previous experiments results as well as the used photo;
5. Choose which and how many patches to use in the color correction algorithm;
6. Visualize the final result of the analysis (sulfonamides concentration);

A complete flowchart describing the flow between all the applications screens and functionalities is displayed in the Fig. 7.

For the development of the application we used Flutter, which is an open-source cross-platform mobile app development framework which allows to build a native application for both iOS and Android with a single codebase. Flutter allows faster development of cross-platform applications and reduce their cost of production. In terms of differentiation, Flutter moves away from the other frameworks since it does not require web browser technology using instead its own high-performance rendering engine improving the overall performance of the applications. Below we enumerated the main advantages of this framework:

- **Highly Productive:** Flutter use just one base code to build applications for both iOS and Android which can save time to the developers;
- **Performance:** as it does not need any Java Script Bridge it is highly reliable and can achieve great scores in terms of performance;
- **Easy handling:** easy handling as it does not requires bundles of code, rules or any kind of regulations;
- **Compatibility with TensorFlow Lite models:** has two different libraries capable of handling these models, ensuring maximum performance.

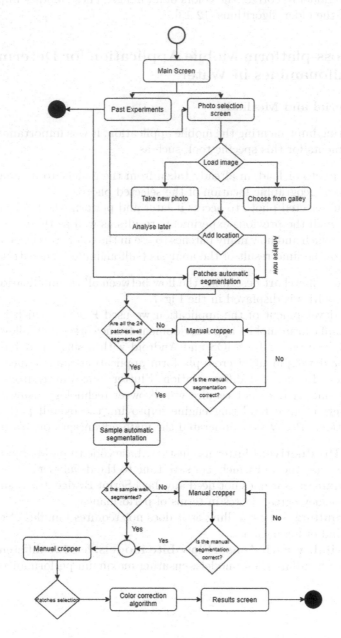

Fig. 7. Mobile application flowchart.

3.2 The Mobile Application

The application **main screen** is very simple, having a button at the center of the screen with a wave effect around it (see Fig. 8a). From here the user can navigate to: **Photo selection screen** by pressing the central button with the text "ANALYSE"; **Past experiments screen** by pressing the menu button in the *AppBar* followed by selecting the option "Past Experiments" it will take the user to the past experiments screen.

The **photo selection screen** allows the user to select the photo of the sample to analyse, add a location and start an analysis immediately or saving it for later. This screen consists in the following elements: **Camera Button:** The button with the camera logo when pressed allow the user to choose a photo from the gallery or take a new one using the mobile phone camera. **Location InputTextBox:** In this text the user must add a location where the photo belongs; **Analyse Now Button:** When pressed, an analysis of the selected photo starts instantly. If no photo is selected or the location field is empty, the user is warned and the analysis does not start; **Analyse Later Button:** When pressed the selected photo and location are saved and sent to the past experiments to be analysed later; **Discard Button:** The selected photo and location are deleted. Figure 8b shows the photo selection screen when neither the photo or the location have been added and Fig. 8c shows the photo selection screen after adding both of them.

(a) Application main screen

(b) Photo selection screen

(c) Photos election screen filled

Fig. 8. Main screen and photo selection screen

The first step of the entire process is the **segmentation of the patches**. The segmentation is made using the algorithm described in Sect. 2.2. Figure 9a shows the patches segmentation screen where is possible to visualize the selected photo with the bounding boxes around the 24 patches. If the automatic algorithm does not produce the correct segmentation, the user can activate a cropper tool to manually crop the patches. Before the cropper appears, a dialog box is displayed giving the user guidance on how to crop the patches (see Fig. 9b and c for Android and iOS, respectively. Figure 9d and e represent the cropper aspect in iOS and Android, respectively. After the user manually crops the patches, a new screen appears, allowing the user to verify if it is correct (see Fig. 9f).

(a) Patches segmentation screen

(b) Patches crop tutorial dialog box in Android

(c) Patches crop tutorial dialog box in iOS

(d) Patches manual cropper in iOS

(e) Patches manual cropper in Android

(f) Patches crop verification screen

Fig. 9. Patches segmentation screen

The next step is the **segmentation of the sample**, which is identical to the patches segmentation screen and the user can also choose to manually segment the sample, if the algorithm does not produce the correct segmentation (see

Sect. 2.3 (see Fig. 10a). The dialog box with guidance on how to crop the sample are displayed in the Fig. 10b and c for Android and iOS devices, respectively. Figure 10d and e shows the cropper for Android and iOS devices, respectively. After the user manually crops the sample, a new screen appears, allowing the user to verify if it is correct (see Fig. 9f).

(a) Sample segmentation screen

(b) Sample crop tutorial dialog box in Android

(c) Sample crop tutorial dialog box in iOS

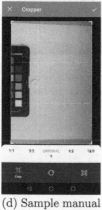

(d) Sample manual cropper in Android

(e) Sample manual cropper in iOS

(f) Sample crop verification screen

Fig. 10. Sample segmentation screen

The color correction algorithms uses by default all the 24 patches of the color chart; however, it can also work with any combination of patches, which can be selected in the **patches selection screen**. In this screen the user has the possibility to choose the following options: **Templates: T24 or t24:** To use all the 24 patches in the color correction algorithm; **T13 or t13:** To use a pre-defined set of the best 13 patches [13]. **Manual insertion:** The user has also the possibility to select any combination of patches. The patches are numbered according to the grid visible in Fig. 11a.

The **results screen** is where the sample concentration is displayed to the user along with the final color of the sample after applying the color correction algorithm (see Sect. 2.4) (see Fig. 11b). The concentration value is obtained by applying a linear regression presented in [12].

In the **past experiments screen** all application usage records can be viewed. Two types of registries are stored here: **Already analysed samples:** The result of an already performed analysis is displayed along with the date and a photo of the sample (see Fig. 11c). **Samples to be analysed:** It is presented all the samples to be analysed (see Fig. 11d).

(a) Patches selection screen

(b) Results screen

(c) Past experiments screen with complete entry

(d) Past experiments screen with sample to be analysed

Fig. 11. Patches selection, results and past experiments screens

3.3 Validation

An usability questionnaire was made with the propose to evaluate and measure the usability of the application. The target audience of this dissertation was a group of researchers specialized in chemistry working in the area of sulfonamides detection in water. Each question was rated on a scale from 1 to 5, where 5 is the most positive possible and 1 the most negative, inspired in a well established questionnaire [18], composed by the following questions:

1. **Is it appropriate for the main task?**
2. **Does the user have control over the interface?**

3. **Is it explicit?**
4. **Is it consistent?**
5. **Does it fulfill the user expectations?**
6. **Is it easy to use?**

It was possible to obtain a total of five different answers. In Table 1, it is possible to observe the average of the ratings assigned to each question as well as their standard deviation.

Table 1. Usability questionnaire results

Question	1	2	3	4	5	6
Average	4.8	4.6	5.0	5.0	4.8	4.4
Standard deviation	0.4	0.5	0.0	0.0	0.4	0.5

3.4 Summary

We cannot deny that smartphones are powerful, in fact, some of them are even more powerful than old and new generation laptops. With this in mind, the solution we worked on consists of a mobile application that leverages the computing power and ease of use of these devices. The mobile application answered to the specific requirements and enabled the ability of in situ analysis, without requiring an Internet connection or specified heavy equipment.

4 Conclusion

The use of antibiotics in the treatment of humans and animals is highly recurrent and with an increasing trend. Despite the benefits it can pollute aquatic environments. The presence of these compounds in environmental waters promote the occurrence of antibiotic-resistant bacteria, which will make intractable diseases more likely to appear.

The determination of the concentration of sulfonamides in water samples requires the use of laboratory equipment, which is not suitable to the analysis in situ. This work provides an affordable and practical method to this problem. Using digital image colorimetry and the application of machine learning algorithms for detection and segmentation, the application developed within the scope of this work allows, to quickly and accurately determine the concentration of these compounds in a water sample without the need of any internet connection or resorting to a laboratory making local analyzes a reality. The application is cross-platform being able for both Android and iOS users.

References

1. Thornton, P.K.: Livestock production: recent trends, future prospects. Trans. R. Soc. B **365**, 2853–2867 (2010). https://doi.org/10.1098/rstb.2010.0134
2. Hruska, K., et al.: Sulfonamides in the environment: a review and a case report. Vet. Med. **57**(1), 1–35 (2012). https://doi.org/10.17221/4969-VETMED
3. Baran, W., Adamek, E,. Ziemiańska, J., Sobczak, A.: Effects of the presence of sulfonamides in the environment and their influence on human health. J. Hazard Mater. **196**, 1–15 (2011). https://doi.org/10.1016/j.jhazmat.2011.08.082
4. Ventola, C.L.: The antibiotic resistance crisis: part 1: causes and threats. Pharm. Ther. **40**(4), 277–83 (2015)
5. European Centre for Disease Prevention and Control: Antimicrobial resistance in the EU/EEA (EARS-Net). Annual Epidemiological Report 2019
6. U.S Department of Health and Human Services: Antibiotic Resistance Threats in the United States. Centers for Disease Control and Prevention (2019)
7. Gbylik-Sikorska, M., Posyniak, A,. Sniegocki, T., Zmudzki, J.: Liquid chromatography-tandem mass spectrometry multiclass method for the determination of antibiotics residues in water samples from water supply systems in food-producing animal farms. Chemosphere **119**, 8–15 (2015). https://doi.org/10.1016/j.chemosphere.2014.04.105
8. Hoff, R., Kist, T.B.: Analysis of sulfonamides by capillary electrophoresis. J. Sep. Sci. **32**(5–6), 854–66 (2009). https://doi.org/10.1002/jssc.200800738
9. Bilandžić, N., et al.: Veterinary drug residues determination in raw milk in Croatia. Food Control **22**(12), 1941–1948 (2011). https://doi.org/10.1016/j.foodcont.2011.05.007
10. El-Dien, F.A.N., Mohamed, G.G., Khaled, E., Frag, E.Y.Z.: Extractive spectrophotometric determination of sulphonamide drugs in pure and pharmaceutical preparations through ion-pair formation with molybdenum(V) thiocyanate in acidic medium. J. Adv. Res. **1**(3), 215–220 (2010). https://doi.org/10.1016/j.jare.2010.05.005
11. Li, C., et al.: A class-selective immunoassay for sulfonamides residue detection in milk using a superior polyclonal antibody with broad specificity and highly uniform affinity. Molecules **24**(3), 443 (2019). https://doi.org/10.3390/molecules24030443
12. Carvalho, P.H., Bessa, S., Silva, A.R.M., Peixoto, P.S., Segundo, M.A., Oliveira, H.P.: Estimation of sulfonamides concentration in water based on digital Colourimetry. In: Morales, A., Fierrez, J., Sánchez, J.S., Ribeiro, B. (eds.) IbPRIA 2019. LNCS, vol. 11867, pp. 355–366. Springer, Cham (2019). https://doi.org/10.1007/978-3-030-31332-6_31
13. Carvalho, P.H., Rocha, I., Azevedo, F., Peixoto, P.S., Segundo, M.A., Oliveira, H.P.: Cost-efficient color correction approach on uncontrolled lighting conditions. In: Tsapatsoulis, N., Panayides, A., Theocharides, T., Lanitis, A., Pattichis, C., Vento, M. (eds.) CAIP 2021. LNCS, vol. 13052, pp. 90–99. Springer, Cham (2021). https://doi.org/10.1007/978-3-030-89128-2_9
14. Reis, P., Carvalho, P.H., Peixoto, P.S., Segundo, M.A., Oliveira, H.P.: Mobile application for determining the concentration of sulfonamides in water using digital image colorimetry. In: Antona, M., Stephanidis, C. (eds.) HCII 2021. LNCS, vol. 12769, pp. 468–484. Springer, Cham (2021). https://doi.org/10.1007/978-3-030-78095-1_34
15. Huang, J., et al.: Speed/accuracy trade-offs for modern convolutional object detectors. In: 2017 IEEE Conference on Computer Vision and Pattern Recognition (CVPR), pp. 3296–3297 (2017)

16. Lin, T., et al.: Feature pyramid networks for object detection. In: 2017 IEEE Conference on Computer Vision and Pattern Recognition (CVPR), Honolulu, HI, USA, pp. 936–944 (2017). https://doi.org/10.1109/CVPR.2017.106
17. Finlayson, G.D., Mackiewicz, M., Hurlbert, A.: Color correction using rootpolynomial regression. IEEE Trans. Image Process. **24**(5), 1460–1470 (2015)
18. Park, K.S., Lim, C.H.: A structured methodology for comparative evaluation of user interface designs using usability criteria and measures. Int. J. Industr. Ergono. **25**(3), 379–389 (1999). https://doi.org/10.1016/S0169-8141(97)00059-0

Image Processing

Visual Semantic Context Encoding for Aerial Data Introspection and Domain Prediction

Andreas Kriegler[1,2](✉)[iD], Daniel Steininger[1][iD], and Wilfried Wöber[3,4][iD]

[1] Vision Automation and Control, Austrian Institute of Technology,
1210 Vienna, Austria
andreas.kriegler@ait.ac.at
[2] Visual Computing and Human-Centered Technology,
TU Wien Informatics, 1040 Vienna, Austria
[3] Industrial Engineering, UAS Technikum Wien, 1200 Vienna, Austria
[4] Integrative Nature Conservation Research, University of Natural Resources
and Life Sciences, 1180 Vienna, Austria

Abstract. Visual semantic context describes the relationship between objects and their environment in images. Analyzing this context yields important cues for more holistic scene understanding. While visual semantic context is often learned implicitly, this work proposes a simple algorithm to obtain explicit priors and utilizes them in two ways: Firstly, irrelevant images are filtered during data aggregation, a key step to improving domain coverage especially for public datasets. Secondly, context is used to predict the domains of objects of interest. The framework is applied to the context around airplanes from *ADE20K-SceneParsing*, *COCO-Stuff* and *PASCAL-Context*. As intermediate results, the context statistics were obtained to guide design and mapping choices for the merged dataset *SemanticAircraft* and image patches were manually annotated in a one-hot manner across four aerial domains. Three different methods predict domains of airplanes: An original threshold-algorithm and unsupervised clustering models use context priors, a supervised CNN works on input images with domain labels. All three models were able to achieve acceptable prediction results, with the CNN obtaining accuracies of 69% to 85%. Additionally, context statistics and applied clustering models provide data introspection enabling a deeper understanding of the visual content.

Keywords: Context encoding · Domain prediction · Aerial scenes

1 Introduction

Humans intuitively incorporate contextual information when trying to understand the environment they perceive. Objects appearing in an unfamiliar semantic context or out-of-context objects [7], such as airplanes on a highway, attract

© Springer Nature Switzerland AG 2022
A. J. Pinho et al. (Eds.): IbPRIA 2022, LNCS 13256, pp. 433–446, 2022.
https://doi.org/10.1007/978-3-031-04881-4_34

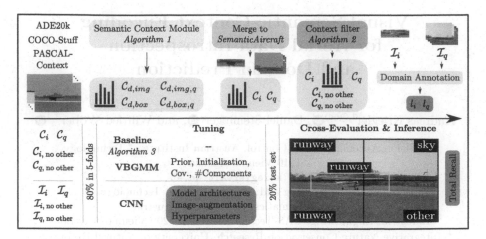

Fig. 1. Upper half: For every source dataset d, context vectors $\mathcal{C}$ on images (img) and object bounding boxes (box) as well as the respective quadrants q are obtained. Merging leads to $\mathcal{C}_i$ and $\mathcal{C}_q$ for instances and quadrants, removal of other-patches yields $\mathcal{C}_{i,no\_other}$ and $\mathcal{C}_{q,no\_other}$. All image patches $\mathcal{I}_i$ and $\mathcal{I}_q$ were then annotated with domain labels l_i and l_q. Lower half: Either the set of context vectors or images with annotations were finally used for domain prediction.

the observer's attention since they are typically related to other scenes. Incorporating this kind of prior information has the potential to improve computer vision (CV) models by assigning meaning to objects and actions, enabling "visual common-sense" and is essential for solving upcoming challenges in scene understanding [1]. In particular, autonomous systems operating in the real world struggle to stay robust when traversing multiple environments, or the surroundings look significantly different due to weather, atmospheric effects, or time of day. Semantics of images or semantic parsing in the field of CV refers to the recognition and understanding of the relationship between objects of interest other objects and their environment [7]. Natural occurrences of objects and corresponding environments are analyzed to transfer this information into a logical-form representation, understandable for machine vision systems. This context can be understood as a statistical property of our world [19]. On a micro level semantic segmentation yields information regarding both foreground objects, commonly referred to as *things* [17] and background scenery, known as *stuff* [4]. Following this segmentation and applying ideas from natural language processing semantic relations between *things* and *stuff* can be formulated. It is well known that cues stem from the *semantic context* surrounding objects and this visual context is therefore a necessity for more complete scene understanding [5]. The models developed in this work are evaluated in the field of avionics, specifically on images showing airplanes. Therefore, the related concept of *domains* in this work holds two specific yet congruent meanings: In the applied sense, a domain describes the local real world surroundings of airplanes, in the more formal sense a domain can be understood as a collection of characteristic classes. The syntactic evolution from semantic context to domains is natural, when considering the focus lies on a person, object or autonomous agent around which context is formulated. Domains can therefore be understood as a result of the analysis of an objects semantic context, placing *things* into a distinctive domain. Following the analysis of this visual semantic context, characteristic statistics can give an understanding of datasets which can in turn be used to guide data-aggregation strategies. Further along the learning pipeline, due to the convolutional kernel in convolutional neural network (CNN), contextual information is usually learned implicitly regardless of the learning task at hand [2,26]. Contextual information is embedded in the feature space and the learned kernel-parameters lead to the well-known bias/variance dilemma [12]. An explicit representation of context in the form of distinct domains might allow intelligent systems to swap between model parameters in a mixture-of-experts fashion. To this end the preliminary step of predicting domains is studied (see Fig. 1 for an overview of the proposed methods). To summarize, this work makes the following contributions:

1. We propose a simple method to encode semantic context from segmentation masks providing context insights for images from the aerial domain.
2. We present the merged dataset *SemanticAircraft* and filter images using context statistics. Additionally, we provide over 17k domain annotations for *SemanticAircraft*.
3. We use unsupervised clustering algorithms for data introspection revealing further information relevant to avionic applications. Finally, we reinterpret clustering results for domain prediction, propose a novel, fast and interpretable prediction algorithm as baseline and compare these results to domain classification results using deep supervised CNNs.

The remainder of the paper is structured as follows: Sect. 2 provides related works. Section 3 details the encoding algorithm, shows context results and explains the domain annotation process for *SemanticAircraft*. Section 4 introduces the three domain prediction models and provides inference results.

2   Related Works

Following classification, object detection and semantic segmentation, a clear trend towards more complex representations is noticeable [16,32]. Before CNNs, semantic segmentation used either conditional random fields (CRFs) or tree models similar to Markov networks [22,23], although with limited accuracy.

2.1   Semantic Segmentation with Deep Learning

In a similar vein to works using CRFs is Wang *et al.* [29]'s multiple-label classification on *NUS-WIDE* [8], *COCO* [17] and *VOC* [10]. They combine a VGG [28] CNN to embed visual features with a LSTM-RNN for label information in a joint space. Zhang *et al.* [34] pose the question whether capturing contextual information with a CNN is the same as simply increasing the receptive field size but perhaps more accurately the question should be how much one can increase the receptive field size and still capture relevant contextual information. In a similar vein Fu *et al.* [11] state that the method with which to effectively capture pixel or region-aware context is still an open and unresolved research question. While these works provide models for obtaining semantically segmented images, neither the learned features nor final masks constitute an explicit context representation that is in line with our concept of visual domains. The literature on domain adaptation techniques [9,30,31] on the other hand is predominately concerned with domain adaptation between synthetic and real images for transfer learning. For the purpose of making a segmentation network robust across multiple domains, Chen *et al.* [6] propose to treat different cities as distinct domains and go on to learn both class-wise and global domain adaptation in an unsupervised manner. The concept of domains are treated as a means-to-the-end for boosting segmentation accuracy which is a common approach. Similarly, Sakaridis *et al.* [24] use the idea of guided curriculum model adaptation for improving semantic segmentation of nighttime images for advanced driver assistance systems (ADASs). Having captured the same scene at daytime, twilight and night using labeled synthetic stylized and unlabeled real data, models are transferred from daytime to night with twilight as an intermediate domain, using the Dark Zürich dataset. In a similar vein are the works of Zhang *et al.* [36] and their follow-up paper [35]. They learn global label distributions over images and local distributions over landmark superpixels and feed those into a segmentation network to boost semantic segmentation performance.

2.2   Domains for Context Generalization

While these works provide a foundation for capturing context in a deep learning (DL) manner, and also tackle the problem of domain generalization, it stands to reason that the semantically segmented output masks are much lower-level in their representation of the context than desirable. It could be argued that pixel-wise classification as final model output is less representative of the actual content of an image than our conceptual usage of domains around target objects, at least for object-centric tasks. To this end the work of Sikirić *et al.* [27] is related. The task is image-wide classification of images captured in various traffic scenes in Croatia. Their treatment of different traffic scenes is similar to the idea of domains in this work: As a concept to describe the environment for scene parsing. Although the methods developed in our work are kept as general as possible to allow the application in multiple domains, we will focus on one domain in particular, the aerial domain. While methods for ADAS applications have gotten strong interest, the aerial domain is much less studied in contemporary literature.

2.3   Public Aerial Datasets

Publicly available datasets providing semantically segmented images are fairly numerous, around 10–15 according to [14]. Nevertheless, no semantically-segmented dataset specifically created for airplanes exists. Three semantic public datasets that feature some images of airplanes are accessible: A derivative of the *ADE20K* dataset for scene parsing [37] referred to as *ADE20K-SceneParsing*. An extension to the *MS COCO* [17] annotation for *stuff* classes [4] denoted as *COCO-Stuff*. And the semantic extension to *PASCAL-VOC* [10], *PASCAL-Context* [19]. For brevity these special derivations will be referred to as *ADE*, *COCO* and *PASCAL*. These three datasets from the basis for our following context analysis.

3   Semantic Context and *SemanticAircraft*

In this section we outline our taxonomy of aerial domains, detail the data aggregation process, introduce our algorithm to compute context vectors and use context statistics and a context filter to merge images to *SemanticAircraft*, for which we finally provide domain label annotations.

3.1   Aerial Domains and Aggregation of Airplane Images

Considering the environment airplanes traverse, three domains can be identified:

Apron: In aviation the area where airplanes are usually parked, loaded or unloaded with goods, boarded or refueled is referred to as apron. A large variety of partially occluded objects, persons and unusual vehicles such as mobile loading ramps, taxiing vehicles and moving stairways, is common.

Runway: The strip of asphalt or concrete used primarily for takeoff and landing of the airplanes is referred to as runway. It is usually enclosed by grass or other types of soil, with more vegetation such as bushes and trees appearing to the sides. Neither vehicles nor persons are usually encountered in this domain.

Sky: Sky is typically a smooth blue or grey background to the airplane, but clouds and time of day can significantly alter its appearance. The elevation angle of the capturing camera plays an important role.

Other: Finally we use a fourth domain, other, for out-of-context airplanes.

Images from *ADE*, *COCO* and *PASCAL* featuring at least one airplane pixel in their semantic masks were aggregated and the following observations made:

ADE20K-SceneParsing: *ADE* features 146 images with airplanes in total, where 33 of the 150 classes are of interest. Images on average are around 600×600 in size. Two pairs of duplicate images exist where only one image is kept.

COCO-Stuff: *COCO* is the largest of the three datasets with 3079 images featuring airplanes. *COCO-Stuff* has 171 classes with 41 being relevant. Average

---

Algorithm 1. Obtaining semantic context

Requisites: A set $\mathcal{M}$ of masks $\mathbf{m}$ mapping from pixels to the list of class ids $\mathbf{x} = \{0, \ldots, N\}$, in particular t for the target class and label v for void pixels.

1: **function** GETCONTEXT($\mathcal{M}, \mathbf{x}$)
2:    **for** $\mathbf{m} \in \mathcal{M}$ **do**   ▷ For every segmantic mask
3:       $\mathbf{m}_{\text{ext}} \leftarrow$ dilate($\mathbf{m}, t, 1, 5$)   ▷ Dilate the mask to deal with void pixels
4:       **for** x in $\mathbf{x}$ **do**   ▷ For every class in consideration
5:          **if** $x \neq t, v$ **then**   ▷ Ignore specific classes
6:             $c_{\mathbf{m}, x} \leftarrow \sum_{i=0, j=0}^{i=W, j=H} (\mathbf{m}_{\text{ext}\,i,j} == x)$ ▷ Count class-specific pixels
7:       $\forall x \in \mathbf{x} : \mathbf{c} \leftarrow 100 \times \frac{c_{\mathbf{m}, x}}{\sum(c_{\mathbf{m}})}$ ▷ Normalize to obtain a $(0, 1]$ squashed vector
8:       **return** $\mathbf{c}$

---

image size is around 640×480. Besides out-of-context airplanes, *COCO* also features synthetic images.

PASCAL-Context: The total number of images with airplanes is 597. It has 456 classes in total of which around 30 are applicable. Here, many variations for building and soil exist. The average image size is around 470×386.

Every image featuring at least one airplane pixel is considered. If the dataset provides bounding box (BBox) annotations they are used for extraction of instances, otherwise an algorithm iteratively extends rectangles encompassing 1 pixel at the start to include all pixels of the same target class touching any already included pixels.

3.2   Encoding Visual Semantic Context

The method used for obtaining semantic context is similar to the concept of label occurrence frequency presented by Zhang *et al.* [36]. In this work we extend the algorithm to handle semantic uncertainty boundaries, exclude undesired classes and apply the method to a finer granularity of image regions. The semantic context module described in Algorithm 1 extracts label frequency for a set $\mathcal{M}$ of masks $\mathbf{m}$. Besides the masks, the list of classes $\mathbf{x} \in \mathbb{R}^{u \times 1}$ is required. As a first step, to deal with void pixels at label transitions, the boundaries of the target instance are expanded by five pixels in every direction, i.e. the instance gets dilated by 1 for five times. Then for every class in $\mathbf{x}$ the number of pixels in a certain patch of $\mathbf{m}$ is obtained and normalized with the total number of pixels in that patch. For void and airplane pixels, they can optionally be ignored. As a result, a number of context distribution vectors $\mathbf{c}$ are obtained in every dataset d. Vector $\mathcal{C}_{d, img, q=II}$ then for example gives the context in dataset d across the second image quadrants. Quadrant-I is the top-right image quadrant counting counter-clockwise. Entries in $\mathbf{c}$ are also sorted by magnitude. We deal with instances and image quadrants separately, since a downstream tracking framework would benefit from this granularity. The module was applied to the datasets *ADE*, *COCO* and *PASCAL* to obtain first statistical context measures.

Algorithm 2. Semantic context filter

Requisites: Set $\mathcal{C}$ of vectors $\mathbf{c}$ holding statistics for every patch to be filtered. Labels $\mathbf{x}$ that shall get filtered with quantile percentages $\mathbf{p} = (0.93, 0.93)$.

1: **function** FILTERCONTEXT($\mathcal{C}, \mathbf{x}, \mathbf{p}$)
2:     **for** (x, p) in $(\mathbf{x}, \mathbf{p})$ **do**   ▷ For every filter label
3:         $q_{i,x} \leftarrow$ quantile($\mathcal{C}_{(i,x)}, p$)   ▷ Calculate quantiles in instances
4:         $q_{q=I...IV,x} \leftarrow$ quantile($\mathcal{C}_{(q=I...IV,x)}, p$)   ▷ And image quadrants
5:         $\mathcal{S}_{i,x} \leftarrow \mathcal{C}_{i,x} < q_{i,x}$ ▷ Filter instances using the threshold
6:         **for** $i \leftarrow I$ to IV **do**
7:             **if** $i \leq II$ **then**
8:                 $\mathcal{S}_{q=i,x} \leftarrow \mathcal{C}_{i,x} < mean(q_{(q=I,x)}, q_{(q=II,x)})$ ▷ Or mean for quadrants
9:             **else**
10:                 $\mathcal{S}_{q=i,x} \leftarrow \mathcal{C}_{i,x} < mean(q_{(q=III,x)}, q_{(q=IV,x)})$
11:     **return** $\mathcal{S} \setminus (\mathcal{S}_{\text{indoor}} \cup \mathcal{S}_{\text{void}})$ ▷ Let both constraints apply for the final set

---

3.3 Aggregation and Context of *SemanticAircraft*

When training a framework on public datasets a single pass of data aggregation does not yield optimal data in terms of domain coverage, redundancy, object size and especially class consistency. Therefore, following inspection of context statistics, semantically similar classes were merged to superclasses for *SemanticAircraft*. Bounding boxes were increased by thirty percent, which strikes a good balance of incorporating distant elements while leaving out largely irrelevant features. First context statistics have shown that many images and instances feature undesirable context traits, e.g. a high-percentage of void and indoor pixels. Since semantic context yields a high-level understanding of the scene, it can be used to filter patches where the context in specific classes is higher than a desired value (Algorithm 2). To be precise, using the set $\mathcal{C}$ of context vectors for any patch, obtain the quantile value q_p at the threshold p and remove all patches with context above q_p, while using the mean for quadrants-I/II and III/IV. Our motivation for using the mean of the upper vs. lower image is, that the majority of images showing airplanes feature a clear horizon (top) or ground (bottom) separation, if the airplane is on the ground. While visual content in the lower left quadrant might be dissimilar to content of the lower right for a limited number of images, assuming a dataset of infinite size, the content of the two quadrants becomes equal. The same reasoning holds for the upper two quadrants. Therefore, to reduce dataset bias we take the mean of the upper and lower quadrants. Following the heuristic of filtering patches showing at least a small amount of indoor pixels, the percentages were determined empirically as $\mathbf{p} = (0.93, 0.93)$. This removes indoor images while also removing samples with excessive void pixels. As a last filtering step, low-level filters were applied: All instances with width or height shorter than 60 pixels and aspect ratio larger than 6:1 were discarded. This leaves 3854 instances and 13265 image-quadrants. Although not visualized in this paper, the filtered out-of-distribution patches include toy-airplanes, airplanes in magazines, LEGO-airplanes and many airplanes in museums and

Fig. 2. A random selection of instances (top) and quadrants (bottom) from *SemanticAircraft*. Clutterness in quadrants is lower than instances, across all domains. Although looking like a sky image, the fourth image in the other row of the upper half actually shows a computer generated image. Images with red outline were misclassified by *any* of the three prediction models (see Sect. 4). (Color figure online)

exhibitions. Without the exclusion, these samples would bring unwanted noise into the data for training domain prediction models. Figure 2 provides example images from the resulting dataset *SemanticAircraft*. Final semantic context statistics were obtained and can be seen in Table 1.

4   Domain Prediction on *SemanticAircraft*

In this section we propose the application of semantic context for the task of domain prediction. Three distinct approaches were chosen for this task:

Baseline: First, define domains as set of superclasses. Using the context vectors C_i and C_q run a threshold algorithm with defined ranges and weights.

Unsupervised: Using the set of context features C_i and C_q, fit an unsupervised machine learning (ML) model to predict the domain for unseen context vectors, i.e. interpret label statistics per patch as features and use unsupervised learning for clustering – reinterpret the clusters for a classification setting.

Table 1. Visual percentage-wise context for the *SemanticAircraft* dataset showing dominant sky-context. Context across all four quadrants was merged.

	Building	Elevation	Object	Pavement	Person	Plant	Sky	Soil	Vehicle	Waterbody
Instances	**7.5**	**3.2**	**1.5**	15.8	**1.2**	**5.1**	57.2	6.3	0.9	1.3
Quadrants	4.2	2.8	1.2	**17.4**	1.1	4.0	**58.6**	**7.8**	**1.0**	**1.9**

Supervised: Instances and quadrants from *SemanticAircraft* with their respective domain labels are used for supervised classification with a CNN.

For the purpose of domain prediction, classification accuracy (recall) is the primary goal, although unsupervised mixture models used other metrics for parameters tuning. The parameters of the baseline algorithm were not tuned, instead were set once using human-expert knowledge. The dataset *SemanticAircraft* consistes of a set of 3854 instance and 13265 quadrant triplets: RGB input images, corresponding context vector $c_{i/q}$, and ground truth (GT) domain label. After setting 20% of data aside for the test set, the experiment took place in two phases. In the first phase hyperparameters and architectural designs were tuned following the evaluation on the validation portion of the remaining 80% using method-specific metrics. For the second phase two separate versions of *SemanticAircraft* were used. The first consists of all remaining 20% of instances and quadrants. For the second all samples with the GT domain label other were removed. The final prediction results of phase 2 can be observed in Table 2.

4.1   Baseline Threshold Model

Algorithm 3 proposed in this subsection serves as the baseline for domain prediction evaluation. The basic premise of the baseline was to develop an algorithm that works similar to human intuition: The relative pixel amount of every context-class contributes towards a certain domain-belief with a set strength if it is as-expected for any domain. For example, apron samples are commonly expected to feature vehicles while runway and sky are not. Observing the context for any patch, e.g. $c_{vehicle} = 0.4$ meaning forty percent of pixels are vehicle, the ranges r of expected vehicle context for all three domains are checked and weights w added for every domain with bounds including 0.4. This cummulative score s signifies the level of distinction all context classes provides. This is done for every superclass with scores adding up to the domain score d. For classification, an image patch then has to reach a configurable score-threshold th. While simple to configure, this algorithm shows multiple shortcomings: 1) It is parameter-heavy, making tuning for a set of domains and extension to other domains difficult, 2) All parameters are partly dependent on expert-knowledge, informed by previous dataset-wide semantic context analysis, 3) Equal domain-scores lead to ambiguity – in this case, this ambiguity was solved with random tie-breaks 4) Patches not meeting the threshold signify high uncertainty in the context or out-of-context patches and it is unclear how this should be resolved. Despite these drawbacks, once set up for a set of domains and datasets, results are reproducible due to the deterministic nature and inference time is negligible.

Algorithm 3. Thresholding domain prediction

Requisites: Set of context vectors $C_{i/q}$. Set of domains **d** and for every domain and superclass **s** consisting of classes c a certain range $\mathbf{r}_{x,y}$ and weight $\mathbf{w}_{x,y}$. Domain-prediction threshold of th and a decrease th_d.

1: **function** $\mathrm{TDP}(C_{i/q}, \mathcal{D}, \mathcal{S}, \mathcal{R}, \mathcal{W}, th, th_d)$
2:     **for** c in $\mathcal{C}$ **do**   ▷ For every context vector
3:         d_s $\leftarrow$ **0** : **0** $\in \mathbb{R}^{n \times 1}$ ▷ Initialize the domain score
4:         **for** d in $\mathcal{D}$ **do**   ▷ And for every dataset
5:             **for** s in $\mathcal{S}$ **do**   ▷ And superclass in that dataset
6:                 s_s $\leftarrow \sum_i c_i, \forall c \in$ **s**   ▷ Aggregate context of all classes
7:                 **if** s_s $\in [\mathbf{r}_{d,s,l}, \mathbf{r}_{d,s,u}]$ **then**   ▷ Check if score is in range
8:                     d_$s_d \leftarrow$ d_$s_d + \mathbf{w}_{d,s}$ ▷ Add a weight to the domain score
9:         **if** $\max($d_s$) > th$ **then**   ▷ Take the top-1 domain
10:            $\mathbf{l_c} \leftarrow argmax(d_s)$ ▷ And assign the domain label
11:        **else**
12:            $th \leftarrow th - th_d$ ▷ Or decrease threshold until domain is found
13:    **return** l   ▷ Return domain labels for every image patch

---

4.2   Unsupervised Clustering and Mixture Models

The mathematical foundations of the unsupervised clustering and mixture models are detailed by [3]. It should be noted, that any created cluster are an internal mathematical construct and do not resemble the set of predefined domains. This makes interpretation in a classification setting not as straightforward as with CNNs. The scikit-learn python library [21] was used for implementation. The clustering model of choice was the variational Bayesian Gaussian mixture model (VBGMM), which provides a larger flexibility than popular K-Means or regular Gaussian mixture models. The optimal hyperparameters are those, where the silhouette-coefficient [15] is at a maximum in $[0, 1]$. Tuned parameters include the distribution prior (Dirichlet process vs. Dirichlet distribution), covariance type (full, diagonal etc.), initialization (K-Means, random) and number of active components to model the data. Parameters were tuned in a grid-search, the highest achieved coefficients are 0.702 and 0.766 for instances and quadrants respectively, only one cluster can be assigned at any time. It should be noted, that in some experiments, the number of clusters was not set, allowing the VBGMM to cluster the context vectors however it sees fit. This would result in up to 13 and 9 clusters for instances and quadrants respectively, a hint that the restraint to 3 or 4 clusters does not fully explain the distributions that created the context vectors. Finally, it should be noted hat context vectors were not assigned to any of the 3 or 4 domains, but only to an equal number of clusters. This means the method is truly unsupervised at the cost of domain prediction accuracy. To obtain classification accuracy (recall), the best-performing permutation of possible cluster-assignments had to be found, since the clusters are not directly related to the defined set of domains. This was done by expressing the confusion matrix across clusters as a cost matrix and obtaining the minimization over all permutations of possible row/column combinations [18], 4! when including other, 3! otherwise. This yields the permutation of the confusion matrix with the highest diagonal sum.

4.3 Supervised Convolutional Neural Networks

For the CNN, not the set of context vectors are used as input data, but instead the images $\mathcal{I}_{RGB}/\mathcal{Q}_{RGB}$ and corresponding class-labels $\mathcal{I}_l/\mathcal{Q}_l$ obtained from domain annotation. The PyTorch library [20] was used for implementation. All samples were resized to 256×256 pixels to meet the input requirements. The tuned parameters included model-type and size (ResNet50 to ResNet34, ResNet18 and various DRN [33] and MobileNet [25] architectures), image augmentation strength, batch-size and the manual addition of a dropout layer. Parameters were tuned using 5-fold cross validation. The best performing models on both instances and quadrants turned out to be ResNet18 [13] variants. For both models, only light image-augmentation yielded the best results. A key difference between the models is the presence of dropout ($p = 0.5$) for the quadrants model. Thus the problem of model overfitting could be addressed by reducing model-size and adding a dropout layer to the architectures. With hyperparameter tuning of the models concluded, they can now be directly compared against another on the held-out test set.

4.4 Domain Prediction Results

Since this can be seen as a classification task, classification accuracy or recall was used. Table 2 shows the final obtained results.

It should be made clear that all three models serve different purposes and only quantifying their usefulness regarding inference accuracy does not give the full picture: While the CNN has given the best prediction performance, the requirement of annotations for the domains are a significant drawback, limiting its application in entirely new domains. Nevertheless, for an applied system that observes airplanes around airports the CNN model would be the preferred choice. For the baseline, the parameterization makes tuning and extension to other domains difficult. At the same time, since context statistics need to be analyzed for choices in data aggregation anyway (at least when dealing with a new dataset from the wild), the parameterization naturally evolves from these ideas and requires little more effort. While the variational Bayesian gaussian mixture model (VBGMM) performs the worst in terms of prediction accuracy, the insights the model can provide into the dataset structure, hinting at possible other subdomains besides apron, runway, sky and other, are noteworthy. If the number of clusters was not specified as it was the case in some VBGMM experiments, more than three or four clusters were created. The results indicate, that the limitation to apron, runway and sky was perhaps too strict. In future work, further analysis with clustering algorithms could provide important insights.

Thus, all three models have benefits and drawbacks but for the explicit task of domain prediction, the supervised CNN performed the best. The exclusion of

Table 2. Accuracy of all three models predicting domains of airplane instances and quadrants from *SemanticAircraft*.

	Instances		Quadrants	
	Including other	Excluding other	Including other	Excluding other
Baseline	0.588 ± 0.015	0.796 ± 0.011	0.639 ± 0.017	**0.799 ± 0.006**
VBGMM	0.586 ± 0.048	0.712 ± 0.06	0.539 ± 0.029	0.637 ± 0.083
ResNet18	**0.716 ± 0.015**	**0.854 ± 0.011**	**0.692 ± 0.013**	0.778 ± 0.006

any other samples does improve all model's performance, most significantly the baseline. The importance of filtering out-of-context samples is not apparent but it stands to reason that without the explicit context representation and filtering using context the CNN would perform worse. Finally the somewhat underwhelming accuracies can be broadly explained due the manual annotation of domain labels. Even for human experts the distinction between apron and runway was hard, especially for image quadrants, and the most common prediction errors was between these two domains. A more principled approach, perhaps using context statistics themselves to assign labels to image patches, could prove more fruitful.

5 Conclusion

With the proposed semantic context module context vectors were extracted from semantically segmented masks. These context vectors were used for improved data aggregation and domain prediction of images in the merged dataset *SemanticAircraft*. Images were further manually annotated with domain labels. Results show that all three domain prediction models, a novel baseline, unsupervised clustering model, and the CNN were capable of predicting domains with acceptable accuracy, although only inferences with the ResNet18 CNN are accurate enough to guide potential downstream models. For baseline and mixture models the fact that they do not require annotations is a significant benefit. The clustering method additionally provides data introspection. In future works, improved domain prediction results could be used to guide parameter-selection for downstream models fine-tuned on specific domains. Finally clusters created with the unsupervised models could be further analyzed for deeper insights into the visual context of the datasets.

References

1. Aditya, S., Yang, Y., Baral, C., Fermuller, C., Aloimonos, Y.: Visual commonsense for scene understanding using perception, semantic parsing and reasoning. In: Logical Formalizations of Commonsense Reasoning - Papers from the AAAI Spring Symposium, Technical Report, pp. 9–16. AAAI Spring Symposium - Technical Report, AI Access Foundation (2015)

2. Bach, S., Binder, A., Montavon, G., Klauschen, F., Müller, K.R., Samek, W.: On pixel-wise explanations for non-linear classifier decisions by layer-wise relevance propagation. PloS one **10**(7) (2015)
3. Bishop, C.M.: Pattern Recognition and Machine Learning. Springer, New York (2006)
4. Caesar, H., Uijlings, J., Ferrari, V.: Coco-stuff: thing and stuff classes in context. In: 2018 IEEE/CVF Conference on Computer Vision and Pattern Recognition, pp. 1209–1218 (2018)
5. Chen, X.: Context Driven Scene Understanding. Ph.D. thesis, University of Maryland (2015)
6. Chen, Y.H., Chen, W.Y., Chen, Y.T., Tsai, B.C., Frank Wang, Y.C., Sun, M.: No more discrimination: Cross city adaptation of road scene segmenters. In: 2017 IEEE International Conference on Computer Vision, pp. 2011–2020 (2017)
7. Choi, M.J., Torralba, A., Willsky, A.S.: Context models and out-of-context objects. Pattern Recogn. Lett. **33**(7), 853–862 (2012)
8. Chua, T.S., Tang, J., Hong, R., Li, H., Luo, Z., Zheng, Y.: Nus-wide: a real-world web image database from national university of Singapore. In: Proceedings of the ACM International Conference on Image and Video Retrieval, pp. 1–9 (2009)
9. Csurka, G.: A comprehensive survey on domain adaptation for visual applications. In: Csurka, G. (ed.) Domain Adaptation in Computer Vision Applications. ACVPR, pp. 1–35. Springer, Cham (2017). https://doi.org/10.1007/978-3-319-58347-1_1
10. Everingham, M., Van Gool, L., Williams, C.K., Winn, J., Zisserman, A.: The pascal visual object classes (voc) challenge. Int. J. Comput. Vision **88**(2), 303–338 (2010)
11. Fu, J., et al.: Adaptive context network for scene parsing. In: 2019 IEEE/CVF International Conference on Computer Vision (ICCV), pp. 6747–6756 (2019)
12. Geman, S., Bienenstock, E., Doursat, R.: Neural networks and the bias/variance dilemma. Neural Comput. **4**(1), 1–58 (1992)
13. He, K., Zhang, X., Ren, S., Sun, J.: Deep residual learning for image recognition. In: 2016 IEEE Conference on Computer Vision and Pattern Recognition (CVPR), pp. 770–778 (2016)
14. Huang, L., Peng, J., Zhang, R., Li, G., Lin, L.: Learning deep representations for semantic image parsing: a comprehensive overview. Front. Comp. Sci. **12**(5), 840–857 (2018). https://doi.org/10.1007/s11704-018-7195-8
15. Kaufman, L., Rousseeuw, P.J.: Finding Groups in Data: An Introduction to Cluster Analysis. John Wiley & Sons (2009). https://doi.org/10.1002/9780470316801
16. Kendall, A.G.: Geometry and Uncertainty in Deep Learning for Computer Vision. Ph.D. thesis, University of Cambridge (2019)
17. Lin, T.-Y., et al.: Microsoft COCO: common objects in context. In: Fleet, D., Pajdla, T., Schiele, B., Tuytelaars, T. (eds.) ECCV 2014. LNCS, vol. 8693, pp. 740–755. Springer, Cham (2014). https://doi.org/10.1007/978-3-319-10602-1_48
18. Morbieu, S.: Accuracy: from classification to clustering evaluation (2019). https://smorbieu.gitlab.io/accuracy-from-classification-to-clustering-evaluation/. Accessed 29 May 20
19. Mottaghi, R., et al.: The role of context for object detection and semantic segmentation in the wild. In: 2014 IEEE Conference on Computer Vision and Pattern Recognition (2014)

20. Paszke, A., et al.: Pytorch: an imperative style, high-performance deep learning library. In: Wallach, H., Larochelle, H., Beygelzimer, A., d'Alché-Buc, F., Fox, E., Garnett, R. (eds.) Advances in Neural Information Processing Systems 32, pp. 8024–8035. Curran Associates, Inc. (2019). http://papers.neurips.cc/paper/9015-pytorch-an-imperative-style-high-performance-deep-learning-library.pdf
21. Pedregosa, F., et al.: Scikit-learn: Machine learning in python. J. Mach. Learn. Res. **12**, 2825–2830 (2011)
22. Rabiner, L., Juang, B.: An introduction to hidden Markov models. IEEE ASSP Magaz. **3**(1), 4–16 (1986)
23. Richardson, M., Domingos, P.: Markov logic networks. Mach. Learn. **62**(1–2), 107–136 (2006)
24. Sakaridis, C., Dai, D., Van Gool, L.: Guided curriculum model adaptation and uncertainty-aware evaluation for semantic nighttime image segmentation. In: 2019 IEEE/CVF International Conference on Computer Vision (ICCV), pp. 7373–7382 (2019). https://doi.org/10.1109/ICCV.2019.00747
25. Sandler, M., Howard, A., Zhu, M., Zhmoginov, A., Chen, L.C.: Mobilenetv 2: inverted residuals and linear bottlenecks. In: 2018 IEEE/CVF Conference on Computer Vision and Pattern Recognition, pp. 4510–4520 (2018). https://doi.org/10.1109/CVPR.2018.00474
26. Selvaraju, R.R., Cogswell, M., Das, A., Vedantam, R., Parikh, D., Batra, D.: Gradcam: Visual explanations from deep networks via gradient-based localization. Int. J. Comput. Vis. **128**(2), 336–359 (2019). http://dx.doi.org/10.1007/s11263-019-01228-7
27. Sikirić, I., Brkić, K., Bevandić, P., Krešo, I., Krapac, J., Šegvić, S.: Traffic scene classification on a representation budget. IEEE Trans. Intell. Transp. Syst. **21**(1), 336–345 (2020). https://doi.org/10.1109/TITS.2019.2891995
28. Simonyan, K., Zisserman, A.: Very deep convolutional networks for large-scale image recognition. CoRR abs/1409.1556 (2015)
29. Wang, J., Yang, Y., Mao, J., Huang, Z., Huang, C., Xu, W.: CNN-RNN: a unified framework for multi-label image classification. In: 2016 IEEE Conference on Computer Vision and Pattern Recognition (CVPR), pp. 2285–2294 (2016)
30. Wang, M., Deng, W.: Deep visual domain adaptation: a survey. Neurocomputing **312**, 135–153 (2018)
31. Weiss, K., Khoshgoftaar, T.M., Wang, D.D.: A survey of transfer learning. J. Big Data **3**(1), 1–40 (2016). https://doi.org/10.1186/s40537-016-0043-6
32. Xiao, T., Liu, Y., Zhou, B., Jiang, Y., Sun, J.: Unified perceptual parsing for scene understanding. In: Proceedings of the European Conference on Computer Vision (ECCV) (2018)
33. Yu, F., Koltun, V., Funkhouser, T.: Dilated residual networks. In: 2017 IEEE Conference on Computer Vision and Pattern Recognition (CVPR), pp. 636–644 (2017)
34. Zhang, H., et al.: Context encoding for semantic segmentation. In: 2018 IEEE/CVF Conference on Computer Vision and Pattern Recognition, pp. 7151–7160 (2018)
35. Zhang, Y., David, P., Foroosh, H., Gong, B.: A curriculum domain adaptation approach to the semantic segmentation of urban scenes. IEEE Trans. Pattern Anal. Mach. Intell. **42**, 1823–1841 (2020)
36. Zhang, Y., David, P., Gong, B.: Curriculum domain adaptation for semantic segmentation of urban scenes. In: 2017 IEEE International Conference on Computer Vision (ICCV), pp. 2039–2049 (2017)
37. Zhou, B., et al.: Semantic understanding of scenes through the ade20k dataset. Int. J. Comput. Vision **127**(3), 302–321 (2016)

An End-to-End Approach for Seam Carving Detection Using Deep Neural Networks

Thierry P. Moreira[1] , Marcos Cleison S. Santana[1] , Leandro A. Passos[2](✉) ,
João Paulo Papa[1] , and Kelton Augusto P. da Costa[1]

[1] Department of Computing, São Paulo State University, Av. Eng. Luiz Edmundo
Carrijo Coube, 14-01, Bauru 17033-360, Brazil
{joao.papa,kelton.costa}@unesp.br
[2] CMI Lab, School of Engineering and Informatics, University of Wolverhampton,
Wolverhampton, UK
l.passosjunior@wlv.ac.uk

Abstract. Seam carving is a computational method capable of resizing
images for both reduction and expansion based on its content, instead
of the image geometry. Although the technique is mostly employed to
deal with redundant information, i.e., regions composed of pixels with
similar intensity, it can also be used for tampering images by inserting
or removing relevant objects. Therefore, detecting such a process is of
extreme importance regarding the image security domain. However, recognizing seam-carved images does not represent a straightforward task
even for human eyes, and robust computation tools capable of identifying such alterations are very desirable. In this paper, we propose an
end-to-end approach to cope with the problem of automatic seam carving
detection that can obtain state-of-the-art results. Experiments conducted
over public and private datasets with several tampering configurations
evidence the suitability of the proposed model.

Keywords: Seam carving · Convolutional neural networks · Image
security

1 Introduction

Seam Carving is an image operator created for content-aware image resizing,
and it works by searching for the least relevant pixel paths (seams) in the image
so that, when they are removed or duplicated, the figure size is changed without

The authors are grateful to São Paulo Research Foundation (Fapesp) grants
#2021/05516-1, #2013/07375-0, #2014/12236-1, #2019/07665-4, to the Petrobras
grant #2017/00285-6, to the Brazilian National Council for Research and Development (CNPq) #307066/2017-7 and #427968/2018-4, as well as the Engineering and
Physical Sciences Research Council (EPSRC) grant EP/T021063/1 and its principal
investigator Ahsan Adeel.

A. J. Pinho et al. (Eds.): IbPRIA 2022, LNCS 13256, pp. 447–457, 2022.
https://doi.org/10.1007/978-3-031-04881-4_35

distortion or relevant information loss. Roughly speaking, seams are defined as a connected line of pixels that crosses the entire image vertically or horizontally.

The general effect for the human eye is that the image was already captured that way, i.e., without being "distorted". Seam carving has good uses for personal editing, but fake images may be used for illegal purposes too, such as perjury, counterfeiting, or other types of digital forgery. This scenario fostered the need to ensure the image integrity, primarily, but not exclusively, for forensic analysis. Therefore, developing efficient methods for its detection became essential.

Avidan and Shamir [1] originally employed the concept of object removal using seam carving. In a nutshell, the idea consists in driving all seams to pass through the object of interest. One may reapply seam carving to enlarge the image to its original size. Seam carving may highlight some texture artifacts in the resulting image as well [10,15,16], which can be considered for further automatic detection. Although these artifacts are mostly imperceptible to the human eye, they can be detected by computer vision techniques. This effect was explored by Yin et al. [25], which applied a set of descriptors directly on the LBP (Local Binary Descriptors) image [24]. The authors employed eighteen energy information and noise-based features from [21], and defined six more features based on statistical measures from the seams.

Zhang et al. [26] proposed an approach that consists of extracting histogram features from two texture descriptors, i.e., LBP and the Weber Local Descriptor [2]. The methodology first extracts both descriptors from the image, and then their histograms are merged for the further application of the Kruskal-Wallis test for feature selection purposes [22]. The resulting vector is then classified with the well-known Support Vector Machines (SVM) technique [5]. Their results indicated that texture features have a strong descriptive power for the problem at hand.

Avidan and Shamir [1] originally employed the concept of object removal using seam carving. In a nutshell, the idea consists in driving all seams to pass through the object of interest. One may reapply seam carving to enlarge the image to its original size. Seam carving may highlight some texture artifacts in the resulting image as well [10,15,16], which can be considered for further automatic detection. Although these artifacts are mostly imperceptible to the human eye, they can be detected by computer vision techniques. This effect was explored by Yin et al. [25], which applied a set of descriptors directly on the LBP (Local Binary Descriptors) image [24]. The authors employed eighteen energy information and noise-based features from [21], and defined six more features based on statistical measures from the seams.

While seam carving detection has been approached using texture features, in this work we propose an end-to-end approach based on deep neural networks. We observe that feeding neural networks with the original images allows for better accuracies when the severity of the tampering is quite subtle to be detected. We show results that outperform the recent ones presented by Cieslak et al. [4] on the smallest rates of change. Therefore, the main contribution of this work is to propose an end-to-end approach for seam carving detection on unprocessed images. This approach is more effective on finding small and subtle tempering.

The remainder of this paper is organized as follows. Section 2 presents a theoretical background related to seam carving, Sect. 3 discusses the proposed approach. Sections 4 and 5 describe the methodology and the experimental section, respectively. Conclusions and future works are stated in Sect. 6.

2 Theoretical Background

In this section, we present the theoretical background concerning the main concepts of seam carving, as well as some energy functions commonly used to compute vertical/horizontal seams.

Seam carving is a content-aware image resizing approach that consists mainly of inserting or removing seams iteratively until the image achieves a desired width or height. Roughly speaking, a seam is defined as a connected line of pixels in completely crossing the image. Seams are constructed using an energy function to ensure that only the low-energy pixels will be removed, keeping the primary structure of the image. These low-energy pixels belong to a low-frequency region in the image where the changes might be imperceptible.

One of the crucial aspects concerning seam carving stands for the concept of "energy", which is used to define the "paths to be carved" in the image. A standard energy function e applied to an image I can be defined as follows:

$$e(I) = \left| \frac{\partial}{\partial x} I \right| + \left| \frac{\partial}{\partial y} I \right|, \tag{1}$$

Such a formulation can also be extended to pixels, i.e., one can compute the energy at pixel $I_{x,y}$. Roughly speaking, a vertical seam is an 8-connected path of adjacent pixels from the top to the bottom of the image, containing only one pixel from each row of the image. An analogous definition can be derived for horizontal seams.

Moreover, the energy function applied to a seam s is defined as follows:

$$E(s) = \sum_{i=1}^{m} e(s_i), \tag{2}$$

where $e(s_i)$ denotes the partial energy computed at the i-th pixel of seam s. The optimal seam, which is the one that minimizes Eq. 2, can be efficiently computed using dynamic programming.

Figure 1 (left) demonstrates the process of seam carving over Fig. 1(a), where Fig. 1(b) shows the seams computed in the vertical axes, and Fig. 1(c) shows Fig. 1(a) after removing those seams. We can also use seams to remove objects from an image, thus restraining the search of seams in a given space. Figure 1 (right) depicts an example of such process[1].

---

[1] Figure 1 is licensed under Creative Commons 0 Public Domain.

Resizing   Object Removal

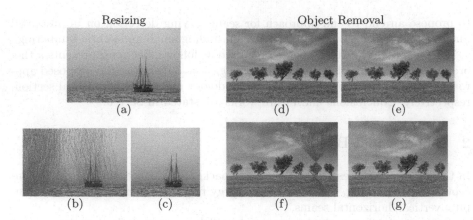

Fig. 1. Left: Seam carving for resizing: (a) original image, (b) original image with seams marked for removing, and (c) resized image using seam carving. Right: Seam carving for object removal: (a) original image, (b) original image with object marked for removal, (c) original image with seams marked for object removal, and (d) desired object removed using seam carving.

Avidan and Shamir [1] propose other energy functions for the problem (for example, Histogram of Oriented Gradients (HoG)) and argue that no single formulation works well for all images at once, but they all may have similar results in the end. The different formulations may vary in the rate at which visual artifacts are introduced.

3   Proposed Approach

As mentioned before, removing or inserting a seam leaves intrinsic local patterns. This way, the inspection of the seam carving traces can be accomplished by exploring the local discontinuities of pixel values. While these discontinuities can be detected by texture-based feature extractors, a deep convolutional model is also capable of identifying such artifacts and distortions. To cope with such issues, we propose an end-to-end deep network training methodology to inspect the use of seam carving manipulation in images.

We opted to employ the well-known deep neural network Xception since it innovates on using separable convolutions as the fundamental operator. The network is composed of three essential parts: (i) the entry flow, (ii) some repeating blocks forming the middle flow, and (iii) the exit flow. The output of exit flow can be used to feed dense layers and later on a classifier, while the entry flow is in charge of extracting low-level features with its residuals obtained from 1×1 convolutions. The middle flow, with its repeating blocks of depthwise convolutions with simple residuals connections, extracts deeper features.

Other architectures could be employed – the choice of the Xception architecture was motivated by the fact the network presents a good trade-off between representation ability and execution time. Such skill is due to its depthwise separable convolutions, which speeds up its execution, even when the number of parameters is kept the same. The advantages of the Xception can be observed in some previous works, such as Giancardo et al. [7], which employed the pretrained model for brain shunt valve recognition on magnetic resonance images, and Liu et al. [11], that applied Xception for facial expression feature extraction.

The model proposed in this work is based upon the original Xception architecture but with its original top layers (fully connected - FC) replaced by two new ones. The first FC layer reduces the dimensionality from 2,048 to 512, while the second is in charge of classifying the input image as seam carved or untouched. Figure 2 depicts the architecture proposed in this work.

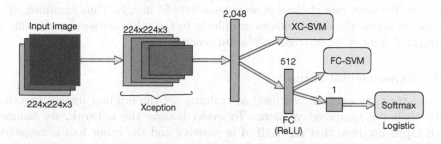

Fig. 2. Proposed deep network architecture: two fully-connected layers are added right after the original Xception model to perform fine-tuning for seam carving recognition purposes.

Training a CNN to a classification task opens the possibility of using intermediate CNN layers as a feature extractor to feed other classifiers. In our experiments, the aforementioned Xception architecture is stacked with fully-connected layers for fine-tuning purposes and weights learned over the well-know ImageNet dataset [9]. When the training process converges, the last FC layer acts as a classifier, but one can remove a number of layers from the stack and use intermediate representations with other classification algorithms.

The model depicted in Fig. 2 allows us to observe three distinct descriptions from the same network: (i) the Xception output with size as of 2,048, (ii) the first fully-connected output with size as of 512, and (iii) the classification output, which stands for a softmax layer activated with a Sigmoid function to scale the output between 0 (not tampered image) and 1 (seam-carved image). The loss function adopted in this work concerns the well-known "binary cross-entropy".

4   Methodology

In this section, we present the datasets and the experimental setup used to validate the robustness of the proposed approach for seam carving detection using deep networks.

4.1 Datasets

In this paper, we used two datasets, being the former composed of natural images, from now on called "SC-Image", and provided by Liu et al. [12]. Such a dataset contains 5,150 images cropped to 224 × 224 pixels and compressed with JPEG using a 75% compression rate. Each image has 10 modified counterparts, obtained by varying the rate of change by 3%, 6%, 9%, 12%, 15%, 18%, 21%, 30%, 40%, and 50%. The higher the rate of change, the easier for an algorithm to detect the seam carving tampering. Evaluation is performed using a 5-fold cross-validation scheme.

The other dataset is composed of images obtained from off-shore well drilling rigs during operation of Petrobras SA, a Brazilian oil and gas company. The dataset, hereinafter called "Petrobras" [23], is private and therefore not available publicly. For each rate of change, we generated 564 images, thus summing up a dataset composed of 6,204 images. Similarly to the other dataset, evaluation is performed using a 5-fold cross-validation scheme.

4.2 Experimental Setup

To fine-tune the network, we used as training set the original images together with their ten tampered versions. To avoid biasing the network, we balance each input batch so that one half of is positive and the other half is negative. Therefore, the resulting network is less invariant to the proportion of changes.

We considered four approaches for data augmentation purposes, each with a 50% chance of occurrence for a given image: (i) horizontal flip, (ii) vertical flip, (iii) Gaussian noise of zero mean and standard deviation as of 0.2 (the original pixel values are scaled to $[-1, 1]$), and (iv) a random black column with width of 10 pixels.

The training process was performed using the Stochastic Gradient Descent with Restart (SGDR) optimization algorithm [13]. In this paper, the learning rate varied as a function of the cosine with an abrupt restart at the end of each cycle. The initial maximum learning rate was $lr_{max} = 10^{-2}$, lr_{max} value, decayed by 30% at each cycle, with lower limit $lr_{min} = 10^{-5}$. The initial period was composed of 10 epochs. The training procedure was implemented with Keras 2.2.4 [3] and Tensorflow 1.12.0 as a backend.

As depicted in Fig. 2, we considered FC-SVM with 512 features, XC-SVM with 2,048 features, and the standard output of the CNN with a softmax layer, hereinafter called SM-CNN, for comparison purposes. Additionally, concerning dataset SC-Images, we also compared the proposed approaches against the works by Cieslak et al. [4] and Zhang et al. [26]. Regarding the SVM implementation, we used the library provided by Fan et al. [6] within SciKit-Learn [20] package.

5 Experiments

In this section, we present the results obtained over SC-Image and Petrobras datasets, as well as a more in-depth discussion about the robustness of the proposed approaches.

5.1 SC-Image Dataset

Table 1 presents the accuracy results concerning seam carving detection over the SC-Image dataset, where the best values are in bold. For lower levels of seam carving (i.e., rates of change), the proposed approaches obtained the best results with SM-CNN and XC-SVM models. Such low levels of distortion make it difficult for the network to discriminate between normal and tampered images since the differences are subtle, i.e., few and sparse information distinguish them. However, the proposed approaches were able to detect seam-carved images with suitable accuracy.

Table 1. Accuracies obtained over SC-Image dataset.

Rate of change	Accuracy (%)				
	SM-CNN	FC-SVM	XC-SVM	Cieslak et al. [4]	Zhang et al. [26]
3%	84.03	85.05	**85.19**	81.46	68.26
6%	87.96	87.76	**88.02**	83.50	84.99
9%	**90.58**	90.02	90.06	81.72	89.86
12%	**92.86**	91.90	91.86	84.21	92.72
15%	93.98	93.21	93.42	87.51	**95.03**
18%	95.15	94.15	94.40	89.73	**95.94**
21%	94.66	94.63	94.94	88.91	**96.02**
30%	95.49	96.18	96.51	91.23	**97.58**
40%	95.73	97.23	97.71	95.12	**98.64**
50%	95.73	97.65	98.10	98.90	**99.20**

For intermediate values of tampering (i.e., 9%–18%), the SM-CNN approach obtained the best results with a small margin regarding XC-SVM. Higher levels of tampering are easier to detect since the seam removal artifacts are present in abundance. For 50% of reduction level, SM-CNN obtained 95.73% of classification rate, while XC-CNN yielded 98.10% of accuracy. Although our approach obtained smaller accuracies on higher rates of change compared with Zhang et al. [26], it achieved much higher results on more subtle tempering.

To explore the decision mechanism that led to the results, we employed dimensionality reduction on the output features. The reduced space was computed using the Uniform Manifold Approximation and Projection (UMAP) technique [14] using 20 neighbors for further projecting the output onto a 2D space. This projection method is based on Riemannian geometry and algebraic topology. Figure 3 (left) illustrates projections for SC-Image dataset images concerning different classifiers and rates of change.

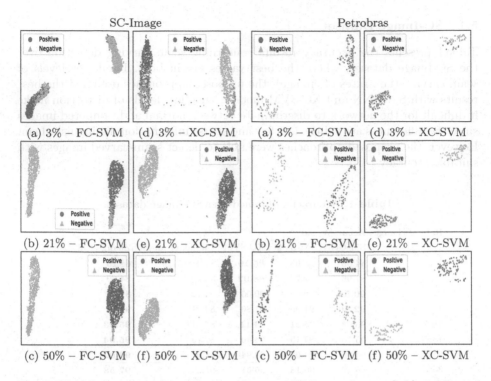

Fig. 3. UMAP dimensionality reduction plots over SC-Image dataset (left) and Petrobras dataset (right), where "positive" means that images were tampered, and "negative" means that images were not attacked.

All plots present two clusters as a consequence of the training of the network as a binary classifier. For images with a higher level of tampering, the clusters present smaller overlap among points from different classes. In such condition, the network presents a better discriminative power provided by the artifacts left by seam carving in the image. Such discriminative power is reduced with the decreasing of the tampering level, resulting in clusters with more misclassified samples. Besides, there are also points far away from clusters, which may reflect in a challenger situation for the classifiers.

5.2   Petrobras Dataset

Table 2 presents the accuracy concerning the experiments over the Petrobras dataset for different levels of tampering, where the best results are in bold. Similarly to the previous dataset, images with lower levels of tampering exhibit smaller accuracy. Among the proposed approaches, XC-SVM obtained the best results in almost every distortion rates, except for 18% and 30%. Regarding a change of rate as of 3%, XC-SVM achieved the accuracy of 72.64%, followed by FC-SVM with 71.75% of recognition rate. As the tampering level is intensified, all approaches also increase their accuracies. For images with 50% of tampering, XC-SVM could reach the best accuracy of 95.45%. Such results corroborate with those from the other dataset.

The networks decision behavior can be easily visualized in Fig. 3 (right), depicting UMAP projections onto a bidimensional space of the output features from the XC-SVM and FC-SVM models. The projection output for 3% of tampering presents two main clusters, where there is a moderate presence of mixing between classes. Such situation illustrates the confusion made by the network when images with a low level of tempering are evaluated. On the other hand, the projection of the 50% carved images shows clearer cluster separation between classes. Although Petrobras and SC-Image datasets are different, their projections present strong similarities.

Table 2. Accuracies obtained over Petrobras dataset.

Rate of change	Accuracy (%)		
	SM-CNN	FC-SVM	XC-SVM
3%	67.86	71.75	**72.64**
6%	66.96	75.58	**75.85**
9%	76.79	77.90	**77.99**
12%	79.91	79.41	**80.04**
15%	81.25	82.27	**83.78**
18%	**83.04**	82.27	82.53
21%	83.04	84.58	**85.03**
30%	84.82	**90.02**	89.93
40%	85.27	92.16	**92.52**
50%	85.71	94.57	**95.45**

6 Conclusions

Seam carving is a resizing method widely used in image manipulation with minimal distortions aiming at preserving essential elements of the image's content. Due to this characteristic, seam carving can be used to tamper with images by removing desired elements from the image.

In this paper, we proposed to detect image tampering using an end-to-end deep learning network. Experiments demonstrated that the proposed approach is suitable to be an auditing tool to identify seam-carved images since it can classify images with a low level of tampering with good accuracy, besides having an excellent performance for images with aggressive tampering.

The projection of the output of the deep neural network features onto a bidimensional space indicates that the obtained clustered space might be suitable to build a hybrid-approach, i.e., we can attach another layer of classifiers right after the mapped space. The proposed approach opened new possibilities to explore new hybrid architectures.

Concerning future works, we intend to evaluate the Optimum-Path Forest classifier [17–19] in the context of seam carving detection since it is parameterless and fast for training. Besides, we are considering employing deep autoencoders and subspace clustering [8] for unsupervised tempering detection as well.

References

1. Avidan, S., Shamir, A.: Seam carving for content-aware image resizing. ACM Trans. Graph. **26**(3), July 2007
2. Chen, J., Shan, S., He, C., Zhao, G., Pietikainen, M., Chen, X., Gao, W.: Wld: a robust local image descriptor. Pattern Anal. Mach. Intell. **32**(9), 1705–1720 (2010)
3. Chollet, F., et al.: Keras (2015). https://keras.io
4. Cieslak, L.F.S., Costa, K.A.P., Papa, J.P.: Seam carving detection using convolutional neural networks. In: IEEE 12th International Symposium on Applied Computational Intelligence and Informatics (SACI), pp. 195–200, May 2018
5. Cortes, C., Vapnik, V.: Support vector networks. Mach. Learn. **20**, 273–297 (1995)
6. Fan, R.E., Chang, K.W., Hsieh, C.J., Wang, X.R., Lin, C.J.: Liblinear: A library for large linear classification. J. Mach. Learn. Res. **9**, 1871–1874 (2008)
7. Giancardo, L., Arevalo, O., Tenreiro, A., Riascos, R., Bonfante, E.: Mri compatibility: automatic brain shunt valve recognition using feature engineering and deep convolutional neural networks. Sci. Rep. **8**, December 2018. https://doi.org/10.1038/s41598-018-34164-6
8. Ji, P., Salzmann, M., Li, H.: Efficient dense subspace clustering. In: IEEE Winter Conference on Applications of Computer Vision, pp. 461–468, March 2014. https://doi.org/10.1109/WACV.2014.6836065
9. Krizhevsky, A., Ilya, S., Hinton, G.E.: Imagenet classification with deep convolutional neural networks. In: Advances in Neural Information Processing Systems, pp. 1097–1105. Curran Associates, Inc. (2012)
10. Li, Y., Xia, M., Liu, X., Yang, G.: Identification of various image retargeting techniques using hybrid features. J. Inf. Secur. Appl. **51**, 102459 (2020)
11. Liu, N., Fang, Y., Guo, Y.: Enhancing feature correlation for bi-modal group emotion recognition. In: Hong, R., Cheng, W.-H., Yamasaki, T., Wang, M., Ngo, C.-W. (eds.) PCM 2018. LNCS, vol. 11165, pp. 24–34. Springer, Cham (2018). https://doi.org/10.1007/978-3-030-00767-6_3
12. Liu, Q., Cooper, P.A., Zhou, B.: An improved approach to detecting content-aware scaling-based tampering in jpeg images. In: IEEE China Summit International Conference on Signal and Information Processing (ChinaSIP), pp. 432–436, July 2013
13. Loshchilov, I., Hutter, F.: Sgdr: stochastic gradient descent with warm restarts. arXiv preprint arXiv:1608.03983 (2016)
14. McInnes, L., Healy, J.: Umap: uniform manifold approximation and projection for dimension reduction. arXiv preprint arXiv:1802.03426 (2018)
15. Nam, S.H., Ahn, W., Yu, I.J., Kwon, M.J., Son, M., Lee, H.K.: Deep convolutional neural network for identifying seam-carving forgery. IEEE Trans. Circuits Syst. Video Technol. **31**, 3308–3326(2020)

16. Nataraj, L., Gudavalli, C., Manhar Mohammed, T., Chandrasekaran, S., Manjunath, B.S.: Seam carving detection and localization using two-stage deep neural networks. In: Gopi, E.S. (ed.) Machine Learning, Deep Learning and Computational Intelligence for Wireless Communication. LNEE, vol. 749, pp. 381–394. Springer, Singapore (2021). https://doi.org/10.1007/978-981-16-0289-4_29
17. Papa, J.P., Falcão, A.X., Albuquerque, V.H.C., Tavares, J.M.R.S.: Efficient supervised optimum-path forest classification for large datasets. Pattern Recogn. **45**(1), 512–520 (2012)
18. Papa, J.P., Falcão, A.X., Suzuki, C.T.N.: Supervised pattern classification based on optimum-path forest. Int. J. Imaging Syst. Technol. **19**(2), 120–131 (2009)
19. Papa, J.P., Fernandes, S.E.N., Falcão, A.X.: Optimum-path forest based on k-connectivity: theory and applications. Pattern Recogn. Lett. **87**, 117–126, 102459 (2017)
20. Pedregosa, F., et al.: Scikit-learn: machine learning in Python. J. Mach. Learn. Res. **12**, 2825–2830, 102459 (2011)
21. Ryu, S., Lee, H., Lee, H.: Detecting trace of seam carving for forensic analysis. IEICE Trans. Inf. Syst. E97.D(5), 1304–1311 (2014)
22. Saeys, Y., Inza, I., Larrañaga, P.: A review of feature selection techniques in bioinformatics. Bioinformatics **23**(19), 2507–2517 (2007)
23. Santana, M.C., Passos, L.A., Moreira, T.P., Colombo, D., de Albuquerque, V.H.C., Papa, J.P.: A novel siamese-based approach for scene change detection with applications to obstructed routes in hazardous environments. IEEE Intell. Syst. **35**(1), 44–53 (2019)
24. Wang, L., He, D.: Texture classification using texture spectrum. Pattern Recogn. **23**(8), 905–910 (1990)
25. Yin, T., Yang, G., Li, L., Zhang, D., Sun, X.: Detecting seam carving based image resizing using local binary patterns. Comput. Secur. **55**, 130–141, 102459 (2015)
26. Zhang, D., Li, Q., Yang, G., Li, L., Sun, X.: Detection of image seam carving by using weber local descriptor and local binary patterns. J. Inf. Secur. Appl. **36**(C), 135–144 (2017)

Proposal of a Comparative Framework for Face Super-Resolution Algorithms in Forensics

Antonio Salguero-Cruz, Pedro Latorre-Carmona[✉],
and César Ignacio García-Osorio

Departamento de Ingeniería Informática, Universidad de Burgos,
Avda. de Cantabria s/n, 09006 Burgos, Spain
{plcarmona,cgosorio}@ubu.es

Abstract. The number of images acquired by electronic devices is growing exponentially. This is partly due to an easier and more extended access to high-tech portable acquisition devices. These devices could record people of interest, either because they were involved in some type of criminal act, or they could be missing people, so their identification (and specifically the identification of their faces) is very relevant. On the other hand, image quality sometimes does not allow a positive identification of these individuals. This paper presents a framework for the comparative analysis of face super-resolution (*hallucination*) algorithms in forensics. The *super-resolved* images could be used to help identify a person of interest, and subsequently be admitted by the competent judicial authority in a criminal process, maintaining, like the rest of evidence, the so-called *chain of custody*, as required by the Spanish law and other similar legal systems.

Keywords: Face hallucination · Perceptual image quality ·
Identification · Forensics

1 Introduction

Face super resolution (*face hallucination*) improves the identification potential of an image. To the best of our knowledge, these techniques have not yet been used to identify people involved in a legal proceeding. On the other hand, other types of techniques (considered as innovative tools when first implemented, i.e., by 1891) allow for the acquisition of fingerprints on diverse surfaces, and are now accepted by judicial authorities as an identification method. This is usually made by comparing the so-called *doubtful* samples with the corresponding *undoubtful* ones. The latter must be obtained under controlled conditions, and would allow for an *undoubtful* assignment of fingerprints to a particular person (for example, when dealing with a *police profile*).

In fingerprint development, all the steps to be carried out are well defined and standardized. The revealing product is standardized as well (we might consider *black powder*, *iodine vapour*, or *silver nitrate*, to cite a few cases). On the

© Springer Nature Switzerland AG 2022
A. J. Pinho et al. (Eds.): IbPRIA 2022, LNCS 13256, pp. 458–469, 2022.
https://doi.org/10.1007/978-3-031-04881-4_36

other hand, there is no standardized method (to the best of our knowledge) to assess the reconstruction quality of the super resolution algorithms themselves, at least for faces. The existence of such a methodology would allow these methods to be scientifically validated, as it has been done with DNA or fingerprint information [12]. The most important problem to deal with would be the low quality of some of the *doubtful* images, and their quality would be a key factor for its potential identification application, as shown in [10]. A similar situation appears when we get only partial fingerprints from a crime scene. Fortunately, standardized procedures applied to fingerprint analysis may make the partial *doubtful* fingerprints to be valid, on condition that they may have a minimum number of (good enough) characteristic points.

The application of face hallucination (FH) algorithms on low quality face images would allow to generate improved images which might have enough level of detail to compare them against *undoubtful* images in order to get a positive identification. This positive identification might not have been obtained otherwise. However, from our point of view, the most effective way to apply the aforementioned strategy would be to create a legal *framework* for those *improved* images, to be used in a legal procedure. This strategy could have become a capital tool during the terrorist attack that occurred in the 2013 Boston Marathon. The Carnegie-Mellon University CyLab Biometrics Center [3] presumably applied a type of FH algorithm to these images, but this occurred after the FBI had already identified the suspects. We can find an example in [27] about how an FH algorithm is used aiming at improving video quality recording obtained by a surveillance camera. One potential way to assess super-resolution performance would be by locating facial landmarks [2,7] and comparing their (x, y) spatial coordinate positions. Facial landmarks are mostly located in the so-called *Penry's area*. In fact, the assessment of features with identifying potential on the FH resolved images might represent an alternative tool when comparing that FH image with an undoubtful image (as shown in [13]) with the aim of achieving its identification. However, this idea was dismissed because it would be considerably difficult to find exact matched images in a real case scenario to compare the location of facial landmarks.

The aim of this paper is two-fold: (a) To validate the use of FH algorithms in a legal (forensic) procedure (normally of criminal nature), and (b) To objectively analyze and assess the results obtained by those algorithms, using the *merit figures* developed in [6], allowing to discern whether their performance is significantly different or not (from a statistical point of view), using the methodology proposed in [9]. The rest of the paper is divided as follows: Sect. 2 presents a comparative analysis of the different FH methods that were originally proposed in [19], including a brief description (Sect. 2.1), the image datasets used in this paper (Sect. 2.2), the creation of the images that are *fed into* the FH algorithms (Sect. 2.3), and the application of the FH algorithms to the input low resolution images. Section 3 shows, on the one hand, the analysis of the results obtained by applying the *merit figures* proposed in [6], and a statistical analysis of the performance of each one of them, on the other. Conclusions are presented in Sect. 4.

2 Methodology

This Section describes the process to validate the FH algorithms for identification purposes, trying to work as closely as possible to the conditions given in a real environment/scenario. In order to assess the performance of the different FH algorithms, the *merit figures* proposed in [6] will be used. These *figures* do not need an original image (also called *groundtruth*) with which to make comparisons. They give an estimation of the perceptual quality (from a *human vision* point of view) of those images without the need for a reference image. This is particularly beneficial in our case, because it is usually uncommon to have a reference image with which to evaluate the quality of an image.

Subsequently, once the different FH algorithms have been applied to the images and the corresponding *merit figures* have been obtained, a descriptive, as well as a statistical analysis of the obtained results, will be made. This statistical analysis will be made by applying some of the procedures mentioned in [9], which will make it possible to determine whether the results obtained by the different methods are statistically (significantly) different or not.

2.1 Face Hallucination Methods

In this section, we briefly describe the FH algorithms that will be used in the comparative analysis. For additional details of the methods, the reader is referred to the references that appear in the following descriptions:

- **C-SRIP** [19]: It uses a super-resolution cascade network (a convolutional neural network, or CNN, with 52 layers) and a group of face recognition models that are used as identity priors during the network training process.
- **EDSR** [23]: It is similar to the SRResNet approach [22]. However, a simple modification consisting of removing the batch normalization layers allows the algorithm to save 40% on memory usage during training, which makes it possible to build larger models that improve the results obtained. The final model is a CNN with 32 layers and 256 feature channels.
- **l_p** [20]: It is formed by an image transformation network and a loss network, and it is used to define different types of loss functions (two perceptual loss functions and two simple loss functions).
- **LapSRN** [21]: It is a Laplacian Pyramid Super-Resolution Network based on a cascade CNN. Each convolutional layer is formed by 64 filters.
- **SICNN** [29]: It is a cascade FH network with a recognition network, which extracts identity related features.
- **SRGAN** [22]: It is a Generative Adversarial Network (GAN) with a new type of perceptual loss function. The Mean Squared Error (MSE) loss is substituted by a loss function based on the feature map given by the Visual Geometry Group (VGG) convolutional neural network [25].

All of these methods have been used in [19]. We have selected them because they are easily available. They all have also been trained using the same images, and applied as described in [1]. In all the cases, we will use an ×8 super-resolution factor.

2.2   Image Databases

A series of images were obtained from databases of facial images, in an ad-hoc recording framework, or from a series of video surveillance cameras. Authorization for public dissemination (and in the context of data protection laws, by showing images in this paper we are making them public) has been obtained from non-public databases (as explained below), in order to avoid legal problems related to data protection legislation (including [24] and [18]). The databases from which the images used in this article have been obtained are the following:

- **QMUL-SurvFace** [8]: Public database, containing real images from surveillance cameras. A group of 37, 24 × 24 pixels images was selected.
- **QMUL-TinyFace** [4]: Public database, containing low-resolution images. A group of 8, 24 × 24 pixels images was selected.
- **SN-Flip** [5]: Image database from the university of Notre Dame, containing cropped face images from some camera recorded videos. A group of 40 images was selected. However, their size is not 24 × 24 pixels, because that is not the size they originally are.
- **VBOLO** [5]: Image database from the university of Notre Dame, consisting of a real surveillance camera video. A group of 7 images were selected. Their 24 × 24 pixels size was obtained by cropping them.

None of the images of the previous databases have the corresponding *groundtruth* image their qualities could be assessed against. On the other hand, these images can be considered as similar to those that would be used in a real case scene, obtained for example from *home made* videos in which the person of interest is further away from an ideal acquisition distance, or from low quality security cameras, to cite a few cases. Figure 1 shows some examples of the images in the datasets, in their original 24 × 24 pixels size.

Fig. 1. Some images that the image databases used (sn-flip, VBOLO, QMUL-TinyFace and QMUL-SurvFace) contains.

We would like to emphasize the difficulty of finding databases of face images, which might meet the necessary conditions for our research objectives, specially given the current legislation concerning data protection of personal nature and the impossibility of public dissemination (at least in the European Union, Spain included, as [24] and [18]) without the express and verifiable authorization of each person appearing in them. Nevertheless, the images used in this article are in the public domain, or it is possible to use them in scientific studies after signing an agreement with the database owners (e.g., the University of Notre Dame databases [5] used in our research).

2.3 Low Resolution Image Generation

All the selected images are 24 × 24 pixels in size. There are two reasons for this. First, we are particularly interested in processing small image sizes, as this is more similar to what happens in a real life scenario (for example, a traffic control camera on top of a mast that has acquired the image of a person of interest). Secondly, because that is the input size of all the FH algorithms that are considered in our research. In the case of QMUL-SurvFace and QMUL-TinyFace image databases, the selected images are already of that size. In the case of the VBOLO image database (which is a video from a surveillance camera), the faces used have been cropped to match that size. However, the images in the SN-Flip database, are small, but bigger than 24 × 24, so they have had to be resized to 24 × 24 using a bilinear interpolation approach.

An important point to take into account in a real life situation is whether the image to be used (i.e., the low resolution image) is *authentic* or not. We will assume that all of them are (that is, they are not images that have been previously modified or processed, and they are indeed images obtained by the acquisition device). If there are doubts about their authenticity (they have not been collected directly by the Police, *watermarks* could have been modified, etc.), the authenticity of those images should be checked (see [11]), in order to make sure that the persons in the pictures were there indeed.

3 Results and Discussion

In the quest for a standardized procedure for the use of images obtained from surveillance devices for the identification of subjects, a key step will be the application of techniques to improve their quality. In this context, face hallucination (FH) techniques can be key. In this section we evaluate the most representative FH algorithms using as a criterion the quality of the obtained images, measured with figures of merit that take into account perceptual aspects of the human vision system. The aim is both to determine which algorithm is better and to determine which algorithms provide statistically similar results.

3.1 Application of Perceptual Image Quality Measures

The generation of image quality assessment (IQA) models has been a very active area of recent research. The two most representative are the peak signal to noise ratio (PSNR) and the structural similarity index (SSIM) [26]. However, their correlation with image quality is far from perfect, specially for image super resolution [28]. Besides, these and other models are called full reference (FR) models, for which a reference image is needed.

There are methods that do not need of reference images, the so-called *blind image quality assessment* (BIQA) algorithms. Some of these methods are based on the use of distorted image databases whose quality has been manually annotated by human evaluators. Thus, the values associated with the images include the subjective perceptual aspects that accompany this type of evaluation. The main disadvantage of these methods is that, for the algorithms to give reasonable results, one must have a large number of annotated training images.

The methodolgy defined in [6] defines a series of *perceptual features* associated to each image (see Section III.A in [6] for details), and measures their quality by calculating the standardized Euclidean distance between the feature vector obtained from each image and a multivariate Gaussian model (MVG) learned from the corpus of pristine naturalistic images (Eq. 8 in [6]). Qodu, Q1odu, and Q2odu are the three *merit figures* proposed by Beron et al. in [6] and they differ in the subgroup of features (of the complete feature vector) that are considered in order to characterize each one of the images.

A group of 92 input LR images, and 6 different FH algorithms, were used. The results of each one of the FH algorithms have been assessed using Qodu, Q1odu, and Q2odu. Image quality (according to each *merit figure*) being better the lower the score obtained, taking into account that these *figures* measure the distance to a pristine image, and therefore, that is the reason of the inverse dependence. In order to have more complete information, results have been organized according to the source database of that image. Therefore, from each database, there are three sets of results (one for each of the three *merit figures*), which collect the information of the results given by each algorithm. Figure 2 shows the results obtained, taking into account that the best score obtained is the lowest one.

Figure 2 shows that the Qodu *merit figure* would determine that the best results are obtained by the l_p algorithm, for all the databases. The same tendency appears for the Q1odu *merit figure*. However, Q2odu determines that the l_p algorithm is the best, except for the VBOLO and Tinyface databases, where CSRIP would be the best FH algorithm. This can be seen more clearly by analyzing the results independently of the image database it belongs to. Thus, Fig. 2, last row, shows the same trend as that shown for each one of the *merit figures*.

Figure 3 shows some examples of FH results. The corresponding input low resolution image appears in a red *frame*. The FH image size is ×8 that of the input LR image(s). As a reference, the score values obtained by applying the Q1odu *merit figure* have been used. The results of the best (i.e., lowest) Q1odu value (left side of the pair) can be visually compared to the worst Q1odu score, for the same image (right side of the pair).

3.2   Statistical Analysis

To compare the different FH algorithms, two of the statistical tests proposed in [9] have been used on the values of the *merit figures* for the super resolved images obtained applying each of the FH algorithms. More specifically, first

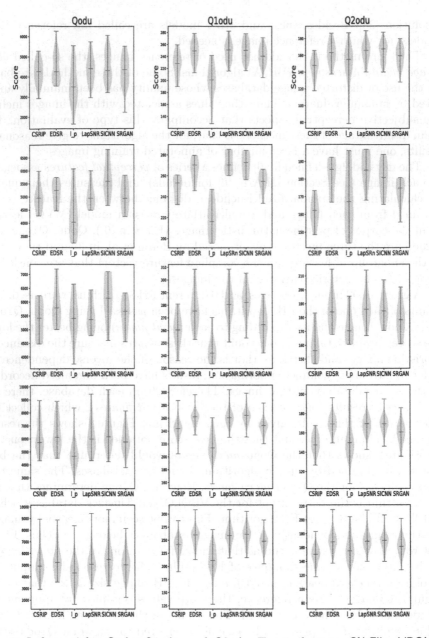

Fig. 2. Left to right: Qodu, Q1odu, and Q2odu. Top to bottom: SN-Flip, VBOLO, QMUL-TinyFace, and QMUL-SurvFace. Last row, all the databases, simultaneously.

Fig. 3. Top to bottom rows: SN-Flip, QMUL-TinyFace, VBOLO, and QMUL-SurvFace image examples. Columns, in pairs: Left, best Q1odu score. Right, worst Q1odu score. The image inside te red box is the original low resolution image (Color figure online)

the Friedman Test, which uses the average ranks to determine whether all the algorithms are statistically equivalent, and once it has been determined that they are not equivalent, the Nemenyi post-hoc test, to determine which of the algorithms are statistically equivalent.

When the Friedman test is used to compare classifiers, the classifiers are ordered according to the results (for example, of precision) obtained for several data sets, taking the average of the ranks obtained in each of the data sets. But since here we are going to evaluate super-resolution algorithms on images, what we will do is use the ranks of the algorithms on each of the images[1], where the algorithms are ordered using the various figures of merits.

The process followed for each of the merit figures is as follows. The FH algorithms are first applied to each low resolution image to obtain the high

---

[1] Images are used regardless of their origin, that is, it is like having a new database of images with the union of all the images in the databases of images described in Sect. 2.2.

resolution image. On the image obtained, the figure of merit is applied and with the ordering of the algorithms based on the results, a rank is assigned to each of the algorithms. Rank 1, if according to the merit figure the algorithm has given the best result, rank 2, if it has been the second best and so on. If two methods tie, they are assigned the average rank (for example, if after assigning the first two positions, two methods dispute the third, the rank assigned to both is $(3 + 4)/2 = 3.5$).

All these rankings obtained for each of the images are finally used for the global average ranking for each FH algorithm, which is the performance measure that will be used to compare the algorithms. Table 1 shows these global measures, where it can be seen that for both Qodu and Q1odu *merit figures*, the best FH algorithms would be l_p, while for the Q2odu *merit figure* the best algorithm would be CSRIP, being l_p the second best.

In order to reject the null hypothesis of the Friedman test ($H_0 \equiv$ *"all algorithms are equivalent"*), the p-value of the test has to be less than 0.05 (which is the value for the α of the confidence interval under consideration). As can be seen in Table 2, the p-values are less than 0.05, so that for all merit figures the Friedman test concludes that the methods are statistically different.

Once H_0 is rejected, the Nemenyi post-hoc test is applied. This post-hoc test would allow us to know, for each *merit figure*, which methods are statistically equivalent. For this, the so-called *critical distance* (CD) is calculated. For all the *merit figures*, the CD value is 0.786. Therefore, algorithms whose average rank (Table 1) difference is not greater than this value can be considered statistically equivalent (this is what happens with EDSR and LapSRN for the Qodu merit function, as $4.304 - 3.685 < 0.786$).

These results can be seen visually in Fig. 4. This figure shows the average rank for each method, where the methods that are not statistically different are *grouped together* by a horizontal segment, representing the CD value. Thus, the main conclusions we may obtain from Fig. 4 would be the following:

- SICNN is not significantly different from EDSR, but it is from LapSRN, CSRIP, SRGAN and l_p.
- EDSR is not significantly different from SICNN and LapSRN, but it is from CSRIP, SRGAN and l_p.
- LapSRN is not significantly different from EDSR, CSRIP and SRGAN, but it is from SICNN, and l_p.
- CSRIP is not significantly different from LapSRN and SRGAN, but it is from SICNN, EDSR, and l_p.
- SRGAN is not significantly different from LapSRN, and CSRIP, but it is from SICNN, EDSR and l_p.
- l_p is significantly different from all other FH algorithms used.

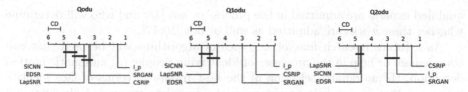

Fig. 4. Nemenyi post-hoc test results

Table 1. Average ranks of the FH algorithms for each of the figures of merit (the best ones highlighted in bold).

Merit figure	FH method					
	l_p	CSRIP	SICNN	SRGAN	EDSR	LapSRN
Qodu	**1.500**	3.357	4.880	3.272	4.304	3.685
Q1odu	**1.076**	2.163	5.380	2.989	4.478	4.913
Q2odu	2.391	**1.489**	4.685	3.011	4.619	4.804

Table 2. *p_value* results

	Qodu	Q1odu	Q2odu
p_value	4.185×10^{-56}	2.667×10^{-54}	4.320×10^{-46}

4 Conclusions

In this paper, we have analysed the potential capability to use face super resolution algorithms as image quality enhancement tools in a forensics context, either to identify suspects, or to search for missing individuals. The performance evaluation of these algorithms is essential so that they can be legally accepted as a piece of evidence in a judicial process (similarly to the techniques used to obtain latent and partial fingerprints). A real case scenario was considered by using images from databases of facial images, in an ad-hoc recording framework or from video surveillance cameras, being all of them very low spatial resolution images. The objective criterion for the assessment of the quality improvement has been the application of three perceptual quality measures, no reference *merit figures*. A statistical analysis was subsequently applied to determine whether those SR methods were significantly different from each other or not.

We must emphasize that any processing protocol aimed at obtaining identification results must be perfectly standardized so that it can be used in a legal proceeding. Furthermore, we believe that this procedure is totally new in the context of its application as part of a criminal process and therefore we could consider it would fall under one of the legal assumptions reflected in the Law referring to the creation of an expert report [14]. We must also stress that, at least in Spain, it will be the *judicial authority* who will determine whether the qualified experts are admitted in the process or not [15] and who will determine whether these works are admitted as well or not [16,17].

As a future research line, other types of algorithms will be used that can complement or help in the processes of identifying people: (a) automatic motion detection, (b) automatic selection of the best frames in terms of face identification, so that it may help to automate the complete processing pipeline, and therefore provide a controlled and audited framework that might give base for the judicial use of this type of super resolution algorithms, as it is already the case with fingerprints.

Acknowledgments. This work was supported by the Junta de Castilla y León under project BU055P20 (JCyL/FEDER, UE) and by the Ministry of Science and Innovation under project PID2020-119894GB-I00, co-financed through European Union FEDER funds.

References

1. Face hallucination using cascaded super-resolution and identity priors - Laboratory for Machine Intelligence. https://lmi.fe.uni-lj.si/en/research/fh/. Accessed 07 Dec 2021
2. Facial landmarks with dlib, OpenCV, and Python - PyimageSearch. https://www.pyimagesearch.com/2017/04/03/facial-landmarks-dlib-opencv-python/. Accessed 07 Dec 2021
3. Hallucinating a face, new software could have ID'd Boston bomber – Ars Technica. https://arstechnica.com/information-technology/2013/05/hallucinating-a-face-new-software-could-have-idd-boston-bomber/. Accessed 07 Dec 2021
4. Tinyface: Face recognition in native low-resolution imagery. https://qmul-tinyface.github.io. Accessed 07 Dec 2021
5. Barr, J.R., Cament, L.A., Bowyer, K.W., Flynn, P.J.: Active clustering with ensembles for social structure extraction (2014)
6. Beron, J., Benitez-Restrepo, H.D., Bovik, A.C.: Blind image quality assessment for super resolution via optimal feature selection. IEEE Access **8**, 143201–143218 (2020). https://doi.org/10.1109/ACCESS.2020.3014497
7. Bulat, A., Tzimiropoulos, G.: Binarized convolutional landmark localizers for human pose estimation and face alignment with limited resources. Technical report https://arxiv.org/abs/1703.00862
8. Cheng, Z., Zhu, X., Gong, S.: Surveillance face recognition challenge. arXiv preprint arXiv:1804.09691 (2018)
9. Demšar, J.: Statistical comparisons of classifiers over multiple data sets. J. Mach. Learn. Res. **7**, 1–30 (2006)
10. on Digital Evidence, S.W.G.: SWGDE Best Practices for Image Content Analysis (2017)
11. on Digital Evidence, S.W.G.: SWGDE Best Practices for Image Authentication (2018)
12. on Digital Evidence, S.W.G.: SWGDE Establishing Confidence in Digital and Multimedia Evidence Forensic Results by Error Mitigation Analysis (2018)
13. on Digital Evidence, S.W.G.: SWGDE Technical Overview for Forensic Image Comparison (2019)

14. España: Real Decreto de 14 de septiembre de 1882, aprobatorio de la Ley de Enjuiciamiento Criminal. Boletín Oficial del Estado Title V, Chapter VII of the Informe Pericial, article 456 and following (1882)
15. España: Real Decreto de 14 de septiembre de 1882, aprobatorio de la Ley de Enjuiciamiento Criminal. Boletín Oficial del Estado Title V, Chapter VII of the Informe Pericial, article 473 (1882)
16. España: Real Decreto de 14 de septiembre de 1882, aprobatorio de la Ley de Enjuiciamiento Criminal. Boletín Oficial del Estado Title V, Chapter VII of the Informe Pericial, article 741 (1882)
17. España: Real Decreto de 14 de septiembre de 1882, aprobatorio de la Ley de Enjuiciamiento Criminal. Boletín Oficial del Estado Title V, Chapter VII of the Informe Pericial, article 973 (1882)
18. España: Ley orgánica 3/2018, de 5 de diciembre, de protección de datos personales y garantía de los derechos digitales. BOE (2018)
19. Grm, K., Scheirer, W.J.: Face hallucination using cascaded super-resolution and identity priors. IEEE Trans. Image Process. **29**, 2150–2165 (2019)
20. Johnson, J., Alahi, A., Fei-Fei, L.: Perceptual losses for real-time style transfer and super-resolution, Mar 2016. http://arxiv.org/abs/1603.08155
21. Lai, W.S., Huang, J.B., Ahuja, N., Yang, M.H.: Deep laplacian pyramid networks for fast and accurate super-resolution, pp. 624–632 (2017). http://vllab1.ucmerced.edu/wlai24/LapSRN
22. Ledig, C., et al.: Photo-Realistic Single Image Super-Resolution Using a Generative Adversarial Network, pp. 4681–4690 (2017)
23. Lim, B., Son, S., Kim, H., Nah, S., Lee, K.M.: Enhanced Deep Residual Networks for Single Image Super-Resolution, pp. 136–144 (2017)
24. Parliament, E.: Regulation (eu) 2016/679 of the european parliament and of the council of 27 april. Eur-LEX (2016)
25. Simonyan, K., Zisserman, A.: Very deep convolutional networks for large-scale image recognition. In: International Conference on Learning Representations, ICLR, September 2015
26. Timofte, R., Gu, S., Wu, J., Gool, L.V., Zhang, L., Yang, M.: Ntire 2018 challenge on single image super-resolution: Methods and results. IEEE (2018)
27. Villena, S., Vega, M., Mateos, J., Rosenberg, D., Murtagh, F., Molina, R., Katsaggelos, A.K.: Image super-resolution for outdoor digital forensics. Usability and legal aspects. Comput. Ind. **98**, 34–47 (2018). https://doi.org/10.1016/j.compind.2018.02.004
28. Yang, C.-Y., Ma, C., Yang, M.-H.: Single-image super-resolution: a benchmark. In: Fleet, D., Pajdla, T., Schiele, B., Tuytelaars, T. (eds.) ECCV 2014. LNCS, vol. 8692, pp. 372–386. Springer, Cham (2014). https://doi.org/10.1007/978-3-319-10593-2_25
29. Zhang, K., Zhang, Z., Cheng, C.W., Hsu, W.H., Qiao, Y., Liu, W., Zhang, T.: Super-Identity Convolutional Neural Network for Face Hallucination (2018)

On the Use of Transformers
for End-to-End Optical Music
Recognition

Antonio Ríos-Vila(✉), José M. Iñesta, and Jorge Calvo-Zaragoza

U.I for Computer Research, University of Alicante, Alicante, Spain
{arios,inesta,jcalvo}@dlsi.ua.es

Abstract. State-of-the-art end-to-end Optical Music Recognition
(OMR) systems use Recurrent Neural Networks to produce music transcriptions, as these models retrieve a sequence of symbols from an input
staff image. However, recent advances in Deep Learning have led other
research fields that process sequential data to use a new neural architecture: the Transformer, whose popularity has increased over time. In
this paper, we study the application of the Transformer model to the
end-to-end OMR systems. We produced several models based on all the
existing approaches in this field and tested them on various corpora with
different types of encodings for the output. The obtained results allow
us to make an in-depth analysis of the advantages and disadvantages of
applying this architecture to these systems. This discussion leads us to
conclude that Transformers, as they were conceived, do not seem to be
appropriate to perform end-to-end OMR, so this paper raises interesting
lines of future research to get the full potential of this architecture in
this field.

Keywords: Optical Music Recognition · Transformers · Connectionist
Temporal Classification · Image-to-sequence

1 Introduction

Optical Music Recognition (OMR) is the research field that studies how to computationally read music scores [4], in the same way Optical Character Recognition (OCR) and Handwritten Text Recognition (HTR) are applied to extract
content from text images. Despite OMR being sometimes labelled as "OCR for
music", the two-dimensional nature of music compositions and the many ubiquitous contextual dependencies between symbols on a staff differentiate OMR
from other document recognition areas, such as those mentioned above [3].

Most OMR literature is framed in a multi-stage workflow, with several steps
involving image pre-processing, symbol classification, and notation assembly [15].

This paper is part of the project I+D+i PID2020-118447RA-I00 (MultiScore),
funded by MCIN/AEI/10.13039/501100011033. The first author is supported by grant
ACIF/2021/356 from "Programa I+D+i de la Generalitat Valenciana".

© Springer Nature Switzerland AG 2022
A. J. Pinho et al. (Eds.): IbPRIA 2022, LNCS 13256, pp. 470–481, 2022.
https://doi.org/10.1007/978-3-031-04881-4_37

Recent advances in machine learning field, specifically those related with deep neural networks, have led OMR technologies to evolve towards alternatives to the traditional workflow. Currently, there are two modern OMR approaches: segmentation-based and end-to-end.

The former group replaces complex symbol isolation pipelines with Region-based Convolutional Neural Networks, which directly identify the symbols within a music score image [13]. The latter one, which is the topic of interest of this paper, combines techniques from the Computer Vision (CV) and the Natural Language Processing (NLP) fields to obtain a model that receives an input image —namely a complete music staff— and output its complete transcription. These systems can be implemented in a number of ways, but the most popular ones in the OMR literature are the Image-to-Sequence (Img2Seq) approaches [2], which follow a Sequence-to-Sequence (Seq2Seq) process as it is done in Machine Translation, and those trained with the Connectionist Temporal Classification (CTC) loss strategy [10] which learn how to distribute predictions within the recognized document.

Despite being labeled as different strategies, both approaches share techno-logical features in their implementation, as CNNs are used to differentiate rele-vant image features for recognition and Recurrent Neural Networks (RNN) model temporal dependencies between the recognized symbols within the document. However, recent advances in NLP have evolved towards the use of non-recurrent systems to model these dependencies while avoiding the issues that RNNs have (such as computational time cost or the modeling of long-term dependencies). In particular, these advances have led to the so-called attention-based systems, where the network learns what information is relevant within a sequence to make a specific prediction. The most relevant model of this kind is the Trans-former [20], which is currently the state-of-the-art in many NLP tasks aside from Machine Translation, where it was created. These are, for example, Document Summarization or Text Generation [9]. This model has been extended to other areas, such as CV [12], with satisfactory results.

However, unlike in OMR-analogous fields such as HTR [11], there are no similar studies in the literature about the application of Transformer models to OMR, specifically in the end-to-end recognition approaches. We believe it is important to carry out this study, since it is necessary to explore the applicability of the new general trends to create a foundation for the improvement of state-of-the-art recognition systems. Therefore, in this paper, we explore the use of the Transformer architecture in end-to-end OMR, both for the Img2Seq and the CTC-based approaches, and discuss the advantages and disadvantages of bringing this technology as an option to implement those systems.

The rest of the paper is organized as follows: in Sect. 2 we formally define the end-to-end approaches and describe the specific implemented systems used to perform our experiments. In Sect. 3, we explain our experimentation environment for the sake of replicability; in Sect. 4 we present and discuss the obtained results regarding the comparison between different alternatives; and, we conclude our work in Sect. 5.

2   Methodology

In this section, we present the end-to-end approaches and describe the implementation strategies of the systems used for this experimentation.

First, we formally describe our recognition system. End-to-end approaches consist of a system that takes an input image and produces its corresponding sequence of transcribed symbols. The specific encoding used in this experimentation is described in Sect. 3. Given an input image $x \in \mathcal{X}$ that contains image notation, f_g seeks for the sequence $\hat{s}$ of symbols from the alphabet Σ_a such that

$$\hat{s} = \arg \max_{s \in \Sigma_a^*} P(s \mid x). \tag{1}$$

This procedure can be approached in two ways, as follows.

The first one is the Image-to-Sequence approach, which implements a Seq2Seq model [18] which consists of two modules: the encoder and the decoder. The encoder receives an image and creates an intermediate representation of its visual features in a multidimensional vector, which is usually referred to as the *context vector*. Then, the decoder predicts all the corresponding tokens to the transcription depending on this generated context vector and the previous predictions (as transcribed tokens are predicted one by one), adapting the definition above to

$$\hat{s}_t = \arg \max_{s_t \in \Sigma_a} P(s_t \mid \mathbf{c}, s_{1,t-1}). \tag{2}$$

being $\hat{s}_t$ the current token prediction, $\mathbf{c}$ the context vector generated by the encoder and $s_{1,t-1}$ the subsequence of previous predictions of the model.

The second approach is the CTC-based implementation, where only an encoder is used. The network is trained to make a complete sequence prediction and the loss is computed following a CTC strategy, which allows the network to align its predictions with the available temporal steps (namely, column-wise frames in the input image) to produce an output that correlates in length with the input. The network produces a posteriorgram (as the CTC loss strategy assumes the output of the network is a linear layer with softmax activation) with the probabilities of all the tokens corresponding to Σ_a with an additional void label (ϵ) to indicate time step separations, $\Sigma_a' = \Sigma_a \cup \{\epsilon\}$. In this paper, this posteriorgram is decoded by following a *greedy decoding* approach, where the most probable token is predicted for each time step. We refer to the last layer with the CTC strategy of this network as a decoder module.

2.1   Transformer Modules

The two approaches described above are usually implemented by using RNN layers, both in the encoder and the decoder, to get information about temporal dependencies between the input data. Here, we replace these recurrent units with Transformer modules. This mechanism [20] is based in two main components: first, the model obtains information about the sequence data and its dependencies with the so-called Multi Head Attention (MHA) mechanism. This is an *attention* mechanism [1] that measures the influence of the sequence elements to define their neighbors. This attention can be applied locally in the input sequence (Self-Attention) or between different sequences (Cross-Attention).

The second concept that the Transformer introduced is the modelling of position awareness of the introduced data, as recurrence warranted this concept by nature. The Transformer solves this issue by adding a juxtaposition of sinusoidal signals to the sequence embeddings. The intuition of this idea can be found in [20], from which the Position Encoding applied here has been taken.

2.2 Application of Transformers to OMR Architectures

Now we specify how the Transformer modules are applied in the different architectures that perform music transcription.

Encoder. A CRNN is used to model the sequential data obtained from an input image. Typically, this is composed by two main blocks: first a convolutional block that extracts the relevant image features and then a recurrent layer to obtain the dependency information between frames. Here, we replace the recurrent block with a Transformer Encoder unit, which is composed by stacked MHA layers combined with position-wise linear blocks and layer normalization units. Note that the output from the Convolutional block has to be reshaped before the Transformer Encoder, as these modules cannot handle information in a 2D space. To do so, the output of the convolutional block, (b, w, h, f), is reshaped into $(b, w, h \times f)$. An illustrative example of a CNN Transformer encoder is depicted in Fig. 1. This is applied in both the Img2Seq and CTC approaches.

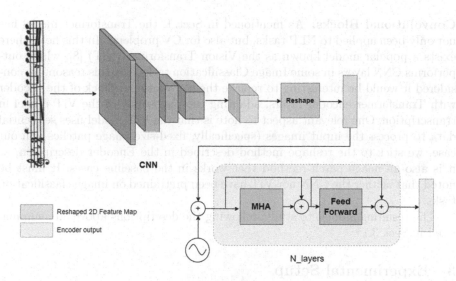

Fig. 1. Encoder architecture using the Transformer module.

Decoder. This part is only applicable to the Img2Seq models, usually consists of a recurrent or stacked recurrent layers that use the encoder output as their initial state. Attention mechanisms are also applied to enhance the decoder predictions [2]. A Transformer decoder is used to replace the recurrent features of this part. This model uses two types of MHA layers. First, it applies Self Attention to its inputs to obtain relevant features about the previously predicted tokens. Then, it uses a simple MHA module which receives both the generated context vector by the Encoder and the previous layer input. A depiction of this implementation can be seen in Fig 2.

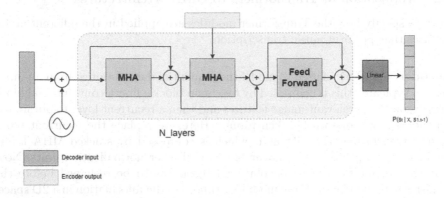

Fig. 2. Decoder architecture with the Transformer module for the Img2Seq.

Convolutional Blocks. As mentioned in Sect. 1, the Transformer model has not only been applied to NLP tasks, but also for CV problems. In this field, there exists a popular model known as the Vision Transformer (ViT) [8], which outperforms CNN layers in some Image Classification tasks. For this reason, we considered it would be interesting to replace the convolutional block of the encoder with Transformer encoder layers, adapting the philosophy of the ViT model in transcription. One relevant aspect to note is that the ViT model uses sequential data to process the input images (specifically, fixed-size image patches). In our case, we stick to the reshape method described in the Encoder description, as it is also an image patch method that works in the baseline cases. It must be noted that neither the CNN nor ViT have been pretrained on image classification tasks.

The resulting models to study, following the descriptions above, are summarized in Sect. 3.1

3   Experimental Setup

In this section, we present our experimental environment to study the application of Transformers to OMR. We specify the implemented models, detail the corpora used, and define the evaluation process considered obtaining the results presented in Sect. 4.

3.1 Studied Models

The models proposed for study are a Cartesian product between all the Encoder and Decoder strategies defined in Sect. 2.1. This product results in the following models:

- CNN Transformer Encoder (CNNT) + CTC
- ViT + CTC
- CNN Transformer Encoder (CNNT) + Transformer Decoder (following the implementation in [11])
- ViT + Transformer Decoder

An additional model has been included in this study, that is the state of the art for end-to-end OMR. This model is the CRNN-CTC architecture proposed in [6]. It will be used as a baseline to compare the performance of the proposed implementations.

It must be noted that all images are resized to 64 pixels height, maintaining the aspect ratio, and all the CNN processed inputs have pooling reductions. Depending on the implementation, the amount of reduction varies.

3.2 Corpora

Four corpora of music score images, with different characteristics in printing style, have been used to carry out this study.

The first corpus is the "Printed Images of Music Staves" (PrIMuS) dataset; specifically, the camera-based version [5]. It consists of 87,678 music incipits[1] from the RISM collection. They consist of music scores in common western modern notation, rendered with digital musicology software and extended with synthetic distortions to simulate the imperfections that may be introduced by taking pictures of music sheets in a real scenario, such as blurring, low-quality resolutions, and rotations. For these experiments, a subset of 30,000 samples was considered for experimentation.

The second corpus is the "Il Lauro Secco" manuscript (denoted as SEILS for the rest of the paper), which corresponds to an anthology of 150 typeset images from the 16th-century Italian madrigals in the so-called mensural notation.

The third corpus is the "Capitan" dataset, which contains a complete ninety-six pages manuscript of the 17th century corresponding to a *missa*, handwritten in mensural notation.

The fourth corpus is a collection of four groups of handwritten score sheets of popular Spanish songs taken from the "Fondo de Música Tradicional IMF-CSIC" (FMT), that is a large set of popular songs manually transcribed by musicologists between 1944 and 1960.

---

[1] Short sequence of notes, typically the first ones, used for identifying a melody or musical work.

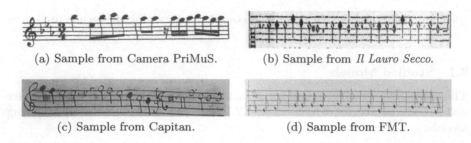

(a) Sample from Camera PriMuS.   (b) Sample from *Il Lauro Secco*.

(c) Sample from Capitan.   (d) Sample from FMT.

Fig. 3. Music excerpts from the used corpora in this paper.

It must be noted that no data augmentation or model pretraining was performed with the corpora, as the baseline for this task did not require such techniques to converge.

Table 1. Details of the corpora used in the experimental setup (Pr. = printed; Hw. = handwritten).

	PrIMuS	SEILS	Capitan	FMT
Engraving	Pr.	Pr.	Hw.	Hw.
Staves	30,000	1,278	737	872
Running symbols	829,978	31,589	17,112	17,093
Vocabulary size (std.)	426	365	320	266
Vocabulary size (split)	91	54	71	75

3.3   Music Encoding

OMR systems typically represent the recognized elements of a music staff sequentially in a standard encoding (Std), sometimes called *agnostic* encoding [5], which is a joint representation of both the symbol shape (from a set of possible glyphs Σ_g) and its position in the staff (from a set of possible positions Σ_p) in a single output code. The size of the output vocabulary in this encoding is the product of the cardinalities of both sets ($|\Sigma_g| \times |\Sigma_p|$). However, recent research has led to an alternative split-sequence (Split) encoding representation [16] taking advantage of the two-dimensional nature of music that improves the recognition performance. In this encoding, each symbol is represented by a pair of tokens representing the glyph and its position. While assuming that output sequences will be larger in this case, the output vocabulary for the network will be significantly reduced, as it consists of the sum of the glyph and position vocabulary sizes ($|\Sigma_g| + |\Sigma_p|$), thus simplifying the task for the classifier since the number of classes is much smaller (see Table 1).

3.4   Evaluation Metrics

One issue that one may find when performing OMR experiments is to correctly evaluate the performance of a proposed model, as music notation has specific features to take into account. However, OMR does not have a such a specific evaluation protocol [4]. The performance of the proposed models have been measured with the Sequence Error Rate (SER) in this paper. This metric computes the edit distance between the hypothesized sequence $\hat{s}$ and its corresponding ground truth, s, and it is divided by the length (in tokens) of s. We chose this metric because it represents accurately the recognition performance of the model and correlates with the effort that a user would have to invest to manually correct the output sequence.

The data sets have been split into fixed partitions, where 50% has been used for training and 25% has been used for both validation and test.

4   Results

The experiments were carried out applying the proposed Transformer-based models and the baseline to the four corpora described above. The results obtained are shown in Table 2.

Table 2. SER (%) obtained for the studied models for the standard encoding output (Std) and the split-sequence encoding (Split). Bold figures highlight the best results obtained for each corpus for both encodings. Underlined figures indicate the best results obtained among all the transformer-based implementations for both encodings.

Model	Encoding	PrIMuS	SEILS	CAPITAN	FMT
CNNT - CTC	Std	3.8	3.9	26.0	25.5
	Split	_3.1_	_2.8_	_19.0_	_21.6_
ViT - CTC	Std	16.4	21.2	89.6	71.4
	Split	17.2	28.2	98.8	77.3
CNNT - Img2Seq	Std	19.0	45.3	44.6	72.7
	Split	15.4	38.5	50.5	64.9
ViT - Img2Seq	Std	20.5	69.7	42.3	92.9
	Split	18.3	76.0	55.9	99.8
CRNN - CTC (baseline)	Std	2.7	2.1	16.5	24.0
	Split	**1.7**	**1.9**	**14.5**	**12.9**

As can be seen at first glance, none of the Transformer-based architectures outperforms the baseline method. This observation is aligned with conclusions in similar areas, such as HTR [11], where Transformers were not able to outperform CTC-based RNN approaches. The poorest results were obtained by the ViT-based approaches (rows 2 and 4). This is probably due to a lack of data, as reported in the literature in its own field [12]. It is worth noting that the ViT model requires a pre-training step with billions of data and then fine-tuning on a specific dataset, something that is unfeasible to do in OMR, where labeled data is usually scarce (specially in historic music manuscripts). In addition, the ViT model was designed for a specific task, which is image classification. This leads us to believe that, even if being able to achieve state-of-the-art results in some CV fields, it is rash to claim that this can be a substitute for CNN layers.

It is observed in this study that, in document transcription, where detailed information about the input must be retrieved to produce good results, CNNs are able to obtain more details, unlike ViT which seems to get more general information (and requires more data to do so). However, there was an atypical result regarding the *Capitan* dataset, where the ViT-Img2Seq (row 4) obtained better results than the CNN-based implementation. We consider this was a circumstantial situation, where the encoder was able to generate enough generic features of the image (note the *Capitan* distinctive features in Fig. 3a) and the decoder was able to converge with this little information. However, this case is not representative of the global tendency of the ViT-based techniques performance, as the rest of the corpora showed that this approach did not work effectively with the available amount of data, specially for those corpora with less training data available.

Furthermore, we observe that, in general, the CNN-based Img2Seq approach (row 3) did not outperform the CTC-based (rows 1 and 2) implementations, specially for the biggest corpora (*PrIMuS* and *SEILS*). We believe this is due to the ease of alignment the CTC loss strategy brings, as it forces the network to predict in harmony within the available temporal steps (column-wise image frames). The Transformer decoder in the Img2Seq implementation has to inherently learn to align its predictions within the document frames directly from a sequential categorical cross-entropy loss, which we believe is significantly harder to perform. Indeed, this difficulty is clearly visible for the PrIMuS dataset, where the performance was 5 times worse than its CTC counterpart. This tendency is consistent when we analyze the rest of the datasets, where there is much less training data to adjust the models (see Table 1).

Finally, the last point to observe is the comparison between the CNN with CTC approaches when using a Transformer or an RNN (rows 1 and 5). Results are very similar independently on the availability of the data, being always better for the split-sequence encoding. That is, modelling sequential data from a previous convolutional block seems to report very similar performance between the available models. However, we can observe that the baseline recurrent model (specially with the split encoding) obtains better performance than the Transformer-based approach. These results are probably correlated to the sequence length that both architectures have to deal with. Specifically, the length of the reduced input images can be easily handled by both architectures. However, the Transformer, being a model that is known to require much more data to converge than an RNN, understandably obtains slightly worse results in the same conditions.

We believe that an advantage that RNNs have compared to Transformers is the way of processing data. A recurrent architecture process information sequentially (as the information of a current time step cannot be processed without having the results of the previous ones). This feature makes the model inherently learn how to read information from left to right, as information propagates in that direction. However, the attention matrix generated in the MHA block is not linked to the positions of the tokens received in its inputs, since they are all processed within the same linear operation. That is, the Transformer needs to learn the position in which these tokens are located in order to process the information sequentially, which is an additional challenge that RNNs tackle by definition. This mainly explains the results obtained in our experimentation and leads us to conclude that, until the Transformer is adapted to perform a computational reading, RNN remains the most viable option to perform this task.

5   Conclusions

In this paper, we have studied the application of the Transformer architecture to current end-to-end recognition architectures for Optical Music Recognition. We implemented several models based on the state-of-the-art technologies for OMR and implementations made in writing recognition. From the analysis of the results, several conclusions for this study can be extracted.

Firstly, the Transformer, even being a promising Deep Learning neural architecture, did not yield better results than the recurrent models for music transcription. However, we have observed that, in some cases, specially when there were a large amount of training data, the performance was very similar to that produced by the state of the art. But, when the amount of data is not vast, it cannot be considered as a replacement option for the recurrent models (as it happens in other deep learning tasks).

This way, it seems that the main problem is the lack of data. By definition, the Transformer is a data-devouring architecture that is usually pretrained on large datasets and then fine-tuned with specific corpora. However, addressing this issue is not as easy in OMR as it may be in other tasks, since there are few labelled datasets from music archives, specially for historic manuscripts. This makes it hard to perform a pretraining step, as it is mandatory to have labeled data to do so. This leads research to focus more on architectures and training techniques rather than data labeling and augmentation. One possible solution is to use what are known as data-efficient Transformers [19]. Another way to address this issue could be to use Self-Supervised Learning (SSL), which is a well-known approach that has been applied successfully in the NLP field, giving birth to popular models such as BERT or GPT [7,14]. In this branch, pretext tasks are applied to the model in order to pretrain it before adjusting it to a given dataset. However, the relevant feature of this pretraining is that it does not need labelled data, as the pretext tasks ensure that the labels come from the input itself (for instance, one pretrain task could be to reorder input sequence or detect missing tokens). One interesting approach would be to pretrain the model in a specific pretext task where no labels are required, and then fine-tune it to a specific corpus. This is a powerful tool that OMR can take advantage of, as there are many music scores without their transcription. However, as promising as this proposition may be seen, currently SSL for CV is, to our knowledge, not as developed as it is in the NLP one, as performing pretext tasks adapted to the image domain is challenging.

Another issue to work on in the future is how the Transformer encoder can learn to read data in a specific direction (in this case, from left to right), something that RNNs do not need to. This is a challenging task, as one of the key factors of the Transformer model is that it computes a Linear Application that processes all sequence data in a parallel way. This issue has to be studied, as we believe research about Attention matrices regularization by masking or the evaluation of specific constraints do not allow the model to process data in other ways may boost performance, as models could learn to read naturally.

Aside from these conclusions, the scope of this study has been on what currently is being done in OMR: the recognition of single-staff images (which we assume they are extracted in previous steps). However, we believe that end-to-end recognition has to evolve into full page recognition, where several challenges may appear and the current state-of-the-art models (specially those based on CTC) may have scalability issues. This may be the key opportunity for the studied models to be applied more successfully, as it is starting to happen in HTR field [17]. The knowledge extracted from this study can be applied to this case, and some proposed techniques could be considered to extend the analysis and obtain better recognition systems.

In summary, we believe Transformer models are a promising neural architecture that, in its current state, cannot be exploited to its best for existing OMR tasks. However, maybe further research on architecture adaptations and data-efficient techniques on this model might be the key to produce the next level of recognition models.

References

1. Bahdanau, D., Cho, K., Bengio, Y.: Neural machine translation by jointly learning to align and translate. arXiv preprint arXiv:1409.0473 (2014)
2. Baró, A., Badal, C., Fornés, A.: Handwritten historical music recognition by sequence-to-sequence with attention mechanism. In: 2020 17th International Conference on Frontiers in Handwriting Recognition (ICFHR), pp. 205–210 (2020)
3. Byrd, D., Simonsen, J.: Towards a standard testbed for optical music recognition: definitions, metrics, and page images. J. New Music Res. **44**, 169–195 (2015)
4. Calvo-Zaragoza, J., Hajič, J., Jr., Pacha, A.: Understanding optical music recognition. ACM Comput. Surv. **53**(4), 1–35 (2020)
5. Calvo-Zaragoza, J., Rizo, D.: Camera-PrIMuS: neural end-to-end optical music recognition on realistic monophonic scores. In: Proceedings of the 19th International Society for Music Information Retrieval Conference, ISMIR 2018, Paris, France, September 23–27, 2018, pp. 248–255 (2018)

6. Calvo-Zaragoza, J., Toselli, A.H., Vidal, E.: Handwritten music recognition for mensural notation with convolutional recurrent neural networks. Pattern Recogn. Lett. **128**, 115–121 (2019)
7. Devlin, J., Chang, M., Lee, K., Toutanova, K.: BERT: pre-training of deep bidirectional transformers for language understanding. CoRR abs/1810.04805 (2018)
8. Dosovitskiy, A., et al.: An image is worth 16x16 words: transformers for image recognition at scale. CoRR abs/2010.11929 (2020)
9. Floridi, L., Chiriatti, M.: Gpt-3: Its nature, scope, limits, and consequences. Mind. Mach. **30**(4), 681–694 (2020)
10. Graves, A., Fernández, S., Gomez, F.J., Schmidhuber, J.: Connectionist temporal classification: labelling unsegmented sequence data with recurrent neural networks. In: Proceedings of the Twenty-Third International Conference on Machine Learning, (ICML 2006), Pittsburgh, Pennsylvania, USA, June 25–29, 2006, pp. 369–376 (2006)
11. Kang, L., Riba, P., Rusiñol, M., Fornés, A., Villegas, M.: Pay attention to what you read: Non-recurrent handwritten text-line recognition. CoRR abs/2005.13044 (2020)
12. Khan, S.H., Naseer, M., Hayat, M., Zamir, S.W., Khan, F.S., Shah, M.: Transformers in vision: A survey. CoRR abs/2101.01169 (2021)
13. Pacha, A., Hajič, J., Jr., Calvo-Zaragoza, J.: A baseline for general music object detection with deep learning. Appl. Sci. **8**(9), 1488 (2018)
14. Radford, A., Wu, J., Child, R., Luan, D., Amodei, D., Sutskever, I.: Language Models are Unsupervised Multitask Learners (2019)
15. Rebelo, A., Fujinaga, I., Paszkiewicz, F., Marçal, A.R.S., Guedes, C., Cardoso, J.S.: Optical music recognition: state-of-the-art and open issues. Int. J. Multim. Inf. Retr. **1**(3), 173–190 (2012)
16. Ríos-Vila, A., Calvo-Zaragoza, J., Iñesta, J.M.: Exploring the two-dimensional nature of music notation for score recognition with end-to-end approaches. In: 2020 17th International Conference on Frontiers in Handwriting Recognition (ICFHR), pp. 193–198 (2020)
17. Singh, S.S., Karayev, S.: Full page handwriting recognition via image to sequence extraction. arXiv preprint arXiv:2103.06450 (2021)
18. Sutskever, I., Vinyals, O., Le, Q.V.: Sequence to sequence learning with neural networks. In: Advances in Neural Information Processing Systems, pp. 3104–3112 (2014)
19. Touvron, H., Cord, M., Douze, M., Massa, F., Sablayrolles, A., Jegou, H.: Training data-efficient image transformers & distillation through attention. In: Proceedings of the 38th International Conference on Machine Learning, vol. 139, pp. 10347–10357 (2021)
20. Vaswani, A., et al.: Attention is all you need. CoRR abs/1706.03762 (2017)

Retrieval of Music-Notation Primitives via Image-to-Sequence Approaches

Carlos Garrido-Munoz, Antonio Ríos-Vila[✉], and Jorge Calvo-Zaragoza

U.I for Computer Research, University of Alicante, Alicante, Spain
{carlos.garrido,antonio.rios,jorge.calvo}@ua.es

Abstract. The structural richness of music notation leads to develop specific approaches to the problem of Optical Music Recognition (OMR). Among them, it is becoming common to formulate the output of the system as a graph structure, where the primitives of music notation are the vertices and their syntactic relationships are modeled as edges. As an intermediate step, many works focus on locating and categorizing the symbol primitives found in the music score image using object detection approaches. However, training these models requires precise annotations of where the symbols are located. This makes it difficult to apply these approaches to new collections, as manual annotation is very costly. In this work, we study how to extract the primitives as an image-to-multiset problem, where it is not necessary to provide fine-grained information. To do this, we implement a model based on image captioning that retrieves a sequence of music primitives found in a given image. Our experiments with the MUSCIMA++ dataset demonstrate the feasibility of this approach, obtaining good results with several models, even in situations with limited annotated data.

Keywords: Optical music recognition · Music notation · Image to sequence

1 Introduction

Optical Music Recognition (OMR) is the computational process of reading music notation from documents. Capturing all the richness and variety of music notation is a challenging task for computer vision [1].

In this field, the configuration of the different primitives and musical structures is key to its understanding. In addition, long-range dependencies between symbols and structures make this interpretation even more complicated. Given

Project supported by a 2021 Leonardo Grant for Researchers and Cultural Creators, BBVA Foundation. The BBVA Foundation accepts no responsibility for the opinions, statements and contents included in the project and/or the results thereof, which are entirely the responsibility of the authors. The second author is supported by grant ACIF/2021/356 from "Programa I+D+i de la Generalitat Valenciana".

© Springer Nature Switzerland AG 2022
A. J. Pinho et al. (Eds.): IbPRIA 2022, LNCS 13256, pp. 482–492, 2022.
https://doi.org/10.1007/978-3-031-04881-4_38

the nature of this domain, it is necessary to capture these complex relationships as accurately as possible. The common workflow of symbol detection plus classification is insufficient to solve OMR effectively, as it does not retrieve the relationships among the different components [2].

Some authors suggest that a suitable representation for music notation is a graph [3]. On this basis, the problem we face is to convert a music score into a graph-alike structure, where primitives (basic notation units) are modeled as vertices and their relationships are modeled as edges. Therefore, a promising avenue to solve OMR is to address the problem as an image-to-graph task.

There exist certain works that have pointed toward that direction. Most of them focus on identifying the primitives via object detection, with both general [5,6] and music-specific approaches [7]. From this, a graph is eventually retrieved by inferring the relationships between pairs of primitives, as in the work of Pacha et al. [4]. However, these approaches have a disadvantage, as they need fine-grained labeled data to work; that is, the ground-truth must provide the exact location of the primitives in the image. Unfortunately, this is not scalable in practice and might explain the lack of OMR datasets for this formulation.

Unlike previous approaches, our goal is to look for strategies that can learn from weakly annotated data. In a first step towards this end, our work proposes to solve the problem of retrieving the primitives following an image-to-multiset approach.[1] We will use excerpts of the MUSCIMA++ [3], a corpus that models music notation as graphs, and try to retrieve the primitives directly without providing geometric information on where they are within the image.

We address our image-to-multiset formulation using different image-to-sequence techniques, which are common in task such as image captioning. Our experiments show that the approach is promising and that accurate results are attainable. Although this work does not solve the global problem of retrieving a graph, our results open the possibility for further research in the field of image-to-graph OMR with weakly annotated data.

The rest of the work is organized as follows: in Sect. 2, we introduce the problem formulation and the techniques used; the experimental setup, corpora and metrics are described in Sect. 3; the results obtained are presented and discussed in Sect. 4 and, finally, the main conclusions of this work are summarized in Sect. 5.

2   Methodology

In this section, we formally define the problem that this paper addresses and describe the solution proposed to deal with it.

---

[1] A *multiset* is a generalization of the concept of *set*, which allows for repeated elements. This is our case, given that the same primitive category can appear many times in a music score image.

2.1 Formulation

As aforementioned, our goal in this work is to extract the multiset of symbol primitives that conform musical structures present in music score images.

The formal definition of the problem is as follows. Let Σ represent the vocabulary of music-symbol primitives (e.g., notehead, stem, flags, ...) and let $\mathcal{P}(\Sigma)$ denote the set of all possible multisets using elements from Σ. Given an input image $\mathbf{x}$, we seek for the multiset $\hat{S} = \{s_1, s_2, \ldots, s_n\}$ such that

$$\hat{S} = \arg \max_{S \in \mathcal{P}(\Sigma)} P(S \mid \mathbf{x}) \tag{1}$$

To solve Eq. 1, we reformulate the problem as a sequence prediction, given that a sequence can be converted into a multiset by just ignoring the order of its elements. Therefore, this is rewritten as:

$$\hat{S} = \varphi \left(\arg \max_{s \in \Sigma^*} P(\mathbf{s} \mid \mathbf{x}) \right) \tag{2}$$

where φ is a function that converts a sequence into its corresponding multiset.

Equation 2 can be addressed by existing image-to-sequence approaches from computer vision, that deal with challenges that are formulated alike. These consider the combined use of Convolutional Neural Networks plus Recurrent Neural Networks, which represent an appropriate solution to deal with both the image and sequential nature of the problem. Below, we describe the approach of the image-to-sequence models considered.

2.2 Image-to-Sequence Models

To implement an image-to-sequence approach, we resort to image captioning models, which take an image as input and output a sequence that describe the elements within it.

The model, at each timestep t, aims to maximize the likelihood of a symbol (or *token*) given a set of features present in the input image, along with the sequence generated until t. This solution is expressed as $\hat{s}_t = P(s_t \mid \mathbf{x}; \mathbf{s}_{0,t-1})$, where $\mathbf{x}$ represents the input image, $\mathbf{s}_{0,t-1}$ the tokens predicted previously, and s_t the symbol to predict in the current timestep.

Since we want our model to output a sequence conditioned to the input, we need to introduce image features to the computation of the sequential model. This feature extraction can be done in several ways, being the most common the one that uses CNN. Then, an RNN is used to model time dependencies between the features and the network's previous outputs to predict a given token in a specific timestep. In this work, we implement this approach by two means: a simple CNN+RNN and a visual-attention-based CNN+RNN.

CNN + RNN. The first way to implement the image captioning model is by directly combining a CNN and an RNN [8]. The CNN, usually referred to as the *encoder*, processes the input image and learns to project its relevant features into a feature map. This data structure is then reshaped and fed as an initial state to an RNN, known as the *decoder*. This part also receives the previous predictions of the network as an input. From this data, the RNN predicts the token that corresponds to each timestep.

Visual-Attention-Based CNN + RNN. As a second approach, we use an attention-based model to condition the RNN to the image features extracted by the encoder, as in the work of Xu et al. [9]. Note that, in this case, the image features are extracted directly from one of the feature maps generated through the CNN. Figure 1 shows an example of how this is done.

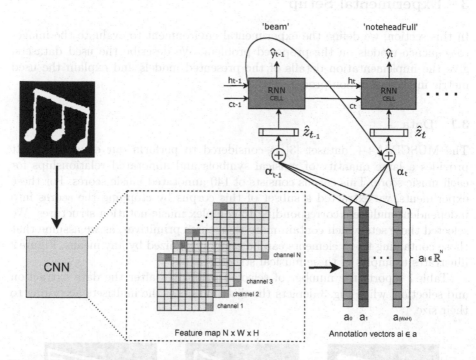

Fig. 1. General schema of the methodology. The annotation extraction process is depicted in the figure. Each annotation a_i is composed of the features of the pixel i along all the N channels extracted from the CNN feature map.

In this case, the decoder receives a composed input, with the encoder output as an initial state. This new input can be expressed as $[\hat{y}_{t-1}, \hat{z}_t]$, where y_{t-1} is the previously predicted token and $\hat{z}_t$ is a context vector, which contains filtered information from the encoder at the current timestep. This vector is produced by an attention mechanism, which filters the extracted feature maps by means of an attention mask. This mask is learned by a Multi-Layer Perceptron, which receives as an input the concatenated feature map along with the previously predicted tokens. Generally speaking, the network learns to select which information is relevant from the input image in order to enhance the decoder predictions.

One detail about this model is that its loss function—a cross-entropy loss—is regularized with a penalty term as follows:

$$L(\mathbf{x}, \mathbf{s}) = -\sum_{t=1}^{N} \log p_t(s_t) + \lambda \sum_{i}^{L} (1 - \sum_{t}^{c} \alpha_{ti})^2 \tag{3}$$

This loss can be seen as a way to "force" the model to pay attention to every part of the image at some point in the prediction of the entire sequence.

3   Experimental Setup

In this section, we define the experimental environment to evaluate the image-to-sequence models on the proposed problem. We describe the used datasets, give the implementation details of the presented models and explain the used metric in the evaluation process.

3.1   Data

The MUSCIMA++ dataset [3] is considered to perform our experiments. It provides a large quantity of musical symbols and annotated relationships for each music score. This corpus consists of 140 annotated music scores. For these experiments, we extracted a subset of this corpus by cropping the scores into independent multisets corresponding to complex music-notation structures. We selected those sets which contain more than three primitives, as we assume that those containing fewer elements can be easily recognized by any means. Figure 2 illustrates examples of these musical structures.

Table 1 reports the number of selected primitives after the data extraction and selection, while Fig. 3 depicts the distribution of the multisets according to their size.

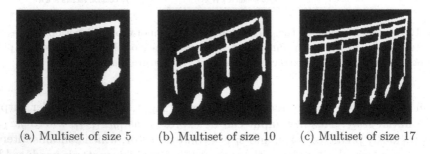

(a) Multiset of size 5   (b) Multiset of size 10   (c) Multiset of size 17

Fig. 2. Examples of different musical structures selected for our experiments.

Table 1. Statistics as regards the number of symbol primitives in the selected music-notation structures from MUSCIMA++ dataset.

Primitive	
Full notehead	4,868
Empty notehead	213
Stem	13,843
Beam	6,575
8th flag	698
16th flag	487
Total	36,684

3.2 Implementation Details

In this paper, we implement both a "classical" and a visual-attention image captioning models, which were described in Sect. 2. For the encoder (CNN), both share the same architecture with minor differences: 4 layers of alternating convolutions and poolings, with 64, 128, and 256 filters. In the case of the visual-attention model, this last layer produces 512 filters. Poolings are 2×2, except for the last layer of the simple model, which is increased to 4×4 to reduce the eventual features before the RNN. To reshape the encoder output in order to fed an initial state to the decoder, we implement a flattening layer in the "classical" model, while in the visual-attention one features are extracted as described in Fig. 1.

For the decoder (RNN), we make use of Long Short-Term Memory (LSTM) units. In the case of the simple model, the hidden state consists of 1024 dimensions. The input at each timestep is the token predicted in the previous one.[2] This model receives the flattened features of the encoder image with a size of 1024 dimensions and is introduced as the initial hidden state

In the visual-attention model, we add the attention mechanism described in Sect. 2.2. This model receives as input the concatenation of the input token and the context vector $\hat{z}_t$. Features from the encoder in this case are the annotations extracted from 512 feature maps of 8×8, which give a total of 64 vectors with 512 dimensions.

Finally, both models have a linear layer with as many units as the size of the vocabulary, and a softmax operation to take the most probable token at each timestep for the prediction. Both models are trained with the Adam optimizer with a learning rate of 0.001.

---

[2] During training, we use a *teacher forcing* methodology in both models.

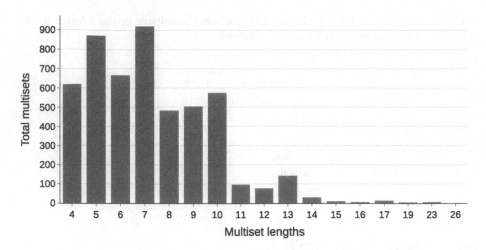

Fig. 3. Size distribution of the multisets after data selection.

3.3 Metrics

Given that we aim to measure the quantity of symbols detected on a multiset, the main metric used was the Intersection-Over-Union (IoU). Given two sets, A and B, we compute the IoU as in the Eq. (4).

$$\mathrm{IoU}(A, B) = \frac{|A \cap B|}{|A \cup B|} = \frac{|A \cap B|}{|A| + |B| - |A \cap B|} \tag{4}$$

Note that by definition $0 \leq \mathrm{IoU}(A, B) \leq 1$. Since multisets can have repeated symbols, we first enumerate symbols as

$$S_1 = \{a, b, c, a\}, S_2 = \{a, b, a, c\} \rightarrow S_1' = \{a_1, b_1, c_1, a_2\}, S_2' = \{a_1, b_1, a_1, c_2\}$$

and then the IoU between S_1', S_2' is computed.

3.4 Evaluation Scenarios

In addition to the two approaches for image captioning, we include some factors to be considered in the evaluation. Specifically, these are:

1. **Scarcity of data.** Studying performance under conditions of data limitations will allow us to understand the influence of these constraints on our formulation. This could be of special relevance in the case of handwritten scores, which are not easy to annotate. To simulate different conditions, we will restrict the data used for training to varying percentages (1, 5, 10, 25, 50, 75, and 100).

2. **Order in sequences.** Let us recall that we are implementing image-to-sequence models but the actual data comes in the form of a multiset. Therefore, we have to define a consistent ordering of the primitives from the excerpts to train the sequential neural approaches. Here, we consider the following ones. (i) To sort the primitives by the topology of the music-notation structure in the image; in this case, we consider primitives ordered from left to right and top to bottom. Hereafter, we will denote this policy as "no-sort". (ii) To˙ sort the primitives alphabetically according to the vocabulary considered. This will also place the primitives of the same category consecutively in the sequence. This option will be denoted as "sort".

3. **Data Augmentation.** The presence of basic data augmentation techniques—rotations up to 45 °C, padding, and vertical and horizontal flips—is included in order to analyze the response of these models and their relation with the performance. Note that the fact of avoiding precise primitive location facilitates the possibility of using these data augmentation techniques in a straightforward way (no need to modify the ground-truth annotation).

The combination of all these factors, along with the two neural approaches, will be studied empirically.

4 Results

In this section, we present and discuss the results obtained after performing a 5-fold cross-validation experiment. Table 2 reports the average performance in terms of IoU of each model configuration. For observing general trends, we also provide Fig. 4.

An initial remark about the results is that all the considered approaches can be considered successful, which indicates that the task is attainable with existing models. The amount of data has a direct relationship with the performance, but it only degrades noticeably when the percentage is very small (less than 10%). Inspecting the results with all the available data, we observe that performance is always around 95 of IoU. To help the reader understand the meaning of these values, Fig. 5 shows one example of perfect recognition and another where one error occurs, using the best possible model (identified below). It can be seen that, despite the large number of primitives to retrieve from the image, some of them (such as beams) with a certain graphic complexity, the approach is capable of operating without the need to specify geometric information about their location during training.

Table 2. Average IoU results on 5-CV with respect to the approach (the use or not of attention mechanisms, data augmentation, or sequence sorting) and the percentage of the training data used. Results in bold indicate the best IoU per % of training data.

Approach			% of training data						
Model	Data Aug.	Seq. sort	1	5	10	25	50	75	100
CNN + RNN	No	No	68.9	67.8	80.4	89.1	92.5	94.1	95.0
	No	Yes	66.1	68.2	71.0	80.9	92.6	88.9	95.8
	Yes	No	66.6	69.7	80.9	89.4	93.0	94.3	95.0
	Yes	Yes	66.5	68.2	70.1	85.8	87.8	94.6	96.0
Visual-attention CNN + RNN	No	No	**72.4**	78.8	82.2	88.0	92.1	93.5	94.2
	No	Yes	68.6	79.7	**86.2**	90.3	**94.4**	95.2	96.2
	Yes	No	67.0	79.5	80.8	88.7	92.0	93.6	94.6
	Yes	Yes	62.3	**80.8**	85.7	**90.6**	93.7	**95.4**	**96.4**

As regards the two neural models, no significant differences are observed when the entire dataset is used. However, the model that includes attention achieves slightly better results, the difference of which varies depending on the specific configuration. For example, while the model without attention gets an IoU of 96 (with data augmentation and alphabetically sorted sequences), the model with attention goes up to 96.4. The difference between the two approaches is more noticeable with the lowest percentages of data. For instace, with only 5% of data, the model without attention only reaches 69.7 of IoU, while the model with attention manages to reach 80.8 IoU.

Furthermore, the other two factors studied (data augmentation and the way of sorting the sequences) do not seem to significantly affect performance. The data augmentation fails to improve noticeably the results, even in conditions of scarce data. In fact, sometimes it is detrimental, as occurs in the model with attention with only 1 % of data (from 72.4 to 67.0 and from 68.6 to 62.3 of IoU, without and with sequence sorting, respectively). This may be due to the fact that the generic data augmentation introduced inconsistencies that hinder the learning process. Concerning the way of ordering the sequences, we did not observe significant changes either, and its adequacy depends on the rest of the configuration. In general, however, a regularity is observed: the model without attention prefers the topological sort, whereas the model with attention performs better when the primitives must appear alphabetically.

With all of the above, we can claim that, in general, the models considered are capable of dealing with the problem at issue. While it is true that there are other factors to take into account, they do not seem decisive for the operation of the models. The exception to this is the amount of data, since a clear correlation can be observed between the size of the training set and the performance of the models (see Fig. 4).

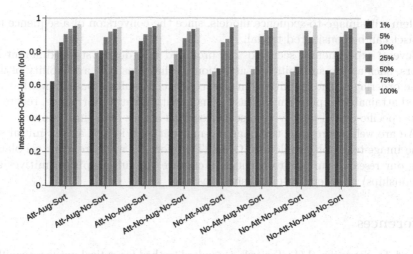

Fig. 4. Results for each model measured in terms of the Intersection-Over-Union (IoU) metric. The chart shows the average results obtained in a 5-CV. Each bar represents the performance according to the percentage of data used for training, while x-axis indicate the specific configuration.

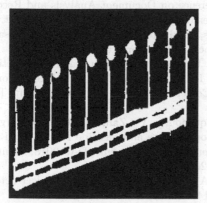 

(a) Test sample for which the best model retrieves all primitives.   (b) Test sample for which the best model misses one primitive.

Fig. 5. Examples of different predictions obtained with the best model. (a) This example corresponds to an image with 23 symbols: 10 full noteheads, 10 stems and 3 beams. The IoU obtained is 100. (b) This example contains 16 symbols: 6 full noteheads, 6 stems and 4 beams. The model misses one "beam". The IoU obtained is 94%.

5   Conclusion

In this work, we study the direct extraction of music primitives from different music structures present in the MUSCIMA++ dataset without using any object detection technique. We approach this retrieval task as an image-to-multiset problem with image-to-sequence models, since the conversion of a sequence to a multiset can be considered trivial.

Several experimental scenarios are proposed in order to study different key factors, such as the suitability of attention mechanisms, the availability of data or the ordering of the primitives in the sequence. Results report show that it is indeed attainable to perform the task at hand with great performance, regardless of the specific model and even in scenarios with severe data limitations.

We are well aware that the image-to-multiset task is just a very initial step of the image-to-graph problem in OMR. Therefore, as a future work we plan to move our research further to eventually capture the full graph (primitives and relationships) in an end-to-end fashion, as well.

References

1. Byrd, D., Simonsen, J.G.: Towards a standard testbed for optical music recognition: definitions, metrics, and page images. J. New Music Res. **44**(3), 169–195 (2015)
2. Calvo-Zaragoza, J., Jr., J.H., Pacha, A.: Understanding optical music recognition. ACM Comput. Surv. **53**(4), 77:1–77:35 (2020)
3. Jan Hajič, j., Pecina, P.: The MUSCIMA++ Dataset for Handwritten Optical Music Recognition. In: 14th International Conference on Document Analysis and Recognition, ICDAR 2017, Kyoto, Japan, November 13 - 15, 2017. pp. 39–46. Dept. of Computer Science and Intelligent Systems, Graduate School of Engineering, Osaka Prefecture University, IEEE Computer Society, New York, USA (2017)
4. Pacha, A., Calvo-Zaragoza, J., Jan Hajič, J.: Learning notation graph construction for full-pipeline optical music recognition. In: Proceedings of the 20th International Society for Music Information Retrieval Conference, pp. 75–82. ISMIR, Delft, The Netherlands (Nov 2019)
5. Pacha, A., Choi, K., Coüasnon, B., Ricquebourg, Y., Zanibbi, R., Eidenberger, H.: Handwritten music object detection: open issues and baseline results. In: 13th IAPR International Workshop on Document Analysis Systems, DAS 2018, Vienna, Austria, 24–27 April, 2018, pp. 163–168 (2018)
6. Pacha, A., Hajič, J., Calvo-Zaragoza, J.: A baseline for general music object detection with deep learning. Appl. Sci. **8**(9), 1488 (2018)
7. Tuggener, L., Elezi, I., Schmidhuber, J., Stadelmann, T.: Deep watershed detector for music object recognition. In: Proceedings of the 19th International Society for Music Information Retrieval Conference, ISMIR 2018, Paris, France, 23–27 September, 2018, pp. 271–278 (2018)
8. Vinyals, O., Toshev, A., Bengio, S., Erhan, D.: Show and tell: a neural image caption generator. 2015 IEEE Conference on Computer Vision and Pattern Recognition (CVPR), pp. 3156–3164 (2015)
9. Xu, K., et al.: Show, attend and tell: Neural image caption generation with visual attention. In: Bach, F., Blei, D. (eds.) Proceedings of the 32nd International Conference on Machine Learning. Proceedings of Machine Learning Research, vol. 37, pp. 2048–2057. PMLR, Lille, France, 07–09 July 2015

Digital Image Conspicuous Features Classification Using TLCNN Model with SVM Classifier

Swati Rastogi(✉) ⓘD, Siddhartha P. Duttagupta, and Anirban Guha

Indian Institute of Technology Bombay, Mumbai 400076, India
swatirastogi.iitb@gmail.com

Abstract. We present a transfer learning convolutional neural network (TLCNN) model in this study that permits classification of noticeable noise characteristics from degraded images using dispositional criteria. Various digitally degraded images comprising additive, impulsive, as well as multiplicative noise are used to test the suggested approach for noise type detection techniques. We further show that our representations sum up effectively when applied to additional datasets, achieving best-in-class results. In order to use the TL approach, we selected three CNN models: VGG19, Inception V3, and ResNet50, which are profound for visual recognition in computer vision to detect correct noise distribution. In this perplexing setting, the system's capacity to deal with the degraded image has outperformed human vision in noise type recognition. By getting noise classification performance of 99.54, 95.91, and 99.36%, while observing the nine classes of noises, the author's testing confirmed the constant quality of the recommended noise type's categorization.

Keywords: Conspicuous features · Noise classification · CNN · SVM classifier · Transfer learning

1 Introduction

1.1 Background and Related Work

When it comes to image processing, noise is always a component of the equation. During image transfer, image capture, programming/coding, analyzing or relaying processes, there's a significant chance that noise will persist. In practice, image classification techniques diminish in performance when images become contaminated with noise, despite the fact that these models are trained on pre-processed data. Various customized and classic ML algorithms are available, and they have been proven to be effective for noisy image classification because to their deep layer-wise design to emulate highlights from current information; nonetheless, they suffer from the same noise issue.

Indian Institute of Technology Bombay.

© Springer Nature Switzerland AG 2022
A. J. Pinho et al. (Eds.): IbPRIA 2022, LNCS 13256, pp. 493–504, 2022.
https://doi.org/10.1007/978-3-031-04881-4_39

Noise is the most prevalent occurrence with imaging, according to several research conducted over the last few decades. Several solutions with substantial performance are available for previously state-of-the-art extremely well presented with random valued noise reduction and removal. Noise recognition and identification, on the other hand, have remained a stumbling block and are one of the most important elements to address before proceeding with image denoising. As a result, this research investigates the categorization of digital image noise and questions: *Is it possible to develop a solution that allows us to investigate a better image noise feature classification?*

Denoising images will be exceedingly productive and beneficial once perfect noise detection and classification is accomplished, as no damaged picture elements will go undetected. In recent years, picture element-based ML algorithms have shown to be successful in solving several image processing challenges. As a consequence, the Eq. 1 will be employed in to represent the entire image,

$$n_i(p,q) = i_n(p,q) + o_i(p,q) \tag{1}$$

where, $n_i(p,q)$ is the noisy image function, $i_n(p,q)$ is image noise function, and $o_i(p,q)$ is the original image function.

Image noise typically appears as random speckles on a single layer but can substantially impact image quality. However, sometimes enhancing the perceived sharpness of a digital image may be favorable. The degree of noise typically rises with the duration of exposure, the physical temperature, and the camera's sensitivity setting. The proportion of various forms of image noise present at a specific setting varies between camera models and is also determined by sensor technology.

During acquisition and transfer, noise is commonly incorporated into digital images [3,6,7]. Additive noise, impulse noise, multiplicative noise, such kinds of noises can degrade image quality. These noises are caused by improper memory placement, after-filtration, a short focal length, and other unfavorable environmental factors or image capturing devices. The top to bottom assessment of the noise model and an appropriate denoising approach are essential preparatory measures towards recovering the original image [4,12]. An evaluation of the noise model and a suitable denoising method is required before recovering the original image. The impulse noise detection system extracts the critical data from each image window's feature vectors, resulting in better pixel classification [8]. It incorporates digital image noise categorization and is preceded by several integrated noise detectors and filters for superior performance over standard uniformly applied filters.

An in-depth examination of the noise model and an appropriate denoising approach are essential preparatory measures towards recovering the original image. As an outcome of utilizing robust noise feature extraction parameters and ML techniques, the classification of noisy parameters with an adequately trained classifier, it can be made more effective, and it can also be improved. The current robust model was used since it conveys critical information to machine learning classifiers.

1.2 Our Work and Contribution

We contribute in a variety of ways. The research focuses on developing a robust image categorization system that can cope with any degree of noise. It also avoids the requirement for repeated system training with imagery with varied degrees of noise when dealing with images with different levels of noise.

– To begin, we used transfer learning-based machine learning approaches to cat-egorise noise in our research. To do this, we combined numerous CNN models and employed SVM classifiers to lower the training filter's high processing cost.
– Various forms of digital image noises are investigated, and the system's per-formance is certified for nine noise level classes. The difficult experiment is one in which the classes are produced from imagery that have been damaged by a mixture of several types of noise.
– In the trials, accuracies of 99.54, 95.91, and 99.36% were achieved. It has been proven that the suggested study can exceed cognitive development in recognizing noise types.
– To categorize distorted data, we employed CNNs. In this study, we used VGG-19 and Inception-V3 deep CNN architectures, as well as ResNet CNN models. Gaussian noise, Lognormal noise, Uniform noise, Exponential noise, Poisson noise, Salt and Pepper noise, Rayleigh noise, Speckle noise, and Erlang noise are among the nine types of noise representations classified by CNN.

To our knowledge, neither these distinctions nor the literature has ever dis-cussed the degree of complexity. On the other hand, the majority of relevant research focused on only a few forms of noise spectrum [1,2,11,17–19].

1.3 Paper Organization

Section 2 of this paper outlines the proposed CNN modeling approach, including information on the CNN networks and dataset utilized in this research. While approaching the end of this part, the ML SVM classifier is used in this research. Section 3 goes through the experiment that was carried out utilizing several noise models. The results of these pre-trained models are passed to noise estimators, which determine the digital noise intensities and levels. These estimated noise intensities are then categorized using an ML technique focusing on the SVM classifier. Section 4 of this study discusses the noisy image classification method-ological experiment. Section 5 displays the conclusion.

2 Proposed CNN Modeling Strategy

2.1 ML Classifier for Noise Classification

The feature acquired from the noisy imagery is used by the ML-based classifier to categorize degraded pixels in the image data. The strong statistical properties of the image produced from the images serve as resilient features for ML classifiers, resulting in successful learning and, as a result, thorough analysis. Because they perform better in their particular fields, ML algorithms are built for specialized purposes. The purpose is to use statistical data acquired from the image to build a classification model. The Support Vector Machine (SVM) classifier [10] was used in this investigation since it has shown benefits in terms of accuracy and response time. It's essentially a linear model that may be used to solve classification and regression problems. Furthermore, multiple noise levels may be used to train the SVM, and the learnt model is predicted to function with a wide variety of noise levels. Figure 1 illustrates the whole approach utilized in this literature as an architecture of the proposed image conspicuous features classification TLCNN model. While Fig. 2 depicts the transfer learning model utilized in Fig. 1.

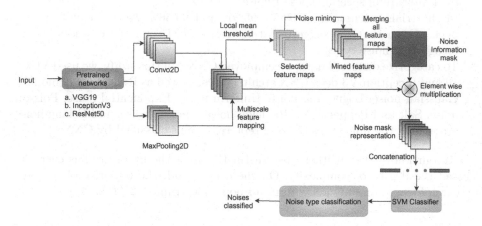

Fig. 1. The architecture of the proposed image conspicuous features classification TLCNN model.

2.2 Convolution Neural Network

CNN is an end-to-end technique that utilizes numerous convolution layers (CL) followed by maxpooling layers (PL) to anticipate the output from digital images. It can extract crucial visual properties using the neural network's link network. To accomplish the same, intermediate layered neurons get information from their succeeding layer. The higher-level properties are then derived by analysing and combining these qualities in successive intermediary levels. To create the finest imperative coding to a considerably localised learning algorithm, weight values are shared across nodes. A weight-sharing set is a convolutional kernel or, more commonly, a filter. Feature maps will be produced as a result of these filters, which will be convolved with the input. The pooling layer is another useful aspect of CNN for lowering processing complexity. The network is often followed by many fully convolutional (FC) levels in order to get the desired result.

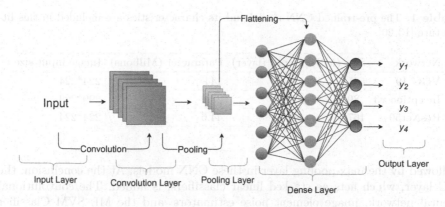

Fig. 2. Transfer learning approach.

CNN architecture arrangement can vary according to the application and can be enhanced by repetitive parameter adjustments. The following are the configurations of the different networks explored in this paper:

VGG19 Architecture. VGG19 [16], which has 19 CL and consists of infinitesimally tiny convolutional filters with three FC-linked layers, was used in this investigation. This model has 48 layers when the input, output, pooling, and activation layers are taken into consideration. This model beat a number of well-known CNNs in classification tests, including AlexNet and the MobileNet reference model.

Inception V3 Architecture. The GoogLeNet Inception-V3 [9] model from Google Inc. is a 42-layer CNN-based model that is similar in difficulty to VGG-19. The model includes several convolutional frames, of which we used 1×1, 3×3, and 5×5, as well as a 3×3 maxpooling in our studies. Before completing the larger convolutions, the generated feature map is processed via a ReLU (5×5 or 3×3). When opposed to executing the larger convolutions first, this method significantly reduces the amount of calculations.

ResNet50 Architecture. The ResNet-50 [5] CNN model, also known as Residual Network, being investigated in the present study. This approach is most commonly used to examine visual images when compared to earlier literary works. The entire model consists of 48 CLs, 1 FC, and one max-pooling layer, resulting in a ResNet50 with 50 layers. There are 3.8 billion floating point operations in the model.

The available pre-trained networks trained on the noisy picture dataset [14, 15], as well as some of its properties, are listed in Table 1. "Network depth" refers to the total number of consecutive convolution layers (CL) and fully-connected (FC) layers on a pathway from the input to the output layer. CL is

Table 1. The pre-trained CNN model and its characteristics are included in this literature [13, 20].

Network	Size (MB)	Depth (layer)	Parameter (Millions)	Image input size
VGG 19	535	19	144	224*224
Inception v3	89	48	23.9	299*299
ResNet50	167	50	44.6	224*224

followed by the max-pooling layer in these CNN models. At the conclusion, the FC layer, which acts as stacked linear classifiers, is added. The convolutional neural network, image element noise estimators, and the ML SVM Classifier are the initial topics covered in this part. The model suggested in this paper is supervised since the SVM classifier is a supervised ML approach that may be utilized for regression and classification tasks.

3   Experimental Work and Discussion

3.1   Dataset Description

We put our suggested approach for categorising picture noise through its paces in order to assess its performance. The database utilised in this study is [14, 15], which is a standard 9-class noisy picture dataset. This dataset was heavily utilized in the research, and various experiments were conducted to evaluate the usefulness of the proposed model for digital noisy picture classification. Each test set has nine folders and is 2.68 GB in size. The dataset that will be worked on in this research has a total of 14000 photos, of which 12000 are considered the training image set and 2000 are considered the test image set, on which the proposed study is based.

3.2   Noise Classification and Noise Models

The objective was to find out which of the nine possible noise distributions is present in a noisy image. As a result, we frame the job as a classification issue, with the goal of determining the kind of noise contained in each random image. ML algorithms are now widely used to solve any classification challenge since they produce far more accurate answers. The two viable logics for dealing with categorization supervised learning issues are training and testing.

Additive noise, multiplicative noise, and impulsive noise are the three main types of noise in digital images. The subcategories of these noises include erlang noise, exponential noise, gaussian noise, lognormal, poisson, rayleigh, salt and pepper, speckle, and uniform noise. We picked these noises because they are the most basic type of digital imagery noise and have been used in a variety of works of literature throughout the years. As stated in Eq. 2, the digital image noisy dataset is randomised into two subsets, the training (t_r) subset and the testing (t_s) subset.

$$t_r \quad \text{and} \quad t_s \subseteq noisy\ digital\ images \tag{2}$$

Both the training and testing subsets of the [14, 15] datasets were used to achieve the findings. Rather than entering all of the training data in one epoch, we used sequential training to train all three CNN models, with the weights being adjusted after each learning matrix is passed through the model. The t_r subset differs from the t_s subset in that it shows how well our model performs when faced with unknown data.

Epochs represent the number of times the whole training subset is measured in order to update the neural network model weights all at once. As a result, the total number of epochs used in this study for all three CNN models is 20. VGG19 and Resnet50 imagery are shrunk to 224X224 pixels, while the InceptionV3 CNN model is resized to 299X299 pixels. Because CNN models are subject to pixel values and will not work on images that do not satisfy the specified resolution, images should be scaled. The training and testing subsets are uploaded from the relevant folders and fed to the scripts once the graphics have been resized.

To acquire the learning curves (LC) for TLCNN's, several Keras and TensorFlow applications, as well as specialised modules on the Python platform, are evaluated. The LC is a graph that shows how a model's learning effectiveness changes over time or with different experiences. For algorithms that learn gradually from a training dataset, LC are a popular analytical tool in machine learning. The model may be assessed on the training dataset and holdout validation dataset after each training update, and graphs of measured performance can be constructed to validate and illustrate LC.

In this study, we created plots for each metric's LC, with each plot displaying two LC's, one for each of the train and validation datasets. LC based on the metric used to improve the model's parameters, i.e., training and validation loss curve is known as Optimization Learning Curves (OLC). While LC based on the measure used to evaluate and select the model, i.e., training and validation accuracy curve is known as Performance Learning Curves (PLC).

4   Results and Discussion

4.1   TLCNN Behaviour

The dataset has been trained utilizing various CNN models, it has been tested and validated for an unknown subset, the LC's have been acquired, and the model behaviour has been determined at this phase of the experiment. The evaluation is conducted using an ML SVM classifier for noise classification in the following steps of the experiment. We looked at nine different forms of digital image noises in our research. The SVM classifier has benefit over the computing time and can more effectively identify the model's parameter, as explained in Sect. 2 of this work. The SVM classifier is used to the learnt model in this study. The required outcomes are listed in Table 3 of the Sect. 4 report noise

Table 2. Classification accuracy for nine types of digital image noise and overall average at last stage of the experiment.

Parameter	VGG19	Inception V3	ResNet 50
Training accuracy	99.54	95.91	99.36
Validation accuracy	97.88	95.91	99.04
Model behavior	Overfit	Underfit	Good fit

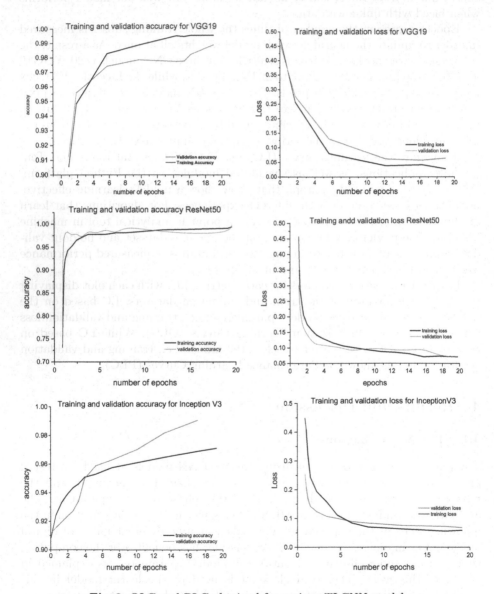

Fig. 3. OLC and PLC obtained for various TLCNN model

level classification accuracies. Figure 3 shows the obtained OLC and PLC for every TLCNN model. Model behavior is tabulated in Table 2 which helped us in determining whether the behaviour is overfit, underfit, or good fit.

4.2   TLCNN Performance

For TLCNN's, OLC and PLC are derived as shown in Table 3. TL VGG-19 and TL Inception V3 accuracy and loss vary in unanticipated and inconsistent ways. TL VGG-19 architecture, on the other hand, gives 99.54% training and validation accuracy after 20 epochs, whilst TL Inception v3 provides 95.91% training and validation accuracy. ResNet50 has a training and validation accuracy of 99.36% on the same set of epochs, showing that the model behaviour is good-fit (see graphical depiction in Fig. 3 (c) and (d)). ResNet50 is the best model in terms of model behaviour and accuracy, according to the data. Later design performed better for the chosen noise class. TL Inception V3 and TL VGG19, on the other hand, provide cause for alarm.

According to the Table 3., the total categorization accuracy has topped 99%, which is greater than the preceding pieces of literature, to the best of our knowledge and studies. When using nine classes of noisy image dataset as our study's benchmark, the following table illustrates the noticeable noise feature classification accuracy for nine types of noise features.

4.3   Conspicuous Noise Feature Classification

Various libraries and modules, including Keras and TensorFlow, are used in the experiment, which is executed on Anaconda Navigator → Spyder (Python 3.8). As mentioned in Fig. 4, the experiment was carried out in four phases in two separate sets. For each of the three TLCNN models, the first portion determines the training and validation accuracy as well as the loss percentage.

In comparison to existing research, the first stage experiment portrays a detailed vision of the TLCNN model with the best accuracy and least loss. As part of future research, the best model will be examined. The trained and validated model is then subjected to an MLSVM classifier, which will categorize one of the nine noises in the image.

As a result, the proposed research aims to evaluate the proposed model's effectiveness in recognising and categorising the type of digital picture noise that exists. The training and validation accuracy and loss for all of the CNN models presented in this study are included in the Table 3 below.

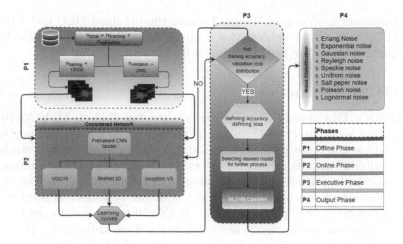

Fig. 4. Experimental description that has been conducted in four phases in two different sets

Table 3. Classification accuracy for nine types of digital image noise and overall average at last stage of the experiment.

Noise type	VGG19	Inception V3	ResNet 50
Gaussian	99.79	94.9	99.41
Lognormal	99.91	95.62	99.36
Uniform	99.65	95.48	99.52
Exponential	99.87	95.86	99.87
Poisson	100	95.12	99.21
Salt and pepper	100	100	100
Rayleigh	99.49	95.12	99.27
Speckle	99.15	95.89	99.62
Erlang	100	95.11	97.98
Classification average accuracy	99.54	95.91	99.36

5   Conclusion

The noise feature classes and classification average accuracy for TLCNNs are shown in Fig. 5 and Fig. 6. The presence of noise in visual data is common. Conventional classification algorithms can readily classify preprocessed noiseless images. Dealing with noisy data directly fed into a supervised classifier, on the other hand, is difficult, and failure to classify is pretty much assured. Several new research pathways have been explored as a result of this work. The design of this study was found to be effective. Stacking layers using optimization approaches may be used in future research to increase performance.

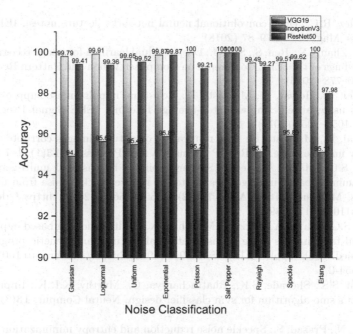

Fig. 5. Conspicuous noise feature classification.

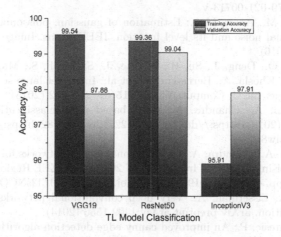

Fig. 6. Classification average accuracy for various TLCNN.

References

1. Ahumada, A.J.: Classification image weights and internal noise level estimation. J. Vis. **2**(1), 8–8 (2002)
2. Ahumada Jr, A.J.: Classification image weights and internal noise level estimation (expanded)
3. Bovik, A.C.: Handbook of image and video processing. Academic press (2010)

4. Gonzalez, R.C.: Deep convolutional neural networks [lecture notes]. IEEE Signal Process. Mag. **35**(6), 79–87 (2018)
5. He, K., Zhang, X., Ren, S., Sun, J.: Deep residual learning for image recognition. In: Proceedings of the IEEE Conference on Computer Vision and Pattern Recognition, pp. 770–778 (2016)
6. Hosseini, H., Hessar, F., Marvasti, F.: Real-time impulse noise suppression from images using an efficient weighted-average filtering. IEEE Signal Process. Lett. **22**(8), 1050–1054 (2014)
7. Hosseini, H., Marvasti, F.: Fast restoration of natural images corrupted by high-density impulse noise. EURASIP J. Image Video Process. **2013**(1), 1–7 (2013)
8. Javed, S.G., Majid, A., Lee, Y.S.: Developing a bio-inspired multi-gene genetic programming based intelligent estimator to reduce speckle noise from ultrasound images. Multimed. Tools Appl. **77**(12), 15657–15675 (2017). https://doi.org/10.1007/s11042-017-5139-2
9. Javed, S.G., Majid, A., Mirza, A.M., Khan, A.: Multi-denoising based impulse noise removal from images using robust statistical features and genetic programming. Multimed. Tools Appl. **75**(10), 5887–5916 (2015). https://doi.org/10.1007/s11042-015-2554-0
10. Keerthi, S.S., Shevade, S.K., Bhattacharyya, C., Murthy, K.R.K.: Improvements to platt's smo algorithm for svm classifier design. Neural Comput. **13**(3), 637–649 (2001)
11. Mehta, N., Prasad, S.: Speckle noise reduction and entropy minimization approach for medical images. Int. J. Inf. Technol. **13**(4), 1457–1462 (2021). https://doi.org/10.1007/s41870-021-00713-y
12. Rakhshanfar, M., Amer, M.A.: Estimation of gaussian, poissonian-gaussian, and processed visual noise and its level function. IEEE Trans. Image Process. **25**(9), 4172–4185 (2016)
13. Russakovsky, O., Deng, J., Su, H., Krause, J., Satheesh, S., Ma, S., Huang, Z., Karpathy, A., Khosla, A., Bernstein, M., et al.: Imagenet large scale visual recognition challenge. Int. J. Comput. Vision **115**(3), 211–252 (2015)
14. Sil, D., Dutta, A., Chandra, A.: CNN based noise classification and denoising of images (2019). https://doi.org/10.21227/3m26-dw82. https://dx.doi.org/10.21227/3m26-dw82
15. Sil, D., Dutta, A., Chandra, A.: Convolutional neural networks for noise classification and denoising of images. In: TENCON 2019–2019 IEEE Region 10 Conference (TENCON), pp. 447–451 (2019). https://doi.org/10.1109/TENCON.2019.8929277
16. Simonyan, K., Zisserman, A.: Very deep convolutional networks for large-scale image recognition. arXiv preprint arXiv:1409.1556 (2014)
17. Stosic, Z., Rutesic, P.: An improved canny edge detection algorithm for detecting brain tumors in MRI images. Int. J. Signal Process. **3** (2018)
18. Tuba, E., Bačanin, N., Strumberger, I., Tuba, M.: Convolutional neural networks hyperparameters tuning. In: Pap, E. (ed.) Artificial Intelligence: Theory and Applications. SCI, vol. 973, pp. 65–84. Springer, Cham (2021). https://doi.org/10.1007/978-3-030-72711-6_4
19. Tuba, E., Tuba, I.: Swarm intelligence algorithms for convolutional neural networks (2021)
20. Zhou, B., Khosla, A., Lapedriza, A., Torralba, A., Oliva, A.: Places: An image database for deep scene understanding. arXiv preprint arXiv:1610.02055 (2016)

Contribution of Low, Mid and High-Level Image Features of Indoor Scenes in Predicting Human Similarity Judgements

Anastasiia Mikhailova[1]([✉]) [iD], José Santos-Victor[1] [iD], and Moreno I. Coco[2] [iD]

[1] Instituto Superior Técnico, Universidade de Lisboa, Lisbon, Portugal
anastasiia.mikhailova@tecnico.ulisboa.pt
[2] Sapienza, University of Rome, Rome, Italy

Abstract. Human judgments can still be considered the gold standard in the assessment of image similarity, but they are too expensive and time-consuming to acquire. Even though most existing computational models make almost exclusive use of low-level information to evaluate the similarity between images, human similarity judgements are known to rely on both high-level semantic and low-level visual image information. The current study aims to evaluate the impact of different types of image features on predicting human similarity judgements. We investigated how low-level (colour differences), mid-level (spatial envelope) and high-level (distributional semantics) information predict within-category human judgements of 400 indoor scenes across 4 categories in a Four-Alternative Forced Choice task in which participants had to select the most distinctive scene among four scenes presented on the screen. Linear regression analysis showed that low-level (t = 4.14, p < 0.001), mid-level (t = 3.22, p< 0.01) and high-level (t = 2.07, p < 0.04) scene information significantly predicted the probability of a scene to be selected. Additionally, the SVM model that incorporates low-mid-high level properties had 56% accuracy in predicting human similarity judgments. Our results point out: 1) the importance of including mid and high-level image properties into computational models of similarity to better characterise the cognitive mechanisms underlying human judgements, and 2) the necessity of further research in understanding how human similarity judgements are done as there is a sizeable variability in our data that it is not accounted for by the metrics we investigated.

Keywords: Image similarity · Scene semantics · Spatial envelope · SVM · Hierarchical regression

This research was supported by Fundação para a Ciência e Tecnologia with a PhD scholarship to AM [SFRH/BD/144453/2019] and Grant [PTDC/PSI-ESP/30958/2017] to MIC.

A. J. Pinho et al. (Eds.): IbPRIA 2022, LNCS 13256, pp. 505–514, 2022.
https://doi.org/10.1007/978-3-031-04881-4_40

1   Introduction

Evaluating the similarity of visual information is an important challenge for computer vision that finds its application in object recognition [1], template matching [2], generalization of robot's movement in space [3,4], reverse search of products or images [5] and many other practical applications.

As image similarity is a difficult computer vision task, humans are still often required to provide their assessment as they can successfully evaluate image similarity even in noisy conditions [6]. But relying on humans is expensive and time-consuming because the number of judgements needed grows quadratically according to the number of evaluated pair of images. However, even though the nature of human judgements is subjective, it may still be possible to frame this problem computationally by considering different types of metrics that could characterise an image. We could identify three types of features on which human similarity judgements may be based: low-level visual information (e.g., pixels, colour), mid-level structural information that is specific to scenes (spatial envelope) and high-level semantic information (e.g., object and scene concepts) [7].

At present, the majority of computational models assessing image similarity rely only on low-level visual image properties (e.g., [1,2]) as it is the easiest information that could be extracted from an image. Thus, other types of information, as those we listed just above, are often not accounted for when calculating their similarity. Hence, to develop more reliable computational models of image similarity, it is important to understand the interplay of other types of mid and high-level information when humans are asked to perform similarity judgements.

Of the existing literature we are aware of, only a few studies are attempting to model the features that drive similarity judgements in humans. The study by [8] is one such example, where authors trained Sparse Positive Similarity Embedding (SPoSE) [9] on more than 1.5 million similarity judgements on images of objects in an odd-one-out task, where participants needed to exclude the most distinct image out of three. Their results, which are based on interpreting the SPoSE dimensions, highlighted that humans rely on both low and high-level features of images to provide their similarity judgments.

The study by Hebart and colleagues' [8], which we briefly touched upon, was done on individual objects and measured their similarities across different semantic categories (e.g., toy vs kettle). Even though it is common to operationalise this task between categories, it is important to frame it also in the within-category context [10] and expand it to the larger scope of naturalistic scenes. A within-category similarity task is, in fact, harder than a between-category and probably require access and use of a variety of features to be accurately performed. Moreover, by looking at naturalistic scenes instead of individual objects, we can also evaluate the role that mid-level features, such as their spatial envelope [11], play in similarity judgements.

In the current study, we precisely aimed to investigate how humans perform a within-category similarity judgement of naturalistic scenes and evaluated the importance of their low, mid and high-level features in such a task. Our goal is to provide a better understanding of the cognitive factors implied in human similarity judgments while evaluating the predictivity of different computational metrics that may be implicated in performing this task.

2   Study of Similarity Judgements

2.1   Image Dataset

We used the ADE20K [12] and SUN [13] databases and selected images following these criteria:

(1) contained full annotation of object labels, which is necessary to compute their high-level semantic similarity (see next section for details on this metric).
(2) did not contain animate objects (e.g., people or animals) as it makes them more distinctive [14].
(3) at least 700 pixels in width or height to ensure a reasonable resolution quality.

According to these criteria, we kept 4 categories with more than 200 images each: bathroom, bedroom, kitchen and living room (1,319 images). These images were further filtered down to 100 images per category for the study to fit within a reasonable time frame (i.e., under one hour), and selected by excluding the most dissimilar images based on low, mid, and high-level measures described below. Thus, the final dataset comprised 400 images in 4 different semantic categories.

2.2   Low, Mid, and High-Level Image Features

As mentioned above, we can identify three possible features on which similarity of images can be estimated: low (e.g., colour) mid (e.g., spatial envelope) and high-level (e.g., semantics).

A very simple measure of low-level similarity between images can be obtained from their colour at the pixel level [15] (Eq. 1). Specifically, images can be represented as three-dimensional matrices in RGB space and obtain a pairwise similarity by simply subtracting matrices. Then, the sum of all differences at the pixel level, raised to the power of two, can be taken as a single pairwise similarity value (sum of square differences, SSD, or l^2-norm) measure.

$$S_{SSD} = \sum (I[n, m, k] - J[n, m, k])^2 \tag{1}$$

where I and J indicate the indices of two images being compared, $[n, m, k]$ points to a pixel position in three dimensional matrix of RBG image and S_{SSD} is the similarity distance of the image I and J according to the SSD measure.

Mid-level information of an image can be obtained from the spatial distribution of low-level spectral information in it. We follow the classic study by [11] and compute the power spectral density of each image divided into 8×8 regions in 8 different orientations. Thus, each image is represented as a vector of 512 values (64 regions × 8 orientations), a spatial envelope, which is also known to convey coarse information about its semantic category [16]. Then, the similarity between images is computed in pairwise fashion as the sum of the square difference of the GIST vectors:

$$S_{GIST} = \sum (GIST_i - GIST_j)^2 \qquad (2)$$

where GIST is a vector of 512 values of GIST descriptor, i and j are indices of two images being compared and S_{GIST} is the similarity distance of the image i and j according to the GIST measure.

Finally, high-level semantic information of a scene can be approached by considering the objects therein as its conceptual building blocks and their co-occurrence statistics as the metric. This approach assumes that the more objects scenes share, the more semantically similar they will be [17] and we utilised the method conceived by Pennington and colleagues [18] (Global Vectors model, GloVe) to operationalise this measure. Specifically, the labels of objects in each image were used to create a co-occurrence matrix of objects across all images in each of four categories. This matrix was transformed using a weighted least squares regression into a vector representation of each object (50 dimensions), and each image was represented as the sum of vectors of objects contained therein. Then, the pairwise similarity between images was calculated as a reverse cosine distance of the GloVe vectors:

$$S_{GloVe} = 1 - cos(GloVe_i, GloVe_j) \qquad (3)$$

where GloVe is a vector of 50 values of GloVe descriptor, i and j are indices of two images being compared and S_{GloVe} is the similarity distance of the image i and j according to the GloVe measure.

These three measures of similarity described above were applied to all images belonging to a certain semantic category and the value for each image represented as its distance to all other images in the given category. The three measures were then normalized using a z-score transformation to make them more directly comparable.

Additionally, we considered average feature maps from VGG-16 neural network trained on Places365 image set [19] from the first and the last convolutional layers as the proxy for low- and high-level feature information of the scenes. The average maps were compared using Kullback-Leibler divergence between activation maps of the images but as it did not lead to significant results, we would not report this data here.

2.3   Similarity Task

Forty participants viewed the stream of images presented four at a time, all from the same semantic category, and asked to click on the one that was, in their opinion, the most different in the set, so performing a Four-Alternative Forced Choice task (see Fig. 2 for the examples of design and human judgements

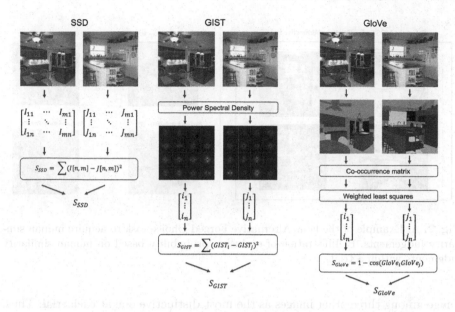

Fig. 1. Illustration of the similarity measures: SSD, GIST and GloVe.

output). For each trial/image combination we collected 10 subjective similarity judgements. To test the predictivity of GIST or GloVe in explaining similarity judgments, the four images in each trial were arranged such that there was one more distinctive (target) selected on the basis of its maximum distance to the other three (foils) images according to GIST or GloVe measures of similarity. This resulted in two different experimental conditions based on either GIST or GloVe, respectively. Additionally, we ensured that all images were properly counterbalanced in their possible combinations by creating two further lists from each condition.

The participants had a time limit of 6 s to provide their judgment, otherwise, a null response was logged in, and the trial was excluded from the analysis. Out of 4,000 trials (25 trials per category × 4 categories × 40 participants), we excluded 14.5% of the trials because of timeout. The Gorilla platform [20] was used to implement the study and collect the data.

2.4 Analysis

The data of this study were analysed in two ways.

The first analysis used linear-mixed effect modelling (using the R package lme4 [21]) to investigate how and to what extent the probability of excluding a target image can be predicted by similarities in their SSD, GIST and GloVe with the foils. The probability of an image to be excluded was computed by averaging similarity judgements across all participants for the trial in which such an image appeared. So, it was defined as the proportion of participants selecting a certain

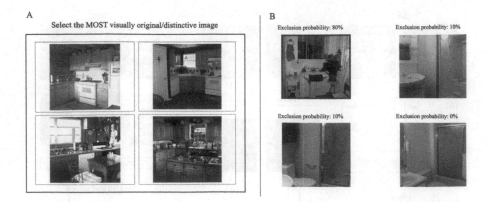

Fig. 2. A. Example of the Four-Alternative Forced Choice task to acquire human similarity judgements. B. Illustration of exclusion probability based on human similarity judgments within a trial.

image among three other images as the most distinctive one at each trial. Thus, a high exclusion probability indicated that the participant found this image to be the most distinctive relative to the other images (see Fig. 2). This measure, the exclusion probability of an image, was used as the dependent variable for the regression analysis as it reflects the agreement or consistency between the participants in their similarity judgment.

In the second analysis, we used Support Vector Machine (SVM) classifiers to examine the predictivity of the similarity metrics in approximating human judgments (binary classification). For this analysis, exclusion probability was expressed as a binary variable whereby a 1 indicated the most selected scene in each trial (i.e., an exclusion probability above 0.25, which is chance level) and 0 otherwise. Four different SVM models were created to predict the exclusion probability using each metric as an independent predictor (GIST, GloVe or SSD), and one more SVM was built as an additive (linear) combination of all predictors (GIST, GloVe and SSD) to explore the joint contribution of low, mid and high level information.

SVMs were trained and tested with a 70/30% ratio on 10-fold cross-validation over 100 iterations. We used the *ksvm* function with default settings from kernlab R package [22] to implement the SVM. The *predict* and *confusionMatrix* functions were used to extract the accuracy, f-score, and other measures that characterised the performance of the SVM models.

We compared the performance of the SVM (all predictors combined: GIST, GloVe and SSD) against the human data by averaging all SVM predictions (i.e., 100 iterations * 10 folds) for each image, and so obtain a dependent measure more directly comparable to the human judgments, and used independent-sample t-test as well as the Kolmogorov-Smirnov test to assess whether differences between human judgments and model predictions were significant.

2.5   Results

To provide a descriptive analysis of the task and measure the consistency in human judgements, we calculated the percentage of inter-observer agreement. On a trial-by-trial basis, we calculated the percentage of observers to pick the most selected image in the trial. We observed a 49% inter-observer agreement, which is much above change (i.e., 25% in this task) but still shows a certain degree of variability among them.

When looking at the linear mixed regression analysis, we found that all three measures GIST, GloVe and SSD had a significant effect on the probability of exclusion (see Table 1), which indicates that humans rely on all three types of features to perform similarity judgements.

Table 1. Results of linear regression analysis, where the effect of GIST, SSD and GloVe on probability of image exclusion in Four-Alternative Forced Choice task is tested.

Predictor	Beta coefficient	t-value	SE	p-value
GIST	0.11	4.14	0.01	<0.001***
SSD	0.08	3.22	0.01	<0.01***
GloVe	0.05	2.07	0.01	<0.04*

Turning onto the SVM predictions, we note that the model combining all 3 similarity measures predicts human similarity judgements best (i.e., 56%) and significantly better than models with single predictors (refer to Table 2).

Table 2. Accuracy of SVM models averaged across 10 fold and 100 random initialisation and the t-test comparison of single predictor models relative to the full model (GIST, GloVe and SSD).

Model	Accuracy	t-score	SE	p-value
GIST + GloVe + SSD	55.9%			
SSD	54.9%	−3.08	0.32	<0.01**
GIST	54.2%	−5.15	0.32	<0.001***
GloVe	53.9%	−6.17	0.32	<0.001***

When comparing the performance of SVM and humans, we observe no significant difference under the t-test (t = 0.64, p = 0.52), which instead is significant under the Kolmogorov-Smirnov test (D = 0.35, p < 0.001) indicating that the distributions of exclusion probability according to humans and SVM are significantly different even though, they have a similar mean.

3   Discussion

Current research shows that humans rely on both low- and high-level image information to judge their similarities [8], even though most computational models seem to be largely based only on low-level image properties [10]. Moreover, most of this research focused on similarity between categories rather than within, which is a much harder task to solve (e.g., assessing the similarities between different kitchens) [10]. The goal of the present study was precisely to evaluate the role played by low, mid and high-level image features on driving within-category human similarity judgements. Our aims were two-fold: expand our understanding of the cognitive processes humans rely on when assessing the similarity of images and evaluate the computational predictivity of low, mid and high-level metrics in approximating human performance.

Our data shows that all types of image features (low, mid and high) contribute to explaining how humans perform similarity judgements. These results corroborate findings by Hebart and colleagues' [8] but in naturalistic scenes, and especially in within-category context, which is not typically done in this research field. To provide a further, mainly qualitative, comparison with this study, we retrained their SPoSE model on our within-category data and found that the embedding dimensions obtained from the model do not seem to have a significant predictive effect on similarity judgements (t = 1.53, p = 0.13). Such results tentatively indicate that even the most recent computational models of image similarity are not well-suited yet to assess more fine-grained within-category similarities. We acknowledge, however, that our results obtained using SPoSE might be limited due to the significantly smaller sample size and call for further research on within-category similarity.

Another theoretical contribution of the current study was to show the importance of mid-level features in this task (i.e., GIST), which are often neglected by previous research. This finding confirms the necessity to incorporate other types of features into computational models of image similarity and so aiming to better approximate the way humans may be solving this task [6]. When looking at the best predictivity of human performance using SVM, we found that all features should be used to achieve the highest performance (i.e., 56%). We note, however, that such accuracy remains still pretty low and so there may likely be other factors, and perhaps also more efficient metrics, that should be used to better model the performance of humans in this task. Crucially, the accuracy of our SVM models remains the same even when using other measures of low- and high-level information. In an exploratory analysis not presented at length in the current study, we also used the average activation map from the first layer of the VGG-16 neural network [19] as a proxy to low-level features and of the last layer as a proxy to high-level features. The pairwise similarity between images, in this case, was computed as Kullback-Leibler divergence between activation maps. SVM model trained on these features still produces a prediction performance of 54%. We acknowledge that there are more modern sophisticated deep-neural network models that could be used to derive metrics of image similarity and so better understand the nature of human judgement, compared to the rather simple metrics this work utilise. This consideration reinforces the call to further research into other metrics to compute image similarity, and perhaps aspects in the task, that may better uncover the cognitive underpinnings of humans similarity judgement. Finally, we also note that the inter-observer agreement is also relatively low (49%) so indicating quite some disagreement between human judgements. We believe that precisely understanding root causes of their disagreements would indirectly help us capturing the processes that would make them agree on their judgements instead.

Finally, we realise that the scope of the current study is restricted to a within-category similarity scenario and considers only few indoor categories; and agree that to evaluate a wider spectrum of low-, mid- and high level differences on similarity judgements outdoor scenes should be considered.

4 Conclusion

First, our results demonstrate the importance of incorporating mid-level spatial and high-level semantic image information along commonly used low-level visual information when modelling similarity judgements. Secondly, we bring attention to the investigation of within-category similarity judgements, as it is an intrinsically harder task for similarity models to solve. Lastly, we point out that more research is needed to understand all aspects of human similarity judgments as our modelling, which incorporated low-mid-high level information, is still not sufficient to account for the overall variability observed across human judgements.

References

1. Sampat, M.P., Wang, Z., Gupta, S., Bovik, A.C., Markey, M.K.: Complex wavelet structural similarity: a new image similarity index. IEEE Trans. Image Process. **18**(11), 2385–2401 (2009)
2. Zhang, Y., Zhang, C., Akashi, T.: Multi-scale Template Matching with Scalable Diversity Similarity in an Unconstrained Environment (2019)
3. Wu, A., Piergiovanni, A.J., Ryoo, M.S.: Model-based behavioral cloning with future image similarity learning. In: Conference on Robot Learning, pp. 1062–1077 (2020)
4. Wang, L., et al.: Image-similarity-based convolutional neural network for robot visual relocalization. Sens. Mater. **32**, 1245–1259 (2020)
5. Bell, S., Bala, K.: Learning visual similarity for product design with convolutional neural networks. In: ACM Trans. Graph. (TOG) **34**(4), 1–10 (2015)
6. Silva, E.A., Panetta, K., Agaian, S.S.: Quantifying image similarity using measure of enhancement by entropy. In: Mobile Multimedia/Image Processing for Military and Security Applications 2007 6579, p. 65790U (2007)
7. Liu, Y., Gevers, T., Li, X.: Color constancy by combining low-mid-high level image cues. Comput. Vision Image Understanding **140**, 1–8 (2015)
8. Hebart, M.N., Zheng, C.Y., Pereira, F., Baker, C.I.: Revealing the multidimensional mental representations of natural objects underlying human similarity judgements. Nat. Hum. Behav. **4**(11), 1173–1185 (2020)

9. Zheng, C.Y., Pereira, F., Baker, C.I., Hebart, M.N.: Revealing interpretable object representations from human behavior. In: International Conference on Learning Representations (2018)
10. Wang, J., et al.: Learning fine-grained image similarity with deep ranking. In: Proceedings of the IEEE Conference on Computer Vision and Pattern Recognition, pp. 1386–1393 (2014)
11. Oliva, A., Torralba, A.: Modeling the shape of the scene: A holistic representation of the spatial envelope. Int. J. Comput. Vision **42**(3), 145–175 (2001)
12. Zhou, B., Zhao, H., Puig, X., Fidler, S., Barriuso, A., Torralba, A.: Scene parsing through ade20k dataset. In: Proceedings of the IEEE Conference on Computer Vision and Pattern Recognition, pp. 633–641 (2017)
13. Xiao, J., Hays, J., Ehinger, K. A., Oliva, A., Torralba, A.: Sun database: large-scale scene recognition from abbey to zoo. In: 2010 IEEE Computer Society Conference on Computer Vision and Pattern Recognition, pp. 3485–3492 (2010)
14. Bylinskii, Z., Isola, P., Bainbridge, C., Torralba, A., Oliva, A.: Intrinsic and extrinsic effects on image memorability. Vision Res. **116**, 165–178 (2015)
15. Ulysses, J. N., Conci, A.: Measuring similarity in medical registration. In: IWSSIP 17th International Conference on Systems, Signals and Image Processing (2010)
16. Oliva, A., Torralba, A.: Building the gist of a scene: the role of global image features in recognition. Progress Brain Res. **155**, 23–36 (2006)
17. Sadeghi, Z., McClelland, J.L., Hoffman, P.: You shall know an object by the company it keeps: an investigation of semantic representations derived from object co-occurrence in visual scenes. Neuropsychologia **76**, 52–61 (2015)
18. Pennington, J., Socher, R., Manning, C. D.: Glove: global vectors for word representation. In: Proceedings of the 2014 Conference on Empirical Methods in Natural Language Processing (EMNLP), pp. 1532–1543 (2014)
19. Simonyan, K., Zisserman, A.: Very deep convolutional networks for large-scale image recognition. In: arXiv preprint arXiv:1409.1556 (2014)
20. Anwyl-Irvine, A.L., Massonnié, J., Flitton, A., Kirkham, N., Evershed, J.K.: Gorilla in our midst: an online behavioral experiment builder. Behav. Res. Methods **52**(1), 388–407 (2019). https://doi.org/10.3758/s13428-019-01237-x
21. Bates, D., Mächler, M., Bolker, B., Walker, S.: Fitting linear mixed-effects models using lme4. J. Stat. Softw. **67**(1), 1–48 (2015)
22. Karatzoglou, A., Smola, A., Hornik, K., Zeileis, A.: kernlab-an S4 package for kernel methods in R. J. Stat. Softw. **11**(9), 1–20 (2004)

On the Topological Disparity Characterization of Square-Pixel Binary Image Data by a Labeled Bipartite Graph

Pablo Sanchez-Cuevas[1]([⊠]), Pedro Real[2], Fernando Díaz-del-Río[2],
Helena Molina-Abril[3], and María José Moron-Fernández[1]

[1] Department of Computer Architecture and Technology,
University of Seville, Seville, Spain
pablo-aries-7@hotmail.com, pabsancue@alum.us.es
[2] Research Institute of Computer Engineering (I3US), Sevilla, Spain
[3] Mathematical Institute of the University of Seville (IMUS), Sevilla, Spain

Abstract. Given an nD digital image I based on cubical n-xel, to fully characterize the degree of internal topological dissimilarity existing in I when using different adjacency relations (mainly, comparing $2n$ or $2^n - 1$ adjacency relations) is a relevant issue in current problems of digital image processing relative to shape detection or identification. In this paper, we design and implement a new self-dual representation for a binary 2D image I, called $\{4, 8\}$-region adjacency forest of I ($\{4,8\}$-RAF, for short), that allows a thorough analysis of the differences between the topology of the 4-regions and that of the 8-regions of I. This model can be straightforwardly obtained from the classical region adjacency tree of I and its binary complement image I^c, by a suitable region label identification. With these two labeled rooted trees, it is possible: (a) to compute Euler number of the set of foreground (resp. background) pixels with regard to 4-adjacency or 8-adjacency; (b) to identify new local and global measures and descriptors of topological dissimilarity not only for one image but also between two or more images. The parallelization of the algorithms to extract and manipulate these structures is complete, thus producing efficient and unsophisticated codes with a theoretical computing time near the logarithm of the width plus the height of an image. Some toy examples serve to explain the representation and some experiments with gray real images shows the influence of the topological dissimilarity when detecting feature regions, like those returned by the MSER (maximally stable extremal regions) method.

Keywords: Hierarchical representation · Digital image · Topological dissimilarity · Parallelism · (4 · 8)-adjacency tree · $\{4,8\}$-adjacency forest

1 Introduction

In 1979, Azriel Rosenfeld presented in [18] a new area of mathematical knowledge, called digital topology, that is dedicated to studying basic topological

© Springer Nature Switzerland AG 2022
A. J. Pinho et al. (Eds.): IbPRIA 2022, LNCS 13256, pp. 515–527, 2022.
https://doi.org/10.1007/978-3-031-04881-4_41

properties of digital images. This work was later continued in [8]. Rosenfeld's framework uses a dual pair of adjacencies to get rid of connectivity paradoxes. This fact leads to ambiguities: depending on the chosen dual pair of adjacencies, the results may be different, even for the most elementary digital images processing algorithms. To overcome this problem, Latecki, Eckhardt and Rosenfeld [10] introduced in 1995 a new concept of $2D$ sets free from topological paradoxes, called *well-composed sets*. The connected components (CC, for short) of these sets (and of their complements) do not depend on the chosen connectivity. In others words, in the case of rectangular pixels, a set $D = I[1] \subset I$ is well-composed if and only if 8-adjacency (vertex-connectedness) implies 4-adjacency (edge-connectedness). It is clear that the two patterns consisting of four mutually 8-adjacent square pixels having four (Foreground or Background) 4-CCs are the only ones which are forbidden in well-composedness.

On the other hand, starting from a binary digital image I defined on a square grid, Rosenfeld [17] appropriately redefined the classical region adjacency tree representation of a continuous binary image (a tree-based representation of the nested relationship between the connected components in the image). Then the asymmetric concept of (a, b)-adjacency tree arose (a-adjacency for black/foreground/FG pixels and b-adjacency for white/background/BG pixel, such that $a, b \in \{4, 8\}$ with $a \neq b$).

In this paper, the problems of classification of 2D images in terms of degree of topological disparity or well-composedness and adjacency tree representations are both dealt with here. We use a complete parallel approach based on the HSF framework [14, 15], allowing an efficient computation of all the needed structures and measures.

Section 2 contains a brief introduction to related works. In Sect. 3 some intuitive global topological discrepancy measures are defined. Section 4 explains the generation of the labeled bipartite graph $\{4, 8\}$-RAF, and in Sect. 5 the potential importance of local disparity measures derived from that representation in evaluating the topological robustness of the well-know feature detector MSER is uncovered. The paper concludes in Sect. 6.

2   Related Work

From a theoretical point of view, in this paper we propose a generalization of the classical topological Region Adjacency Tree (AdjT) notion to compute disparity-related properties. More concretely, these properties are based on the computing of topological Euler numbers and derived local and global homological information for adjacency graphs built with $(4, 8)$ and $(8, 4)$ adjacency pairs. The calculation of Euler numbers is well-known in literature, and several papers provide algorithms to make fast calculations on large images, as can be seen in [3, 6, 20], not only for 2D-images, but also for 3D-images [19]. Moreover, there exists a strong relation between these properties and the "well-composedness" problem in image analysis [1, 2].

From a computational point of view, topology is the ideal mathematical scenario for promoting parallelism in a natural way, although it drives to less classical parallelism approaches. The nature of the topological properties is essentially qualitative and local-to-global, having the additional advantage that its magnitudes are robust under deformations, translations and rotations. Nevertheless, the results in the literature in that sense are rare. Up to now, the only topological invariant that has been calculated using a fully parallel computation is the Euler number [5]. Other authors have recently proposed parallel algorithms that compute some aspects of the homological properties of binary images [12]. In [16], a digital framework for parallel topological computation of 2D binary digital images based on a sub-pixel scenario was developed, modeling the image as a special abstract cell complex [9], in order to facilitate the generalization of this work to higher dimensional images. In addition, some software libraries of flexible C++ (RedHom, [13]) have appeared for the efficient computation of the homology of sets. These libraries implement algorithms based on geometric and algebraic reduction methods.

3   Topological Discrepancy of Binary Images Defined on a Square Grid

The topological duality properties of the (a, b)-adjacency tree are remarkable: (a) each a-connected component of the set $D = I[0]$ of FG pixels of I can be identified with exactly one 1-dimensional hole of its complement D^c; (b) and each b-connected component of the set D^c can be identified with exactly one 1-dimensional hole of D. Moreover, from an (a, b)-adjacency tree, we can straightforwardly extract an a-connected component labeling (CCL, for short) for D and a b-CCL for D^c.

Given the (a, b)-adjacency tree of I and in order to create a self-dual tool (that is, a model that coincides for I and for I^c), the first candidate is, without hesitation, the labeled forest formed by the two asymmetric adjacency trees $\{T_{(4,8)}(I), T_{(8,4)}(I)\}$ of I. Using as labeling set of FG (resp. BG) nodes in both trees the set of labels obtained in the 8-CCL of D (resp. of D^c), the resulting labeled forest, called $\{4, 8\}$-region adjacency forest ($\{4, 8\}$-$RAF(I)$, for short), will be the representation model from which intra and inter topological dissimilarity measures can be easily derived. It is immediate to see that the $\{4, 8\}$-$RAF(I)$ of I can be also described as a bipartite graph. Note that if I is a well-composed image, there is no internal topological dissimilarity. Due to the fact that in this case the $\{4, 8\}$-$RAF(I)$ is reduced to $T_{(4,8)}(I) = T_{(8,4)}(I)$, the local and global dissimilarity measures and descriptors of a non well-composed image must be constructed comparing invariants or topological features that are present in both trees. In this paper, we restrict our attention to global disparity characteristics.

Being I a non well-composed binary image and D the set of FG pixels of I, a first global topological discrepancy characteristic $dsm(\ )$ is based on the Euler number:

$$dsm(\aleph)(I) = |\aleph_8(D) - \aleph_4(D)| = |\aleph_8(D^c) - \aleph_4(D^c)|.$$

This is due to the fact that $\aleph_8(D \cup D^c) = \aleph_4(D \cup D^c)$ where $\aleph_a(C)$ $(a \in \{4, 8\})$ is the classical Euler number of the set C of pixels, considering a-adjacency between them.

This heterogeneity number $dsm(\aleph)(I)$ can be obtained from the $\{4, 8\}$-$RAF(I) = \{T_{(4,8)}(I), T_{(8,4)}(I)\}$ without difficulty. For instance, $\aleph_8(D)$ is the number of FG nodes in $T_{(8,4)}(I)$ minus the sum of the number of children of each FG node in $T_{(8,4)}(I)$. This is due to the fact that $\aleph_8(D)$ is globally defined in homological terms as the number of 8-CCs of D minus the number of 1-dimensional holes detected on the digital object D with 8-adjacency. An equivalent topological invariant of I to the "Euler" 4-dimensional vector: $(\aleph_8(D), \aleph_8(D^c), \aleph_4(D), \aleph_4(D^c))$ is the following one, called *Betti's vector of I*:

$$\beta(I) = (8CC(D), 8CC(D^c), 4CC(D), 4CC(D^c)).$$

where $a - CC(X)$ (resp. $a - HL(X)$) specify the number of $a - CCs$ (resp. of 1-dimensional holes) the set of pixels X has $(a \in \{4, 8\})$. Let us call *Betti's vector of I* to $\beta(I)$. This definition takes advantage of well-know homological duality properties between a-connected component and 1-dimensional b-holes $(a \neq b$ with $a, b \in \{4, 8\})$.

Finally, given two binary images I and I' defined on a square grid, let us define the *global {4,8}-homological discrepancy* between I and I' as:

$$dsm(\beta)(I) = \sum_{i=1}^{4} |\beta(I)[i] - \beta(I')[i]|,$$

where $\beta(J)[k]$ means the k-component of the vector $\beta(J)$ $(k = 1, \ldots, 4)$ of the binary image J.

In the next section, the programming approach to compute the labeled forest $\{4, 8\}$-RAF and derived local and global dissimilarity measures is shown.

4   Generation of the Bipartite Graph

In order to tackle the disparity between Adjacency Trees when employing $(4, 8)$ and $(8, 4)$-adjacency, a new hierarchical representation called $\{4, 8\}$-$RAF(I)$ is developed here. This representation is computed by means of a bipartite graph. We consider the union of the $(4, 8)$ and $(8, 4)$-AdjTs and then add edges connecting two nodes of the same color FG or BG), one of each AdjT. That is, 4-CC nodes are connected to the 8-CC nodes containing them. In other words the $\{4, 8\}$-$RAF(I)$ defines the inclusion relationships among regions from the AdjTs for both $(4, 8)$ and $(8, 4)$-adjacencies.

This computation is based on the algorithm $CCLT$ (Connected Component Labelling Tree) for labelling connected components published in [15]. An example of this (non-unique) $CCLT$ representation is shown in Fig. 1.

This $CCLT$ tree contains all the information related to the AdjT, having a unique root on the left bottom corner. This tree can be divided into rooted sub-trees. Within each sub-tree, the root (called attractor) appears when an edge

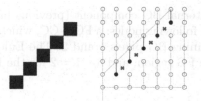

Fig. 1. Left: a synthetic image with 5 points in diagonal (using 4 adjacency criterion). Right: its double rooted tree. Blue crosses are added to represent the 8-connectivity of white pixels (crosses of the chessboard).

"touches" two different colors. This simplified representation implies a more efficient parallel topological computation [14, 15]. From now on, the $CCLT$ will be the underlying topological encoding for all the digital image structures used in this paper.

In order to explain the consequences of the different adjacency criterion on our double rooted tree, Fig. 1, left, shows a simple but very common case when binarizing color images: the apparition of "thin" oblique lines. Using 8-adjacency criterion for FG pixels (and 4 for the BG ones) would evidently produce a single FG CC surrounded by the BG. But the reverse criterion drives to a very different composition (see Fig. 1, right): a set of 5 FG CCs of only one pixel each. Whereas BG pixels are really adjacent (blue crosses have been inserted to denote this adjacency) the tree for the BG regions "chose" the edge that runs parallel to the 5 FG pixels. Another tree building could have "decide" to cross among them, by inserting green edges along the blue crosses. These blue crosses represent alternatively the frontier between two FG regions. A fast measure of the disparity between these two adjacency criteria is computed from the Euler number. In the first case, this topological magnitude would value 1 for the FG pixels and $(1 - 1) = 0$ for the BG ones. On the other hand, the opposite criterion (4-adjacency for the FG and 8 for the BG) would produce an Euler number of 5 for the FG pixels, and $(1 - 5) = -4$ for the BG ones (it has 5 holes, that are the FG pixels). Obviously, Euler number for the whole image remains to be 1 in both cases.

Figure 2 contains a synthetic image with a 4×4 chessboard that is surrounded by a monochrome background. The double rooted tree has been depicted with green edges for the tree connecting BG pixels and red edges ones for that connecting the FG ones. In addition, blue crosses have been inserted to represent the 8-connectivity of white pixels (crosses of the chessboard). These crosses represent the frontier among the FG components. The $(4, 8)$-AdjT in this case is composed of a unique (8-connected) BG component that includes up to eight (4-connected) FG components. Note that the BG component is the root of the complete tree that ends in the left bottom pixel.

On the contrary, when using the opposite adjacency criterion, the very same image would have only two BG isolated 4-CC (marked with red numbers 1 and 2), and only one 8-adjacent FG region. The $(8, 4)$-AdjT in this case would be composed of an external BG component (previous background plus six BG pixels), containing the following nodes: a FG 8-CC, which includes, in turn, only the two internal remaining BG pixels 1 and 2. The Euler numbers would then be: $1 - 2 = -1$ for the FG region, and $3 - 1 = 2$ for the BG one.

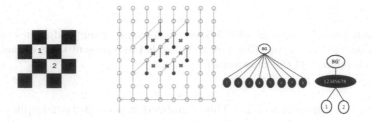

Fig. 2. Left: a synthetic image with a 4×4 chessboard surrounded by a monochrome background. Central left: its $CCLT$ for 8-connected BG (4-connected FG) –Green edges: BG tree; red edges: FG tree; blue crosses: the 8-connectivity of white pixels (crosses of the chessboard)–. Central right: $(4,8)$-AdjT. Right: $(8,4)$-AdjT.

Consequently, we propose an algorithm which firstly computes the $(4,8)$ and $(8,4)$-AdjTs and, then, adds the edges between pairs that follow the mentioned inclusion relationship. The complete pipeline is portrayed in Fig. 3.

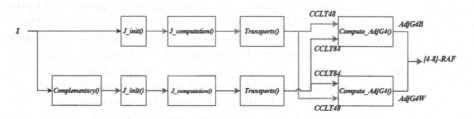

Fig. 3. Pipeline of the method for computing the final labeled bipartite graph $\{4-8\}$-RAF from an image I. $Compute\_AdjG4()$ function is defined in Fig. 7.

Indeed, the capability of the $CCLT$ to extract the topological information of the given image in the form of its AdjT in an efficient, parallel manner is here exploited. We can summarize the $CCLT$ algorithm in the following phases:

1. $J\_init()$: A jump matrix J of the same size as I is initialized, where each FG or BG pixel points into an 8-adjacent FG or a 4-adjacent BG pixel resp.
2. $J\_computation()$: Jumps/distances are propagated, where each pixel will eventually point into another of the same FG or BG region whose jump value is equal to zero. These pixels are provisional "attractors".
3. $Transports()$: Connectivity inconsistencies are solved, where the same region could be composed of different attractors. Hence, "false" attractors end up pointing into the "true" attractor of its FG or BG region.

Finally, $AdjT(I)$ has been implicitly defined in J once theses phases are performed, since the color-opposite region adjacent to each attractor is the region that surrounds the region of the attractor (which is a hole). Hence, this inclusion relationship builds up the AdjT of the image, with $(4,8)$-adjacency.

Using this method, a straightforward solution for building the $(8,4)$-AdjT of a given binary image I is as simple as computing the $(4,8)$-AdjT of its complement I^c. This approach is feasible, by considering attractors as pixels not related to its color (FG or BG), using 8-adjacency for FG pixels and 4-adjacency for BG pixels. Thus, the relationship between attractors of both AdjTs is established.

Different 4-adjacent attractors of one AdjT are contained in the region represented by one 8-adjacent attractor of the opposite AdjT and, therefore, the bipartite graph $\{4-8\}$-RAF is formed. The proposed algorithm for such task is fully described in Fig. 7. Its execution consists of searching for 4-adjacent attractors (pixels with value equal to zero in the jump matrix J) via matrix scan and checking if that pixel is also an 8-adjacent attractor or, on the contrary, it has a jump value pointing into another 8-adjacent attractor. In the later case, a new edge in the bipartite graph is detected, and hence it is stored in a table containing all the 4 to 8-adjacency inclusion relations.

One example of the application of the proposed algorithm is shown in Fig. 4. Considering the original binary image and its complement, the results of computing the $CCLT$ algorithm are summarized in table of Fig. 5, where all the regions are stored with the linear indexes of their respective attractors. Finally, a bipartite graph is fully computed when checking the inclusion relations between attractors, and in this case the results are written as sets of 4-adjacent regions in parenthesis. Its graphical representation is portrayed in Fig. 6.

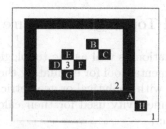

Fig. 4. A synthetic 11×12 binary image showing several cases where 4 and 8 adjacency criteria produce different adjacency trees. 4-regions are labelled with numbers and letters for BG and FG regions resp.

Once the $\{4,8\}$-$RAF(I)$ is constructed, extracting local and global discrepancy measures (like $dsm(\aleph)(I)$ and $dsm(\beta)(I)$ defined in Sect. 3) is straightforward from $\{4,8\}$-RAF (tables of Fig. 5). For the image in Fig. 2, the last two elements of $dms(\beta)$ are respectively the number of rows of each table; and the first two elements are computed as the number of labels that appear in the last column of each table, that is: $dsm(\beta)(I) = (3, 2, 8, 3)$.

	AdjG4W (from CCLT48)			AdjG4W
Column indexes:	0	1	2	3
Name of 4-white regions	Linear indexes of 4-white regions	Surrounding 8-black region	Surrounding 4-white region (of each 8-black region)	8-white region
1	131=11×12-1	(-)	(-)	131 (1)
2	92=7×12+8	105 (AH)	131 (1)	92 (23)
3	64=5×12+4	77 (DEFG)	92 (2)	92 (23)

	AdjG4B (from CCLT84)			AdjG4B
Column indexes:	0	1	2	3
Name of 4-black regions	Linear indexes of 4-black regions	Surrounding 8-white region	Surrounding 4-black region (of each 8-white region)	8-black region
A	105=8×12+9	131 (1)	(-)	105 (AH)
B	43=3×12+7	92 (23)	105 (A)	56 (BC)
C	56=4×12+8	92 (23)	105 (A)	56 (BC)
D	64=5×12+4	92 (23)	105 (A)	77 (DEFG)
E	53=4×12+5	92 (23)	105 (A)	77 (DEFG)
F	66=5×12+6	92 (23)	105 (A)	77 (DEFG)
G	77=6×12+5	92 (23)	105 (A)	77 (DEFG)
H	118=9×12+10	131 (1)	(-)	105 (AH)

Fig. 5. Tables containing the $\{4,8\}$-RAF (and CCL48 and CCL84 trees) of Fig. 4. The highest index of the 4-regions composing an 8-regions is considered to be an attractor for this image (which may fall elsewhere within the 8-region).

Moreover, our method allows to compute at the same time (without any additional time complexity) more information of the regions and contours, like areas and perimeters, which might be useful for further processing (see Sect. 6). Additionally, no processing step is done in a sequential manner. This allows to maintain the theoretical time complexity of the whole process near the logarithm of the width plus the height of the initial image.

5 $\{4,8\}$-RAF and Topological Feature Detectors

The $\{4,8\}$-RAF representation as well as the topological discrepancy measures proposed here are a fundamental tool for evaluating the robustness of image processing algorithms dealing with analytical or geometrical measures that strongly depend on the type of connectivity used for their calculation (area, perimeter, boundaries, curvatures, etc.).

In this Section, trying to illustrate this issue, we present a preliminary study about evaluating the robustness of the stability criteria used in the well known features detector algorithm MSER (Maximal Stable Extremal Region, [11]).

It is worth mentioning that the effectiveness of MSER for gray images is limited in its capability to detect regions of interest in extreme conditions, such as high contrast, low luminance, high light reflection, etc. [7]. Likewise, as the authors of [4] note, MSER is sensitive to blurring. Specifically, in blurred images, the values of intensity in the boundary of the regions change more slowly and this issue impact on the stability criteria on which the MSER is based. The limitations on blurred and/or textured images are related to image scaling, since blurring could be equivalent to image downscaling. Analogously, shapes

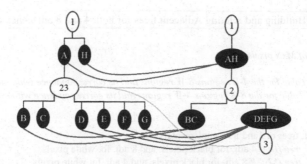

Fig. 6. $(4, 8)$-AdjT Right (Left) and $(8, 4)$-AdjT (Right) of Fig. 4. Relations between both trees (detailed in the tables of Fig. 5.) determine the Bipartite Graph.

associated with fine textures can vary fitfully in response to changes of scale. Specifically, the jutting fragments of non-convex MSERs are especially impacted by scaling changes [21]. It can be supposed that these issues could be related to the fact that the computing of the areas of the regions is carried out using 4-adjacency. Specifically, at [11], it is specified that the adjacency relation used is 4-adjacency. In fact, a small transformation that modifies the pixels diagonally, affecting connectivity at the 4-adjacency level, practically imperceptible in the image, causes an alteration in the number and location of the detected regions. The Figs. 8 and 9 illustrate this effect.

On the one hand, in Fig. 8 the MSER regions for the original image are shown using the *detectMSERFeatures()* MATLAB function. On the other hand, Fig. 9 shows the original image slightly modified to include additional 4-connected pixels by setting each horizontal couple with the same grey tone (that of the arithmetic mean of the values of both pixels). In addition the MSER regions for this modified image are also shown using the same MSER detector.

Note that both images are almost identical, but the ellipses that mark the area and orientation of regions have changed considerably. This is noticeable for the regions on the border of the most bottom coin: the slight change has introduced two new regions on the bottom, and other previous region on the left have disappeared. Likewise, a similar analysis can be discovered for the border regions of the most left coin.

Going further, the changes on the coin on the right of this last one are noticeable: instead of the seven original regions, the second image presents only four regions on this coin. The rest of coins have also suffered relevant changes on their MSER regions.

Additional tests have been done to slightly disturb the image by introducing 4 or 8 additional connected pixels. Specifically, it can be verified that the number of MSER regions change from the 60 of the original image to 51 for the previous Fig. 8, or 54 if the couple is done vertically. If 8-connected pixels were modified using the same arithmetic mean of two diagonal pixels, number of MSER regions would fall up to 47 for a left to right diagonal and 41 for the right to left one. Although the total number of detected MSER regions vary from 10 up to 32%,

---

Algorithm 1. [Building and relating Adjacent trees for both 4 and 8 adjacency criteria]

Input:
I: binary image of MxN pixels.
Output:
AdjG4B: CCLT Table for the 4-B regions/8-W regions and its correspondence with and 8-B regions
AdjG4W: CCLT Table for the 8-B regions/4-W regions and its correspondence with and 8-W regions

begin
//computing CCL trees of *I* and its complementary *! I*
CCLT48 = CCL_tree (I); // 4 adj. for black pixels, and 8 adj. for white pixels,
CCLT84 = CCL_tree (! I); // 8 adj. for black pixels, and 4 adj. for white pixels,

(n_4Black_regions , AdjG4B) = compute_ AdjG4(CCLT48, CCLT84);
(n_4White_regions , AdjG4W) = compute_ AdjG4(CCLT84, CCLT48);
end

function (idx_4, AdjG4) = compute_ AdjG4(CCLT1, CCLT2)
  *idx_4 = 0; //*index for the table of 4-black regions
  // matrixes *CCLT48* and *CCLT84* are indexed with a linear index running from 0 to *M×N-1*
  for *l* = 0: *M×N-1* **do**
    // running along the CCL tree to extract the critical cells, that is, the representative of each region
    if *(CCLT1 (l) == 0)* **then**
      AdjG4 (idx_4)(0) = 1;
      opp_idx = CCLT1 (l+1) ; // (l+1) points to a pixel of inverse color in the *CCL_tree (I)*
                               // Only for the last pixel M×N-1 this indexation is not valid
      AdjG4 (idx_4)(1) = (l+opp_idx);
      idx8 = CCLT1 (l+opp_idx);
      AdjG4 (idx_4)(2) = idx8;

      // looking in the *CCL_tree (!I)* of the complement image
      idx8_2 = CCLT2 (l);
      AdjG4 (idx_4)(3) = (l+idx8_2); // this is the linear index of the 8 CC
      idx_4 = idx_4+1;
    end if
  end for
  // finally *idx_4* contains the number of 4-black regions,
  // and the representative of the regions in the tables are ordered by the linear indexes
  // theoretical time complexity is very near *O(1)* because the only loop-carried
  // dependences among iterations is thus of the index *idx_4* that fill the table.

Fig. 7. Algorithm 1. [Building and relating the $(4,8)$ and $(8,4)$-AdjT]

as previously mentioned, most regions (around 60%) change significantly from position, area or orientation, mainly those situated near the coin borders. Only those large evident regions, like that embedding a whole coin, are insensitive to these small changes.

It is worth to mention that this issue occurs in most gray images; we have simply chosen the coin examples because they clearly show the effect in the alteration of the MSER regions. This simple test illustrates the relevance of the

Fig. 8. Left: classical image of coins showing the MSER regions, with an ellipse that denotes their orientation and area. Right: same regions coloured.

Fig. 9. Left: classical image of coins, slightly modified by adding additional 4-connected pixels, that includes the MSER regions, with an ellipse that denotes their orientation and area. Right: same regions coloured.

connectivity criterion when processing areas of images resulting from a threshold operation.

From this short analysis, the following starting hypothesis is plausible: Having at hand an efficient software for computing topological dissimilarity measures (like $dms(\beta)$) extracted from the $\{4,8\}$-RAF, constitutes a breakthrough in evaluating digital image processes that are based on the notion of topological region.

6   Conclusions

We propose here a parallel computational framework of a new hierarchical representation called $\{4,8\}$-RAF for digital images based on a square grid. As a conclusion from a basic analysis about the robustness of the stability criterion of the MSER feature detector, it can be said that an efficient software for computing (mainly, local) topological discrepancy measures derived from the RAF within the 2D and 3D context, could have a considerable influence on the fast and correct detection of characteristics in feature detectors that are based on the topological notion of region. In a near future, we plan to advance in the following questions: (a) to have a fully operative software for calculating the $\{4,8\}$-RAF and derived local and global measures; (b) to design and implement a topologically improved MSER method; (c) to obtain the $\{4,8\}$-RAF (or, an equivalent graph structure) directly from the region contours via the Homological Spanning Forest model; and (d) to build a topologically robust computational framework of topological 3D disparity.

Acknowledgement. This study was funded by the research project of Ministerio de Economía, Industria y Competitividad, Gobierno de España (MINECO) and the Agencia Estatal de Investigación (AEI) of Spain, cofinanced by FEDER funds (EU): Par-HoT (Parallel Data Processing based on Homotopy Connectivity: Applications to Stereoscopic Vision and Biomedical Data, PID2019-110455GB-I00) and CIUCAP-HSF: US-1381077.

References

1. Boutry, N., González-Díaz, R., Jiménez, M., Paluzo-Hidalgo, E.: Strong euler well-composedness. J. Comb. Optim. (2021)
2. Boutry, N., Géraud, T., Najman, L.: A tutorial on well-composedness. J. Math. Imaging Vision **60** (2018)
3. Chao, Y., Kang, S., Yao, B., Zhao, X., He, L.: An efficient euler number computing algorithm. In: 2015 IEEE International Conference on Information and Automation (2015)
4. Chen, S.y., Cai, H., Wang, X., Xia, M., Wang, Y.: Entropy-based maximally stable extremal regions for robust feature detection. Math. Probl. Eng. **2012**, 857210 (2012)
5. Chiavetta, F., Di Gesù, V.: Parallel computation of the euler number via connectivity graph. Pattern Recogn. Lett. **14**(11), 849–859 (1993)
6. He, L., Chao, Y.: A very fast algorithm for simultaneously performing connected-component labeling and euler number computing. IEEE Trans. Image Process. **24**(9), 2725–2735 (2015)
7. Kang, S., Cha, D., Kim, Y., Han, D.: Text region extraction in high contrasting image. Int. J. Future Comput. Commun. **6**, 106–109 (2017)
8. Kong, T., Ronsenfeld, A.: Digital topology: introduction and survey. Comput. Vis. Graph. Image Process. **48**(3), 357–393 (1989)
9. Kovalevsky, V.: Algorithms in digital geometry based on cellular topology. In: Klette, R., Žunić, J. (eds.) IWCIA 2004. LNCS, vol. 3322, pp. 366–393. Springer, Heidelberg (2004). https://doi.org/10.1007/978-3-540-30503-3_27
10. Latecki, L., Eckhardt, U., Rosenfeld, A.: Well-composed sets. Comput. Vis. Image Underst. **61**(1), 70–83 (1995)
11. Matas, J., Chum, O., Urban, M., Pajdla, T.: Robust wide-baseline stereo from maximally stable extremal regions. Image Vis. Comput. **22**(10), 761–767 (2004)
12. Murty, A., Natarajan, V., Vadhiyar, S.: Efficient homology computations on multicore and manycore systems. In: 20th Annual International Conference on High Performance Computing, pp. 333–342 (2013)
13. REDHOM: http://redhom.ii.uj.edu.pl/. Institute of Computer Science, Jagiellonian University (2015)

14. Diaz-del Rio, F., et. al.: Computing the component-labeling and the adjacency tree of a binary digital image in near logarithmic-time. CTIC, pp. 82–95 (2019)

15. Diaz-del Rio, F., et. al.: Parallel connected-component-labeling based on homotopy trees. Pattern Recogn. Lett. **131**, 71–78 (2020)

16. Diaz-del Rio, F., Real, P., Onchis, D.: A parallel homological spanning forest framework for 2d topological image analysis. Pattern Recogn. Lett. **83**, 49–58 (2016)

17. Rosenfeld, A.: Adjacency in digital pictures. Inform. Control **26**, 24–33 (1974)

18. Rosenfeld, A.: Digital topology. Amer. Math. Monthly **86**, 621–630 (1979)

19. Toriwaki, J., Yonekura, T.: Euler number and connectivity indexes of a three dimensional digital picture. Forma **17**, 183–209 (2002)

20. Yao, B., He, L., Kang, S., Chao, Y., Zhao, X.: A novel bit-quad-based euler number computing algorithm. Springerplus **4**, 735–735 (2015)

21. Śluzek, A.: Improving performances of mser features in matching and retrieval tasks. In: Hua, G., Jégou, H. (eds.) ECCV 2016. LNCS, vol. 9915, pp. 759–770. Springer, Cham (2016). https://doi.org/10.1007/978-3-319-49409-8_63

Learning Sparse Masks
for Diffusion-Based Image Inpainting

Tobias Alt$^{(\boxtimes)}$, Pascal Peter, and Joachim Weickert

Mathematical Image Analysis Group, Faculty of Mathematics and Computer Science,
Campus E1.7, Saarland University, 66041 Saarbrücken, Germany
{alt,peter,weickert}@mia.uni-saarland.de

Abstract. Diffusion-based inpainting is a powerful tool for the reconstruction of images from sparse data. Its quality strongly depends on the
choice of known data. Optimising their spatial location – the inpainting
mask – is challenging. A commonly used tool for this task are stochastic
optimisation strategies. However, they are slow as they compute multiple
inpainting results. We provide a remedy in terms of a learned mask generation model. By emulating the complete inpainting pipeline with two
networks for mask generation and neural surrogate inpainting, we obtain
a model for highly efficient adaptive mask generation. Experiments indicate that our model can achieve competitive quality with an acceleration
by as much as four orders of magnitude. Our findings serve as a basis for
making diffusion-based inpainting more attractive for applications such
as image compression, where fast encoding is highly desirable.

Keywords: Image inpainting · Diffusion · Partial differential
equations · Data optimisation · Deep learning

1 Introduction

Inpainting is the task of restoring an image from limited amounts of data. Diffusion processes are particularly powerful for reconstructions from sparse data;
see e.g. [32]. By solving a partial differential equation (PDE), they propagate
information from a small known subset of pixels, the inpainting mask, to the
missing image areas. Inpainting from sparse data is successful in applications
such as image compression [13,27,29], adaptive sampling [9], and denoising [1].

Optimising the inpainting mask is essential for a good reconstruction. However, this is a challenging combinatorial problem. While there are theoretical
results on optimal masks [5], practical implementations are often qualitatively
not that convincing albeit highly efficient. On the other hand, stochastic mask
optimisation strategies [15,22] produce high quality masks, but are computationally expensive.

This work has received funding from the European Research Council (ERC) under the
European Union's Horizon 2020 research and innovation programme (grant agreement
no. 741215, ERC Advanced Grant INCOVID).

In the present paper, we combine efficiency and quality of mask optimisation for PDE-based inpainting with the help of deep learning. To this end, we design a hybrid architecture which, to the best of our knowledge, constitutes the first instance of learned sparse masks for PDE-based inpainting.

Our Contribution. We present a model for learning sparse inpainting masks for homogeneous diffusion inpainting. This type of inpainting shows good performance for optimised masks [22], and does not depend on any free parameters. We employ two networks: one which generates a sparse inpainting mask, and one which acts as a surrogate solver for homogeneous diffusion inpainting. By using different loss functions for the two networks, we optimise both inpainting quality and fidelity to the inpainting equation.

The use of a surrogate solver is a crucial novelty in our work. It reproduces results of a diffusion-based inpainting process without having to perform backpropagation through iterations of a numerical solver. This replicates the full inpainting pipeline to efficiently train a mask optimisation model.

We then evaluate the quality of the learned masks in a learning-free inpainting setting. Our model combines the speed of instantaneous mask generation approaches [5] with the quality of stochastic optimisation [22]. Thus, we reach a new level in sparse mask optimisation for diffusion-based inpainting.

Related Work. Diffusion-based inpainting plays a vital role in image and video compression [3,13,29], denoising [1], and many more. A good inpainting mask is crucial for successful image inpainting. Current approaches for the spatial optimisation of sparse inpainting data in images can be classified in four categories.

1. *Analytic Approaches.* Belhachmi et al. [5] have shown that in the continuous setting, optimal masks for homogeneous diffusion inpainting can be obtained from the Laplacian magnitude of the image. In practice this strategy is very fast, allowing real-time inpainting mask generation by dithering the Laplacian magnitude. However, the reconstruction quality is lacking, mainly due to limitations in the quality of the dithering operators [15,22].
2. *Nonsmooth Optimisation Strategies.* Several works [6,7,15,25] consider sophisticated nonsmooth optimisation approaches that offer high quality, but do not allow to specify the desired mask density in advance. Instead one influences it by varying a regularisation parameter, which requires multiple program runs, resulting in a slow runtime. Moreover, adapting the model to different inpainting approaches is not trivial.
3. *Sparsification Methods.* They successively remove pixel data from the image to create an adaptive inpainting mask. For example, the *probabilistic sparsification (PS)* of Mainberger et al. [22] randomly removes a set of points and reintroduces a fraction of those points with a high inpainting error. Sparsification strategies are generic as they work with various inpainting operators such as diffusion-based ones [15,22] or interpolation on triangulations [11,23]. Moreover, they allow to specify the desired mask density in advance. However,

they are also computationally expensive as they require many inpaintings to judge the importance of individual data points to the reconstruction. Due to their simplicity and their broad applicability, sparsification approaches are the most widely used mask optimisation strategies.

4. *Densification Approaches.* Densification strategies [8,10,19] start with empty or very sparse masks and successively populate them. This makes them reasonably efficient, while also yielding good quality. They are fairly easy to implement and work well for PDE-based [8,10] and exemplar-based [19] inpainting operators. Still, they require multiple inpainting steps in the range of 10 to 100 to obtain a sufficiently good inpainting mask.

In order to escape from suboptimal local minima, the Categories 3 and 4 have been improved by *nonlocal pixel exchange* (NLPE) [22], at the expense of additional inpaintings and runtime. Moreover, it is well-known that optimising the grey or colour values of the mask pixels – so-called tonal optimisation – can boost the quality even further [15,22]. Also the approaches of Category 2 may involve tonal optimisation implicitly or explicitly.

Qualitatively, carefully tuned approaches of Categories 2–4 play in a similar league, and are clearly ahead of Category 1. However, their runtime is also substantially larger than Category 1, mainly due to the many inpaintings that they require. Last but not least, all aforementioned approaches are fully model-based, in contrast to most recent approaches in image analysis that benefit from deep learning ideas.

The goal of the present paper is to show that the incorporation of deep learning can give us the best of two worlds: a real-time capability similar to Category 1, and a quality similar to Categories 2–4. In order to focus on the main ideas and to keep things simple, we restrict ourselves to homogeneous diffusion inpainting and compare only to probabilistic sparsification without and with NLPE. Also tonal optimisation is not considered in our paper, but is equally possible for our novel approach. More refined approaches and more comprehensive evaluations will be presented in our future work.

Learning-based inpainting has also been successful in recent years. Following the popular work of Xie et al. [33], several architectures and strategies for inpainting have been proposed; see e.g. [18,21,26,34,35]. However, inpainting from sparse data is rarely considered. Vašata et al. [31] present sparse inpainting based on Wasserstein generative adversarial networks. Similarly, Ulyanov et al. [30] consider inpainting from sparse data without mask generation. Dai et al. [9] present a trainable mask generation model from an adaptive sampling viewpoint. Our approach is the first to combine deep learning for mask optimisation for PDE-based inpainting in a transparent and efficient way.

Organisation of the Paper. In Sect. 2, we briefly review diffusion-based inpainting. Afterwards in Sect. 3, we introduce our model for learning inpainting masks. We evaluate the quality of the learned masks in Sect. 4 before presenting our conclusions in Sect. 5.

2   Review: Diffusion-based Inpainting

The goal of inpainting is to restore missing information in a continuous greyscale image $f : \Omega \to \mathbb{R}$ on some rectangular domain Ω, where image data is only available on an inpainting mask $K \subset \Omega$. In this work we focus on homogeneous diffusion inpainting, which computes the reconstruction u as the solution of the PDE

$$(1 - c)\,\Delta u - c\,(u - f) = 0 \tag{1}$$

with reflecting boundary conditions. Here, a confidence measure $c : \Omega \to \mathbb{R}$ denotes whether a value is known or not. Most diffusion-based inpainting models consider binary values for c: A value of $c(\boldsymbol{x}) = 1$ indicates known data and thus $u = f$ on K, while $c(\boldsymbol{x}) = 0$ denotes missing data, leading to homogeneous diffusion [17] inpainting $\Delta u = 0$ on $\Omega \backslash K$, where $\Delta = \partial_{xx} + \partial_{yy}$ denotes the Laplacian. However, it is also possible to use non-binary confidence measures [16], which we will exploit to our advantage.

We consider digital greyscale images $\boldsymbol{u}, \boldsymbol{f} \in \mathbb{R}^{n_x n_y}$ with dimensions $n_x \times n_y$ and discretise the inpainting Eq. (1) by means of finite differences. Then a numerical solver for the resulting linear system of equations is used to obtain a reconstruction $\boldsymbol{u}$. For a good inpainting quality, optimising the binary mask $\boldsymbol{c} \in \{0, 1\}^{n_x n_y}$ is crucial. This problem is constrained by a desired mask density d which measures the percentage of mask pixels w.r.t. the number of image pixels.

One strategy for mask optimisation has been proposed by Belhachmi et al. [5]. They show that an optimal mask in the continuous setting can be obtained from the rescaled Laplacian magnitude of the image. However, transferring these results to the discrete setting often suffers from suboptimal dithering strategies. While being highly efficient, reconstruction quality is not fully satisfying.

Better quality can be achieved with the popular stochastic strategies of Mainberger et al. [22]. First, one employs *probabilistic sparsification (PS)*: Starting with a full mask, one removes a fraction p of candidate pixels and computes the inpainting. Then one reintroduces a fraction q of the candidates with the largest local inpainting error. One repeats this step until reaching a desired mask density d.

Since sparsification is a greedy local approach, it can get trapped in bad local minima. As a remedy, Mainberger et al. [22] also propose a *nonlocal pixel exchange (NLPE)*. Pixel candidates in a sparsified mask are exchanged for an equally large set of non-mask pixels. If the new inpainting result improves, the exchange is kept, otherwise it is discarded. In theory, NLPE can only improve the mask, but in practice convergence is slow.

The use of PS and NLPE requires to solve the inpainting problem numerous times, leading to slow mask optimisation. To avoid this computational bottleneck, we want to reach the quality of stochastic mask optimisation with a more efficient model based on deep learning.

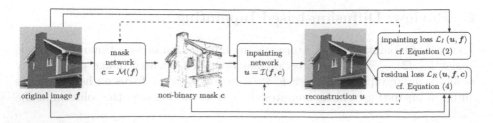

Fig. 1. Overview over our model structure. Solid lines denote forward passes, dashed lines denote backpropagation.

3   Sparse Masks with Surrogate Inpainting

Our model consists of two equally shaped U-nets [28] with different loss functions. By optimising both inpainting quality and fidelity to the inpainting equation, we obtain masks with good reconstruction quality for the inpainting problem.

3.1   The Mask Network

The *mask network* takes an original image f and transforms it into a mask c. We denote the forward pass through the mask network by $\mathcal{M}(\cdot)$, i.e. the mask is computed as $c = \mathcal{M}(f)$.

The mask entries lie in the interval $[0, 1]$. Permitting non-binary values allows for a differentiable network model. To obtain mask points in the desired range, we apply a sigmoid function to the output of the network. Moreover, the mask network is trained for a specific mask density d. To this end, we rescale the outputs of the network if they exceed the desired density. We do not require a lower bound, since the loss function incites a sufficiently dense mask.

The mask network places the known data such that the inpainting error between the reconstruction u and the original image f is minimised. This yields the *inpainting loss*

$$\mathcal{L}_I(u, f) = \frac{1}{n_x n_y} \|u - f\|_2^2 \tag{2}$$

as its objective function where $\| \cdot \|_2$ is the Euclidean norm. Its implicit dependency on the inpainting mask links the learned masks to the reconstructions.

We found that the mask network tends to get stuck in local minima with flat masks which are constant at every position, yielding a random sampling. To avoid this, we add a regularisation term $\mathcal{R}(c)$ to the inpainting loss $\mathcal{L}_I(u, f)$. It penalises the inverse variance of the mask via $\mathcal{R}(c) = \left(\sigma_c^2 + \epsilon\right)^{-1}$ where a small constant ϵ avoids division by zero. The variance of a mask describes how strongly the confidence measures of the individual pixels differ from the mean probability. Thus, the regulariser serves two purposes: First, it lifts the bad local minima for flat masks by adding a strong penalty to the energy. Second, it promotes probabilities closer to 0 and 1, as this maximises the variance. The impact of the regularisation term is steered by a positive regularisation parameter α.

3.2 The Inpainting Network

The second network is called the *inpainting network*. Its task is to create a reconstruction u which follows a classical inpainting process. In [2], it has been shown that U-nets realise an efficient multigrid strategy at their core. Thus, we use a U-net as a surrogate solver which reproduces the results of the PDE-based inpainting. The inpainting network takes the original image f and the mask c and creates a reconstruction $u = \mathcal{I}(f, c)$. This result should solve the discrete version of the inpainting Eq. (1) which reads

$$(I - C)\, Au - C\,(u - f) = 0. \tag{3}$$

Here, A is a discrete implementation of the Laplacian Δ with reflecting boundary conditions, and $C = \mathrm{diag}(c)$ is a matrix representation of the mask. To ensure that the reconstruction u approximates a solution to this equation, we minimise its residual, yielding the *residual loss*

$$\mathcal{L}_R(u, f, c) = \frac{1}{n_x n_y}\| (I - C)\, Au - C\,(u - f) \|_2^2. \tag{4}$$

As the residual loss measures fidelity to the PDE-based process, an optimal network approximates the PDE solution in an efficient way that allows fast backpropagation. This strategy has been proposed in [2] and is closely related to the idea of deep energies [14].

Figure 1 presents an overview of the full model structure. Note that the inpainting network receives both the mask and the original image as an input. Thus, this network is not designed for standalone inpainting. However, this allows the network to easily minimise the residual loss by transforming the original into an accurate inpainting result, given the mask as side information.

3.3 Practical Application

After training the full pipeline in a joint fashion, the mask network can be used to generate masks for homogeneous diffusion inpainting. To this end, we apply the mask network to an original image and obtain a non-binary mask. This mask is then binarised: The probability of a pixel belonging to a mask is given by its non-binary value. At each position, we perform a weighted coin flip with that probability. Afterwards, the binarised masks are fed into a numerical solver of choice for homogeneous diffusion inpainting.

While binarising the mask is not necessary in this pure inpainting framework, it is important for compression applications since storing binary masks with arbitrary point distributions is already highly non-trivial [24].

4 Experiments

4.1 Experimental Setup

We train both U-nets jointly with their respective loss function on the BSDS500 dataset [4] which contains a broad selection of natural images. As a training set,

boat   cameraman   house   trui   peppers

Fig. 2. Test images of resolution 256 × 256.

we use 200 cropped grey value images of size 256 × 256 with values in the range [0, 255]. We do not use a validation set as the training process is fully fixed.

Both U-nets employ 5 scales, with 3 layers per scale. On the finest scale, they use 10 channels, and this number is doubled on each scale. Thus, each U-net possesses around 9×10^5 parameters. We use the Adam optimiser [20] with standard settings, a learning rate of $5 \cdot 10^{-4}$, and 4000 epochs. As a regularisation parameter we choose $\alpha = 0.01$. We found this combination of hyperparameters to work well in practice. We train multiple instances of the model for densities between 10% and 1% with several random initialisations.

After training, we binarise the masks and use them with a conjugate gradient solver for homogeneous diffusion inpainting to obtain a reconstruction. Since we aim at the highest quality, we take the best result out of 30 samplings.

Analogously, we generate masks with PS as well as with PS with additional NLPE. In the following, we denote the latter combination by PS + NLPE. In our sparsification we use candidate fractions $p = 0.1$ and $q = 0.05$ as suggested by Mainberger et al. [22], and we take the best result out of 5 runs. For NLPE, we use 30 candidates of which 10 are exchanged. We run NLPE for 10 cycles: In a single cycle, each mask point is exchanged once on average. Moreover, we compare against the strategy of Belhachmi et al. [5]. This approach is realised by taking the Laplacian magnitude of the image, rescaling it to obtain a desired density, and dithering the result with a binary Floyd–Steinberg algorithm [12].

We compare our results on five popular test images (see Fig. 2), since performing PS and NLPE on a large database is infeasible. We measure the quality in terms of peak signal-to-noise ratio (PSNR). Higher values indicate better quality.

4.2   Reconstruction Quality

Figure 3 shows a visual comparison of optimised masks and the corresponding inpainting results. For both test cases, we observe that our learned masks are structurally similar to those obtained by PS with NLPE. This helps to create sharper contours, whereas the inpainting results of Belhachmi et al. suffer from fuzzy edges. The visual quality of the inpainting results for our model and PS+NLPE is indeed competitive.

Figure 4(a) presents a comparison of the reconstruction quality averaged over the test images. Our learned masks consequently outperform the strategy of

original   Belhachmi et al.   PS + NLPE   our model

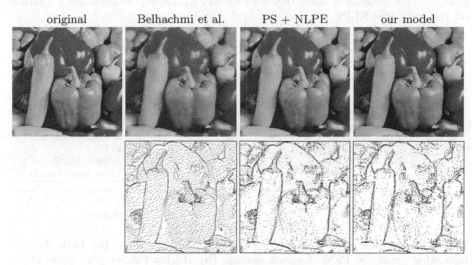

(a) *peppers* with 8% density

original   Belhachmi et al.   PS + NLPE   our model

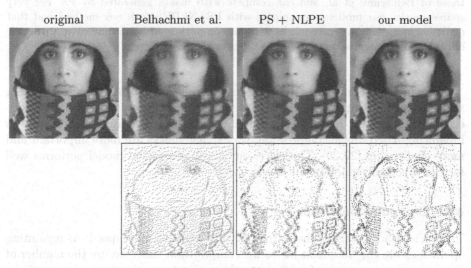

(b) *trui* with 5% density

Fig. 3. Visual comparison of inpainting results on two exemplary images for different mask densities. Mask points are shown in black, and mask images are framed for better visibility. Top rows depict the inpainting results, and bottom rows display the masks, respectively. The learned masks yield inpainting results which are visually comparable to PS with additional NLPE.

Belhachmi et al. Moreover, our model is on par with PS for densities smaller than 5%. For extremely small densities up to 2%, it even outperforms PS and is on par with PS+NLPE.

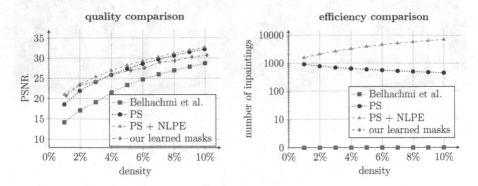

Fig. 4. Comparison of models in terms of quality and efficiency. (a) **Left:** Average inpainting quality in PSNR for each density. (b) **Right:** Efficiency in terms of the number of inpaintings for each density. The learned masks consistently outperform those of Belhachmi et al. and can compete with masks generated by PS. For very sparse masks, our model can compete with PS+NLPE. Both our method and that of Belhachmi et al. generate masks without computing an inpainting. The stochastic optimisation strategies compute up to thousands of inpaintings.

For larger mask densities, the margin between the methods becomes smaller, and our model cannot outperform its stochastic counterparts. Still, all models produce a good reconstruction quality. However, for applications such as inpainting-based image compression, very sparse masks are more important and more challenging [22,29]. Therefore, our mask generation model performs well for the practically relevant mask densities.

4.3   Computational Efficiency

The decisive advantage of the learned mask generation is its speed. As inpainting operations are the dominant factor for computation time, we use the number of inpaintings as a measure for efficiency. In comparison, the forward pass of the mask network is negligible.

Figure 4(b) visualises the average number of inpaintings required to obtain masks of a specific density for the test set. To generate a mask, both our model and that of Belhachmi et al. do not require any inpainting operations. Thus, the efficiency of these mask generation strategies does not depend on the density.

For PS, lower densities require more inpainting operations. Adding NLPE requires even more inpaintings depending on the number of cycles and the mask density. Both strategies trade computational efficiency for inpainting quality.

For example, a single sparsification run for a 3% mask on the *cameraman* image with realistic parameter settings requires 700 steps. On an *Intel Core i7-7700K CPU @ 4.20* GHz, this amounts to 58 s of runtime. The subsequent NLPE optimisation requires another 2000 steps, resulting in more than 3 min of additional runtime. In contrast, the strategy of Belhachmi et al. does not require any inpainting, and a mask can be generated in only 24 ms.

Our model requires only 85 ms for passing a single image through the mask network on the CPU. Thus, it plays in the same league as the strategy of Belhachmi et al., while being on par with the stochastic optimisation in terms of quality. This allows instantaneous high quality mask generation. As a consequence, our learned model can serve as a highly efficient replacement of stochastic mask optimisation.

5   Conclusions

We have proposed the first approach of sparse mask learning for diffusion-based inpainting. It fuses ideas from deep learning with classical homogeneous diffusion inpainting. The key of this strategy is a combination of an inpainting loss for the mask generator and a residual loss for the surrogate inpainting network. Its results are competitive with stochastic mask optimisation, while being up to four orders of magnitude faster. This constitutes a new milestone in mask optimisation for diffusion-based inpainting.

We are currently extending this idea to more sophisticated inpainting operators, as well as to further optimisations of the network architecture. We hope that this will pave the way to overcome the current time-consuming data optimisation strategies and will become an essential component for real-time diffusion-based codecs in hitherto unmatched quality.

References

1. Adam, R.D., Peter, P., Weickert, J.: Denoising by inpainting. In: Lauze, F., Dong, Y., Dahl, A.B. (eds.) Scale Space and Variational Methods in Computer Vision. LNCS, vol. 10302, pp. 121–132. Springer, Cham (2017)
2. Alt, T., Schrader, K., Augustin, M., Peter, P., Weickert, J.: Connections between numerical algorithms for PDEs and neural networks. arXiv:2107.14742v1 [math.NA], July 2021
3. Andris, S., Peter, P., Mohideen Kaja Mohideen, R., Weickert, J., Hoffmann, S.: Inpainting-based video compression in FullHD. In: Elmoataz, A., Fadili, J., Quéau, Y., Rabin, J., Simon, L. (eds.) SSVM 2021. LNCS, vol. 12679, pp. 425–436. Springer, Cham (2021). https://doi.org/10.1007/978-3-030-75549-2_34
4. Arbelaez, P., Maire, M., Fowlkes, C., Malik, J.: Contour detection and hierarchical image segmentation. IEEE Trans. Pattern Anal. Mach. Intell. **33**(5), 898–916 (2011)
5. Belhachmi, Z., Bucur, D., Burgeth, B., Weickert, J.: How to choose interpolation data in images. SIAM J. Appl. Math. **70**(1), 333–352 (2009)

6. Bonettini, S., Loris, I., Porta, F., Prato, M., Rebegoldi, S.: On the convergence of a linesearch based proximal-gradient method for nonconvex optimization. Inverse Probl. **33**(5), 055005 (2017)
7. Chen, Y., Ranftl, R., Pock, T.: A bi-level view of inpainting-based image compression. In: Kúkelová, Z., Heller, J. (eds.) Proceedings 19th Computer Vision Winter Workshop. Křtiny, Czech Republic, Feburary 2014
8. Chizhov, V., Weickert, J.: Efficient data optimisation for harmonic inpainting with finite elements. In: Tsapatsoulis, N., Panayides, A., Theocharides, T., Lanitis, A., Pattichis, C., Vento, M. (eds.) CAIP 2021. LNCS, vol. 13053, pp. 432–441. Springer, Cham (2021). https://doi.org/10.1007/978-3-030-89131-2_40
9. Dai, Q., Chopp, H., Pouyet, E., Cossairt, O., Walton, M., Katsaggelos, A.K.: Adaptive image sampling using deep learning and its application on X-ray fluorescence image reconstruction. IEEE Trans. Multimedia **22**(10), 2564–2578 (2019)
10. Daropoulos, V., Augustin, M., Weickert, J.: Sparse inpainting with smoothed particle hydrodynamics. SIAM J. Appl. Math. **14**(4), 1669–1704 (2021)
11. Demaret, L., Dyn, N., Iske, A.: Image compression by linear splines over adaptive triangulations. Sign. Process. **86**(7), 1604–1616 (2006)
12. Floyd, R.W., Steinberg, L.: An adaptive algorithm for spatial grey scale. Proc. Soc. Inf. Disp. **17**, 75–77 (1976)
13. Galić, I., Weickert, J., Welk, M., Bruhn, A., Belyaev, A., Seidel, H.P.: Image compression with anisotropic diffusion. J. Math. Imaging Vis. **31**(2–3), 255–269 (2008)
14. Golts, A., Freedman, D., Elad, M.: Deep energy: task driven training of deep neural networks. IEEE J. Select. Top. Sign. Process. **15**(2), 324–338 (2021)
15. Hoeltgen, L., et al.: Optimising spatial and tonal data for PDE-based inpainting. In: Bergounioux, M., Peyré, G., Schnörr, C., Caillau, J.P., Haberkorn, T. (eds.) Variational Methods in Imaging and Geometric Control, Radon Series on Computational and Applied Mathematics, vol. 18, pp. 35–83. De Gruyter, Berlin (2017)
16. Hoeltgen, L., Weickert, J.: Why does non-binary mask optimisation work for diffusion-based image compression? In: Tai, X.-C., Bae, E., Chan, T.F., Lysaker, M. (eds.) EMMCVPR 2015. LNCS, vol. 8932, pp. 85–98. Springer, Cham (2015). https://doi.org/10.1007/978-3-319-14612-6_7
17. Iijima, T.: Basic theory on normalization of pattern (in case of typical one-dimensional pattern). Bull. Electrotech. Lab. **26**, 368–388 (1962). in Japanese
18. Isogawa, K., Ida, T., Shiodera, T., Takeguchi, T.: Deep shrinkage convolutional neural network for adaptive noise reduction. IEEE Sign. Process. Lett. **25**(2), 224–228 (2017)
19. Karos, L., Bheed, P., Peter, P., Weickert, J.: Optimising data for exemplar-based inpainting. In: Blanc-Talon, J., Helbert, D., Philips, W., Popescu, D., Scheunders, P. (eds.) ACIVS 2018. LNCS, vol. 11182, pp. 547–558. Springer, Cham (2018). https://doi.org/10.1007/978-3-030-01449-0_46
20. Kingma, D.P., Ba, J.: Adam: A method for stochastic optimization. In: Proceedings of 3rd International Conference on Learning Representations. San Diego, CA, May 2015
21. Liu, H., Jiang, B., Xiao, Y., Yang, C.: Coherent semantic attention for image inpainting. In: Proceedings 2019 IEEE/CVF International Conference on Computer Vision, pp. 4170–4179. Seoul, Korea, October 2017
22. Mainberger, M., et al.: Optimising Spatial and Tonal Data for Homogeneous Diffusion Inpainting. In: Bruckstein, A.M., ter Haar Romeny, B.M., Bronstein, A.M., Bronstein, M.M. (eds.) SSVM 2011. LNCS, vol. 6667, pp. 26–37. Springer, Heidelberg (2012). https://doi.org/10.1007/978-3-642-24785-9_3

23. Marwood, D., Massimino, P., Covell, M., Baluja, S.: Representing images in 200 bytes: Compression via triangulation. In: Proceedings of 2018 IEEE International Conference on Image Processing, pp. 405–409. Athens, Greece, October 2018
24. Mohideen, R.M.K., Peter, P., Weickert, J.: A systematic evaluation of coding strategies for sparse binary images. Sign. Process. Image Commun. **99**, 116424, November 2021
25. Ochs, P., Chen, Y., Brox, T., Pock, T.: iPiano: inertial proximal algorithm for nonconvex optimization. SIAM J. Imaging Sci. **7**(2), 1388–1419 (2014)
26. Pathak, D., Krähenbühl, P., Donahue, J., Darrell, T., Efros, A.A.: Context encoders: feature learning by inpainting. In: Proceedings 2016 IEEE Conference on Computer Vision and Pattern Recognition, pp. 2536–2544. Las Vegas, NV, June 2016
27. Peter, P., Hoffmann, S., Nedwed, F., Hoeltgen, L., Weickert, J.: Evaluating the true potential of diffusion-based inpainting in a compression context. Sign. Process. Image Commun. **46**, 40–53 (2016)
28. Ronneberger, O., Fischer, P., Brox, T.: U-Net: convolutional networks for biomedical image segmentation. In: Navab, N., Hornegger, J., Wells, W.M., Frangi, A.F. (eds.) MICCAI 2015. LNCS, vol. 9351, pp. 234–241. Springer, Cham (2015). https://doi.org/10.1007/978-3-319-24574-4_28
29. Schmaltz, C., Peter, P., Mainberger, M., Ebel, F., Weickert, J., Bruhn, A.: Understanding, optimising, and extending data compression with anisotropic diffusion. Int. J. Comput. Vis. **108**(3), 222–240 (2014)
30. Ulyanov, D., Vedaldi, A., Lempitsky, V.: Deep image prior. In: Proceedings of 2018 IEEE Conference on Computer Vision and Pattern Recognition, pp. 9446–9454. Salt Lake City, UT, June 2018
31. Vašata, D., Halama, T., Friedjungová, M.: Image inpainting using Wasserstein generative adversarial imputation network. In: Farkaš, I., Masulli, P., Otte, S., Wermter, S. (eds.) ICANN 2021. LNCS, vol. 12892, pp. 575–586. Springer, Cham (2021). https://doi.org/10.1007/978-3-030-86340-1_46
32. Weickert, J., Welk, M.: Tensor field interpolation with PDEs. In: Weickert, J., Hagen, H. (eds.) Visualization and Processing of Tensor Fields, pp. 315–325. Springer, Berlin (2006)
33. Xie, J., Xu, L., Chen, E.: Image denoising and inpainting with deep neural networks. In: Bartlett, P.L., Pereira, F.C.N., Burges, C.J.C., Bottou, L., Weinberger, K.Q. (eds.) Proceedings of 26th International Conference on Neural Information Processing Systems. Advances in Neural Information Processing Systems, vol. 25, pp. 350–358. Lake Tahoe, NV, December 2012
34. Yang, C., Lu, X., Lin, Z., Shechtman, E., Wang, O., Li, H.: High-resolution image inpainting using multi-scale neural patch synthesis. In: Proceedings of 2017 IEEE Conference on Computer Vision and Pattern Recognition, pp. 6721–6729. Honolulu, HI, July 2017
35. Yu, J., Lin, Z., Yang, J., Shen, X., Lu, X., Huang, T.S.: Generative image inpainting with contextual attention. In: Proceedings of 2018 IEEE Conference on Computer Vision and Pattern Recognition, pp. 5505–5514. Salt Lake City, UT, June 2018

Extracting Descriptive Words
from Untranscribed Handwritten Images

Jose Ramón Prieto[1]([⊠])(iD), Enrique Vidal[1](iD), Joan Andreu Sánchez[1](iD),
Carlos Alonso[2](iD), and David Garrido[3]

[1] PRHLT Research Center, Universitat Politècnica de València, Valencia, Spain
{joprfon,evidal,jandreu}@prhlt.upv.es
[2] tranSkriptorium AI, Valencia, Spain
[3] HUM313 Research Group, Universidad de Cádiz, Cádiz, Spain

Abstract. Extracting descriptive text from manuscripts to be included
in the manuscript metadata is an important task that is generally performed in archives and libraries by experts with a wealth of knowledge
on the manuscripts contents. Unfortunately, many manuscript collections
are so vast that it is not feasible to rely solely on experts to perform this
task. To our knowledge, this is the first work aiming at automatic extraction of descriptive text from untranscribed text images. To attempt dealing with such a task, a first step would be to transcribe the handwritten
images into text – but achieving sufficiently accurate transcripts is generally unfeasible for large sets of historical manuscripts. We propose new
approaches to automatically extract descriptive words which do not rely
on any explicit image transcripts. They are based on "probabilistic indexing", a relatively novel technology which allows to effectively represent
the intrinsic word-level uncertainty generally exhibited by handwritten
text images. We assess the performance of this approach on samples
of a large collection of complex manuscripts from the *Spanish Archivo
General de Indias*. Since no standard metrics exist for the novel task
considered in this work, we propose two new evaluation measures which
aim at measuring the quality of the detected descriptive words in terms
close to practical usage of these words. Using these metrics we report
promising preliminary results.

Keywords: Descriptive words · Content-based image retrieval ·
Historical manuscripts

1 Introduction

We consider the task of automatically finding a set of words that serve to identify
a file, folder or bundle (hereafter uniquely referred to as a "bundle"[1]) of untranscribed handwritten page images, with respect to other bundles of the collection.
This task has countless applications in libraries and archives where billions of

---

[1] We will nevertheless use the time-honored term "document" when dealing with plain
text rather than text images.

© Springer Nature Switzerland AG 2022
A. J. Pinho et al. (Eds.): IbPRIA 2022, LNCS 13256, pp. 540–551, 2022.
https://doi.org/10.1007/978-3-031-04881-4_43

manuscripts are stored without a sufficiently useful identification and/or digest of their contents. To our knowledge, no previous works have approached this difficult problem. The current commonly accepted wisdom would be to split the process into two sequential stages. First, a handwritten text recognition (HTR) system should be used to transcribe the images into text and, second, traditional text analytics methods would be tried on the resulting text documents.

This approach might be reasonable for simple manuscripts with uniform writing style and good quality images, where automatic handwritten text recognition (HTR) can provide highly precise transcripts with over 90% word recognition accuracy [13]. It can of course be reasonable also for small-scale collections, where manual correction of HTR errors can be affordable.

But this is not an option for countless large historical collections of, say, hundreds of thousands of images. Moreover, for many of these collections, the best available HTR systems can only provide word recognition accuracies as low as 50–70% (see e.g. the ICDAR-2015 benchmark [13]). An example of this situation is the CARABELA collection[2] , considered in this paper. It encompasses more than the 125 000 complex page images [4] and the average word recognition accuracy achieved in optimistic laboratory conditions is 65% [12], dropping to 46% when conditions are closer to real-world usage [4]. Similar cases abound; for instance the Chancery collection [2], the Bentham Papers [14], etc.

Clearly, for these kinds of manuscript collections, the aforementioned two-stage idea is to be ruled out and more holistic approaches should be devised. To the best of our knowledge, this is the first paper proposing, developing and testing these kinds of approaches on a large manuscript dataset.

The approach here proposed strongly relies on the so-called *probabilistic indexing* (PrIx) technology, recently developed to deal with the intrinsic word-level *uncertainty* generally exhibited by handwritten text and, more so, by historical handwritten text images [2,8,11,15,16]. The probabilistic index of a text image can be seen as a "heat-map" image representation which highlights positions of words and "pseudo-word" character sequences which were likely written in the original manuscript. This technology was primarily developed to allow search and retrieval for textual information in untranscribed manuscript collections. In fact, it has recently been successfully applied to allow textual searching in several large collections of untranscribed manuscripts[3] [2,4,14].

But probabilistic indexing can go far beyond search and retrieval applications: Since a probabilistic index of a text image provides a distribution of likely words, it allows to properly estimate statistical expectations of the text features typically required by most plaintext analytics methods.

This is the main idea proposed in this paper. We explore two approaches to extract relevant, distinguishing words from untranscribed text images, based on word frequencies which are estimated using the probabilistic information provided by PrIx's. The first approach consists in a rather direct use of word *information gain*, estimated from PrIx's. The second, more sophisticated method,

---

[2] http://prhlt-carabela.prhlt.upv.es/carabela.

[3] http://prhlt-carabela.prhlt.upv.es/PrIxDemos provides a list of public search interfaces for these collections.

derives the distinguishing words from the weights of a Neural Network classifier trained to predict the bundle text images belong to.

We report empirical results which are still preliminary but clearly suggest the feasibility of automatic extraction of descriptive words from untranscribed sets of text images.

2   Background

Much of the background to this work is provided in [10] and [5]. Here we summarize only the essential concepts and formulation needed to present the methods proposed in this paper.

2.1   Probabilistic Indexing of Handwritten Text Images

In order to deal with the intrinsic word-level uncertainty, usually exhibited in images of historical manuscripts, the Probabilistic Indexing (PrIx) framework was proposed. In PrIx, image elements that can be interpreted as words are detected and stored, along with their *relevance probability* (RP) and location in the image. These text elements are referred to as *"pseudo-word spots"*.

In [11,16], the RP of an image region x for a pseudo-word v is denoted as $P(R = 1 \mid X = x, V = v)$, but for the sake of conciseness, we will write it just as $P(R|x,v)$. RP's can be computed using the approximations discussed in [11,17]. See also [10] for a brief account of concepts and methods of the PrIx framework.

This word indexing approach has proved to be very robust, and it has been used to very successfully index several large iconic manuscript collections, such as the French CHANCERY collection [2], the BENTHAM PAPERS [14], and the Spanish CARABELA collection considered in this paper, among others (See: footnote 3).

PrIx technology is also very flexible. Since it provides a distribution of words likely written in the images, it can be used to compute statistical expectations of the text features required by traditional plaintext analytics techniques. We capitalize on this flexibility to develop the methods proposed in this paper.

2.2   Document Representation and Classification

If a text document is given in some electronic form, its words can be trivially identified as discrete, unique elements, and then the whole field of *text analytics* [1,9] is available to approach many document processing problems, including *document classification* (DC). Most DC methods assume a document representation model known as *vector model* or *bag of words* (BOW) [1,7,9]. In this model, the order of words in the text is ignored, and a document is represented as a *feature vector* indexed by the words in a vocabulary V.

However, conventional DC methods are *not* directly suitable to deal with the word uncertainty underlying historical manuscripts. For this reason, we will rely on the probabilistic information provided by PrIx pseudo-word spots, and follow the feature selection and feature extraction methods proposed in [10].

Feature selection is based on Information Gain (IG), which is computed for all the (pseudo-)words in V. Then a "good" vocabulary V_n, of reasonable size $n < N$, is determined by selecting from V the n words with largest IG.

Using the notation of [10], let t_v be the value of a boolean random variable that is *True* iff, for some random document D in a document collection $\mathcal{D}$, the word v appears in D. So, $P(t_v)$ is the probability that $\exists D \in \mathcal{D}$ such that v is used in D, and $P(\bar{t}_v) = 1 - P(t_v)$ is the probability that *no* document uses v. The IG of a word v is then defined as:

$$
\begin{aligned}
\mathrm{IG}(v) = &- \sum_{c \in C} P(c) \log P(c) \\
&+ P(t_v) \sum_{c \in C} P(c \mid t_v) \log p(c \mid t_v) \\
&+ P(\bar{t}_v) \sum_{c \in C} P(c \mid \bar{t}_v) \log P(c \mid \bar{t}_v)
\end{aligned}
\tag{1}
$$

where $P(c)$ is IE prior probability of class c, $P(c \mid t_v)$ is the conditional probability that a document belongs to class c, given that it contains the word v, and $P(c \mid \bar{t}_v)$ is the conditional probability that a document belongs to class c, given that it does *not* contain v. All these probabilities are estimated as explained in [10], using the RP's provided by the PrIx's of the text images considered.

On the other hand, conventional Tf·Idf (term frequency weighted by inverse document frequency) [9] is used as *feature extraction* to represent each document under the BOW model. Again, as discussed in [10], word and document frequencies needed to compute Tf·Idf are estimated from the probabilistic information provided by the PrIx pseudo-word spots.

Finally, from the classifiers studied in [10], the Plain Multiclass Perceptron (MCP, or MLP-0 in [10]) is used in the present work. As discussed later, the simplicity of this model allows us to adequately interpret the trained weights in order to extract descriptive words.

3   Descriptive Words from IG and MCP Weights

The task considered in this work is to determine, for each image bundle a set of *descriptive words* (DWs); i.e., words which adequately describe the bundle by distinguishing it from other bundles of the collection. This suggests to more or less directly address this problem as a classification problem where each class is associated with a single bundle and "documents" are subsets of page images of the different bundles. Following this idea, two approaches are proposed below, one based on IG and another on the analysis of the weights of an MCP classifier.

3.1   Using Information Gain

A simple approach for DW extraction can be derived from Eq. (1). Note that the last two addends in this expression could not contribute equally to decrease

IG(v). In fact, it can be easily shown that the lesser decrement in IG is obtained when all involved conditional probabilities are equal. This fact can be interpreted in a reverse way, that is, the more a word v appears in a class and the less in the other classes, the large the IG(v) is since it discriminates better the classes in which it appears. These considerations can be used to select a subset of words that better describe a specific class (i.e., a bundle), c. To this end, words included in the "good" vocabulary V_n defined in Sect. 2.2 are sorted according to:

$$S_c(v) \overset{\text{def}}{=} P(t_v)P(c \mid t_v) \log P(c \mid t_v) \tag{2}$$

and the N words with largest values of $S_c(v)$ are selected as DWs of the class c.

3.2   Using Perceptron Weights

The MCP can be trained to minimize the cross-entropy between the desired outputs and the outputs provided by the model. Since each output neuron has a weighted connection to all the inputs, this learning criterion leads to adjusting the weights so that they become high for the input features (words) that are useful for the target class and low, maybe negative, for those that are not useful in that class.

However, the training of MCP weights does not care about the weight associated to a given input (word) v for a class c to be comparable with the corresponding weight of the same input for another class c'. So, some form of weight "shift" is needed. Let w'_{vc} be the MCP weight for input word v and output class c. To shift these weights, we subtract from each w'_{vc} the class-wise average weight for each word v. That is:

$$w_{vc} = w'_{vc} - \frac{1}{C} \sum_{c'=1}^{C} w'_{vc'}, \quad 1 \leq c \leq C, \quad v \in V_n \tag{3}$$

Then, for each class c, the weights w_{vc} are sorted in decreasing order and the words v associated with the N largest weights are selected as DWs of class c.

Now remember that our task is to determine a set of DWs per bundle. Therefore, as discussed at the beginning of this section, each class is associated with a single bundle and the MCP is trained with samples that are Tf·Idf representations of small sets of page images of the collection, each set class-labeled with the identifier of the bundle the pages belong to.

It is worth noting that here we are not interested in using such a classifier to actually classify images into bundles – even though such an application might actually have a great interest in some scenarios related with the organization of unstructured manuscript collections in archives and libraries. Instead, as explained above, we will use the weights of an MCP trained with all the images of a collection in order to find, for each bundle, words that convey enough "semantic" information to "describe" it.

4   Using Descriptive Words for Bundle Retrieval

A set of DWs for each bundle can be used for many purposes. The most obvious one, mainly considered in this paper, is to use them as metadata to help identifying the content of image bundles. As discussed later, objectively assessing performance in such an application is fairly elusive. Another use of the DWs is searching for specific bundles in a large manuscript collection. To this end we can resort to the primary use of PrIx's and search for images which contain specific words. If we use enough words from the DWs of an image bundle X for this purpose, we expect that X will be accurately retrieved, as far as the DWs are actually a good digest of the textual contents of X.

Apart from the interest of this scenario to provide simple access to handwritten information in archives and libraries, it offers a way to objectively assess the quality of a set of DWs, as discussed later.

To develop this idea, let V_{X_N} the set of N best DWs of X. We can issue an AND query [15], $q(V_{X_N})$, with all the words in V_{X_N}, extended to all the images over the whole bundle collection $\mathcal{X}$. The relevance probability of a bundle $X \in \mathcal{X}$ for the AND query $q(V_{X_N})$ can be approximately computed, like in [15], as the minimum over all the words in V_{X_N} of the relevance probability of X for v. Also as in [15], single-word relevance probabilities over a bundle X can be approximated as the maximum of the relevance probability of each $x \in X$ for v. Wrapping it all up:

$$P(R \mid X, q(V_{X_N})) \approx \min_{v \in V_{X_N}} \max_{x \in X} P(R \mid x, v) \qquad (4)$$

It should be taken into account, however, that in contrast with the size (number running words) of individual page images, which tends to be relatively uniform across a given collection, the size of bundles can vary dramatically (from 5 page images to more than 2000 in the collection considered in this work). Therefore, the simple $\max - \min$ approximation outlined above needs to be tweaked to take into account that it is more likely to find words in a large bundle than in another having only a few page images. For the experiments presented in Sect. 6.1, the following simple heuristic "normalization" is applied:

$$\hat{P}(R \mid X, q(V_{X_N})) = \exp\left(-\frac{\hat{n}(X)}{\hat{n}}\right) P(R \mid X, q(V_{X_N})) \qquad (5)$$

where $\hat{n}(X)$ is the estimated number of running words in X and $\hat{n}$ is the average estimated number of running words per bundle (34 890, see Table 1).

5   Dataset, Empirical Settings and Assessment Metrics

Here we provide details of the setting adopted for the experiments, including the dataset and the metrics used to assess the proposed approaches.

5.1   Dataset

The images used in the present work correspond to bundles from the fabled Spanish *Archivo General de Indias* (AGI) and *Archivo Histórico Provincial de Cádiz* (AHPC). These massive archives hold key information about Spanish travels and naval commerce during the XV-XIX centuries.

The CARABELA collection [4] consists of selected bundles from these archives, with more than 125 000 images of manuscripts of interest to underwater archaeology. This collection has been considered one of the most challenging sets of historical manuscripts due to the poor quality of the scanned images, the heavy use of archaisms and non-standard abbreviations, etc. [4,12]. In this work, only the AGI part is considered, mainly because most its bundles are accompanied with some descriptive metadata which, to some extent, can be used for evaluation purposes. Table 1 (upper, left) shows basic statistics of this corpus.

Many AGI bundles are very small (just a few pages, several of which contain no text). Therefore, only those bundles which have a minimum amount of text have been selected for the experiments of this work. This resulted in a much smaller number of useful bundles; namely only 165, each having at least 10 handwritten page images. As mentioned above, this means that the number of classes for computing IG and training the MCP model is 165.

Table 1 (right) also shows relevant features of this selected dataset. It is worth noting that the selected bundles do preserve most of the substance of AGI; both in terms of handwritten text images and estimated number of running words.

As discussed in Sect. 3.2, the MCP classifier has to be trained to detect which bundle an image or small set of images belongs to. To this end, after a few tentative tests, the images of each bundle have been split into sets of sizes ranging from 2 to 5 page images. Features of these small image sets are also shown in the lower part of Table 1.

Table 1. Features of both the AGI corpus and the finally selected set of 165 bundles.

	Full AGI	Selected
Number of bundles	328	165
Total number of images	30 765	30 134
Average number of images per bundle	94	183
Estimated number of running words (RW)	5 879 328	5 756 948
Estimated avg. num. of RW per bundle	17 924	34 890
Estimated avg. num. of RW per image	191	191
Number of small sets of (2–5) images	–	6 142
Estimated avg. num. of RW per image set	–	937
min/std. deviation/max	–	5/477/3/817

Most AGI bundles come accompanied with metadata files which often include a short free-text description of the bundle contents. We capitalize on these metadata in one of the assessment metrics discussed below. These metadata were compiled long time ago by the Spanish Archives Portal (PARES) archivists and are available when performing a thematic search on the PARES interface.[4] Although such metadata provide subjective, maybe biased or incomplete information, many of the words included are actually useful for bundle description.

We have used a few of these descriptions to obtain some results using the second metric we propose in Sect. 5.3, below. For the present preliminary results, 6 bundles where selected, according to availability of clear metadata and diversity of contents, bundle sizes, and writing styles in the manuscripts. Descriptive text was manually extracted from each metadata file. Finally, the human-produced description texts for the 6 selected bundles contain 919 words in total, or 723 "useful" words, considering only words with more than 2 characters.

5.2   Experimental Settings

As mentioned earlier, 165 classes were defined, one per bundle. The machine learning task thus consists in training a MCP to classify groups of pages into one of these classes ($C = 165$).

Typically, PrIx vocabularies are huge because they are composed of large amounts of (pseudo-)word hypotheses. Many of these hypotheses have low relevance probability and most of the low-probability pseudo-words are not real words. Therefore, as a first step, we pruned out the huge PrIx vocabulary, avoiding words with less than three characters, as well as words with estimated document frequency lesser than 1.0. This resulted in a vocabulary of 230 704 pseudo-words. Further, pseudoword hypotheses with a relevance probability lower than 0.03 have been removed.

After performing some classification tests, it has been decided to use 2 048 input features (words), selected by IG, as mentioned in Sect. 2.2. Larger number of words did not show clear improvements.

The output of MCP classifier is a softmax layer with $C = 165$ units. Its parameters were initialized following [6] and trained by backpropagation according to the cross-entropy loss for 50 epochs. The ADAM optimizer [3] was used, with a learning rate of 0.1, $\beta_1 = 0.5$, $\beta_2 = 0.999$.

As explained in Sect. 3.2, we remind that our target task is *not* to actually classify page images into bundles. Therefore it makes not sense to split the dataset into training and test. Instead, the MCP is trained with all the data available. Clearly, we are not explicitly interested in using the model as an actual classifier, but in the most important words upon which the model would relay for classification. Hopefully, the model will learn to relay on the "most descriptive" words of the training documents.

---

[4] http://pares.mcu.es/ParesBusquedas20/catalogo/search.

5.3   Assessment Measures

Assessment Based on Bundle Retrieval Using DW Queries. As commented in Sect. 4, search and retrieval tests using DWs can provide information about the quality of automatically extracted DWs. If retrieval accuracy is high, DWs are considered good, the smaller number of DWs needed, the better. This metric directly aims at assessing the *distinguishing* capabilities of DWs.

Retrieval accuracy can be measured in many ways. Here we will just use the *retrieval error rate* for increasing number N of DWs and for a few amounts K of bundle hypotheses. That is, given an AND query using a number N of descriptive words extracted from a target bundle X_t, we obtain the K-best bundle hypotheses, $X_1, \ldots, X_k$. Then a retrieval error occurs if $X_t \neq X_i$, $1 \leq i \leq K$.

Assessment Based on Human-Produced Metadata Descriptions. Another (daring) idea to assess the quality of a set of DWs is to compute how many automatically extracted DWs were also used by archive experts to produce the descriptions included in the metadata of the same bundles (cf. Sect. 5.1). In contrast with the previous metric, this score aims more at assessing the *descriptive* capabilities of DWs.

Let $T(X)$ be the metadata, human-produced description of the bundle X and $W(X)$ the set of (unique) DWs automatically extracted from X. A metric, referred to as *"match"*, can be defined as the number of words in $W(X)$ which appear in $T(X)$ divided by the total number of running words in $T(X)$.

6   Empirical Results

Results for both metrics explained in Sect. 5.3 are reported in this section.

6.1   Accuracy of Bundle Retrieval Using DW AND-Queries

Table 2 shows bundle retrieval error rates of AND queries using DWs extracted directly using IG *per class* and IG–MCP weights (as discussed in Sect. 3), trained on the whole set of 165 AGI bundles.

Using only 8 DWs for each AND query, the IG *per class* approach achieves almost perfect 1-best accuracy. If 3-best bundle hypotheses are considered, results are perfect with 8 DWs and almost perfect with only 4 DWs per query. The IG–MCP approach also provides reasonably good results but IG *per class* clearly excels in this test – which mainly aims at measuring the *distinguishing* capabilities of DW.

For comparison, we also made an experiment where each set of N DWs was selected by random sampling the set of 2 048 words used to obtain the input feature vectors of the MCP. In this case, for all N, 1-best and 10-best retrieval errors were higher than 95% and 90%, respectively.

Table 2. AGI bundle retrieval error rates (in %) corresponding to AND queries using an increasing number (N) of DWs automatically extracted by two methods.

	IG–MCP				IG *per class*			
N	1-best	3-best	5-best	10-best	1-best	3-best	5-best	10-best
1	97.6	90.9	78.2	63.0	87.3	57.0	45.5	20.0
2	73.3	54.5	48.5	33.9	33.9	7.3	4.2	1.2
4	43.6	29.1	20.6	12.7	6.1	0.6	0.6	0.6
8	27.3	9.1	5.5	2.4	0.6	0.0	0.0	0.0
16	7.3	1.8	1.8	1.2	0.6	0.0	0.0	0.0
32	3.0	1.8	1.2	0.0	0.6	0.0	0.0	0.0
64	2.4	0.0	0.0	0.0	0.0	0.0	0.0	0.0

6.2 Evaluation by Metadata Human Description Matching Scores

For a few bundles, Table 3 reports metadata matching scores for DWs extracted by three methods: Random selection as in the previous section, IG *per class* as explained in Sect. 3.1, and use of IG–MCP trained weights, as explained in Sect. 3.2, For the random "method", the table shows the (rounded) average of 100 different random selections. Results are shown for the best 25, 50, 100 and 300 DWs automatically extracted by each method. The IG–MCP approach is clearly superior in this test – which is more aimed at measuring the *descriptive*, "semantic" quality of DWs.

Figure 1 illustrates how automatically extracted DWs using IG–MCP match human-produced metadata descriptive text for a specific bundle.

Table 3. How descriptive words automatically extracted from 6 image bundles match human-produced bundle metadata descriptive texts. *"match"* is the total number of matching words and the percentage is referred to the total of number of "useful" words in the selected bundles (723 running words). Higher percentage is better.

Number of DWs	25		50		100		300	
	match	(%)	match	(%)	match	(%)	match	(%)
IG + random	5	0.7	10	1.4	20	2.8	60	8.3
IG *per class*	10	1.4	27	3.7	30	4.1	127	17.6
IG–MCP weights	71	9.8	99	13.7	145	20.1	212	29.3

```
CARTA DE DOMINGO DE ZABALBURU GOBERNADOR DE FILIPINAS REPITIENDO LA
NOTICIA DADA EN 1705 EN EL GALEON QUE SE PERDIO SOBRE HABER SACADO
CUATRO PIEZAS DE ARTILLERIA EN LAS ISLAS MARIANAS Y DISPOSICIONES
DADAS PARA SACAR EL RESTO

ACOMPAÑA
CERTIFICACION DE LOS OFICIALES REALES DE HABERSE SACADO 17 PIEZAS DE
ARTILLERIA DE BRONCE Y 8 ANCLAS  MANILA 22 DE MARZO DE 1708  CARTA
DE DOMINGO DE ZABALBURU SOBRE HABER SACADO DE LA SARPANA GRANDE EN LAS
ISLAS MARIANAS 18 PIEZAS DE ARTILLERIA DE BRONCE DE LA NAO NUESTRA
SEÑORA DE LA CONCEPCION QUE SE PERDIERON EL AÑO 1638 MANILA 8 DE
JUNIO DE 1707
```

Fig. 1. The 300 most descriptive words extracted from the bundle FILIP,129,N.13 through IG–MCP weights are matched with the human-produced metadata descriptive text of the same bundle. The matching words are marked in red. (Color figure online)

7   Conclusion

Based on Probabilistic Indexes (PrIx's) automatically computed for each image of a large collection, two approaches have been proposed to extract descriptive, distinguishing words from bundles of untranscribed handwritten text images. These words allow finding a bundle by means of simple AND queries and can be used to automatically produce metadata information without image transcription or human intervention.

To assess the quality of these methods two new metrics are proposed, which aim at measuring the usefulness of the extracted words for practical applications. Preliminary good results using these metrics encourages us to follow this line of research in upcoming projects.

Acknowledgments. Work partially supported by the research grants: Ministerio de Ciencia Innovación y Universidades "DocTIUM" (RTI2018-095645-B-C22), Generalitat Valenciana under project DeepPattern (PROMETEO/2019/121) and PID2020-116813RB-I00a funded by MCIN/AEI/ 10.13039/501100011033.

The first author's work was partially supported by the Universitat Politècnica de València under grant FPI-I/SP20190010.

References

1. Aggarwal, C.C., Zhai, C.: Mining Text Data. Springer Science & Business Media (2012)
2. Bluche, T., et al.: Preparatory KWS experiments for large-scale indexing of a vast medieval manuscript collection in the HIMANIS project. In: 14th IAPR International Conference on Document Analysis and Recognition (ICDAR), vol. 01, pp. 311–316, November 2017
3. Diederik P., K., Ba, L.: Adam: a method for stochastic optimization. In: AIP Conference Proceedings, vol. 1631, pp. 58–62 (2014)
4. Vidal, E., et al.: The carabela project and manuscript collection: large-scale probabilistic indexing and content-based classification. In: Proceedings of 16th ICFHR (2020)

5. Flores, J.J., Prieto, J.R., Alonso, C., Garrido, D., Vidal, E.: Classification of untranscribed handwritten notarial documents by textual contents. In: Proceedings of the 2022 Iberian Conference on Pattern Recognition and Image Analysis (IbPRIA) (2022)

6. Glorot, X., Bengio, Y.: Understanding the difficulty of training deep feedforward neural networks. J. Mach. Learn. Res. **9**, 249–256 (2010)

7. Ikonomakis, M., Kotsiantis, S., Tampakas, V.: Text classification using machine learning techniques. WSEAS Trans. Comput. **4**(8), 966–974 (2005)

8. Lang, E., Puigcerver, J., Toselli, A.H., Vidal, E.: Probabilistic indexing and search for information extraction on handwritten German parish records. In: 2018 16th International Conference on Frontiers in Handwriting Recognition (ICFHR), pp. 44–49, August 2018

9. Manning, C.D., Raghavan, P., Schtze, H.: Introduction to Information Retrieval. Cambridge University Press, New York, NY, USA (2008)

10. Prieto, J.R., Bosch, V., Vidal, E., Alonso, C., Orcero, M.C., Marquez, L.: Textual-content-based classification of bundles of untranscribed manuscript images. In: 2020 25th International Conference on Pattern Recognition (ICPR), pp. 3162–3169. IEEE (2021)

11. Puigcerver, J.: A Probabilistic Formulation of Keyword Spotting. Ph.D. thesis, Univ. Politècnica de València (2018)

12. Romero, V., Toselli, A.H., Vidal, E., Sánchez, J.A., Alonso, C., Marqués, L.: Modern vs diplomatic transcripts for historical handwritten text recognition. In: Cristani, M., Prati, A., Lanz, O., Messelodi, S., Sebe, N. (eds.) ICIAP 2019. LNCS, vol. 11808, pp. 103–114. Springer, Cham (2019). https://doi.org/10.1007/978-3-030-30754-7_11

13. Sánchez, J.A., Romero, V., Toselli, A.H., Villegas, M., Vidal, E.: A set of benchmarks for handwritten text recognition on historical documents. Pattern Recogn. **94**, 122–134 (2019)

14. Toselli, A., Romero, V., Vidal, E., Sánchez, J.: Making two vast historical manuscript collections searchable and extracting meaningful textual features through large-scale probabilistic indexing. In: 2019 15th IAPR International Conference on Document Analysis and Recognition (ICDAR) (2019)

15. Toselli, A.H., Vidal, E., Puigcerver, J., Noya-García, E.: Probabilistic multi-word spotting in handwritten text images. Pattern Anal. Appl. **22**(1), 23–32 (2018). https://doi.org/10.1007/s10044-018-0742-z

16. Toselli, A.H., Vidal, E., Romero, V., Frinken, V.: HMM word graph based keyword spotting in handwritten document images. Inf. Sci. **370–371**, 497–518 (2016)

17. Vidal, E., Toselli, A.H., Puigcerver, J.: A probabilistic framework for lexicon-based keyword spotting in handwritten text images. Tech. rep, UPV (2017)

Other Applications

GMM-Aided DNN Bearing Fault Diagnosis Using Sparse Autoencoder Feature Extraction

Andrei Maliuk, Zahoor Ahmad, and Jong-Myon Kim[✉]

University of Ulsan, Ulsan 44610, South Korea
jmkim07@ulsan.ac.kr

Abstract. Deep learning techniques are gaining popularity due to their ability of feature extraction, dimensionality reduction, and classification. However, one of the biggest challenges in bearing fault diagnosis is reliable feature extraction. When using the bearing fault vibration spectrum, the deep neural network (DNN) model can learn the relationships in data that are unrelated to the task. In this work, a simple approach to bearing fault diagnosis using the elimination of unrelated data artifacts for DNN is proposed. The proposed fault diagnosis pipeline is explained and the comparison with popular fault diagnosis methods is performed.

Keywords: Fault diagnosis · Bearing · Vibration · Deep learning

1 Introduction

The rolling-element bearings are the most common industrial equipment which is necessary for rotary motion machines. In electric motors, bearings are mounted on both ends of the rotor shaft to provide smooth rotation. Especially in traction motors, bearings have very high requirements for reliability due to the sensitive areas of applications where a failure of the electric motor can lead to damage of the plant or even human casualties. Electric motors are gaining attraction over internal combustion engines due to their simple construction, however, a mechanical fault in the electric motor can result in catastrophic failure. Specifically, the rolling-element bearing faults are responsible for 45% of all the electric motor faults occurrences [1]. For this reason, condition monitoring of the bearings in electric motors is of primary importance.

The most common types of faults in bearing are outer and inner race faults. When a bearing is in the process of operation, rolling elements pass the damaged areas on the surface of the bearing race and create vibration impulses with a certain rate defined as fundamental defect frequency. In the case of the outer or inner race faults, the frequencies are the ball pass frequency of the outer race (BPFO) and the ball pass frequency of the inner race (BPFI) [2]. The definitions of these frequencies are given in Eqs. (1–2).

$$BPFO = N \times \left(\frac{S_{sh}}{2} \right) \left(1 - \frac{d_r}{D_p} \cos \phi \right), \tag{1}$$

$$BPFI = N \times \left(\frac{S_{sh}}{2} \right) \left(1 + \frac{d_r}{D_p} \cos \phi \right), \tag{2}$$

© Springer Nature Switzerland AG 2022
A. J. Pinho et al. (Eds.): IbPRIA 2022, LNCS 13256, pp. 555–564, 2022.
https://doi.org/10.1007/978-3-031-04881-4_44

where S_{sh} is a shaft speed in RPMs, d_r is the diameter of the rolling element, D_p is the pitch diameter.

In recent years bearing fault diagnosis methods can mostly be described as an application of Machine Learning techniques to the bearing operation history data such as vibration, current or acoustic emission data. The usual pipeline for these methods is signal preprocessing, feature extraction, feature selection, and classification. Signal processing methods commonly used for bearing fault diagnosis are envelope analysis, which has proven to be useful in analyzing low-amplitude high-frequency broadband signals containing bearing fault characteristics, Spectral kurtosis [3], which is capable of extracting the transient components masked by a noisy signal and can find the locations of transients in the frequency domain. Another technique is Wavelet Packet Transform based on Wavelet Transform and is used for signal multi-band filtering and denoising [4].

Following the signal preprocessing step, the feature extraction step allows to represent a time, frequency, or time-frequency domain signal using several statistical parameters of the signal to facilitate the work of the classifier. The feature selection process in its turn is used to choose the best features out of the feature pool and lower the feature space dimensionality. Xie and Zhang performed a comparative study of the most popular feature selection techniques for fault classification tasks [5]. Separately using Principal Component Analysis (PCA) and Linear Discriminant Analysis (LDA) with SVM classifier the performance of two was compared and both significantly reduced data dimensions and showed improvement for classification accuracy, however, LDA had better results due to consideration of interclass and intraclass correspondence.

After the feature pool is constructed, it is provided to a machine learning classification algorithm with labels. Classification algorithms such as Support Vector Machine, Decision trees, and k-Nearest Neighbour are of the highest prevalence in rotating machinery fault diagnosis field these days [6–8].

Gaining huge popularity during the last decade, Deep learning is a technology that can automatically learn representative features from the data and perform classification tasks. It significantly reduced the dependency on manual feature selection. Deep Neural Network architecture can acquire hidden relationships contained in the original data and amplify the meaningful interclass differences while suppressing irrelevant information that can cause interference.

However, working with real data it is impossible to guarantee the absence of discriminant fault-unrelated information, which helps in the improvement of classification performance. This redundant information in the data may have no relationship to the observed phenomena and in other terms can be referred to as artifacts. Thus, expert data preprocessing for intrinsic fault-related information extraction is of primary concern. This is true for both conventional feature extraction and DNN methods.

The diagnosis of the bearing faults traditionally relies on the analysis of the bearing characteristic frequencies which appear at certain stages of the bearing fault formation. These four main stages of bearing fault formation are presented in Fig. 1. In Stage 1 the emerging subsurface microcracks start to appear at the ultrasonic frequencies from 20 kHz to 350 kHz in the range D. The progressing wear results in ringing natural frequencies that start to appear in the range C at 500–2000 Hz together with higher signal amplitude in the range D. Bearing defects become visible in Stage 3 when fundamental frequencies, accompanied by the well-formed sidebands, start to appear in the frequency range B. At this stage, the fault diagnosis can be performed, and the damaged component can be found. Stage 4 is characterized by decreased amplitude of characteristic frequencies and the presence of broadband random vibration. It is caused by the increased rotor vibration which is a result of the growth of the bearing damaged area. At Stage 4 the bearing has to be replaced immediately.

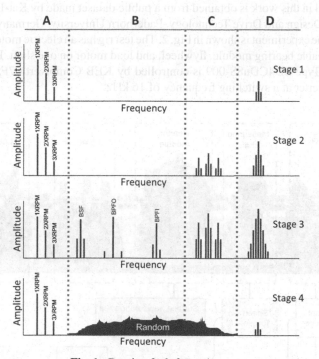

Fig. 1. Bearing fault formation stag.

From the above, it is possible to conclude that for bearing fault diagnosis the amount of useful vibration spectrum information can be scarce. So, for the successful analysis and classification of the bearing fault, this information must be precisely selected to improve the performance of the classification algorithms.

In this paper, a solution for this problem is proposed as a supervised learning model using expert-selected frequency bands. Here, from the full envelope frequency spectrum of vibration signal, fault characteristic frequency bands are isolated and selected using Gaussian windows. It is done to neutralize the chances for DNN to learn the bearing resonance frequencies or normal frequency component information that is present in the envelope spectrum as meaningful information. Feature extraction is performed using sparse autoencoders trained specifically for each selected band and the features from each autoencoder bottleneck layer are used to create the feature pool. After that, the obtained feature pool is provided to Deep Neural Network for classification.

The rest of the paper is organized as follows. The experimental setup and data collection are described in Sect. 2. The proposed method is described in Sect. 3. Section 4 contains results and discussion. The conclusion is made in Sect. 5.

2 Experimental Setup

The data used in this work is obtained from a public dataset made by Kat-DataCenter of the Chair of Design and Drive Technology, Paderborn University, Germany [9]. The test rig used for the experiment is shown in Fig. 2. The test rig has an electric motor, measuring shaft, replaceable bearing module, flywheel, and load motor on one shaft. PMSM 425W drive motor Type SD4CDu8S-009 is controlled by KEB Combivert 07F5E 1D-2B0A industrial inverter at a switching frequency of 16 kHz.

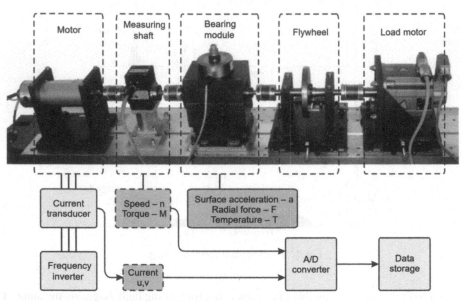

Fig. 2. Modular test rig.

Vibration data was collected using an accelerometer (Model No. 336C04, PCB Piezotronics, Inc.) and a charge amplifier with a 30 kHz Low-Pass filter. The sensor was attached to the top end of the test rig bearing module. The signal was digitalized and saved with a sampling rate of 64 kHz.

In the experiment, healthy and faulty bearings were used. The damages of the bearing were generated by accelerated lifetime tests. In this work in total 17 different bearing signals were used, where 6 bearings are healthy, 5 have a fault in the outer ring and 6 have a fault in the inner ring. The dataset structure used in this work with bearing codes, types of faults and labels are presented in Table 1.

Table 1. Bearing state codes.

Bearing code	Type of fault	Extent of damage	Class label
K001	Healthy	Run-in period > 50 h	0
K002	Healthy	Run-in period = 19 h	0
K003	Healthy	Run-in period = 1 h	0
K004	Healthy	Run-in period = 5 h	0
K005	Healthy	Run-in period = 10 h	0
K006	Healthy	Run-in period = 16 h	0
KA04	Outer ring	1	1
KA15	Outer ring	1	1
KA16	Outer ring	2	1
KA22	Outer ring	1	1
KA30	Outer ring	1	1
KI04	Inner ring	1	2
KI14	Inner ring	1	2
KI16	Inner ring	3	2
KI17	Inner ring	1	2
KI18	Inner ring	2	2
KI21	Inner ring	1	2

Besides the different bearing damage extent, for higher reliability of methods developed with this data, the test rig was operated at 4 different operational conditions with changing rotational speed, load torque, and radial force. The operating parameters of the test rig are shown in Table 2. Bearing vibration data with all operation parameters are included in the dataset used in this research work. Time-domain plots of vibration signal of healthy bearing, bearing with outer ring fault, and bearing with inner ring fault are presented in Fig. 3.

Table 2. Test rig operation parameters.

No.	Rotational speed [rpm]	Load torque [Nm]	Radial force [N]
0	1500	0.7	1000
1	900	0.7	1000
2	1500	0.1	1000
3	1500	0.7	400

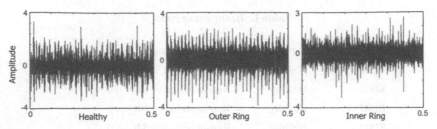

Fig. 3. Time-domain vibration signal plots.

3   Proposed Method

The pipeline of the proposed methodology is shown in Fig. 4. As it can be seen from Fig. 4, in the first step time-domain bearing vibration signals are going through envelope analysis and FFT to obtain the envelope frequency spectrum of the vibration signal. It allows extracting the fault frequencies which are amplitude-modulated to the high-frequency region that cannot be distinguished in the raw frequency spectrum.

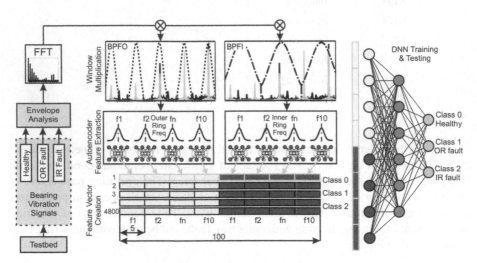

Fig. 4. A pipeline of the proposed method.

Secondly, the vibration envelope spectrum of every one-second sample is multiplied with the Gaussian windows focused on the characteristic frequency components of the outer ring and inner ring faults. The fault bands around inner ring characteristic frequency harmonics tend to be wider than in the case with outer ring faults due to the inner ring rotation at the shaft speed. Therefore, Gaussian window shapes depend on the rotational speed and the expected shapes of the fault signature frequency band. The Gaussian windows means are placed at the harmonics of fault characteristic frequencies which values are calculated using Eqs. 1 and 2. Ten windows are deployed for inner ring fault harmonics selection and in a similar way ten windows are deployed for outer ring harmonics.

Following that, for each of 20 selected bands, a unique three-layer sparse autoencoder is trained in an unsupervised manner for feature extraction with the dimension of the bottleneck layer equal to 5, while the input and output dimensions depend on the width of each frequency band. Consequently, when the feature extraction process is done, the data from each autoencoders bottleneck layer is gathered in the feature vector, where each of the 10 bands obtained from multiplication with IR windows and each of the 10 bands obtained from the multiplication with OR windows is characterized by 5 features from autoencoder. As a result, each envelope spectrum of a 1-s vibration signal sample is now represented by 100 features.

This number of features can be too high and lead to classification performance decrease for conventional Machine Learning techniques, therefore in this work Deep Neural Network architecture was chosen to perform the classification task. A DNN with a 100-neurons input layer, one 25-neurons hidden layer, and a 3-neuron output layer is constructed for classification.

Before training the dataset is balanced among three classes using the stratified sampling technique and is normalized between zero and one. The activation function for the input layer and a hidden layer of the DNN is ReLu and the output layer activation function is SoftMax. Cross entropy loss function with Adam optimizer are used for DNN training.

4 Experimental Results and Analysis

Figure 5 shows network loss and accuracy plot for one training iteration. Here the network was trained for 51 epochs at which the loss and the accuracy tended to converge, and the training process was automatically early stopped with the training accuracy reaching 0.9976, and loss reaching 0.099.

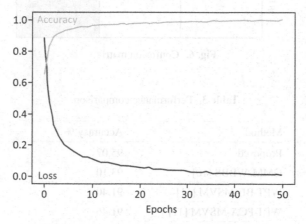

Fig. 5. Loss-accuracy plot for DNN training.

To evaluate the performance of the proposed method, a holdout validation approach was chosen. For this, prior to training the network, the dataset was randomly split into training and testing parts in 80/20 manner with the same proportion of the samples for each class. The network was trained using the training data and then tested using previously unseen testing data. To verify the stable performance of the proposed method, the neural network was trained and tested 10 times using the unique data split for each iteration. Figure 6 shows the confusion matrix obtained by averaging of the results from 10 training and testing iterations. The accuracy of the proposed method is also calculated as the average of 10 iterations.

Performance comparison with other bearing fault diagnosis methods trained and tested using Paderborn bearing vibration data is given in Table 3.

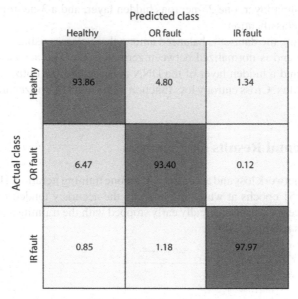

Fig. 6. Confusion matrix.

Table 3. Performance comparison.

Method	Accuracy %
Proposed	95.07
GMM-WBBS [10]	93.10
WPT-BE-MSVM [11]	91.40
WPT-PCA-MSVM [11]	91.46

The comparison shows that the accuracy of the proposed method is generally higher than the accuracy of the comparison methods. This difference can be mainly explained by the usage of the Sparse Autoencoder for feature extraction due to its capability to create an effective characteristic representation of the data.

The Sparse Autoencoder and DNN parts were implemented using TensorFlow with Python 3.8. All signal processing was performed in MATLAB R2021b.

5 Conclusions

In this work, a novel bearing fault diagnosis method was proposed using sparse autoencoder for feature extraction and Deep Neural Network for classification. The workflow of the process was described and on an overall idea of the fault diagnosis method and its structure was given. Method performance was compared to a set of state-of-the-art methods.

The GMM-based spectrum selection for fault characteristic frequency bands was intended to neglect the possible negative effects of the model learning bearing resonance frequencies and bearing normal vibration components along with gear mesh frequency components as related to the bearing fault. Thanks to this, the proposed approach allows reducing the possibility of the negative influence of the data artifacts.

Although the proposed bearing fault diagnosis algorithm had access exclusively to narrowband meaningful data instances selected using Gaussian window series, it showed very high accuracy results compared with techniques that utilized the whole range of the data thanks to the highly effective feature selection performance of Sparse Autoencoders.

Acknowledgements. This work was supported by the Korea Institute of Energy Technology Evaluation and Planning (KETEP) and the Ministry of Trade, Industry & Energy (MOTIE) of the Republic of Korea (No. 20192510102510). This work was also supported by the Technology development Program (S3126818) funded by the Ministry of SMEs and Startups (MSS, Korea).

References

1. Bazurto, A.J., Quispe, E.C., Mendoza, R.C.: Causes and failures classification of industrial electric motor. In: 2016 IEEE ANDESCON, Arequipa, Peru, pp. 1–4, October 2016. https://doi.org/10.1109/ANDESCON.2016.7836190
2. Nandi, A.K., Ahmed, H.: Condition Monitoring with Vibration Signals: Compressive Sampling and Learning Algorithms for Rotating Machines. Wiley-IEEE Press, Hoboken, NJ, USA (2019)
3. Sawalhi, N., Randall, R.B., Endo, H.: The enhancement of fault detection and diagnosis in rolling element bearings using minimum entropy deconvolution combined with spectral kurtosis. Mech. Syst. Sign. Process. **21**(6), 2616–2633 (2007). https://doi.org/10.1016/j.ymssp.2006.12.002
4. Rajeswari, C.: Bearing fault diagnosis using wavelet packet transform, hybrid pso and support vector machine. Procedia Eng. **97**, 12 (2014)
5. Xie, Y., Zhang, T.: A fault diagnosis approach using SVM with data dimension reduction by PCA and LDA method, p. 6

6. Soualhi, A., Medjaher, K., Zerhouni, N.: Bearing health monitoring based on hilbert-huang transform, support vector machine, and regression. IEEE Trans. Instrum. Meas. **64**(1), 52–62 (2015). https://doi.org/10.1109/TIM.2014.2330494
7. Demetgul, M.: Fault diagnosis on production systems with support vector machine and decision trees algorithms. Int. J. Adv. Manuf. Technol. 12 (2013)
8. Lee, C.-Y., Huang, Y.K.-Y., Shen, X., Lee, Y.-C.: Improved weighted k-nearest neighbor based on PSO for wind power system state recognition. Energies **13**(20), 5520 (2020). https://doi.org/10.3390/en13205520
9. Lessmeier, C., Kimotho, J.K., Zimmer, D., Sextro, W.: Condition Monitoring of Bearing Damage in Electromechanical Drive Systems by Using Motor Current Signals of Electric Motors: a Benchmark Data Set for Data-Driven Classification, p. 17 (2016)
10. Maliuk, A.S., Prosvirin, A.E., Ahmad, Z., Kim, C.H., Kim, J.M.: Novel bearing fault diagnosis using Gaussian mixture model-based fault band selection. Sensors **21**(19), 6579 (2021). https://doi.org/10.3390/s21196579
11. Rapur, J.S., Tiwari, R.: Experimental fault diagnosis for known and unseen operating conditions of centrifugal pumps using MSVM and WPT based analyses. Measurement **147**, 106809 (2019). https://doi.org/10.1016/j.measurement.2019.07.037
12. Xu, W.: Research on bearing fault diagnosis base on deep learning. In: 2021 4th International Conference on Artificial Intelligence and Big Data (ICAIBD), Chengdu, China, pp. 261–264, May 2021. https://doi.org/10.1109/ICAIBD51990.2021.9459073

Identification of External Defects on Fruits Using Deep Learning

Henrique Tavares Aguiar$^{(\boxtimes)}$ and Raimundo C. S. Vasconcelos$^{(\boxtimes)}$

Federal Institute of Brasilia, Taguatinga DF CEP 72146-050, Brazil
henrique.aguiar@estudante.ifb.edu.br, raimundo.vasconcelos@ifb.edu.br
https://www.ifb.edu.br/index.php

Abstract. The quality of fruit plays a fundamental role in their marketing and is mainly defined by its shape, color, and size. The classification process is traditionally done manually and takes time. The use of image processing techniques can help this task. Some methodologies for image classification are presented, using deep neural networks. A set of combinations between Convolution Neural Networks (CNN), deep neural networks (DNN) using Gabor filter, over RGB and grayscale images, extracting texture properties of a GLCM (Gray Level Co-occurrence Matrices) is used in this project. Background segmentation, contrast enhancement, and data augmentation are also used to improve generalization and minimize overfitting. Applying it to a set of tropical fruits resulted in an excellent set of results, above 95% on average.

Keywords: Defect detection · Feature extraction · Image processing

1 Introduction

Fruit selection is a highly important economic activity. The traditional method of selection is made by human labor. During postharvest, some people are responsible for visually inspect fruits and classify them according to their color, weight, maturity, diseases, deformations, among other attributes.

An alternative to the traditional method is to automatize the process with the help of fruit selecting machines. They can have several conveyor belts available for carrying fruits towards their sensors. On them, data is obtained, and through software, the destination of each fruit is decided. Upon determining their destinations, the software acts on the conveyor belts to ensure that those fruits arrive correctly.

Automizing this process comes with three benefits:

Error reduction—Given its repetitive nature, humans are more prone to errors when compared to machines;

© Springer Nature Switzerland AG 2022
A. J. Pinho et al. (Eds.): IbPRIA 2022, LNCS 13256, pp. 565–575, 2022.
https://doi.org/10.1007/978-3-031-04881-4_45

Cost-effectiveness—Classifying fruit fast and correctly demands a lot of training. Training people takes time, needs to be done for every new employee, and it is not certain that newly trained employees will perform the same way as more experienced employees. On the other hand, to replicate a selecting machine, building a new one and loading the same software is enough;

Higher rate of selection—There is a maximum rate of how many fruits a human can classify per minute. A machine can easily go beyond that limit.

A more intuitive solution is to use traditional cameras to photograph fruits, while the software decides those fruits' destinations based on those images. Under this perspective, the challenges of building those fruit selecting machines involve computer vision.

To enhance the quality and quantity of the agriculture product, there is a need to adopt the new technology. Image processing approach is a non-invasive technique, which provides consistent, reasonably accurate, less time consuming and cost effective solution.

Machine learning suggests the usage of deep neural networks to approximate functions of all types of complexities. The functions are built based on the union of simple computational units, and they are capable of recognizing relevant patterns to complete a task.

Therefore, the usage of machine learning is already established and offers promising results regarding the problem of detecting external defects on fruit. Pandey et al. [8] revise the literature on automatic fruit selection. The goal of this paper is to develop a computer vision algorithm as good as humans in identifying external defects on fruit with the help of deep learning.

2   Related Work

Image processing and computer vision techniques have been growing important for the fruit industry over the years, especially when applied to quality inspection tasks and defect classification.

It is presented in [10] an effort in recognizing defects on apples using a machine vision system based on three colors cameras. The apple image is segmented from its background by using a multiple threshold method, then, defect detection and counting are made upon the apple image.

Various techniques are analyzed in [3] for automatic fruit inspection with the help of computer vision. The analyzed fruits are apples and citric ones. The authors revised possible algorithms for defects detection on those fruits. The solutions are different and well suited for each type of fruit.

The proposed method in [1] uses pre-processing, segmentation, border detection, and features extraction to classify a fruit as either defective or fresh.

Fruit peel color is an important factor in identifying defects on fruit in [7]. The procedures adopted are background segmentation, identification of spots for a quality calculation, and a classification process.

A fruit defect detection in [11] is made by using some local features such as energy, homogeneity, contrast, and correlation, all from segmented and filtered images.

The purpose of the work made in [9] is to identify whether mangos are mature or not. The fruit images are obtained, then the fruits are harvested if mature. After the fruits' backgrounds are removed, features such as shape, color, and size are extracted. The result is a classification of whether the fruit is mature or not.

Another fruit external defects classification is made in [4]. The images are collected, segmented, and then a color histogram, global correlation, and local binary pattern are calculated in order to feed an SVM classifier.

The work in [2] consists of analyzing orange images and classifying them. Features such as color, texture, and shape are extracted, and then an RBPNN (Radial Basis Probabilistic Neural Network) is used to perform the classification.

Evaluated articles use some kind of pre-processing (e.g., background removal and segmentation). Direct features (color, texture, shape, size) or secondary features (color histogram, global correlation, energy, homogeneity, contrast, and correlation), or both are extracted. Aside from the pre-processing, they also use different classification techniques: algorithms based on thresholds, Support Vector Machines or Neural Networks.

3   Materials

3.1   Hardware

This work was made by using the Google Colab environment. Google Colab is a free cloud storage service in which it is possible to write and run Jupyter notebooks with support to GPU and TPU.

3.2   Software

As Google Colab uses Jupyter notebooks, the scripts were written in Python. As for the libraries, we used scikit-image (ver. 0.19.x) that provides tools to perform image pre-processing like background segmentation and Tensorflow (ver. 2.x) to develop the Neural Networks.

4   Proposed Methodology

We chose the following fruits for this paper: abiu (*pouteria caimito*) and siriguela (*spondias purpurea*). Both of them are tropical fruits usually found in Latin America. Some characteristics were taken into account, such as availability, size, shape, and colors.

Figure 1 summarizes the process developed throughout this work can be summarized in four main steps:

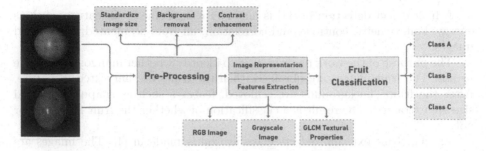

Fig. 1. Flowchart representing the whole process of fruit classification, divided into three parts: pre-processing, image representation, and the classification itself.

1. Manual imagery acquisition and labeling
2. Image pre-processing and segmentation
3. Image representation and feature extraction
4. Fruit classification

5   Segmentation and Pre-processing

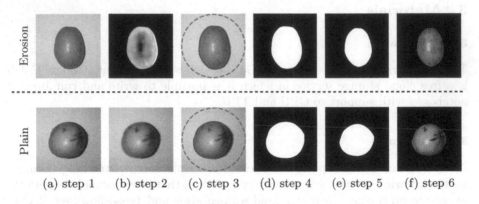

Fig. 2. Fruit images pre-processing, which includes color and contrast enhancement and background removal. Both lines in (c) were drawn on the image from the first step only for visualization purposes.

We took batches of photographs of all the fruits using the same camera, chose the best ones, and then a group of 20 students divided them into three classes (A, B, and C) accordingly to their quality. Class A represents the best fruits, almost stainless with minor defects. Class B represents mid-term fruits with a few significant defects limited only to the fruit peel, so the fruits remain edible and practically fresh inside. And Class C represents the ones with severe defects and are considered to be rotten.

Figure 2 presents the first process, background segmentation. We chose this approach because the background contains irrelevant information regarding the fruit. Thus, removing it will lead to better results.

Figure 2a shows images resized to 512 × 512 pixels. Figure 2b shows the phase in which images are being prepared for a contour detection algorithm. This phase was divided into two separate ways, the first one called plain method smooths the image using a median filter, the other one called erosion method is more complex. It performs a morphological reconstruction by erosion on the image, in other words, low-intensity values replace high-intensity values and are limited to a minimum value. Alternatively, this reconstruction can be seen as a way to isolate connected regions of an image, in this case, the fruit and the background. Also, a median filter is applied.

In general, the erosion method is a faster but slightly less accurate method when compared to the plain method. When the background is too noisy, the erosion method most of the time is unable to correctly find the contours. However, when the shape of a fruit is too irregular, the contour found by using the plain method may not exactly fit the actual shape of that fruit. The erosion method is preferred over the plain method due to its time and computational efficiency.

The Active Contour Model [6] is used as shown in Fig. 2c to find the contours of a fruit in an image. It is applied to the image obtained from the previous step. It works by minimizing the energy of an initial spline, seen as the red dotted line in Fig. 2c, guided by both the image and shape of the spline, fitting onto nearby lines and edges. The result can be seen as the blue line in Fig. 2c.

The image in Fig. 2d is simply a mask image built from the contour calculated previously, filling its area. Alternatively, it is the shape of the fruit.

After that, in Fig. 2e, a basic morphological erosion is applied on the mask image to reduce the shape's area, aiming to discard possible shadows and a portion of its area on the edges that may be misleading due to illumination differences.

Finally, in Fig. 2f, the mask image is applied to the image from the first step. Also, a few improvements are made to the image: gamma adjustment, intensity rescaling and contrast enhancement.

5.1   Image Representation and Features Extraction

There are several ways to represent an image and, for this work, the following ones were chosen: RGB image, grayscale image, and image representation through features.

The first method to represent the images is in the form of an RGB image. It can be considered the default method, since it does not require any additional computation after the pre-processing phase.

The second one transforms an image into a grayscale image, as the colors may not only be irrelevant but also misleading, and the difference in lighting between a healthy and a rotten fruit peel may be enough to classify the fruits correctly.

The third and last one represent an image through some textural properties from a Gray Level Co-occurrence Matrix (GLCM) [5]. A GLCM is a matrix defined over an image, and it represents the distribution of co-occurring gray values (intensity) at a given offset. In other words, it calculates the spatial distribution of intensity values in a neighborhood from a one-color channel image, like a grayscale image, which is why a GLCM is good to obtain textural information about an image.

Concerning the GLCM method, to calculate the textural information necessary to classify the fruits, the following statistical descriptors will be used: Contrast, Dissimilarity, Homogeneity, Angular Second Moment (ASM), Energy and Correlation.

Considering that a GLCM uses gray levels values or intensity values, an RGB image cannot be used as an input, neither a grayscale image, as it loses relevant information about each color channel. So far, 3 GLCMs are needed, one for each channel in the RGB color scheme. In addition, as mentioned above, the distance used is one pixel, moreover, four angles were used: 0°, 45°, 90°, and 135°. A combination between an angle and a distance makes an offset. Now, 4 GLCMs are needed for each color channel. Therefore, in total, 12 GLCMs need to be built. Then, as 6 textural descriptors are calculated. To represent a single image, 72 textural descriptors are calculated.

Moreover, for every method discussed above, a Gabor filter is applied. If it is a grayscale image, the Gabor filter is applied after the image is converted into grayscale. If a GLCM is used, the filter is applied to the image before the GLCM is calculated.

A Gabor filter can be defined as a sinusoidal signal of a particular frequency and orientation, modulated by a Gaussian wave. In other words, it allows certain frequencies and rejects others. When a Gabor Filter is applied to an image, it emphasizes the edges and points where texture changes accordingly to the filter pattern, i.e., its parameters. Therefore, these Gabor Filters are well suited for texture analysis. Mathematically, a Gabor Filter has a real and imaginary component representing orthogonal directions. For this work, only the real component was used.

5.2   Proposed Neural Network Models

There were 120 samples of both fruits, i.e., 240 images in total, and all the fruit images were split equally into those three classes. In other words, 40 images for each class (A, B, or C). 80% of fruit images were used to train the models and the remaining 20% for the validation step. Since there were defined very distinct ways to represent the images, it demands different neural networks architectures to fit better the data fed into them.

The architecture seen in Fig. 3 represents the structure of a Convolution Neural Network (CNN). A CNN is mainly composed of four different layers: Convolution, Pooling, Flattening, and Classification.

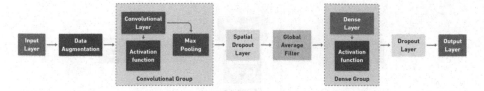

Fig. 3. Represents the overall structure of the convolutional neural network models used.

Convolutional layers allow one to systematically create and apply a series of filters over an array input (e.g., an image). These filters act like feature detectors, and then the layer outputs a feature map that summarizes the features detected (e.g., lines and edges) by the filters. Although the Convolutional layers have their limitations, they are translation-invariant, because they are capable of detecting patterns locally rather than globally. It means that if a Convolution layer leaned a certain pattern, and if the same pattern occurs again, but in a different location, they would still be able to detect that same pattern.

A Pooling Layer is used to perform a downsample on feature maps, which reduces their dimensions, but they still carry the main elements. It works by aggregating a group of pixels and applying a pooling operation. The two most commons are Average Pooling, which calculates the average of each group, and Max Pooling, which calculates the maximum value in each group. The purpose of using a Pooling Layer is to decrease the computational power needed to process the data. And as a side-effect, it helps to extract dominant features, making the training process more efficient.

Before the Pooling layer, an activation function (e.g., ReLU (Rectified Linear Unit)) is used upon the features maps. This function usually breaks the linearity and adds complexity, much like it is done for a Dense Layer.

The combination between a Convolutional Layer, a Pooling Layer, and an activation function makes a Convolution Group, seen in Fig. 3. When stacking these Convolutional Layers, instead of detecting simple features (e.g., lines and edges), deeper layers learn to detect more abstract features, like shapes or specific objects.

After that, a Flattening Layer is used because the output of the combination of multiple Convolutional Groups is multi-dimensional, composed of several features maps. So, it needs to be flattened.

Finally, the Classification part is composed of a sequence of Dense Groups. A Dense Group is made of Dense Layers, also known as Fully Connected Layers, followed by an activation function. Before the Output Layer, a Dropout Layer is used to help reducing overfitting while training (useful for a small dataset). It works by randomly deactivating a percentage of the neurons from the previous layer. Then for the Output Layer, which is also a Dense Layer, the activation function chosen is often different from the ones used in the hidden layers.

The other architecture, DNN, seen in Fig. 4, is a simple deep neural network composed of a Flattening Layer at the beginning, a sequence of Dense Groups,

Fig. 4. Represents the overall structure of the deep neural network models used.

and a Dropout Layer right before the Output Layer. This architecture may look very similar to the one used in the Flattening and Classification part inside the CNN. However, in reality, a CNN uses a combination of a convolutional technique to extract useful information from images and passes it as input to a fully connected deep neural network.

Data Augmentation technique was used at the beginning of the CNN seen in Fig. 3. Since there was a total of 120 images for each fruit, and each model was trained for only one type of fruit, meaning the models were on a small dataset. It helps by reducing overfitting by generating new images from existing ones by doing simple modifications such as random horizontal and vertical flipping with a random rotation. For the same reason, a Spatial Dropout Layer was used right before the Flattening part. It works similarly to a regular Dropout Layer. However, it is applied to 2D feature maps instead of individual elements.

Regarding the Flattening Layer in the CNN, a Global Average Filter was used. It uses an average pooling to reduce the size of feature maps, and then it flattens and passes it as input to the next part. This Global Average Filter is important because its input came from either RGB or Grayscale image methods. Thus, if it were simply flattened, there would be too much information for the dense neural network, i.e., the Classification part, to process and classify efficiently. That was unnecessary for the DNN since its input came from the GLCM textural properties, and it consisted of only 72 elements, the GLCM descriptors when flattened.

Inside the Convolutional Groups, the activation functions used were the ReLU or the swish function, which behaves similarly to the ReLU. Concerning the Dense Groups, they use either the swish or the ELU (Exponential Linear Unit) function, which is also similar to the ReLU.

The softmax function is used at the Output Layer because it converts its input into a probability distribution. Useful for classification tasks when there are more than two classes.

6   Results

In this section, we will present the results of the combination between the various imaging representation methods, their appropriate neural network architectures, and a Gabor Filter. Experiments were made using both fruits separately, they can be seen Figs. in 5 and 6.

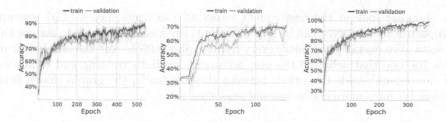

Fig. 5. Training curves for the models trained on *abius* images. ((a) over RGB images. (b) over grayscale images. (c) using texture properties from a GLCM

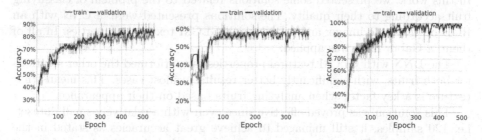

Fig. 6. Training curves for the models trained on *siriguelas* images. (a) over RGB images. (b) over grayscale images. (c) using texture properties from a GLCM

Across all the graphs, the y-axis represents the accuracies (training and validation) achieved by the model, while the x-axis represents the epochs from when the models were being trained. Those models were trained for different epochs, due to a technique called Early Stopping, which stops the training when there is no improvement in loss, which is a measure of how wrong a model is, for a number of consecutive epochs.

Analyzing the graphs, it is clear that the GLCM textural properties-based models outran the image-based ones. Between the image-based models, the ones that used RGB images managed to be slightly better in general. It is also visible a strange behavior on various graphs. It tends to occur more often in the GLCM based models. And this strange behavior is when a model achieved better results on the validation set. A possible explanation is when the validation set is too easy, which is completely plausible due to the small datasets used.

Table 1. Train and validation accuracy results for models trained on both fruits. The bold values show the best configuration

	Abius		Siriguelas	
	Train	Validation	Train	Validation
RGB	86.74%	84.62%	79.36%	84.94%
Grayscale	71.79%	68.38%	59.13%	57.50%
GLCM	**99.01%**	**95.83%**	**99.42%**	**98.10%**

A closer look at the best results across all the fruits, i.e., GLCM textural properties with a DNN, can be seen in Table 1. First of all, it is safe to assume the results were overall great, since every result is above 95%. And the *siriguela* fruit was the one that achieved the best accuracies on both training and validation environments. Aside from that, between the two fruits, the graphs drawn from *siriguelas* have more consistent and smoother lines.

7   Conclusion

In this work, we presented some solutions related to the problem of classifying fruit according to their quality. The solutions presented were: a CNN with an RGB or grayscale image; and a DNN with GLCM textural properties. In all of them, a Gabor filter was applied.

The DNN with GLCM textural properties outperformed the other solutions for both fruits, with significantly better results in most cases. This means that texture is a key factor when analyzing fruits based on their appearance.

This solution was proved effective since even with a relatively small dataset, i.e. 120 samples, it still managed to achieve great accuracies as stated in the previous section. Further experimentations can be done by using larger datasets and testing even more possible solutions.

Acknowlegements. This study was financed in part by the Conselho Nacional de Desenvolvimento Científico e Tecnológico - CNPq, and Fundação de Apoio à Pesquisa do Distrito Federal - FAPDF

References

1. Arunachalam, S., Kshatriya, H.H., Meena, M.: Identification of defects in fruits using digital image processing. Int. J. Comput. Sci. Eng. **6**, 637–640 (2018). https://doi.org/10.26438/ijcse/v6i10.637640
2. Capizzi, G., Sciuto, G., Napoli, C., Tramontana, E., Woźniak, M.: A novel neural networks-based texture image processing algorithm for orange defects classification. Int. J. Comput. Sci. Appl. **13**, 45–60 (2016)
3. Devi, P., Vijayarekha, K.: Machine vision applications to locate fruits, detect defects and remove noise: a review. Rasayan J. Chem. **7**, 104–113 (2014)
4. Dubey, S.R., Jalal, A.S.: Adapted approach for fruit disease identification using images. Int. J. Comput. Vis. Image Processing (IJCVIP) **2**, 44–58 (2012). https://doi.org/10.4018/ijcvip.2012070104
5. Haralick, R.M., Shanmugam, K., Dinstein, I.: Textural features for image classification. IEEE Trans. Syst. Man Cybern. SMC **3**(6), 610–621 (1973). https://doi.org/10.1109/TSMC.1973.4309314
6. Kass, M., Witkin, A., Terzopoulos, D.: Snakes: active contour models. Int. J. Comput. Vis. **1**(4), 321–331 (1988). https://doi.org/10.1007/BF00133570
7. Moradi, G., Shamsi, M., Sedaghi, M.H., Alsharif, M.R.: Fruit defect detection from color images using ACM and MFCM algorithms. In: 2011 International Conference on Electronic Devices, Systems and Applications (ICEDSA), pp. 182–186 (2011). https://doi.org/10.1109/ICEDSA.2011.5959033

8. Pandey, R., Naik, S., Marfatia, R.: Image processing and machine learning for automated fruit grading system: a technical review. Int. J. Comput. Appl. **81**, 29–39 (2013). https://doi.org/10.5120/14209-2455

9. Sahu, D., Potdar, R.M.: Defect identification and maturity detection of mango fruits using image analysis. Am. J. Artif. Intell. **1**, 5–14 (2017)

10. Xul, Q., Zou, X., Zhao, J.: On-line detection of defects on fruit by machinevision systems based on three-color-cameras systems. In: Zhao, C., Li, D. (eds.) Computer and Computing Technologies in Agriculture II, vol. 3, pp. 2231–2238. Springer, US, Boston, MA (2009)

11. Yogesh, Dubey, A.K.: Fruit defect detection based on speeded up robust feature technique. In: 2016 5th International Conference on Reliability, Infocom Technologies and Optimization (Trends and Future Directions) (ICRITO), pp. 590–594 (2016). https://doi.org/10.1109/ICRITO.2016.7785023

Improving Action Quality Assessment Using Weighted Aggregation

Shafkat Farabi[1](✉)[iD], Hasibul Himel[1][iD], Fakhruddin Gazzali[1][iD],
Md. Bakhtiar Hasan[1][iD], Md. Hasanul Kabir[1][iD], and Moshiur Farazi[2][iD]

[1] Islamic University of Technology, Gazipur, Bangladesh
{shafkatrahman,hasibulhaque,fakhruddingazzali,bakhtiarhasan,
hasanul}@iut-dhaka.edu
[2] Data61-CSIRO, Canberra, Australia
moshiur.farazi@data61.csiro.au

Abstract. Action quality assessment (AQA) aims at automatically judging human action based on a video of the said action and assigning a performance score to it. The majority of works in the existing literature on AQA divide RGB videos into short clips, transform these clips to higher-level representations using Convolutional 3D (C3D) networks, and aggregate them through averaging. These higher-level representations are used to perform AQA. We find that the current clip level feature aggregation technique of averaging is insufficient to capture the relative importance of clip level features. In this work, we propose a learning-based weighted-averaging technique. Using this technique, better performance can be obtained without sacrificing too much computational resources. We call this technique Weight-Decider(WD). We also experiment with ResNets for learning better representations for action quality assessment. We assess the effects of the depth and input clip size of the convolutional neural network on the quality of action score predictions. We achieve a new state-of-the-art Spearman's rank correlation of 0.9315 (an increase of 0.45%) on the MTL-AQA dataset using a 34 layer (2+1)D ResNet with the capability of processing 32 frame clips, with WD aggregation.

Keywords: Action quality assessment · Aggregation · MTL-AQA

1   Introduction

Action quality assessment (AQA) addresses the problem of developing a system that can automatically judge the quality of an action performed by a human. This is done by processing a video of the performance and assigning a score to it. The motivation to develop such a system stems from its potential use in applications such as health care [13], sports video analysis [16], skill discrimination for a specific task [14], assessing the skill of trainees in professions such as surgery [3], etc.

© Springer Nature Switzerland AG 2022
A. J. Pinho et al. (Eds.): IbPRIA 2022, LNCS 13256, pp. 576–587, 2022.
https://doi.org/10.1007/978-3-031-04881-4_46

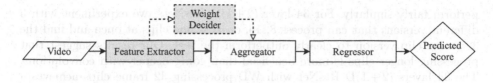

Fig. 1. Overview of a general AQA pipeline and our improvement over it.
Generally, the input video is divided into clips. A feature extractor extracts features
from these clips. These features are then aggregated into a video level feature vector. A
linear-regressor predicts action quality scores based on this feature vector. We introduce
a Weight-Decider module to this architecture, which proposes weights based on the
clip level features for better aggregation. Additionally, we use a ResNet instead of the
commonly used C3D.

Almost all existing works have treated AQA as a regression problem [14,15,
18]. As shown in Fig. 1, most approaches boil down to dividing an RGB video
of the action in multiple clips, extracting higher-level features from each clip,
aggregating them, and then training a linear-regressor to predict a score based on
these aggregated features. Most of these works [15,18,22] utilize a convolutional
neural network [11] to extract complex higher-level features and simple averaging
to aggregate the features. The best performing models [15,16,18] make use of
the Convolutional-3D (C3D) network [20] with average aggregation.

The majority of works in AQA [15,16,18,22] aggregate clip level features
into a video level feature vector by simply averaging them. We think this fails
to preserve temporal information present in the data. We hypothesize that
a more sophisticated method would improve the overall performance of the
pipeline. We propose one such technique by introducing a module called Weight-
Decider(WD). This module inspects the feature vectors extracted from individual clips and proposes a corresponding weight vector. Finally, these weight vectors can be used to calculate a weighted average of the clip-level feature vectors.
In this way, the final video level feature vector contains more contributions from
the important features of each clip. This is similar to how real-world judges base
their final scoring on key mistakes/skills of the performer and not on an average
of all the moments in the action. We design the WD as a shallow neural network
that can be trained along with the rest of the AQA pipeline. In our experiments,
we show the performance of the AQA pipeline to improve when using WD as
aggregation.

Spatio-temporal versions of ResNets capable of processing videos have been
proposed by [5,21]. These approaches have achieved state-of-the-art results in
the task of action recognition and outperform the C3D network [5]. Hence, we
decide to use spatio-temporal ResNets as feature extractors for performing AQA.
We experiment with various 3D and $(2+1)$D ResNet feature extractors on the
MTL-AQA dataset [16]. We find that 3D and $(2+1)$D ResNets of depth 34 and
50 with pretraining on large-scale action recognition datasets have performance
comparable to the state-of-the-art. We see that $(2+1)$D and 3D convolutions

perform fairly similarly. For 34 layer $(2+1)$D ResNets, we experiment with 3 different versions that can process 8, 16, or 32 frame clips at once and find the 32 frame clip version to clearly outperform the rest. Our results suggest that processing longer clips is more beneficial than going deeper with convolutions. The 34 layers $(2+1)$D ResNet with WD processing 32 frame clips achieves a Spearman's rank correlation of 0.9315 on the MTL-AQA dataset, achieving a new state-of-the-art.

Contributions:

- We propose a novel learning-based light-weight aggregation technique called Weight-Decider and demonstrate that it can improve the AQA pipeline's performance.
- To the best of our knowledge, this is the first work to do a comparative analysis of the effect of the depth, convolution type, and input clip size of the ResNet feature extractor on the final quality of the scores predicted in AQA.
- One of our approaches outperforms all the previous works in AQA on the MTL-AQA dataset.

2   Related Work

Pirsiavash et al. [18] proposed a novel dataset containing videos of Diving and Figure-skating annotated with action quality scores by expert human judges. They provided Discrete Cosine Transform (DCT) and extracted human pose as inputs to a Support Vector Regressor, which predicted the score.

More recent works have utilized the Convolutional 3D (C3D) network [20] as a feature extractor. Parmar and Morris [15] proposed three architectures, C3D-SVR, C3D-LSTM, C3D-LSTM-SVR, all of which used features extracted from short video clips using C3D network, and later aggregated them and predicted an action score using Support Vector Regressor (SVR) and Long Short-Term Memory (LSTM). In a later work, Parmar and Morris [16] took a multitask approach towards action quality assessment. They released a novel AQA dataset called MTL-AQA and proposed a multi-task learning-based C3D-AVG-MTL framework that extracted features using the C3D network and aggregated these through averaging. They trained these features to do well in score prediction, action classification, and generating captions. Tang et al. [19] took a probabilistic approach (MUSDL). They divided 103 frame videos into 10 overlapping 16 frame clips, used I3D [1] architecture to extract clip level features, averaged as aggregation, and finally predicted parameters of a probabilistic distribution from which the final score prediction was sampled. The authors calculated 7 different scores corresponding to 7 judges for Olympic scoring and summed up the 5 scores in the middle. With this advantage over simple regression-based methods which directly predict the score, this approach achieved a SOTA spearman's correlation of 0.9273. Diba et al. [2] used a method called "STC Block" for action recognition. This is similar to our proposed aggregation method. However, they utilize this on spatial and temporal features separately after each convolution layer for action recognition, whereas our method is applied to the output of the CNN to aggregate clip level spatiotemporal features for performing AQA.

Our proposed approach differs from these works in that we use 3D and (2+1)D ResNets as the feature extractor and we aggregate these features using the WD network, which is a lightweight and learning-based feature aggregation scheme.

3 Our Approach

3.1 General Pipeline Overview

Let, $V = \{F_t\}_{t=1}^{L}$ be the input video containing L frames, where F_t denotes the tth frame. It is divided into N non-overlapping clips, each of size $n = \lceil \frac{L}{N} \rceil$. Thus we define the ith clip as $C_i = \{F_j\}_{j=i\times n}^{i\times(n+1)-1}$. The feature extractor takes in a clip C_i and outputs a feature vector f_i. For the feature extractor, we experiment with 3D ResNets[6] and (2+1)D ResNets [21] with varying depth and input clip size. Next, we aggregate these clip level features to obtain a global video level representation. Finally, a linear-regressor is trained to predict the score from the video level feature representation. Following the majority of previous works [14–16,18], we model the problem as linear regression. This makes sense as the action quality score is a real number. To experiment with the relation of the ResNet feature extractor's depth with the AQA pipeline's ability to learn, we experiment with 3 different depths:

- **34 layer:** We experiment with both 34 layer 3D and (2 + 1)D ResNets. The only difference being 3D ResNets use $3 \times 3 \times 3$ convolution kernels, on the other hand (2 + 1)D ResNets use a $1 \times 3 \times 3$ convolution followed by $3 \times 1 \times 1$ convolution [21]. We take the final average-pool layer output (512 dimensional) and pass it through 2 fully-connected layers having 256 and 128 units. The 3D ResNet takes input 16 frame clips. On the other hand, we experiment with 3 different variations of (2 + 1)D 34-layer ResNet, processing 8, 16, and 32 frame clips using available pre-trained weights,
- **50 layer:** We experiment with 50 layer 3D and (2 + 1)D ResNets. The final average-pool layer outputs a feature vector of size 2048. We take this feature vector and input it into 3 fully-connected layers having 512, 256, and 128 units. Only 16 frame clips are processed.
- **101 layer:** We experiment with 101 layer 3D ResNet. The remaining details about the input clip size and output feature vector processing are identical to the 50 layer ResNets.

3.2 Feature Aggregation

Most of the previous works dealing with AQA process the entire input video by first dividing it into multiple smaller clips of equal size, due to memory and computational budget. Most CNNs are designed to process 8, 16, or 32 frames at once. Then the features extracted by the CNN are aggregated to form a video

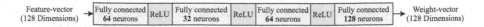

| Feature-vector (128 Dimensions) | → | Fully connected 64 neurons | ReLU | Fully connected 32 neurons | ReLU | Fully connected 64 neurons | ReLU | Fully connected 128 neurons | → | Weight-vector (128 Dimensions) |

Fig. 2. The architecture of the WD network.

level feature description. Next, a linear-regressor predicts the final score based on this feature description. The best performing works aggregated the clips by simply averaging them [15,16,19]. Some other works [14,15] aggregated using LSTMs [7]. However, LSTM networks, although make sense in theory because of their ability to handle time sequences, perform worse due to the lack of big-enough datasets dedicated to AQA.

We propose that simply averaging the clip-wise features is an ineffective measure. It should not be able to preserve the temporal information available in the data. This follows from the fact changing the order of the clip level features will generate the same average and hence the same score prediction. Furthermore, expert human judges focus more on mistakes and deviations of the performers and these have a bigger impact on the score. Hence we think, a weighted averaging technique might be more suitable, as the linear-regressor will be able to base its decision on features more important from each clip.

More concretely, if the feature vector extracted from clip C_i is f_i, we propose the video level feature vector as

$$f_{video} = \sum_{i=1}^{N}(f_i \odot w_i) \tag{1}$$

where w_i is a weight vector corresponding to the feature vector f_i and $\odot$ represents Hadamard Product or elementwise multiplication. w_i is of the same dimensions as f_i and learned using a small neural network of 4 layers. This smaller neural network takes as input 128-dimensional feature vector f_i and runs it through fully connected layers containing 64, 32, 64, and 128 neurons. All but the final layers employ a ReLU activation function. The architecture is explained in Fig. 2. Finally, to ensure the weights corresponding to the same element of different weight vectors sum up to one, a softmax is applied along with the corresponding elements of all the weight vectors. We call this shallow neural network Weight-Decider(WD).

$$w_i' = WD(f_i) \tag{2}$$

$$[w_0 \quad w_1 \quad \ldots \quad w_N] = softmax([w_0' \quad w_1' \quad \ldots \quad w_N']) \tag{3}$$

Finally, the linear-regressor can predict the score using the feature vector f_{video} as proposed by Eq. 1.

The proposed WD module can be used with any of the feature extractors we planned to use in our experiments. WD replaces averaging as aggregation within the typical AQA pipeline (Fig. 3). It can be trained with the rest of the AQA model in an end-to-end manner.

Adding WD to the AQA pipeline to replace averaging as aggregation does not cost much in terms of computational resources. To see this, recall that the

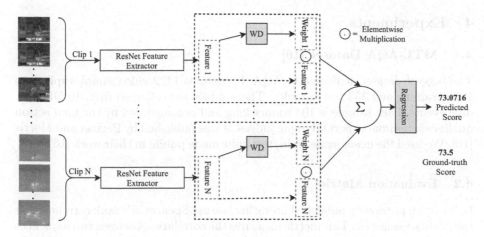

Fig. 3. The video is divided into clips and each clip is processed by a ResNet feature extractor. The ResNet can have depths 34, 50, or 101. Elements with a solid border indicate trainable modules. The solid bordered white modules are initialized with pretrained weights. The solid bordered green modules are trained from scratch. Best viewed in color. (Color figure online)

WD module has 3 hidden layers and an output layer, consisting of 64, 32, 64, and 128 neurons in that order. Each of these layers contains a weight matrix and a bias. The first hidden layer takes 128 inputs and has 64 neurons. Thus the weight matrix is of dimensions 64×128. Hence, number of trainable parameters in this layer (including the bias term) is $(64 \times 128 + 1) = 8193$. Similarly, the second hidden layer, third hidden layer, and the output layers contain 2049, 2049, and 8193 trainable parameters correspondingly. By summing up all the trainable parameters in each layer, we can see that WD only contains 20484 trainable parameters. Spatio-temporal ResNet feature extractors have millions of trainable parameters. Hence, additional resources required to train the WD module used in an AQA pipeline are not much. As an example, the 3D and (2+1)D ResNet-34 feature extractors we are using contain 63.6 million trainable parameters [12]. Thus, using the WD module on top of ResNet-34 would increase the number of trainable parameters by approximately 0.03%. The ratio would be smaller for a deeper ResNet. Hence, our proposed WD does not require many computational resources to be incorporated into the pipeline. Our experiments support this. We found the training time for 32-frame ResNet(2+1)D-34 to be 4557 s per epoch, and the inference time to be 604 s per epoch. The introduction of WD increases the training time per epoch by 24 s (0.52%) and the inference time by 21 s (3.47%).

4   Experiments

4.1   MTL-AQA Dataset [16]

The biggest dataset dedicated to AQA. It contains 1412 video samples split into 1059 training and 353 test samples. The samples are collected from 16 Olympic dive events. Each sample is 103 frames long and accompanied by the final action quality scores from expert Olympic judges. It was published by Parmer and Morris [16]. We used the exact same train/test split made public in their work [16].

4.2   Evaluation Metric

In line with previously published literature, we use Spearman's rank correlation as the evaluation metric. This metric measures the correlation between two sequences containing ordinal or numerical data. It is calculated using the equation:

$$\rho = 1 - \frac{6 \sum d_i^2}{n(n^2 - 1)} \qquad (4)$$

ρ = Spearman's rank correlation
d_i = The difference between the ranks of corresponding variables
n = Number of observations

4.3   Implementation Details

We implemented our proposed methods using PyTorch [17]. All the 3D ResNets and (2+1)D ResNets processing 16 frame clips were pre-trained on Kinetics-700 [9] dataset[1]. The $(2+1)$D Resnets processing 8 and 32 frame clips were pre-trained on IG-65M dataset [4] and fine tuned on Kinetics-400 [9] dataset[2].

For each ResNet, we separately experimented using both averaging and WD as feature aggregation. We temporally augmented by randomly picking an ending frame from the last 6 and chose the preceding 96 frames for processing. The frames were resized to 171×128 and center cropped to 112×112. Random horizontal flipping was applied. Batch-normalization was used for regularization. We defined the loss function as a sum of L2 and L1 loss between the predicted score and ground-truth score as Parmar and Morris [16] suggested. We trained the network using the ADAM optimizer [10] for 50 epochs. We used a learning rate of 0.0001 for modules with randomly initialized weights and 0.00001 for modules with pretrained weights. Train and Test batch sizes were 2 and 5.

4.4   Results on MTL-AQA Dataset

In Table 1a, we present the experiment results of varying the depth of the ResNet feature extractor and the aggregation scheme. We can see that 34 layer $(2+1)$D

---

[1] Weights available at: https://github.com/kenshohara/3D-ResNets-PyTorch.
[2] Weights available at: https://github.com/moabitcoin/ig65m-pytorch.

Table 1. Performance comparison of the various types of ResNets as feature extractors and varying clip length in our pipeline

(a) Effect of various types of ResNets

Depth	Convolution	Aggregation	
		Average	WD
ResNet-34	3D	0.8982	0.8961
	(2+1)D	0.8932	0.8990
ResNet-50	3D	0.8880	0.8935
	(2+1)D	0.8818	0.8814
ResNet-101	3D	0.6663	0.6033

(b) Effect of varying input clip size in ResNet(2+1)D-34

Clip length	Aggregation	
(Input Frames)	Average	WD
8	0.8579	0.8853
16	0.8932	0.8990
32	0.9289	0.9315

ResNet with WD as aggregation performs the best with a Spearman's correlation of 0.8990. Increasing the depth to 50 layers somewhat decreases Spearman's correlation. However, the results are still competitive. At 101 layer depth, even when initialized with pretrained weights from Kinetics [9], overfitting occurs fairly quickly. The overfitting is also evident from the train/test curves presented in Fig. 4. The likely reason behind this is the increased number of parameters in the feature extractor. This leads us to establish that the current biggest AQA dataset has enough data to train a 34-layer and 50-layer ResNet feature extractor with generalization, however it overfits the 101-layer ResNet feature extractor.

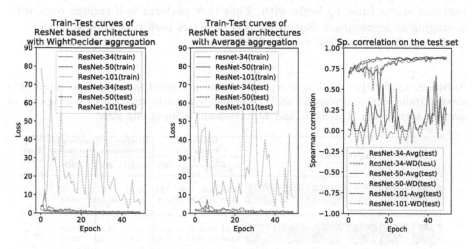

Fig. 4. Train and Test curves obtained from training the pipeline using 3D-ResNet 34, 50, and 101 as feature extractors. Notice that ResNet-101 based architectures show signs of significant overfitting compared to ResNet-34 and ResNet-50 based architectures.

Because of how $(2+1)$D ResNets are designed, they have a similar parameter count to their 3D counterparts [21]. Because the overfitting is occurring due to the high parameter count, we do not repeat the experiment with a $(2+1)$D ResNet-101 feature extractor.

Effect of Clip Length: We check the effect of clip length on the performance. We take the best performing model from Table 1a (ResNet$(2+1)$D-34-WD) and input various clip sizes. We experiment with clip sizes of 8 frames, 16 frames, and 32 frames. We vary the aggregation method as well.

From Table 1b we can see that the performance of the pipeline increases with the number of frames in each clip. We hypothesize that longer clips allow the ResNet to look for bigger patterns in the temporal dimension, which in turn enables the feature descriptors to be more informative. This enables the linear-regressor to better discriminate between similar-looking examples with fine-grained action quality differences. From Table 1b, notice that using WD over simple averaging as aggregation gives a boost in performance. However, this performance boost is quite significant in case of 8 frame clips. We believe the reason behind this improved performance is that increasing clip size reduces the number of clips being averaged. Whatever detrimental effect the averaging might have, it will be more prominent when the number of objects being averaged is larger, and less when this number is smaller. Furthermore, CNNs with longer input clips can look at more frames, this in effect increases their temporal horizon. It follows that the feature vectors extracted would have a better encoding of action patterns across time, to begin with. Thus they perform well enough even with averaging as aggregation. But using WD increases performance nevertheless.

For qualitative results, refer to Table 2.

Table 2. Qualitative results. Every 16th frame processed starting from frame 0 is shown. Italic scores correspond to WD aggregation, plain text scores correspond to average aggregation. The 8, 16, and 32 correspond to input clip sizes.

Input Frames	ResNet-34 Prediction				ResNet-50 Prediction		ResNet-101 Prediction	True Score
	(2+1)D			3D	(2+1)D	3D	3D	
	8	16	32					
	54.84	30.46	8.39	7.29	33.23	34.10	45.22	25.64
	38.76	*18.11*	*16.41*	*22.93*	*38.29*	*29.93*	*52.21*	
	66.94	59.69	47.92	67.92	43.57	58.30	122.20	52.79
	63.85	*40.88*	*53.21*	*63.62*	*52.80*	*52.31*	*76.64*	
	71.46	71.34	69.39	83.34	67.38	80.41	167.60	69.59
	69.85	*64.90*	*70.40*	*81.31*	*67.53*	*74.25*	*132.50*	
	67.13	46.16	27.73	34.25	44.06	46.61	54.28	46.20
	64.54	*32.29*	*42.87*	*39.62*	*49.03*	*47.13*	*51.51*	

Table 3. Comparison with the state of the art on the MTL-AQA dataset

Method	Sp. corr.
Pose+DCT [18]	0.2682
C3D-SVR [15]	0.7716
C3D-LSTM [15]	0.8489
MSCADC-STL [16]	0.8472
MSCADC-MTL [16]	0.8612
USDL-Regression [19]	0.8905
C3D-AVG-STL [16]	0.8960
C3D-AVG-MTL [16]	0.9044
MUSDL [19]	0.9273
Ours C3D-WD	0.9223
Ours ResNet34-(2 + 1)D-WD (32 frame)	**0.9315**
Ours ResNet50-3D-WD (16 frame)	0.8935
Ours ResNet101-3D-AVG (16 frame)	0.6633

The ground truth scores provided in the MTL-AQA [16] are taken from expert Olympic judges during a live broadcast of events. 7 judges independently score the athlete's performance on a scale of 0 (completely failed) to 10 (excellent). The 3 median scores are then added together and multiplied with a predetermined difficulty degree to obtain the final score. This final score is the one provided in the "True Score" column of Table 2. Our various pipelines attempt to predict this final score from the input performance video.

Comparison with the State of the Art: In Table 3, we compare our best performing models of each depth with previous works on the MTL-AQA dataset. For comparison, we combined C3D architecture with WD to test the result. For the C3D-WD model, 16 frame clips were used. The C3D portion of the model was initialized with Sports-1M [8] pretrained weights. We include this result in Table 3 as well. We can see that our $(2 + 1)$D ResNet-34 (32 frame)- WD outperforms all previous works in the literature. This shows the effectiveness of our approach. We can further see that 3D ResNet-50 (16 frame) obtains results comparable to the SOTA. However, the ResNet-101 based approach overfits the dataset and hence performs poorly.

5   Conclusion

In this work, we proposed a ResNet-based regression-oriented pipeline for action quality assessment. We demonstrated experimentally that the MTL-AQA dataset has enough data to train 34 and 50 layer ResNet-based pipelines when initialized with pretrained weights from a related task (like action recognition).

Our experiments suggest processing longer clips is more effective than using deeper ResNets. We also propose a lightweight learning-based aggregation technique called WD to replace simple averaging. Experiments show our methods to be more effective than previous works. In the future, we want to investigate with CNNs that can process longer clips (64 or higher) to see if this translates to better performance.

References

1. Carreira, J., Zisserman, A.: Quo vadis, action recognition? A new model and the kinetics dataset. In: 2017 IEEE Conference on Computer Vision and Pattern Recognition, CVPR 2017, Honolulu, HI, USA, 21–26 July 2017, pp. 4724–4733. IEEE Computer Society (2017). https://doi.org/10.1109/CVPR.2017.502
2. Diba, A., et al.: Spatio-temporal channel correlation networks for action classification. In: Ferrari, V., Hebert, M., Sminchisescu, C., Weiss, Y. (eds.) ECCV 2018. LNCS, vol. 11208, pp. 299–315. Springer, Cham (2018). https://doi.org/10.1007/978-3-030-01225-0_18
3. Funke, I., Mees, S.T., Weitz, J., Speidel, S.: Video-based surgical skill assessment using 3D convolutional neural networks. Int. J. Comput. Assist. Radiol. Surg. **14**(7), 1217–1225 (2019). https://doi.org/10.1007/s11548-019-01995-1
4. Ghadiyaram, D., Tran, D., Mahajan, D.: Large-scale weakly-supervised pre-training for video action recognition. In: IEEE Conference on Computer Vision and Pattern Recognition, CVPR 2019, Long Beach, CA, USA, 16–20 June 2019, pp. 12046–12055. Computer Vision Foundation/IEEE (2019). https://doi.org/10.1109/CVPR.2019.01232
5. Hara, K., Kataoka, H., Satoh, Y.: Learning spatio-temporal features with 3d residual networks for action recognition. In: 2017 IEEE International Conference on Computer Vision Workshops, ICCV Workshops 2017, Venice, Italy, 22–29 October 2017, pp. 3154–3160. IEEE Computer Society (2017). https://doi.org/10.1109/ICCVW.2017.373
6. He, K., Zhang, X., Ren, S., Sun, J.: Deep residual learning for image recognition. In: 2016 IEEE Conference on Computer Vision and Pattern Recognition, CVPR 2016, Las Vegas, NV, USA, 27–30 June 2016, pp. 770–778. IEEE Computer Society (2016). https://doi.org/10.1109/CVPR.2016.90
7. Hochreiter, S., Schmidhuber, J.: Long short-term memory. Neural Comput. **9**(8), 1735–1780 (1997). https://doi.org/10.1162/neco.1997.9.8.1735
8. Karpathy, A., Toderici, G., Shetty, S., Leung, T., Sukthankar, R., Li, F.: Large-scale video classification with convolutional neural networks. In: 2014 IEEE Conference on Computer Vision and Pattern Recognition, CVPR 2014, Columbus, OH, USA, 23–28 June 2014, pp. 1725–1732. IEEE Computer Society (2014). https://doi.org/10.1109/CVPR.2014.223
9. Kay, W., et al.: The kinetics human action video dataset. arXiv preprint arXiv:1705.06950 abs/1705.06950 (2017)
10. Kingma, D.P., Ba, J.: Adam: a method for stochastic optimization. In: Bengio, Y., LeCun, Y. (eds.) 3rd International Conference on Learning Representations, ICLR 2015, San Diego, CA, USA, 7–9 May 2015, Conference Track Proceedings (2015). http://arxiv.org/abs/1412.6980
11. LeCun, Y., Bengio, Y.: Convolutional Networks for Images, Speech, and Time Series, pp. 255–258. MIT Press, Cambridge (1998). https://doi.org/10.5555/303568.303704

12. Leong, M., Prasad, D., Lee, Y.T., Lin, F.: Semi-CNN architecture for effective spatio-temporal learning in action recognition. Appl. Sci. **10**, 557 (2020). https://doi.org/10.3390/app10020557
13. Parmar, P., Morris, B.T.: Measuring the quality of exercises. In: 2016 38th Annual International Conference of the IEEE Engineering in Medicine and Biology Society (EMBC), pp. 2241–2244 (2016). https://doi.org/10.1109/EMBC.2016.7591175
14. Parmar, P., Morris, B.: Action quality assessment across multiple actions. In: IEEE Winter Conference on Applications of Computer Vision, WACV 2019, Waikoloa Village, HI, USA, 7–11 January 2019, pp. 1468–1476. IEEE (2019). https://doi.org/10.1109/WACV.2019.00161
15. Parmar, P., Morris, B.T.: Learning to score Olympic events. In: 2017 IEEE Conference on Computer Vision and Pattern Recognition Workshops, CVPR Workshops 2017, Honolulu, HI, USA, 21–26 July 2017, pp. 76–84. IEEE Computer Society (2017). https://doi.org/10.1109/CVPRW.2017.16
16. Parmar, P., Morris, B.T.: What and how well you performed? A multitask learning approach to action quality assessment. In: IEEE Conference on Computer Vision and Pattern Recognition, CVPR 2019, Long Beach, CA, USA, 16–20 June 2019, pp. 304–313. Computer Vision Foundation/IEEE (2019). https://doi.org/10.1109/CVPR.2019.00039
17. Paszke, A., et al.: Automatic differentiation in PyTorch. In: NIPS 2017 Workshop on Autodiff. Long Beach, California, USA (2017). https://openreview.net/forum?id=BJJsrmfCZ
18. Pirsiavash, H., Vondrick, C., Torralba, A.: Assessing the quality of actions. In: Fleet, D., Pajdla, T., Schiele, B., Tuytelaars, T. (eds.) ECCV 2014. LNCS, vol. 8694, pp. 556–571. Springer, Cham (2014). https://doi.org/10.1007/978-3-319-10599-4_36
19. Tang, Y., et al.: Uncertainty-aware score distribution learning for action quality assessment. In: 2020 IEEE/CVF Conference on Computer Vision and Pattern Recognition, CVPR 2020, Seattle, WA, USA, 13–19 June 2020, pp. 9836–9845. IEEE (2020). https://doi.org/10.1109/CVPR42600.2020.00986
20. Tran, D., Bourdev, L.D., Fergus, R., Torresani, L., Paluri, M.: Learning spatiotemporal features with 3D convolutional networks. In: 2015 IEEE International Conference on Computer Vision, ICCV 2015, Santiago, Chile, 7–13 December 2015, pp. 4489–4497. IEEE Computer Society (2015). https://doi.org/10.1109/ICCV.2015.510
21. Tran, D., Wang, H., Torresani, L., Ray, J., LeCun, Y., Paluri, M.: A closer look at spatiotemporal convolutions for action recognition. In: 2018 IEEE Conference on Computer Vision and Pattern Recognition, CVPR 2018, Salt Lake City, UT, USA, 18–22 June 2018, pp. 6450–6459. IEEE Computer Society (2018). https://doi.org/10.1109/CVPR.2018.00675
22. Xiang, X., Tian, Y., Reiter, A., Hager, G.D., Tran, T.D.: S3D: stacking segmental P3D for action quality assessment. In: 2018 IEEE International Conference on Image Processing, ICIP 2018, Athens, Greece, 7–10 October 2018, pp. 928–932. IEEE (2018). https://doi.org/10.1109/ICIP.2018.8451364

Improving Licence Plate Detection Using Generative Adversarial Networks

Alden Boby[✉] and Dane Brown

Department of Computer Science, Rhodes University, Grahamstown, South Africa
boby.alden128@gmail.com, d.brown@ru.ac.za

Abstract. The information on a licence plate is used for traffic law enforcement, access control, surveillance and parking lot management. Existing licence plate recognition systems work with clear images taken under controlled conditions. In real-world licence plate recognition scenarios, images are not as straightforward as the 'toy' datasets used to benchmark existing systems. Real-world data is often noisy as it may contain occlusion and poor lighting, obscuring the information on a licence plate.

Cleaning input data before using it for licence plate recognition is a complex problem, and existing literature addressing the issue is still limited. This paper uses two deep learning techniques to improve licence plate visibility towards more accurate licence plate recognition. A one-stage object detector popularly known as YOLO is implemented for locating licence plates under challenging situations. Super-resolution generative adversarial networks are considered for image upscaling and reconstruction to improve the clarity of low-quality input. The main focus involves training these systems on datasets that include difficult to detect licence plates, enabling better performance in unfavourable conditions and environments.

Keywords: Object detection · Optical character recognition · Generative adversarial networks

1 Introduction

Detecting licence plates in an unconstrained environment is challenging as real-world data is unideal and can have a low resolution, resulting in reduced image quality, affecting licence plate recognition (LPR) results, and subsequently, character recognition [7]. Existing LPR systems work effectively with 'flawless' input data, i.e. high-resolution images with negligible distortion or occlusion [4]. In practice, most LPR systems are set up to capture images in controlled conditions such as at boom gates. High clarity input is required to recognise licence plates in the wild. A potential solution to the problem is using deep learning techniques to produce a high-resolution image that is perceptually clearer [6].

This work was undertaken in the Distributed Multimedia CoE at Rhodes University.

Licence plates can be used to identify stolen vehicles, automatically bill car owners for parking and passing through toll gates, aiding security and traffic regulation [4,15]. Creating a LPR system that can sustain good accuracy when fed varying quality data can help automate manual licence plate data collection.

Current literature tends towards the application of deep learning to improve the performance of LPR systems. Despite this, existing literature is primarily image processing based. The use of generative adversarial networks (GANs) for image enhancement, in particular, is minimal. The input data is the fundamental problem to overcome in such a system, as unclear data may lead to erroneous output. This paper aims to show the effectiveness of advanced GAN algorithms at improving the input into LPR systems—towards higher character recognition rates. Using specialised cameras to mitigate unclear input is costly as it is hardware-specific. A software-based solution can reduce expensive hardware costs.

2   Related Work

You only look once (YOLO) is a framework for identifying objects within an image in a single pass using a specialised convolutional neural network (CNN) architecture. YOLO is a general-purpose object detector, making it attractive for LPR [3].

Lee et al. [7] trained the first version of YOLO to detect licence plates using general imaging data rather than licence plate-specific data, achieving a total recall of 93.2%.

YOLOv2 improves over its first version by including batch normalisation layers and pruning insignificant convolutional layers. Laroca et al. [5] used YOLOv2 in two stages of their pipeline, first to detect vehicles in an image or frame, followed by detection of the number plates on the vehicles located in the first instance.

YOLOv3 uses a backbone called Darknet-53, which outperforms more complex backbones such as ResNet-101 and ResNet-152 in accuracy whilst providing a 1.5× speedup [11]. YOLOv3 has improved bounding box predictions on small objects and is substantially better at utilising GPU resources. Input images for LPR systems may have cars far from the camera, i.e. a tiny and misaligned region of interest (ROI): thus, YOLOv3 is a good choice.

Lee et al. [7], explored super-resolution using a GAN instead of traditional CNN methods, which produced blurry high-resolution images lacking clarity. A GAN comprises two neural networks: A discriminator and a generator. These networks work against each other, learning through adversarial loss. The resulting images formed by GANs are sharper as they use adversarial loss rather than mean squared error (MSE) loss. MSE methods used by existing deep neural networks produce reasonable high-resolution output [8]. However, they do not utilise the full potential of such networks resulting in blurry images. Furthermore, MSE does not equate to the human perception of image quality and fidelity; an obscure image can have the same MSE as one that is perceptually clearer to the human eye [14]. More applicable to this field of study, a super-resolution generative adversarial network (SRGAN) can be used for image upscaling. The SRGAN is fed low-resolution images and produces high-resolution images; the discriminator's job is to distinguish whether an input image is generated or a true high-resolution image [6]. The enhanced super-resolution generative adversarial network (ESRGAN) further improves the appearance of upscaled images by providing sharper details typically lost with standard super-resolution techniques [13].

Yuan *et al.* [16] argued that an end-to-end mapping of low- and high-resolution images is impractical in the real world as low-resolution images rarely have a corresponding high-resolution counterpart. They aimed to use GANs to achieve super-resolution via unsupervised learning. This is in contrast to Lee *et al.* [7] who utilised end-to-end mapping for training their GAN.

Existing literature use low-resolution images that have been downsampled using the same kernel, creating an artificial consistency within the dataset, contrary to real-world data, which is more erratic [16]. GANs tackle the problem of unsupervised learning as they can visualise data that is not directly present within the input.

Denoising an image typically eliminates minor imperfections whilst keeping the majority of important structural information by using values from neighbouring pixels [10]. Yuan *et al.*'s [16] solution included mapping low-resolution images to denoised images of the same resolution. Subsequently, the clean image was upsampled to a higher resolution using an existing GAN.

The results of the model in their paper were evaluated using peak signal-to-noise ratio (PSNR) and structural similarity index measures (SSIM) which are used to measure the difference between two images.

Brisnello *et al.* [2] considered upscaling images to see its effect on the performance of optical character recognition (OCR). OCR converts text from an image by classifying characters within the image. OCR is not without weaknesses. Some characters are almost indistinguishable from one another, such '1' and '1' and '0' and 'O' [9]. Before applying OCR, they used bicubic interpolation to upscale their images, as it preserves more detail than many alternative methods. Their results demonstrated that resizing an image increased OCR accuracy by up to 20%.

3   Proposed System

The LPR system is divided into three stages. The first stage handled the location and classification of a licence plate.

In the second stage, the input was an image of the ROI from the predicted coordinates supplied by the YOLO model. The ESRGAN upscaled the ROI by 4× to produce a more precise image.

The final stage involved using Pytesseract to analyse the image and produce a string of characters as output. The output of Pytesseract was compared across both the high-resolution and low-resolution images.

3.1 ROI Extraction

The ROI, a licence plate in an input image, was extracted using YOLOv3. The YOLO model was configured to recognise one class—licence plates.

Object Detection

The YOLO model is made up of fifty-three convolutional layers. The architectural structure of the layers in the model is shown in Table 1.

Table 1. Darknet-53 architecture.

Type	Filters	Size	Output
Convolutional	32	3×3	256×256
Convolutional	64	$3 \times 3/2$	128×128
Convolutional	32	1×1	
Residual	64	3×3	128×128
Convolutional	128	$3 \times 3/2$	64×64
Convolutional	64	1×1	
Residual	128	3×3	64×64
Convolutional	256	$3 \times 3/2$	32×32
Convolutional	128	1×1	
Residual	256	3×3	32×32
Convolutional	512	$3 \times 3/2$	16×16
Convolutional	256	1×1	
Residual	512	3×3	16×16
Convolutional	1024	$3 \times 3/2$	8×8
Convolutional	512	1×1	
Residual	1024	3×3	8×8
Global average pooling			
Fully connected	1000		
Softmax			

The YOLO model worked on a set of annotated data that included bounding box coordinates to do feature extraction. These features were stored as weights forming a trained model. The model predicted bounding box coordinates of licence plates based on these weights.

Training Parameters

Bochkovskiy et al. [1] suggested training parameters based on the number of classes the YOLO model is configured to predict. The formulas provided were as follows:

$$\text{maxbatches} = \text{number of classes} \times 2000. \tag{1}$$

$$\text{filters} = (\text{number of classes} + 5) \times 3. \tag{2}$$

The YOLO model configuration file was tuned to the following parameters:

- batchsize = 64
- subdivisions = 16
- maxbatches = 2000
- filters = 18

Classification
Once the YOLO model predicts a bounding box, it classifies the image within the coordinates using independent logistic regression classifiers. Furthermore, the model outputs a confidence value per prediction with a probability between one and zero.

3.2   Image Upscaling

The ESRGAN received an input image containing the ROI from the YOLO model. The image was upscaled to 4× its resolution, producing an image with sharper details than standard non-machine learning upscaling methods, such as bicubic interpolation. The ESRGAN uses a Residual-in-Residual Dense Block (RDDB) to allow for easier training of the model. The ESRGAN also improves the discriminator, judging whether one image is more realistic than the other rather than if an image is generated or authentic; this is what gives the ESRGAN the edge over a standard SRGAN [13].

3.3   Licence Plate Recognition

LPR was done using Pytesseract, a Python wrapper for Tesseract OCR[1]. Tesseract OCR is pre-trained and boasted as an 'out-of-the-box' OCR engine. Adaptive thresholding is applied to convert an input image to binary, followed by connected component analysis to extract character outlines within the binary image. Lastly, Tesseract classifies the text through its training data [9]

4   Experimental Setup

A set of performance metrics were used to evaluate the machine learning models. The YOLO model was evaluated with mean average precision (mAP) and intersection over union (IoU). The images generated by the ESRGAN were evaluated using PSNR and SSIM.

Intersection over Union
IoU is a metric for evaluating an object detection model's prediction capabilities. It represents the ratio between the overlap and the area of intersection between a ground truth bounding box and a predicted bounding box. Higher

---

[1] https://github.com/tesseract-ocr/tesseract.

values represent fewer differences between the predicted bounding box and the ground truth.

$$\text{Intersection over Union (loU)} = \frac{\text{Area of Overlap}}{\text{Area of Union}}. \tag{3}$$

In object detection tasks, IoU is used to calculate precision and recall. A given IoU threshold defines the cut off between false positives and true negatives. If the prediction is below the threshold, it is considered a false positive, and if it is above, it is considered a true positive.

Mean Average Precision

Mean average precision is a standard evaluation metric in computer vision, especially in object detection. It was used to evaluate the YOLO model.

Peak Signal-to-Noise Ratio

The PSNR values were calculated by downsampling a set of images by a scale of four. The downsampled images were upscaled with the ESRGAN and compared with the corresponding ground truth images.

Structural Similarity Index

The SSIM metric was used to calculate the quality of images generated by the ESRGAN, using ground truth images to measure the similarity between the two images.

Character Recognition Rate

The character recognition rate was used to measure the accuracy of predictions made at the OCR stage.

$$\text{character recognition rate} = n/(all + m). \tag{4}$$

where n represents the amount of correctly guessed characters, all represents the total number of characters present in the ground truth, and m represents the number of incorrect predictions or changes required to rectify errors in the prediction [12].

4.1 Datasets

Four datasets were divided into groups. The MediaLAB LPR and Croatian Licence Plate datasets were combined as a training dataset for the YOLO model and ESRGAN. The validation of the YOLO model followed an 80:20 split. One hundred images were used as a training set, and twenty-five pre-annotated images were used as a validation set to rate the performance of the YOLO model. Transfer learning was used to supplement the small pre-annotated data set. All images fed into the YOLO model were downsampled to a resolution of 416×416. MakeML Car Licence Plates and Caltech dataset were used for testing. The following subsections break down the structure of the four datasets.

MediaLAB LPR Dataset
Sixty images were selected from the MediaLAB LPR Dataset to train the YOLO model. The MediaLAB database consists of 12 categorised sets of images. The sets chosen were plates with dirt and shadows to familiarise the object detection model with challenging conditions.

Croatian Licence Plate Dataset
The dataset comprises 500 images containing several different vehicles such as trucks, vans, SUVs and sedans. Images in the dataset are taken from the rear, frontal view and a few at oblique angles. All images in this dataset have a fixed resolution of 640×480.

MakeML Car Licence Plates Dataset
A dataset containing 433 images of cars taken at varying angles. There is inconsistency in the dataset in terms of angle and orientation of the ROI, making it prone to misalignment and consequently challenging. The resolution of the images varies between 400×400 and 600×450.

Caltech Dataset
The Caltech dataset consists of 126 images of vehicles from the Caltech Institute of Technology carpark. All images feature a resolution of 896×592. All licence plates in the dataset are of the rear end of a car. The majority of licence plates are Californian, with few from out of state.

5   Results and Discussion

A set of experiments were created to test the performance of all the stages of the LPR system. Prior to being used in the LPR system, the YOLO model was tested using a validation set of twenty-five images selected from the Media LPR and Croatian licence plate datasets. These images were manually annotated with the correct coordinates of their ROI, which were then compared to the predicted coordinates from the model.

Table 2. Average scores achieved by YOLO object detection.

Metric	Score (%)
Precision	100
Recall	100
f1-score	100
mAP	100
IoU	81.06

5.1 Mean Average Precision

Table 2 shows that the mAP on the validation set was 100%. Several reasons could explain why the model achieves perfect validation precision, such as the size of the validation dataset being insufficient. Moreover, the dataset may not have had enough variance, meaning that all the images in the set were similar, making it easier for the YOLO model to detect. However, these possibilities are only justified if the unseen test accuracies are low.

5.2 IoU

As shown in (the same) Table 2, the average IoU for this model was 81.06%. Figure 1 shows a diagram representing the achieved IoU for this model. The ground truth bounding boxes were very tight around the licence plates; in practice, there is room for a higher tolerance when predicting bounding boxes.

Fig. 1. Illustration of an average IoU of 81.06%.

5.3 Experiment 1

Fifty images were passed through the LPR system. The datasets used for this were the MediaLAB LPR Dataset and Croatian Licence Plates Dataset. The final output from the system was a string of characters representing what was detected on a licence plate. The OCR engine was fed both an original resolution ROI image and a high-resolution counterpart generated by the ESRGAN.

The average accuracy from this experiment was 65.90% for high-resolution images generated by the ESRGAN and 58.44% for images at their original resolution as shown in Table 3. This signifies a 7.46% increase in overall accuracy by using super-resolution techniques to improve the output from the OCR engine.

Table 3. Comparison of system performance between high resolution (HR) and low resolution (LR) input.

Metric	LR (%)	HR (%)
Accuracy	58.44	65.90
Recall	67.59	73.88
Precision	64.22	70.75

The OCR engine generally performed better when given high-resolution images generated by the ESRGAN. In some instances, the OCR engine could not produce any output without upscaling the licence plate. However, when supplied with an upscaled version of the same image generated by the ESRGAN, the OCR engine predicted the text with 100% accuracy. As seen in Fig. 2, the ESRGAN helped recover a significant amount of detail.

(a)   (b)

Fig. 2. (a) Original resolution of the ROI could not be interpreted by Pytesseract. (b) generated by the ESRGAN was interpreted with 100% accuracy.

Conversely, there are instances where the upscaled image performed worse than the input image at its original size. The upscaled version (Fig. 3b) is visually clearer to the human eye than the original image (Fig. 3a). However, the OCR engine misinterpreted the character '1' in the high-resolution image and presented it as a '4'. This result was an outlier, and this occurrence was not consistent throughout the results.

(a) Image 38 at its original resolu-   (b) Upscaled version of image 38.
tion.

Fig. 3. Despite (b) being clearer to the human eye, (a) produced more accurate output at the OCR stage.

(a) Image 29.   (b) Image 45.

Fig. 4. Pytesseract was unable to classify the characters in these images due to their orientation.

The Figs. 4a and b are visually clear to the human eye, but Pytesseract was unable to interpret them. This is due to the orientation of the text in the images, as the OCR engine performs better with a straight line of text rather than skewed or rotated text. This problem is beyond the scope of this paper.

Some results were 100% accurate given the input, but the characters were truncated at the object detection stage. As the truncation resulted from the YOLO model failing, the OCR ground truth values were modified to be compared to the cropped images.

A major problem affecting the LPR system output was ambiguous characters. Typefaces with serifs were correctly predicted as they made the characters easily distinguishable. Contrarily, sans-serif fonts caused problems at the OCR stage, leading to erroneous output whether or not the input to the system was upscaled.

5.4 Experiment 2

The focus of this experiment was to measure the accuracy of the YOLO model and its ability to predict the ROI on an unseen image. This was done in conjunction with Experiment 1.

As shown in Table 4, the YOLO model had a detection rate of 92% on a completely unseen set of data. However, this is not comparable to other systems, as the dataset has not been documented in any other existing literature.

Table 4. YOLO detection rate.

Dataset	Detection rate (%)
MakeML - Car Licence Plates	92
Caltech	75

Examples of the images the model was unable to predict are shown in Fig. 5.

(a)   (b)

Fig. 5. Undetectable by the YOLO model.

5.5   Experiment 3

This experiment aimed to measure the visual clarity of images produced by the ESRGAN. PSNR was used to compare two images. To produce an upscaled image from the dataset, images were downsampled from their original resolution and passed into the ESRGAN. A generated image from the ESRGAN was compared with the original counterpart, and the PSNR was calculated and averaged for each image in the dataset. The images were reduced in resolution by using the bicubic downsampling kernel of OpenCV. The downsampled images were upscaled using the trained ESRGAN model.

The Croatian number plate dataset and Caltech dataset images were reduced to 4× their resolution as the ESRGAN upscales by a factor of four. The output resolution was equal to that of the ground truth images.

Table 5. PSNR results.

Dataset	Upscaling method	PSNR
Car Licence Plates	ESRGAN	17.50
	Bicubic	23.68
Caltech Dataset	ESRGAN	14.92
	Bicubic	20.37

The PSNR values achieved using the ESRGAN are shown in Table 5. The Croatian Licence Plate dataset had an average PSNR value of 17.5 over 50 images, and the Caltech dataset had an average PSNR value of 14.92 over a sample size of 126. The differences in averages are likely due to the sample sizes of the data sets being used. The highest PSNR value achieved for the Croatian dataset was 20.5, and for the Caltech dataset, the highest value was 21.

These PSNR values indicate that the ESRGAN model needs more training data to improve its output. It is important to note the input images were very low quality for this experiment, so perceptually, they appear worse than the images generated in Experiment 1. The ESRGAN struggled to reconstruct the images to match the original images closely. Compared to the low-resolution image, however, the generated image has sharper details but is heavily artifacted. Figure 6a shows the original resolution image, and Fig. 6b shows the same image reconstructed by the ESRGAN from a downsampled version of the same image.

(a)   (b)

Fig. 6. Figure (b) shows output generated from the ESRGAN using a downscaled version of figure (a).

Table 6. Comparison of SSIM results between bicubic upscaling and upscaling via the ESRGAN.

Dataset	Upscaling method	SSIM (%)
Car Licence Plates	Bicubic	76
	ESRGAN	40
Caltech Dataset	Bicubic	56
	ESRGAN	26

SSIM results were calculated for images that were upscaled 4× through bicubic interpolation and are shown in Table 6. The bicubic upscaled images had a higher SSIM value than the generated images from the ESRGAN. This makes sense as the ESRGAN tries to create an image as close to the ground truth as possible by inferring from the available data in a given image. In comparison, the bicubic interpolation preserves the original structure of the image, only slightly altering pixel values. The ESRGAN does not do very well when upscaling extremely low-resolution images.

5.6 Experiment 4

This experiment measured the correlation between the confidence values and the accuracy of the predicted bounding boxes. This experiment used the Caltech dataset. The quality of the predicted bounding box was inspected visually and compared with its confidence value.

The average confidence value from detection on the Caltech dataset was 50%. These results indicate that various features within an image can cause the algorithm to give a low confidence value. While the majority of the time, a low percentage indicated a poor bounding box prediction, at times, the model perfectly extracted the ROI within a given image and classified the licence plate with a sub-fifty confidence value. Conversely, certain predictions with higher confidence values had truncated the information in the bounding box as the box's left and right edges were too small for the ROI.

The YOLO model detected licence plates in 94 out of 126 (75%) images in the Caltech dataset. For the undetected images, the model gave a confidence value of zero. The model was designed to omit predictions with a confidence value below 20% as they likely produce poor output.

6   Conclusion

This study shows that modifying input images and improving their visual quality typically improves the accuracy of LPR systems. A 7.46% increase in accuracy was observed when high-resolution images were used as input over their lower-resolution counterparts. When evaluated with the test data, the YOLO model showed promising results, boasting 100% for all metrics: precision, recall, and mAP. However, this was only true when the model was tested on validation data. In practice, the model was 83.5% accurate, predicting bounding boxes more often than not. A larger dataset would help to improve the performance of the model.

The system, therefore, is not yet commercially ready as it can benefit from future work. A more abundant and varied dataset can advance the machine learning models. More information allows the machine learning model to cater to more diverse situations. A weakness of the proposed system is the performance of the OCR engine, which is an out-of-the-box model designed to cater to a wide range of texts. A fine-tuned OCR engine would perform better as it would better suit the given task.

References

1. Bochkovskiy, A., Wang, C.Y., Liao, H.Y.M.: YOLOV4: optimal speed and accuracy of object detection. arXiv preprint arXiv:2004.10934 (2020)
2. Brisinello, M., Grbić, R., Pul, M., Anđelić, T.: Improving optical character recognition performance for low quality images. In: 2017 International Symposium ELMAR, pp. 167–171. IEEE (2017)
3. Du, J.: Understanding of object detection based on CNN family and YOLO. In: J. Phys. Conf. Ser. **1004**, 012029 (2018). IOP Publishing
4. Du, S., Ibrahim, M., Shehata, M., Badawy, W.: Automatic license plate recognition (ALPR): a state-of-the-art review. IEEE Trans. Circuits Syst. Video Technol. **23**(2), 311–325 (2012)
5. Laroca, R., et al.: A robust real-time automatic license plate recognition based on the YOLO detector. In: 2018 International Joint Conference on Neural Networks (IJCNN), pp. 1–10. IEEE (2018)
6. Lee, D., Lee, S., Lee, H., Lee, K., Lee, H.J.: Resolution-preserving generative adversarial networks for image enhancement. IEEE Access **7**, 110344–110357 (2019)
7. Lee, Y., Yun, J., Hong, Y., Lee, J., Jeon, M.: Accurate license plate recognition and super-resolution using a generative adversarial networks on traffic surveillance video. In: 2018 IEEE International Conference on Consumer Electronics-Asia (ICCE-Asia), pp. 1–4. IEEE (2018)

8. Lucas, A., Lopez-Tapia, S., Molina, R., Katsaggelos, A.K.: Generative adversarial networks and perceptual losses for video super-resolution. IEEE Trans. Image Process. **28**(7), 3312–3327 (2019)
9. Patel, C., Patel, A., Patel, D.: Optical character recognition by open source OCR tool tesseract: a case study. Int. J. Comput. Appl. **55**(10), 50–56 (2012)
10. Qian, Y.: An improved restoration algorithm for blurred images based on complete blind convolution and non-local means filtering. In: 2019 Chinese Automation Congress (CAC), pp. 743–747 (2019). https://doi.org/10.1109/CAC48633.2019.8997281
11. Redmon, J., Farhadi, A.: YOLOv3: an incremental improvement (2018)
12. Shen, M., Lei, H.: Improving OCR performance with background image elimination. In: 2015 12th International Conference on Fuzzy Systems and Knowledge Discovery (FSKD), pp. 1566–1570. IEEE (2015)
13. Wang, X., et al.: ESRGAN: enhanced super-resolution generative adversarial networks. In: The European Conference on Computer Vision Workshops (ECCVW), September 2018
14. Wang, Z., Bovik, A.C.: Mean squared error: love it or leave it? A new look at signal fidelity measures. IEEE Sig. Process. Mag. **26**(1), 98–117 (2009)
15. Xie, L., Ahmad, T., Jin, L., Liu, Y., Zhang, S.: A new CNN-based method for multi-directional car license plate detection. IEEE Trans. Intell. Transp. Syst. **19**(2), 507–517 (2018)
16. Yuan, Y., Liu, S., Zhang, J., Zhang, Y., Dong, C., Lin, L.: Unsupervised image super-resolution using cycle-in-cycle generative adversarial networks. In: Proceedings of the IEEE Conference on Computer Vision and Pattern Recognition Workshops, pp. 701–710 (2018)

Film Shot Type Classification Based on Camera Movement Styles

Antonia Petrogianni[1], Panagiotis Koromilas[2](✉) ⓘ,
and Theodoros Giannakopoulos[2]

[1] University of Thessaly, Volos, Greece
apetrogianni@uth.gr
[2] National Center for Scientific Research - Demokritos, Athens, Greece
{pakoromilas,tyianak}@iit.demokritos.gr

Abstract. Visual information contains the most important characteristics of a movie regarding the related content and filming techniques. Especially the way the camera moves to capture the scene is vital to define the director's aesthetics. However, most of the machine learning tasks existing in the literature treat the movie as shallow content, rather than as an artistic work, and therefore focus on detecting objects and faces, recognizing activities and extracting plot-related topics. On the other hand, cinematography is closely connected to the choice of different ways to handle the camera, and thus camera movements include information that is useful in order to analyse the artistic style of a movie. In this work we present an original, publicly available (https://github.com/magcil/movie_shot_classification_dataset) dataset for film shot type classification that is associated with the distinction across 10 types of camera movements that cover the vast majority of types of shots in real movies. In addition, two different methods are evaluated on the new dataset, one static that is based on feature statistics across frames, and one sequential that tries to predict the target class based on the input frame sequence using LSTMs. Based on the evaluation process it is inferred that the sequential method is more suited for modeling the camera movements.

Keywords: Shot classification · Camera movement classification · Movie analysis

1 Introduction

Machine learning methodologies have been greatly applied in order to conduct movie analysis. Extensive research has been published on different kinds of problems such as movie recommendation [8,9,16,29], movie summarization [25,30,31], movie genre classification [10,27,34] or movie scene detection [2,12,24]. These problems have been faced using different types of approaches (e.g. traditional machine learning or deep learning architectures) or different types of representations (e.g. multimodal recommendation [5] or summarization [22]).

A. J. Pinho et al. (Eds.): IbPRIA 2022, LNCS 13256, pp. 602–615, 2022.
https://doi.org/10.1007/978-3-031-04881-4_48

Most existing tasks in computer vision primarily focus on content rather than style understanding. That is the analysis of movies, which are a work of art, cannot be properly conducted when excluding cinematographic techniques used by the director. Shot type classification could enrich movie analysis techniques and result in a more inclusive automated movie understanding.

The literature on shot classification is mostly centered around machine learning methods in sports events (e.g. [20]) that classify the corresponding shots into one of the following basic categories: long, medium, close and out-field respectively. Most of the movie shot analysis methods adopt this type of categorization in order to form the shot scale classification problem, where a shot is classified based on the apparent distance of the camera from the main subject of a scene. Most of the works (e.g. [3,7]) examine the three-class problem of long, medium and close-up shots. Recent methods have employed Deep Learning architectures [26] and others enrich their overall architecture with the use of semantic segmentation [1]. However, in these methods, the shot analysis is on frame level rather than on video level, and thus the Deep Learning architecture does not include the temporal dimension.

One of the first works to introduce a cinematographic shot taxonomy based on camera movements is [33], which ends up with the following classes: stationary, contextual-tracking, focus-tracking, focus-in, focus-out, intermittent/planing establishment and chaotic. This taxonomy is followed by [13] where a new dataset of 5226 shots is created. A method for videography-based video analysis is introduced in [17], where camera operation classification is used as a module on the proposed architecture in order to classify the shot among four categories, namely static, pan, tilt and zoom.

The so far mentioned works on camera movement classification and their predecessors (e.g. [21]) evaluate their respective approaches on their own private collections, which are not made available. The first publicly available dataset is introduced in [4] with shots categorized as aerial, bird eye, crane, dolly, establishing, pan, tilt, and zoom. However this collection is rather small, consisting of just 263 shots, where most of them are not from movies. The most relevant work is that of [23] MovieShot, a publicly available dataset that contains classes for both scale and camera movement. Still, the proposed camera movement types are rather generic including only static, motion, push shot (zoom in) and pull shot (zoom out).

To sum up, the existing datasets are either private or do not focus on camera movements, where some of them are not movie-oriented since they include other types of videos. However, camera movement is important for expressing mood and style in a movie, and as such it is crucial to be considered as an attribute in any movie analysis system: movie indexing and search, movie visualization and, of course, movie recommendation systems. To put it simple, the way the directors decide to move their cameras makes us, the viewers, like or not a movie, among other characteristics. In this work, we propose a publicly available dataset that contains shots coming strictly from movies. The proposed collection is concerned about camera movements and proposes the categorization of shots in the following extensive 10 classes: static, panoramic, zoom-in, travelling-out, vertical movement, aerial, travelling-in, tilt, handheld, panoramic lateral. We report classification metrics on our dataset based on *(i)* a static method which is based on aggregated statistics on the feature sequence, and *(ii)* a sequential method where an LSTM is applied directly on the sequential features which is not the usual case in the literature since most of the existing works depend on feature aggregations [23].

The rest of the paper is organized as follows. In Sect. 2 the dataset compilation along with the annotation procedure is discussed. The feature extraction process, as well as the proposed methodologies are explained in Sect. 3. Experimental results are provided in Sect. 4, where Sect. 5 concludes our work and proposes some future work directions.

2   Dataset Compilation

2.1   Video Shot Generation

Video *shots* are basic structural elements in film-making and video production that contain a series of frames and run for an uninterrupted period of time [6]. Types of shots can be characterized (among others) with regards to the respective camera movement. Camera movement in film-making is a technique that causes a change in frame or perspective through the movement of the camera. The types of camera movement are crucial in the direction film process, since through these, directors can cause separate feelings to the audience: for example, fast camera movements can cause the viewer anxiety or irritation.

In this work, our goal is to classify the types of shots based on the cinematic aesthetics of camera movements. Towards this end, we have created a manually annotated dataset. To our best knowledge, there is no corresponding dataset that represents a wide range of camera movement styles that is able to cover the vast majority of shot types in any movie. The dataset is created using video shots from 48 films in which a basic shot detection algorithm was applied to generate successive shots. The generated detected video shots were randomly sampled and then led to a multi-annotator pipeline, leaning to the final annotated dataset.

In order to detect shots from movies, we adopted a very basic and fast shot-detection algorithm that is based on three thresholding criteria. In particular, as a first step, the following features are computed on a frame basis:

- Average value of magnitudes of optical flow vectors (*mag_mu*). Optical flow vectors are modeling local movements of video blocks in successive frames [19]. In other words, a flow vector indicates how a particular block from a frame will "move" into the next frame. The rationale behind using this feature is that, if the value of magnitudes of flow vectors changed abruptly from the current frame to the next, then it would probably occur a shot change.
- Proportion of pixels with high absolute differences between two successive frames (*gray_diff*): in particular, we count pixels whose differences is over 50 between two successive frames.

– Average absolute diffs in the histograms of the gray values between two successive video frames (f_diff).

In order to select the aforementioned parameters, we created a small dataset of manually annotated shots (in terms of segment endpoints) and selected the parameters values $mag\_mu = 0.08$, $gray\_diff = 0.65$ and $f\_diff = 0.02$ which resulted to an almost 80 macro F1 performance in the binary task of endpoint shot detection, with a tolerance of 1 s (i.e. the allowed time distance from the ground truth shot endpoints). The aforementioned method has been implemented in the GitHub repository that also implements the basic feature extraction adopted in this work, called *multimodal_movie_analysis*[1].

As soon as this parameter setting of the shot detection algorithm was completed, we executed the algorithm on 48 movies from various genres and eras. This process resulted in around 80,000 detected shots. We then discarded shots shorter than 2 s and randomly selected 4000 shots as the final "annotation poo". In the next Sections we describe the annotation process of these 4000 shots and the way we handled the inter-annotator agreement to form the final ground-truth.

2.2 Video Shot Taxonomy

In order to define the set of classes in which the shots have been labelled, we collaborated with a professional director and at the same time we focused in having a minimum set of classes that covers as many shots as possible from all movies. Our basic criterion for defining the classes was the *type of camera movement*. Based on this, the following classes have been defined:

1. **Static**, The camera is locked on a tripod or pedestal and remains still. Among other types of scenes, it is commonly used in dialogues. A static camera does not necessarily indicate a static scene. Actors and even the background can move while the camera remains still. (Link-for-static-video)
2. **Vertical movement**, of the camera lens while the camera remains locked on a tripod. It is equivalent to someone tilting his head up/down. (Link-for-vertical-movement-video)
3. **Tilt**, Moving the entire camera up or down without moving its lens. Tilting up is like one is moving up his entire body from a sitting position. (Link-for-Tilt-video)
4. **Panoramic**, Lateral movement of the camera lens while the camera remains locked down on its tripod or pedestal. It is like someone is moving his head from one side to another. (Link-for-Panoramic-video)
5. **Panoramic Lateral**, The camera follows the action moving parallel to characters. Specifically, the camera captures the lateral movement of the subject, for example the camera moves parallel to a person walking down the street to keep them in the frame. (Link-for-Panoramic-lateral-video)

---

[1] https://github.com/tyiannak/multimodal_movie_analysis.

6. **Travelling in**, in this type of shot, the camera moves forward, pushes in a character or follows a character. (Link-for-Travelling-in-video)
7. **Travelling out**, in this type of shot, the camera pulls out, moving away from the subject and revealing the surroundings. (Link-for-Travelling-out-video)
8. **Zoom in,** In this type of shot, the camera lens are adjusted so that the image gradually appears larger and closer. (Link-for-Zoom-in-video)
9. **Aerial**, the camera is flown above the action, using a helicopter, drone or a plane. (Link-for-Aerial-video)
10. **Handheld**, the camera is moving throughout the filming set, while the camera operator is physically holding it. These camera shots are shaky and create a hectic feeling. (Link-for-Handheld-video)

Initially we used five more classes that have been found in the literature, but these have been rarely found in the annotation process (in particular they gathered less than 30 annotations after the aggregation procedure described in the next Section). These classes are: (1) **Car Front Windshield**, the camera is mounted on the front windshield, (2) **Car Side Mirror**, the camera is mounted on the side mirror and the viewer can see the driver (and co-driver) from the side (3) **Zoom out**, the entire image appears much smaller and further away (4) **Vertigo**, a combination of travelling and zoom and (5) **Panoramic 360**, a semicircular movement of the camera. These types of shots appeared in less than 30 annotations (i.e. in less than 0.75% of the total data), and therefore their recognition would be of low significance in a real-world scenario of movie analysis. Furthermore, any supervised procedure would fail to model these classes with such few data points.

2.3 Annotation Process

As soon as the 4000 shots had been selected from the 48 films, we began the annotation procedure. Towards this end, we used a simple Python-based video annotation web tool[2]. Figure 1(a) presents a screenshot of the tool: the user can simply view the video and assign it a video shot label. Note that the users had also the ability to annotate a video as "N/A": this label was used for cases that corresponded to either corrupted videos, or to videos with a non clear and fuzzy type of camera movement, or even to videos with more than one types of camera movements. Video shots that have been annotated as "N/A" were discarded from the final dataset.

This process has been carried out by 17 human annotators that annotated an arbitrary number of randomly selected shots. In total, 7500 individual annotations have been made. Then we proceeded to applying a simple aggregation rule: *for each video shot, two minimum annotations (by two different annotators) were required, along with a minimum annotator agreement of 60%.*

---

[2] https://github.com/theopsall/video_annotator.

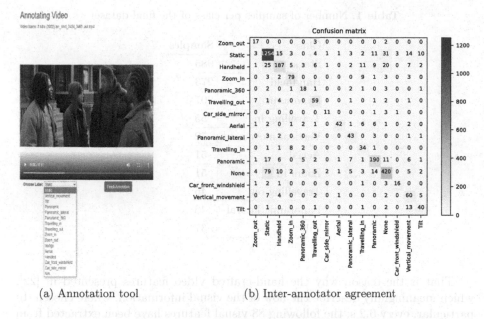

(a) Annotation tool   (b) Inter-annotator agreement

Fig. 1. Web tool and annotation agreement

This means that if a shot was annotated by two persons, this shot will be used in the final ground truth only if both annotators agreed on the label. Similarly, if three persons annotated a video shot, this would be used in the final ground truth if at least two of the annotators agreed on the label (higher than 60% agreement). Figure 1(b) visualizes the agreement between the individual annotators for all adopted classes. The overall inter-annotator agreement was above 80%.

After this process, 1877 videos from all 15 classes "survived" with a confident aggregated label. However, for 5 of these classes, the number of samples was below 30 and therefore have not been used in the final dataset. Given that, the final dataset size was 1803. Table 1 shows the final aggregated counts per final class (for the 10 adopted classes).

3 Video Shot Classification Methods

3.1 Feature Extraction

Image classification methods are usually based on features extracted from well-established Deep Learning computer vision architectures, such as the VGG [28] and ResNet [14] networks. However, in the task of Video Shot Classification we are more interested in capturing the image flow rather than the actual visual information.

Table 1. Number of samples per class of the final dataset

Class	Samples
Static	985
Handheld	273
Panoramic	207
Travelling in	55
Vertical movement	52
Aerial	51
Zoom in	51
Travelling out	46
Panoramic lateral	46
Tilt	37

That is the reason why the hand-crafted video features presented in [22], which meaningfully describe the flow of the visual information, where chosen. In particular, every 0.2 s, the following 88 visual features have been extracted from the corresponding frame:

- Color - related features (45 features):
  - 8-bin histogram of the red values
  - 8-bin histogram of the green values
  - 8-bin histogram of the blue values
  - 8-bin histogram of the grayscale values
  - 5-bin histogram of the max-by-mean-ratio for each RGB triplet
  - 8-bin histogram of the saturation values
- Average absolute difference between two successive frames in grey scale (1 feature)
- Facial features (2 features): The Viola-Jones [32] OpenCV implementation is used to detect frontal faces and the following features are extracted per frame:
  - number of faces detected
  - average ratio of the faces' bounding boxes areas divided by the total area of the frame
- Optical-flow related features (3 features): The optical flow is estimated using the Lucas-Kanade method [19] and the following 3 features are extracted:
  - average magnitude of the flow vectors
  - standard deviation of the angles of the flow vectors
  - a hand-crafted feature that measures the possibility that there is a camera tilt movement – this is achieved by measuring a ratio of the magnitude of the flow vectors by the deviation of the angles of the flow vectors.
- Current shot duration (1 feature): a basic shot detection method is implemented in this library. The length of the shot (in seconds) in which each frame belongs to, is used as a feature.

– Object-related features (36 features): We use the Single Shot Multibox Detector [18] method for detecting 12 categories of objects. For each frame, as soon as the object(s) of each category are detected, three statistics are extracted: number of objects detected, average detection confidence and average ratio of the objects' area to the area of the frame. So in total, $3 \times 12 = 36$ object-related features are extracted. The 12 object categories we detect are the following: person, vehicle, outdoor, animal, accessory, sports, kitchen, food, furniture, electronic, appliance and indoor.

For extracting these visual features, the *multimodal_movie_analysis* library[3] has been used in order to obtain features representing visual characteristics of a video.

The aforementioned features provide a wide range of low (simple color aggregates), mid (optical flows) and high (existence of objects and faces) representation levels. The rationale behind the selection of this wide range of types of features lies in the fact that our goal is to cover a great variety of flow information that may possibly be correlated to the camera movements.

In order to approach the problem from two different perspectives, we create two different types of features:

– **Static features**

The following video-level statistics are calculated for each frame sequence:
- six (6) video-level statistics of the non-object features. In particular, mean, standard deviation (std), median by std ratio, top-10 percentile, mean of the delta features and std of the delta features
- for the object detection, the frame-level predictions are post processed under local time windows with two different ways: (i) the object frame-level confidences are smoothed across time windows in order to increase the accuracy of the predictions and (ii) every object that is not present to at least a minimum number (threshold) of subsequent frames, is excluded from the final feature vector. However, this smoothing procedure is the only post-processing performed on the object-related features: no other statistics are extracted for the whole video, other than the object features' simple averages.

This process therefore results to $52 \times 6 + 36 = 348$ feature statistics that describe the whole video in a static feature vector.

– **Sequential features**

A post processing procedure is applied on the feature sequence in order to smooth the representation across successive frames. Median-filtering with kernel size of 4 (i.e. 0.8 s) were chosen. The processed sequence can be used in a temporal modeling.

3.2 Baseline Classification

In this section we propose two distinct classification methods that can give an intuition about the separability of the introduced classes. These methods will

---

[3] https://github.com/tyiannak/multimodal_movie_analysis.

serve in order to report baseline classification metrics and understand the latent correlations across classes.

Static Method. The first method is based on the static video representation described in Sect. 3.1. It adopts an SVM algorithm with the appropriate data normalization and parameter tuning. This approach aims on finding significant correlations between statistical frame representations and the target shot styles. Such methods can learn to separate continuous camera movements from static shots, but may under-perform in cases of abrupt shot style changes.

Sequential Method. This method is aimed to predict the shot style from the temporal feature sequence. In such a case it would be easier to model different kinds of camera movements since the approach is expected to capture the temporal frame dependencies.

An LSTM architecture was chosen, since it is a well established Recurrent Neural Network that can capture long-term temporal information [11]. Batch-normalization [15] and linear layers where used along with ReLU activation functions in order to perform the classification task. Temporal standard scaling was applied to the input sequence, that is each instance of a feature along the sequence were standarized using the same statistical values (mean value and standard deviation).

4 Experiments

In this section, experimental results are presented for both static and sequential methods.

4.1 Classification Tasks

By combining different shot categories, four classification tasks, one binary and three multi-label, are defined.

Binary Classification Task. The binary task includes the *static* and *non-static* classes. The former consists of shots that have been annotated as **static**, while the latter contains all the classes from the original dataset that are associated with any type of camera movement. That is the corresponding sub-classes are **Panoramic Lateral, Vertical Movement, Zoom-in, Handheld, Aerial, Tilt, Panoramic, Travelling-in, Travelling-out**.

Multi-label Classification Tasks. As it will be described in the results section, predicting the type of camera movement is not a trivial task. At the same time, as explained in the introduction, it is a task that can be mostly used as an automated attribute movie extractor, that is used in the context of a more complex system (such as movie recommendation). Therefore, it is not a task that can form a standalone application or product. Given that, it makes sense to analyze the movie in various levels of detail with regards to its camera movement styles. In this paper, apart from the initial 10-class classification task, we have defined two more classification tasks, by merging classes of similar camera movement (e.g. all vertical movements).

So in total, we define three multi-label classification tasks in which the static class is the original annotated class, while the others are merges from other class combinations.

- **3-class** The corresponding classes are *Zoom, Static* and *Vertical & Horizontal Movements*. The **Zoom** class consists of the *Zoom-in, Travelling-in and Travelling-out* sub-classes, which all contain shots in which the perimeter image changes at very fast intervals, while the centre image remains static or changes at a slower rate. The **Vertical & Horizontal Movements** class consists of the *Vertical Movement, Tilt, Panoramic and Panoramic Lateral* sub-classes from the original dataset, where the position of the camera is moving either in a vertical or in a horizontal way.
- **4-class** In this task, the **Static** and **Zoom** classes of the 3-class problem were kept, while the Vertical & Horizontal Movements class was separated into 2 sub-classes, **Tilt**, which includes all vertical movements and consists of the *Vertical Movement* and *Tilt* original classes, and **Panoramic** that contains shots with lateral movements and consists of the *Panoramic* and *Panoramic Lateral* original classes.
- **10-class** This task includes all provided class from the original dataset, namely: **Static, Panoramic, Zoom-in, Travelling-out, Vertical Movement, Aerial, Travelling-in, Tilt, Handheld** and **Panoramic Lateral**.

4.2 Experimental Set-Up

With regard to the *static* method, an SVM classifier with rbf kernel were initialized. After the data has been scaled using a zero-mean normalization, parameter tuning was performed over the 'C' parameter.

Concerning the *sequential* method, we initially performed a sequential zero-mean normalization. That is, by treating each frame as a different instance, the mean and standard deviation of all features across all frames of the training set of the dataset was calculated and then used to normalize each frame of the dataset, resulting in normalized feature sequences. The LSTM architecture was tuned for each of the four tasks in order to achieve the optimal dropout, weight decay, learning rate and LSTM's hidden dimension. For the binary task, the binary cross entropy loss was used, while for the rest, cross entropy was adopted. Adam was chosen to be the optimizer with the integration of a reduce-on-plateau learning rate scheduler, while the best model was chosen using early stopping.

In order to evaluate the classifiers a cross validation procedure of totally 10 iterations was performed. For each iteration, the shots of the dataset were randomly split in a stratified manner (i.e. keep the initial class distribution) into 20% test set for testing, 9.6% validation set for parameter tuning and 70.4% train set for training.

4.3   Results

Table 2. Macro-averaged F1 of (a) all classes and (b) the non-neutral classes (i.e. excluding static class) calculated on the aggregated predictions across all 10 cross-validated iterations

<table>
<tr><td colspan="3">(a)</td><td colspan="3">(b)</td></tr>
<tr><td>Classification task</td><td>Static</td><td>Sequential</td><td>Classification task</td><td>Static</td><td>Sequential</td></tr>
<tr><td>2-class</td><td>73.9%</td><td>79%</td><td>3-class</td><td>35.7%</td><td>39.8%</td></tr>
<tr><td>3-class</td><td>50.6%</td><td>53%</td><td>4-class</td><td>24.8%</td><td>27.8%</td></tr>
<tr><td>4-class</td><td>39.5%</td><td>40%</td><td></td><td></td><td></td></tr>
<tr><td>10-class</td><td>15.7%</td><td>15.4%</td><td></td><td></td><td></td></tr>
</table>

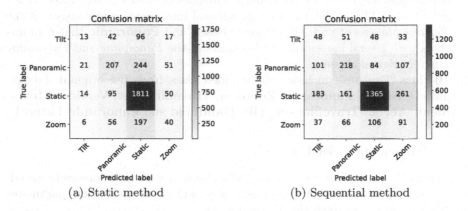

(a) Static method   (b) Sequential method

Fig. 2. Confusion matrices based on aggregated predictions for the 4-class task

The resulted evaluation metrics are listed in Table 2, while the corresponding, aggregated from all iterations, confusion matrices are presented in Fig. 2 for the 3-class and 4-class problems.

As can be clearly seen from the results, the sequential method completely outperforms the static one in terms of the binary task, by scoring a 7% relatively higher score. That is an expected result since the basic distinction of the static and non-static classes is the camera movement, which is easier to capture by a sequential method. With regard to 3-class and 4-class problems, the sequential method achieves a higher by 4.7% and 1.3% relative performance respectively, while for the 10-class task the static method is 1.9% relatively better. It is observed that the more the classes are in the tasks, the less the difference between the two methods is. That is mostly due to the fact that some of the 10 classes are underrepresented and thus a deep learning architecture is harder to learn from that amount of data compared to a traditional machine learning algorithm.

However, in such tasks where the neutral class (static class) is dominant in terms of number of instances, the macro-averaged F1 metric cannot properly represent the performance of the methods. By examining the confusion matrices of Fig. 2, it can be clearly seen that the sequential method achieves to adequately predict the non-neutral classes (tilt, panoramic and zoom). More specifically, as illustrated in Table 2b, which includes the macro-averaged F1 metrics based only on the non-neutral classes (i.e. excluding the static class), the sequential method is 11.5% and 12.1% relatively better for the 3 and 4-class problem respectively. Thus it can be inferred that including temporal information in the classifier leads to better distinction across camera movements.

It needs to be noted that the selection of the features used in our method was not random. More specifically, our pipeline was evaluated by training and testing the LSTM on feature sequences extracted from the first linear layer of the pre-trained VGG16 network. The results showed that when the LSTM is trained on our features (produced by the *multimodal_movie_analysis* library) it performs 21.5%, 23.6%, 24% and 7.7% better for the 2, 3, 4 and 10-class tasks respectively, compared to the model trained on the *VGG*16 features. This outcome is reasonable since while the VGG features contain better image-related information, our features achieve to better model the camera movement, an attribute that is crucial for shot type classification.

5   Conclusion and Future Work

In this work, a new publicly available[4] dataset for film shot classification with annotations in the camera movement styles was presented. This data collection fills the gap in the literature of cinematographic style analysis, since the existing datasets are either private or are not concerned about camera movements in an extensive manner. Additionally, two different methods were used as baseline evaluation, one static and one sequential showing that camera movement classification is better performed when modeling sequential information. Models trained on this dataset can be used in a future work in order to create artistic profiles of either directors or movies that can describe in an informative way the correspond-ing work. In addition, this can provide a new way of representing the movies in recommender systems, so that the extracted recommendations are content-driven and, in particular, driven by the actual underlying visual aesthetics of the movies.

---

[4] https://github.com/magcil/movie_shot_classification_dataset.

References

1. Bak, H.Y., Park, S.B.: Comparative study of movie shot classification based on semantic segmentation. Appl. Sci. **10**(10), 3390 (2020). https://doi.org/10.3390/app10103390, https://www.mdpi.com/2076-3417/10/10/3390
2. Baraldi, L., Grana, C., Cucchiara, R.: Shot and scene detection via hierarchical clustering for re-using broadcast video. In: Azzopardi, G., Petkov, N. (eds.) CAIP 2015. LNCS, vol. 9256, pp. 801–811. Springer, Cham (2015). https://doi.org/10.1007/978-3-319-23192-1_67
3. Benini, S., Svanera, M., Adami, N., Leonardi, R., Kovács, A.B.: Shot scale distribution in art films. Multimedia Tools Appl. **75**(23), 16499–16527 (2016). https://doi.org/10.1007/s11042-016-3339-9
4. Bhattacharya, S., Mehran, R., Sukthankar, R., Shah, M.: Classification of cinematographic shots using lie algebra and its application to complex event recognition. IEEE Trans. Multimedia **16**(3), 686–696 (2014)
5. Bougiatiotis, K., Giannakopoulos, T.: Enhanced movie content similarity based on textual, auditory and visual information. Expert Syst. Appl. **96**, 86–102 (2018)
6. Braudy, L.: Film: an international history of the medium. Film Q. (ARCHIVE) **48**(3), 59 (1995)
7. Canini, L., Benini, S., Leonardi, R.: Classifying cinematographic shot types. Multimedia Tools Appl. **62**(1), 51–73 (2013)
8. Choi, S.M., Ko, S.K., Han, Y.S.: A movie recommendation algorithm based on genre correlations. Expert Syst. Appl. **39**(9), 8079–8085 (2012)
9. Diao, Q., Qiu, M., Wu, C.Y., Smola, A.J., Jiang, J., Wang, C.: Jointly modeling aspects, ratings and sentiments for movie recommendation (JMARS). In: Proceedings of the 20th ACM SIGKDD International Conference on Knowledge Discovery and Data Mining, pp. 193–202 (2014)
10. Ertugrul, A.M., Karagoz, P.: Movie genre classification from plot summaries using bidirectional LSTM. In: 2018 IEEE 12th International Conference on Semantic Computing (ICSC), pp. 248–251. IEEE (2018)
11. Greff, K., Srivastava, R.K., Koutník, J., Steunebrink, B.R., Schmidhuber, J.: LSTM: a search space odyssey. IEEE Trans. Neural Networks Learn. Syst. **28**(10), 2222–2232 (2016)
12. Haq, I.U., Muhammad, K., Hussain, T., Kwon, S., Sodanil, M., Baik, S.W., Lee, M.Y.: Movie scene segmentation using object detection and set theory. Int. J. Distrib. Sens. Networks **15**(6), 1550147719845277 (2019)
13. Hasan, M.A., Xu, M., He, X., Xu, C.: CAMHID: camera motion histogram descriptor and its application to cinematographic shot classification. IEEE Trans. Circ. Syst. Video Technol. **24**(10), 1682–1695 (2014). https://doi.org/10.1109/TCSVT.2014.2345933
14. He, K., Zhang, X., Ren, S., Sun, J.: Deep residual learning for image recognition. In: Proceedings of the IEEE Conference on Computer Vision and Pattern Recognition, pp. 770–778 (2016)
15. Ioffe, S., Szegedy, C.: Batch normalization: accelerating deep network training by reducing internal covariate shift. In: International Conference on Machine Learning, pp. 448–456. PMLR (2015)
16. Lekakos, G., Caravelas, P.: A hybrid approach for movie recommendation. Multimedia Tools Appl. **36**(1), 55–70 (2008)
17. Li, K., Li, S., Oh, S., Fu, Y.: Videography-based unconstrained video analysis. IEEE Trans. Image Process. **26**(5), 2261–2273 (2017). https://doi.org/10.1109/TIP.2017.2678800

18. Liu, W., et al.: SSD: single shot multibox detector. In: Leibe, B., Matas, J., Sebe, N., Welling, M. (eds.) ECCV 2016. LNCS, vol. 9905, pp. 21–37. Springer, Cham (2016). https://doi.org/10.1007/978-3-319-46448-0_2

19. Lucas, B.D., Kanade, T., et al.: An iterative image registration technique with an application to stereo vision. Vancouver, British Columbia (1981)

20. Minhas, R.A., Javed, A., Irtaza, A., Mahmood, M.T., Joo, Y.B.: Shot classification of field sports videos using AlexNet convolutional neural network. Appl. Sci. **9**(3), 483 (2019)

21. Park, S.C., Lee, H.S., Lee, S.W.: Qualitative estimation of camera motion parameters from the linear composition of optical flow. Pattern Recogn. **37**(4), 767–779 (2004)

22. Psallidas, T., Koromilas, P., Giannakopoulos, T., Spyrou, E.: Multimodal summarization of user-generated videos. Appl. Sci. **11**(11), 5260 (2021). https://doi.org/10.3390/app11115260, https://www.mdpi.com/2076-3417/11/11/5260

23. Rao, A., et al.: A unified framework for shot type classification based on subject centric lens. In: Vedaldi, A., Bischof, H., Brox, T., Frahm, J.-M. (eds.) ECCV 2020. LNCS, vol. 12356, pp. 17–34. Springer, Cham (2020). https://doi.org/10.1007/978-3-030-58621-8_2

24. Rasheed, Z., Shah, M.: Scene detection in Hollywood movies and tv shows. In: 2003 IEEE Computer Society Conference on Computer Vision and Pattern Recognition. Proceedings, vol. 2, p. II-343. IEEE (2003)

25. Sang, J., Xu, C.: Character-based movie summarization. In: Proceedings of the 18th ACM International Conference on Multimedia, pp. 855–858 (2010)

26. Savardi, M., Signoroni, A., Migliorati, P., Benini, S.: Shot scale analysis in movies by convolutional neural networks. In: 2018 25th IEEE International Conference on Image Processing (ICIP), pp. 2620–2624. IEEE (2018)

27. Simões, G.S., Wehrmann, J., Barros, R.C., Ruiz, D.D.: Movie genre classification with convolutional neural networks. In: 2016 International Joint Conference on Neural Networks (IJCNN), pp. 259–266. IEEE (2016)

28. Simonyan, K., Zisserman, A.: Very deep convolutional networks for large-scale image recognition. arXiv preprint arXiv:1409.1556 (2014)

29. Subramaniyaswamy, V., Logesh, R., Chandrashekhar, M., Challa, A., Vijayakumar, V.: A personalised movie recommendation system based on collaborative filtering. Int. J. High Perform. Comput. Networking **10**(1–2), 54–63 (2017)

30. Tsai, C.M., Kang, L.W., Lin, C.W., Lin, W.: Scene-based movie summarization via role-community networks. IEEE Trans. Circ. Syst. Video Technol. **23**(11), 1927–1940 (2013)

31. Ul Haq, I., Ullah, A., Muhammad, K., Lee, M.Y., Baik, S.W.: Personalized movie summarization using deep CNN-assisted facial expression recognition. In: Complexity 2019 (2019)

32. Viola, P., Jones, M.: Rapid object detection using a boosted cascade of simple features. In: Proceedings of the 2001 IEEE Computer Society Conference on Computer Vision and Pattern Recognition. CVPR 2001, vol. 1, p. I-I. IEEE (2001)

33. Wang, H.L., Cheong, L.F.: Taxonomy of directing semantics for film shot classification. IEEE Trans. Circ. Syst. Video Technol. **19**(10), 1529–1542 (2009). https://doi.org/10.1109/TCSVT.2009.2022705

34. Zhou, H., Hermans, T., Karandikar, A.V., Rehg, J.M.: Movie genre classification via scene categorization. In: Proceedings of the 18th ACM International Conference on Multimedia, pp. 747–750 (2010)

The CleanSea Set: A Benchmark Corpus for Underwater Debris Detection and Recognition

Alejandro Sánchez-Ferrer, Antonio Javier Gallego[✉] (iD), Jose J. Valero-Mas(iD), and Jorge Calvo-Zaragoza(iD)

Department of Software and Computing Systems, University of Alicante, Carretera San Vicente Del Raspeig S/n, 03690 Alicante, Spain
asf62@alu.ua.es, {jgallego,jjvalero,jcalvo}@dlsi.ua.es

Abstract. In recent years, the large amount of debris scattered throughout the ocean is becoming one of the major pollution problems, causing extinction of species and accelerating the degradation of our planet, among other environmental issues. Since the manual treatment of this waste represents a considerably tedious task, autonomous frameworks are gaining attention. Due to their reported good performance, such frameworks generally rely on Deep Learning techniques. However, the scarcity of data coupled with the inherent difficulties of the field—debris with different shapes and colors due to long-lasting exposure to the ocean, illumination variability or sea conditions—makes detecting underwater objects a particularly challenging task. The contribution of this work to the field is double: on the one hand, we introduce a novel data collection for supervised learning—the CleanSea corpus—annotated at both the bound box and contour levels of the objects to contribute with the research and progress in the field; on the other hand, we devise and optimize a recognition model based on the reference Mask Object-Based Convolutional Neural Network for this set to establish a benchmark for future comparison and assess its performance in both simulated and real-world scenarios. Results show the relevance of the contributions as the devised model is capable of properly addressing the detection and recognition of general debris when trained with the introduced CleanSea corpus.

Keywords: Underwater debris detection · Marine pollution · Deep neural networks · Object detection

1 Introduction

According to United Nations, it is estimated that 6.4 million of tons of anthropogenic waste end up in the oceans every year [5]. Out of them, a large percentage corresponds to plastic-based products such as bags, cans, bottles or fishing

Work supported by the Pattern Recognition and Artificial Intelligence Group (PRAIg) from the University of Alicante, Spain. The third author is supported by grant APOSTD/2020/256 from "Programa I+D+i de la Generalitat Valenciana".

A. J. Pinho et al. (Eds.): IbPRIA 2022, LNCS 13256, pp. 616–628, 2022.
https://doi.org/10.1007/978-3-031-04881-4_49

nets [16]. Due to their durability, this particular type of debris constitutes one of the major threats to biodiversity in marine environments as well as a concerning polluting problem [6]. Hence, a large number of initiatives and human campaigns have been devoted to mitigate that situation, such as The Ocean Cleanup [22], The Big Blue Ocean Cleanup [20] or TrashTag [17].

Nevertheless, these efforts have been reported as insufficient given the overwhelming amount of marine waste. Hence, there exists a great interest in leveraging the current technological developments for fostering automatic recognition and collection mechanisms of these residues [3] by means of Autonomous Underwater Vehicles (AUV) which may eventually perform these cleaning tasks with reduced human supervision and interaction.

Most proposals devoted to the automatic classification and detection of marine debris resort to Deep Learning methods due to the competitive performance reported in the general computer vision field [13]. However, given that such schemes generally require large amounts of data, the general lack of such corpora has hindered their development and practical application, at least when considering supervised paradigms [4]. Note that, due to long-lasting exposure to the ocean, most of these elements present different shapes and colors, altering their original appearance. This, together with the great variability of light or sea conditions, makes detecting underwater objects a particularly challenging task.

Some of the scarce publicly available corpora in the related literature are: Trash Annotation in Context (TACO) dataset [2] which comprises 4,784 object annotations within 1,500 images of 60 different categories, the TrashNet collection of 2,527 images of garbage items categorized into 6 different categories— glass, paper, cardboard, plastic, metal, and trash—, the Marine Debris Segmentation corpus [21] of 1,868 images labeled after 11 different object classes—bottle, can, chain or hook, among others—, the Trashcan set [9] of 7,212 images with specific semantic segmentation labels of 4 different types, or the Marine Debris Archive (MARIDA) [11], which represents a benchmark corpus for litter recognition of 1,381 patches annotated at the pixel level within 15 possible categories comprising waste categories as well as ships, waves or organic elements, among others. Note that, while most of these collections consider annotations only at the class or the bounding box level, such annotation level is insufficient for real-world applications since autonomous cleaning systems require tracking the shape of the object to adequately perform the task. Hence, the creation of public corpora annotated at the contour level is of remarkable for the community.

In this context, the present work introduces a novel corpus of marine debris for supervised learning tasks. This data collection, namely the CleanSea dataset, comprises close to 2,000 objects labeled after 19 different categories in more than 1,200 pictures. Oppositely to most research corpora in this field, these annotations are provided at two different levels: a first one considering the bounding boxes of the objects and a second one based on their contour. In addition, this corpus is evaluated using a reference object detection architecture—Mask Region-Based Convolutional Neural Network or Mask R-CNN—specifically tuned for the case at issue—properly adjusting the hyperparameters of the model, the training strategy, and the data augmentation policy— to provide a baseline performance for further research and comparative purposes.

The rest of the paper is structured as follows: Sect. 2 describes the CleanSea corpus; Sect. 3 presents the proposed approach for debris recognition; Sect. 4 describes the experimental procedure devised and discusses the results obtained; finally, Sect. 5 concludes the work and poses future lines to address.

2   CleanSea Dataset

This section presents the novel CleanSea corpus for marine debris recognition tasks. This dataset is expect to expand the aforementioned scarce amount of labeled data present in the field, hence contributing to the research and development of AUV devoted to cleaning tasks. This data collection may be downloaded from https://www.dlsi.ua.es/~jgallego/datasets/cleansea.

The CleanSea corpus was created using the data from the Japan Agency for Marine-Earth Science and Technology (JAMSTEC) collection [19], which mainly comprises images of marine fauna and flora. Within that dataset, we selected 1,223 images—with resolutions between 5616 × 3744 pixels and 640 × 480 pixels—that contained some type of debris. Specifically, the taxonomy developed comprises the following 19 categories: basket, bottle, bumper, can, fishing net, glove, metal chain, metal waste, pipe, plastic bag, plastic waste, rope, shoe, squared can, tire, towel, washing machine, wood, and wrapper.

The debris within these images were manually labeled at two different levels: a first one by annotating the bounding box of the object and a second one by marking their respective contours at the pixel level [18]. Note that many of the images contain more than a single litter element, being each of them labeled separately and indicating their respective classes, hence constituting a total of 1,994 labeled objects. Table 1 summarizes the distribution of those objects within the previously introduced taxonomy. Additionally, Fig. 1 shows some examples of the selected images and their respective annotations.

Table 1. Description of the CleanSea in terms of the number of samples per label.

Label	Samples	Label	Samples	Label	Samples
Plastic bag	650	Bumper	68	Metal waste	20
Can	339	Wood	62	Tire	16
Packaging	202	Towel	54	Shoe	8
Bottle	143	Glove	46	Metal chain	5
Fishing net	114	Pipe	42	Washer	3
Plastic waste	86	Basket	42		
Rope	79	Square can	20		

Fig. 1. Examples of images from the CleanSea corpus. The top row shows three pictures from the collection while the bottom one shows their respective ground-truth annotations.

3   Methodology

To benchmark the introduced CleanSea corpus, we consider an instance segmentation neural model as it allows the location and classification of the individual objects within the image. More precisely, we contemplate the state-of-the-art Mask Region-Based Convolutional Neural Network (Mask R-CNN) model [8] which retrieves the individual bounding box for each instance in the image, their associated classes, and a binary pixel-level mask for each of the detected objects. Figure 2 shows a graphical representation of the architecture.

Conceptually, Mask R-CNN performs two stages: a first phase, known as Region Proposal Network, which proposes multiple locations of possible objects within the image; and a second phase which, based on the results of the former process, predicts the individual classes of the objects, the actual bounding boxes, and calculates the binary mask for each Region of Interest (ROI).

3.1   Model Optimization Process

This section introduces the optimization carried out in the adjustment of the hyperparameters and the training procedure to optimize the performance of the considered Mask R-CNN model for the CleanSet corpus.

Regarding the first devised adjustment, given that the aim of the work is detecting every possible debris, we modified the confidence acceptance threshold from 70% to 50% to bias the model towards an overestimation of elements in the images. Note that, while such procedure effectively reduces the number of false negatives, it comes along with a slight increase in the number of false positives. However, in this context of marine cleaning, given that these frameworks assume the presence of a human supervisor, it is preferable to detect, and collect if possible, all the waste with some extra items which may not be debris rather than missing any of them.

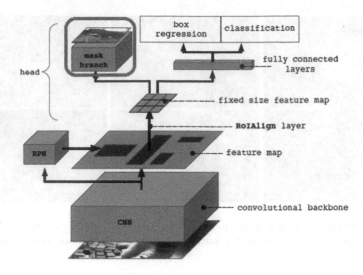

Fig. 2. Outline of the Mask R-CNN object detection and recognition architecture [8].

The influence of the learning rate was studied in preliminary experimentation, eventually establishing the value of this parameter to 0.001. We also considered a weight decay regularization of 0.0001 with a warm restart learning rate policy after 150 iterations [15] to avoid a possible stagnation at local minima. Stochastic Gradient Descent (SGD) [1] was selected as optimization function and the training lasted a maximum of 1,000 iterations with a mini-batch size of 16 samples.

Regarding the actual data, as commonly considered in the Deep Learning field, we contemplated a Transfer Learning approach for obtaining an adequate set of weights for the tasks. In this regard, the network was pre-trained using the Common Objects in COntext (COCO) corpus [14], a large-scale database for image detection and segmentation, for then applying a fine-tuning process using the proposed CleanSet dataset.

In addition, data augmentation was also used to artificially increase the variability of the data in the training set by randomly applying different types of transformations to the original samples [10]. Note that these processes are directly embedded in the network pipeline so that they are performed in an online fashion. In our case, the applied transformations are randomly selected from the follow set of possible processes:

- Horizontal flips or mirroring effect.
- Gaussian blur with random $\sigma \in [0, 0.5]$.
- Contrast alteration given by $v' = 127 + \alpha \cdot (v - 127)$, where v and v' represent the source and target values of a given pixel and $\alpha \in [0.75, 1.5]$.
- Brightness modification.
- Axis-independent scaling within 80% and 120% of their original size.
- Translation of the images by -20% to 20% on both axes.

- Image rotations in the range $[-25°, 25°]$.
- Image shearing in the range $[-8°, 8°]$.

4 Experimentation and Results

The experimental section of the work is structured as follows: a first part is devoted to the adjustment of the training procedure and the hyperparameter optimization process; after that, the individual results obtained for the CleanSea corpus are analyzed and discussed; a third section compares the results with those of other reference works in the literature; finally, a last section provides links to some demo videos showing the performance of the proposed architecture.

To assess the goodness of the proposal we resort to the standard COCO metrics and, more precisely, to the Average Precision (mAP) figure of merit. This metric is defined as the class-wise average of the individual ratios between the number of correctly estimated objects—cases in which detection and ground truth overlap more than a 50% in a pixel-level assessment—and the amount of objects to be detected.

Regarding the data, it must be highlighted that "Washer" and "Metal chain" were included in the "Metal waste" tag due to their scarce representation in the CleanSea corpus. Images were scaled to a spatial resolution of 512×512 pixels.

We used a stratified n-fold cross validation (with $n = 5$) in all the experiments, since it yields a better Monte-Carlo estimate than when performing the tests solely with a single random partition [12]. The dataset was consequently divided into n mutually exclusive folds preserving the percentage of samples for each class in each fold. We used one partition for test (with 20% of the samples) and the rest for training (80%). A validation subset with 10 % of the training samples was additionally used for the adjustment of the hyperparameters. The training and testing processes were repeated $n = 5$ times, using the different partitions of the dataset, and finally, the average result was calculated.

All experiments were carried out using the Python programming language with the TensorFlow (v. 2.4.1) and Keras (v. 2.4.3) libraries. The machine used consists of an Intel(R) Core(TM) i7-8700 CPU @ 3.20 GHz with 16 GB RAM, a NVIDIA GeForce RTX 2070 with 6 GB GDDR6 Graphics Processing Unit (GPU) with the cuDNN library. For the sake of reproducibility and further development, the code has been publicly released and is available at: https://github.com/asanc199/Mask_RCNN-Cleansea.

4.1 Hyperparameters Optimization and Performance Results

This section presents and discusses the results obtained for the introduced CleanSea corpus with the considered Mask R-CNN model in terms of the mAP metric. To study their influence in the overall performance of the model, we provide the results for different numbers of training epochs and data augmentation procedures.

More precisely, Table 2 shows the mAP figures obtained when varying the number of epochs in the range $[25, 1000]$[1] while also contemplating different data augmentation scenarios devised out of the processes described in Sect. 3.1: a first one, denoted as N/A, in which no process is applied, a *Mild* case which only considers flips and rotations, and a third *Severe* mode which applies all transformations described in the aforementioned section.

Table 2. Results, in terms of mAP, when varying the number of training epochs in different data augmentation scenarios.

Type of data augmentation	Number of epochs	mAP (%)
N/A	100	52.0
	200	53.2
	500	54.3
Mild	200	53.6
	500	55.4
	1,000	57.2
Severe	200	57.5
	500	58.7
	1,000	**60.0**

As it may be observed, the data augmentation process remarkably improves the results obtained: focusing on the case of 500 training epochs, the figure of merit improves from 54.3% of the base case to 55.4% when considering mild distortions, reaching a performance of 58.7% with the severe augmentation. Among the figures obtained, the best-performance case is that when training for 1,000 epochs in the severe data augmentation scenario, which reports an mAP of 60%. Hence, this configuration is kept for the rest of the experimentation.

4.2 Detailed Analysis

As previously shown in Table 2, the considered Mask R-CNN model optimized for the CleanSea corpus reports a maximum score of mAP = 60.0%. However, to adequately understand the actual limitations of both the model and the corpus, this section provides a detailed analysis of those figures by checking the confusion figures among classes as well as providing visual examples of the results obtained.

In this regard, Fig. 3 shows the class confusion matrix, where prediction and ground truth are represented in the horizontal and vertical axes, respectively. An additional *Background* (BG) category is included for depicting the False Positive (first row) and the False Negative (first column) type of errors.

---

[1] For conciseness, we only report the subset of the results in which the highest variation was obtained.

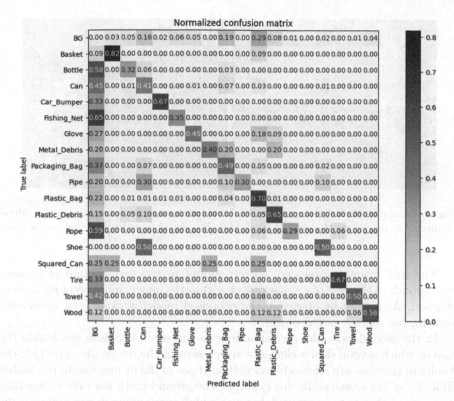

Fig. 3. Results in terms of a normalized confusion matrix obtained for the Mask R-CNN architecture when trained and evaluated with the CleanSea corpus.

Attending to these results, most confusion errors occur in pairs of similar classes as, for instance, "Square can" with "Basket" or "Metal debris". Also, underrepresented labels (e.g., "Shoe") also exhibit large confusion rates given that the model tends to neglect them in favor of the rest of categories.

Another point to highlight is the confusion among the different types of materials. Such task constitutes a remarkable challenge within this marine debris recollection framework given the progressive degradation they suffer, together with the variability in terms of illumination conditions.

A relatively elevated amount of False Positive errors is also observed, especially in images with more than a single object to detect. This is most likely due to a partial labeling of some of the images in the corpora (as will be shown below) as well as the prediction bias considered for the training stage of the model described in Sect. 3.1.

For a visual analysis of the results, Fig. 4 shows two examples of the results obtained. The first row shows the annotated ground truth for reference while the second one depicts the prediction obtained with the proposed model.

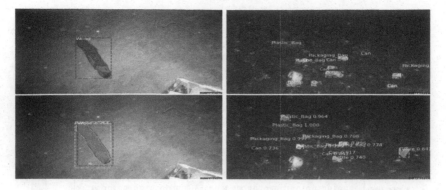

Fig. 4. Examples of prediction results. Each column represents a single example, where the first row represents the ground truth and the second row the obtained prediction.

The first example provided, corresponding to the first column of images, corresponds to the case of a single object within the image, more precisely a piece of wood. As it may be observed, the proposed model is capable of accurately locating the contour of the object.

In the second example, corresponding to the second column, we tackle the case in which several debris elements are present in the image. As expected, the results in this case are relatively worse than those of the former due to the higher difficulty of the scenario. In this example, the ground-truth not only states that there are 13 objects to be tracked, but also a high overlapping degree among some of them. It can be also observed that some False Positive estimations are due to missing annotations in the ground truth. Note that such incomplete annotation is not due to a bad labeling campaign but to the inherent difficulty of the task: even for an expert human agent it may be sometimes difficult to distinguish whether some objects in the images actually correspond to some type of debris or they are environmental elements such as rocks, plants, shades due to the sand, etc. Note that a possibility to tackle such issue could be using the predictions of the recognition model to revise the annotations of the corpus, which may point out problems such as mislabeled elements or non-annotated objects.

4.3   Comparison with State-of-the-art Approaches

Having studied the performance of the considered method, we now comparatively assess the performance achieved by other reference approaches from the literature with the CleanSea corpus. In this regard, Table 3 presents the results obtained by two state-of-the-art approximations: a region-based one which detects the bounding boxes of the objects in the images and a contour-based method which tracks the actual shape of the object to detect. Note that both models are provided in the work by Hong et al. [9].

The evaluation is carried out following two criteria (reflected in the columns of the table): a first object-level evaluation (denoted as "Instance") which comprises the categories "can", "bottle", 'plastic bag", "fishing net", "packaging bag", "rope", "plastic debris", "pipe", and "shoe"; and a second one which focuses on the type of material, considering the classes "wood", "plastic", "metal", and "paper" (including "cardboard"). Metric mAP is considered for all cases.

Table 3. Comparison, in terms of mAP, of the proposed method against the state-of-the-art strategies by Hong et al. [9] for the CleanSea corpus. "Instance" and "Material" depict whether the evaluation is based in the shape or composition labels of the taxonomy. The best performance figures are highlighted in bold type.

Method	mAP (%)	
	Instance	Material
Bounding box detection by Hong et al. [9]	55.4	51.2
Contour segmentation by Hong et al. [9]	55.3	54.0
Our proposal	**59.7**	**67.1**

As it may be observed, the results obtained with the proposed optimized model outperform those obtained by the state-of-the-art architectures by Hong et al. [9], in both evaluation scenarios considered.

4.4 Video Detection in Real and Simulated Environments

To further show the capabilities of the presented method in a practical scenario, we now apply the recognition proposal to two underwater video recordings containing different types of debris[2]:

- A controlled scenario arranged by the authors of the work in which some litter has been introduced in a fishbowl. The results of the experiment may be checked at: https://youtu.be/nQbFYz0dRno.
- A real-world seabed recording provided by the University of Minnesota under the Creative Commons license [9]. The results may be checked at: https://youtu.be/Djy6grfzN1Y.

As it may be observed, while the proposed method is capable of successfully performing the considered recognition task in both scenarios, the results in the first case—the simulated scenario—-are remarkably better than those of the latter one. Such effect is mostly due to the fact that the fishbowl case constitutes a simpler and controlled scenario with almost static and distinguishable elements while the latter depicts moving elements, great variability in the recording distances, and illumination conditions.

---

[2] Note that the debris in the videos may not necessarily match that of the corpus.

Ideally, to improve the overall performance of the method in the provided cases, it would be desirable to increase the size of the CleanSea corpus with larger amounts of varied debris images. However, given the aforementioned cost of the labeling process, one could resort to Domain Adaptation methods [7], which allow adapting a learning-based system trained with a certain data distribution to an alternative but similar one, as long as both sets of data share the same label taxonomy.

5 Conclusions

Anthropogenic marine debris constitutes one of the major threats to biodiversity as well as a remarkable pollution problem nowadays. While some initiatives have been considered for tackling such situation, the vast amount of waste requires leveraging current technological developments for devising autonomous systems capable of performing such task.

However, most of the current methods rely on state-of-the-art computer vision techniques based on Deep Learning. While such approaches are generally related to high-performance rates, they require large amounts of data for being trained. In this regard, the related literature of autonomous marine litter detection and classification points a general lack of annotated corpora for supervised learning purposes which hinders the development of such automated systems.

This work introduces a novel labeled data collection—namely CleanSea corpus—for marine debris detection and classification. This set contains 1,994 objects from 19 different categories distributed in 1,223 images annotated at two levels: the bounding box of the piece and its contour. Additionally, for benchmarking purposes, this corpus is evaluated using a Mask Region-Based Convolutional Neural Network for object detection and classification adequately optimized for this task in terms of training procedure, hyperparameter adjusting, and data augmentation. The results obtained with a thorough hyperparameter process outperform those achieved by state-of-the-art architectures using the same data collection. Moreover, its operation has been demonstrated in image, video, and real-time detection, which makes this model a solid alternative for its implementation in autonomous underwater robots.

Future work considers the expansion of the CleanSea corpus, both in terms of the number of images and categories. We are also interested in performing a curation process of the already annotated data to refine the annotations. Additionally, we aim to evaluate this data collection using other neural architectures for object detection to assess the achievable performances in those cases. Finally, we further consider the exploration of Domain Adaptation techniques for reducing, or even removing, the need for annotated data when tackling scenarios with debris different to that considered when training the method.

References

1. Bottou, L.: Large-scale machine learning with stochastic gradient descent. In: Lechevallier, Y., Saporta, G. (eds.) Proceedings of COMPSTAT 2010. Physica-Verlag HD, pp. 177–186. Springer, Cham (2010). https://doi.org/10.1007/978-3-7908-2604-3_16

2. Cormier, R., Elliott, M.: SMART marine goals, targets and management-is SDG 14 operational or aspirational, is 'life below water' sinking or swimming? Mar. Pollut. Bull. **123**(1–2), 28–33 (2017)

3. Córdova, M., et al.: Litter detection with deep learning: a comparative study. Sensors **22**(2), 548 (2022). https://doi.org/10.3390/s22020548

4. Fulton, M., Hong, J., Islam, M.J., Sattar, J.: Robotic detection of marine litter using deep visual detection models. In: 2019 International Conference on Robotics and Automation (ICRA), pp. 5752–5758. IEEE (2019)

5. Galgani, L., Beiras, R., Galgani, F., Panti, C., Borja, A.: Impacts of marine litter. Front. Mar. Sci. **6**, 208 (2019)

6. Gall, S., Thompson, R.: The impact of debris on marine life. Mar. Pollut. Bull. **92**(1), 170–179 (2015). https://doi.org/10.1016/j.marpolbul.2014.12.041

7. Gallego, A.J., Calvo-Zaragoza, J., Fisher, R.B.: Incremental unsupervised domain-adversarial training of neural networks. IEEE Trans. Neural Networks Learn. Syst. **32**(11), 4864–4878 (2021). https://doi.org/10.1109/TNNLS.2020.3025954

8. He, K., Gkioxari, G., Dollár, P., Girshick, R.: Mask R-CNN. In: 2017 IEEE International Conference on Computer Vision (ICCV), pp. 2980–2988 (2017). https://doi.org/10.1109/ICCV.2017.322

9. Hong, J., Fulton, M., Sattar, J.: Trashcan: A semantically-segmented dataset towards visual detection of marine debris. CoRR abs/2007.08097 (2020)

10. Jung, A.B., et al.: Imgaug (2020). https://github.com/aleju/imgaug. Accessed 20 Jan 2022

11. Kikaki, K., Kakogeorgiou, I., Mikeli, P., Raitsos, D.E., Karantzalos, K.: MARIDA: a benchmark for marine debris detection from sentinel-2 remote sensing data. PloS One **17**(1), e0262247 (2022)

12. Kohavi, R.: A study of cross-validation and bootstrap for accuracy estimation and model selection. In: Proceedings IJCAI. IJCAI 1995, vol. 2, pp. 1137–1143. Morgan Kaufmann Publishers Inc., San Francisco (1995)

13. LeCun, Y., Bengio, Y., Hinton, G.: Deep learning. Nature **521**(7553), 436–444 (2015)

14. Lin, T.-Y., et al.: Microsoft COCO: common objects in context. In: Fleet, D., Pajdla, T., Schiele, B., Tuytelaars, T. (eds.) ECCV 2014. LNCS, vol. 8693, pp. 740–755. Springer, Cham (2014). https://doi.org/10.1007/978-3-319-10602-1_48

15. Loshchilov, I., Hutter, F.: SGDR: stochastic gradient descent with warm restarts. In: 5th International Conference on Learning Representations, ICLR 2017. Toulon, France (2017)

16. Morales-Caselles, C., et al.: An inshore-offshore sorting system revealed from global classification of ocean litter. Nat. Sustain. **4**(6), 484–493 (2021)

17. Reinhold, S.: TrashTag. https://www.trashtag.org/. Accessed 01 Jan 2022

18. Russell, B.C., Torralba, A., Murphy, K.P., Freeman, W.T.: LabelMe: a database and web-based tool for image annotation. Int. J. Comput. Vision **77**(1–3), 157–173 (2008)

19. Sakai, H.: Japan agency for marine-earth science and technology. In: Proc. Shinkai 2000 Kenkyu Symposium 1990 (1990)

20. Sinclair, R.: The Big Blue Ocean Cleanup. https://www.bigblueoceancleanup.org/. Accessed 01 Jan 2022
21. Singh, D., Valdenegro-Toro, M.: The marine debris dataset for forward-looking sonar semantic segmentation. In: Proceedings of the IEEE/CVF International Conference on Computer Vision, pp. 3741–3749 (2021)
22. Slat, B.: The Ocean Cleanup. http:///theoceancleanup.com/. Accessed 01 Jan 2022

A Case of Study on Traffic Cone Detection for Autonomous Racing on a Jetson Platform

Javier Albaráñez Martínez[ID], Laura Llopis-Ibor[ID],
Sergio Hernández-García[✉][ID], Susana Pineda de Luelmo[ID],
and Daniel Hernández-Ferrándiz[ID]

Universidad Rey Juan Carlos, Móstoles, Spain
j.albaranez@alumnos.urjc.es,
{laura.llopis,sergio.hernandez,susana.deluelmo,daniel.hernandezf}@urjc.es

Abstract. Autonomous driving is a growing research line since the future of transportation depends to a great extent on it. Driving is highly dependant on the environment sensing system. Over the last decade, several detection architectures based on neural networks and monocular cameras have been proposed to address this task. However, adapting these proposals to a vehicle with limited resources remains a challenging problem. In our study, we propose a lightweight neural network to perform cone detection from a racing car. We also compare its performance against other popular state-of-the-art proposals on a resource constrained system. From the obtained results, we can conclude that our network outperforms the state-of-the-art works for our use case and it is less resource demanding.

Keywords: Autonomous driving · Object detection · Embedded systems · Deep learning

1 Introduction

Formula Student (FS) is an annual event where students from all over Europe gather to compete designing and driving racing cars. As part of this competition, teams face off against each other to design the fastest and most accurate autonomous vehicle. To successfully complete these tasks, teams must devote a great effort on the perception system in order to detect the track delimiting cones. Since this system must be mounted on the vehicle, it is subject to two limitations: a small size and a limited power supply. These limitations greatly constrain the vehicle's computational capabilities. Thus, a highly efficient solution is required.

Supported by Comunidad de Madrid Y2018/EMT-5062, Ministerio de Ciencia, Innovación y Universidades RTI2018-098743-B-I00 (MICINN/FEDER) and PRE2019-089639 financed by MCIN/AEI/10.13039/501100011033 and FSE+.

A. J. Pinho et al. (Eds.): IbPRIA 2022, LNCS 13256, pp. 629–641, 2022.
https://doi.org/10.1007/978-3-031-04881-4_50

In this paper we propose a computationally efficient neural network for traffic cone detection, SNet-3L (**S**egmentation **Net**work **3 L**evels)[1]. Unlike other detection proposals, our architecture was specifically designed to detect cones, resulting in a network much lighter than off-the-shelf detection networks.

We report the results of a performance comparison between state-of-the-art object detection models and our proposed network. We performed this comparison in the same embedded system that we use in FS. We also provide an analysis of inference times in order to determine which model offers the best performance and real-time detections. Collecting images ranging a variety of weather conditions and cone colors for the training data set is costly and time consuming. Synthetic image generation provides an inexpensive and feature rich source of samples to compensate for the limitations of real data. Thus, we introduce a synthetic data set that contains both real and synthetic images and perform an analysis to show the extent to which this mixed data set is an improvement over the real data set.

This paper is organized as follows: in Sect. 2 we provide a brief overview about object detection. Section 3 describes the hardware, data set and networks used in our comparative. In Sect. 4 we present and analyze the results yielded by the evaluated networks. Finally, our conclusions are drawn in Sect. 5.

2   Related Works

A growing body of literature has examined the object detection problem from different perspectives (e.g. stereo cameras, lidar or radar) [2]. In this work, we are particularly interested in object detection from monocular camera images, since it is the type of device we use in our vehicle. The classical approach to this problem was based on the use of handcrafted features, however, in the last decade research has shifted towards the use of deep learning [10]. Following this approach, Faster R-CNN [15] was proposed as the first end-to-end deep learning solution. It offered near real-time detection while achieving state-of-the-art results. Lin *et al.* [8] proposed a new loss function to address the extreme class imbalance issue between the foreground and background when training dense detectors. They used this novel loss on a new architecture, namely RetinaNet. Recent approaches are capable of reaching real-time detection with improved performance. YOLO [13] is a proposal that used a single network to perform the entire detection task. Following a similar approach, the authors of SSD [11] performed detection at different scales. They defined a series of scales and aspect ratios to generate regions over different levels of feature maps from a CNN. These regions were taken as the basis for the detections, and the network generated displacements to model the positions of the objects of interest.

These latter networks have very low latency, reaching real time. However, this performance is obtained on workstations. For our use case, we are highly limited by the size and energy efficiency of our equipment. Therefore, we need

---

[1] Available at https://github.com/AlbaranezJavier/Object-Detection-with-Segment ation.

more efficient architectures. Motivated by similar constraints, updated versions of previous proposals emerged. YOLOv2 [14] attempted to increase recall and solve the localization problems of its predecessor while increasing its efficiency. Another approach was MobileNet [6], a family of architectures that gives the designer the freedom to choose the architecture that best suits the computational constraints. Unlike the previous works, depthwise separable convolutions are used to reduce the number of model parameters.

Sandler *et al.* [17] also proposed SSDLite, which combined the SSD architecture with MobileNet V2. A recent proposal focused on efficiency is EfficientDet [20]. This architecture is an anchor-based detection network that proposed the use of a weighted bi-directional feature pyramid network to combine features at different resolutions, with EffcientNet [19] as backbone and a scaling factor that controls the dimension of the network.

Taking advantage of the potential of these latest detection architectures and the growing interest in autonomous driving, some works have been published in which object detection is also performed on an embedded system. Dhall *et al.* [3] also participated in the Formula Student competition using YOLOv2 along a Jetson TX2 along a monocular camera in order to detect cones of different colors.

3   Materials and Methods

This section first introduces the hardware used to perform the detection network performance tests. This is followed by a description of the proposed architecture, SNet-3L, and details about the implementations used for the state-of-the-art networks that are compared against it. Finally, detailed information is given on the proposed data set on which all the architectures are trained.

3.1   Hardware

Experiments on this paper were performed on NVIDIA Jetson Xavier NX module (from now on Xavier). This module has a power consumption of 10 W, which is 55 times less energy usage than a desktop PC with an entry level GPU (NVIDIA RTX 3050) [1]. Additionally, Xavier's reduced dimensions ($103 \times 90.5 \times 34.66$ mm) are an another added advantage over a conventional computer (approximately $400 \times 200 \times 500$ mm) in our scenario, i.e. on board a racing vehicle. In terms of processing, Xavier architecture has a Volta-generation GPU with 6 Streaming Multiprocessors (384 CUDA cores), 48 Tensor Cores, and a 64-bit NVIDIA Carmel ARMv8.2 CPU with 6 cores. Moreover, it has a Deep Learning accelerator with 2 NVDLA engines and a vision accelerator with a 7-socket Very Long Instruction Word Vision processor.

3.2   Proposed Cone Detection Network: SNet-3L

Our proposed neural network, SNet-3L, focuses exclusively on cone detection. Since this network only aims at detecting one type of object, we have been able to design a much lighter neural network than the state-of-the-art multi-class networks. This significantly reduces the computational cost in order to meet the requirements of our hardware. SNet-3L performs the detection through an instance segmentation of the image in a two-stage process: first, segmentation is performed followed by a post-processing where the detection is carried through.

Segmentation Stage. The neural architecture proposed is based on U-Net [16] with three resolution levels of $320 \times 180 \times 3$, $160 \times 90 \times 6$ and $80 \times 45 \times 6$. Therefore, we named our network SNet-3L: segmentation network with three resolution levels. This network's input dimensions are 320×180, and the input images are transformed into the YUV color space (we provide the reasoning behind choosing this color space in Appendix). Therefore, the images have to be preprocessed to meet these requirements. Firstly, the edges of the input image are extracted through a convolution with a Laplacian filter and then they are concatenated with the input image. After this, different stages of convolutions are combined and concatenated with fixed bilinear filters [5] with a stride of 2 during the codification and decodification stages. Figure 1 shows the complete architecture and connections between the three kind of operation blocks listed bellow:

- *Trainable convolution block.* Applies a convolution followed by a batch normalization. Additionally, a ReLU activation function is applied to the sum of the *bias* of the convolution layer and the output of the batch normalization.
- *Fixed convolution block.* Applies a convolution with a fixed bilinear or laplacian filter followed by a batch normalization operation.
- *Fixed transposed convolution block.* Applies a transposed convolution with a fixed bilinear filter followed by a batch normalization operation.

The final output of the neural network consist of as many segmentation maps as object classes plus a background map of size $320 \times 180 \times (N + 1)$, where N is the number of classes.

Postprocessing Stage. In this second stage we extract the contours from the segmentation maps to enclose segmented objects within a polygon. Next, we compute a bounding box for each polygon. For these two steps we use the algorithms from [18].

Figure 2 illustrates the complete detection process.

3.3 State-of-the-art Object Detection Neural Networks

We now introduce the implementations used for the three state-of-the-art detection networks that will compete with SNet-3L in solving our problem. For this purpose, all three networks have been trained with the same data set of cones that we present later.

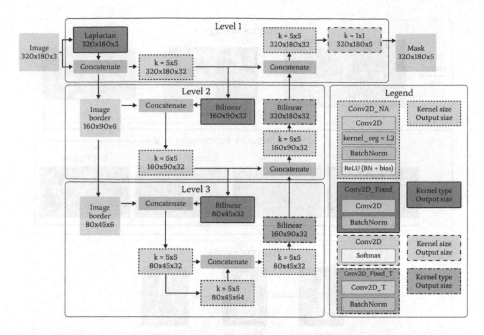

Fig. 1. SNet-3L architecture. It consists of 3 levels of resolutions following a scheme similar to the U-Net. In this architecture fixed kernel convolutions are used to transition between different resolutions. Moreover, these same convolutions are used to extract the edges from the image to be further concatenated with the input image. Additionally, all levels receive initial information in addition to their extracted edges.

EfficientDet-D0 [20]. Among the available EfficientDet networks, we select the smallest architecture, EfficientDet-D0, as it offers the best balance between prediction speed and performance. We use the original implementation.

SSD-MobileNet V2 [17]. We use the Caffe [7] implementation provided in the jetson-inference library [4] pretrained on the COCO dataset [9]. Then, we retrain it using DIGITS [21].

3.4 Synthetic-Real Cones Dataset

We propose a cone data set consisting of 1268 images, 1006 synthetic and 262 real images, containing a total of 9939 cones. We split the data in a set of 1141 images for training formed by 899 synthetic and 242 real and another set of 127 for testing formed by 107 synthetic and 20 real. Synthetic data enlarge our data set by including different scenarios, lighting and weather conditions, whereas real data brings in real-world imperfections, textures and noise. In our experiments, we use three versions of our data set to analyze the contribution of synthetic data. The first version is the one already described, where synthetic and real data are mixed. The second and third versions are entirely composed of real and synthetic data, respectively. We generate all the synthetic images using

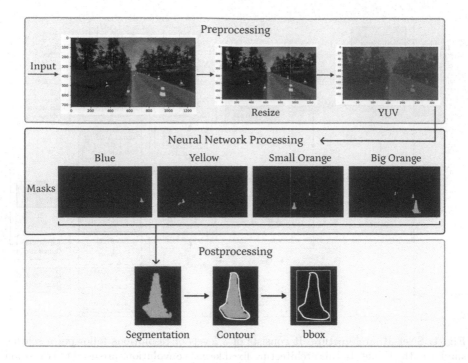

Fig. 2. SNet-3L detection by segmentation process scheme.

Unity3D. The tool employed to generate our synthetic data is available at www.
github.com/AlbaranezJavier/SyntheticConeDatasetGenerator. This tool allows
us to generate cones anywhere in a virtual scene and capture those compositions
along with the cone annotations, which include bounding boxes. These scenes
always include a road and cones have been generated randomly at ground level,
both standing and knocked down. Figure 3 shows some examples of generated
and real data.

3.5 TensorRT

To achieve a more efficient model that uses as little memory as possible without
sacrificing its detection performance, we have optimized the model obtained from
the SNet-3L network using TensorRT. On the other hand, SSD-MobileNet V2
was automatically optimized by the jetson-inference library. Finally, software
and hardware incompatibilities between TensorRT, Tensorflow and Xavier do
not support EfficientDet-D0 optimization.

TensorRT [12] is a library developed by NVIDIA in order to provide faster
inference times on NVIDIA graphics processing units (GPUs). It optimizes network models by layer combination and efficient kernel selection. The optimized
network is able to work with low numerical precision data types, which reduces
the memory requirements and improves the network's efficiency.

Fig. 3. Examples of images from our synthetic-real cones dataset. First two columns contain synthetic images, whereas last two columns contain real images.

4   Experimental Results

In this section we describe our experiments and results. We first analyze the networks performance separately. Next, we compare the FPS that each network delivers when ran on Xavier.

4.1   Individual Network Metrics

In this experiment, we use the intersection over union (IoU) with different thresholds to evaluate the quality of the predictions of each network by measuring how closely the predicted bounding box matches the ground truth counterpart. We measure the recall to assess the network's ability to correctly identify the cones in the image. We also measure the precision to test the network's ability to distinguish between cones and background. Finally, we compute the F1 Score to measure the overall performance of the networks, as this metric combines both recall and precision. We run each object detector in a workstation over the test set of the cones data set described in the previous section.

In Table 1 we show a comparison of the performance results achieved by the reviewed networks. Each network is trained in the following versions of our data set: synthetic, real, and both real and synthetic. We tested them on a scenario that resembles our use case, i.e. testing with real data. All networks obtained worse performance results across all metrics when trained on synthetic images than when trained on a mix of real and synthetic data. Therefore, synthetic data alone is not enough to train reliable networks for real-world cone detection.

EfficientDet-D0 provides the best bounding box estimation for all IoU thresholds when trained with synthetic data, however a very low number of cones are detected (17.62%@0.25, 17.24%@0.50, 8.05%@0.75 of recall). This network also attains the best F1 score (84.19@0.25%, 69.66@0.50%, 32.91@0.75%) while achieving the best IoU (67.58%@0.25, 73.39@0.50%, 84.32@0.75%) when trained on real data. SNet-3L closely follows with a good balance of IoU (66.74%@0.25, 71.05%@0.50, 82.73%@0.75) and F1 score (70.28%@0.25, 59.77%@0.50, 22.33%@0.75) when trained with real data. This network also obtains similar results when trained with synthetic and real data, whereas

Table 1. Detection performance comparison against state-of-the-art detectors. We computed all metrics at three IoU thresholds: 0.25, 0.50 and 0.75. We used three training setups: synthetic data set (S), real data set (R) and a mix of synthetic and real data sets (S+R). Tests were performed on real data. Difference against SNet-3L in brackets.

Network	Train	IoU (%) @0.25	@0.50	@0.75
SNet-3L (Ours)	S	61.35	67.75	80.15
255k parameters	R	66.74	71.05	82.73
	S+R	68.24	72.84	84.33
	Mean	65.44	70.54	82.40
SSD-MobileNet V2	S	62.71 (+1.36)	63.74 (−4.01)	80.95 (+0.80)
15.3M parameters	R	64.49 (−2.25)	64.49 (−6.56)	81.09 (−1.64)
	S+R	66.41 (−1.83)	67.41 (−5.43)	83.32 (−1.01)
	Mean	64.53 (−0.91)	65.21 (−5.32)	81.78 (−0.61)
EfficientDet-D0	S	74.34 (+12.99)	74.97 (+7.22)	83.55 (+3.40)
3.9M parameters	R	67.58 (+0.84)	73.39 (+2.34)	84.32 (+1.59)
	S+R	68.33 (+0.09)	74.66 (+1.82)	85.41 (+1.08)
	Mean	70.08 (+4.64)	74.34 (+3.80)	84.42 (+2.02)

Network	Train	Precision (%) @0.25	@0.50	@0.75
SNet-3L (Ours)	S	28.16	21.36	5.50
255k parameters	R	93.86	79.82	29.82
	S+R	94.22	80.44	31.11
	Mean	65.44	60.54	22.14
SSD-MobileNet V2	S	81.48 (+53.32)	77.78 (+56.42)	7.41 (+1.90)
15.3M parameters	R	100.00 (+6.14)	100.00 (+20.18)	8.82 (−21.00)
	S+R	100.00 (+5.78)	94.29 (+13.84)	20.00 (−11.11)
	Mean	64.53 (−0.91)	90.69 (+30.15)	12.07 (−10.07)
EfficientDet-D0	S	97.87 (+69.71)	95.74 (+74.38)	44.68 (+39.18)
3.9M parameters	R	95.17 (+1.51)	78.74 (−1.08)	37.20 (+7.38)
	S+R	95.27 (+1.05)	78.11 (−2.33)	40.83 (+9.72)
	Mean	70.08 (+4.64)	84.19 (+23.65)	40.90 (+18.76)

Network	Train	Recall (%) @0.25	@0.50	@0.75
SNet-3L (Ours)	S	22.83	17.32	4.46
255k parameters	R	56.17	47.77	17.85
	S+R	55.64	47.51	18.37
	Mean	44.88	37.53	13.56
SSD-MobileNet V2	S	8.43 (−14.4)	8.05 (−9.27)	0.77 (−3.69)
15.3M parameters	R	13.03 (−43.14)	13.03 (−34.74)	1.15 (−16.7)
	S+R	13.41 (−42.23)	12.64 (−34.87)	2.68 (−15.69)
	Mean	11.62 (−33.26)	64.53 (+27.00)	1.53 (−12.03)
EfficientDet-D0	S	17.62 (−5.21)	17.24 (−0.08)	8.05 (+3.49)
3.9M parameters	R	75.48 (+19.31)	62.45 (+12.32)	29.50 (+11.65)
	S+R	61.69 (+6.05)	50.57 (+3.06)	26.44 (+8.07)
	Mean	51.59 (+6.71)	43.42 (+5.89)	21.33 (+7.77)

(continued)

Table 1. (*continued*)

Network	Train	F1 score (%) @0.25	@0.50	@0.75
SNet-3L (Ours)	S	25.22	19.13	4.93
255k parameters	R	70.28	59.77	22.33
	S+R	69.97	59.74	23.1
	Mean	55.16	46.21	16.79
SSD-MobileNet V2	S	15.28 (−9.94)	14.58 (−4.55)	1.39 (−3.54)
15.3M parameters	R	23.05 (−17.23)	23.05 (−36.72)	2.03 (−20.3)
	S+R	23.65 (−46.32)	22.30 (−37.44)	13.64 (−9.46)
	Mean	20.66 (−34.5)	19.97 (−26.23)	5.69 (−11.10)
EfficientDet-D	S	29.87 (+4.65)	29.22 (+10.09)	13.64 (+8.71)
3.9M parameters	R	84.19 (+13.91)	69.66 (+9.89)	32.91 (+10.58)
	S+R	74.88 (+4.91)	61.40 (+1.66)	32.09 (+8.99)
	Mean	62.98 (+7.82)	53.42 (+7.21)	26.21 (+6.42)

Table 2. Jetson Xavier NX power modes.

Power mode	15 W 2 cores	15 W 4 cores	15 W 6 cores	10 W 2 cores	10 W 4 cores	10 W desktop
Power budget (Watts)	15	15	15	10	10	10
# Cores	2	4	6	2	4	4
CPU max frequency (MHz)	1900	1400	1400	1500	1200	1900
GPU max frequency (MHz)	1100	1100	1100	800	800	510

when trained with only synthetic data it results in low F1 score (25.22%@0.25, 19.13%@0.50, 4.93%@0.75). SSD-MobileNet V2 provides a reasonable bounding box estimation but a subpar cone detection. Finally, is worth noting that training all networks on both synthetic and real data improves the bounding box estimation compared to training with real data only. However, this trend generally comes at the expense of a lower F1 score. Looking at the mean results for each metric and network, EfficientDet-D0 offers the best overall performance, closely followed by our proposal.

4.2 Network Throughput Comparison

We carry out a network throughput comparison between SSD-MobileNet V2, EfficientDet-D0 and SNet-3L. We run each detector on Xavier using several power modes (see Table 2) to evaluate the performance of each network under power consumption constraints. For this comparison, we calculate the average inference time of 5 consecutive iterations. We discard the first iteration to avoid wrong measures due to CUDA and GPU warm up.

Table 3. Network throughput for each detector and power mode expressed in frames per second (higher values are better). We use a blue color bar to provide a quick visual comparison of the results. Highlighted cells in orange show an overflow warning in Xavier's capacity. In addition to the state-of-the-art networks, we also show five different configurations on SNet-3L. FP32 and FP16 denote the optimization mode used, 32 and 16-bit floating-point data types respectively. PCO denotes that the network was optimized in a workstation with 2070 RTX GPU. Lastly, XO label denotes that the network was optimized on Xavier itself. All these configurations have been executed on Xavier. Finally, note that EfficientDet-D0 and SNet-3L models were not optimized.

Power mode	15 W 2 cores	15 W 4 cores	15 W 6 cores	10 W 2 cores	10 W 4 cores	10 W desktop
SSD-MobileNet V2	86.81	98.23	96.34	75.87	86.43	52.55
EfficientDet-D0	4.83	4.89	4.92	4.34	4.37	3.59
SNet-3L	0.55	0.47	0.46	0.44	0.38	0.62
SNet-3L FP32 PCO	63.90	79.94	83.61	53.28	66.93	76.80
SNet-3L FP16 PCO	93.46	86.36	86.66	72.05	59.95	76.57
SNet-3L FP32 XO	63.82	77.16	61.65	43.57	52.66	75.82
SNet-3L FP16 XO	47.53	58.96	75.82	52.16	57.80	70.72

To improve the efficiency and memory footprint of our networks, we optimized them with TensorRT. The outputs of this process are optimized models with 32 and 16-bit floating-point data types. When optimizing SSD-MobileNet V2 the FP32 precision was used as no other floating precision was available. SNet-3L was optimized with two different modes, FP32 and FP16, and for two different platforms, a workstation with an NVIDIA 2070 GPU and a Xavier (note that the network throughput is measured on Xavier). EfficientDet-D0 was not optimized with TensorRT due to software and hardware incompatibilities. Therefore, we tested the non-optimized EfficientDet-D0 model on the embedded platform. We also provide a network throughput comparison for the non-optimized version of SNet-3L in Table 3 in order to make a fair comparison between these models.

To address the cone detection problem in real time we set a requirement of 60 FPS for the network throughput. This requirement is met by both SSD-MobileNet V2 and optimized versions of SNet-L3 as seen in Table 3. Furthermore, we discard the configurations that raised a resource overflow warning on Xavier (marked in orange) due to possible system instability. Finally, focusing on the configurations that use the minimum power (10 W) and considering the results from Subsect. 4.2, the best network is SNet-3L FP32 PCO on the 10 W desktop configuration. It is also worth noting that the SNet-3L network obtains better results on the 10 W desktop mode. Considering the frequencies stated in the Table 2, we can observe that in this mode the GPU frequency is lower compared to the other modes and the CPU frequency is higher, benefiting SNet-3L since it has fewer parameters, thus being less GPU bound.

5   Discussion

In this study we have reported the results of an experiment whose purpose was to help us to choose the best cone detection network for a resource constrained system. The chosen network will be used on a racing car to compete in the Formula Student competition, therefore we were limited both in terms of computational resources and energy consumption. Despite having tested leading architectures, the proposed network, SNet-3L, achieves the second best results among the tested networks. EfficientDet-D0 is the best performing network but it requires an order of magnitude more parameters than our proposal (225k against 3.9M). Our problem was limited to a single class, cones, thus complex networks led to low efficiency. Finally, SNet-3L is the best option for our use case because it provides real-time cone detection when used on our embedded platform.

Finally, we also conclude that a mixed data set is a good solution when there is not enough real data. As future work we will investigate the best way to create a fully synthetic data set that simulates better the properties of real images. We will also extend our system to detect cones of different colors and locating them in the environment.

Appendix

We carry out an experiment to validate the selection of YUV as input color space to the neural network. We launch the model over the Synthetic-Real Cones data set using five different color spaces for the input images: HSL, HSV, LAB, YUV and BGR. We use SNet-5L model, which extends SNet-3L model with two further resolution levels at the bottom of the encoder and decoder respectively. Therefore, we have an input level of $1280 \times 720 \times 3$ follow by $640 \times 360 \times 3$, $320 \times 180 \times 3$, $160 \times 90 \times 3$ and $80 \times 45 \times 3$. As shown in Table 4, the YUV color space performs slightly better than the rest.

Table 4. Performance comparison of object detectors on detection and classification in the synthetic-real cones data set.

SNet-5L	BGR	HSL	HSV	LAB	YUV
IoU (%)	75,76	73,91	73,33	76,67	**77,06**
Precision (%)	**89,71**	88,01	87,73	89,09	88,89
Recall (%)	82,97	82,19	81,71	84,61	**85,27**
F1 Score (%)	86,21	85,00	84,61	86,79	**87,04**

References

1. NVIDIA geforce RTX 3050 (2021). https://www.nvidia.com/en-gb/geforce/graphics-cards/30-series/rtx-3050/
2. Arnold, E., Al-Jarrah, O.Y., Dianati, M., Fallah, S., Oxtoby, D., Mouzakitis, A.: A survey on 3D object detection methods for autonomous driving applications. IEEE Trans. Intell. Transp. Syst. **20**(10), 3782–3795 (2019). https://doi.org/10.1109/TITS.2019.2892405
3. Dhall, A., Dai, D., Van Gool, L.: Real-time 3D traffic cone detection for autonomous driving. In: IEEE Intelligent Vehicles Symposium (IV), pp. 494–501 (2019). https://doi.org/10.1109/IVS.2019.8814089
4. Franklin, D.: Hello IA world, NVIDIA Jetson, deploying deep learning (2022). https://github.com/dusty-nv/jetson-inference
5. He, J., Xu, J.: MgNet: a unified framework of multigrid and convolutional neural network. Sci. China Math. **62**(7), 1331–1354 (2019). https://doi.org/10.1007/s11425-019-9547-2
6. Howard, A.G., et al.: MobileNets: Efficient convolutional neural networks for mobile vision applications (2017)
7. Jia, Y., et al.: Caffe: convolutional architecture for fast feature embedding. In: Proceedings of the 22nd ACM International Conference on Multimedia, pp. 675–678 (2014). https://doi.org/10.1145/2647868.2654889
8. Lin, T.Y., Goyal, P., Girshick, R., He, K., Dollár, P.: Focal loss for dense object detection. IEEE Trans. Pattern Anal. Mach. Intell. **42**(2), 318–327 (2020). https://doi.org/10.1109/TPAMI.2018.2858826
9. Lin, T.-Y., et al.: Microsoft COCO: common objects in context. In: Fleet, D., Pajdla, T., Schiele, B., Tuytelaars, T. (eds.) ECCV 2014. LNCS, vol. 8693, pp. 740–755. Springer, Cham (2014). https://doi.org/10.1007/978-3-319-10602-1_48
10. Liu, L., et al.: Deep learning for generic object detection: a survey. Int. J. Comput. Vis. **128**(2), 261–318 (2019). https://doi.org/10.1007/s11263-019-01247-4
11. Liu, W., et al.: SSD: single shot multibox detector. In: Leibe, B., Matas, J., Sebe, N., Welling, M. (eds.) ECCV 2016. LNCS, vol. 9905, pp. 21–37. Springer, Cham (2016). https://doi.org/10.1007/978-3-319-46448-0_2
12. NVIDIA: TensorRT (2018). https://developer.nvidia.com/tensorrt
13. Redmon, J., Divvala, S., Girshick, R., Farhadi, A.: You only look once: Unified, real-time object detection (2016)
14. Redmon, J., Farhadi, A.: YOLO9000: better, faster, stronger. In: Proceedings of the IEEE Conference on Computer Vision and Pattern Recognition, pp. 7263–7271 (2017). https://doi.org/10.1109/CVPR.2017.690
15. Ren, S., He, K., Girshick, R., Sun, J.: Faster R-CNN: towards real-time object detection with region proposal networks. IEEE Trans. Pattern Anal. Mach. Intell. **39**(6), 1137–1149 (2016). https://doi.org/10.1109/ICCV.2015.169
16. Ronneberger, O., Fischer, P., Brox, T.: U-Net: convolutional networks for biomedical image segmentation. In: Navab, N., Hornegger, J., Wells, W.M., Frangi, A.F. (eds.) MICCAI 2015. LNCS, vol. 9351, pp. 234–241. Springer, Cham (2015). https://doi.org/10.1007/978-3-319-24574-4_28
17. Sandler, M., Howard, A., Zhu, M., Zhmoginov, A., Chen, L.C.: MobileNetV 2: inverted residuals and linear bottlenecks. In: Proceedings of the IEEE Conference on Computer Vision and Pattern Recognition, pp. 4510–4520 (2018). https://doi.org/10.1109/CVPR.2018.00474

18. Suzuki, S., et al.: Topological structural analysis of digitized binary images by border following. Comput. Vis. Graph. Image Process. **30**(1), 32–46 (1985). https://doi.org/10.1016/0734-189X(85)90016-7
19. Tan, M., Le, Q.: EfficientNet: rethinking model scaling for convolutional neural networks. In: Chaudhuri, K., Salakhutdinov, R. (eds.) Proceedings of the 36th International Conference on Machine Learning. Proceedings of Machine Learning Research, vol. 97, pp. 6105–6114. PMLR, 09–15 June 2019
20. Tan, M., Pang, R., Le, Q.V.: EfficientDet: scalable and efficient object detection. In: Proceedings of the IEEE/CVF Conference on Computer Vision and Pattern Recognition, pp. 10781–10790 (2020). https://doi.org/10.1109/cvpr42600.2020.01079
21. Yéager, L., Bernauer, J., Gray, A., Houston, M.: DIGITS: the deep learning GPU training system. In: ICML 2015 AutoML Workshop (2015)

Energy Savings in Residential Buildings Based on Adaptive Thermal Comfort Models

Rodrigo Almeida[1,2] , Petia Georgieva[1(✉)] , and Nelson Martins[1]

[1] University of Aveiro, Aveiro, Portugal
{almeida.rodrigo,petia,nmartins}@ua.pt
[2] Bosch Termotecnologia, Cacia, Aveiro, Portugal

Abstract. The residential Heating, Ventilation and Air Conditioning (HVAC) systems nowadays aim to provide tailored solutions, focused on the home occupants' thermal comfort. In that sense, these systems must support the well-being and productiveness throughout the day and the working hours, without compromising the system energetic sustainability.

Relying on user data collected from households in Europe, this work aims to assist residents with optimal behaviours, having in mind the energy efficiency of the system. It seeks a retrofittable solution capable of relying on the limited spectrum of data that these systems are collecting today. To do so, it explores the adaptive comfort theory for residential homes to achieve optimized setpoints without compromising the occupants' thermal needs.

Keywords: Smart home · Smart thermostat · Energy savings · Indoor thermal comfort patterns · Adaptive thermal models

1 Introduction

According to World Health Organization (WHO) indicators, people spend around 90% of their time enclosed in buildings. These indicators are now stressed out by the latest pandemic crisis, as most of the employers and employees had to reinvent their work methodology to perform their responsibilities remotely from their own homes. For this reason, the HVAC system developers aim to provide residences equipped with smart technologies that can enhance the quality of life and promote independent living (i.e. Smart Homes).

Without compromising the thermal comfort of the building occupants, these systems aim to optimize their energy consumption, not only because it would reflect cost reductions for the customer (hence increasing their satisfaction), but also due to the environmental challenges that arise on a global scale. Pointing to the European Union (EU) energy and climate strategies for 2030, a more

This research work is funded by National Funds through the FCT - Foundation for Science and Technology, in the context of the project UIDB/00127/2020.

ambitious reiteration was made, during 2020, by setting the desired reduction of greenhouse emissions target to 55% by 2030, compared to 1990 levels [2]. This goal presents a clear endeavour towards energy sustainability and the usage of cleaner alternatives.

Aligned with the current strategies to reduce the footprint of HVAC systems, in this paper we explore a method to reduce the heating systems energy consumption without loss of comfort to the occupants. For that, it relies on data collected from different European residential homes and the application of regional adaptive thermal comfort models used to establish indoor temperature limits required to grant adequate thermal comfort levels. By analysing the user-defined setpoints, the explored method will propose, when applicable, lower ones that can still grant acceptable thermal comfort levels. Finally, it is proposed an expression to evaluate the relative energy consumption reduction, using as reference real measured datasets.

2   Indoor Thermal Comfort

Human thermal comfort (TC) is achieved in general, when an individual does not feel the need to change the thermal environment or adapt to it. Different standards [3, 9, 13] are available to express indoor human TC perception through empirical models that rely on personal and environmental factors. Two main TC models are recommended to support the current practice of comfort management in buildings: Predicted Mean Vote (PMV) and the adaptive TC model.

2.1   The Predicted Mean Vote

The PMV model of thermal comfort was proposed by Fanger [5] and standardized in EN ISO 7730 [9]. It relates heat balance equations and empirical studies about human factors to express human thermal sensation as the imbalance of the heat transfer between a human body and its surrounding environment [7]. The model explores a set of correlations which can be translated as a function of six dependent variables (Eq. 1): air temperature (t_a), mean radiant temperature (t_{mrt}), air velocity (v), air humidity (p_a), clothing resistance (I_{cl}), and activity level (M) [7].

$$PMV = f(t_a, t_{mrt}, v, p_a, M, I_{cl}) \tag{1}$$

To quantify the model, a thermal discomfort scale was proposed. The scale is comprised between –3 (cold) and +3 (hot), with 0 being the optimal thermal sensation. Based on the PMV index, the Predicted Percentage of Dissatisfied (PPD) indicator can be estimated as well. This last one establishes a quantitative prediction of the percentage of thermally dissatisfied occupants. This relationship is represented by Eq. (2) [9]. The recommended thresholds for an acceptable thermal environment that provides general comfort, correspond to the PMV indexes between –0.5 and 0.5.

$$PPD = 100 - 95 * e^{(-0.3353 * PMV^4 - -0.2179 * PMV^2)} \tag{2}$$

The PMV model has been widely adopted to assess indoor thermal conditions throughout different types of buildings under static or quasi static operating conditions, as the model was initially developed to predict the thermal sensation in buildings with HVAC systems [6]. Table 1 presents the PMV-PPD model boundaries according to the ISO 7730 standard.

Table 1. PMV-PPD model parameter boundaries

Parameters	Value range
Metabolic rate	46–232 W/m^2 (0.8–4 met)
Clothing resistance	0–0.310 m^2 K/W (0–2 clo)
Air temperature indoor	10–30 °C
Mean radiant temperature	10–40 °C
Air velocity	0–1 m/s
Vapor pressure or relative humidity	0–2.7 kPa or 30–70%

Since the introduction of the PMV model, numerous studies on thermal comfort in both real-life situations and climate chambers have been conducted. The PMV model validity and application has been subject to intense study and discussion [15]. It was found that, for naturally ventilated buildings, the indoor comfort temperature increases in warmer climatic zones and decreases in colder ones. This premise led to the development and proposal of the adaptive TC model [15].

2.2   Adaptive Thermal Comfort Model

The adaptive thermal comfort (TC) theory explores the natural tendency that humans have to dynamically adapt and change according to environmental conditions [12]. Adaptation happens both consciously and unconsciously. From behavioural adjustments (e.g., change clothes, open window) to physiological (e.g., acclimatization) and psychological (e.g., routines and habits) [4].

The adaptive TC model relates indoor design temperatures or acceptable temperature ranges, with the outdoor meteorological or climatological parameters. Outdoor environmental factors are set to influence the occupants' clothing, metabolic rate, and the way they interact with the building.

Nicol et al. suggested an adaptive comfort temperature (T_{rm}) based on a running mean outdoor temperature given by Eq. (3) [11,12].

$$T_{rm,n} = (1 - \alpha) \sum_{i=0}^{j} \alpha^i T_{n-1-i} \qquad (3)$$

where:

- α: discount rate to guarantee that the running mean temperature is decreasingly affected by past temperatures.
- T_n: temperature at day n.

Nicol et al. highlight that the recommendations from the standard EN15251 are more focused in European countries, as it is based on data from the Smart Controls and Thermal Comfort (SCATs) European project [11]. From the same project, and based on the same dataset, McCartney et al. derived the adaptive control algorithm for each European country that participated in the SCATs project [10]. Table 2 presents the equations derived by the authors.

From the data of several Belgian houses, Peeters et al. explored the adaptive comfort relation and confirmed that there is a strong dependency between thermal comfort in residences and weather data [14]. For its adaptive control algorithm, they have considered three different zones due to their specific characteristics: bathroom, bedroom, and other rooms. For each zone, the comfort temperature was defined as given in Table 3.

Table 2. Regional adaptive control algorithm from [10].

Country	Adaptive control algorithm	
	$T_{rm} \leq 10\,^{\circ}\mathrm{C}$	$T_{rm} > 10\,^{\circ}\mathrm{C}$
France	$0.049 * T_{rm} + 22.58$	$0.206 * T_{rm} + 21.42$
Greece	N.A	$0.205 * T_{rm} + 21.69$
Portugal	$0.381 * T_{rm} + 18.12$	$0.381 * T_{rm} + 18.12$
Sweden	$0.051 * T_{rm} + 22.83$	$0.051 * T_{rm} + 22.83$
UK	$0.104 * T_{rm} + 22.58$	$0.168 * T_{rm} + 21.63$

Table 3. Adaptive control algorithm for residential zones [14].

Zone	Adaptive control algorithm	
Bathroom	$0.112 * T_{rm} + 22.65,$	$T_{rm} < 11\,^{\circ}\mathrm{C}$
	$0.306 * T_{rm} + 20.32,$	$T_{rm} \geq 11\,^{\circ}\mathrm{C}$
Bedroom	$16,$	$T_{rm} < 0\,^{\circ}\mathrm{C}$
	$0.23 * T_{rm} + 16,$	$0\,^{\circ}\mathrm{C} \leq T_{rm} < 12.6\,^{\circ}\mathrm{C}$
	$0.77 * T_{rm} + 9.18,$	$12.6\,^{\circ}\mathrm{C} \leq T_{rm} < 21.8\,^{\circ}\mathrm{C}$
	$26,$	$T_{rm} \geq 21.8\,^{\circ}\mathrm{C}$
Other rooms	$0.06 * T_{rm} + 20.4,$	$T_{rm} < 12.5\,^{\circ}\mathrm{C}$
	$0.36 * T_{rm} + 16.63,$	$T_{rm} \geq 12.5\,^{\circ}\mathrm{C}$

2.3 Static PMV Versus Adaptive TC Model

Yang et al. reviewed thermal comfort studies and established comparisons between the PMV and adaptive model, particularly from the building energy efficiency point of view. The authors pointed out that, although the static PMV model works well in air-conditioned buildings, it does not have a satisfactory performance when occupants have a more dynamic interaction with the surroundings. Due to the wider range of acceptable comfort temperatures, the adaptive model can provide a significant energy saving when compared with the PMV-PPD. Finally, human factors such as a "pro-environmental" attitude, can impact positively the acceptance of the adaptive model, as they tend to accept a large range of comfort temperatures [16].

As the focus of this work is residential homes, the adaptive model will be explored with finer detail. Moreover, the PMV thermal parameters such as radiant temperature, metabolic rate or clothing insulation might be unfeasible to obtain in an automated and non-intrusive way, limiting its application in real scenarios.

3 Method Description

The usage of programmable schedules to control the HVAC operation is highly convenient, particularly since it avoids the continuous engagement of the user with manual adjustments. Many disengaged users tend to keep the same setpoint configuration and schedule, even when conditions that affect the thermal comfort perception (such as the weather) are changing.

This work explores a strategy that can help to understand when a user-defined setpoint is being over-adjusted. Particularly, when the selected heating setpoint can be replaced with a lower one without compromising the occupants' thermal sensation and feeling. Such strategy can be leveraged by HVAC systems controllers and smart thermostats, having in mind the energy efficiency of the system.

The study starts with the evaluation of regional adaptive thermal comfort models using real data, collected from different European homes. The adaptive thermal comfort models will then be used to calculate the reference comfort temperature that should be met for each user/home. A rule-based algorithm is applied to find user-defined setpoints above the reference thermal comfort temperature value. In such situations, the algorithm must suggest a lower one that can improve the reduction of the system power consumption.

As a final step, the application of the rule-based algorithm is discussed and the potential energy savings are quantified and evaluated.

3.1 Dataset Analysis

An anonymized dataset, that holds information about the overall behaviour of the thermostat and heating system, from several residential homes/users across

Europe, was provided by Bosch Thermotechnology [1]. This dataset is not publicly available and contains data collected and shared by connected smart thermostats as time-series with a sampling rate of one minute.

For the present study, data collection interval from 2020-03-15 to 2021-12-15 was chosen and six users from different European countries (Netherlands, Germany, and United Kingdom) were considered. The following parameters were analysed: i) Indoor temperature (°C); ii) Indoor relative humidity (%); iii) User-defined setpoint (°C); iv) Outdoor temperature (°C).

Since we are dealing with real data, gaps on the time-series or faulty measurements are some of the problems that can arise from power-losses, hardware failures, communication failures, among other unpredictable scenarios. Table 4 summarizes this information.

Table 4. Dataset outliers and missing values.

User id	Country	Dataset length	№ invalid values of outliers and	Overall time-series gaps
user A	NL	994929	18580	18d 02:00
user B	NL	1011326	20858	15d 16:19
user C	DE	986236	39685	11d 05:51
user D	NL	1019992	15941	10d 18:18
user E	UK	997387	13852	27d 14:07
user F	DE	980675	29760	28d 19:32

To improve the dataset characteristics, data cleaning was applied where the following problems were addressed: dataset resampling to remove irrelevant information; outliers filtering; handling missing data through the application of interpolation and filling methods. For missing intervals longer than one hour, none of the methods was considered adequate, therefore, those intervals were discarded. Table 5 outlines the dataset characteristics after the data cleansing.

Table 5. Dataset outliers and missing values after data cleansing.

User id	Dataset length	№ of outliers and invalid values	Overall time-series gaps
user A	59740	0	18d 05:15
user B	59686	0	18d 18:45
user C	60148	0	13d 17:15
user D	60511	0	10d 03:00
user E	58843	0	27d 11:00
user F	58744	0	28d 10:15

Dataset distribution per user is shown in Fig. 1 . The histograms of different temperatures (indoor, outdoor and user-defined setpoint) reveal the data variation range and frequency over the analysed period. Differences between the users are noticeable, and can be mainly explained by the following factors:

- Users/homes are in different geographical areas, they suffer from different meteorological weather conditions.
- Different building constructions lead to different building thermodynamics and insulation.
- Different users follow different approaches to define the desired indoor temperature setpoint.

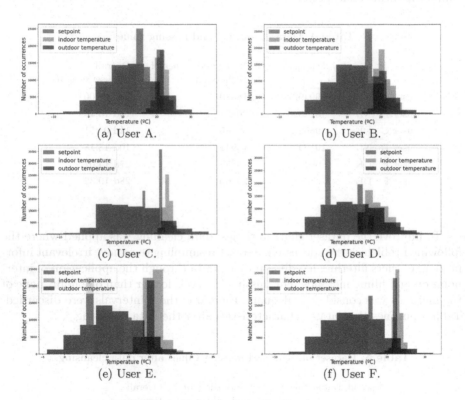

(a) User A.

(b) User B.

(c) User C.

(d) User D.

(e) User E.

(f) User F.

Fig. 1. Temperatures distribution.

Note that in Fig. 1 it is common to have setpoint values targeting rather low temperatures (<16 °C). These low temperature levels cannot be interpreted as the desired thermal comfort value. In most of these cases, the low temperature setpoints are defined when the occupants are absent, as a setback temperature, or even during the summer season when the heating system is switched off.

3.2 Adaptive Control Model for Indoor Thermal Comfort Assessment

Inspired by Ioannou et al. [8], we first verified if the dataset under study fits the adaptive TC model proposed by Peeters et al. [14]. To accomplish this, Eqs. (4) and (5) compute the neutral comfort temperature.

$$T_c = 0.06 \times T_{mean,out} + 20.4, \text{ for } T_{mean,out} < 12.5 \tag{4}$$

$$T_c = 0.36 \times T_{mean,out} + 16.63, \text{ for } T_{mean,out} \geq 12.5 \tag{5}$$

Equation (6) computes the running mean outdoor temperature.

$$T_{mean,out} = \frac{T_{today} + 0.8 \times T_{today-1} + 0.4 \times T_{today-2} + 0.2 \times T_{today-3}}{2.4} \tag{6}$$

Equations (7) and (8) define the acceptable TC boundaries, where: T_c is the comfort temperature, w is the width of the comfort band and α is a constant ($\leq$ 1).

$$T_{upper} = T_c + w \times \alpha \tag{7}$$

$$T_{lower} = max(T_c - w \times (1 - \alpha), 18) \tag{8}$$

The authors of the study pointed that, for 90% acceptability, w should be set to 5 °C. For 80% acceptability, w should be set to 7 °C. The recommended value for the constant α is 0.7.

The adaptive TC model (Eqs. (4)–(8)) was applied to the users/homes under study. The user-defined setpoint temperature and measured indoor temperature were plotted to understand if they would fit within the thermal comfort boundaries provided by the adaptive equations. Figure 2 presents the outcomes.

The visual inspection of Fig. 2, shows that the adaptive TC equations proposed by Peeters et al. [14] present satisfactory results for the different users/homes. It is important to note that, users from different countries were selected for this analysis. User A is from the Netherlands, users C and F are from Germany and user E is from the United Kingdom (see Table 4). All users have a close relation between the user-defined setpoints (assumed as the occupant desired comfort temperature) and the comfort temperatures defined by the adaptive equations. In general, we can see that both the measured indoor temperatures (red dots) and user-defined setpoints (black dots) are fitted within the TC boundaries (zone highlighted as green). Moreover, they follow a similar pattern with the outdoor temperature variation as given by the adaptive TC equation (represented by the blue line). The user-defined setpoints and indoor temperatures below the lower comfort acceptance boundary should be critically analyzed. These values can be representative of periods in which the residents are absent. Lower setpoint values are usually used to maintain the house within a temperature threshold (setback temperature), however, they are not representative of the residents' thermal comfort. Ensuring proper thermal comfort inside a residence is only meaningful when occupied.

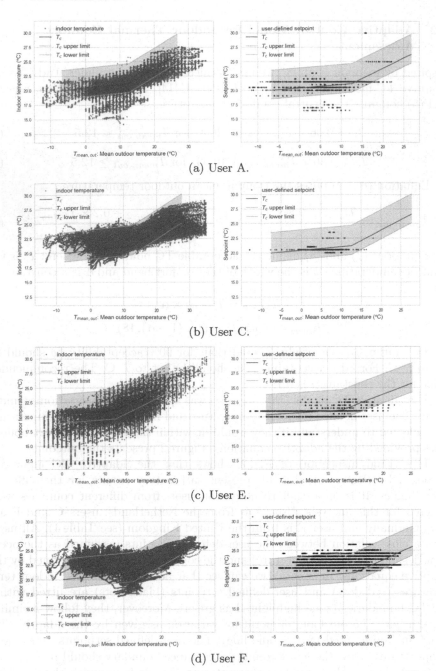

(a) User A.

(b) User C.

(c) User E.

(d) User F.

Fig. 2. Regional adaptive thermal comfort analysis. (Color figure online)

3.3   Heating Setpoint Based on Adaptive TC Model

The energy consumption of an HVAC system can be optimized if the desired heating and cooling setpoints are set according to the occupants' thermal comfort boundaries. As described in the previous section, we consider the thermal comfort limits given by a regional adaptive model which will vary linearly with the running mean outdoor temperature.

Algorithm 1 was implemented to provide suggestions of new setpoint values ($stp_{suggested}$) to improve the system energy consumption while maintaining the thermal comfort limits, as defined by the adaptive TC model. Note that Algorithm 1 will never suggest setpoints above the ones defined by the user. Particularly since the final goal sits on the system energy consumption minimization. Relying on the thermal comfort temperature Tc (Eqs. (4) and (5)) and its upper bound Tc_{upper} (Eq. (7)), each time the indoor temperature is expected to exceed Tc, a more relevant setpoint will be suggested to the user. This rule is based on the assumption that the user will be satisfied to set up the heating setpoint around the estimated TC temperatures. Figure 3 exemplifies the application of Algorithm 1 for one of the users. For better visualization a short time window of 4 days is illustrated. Whenever the indoor temperature (blue curve) goes above the computed TC temperature (red line), the user-defined setpoint (black line) is changed to the suggested setpoint (green line) equal to the TC temperature. Figure 4 provides a way to visualize the potential savings per user. It depicts the relative monthly reduction of the setpoint temperature in case the suggested setpoints are accepted by the users.

Algorithm 1. HVAC system heating setpoint recommendation

---

if Tc level will be reached (or exceeded) in the next time-step **then**
    if $stp > Tc_{upper}$ **then**
        $stp_{suggested} \leftarrow Tc_{upper}$
    else if $stp > Tc$ **then**
        $stp_{suggested} \leftarrow Tc$
    end if
end if

As expected, the proposed data driven method will have a positive impact on the energy consumption reduction. Equation (9) provides a way to quantify the energy savings over time. The difference between the setpoint (user-defined or the recommended by the smart HVAC system) and the outdoor temperature (t_{out}) estimates the heat losses of the residence. Table 6 shows the potential savings per user, computed by Eq. (9). As shown in Fig. 2, most of the user-defined setpoint values for user F are above the referenced TC temperature. Therefore applying the proposed set up rule, user F will get more energy saving without loosing thermal comfort. On the other hand, the proposed methodology may not bring any clear benefit for users who are already conservative while adjusting their setpoints. If a user adjusts its setpoints below the TC temperature, the smart

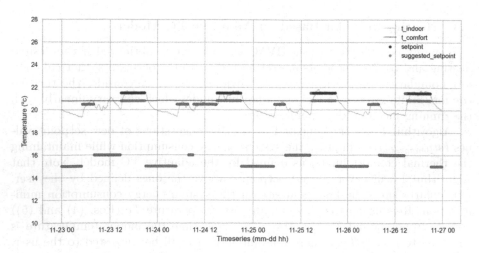

Fig. 3. Heating setpoint recommendation (illustrative example). (Color figure online)

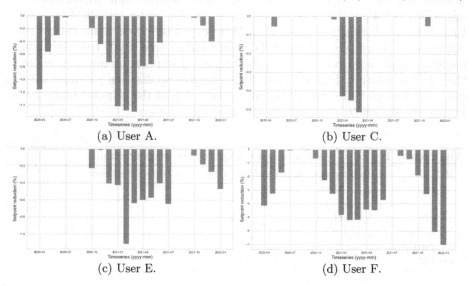

(a) User A.

(b) User C.

(c) User E.

(d) User F.

Fig. 4. Monthly setpoint relative reduction.

HVAC system will not suggest new ones. This effect can be noticed especially in user C, which is the one with the most user-defined setpoints below the TC temperature. As mentioned before, this methodology wants to lead to more conscious energy-driven behaviours by limiting the higher setpoints that can be replaced with lower ones without affecting the occupants thermal comfort.

$$s = 1 - \frac{\sum_{i=0}^{j}(stp_{suggested,i} - t_{out,i}) * \Delta_{t,i}}{\sum_{i=0}^{j}(stp_i - t_{out,i}) * \Delta_{t,i}} \tag{9}$$

Table 6. Energy saving

User id	Savings indicator s [%]
user A	1.507
user C	0.194
user E	2.052
user F	6.500

4   Conclusion and Future Work

This study aims to contribute with innovative advances, in the way that HVAC systems and smart thermostats ensure the customers' thermal comfort and energy savings simultaneously.

The evaluation of regional adaptive models with data collected from real users allowed to explore and confirm the feasibility of these algorithms to assess the thermal comfort level in residential environments.

Relying on the reference comfort temperature that these models provide, a rule-based algorithm was implemented to suggest lower setpoints. The intention is to explore the cases where the user could accept lower setpoints while keeping its thermal balance. The algorithm was applied to different users and an indicator was calculated to understand and quantify the energy savings that each one can get by adopting the suggested setpoints.

The results suggest that the application of regional adaptive models can be a reliable and simpler solution to assess the occupants' thermal comfort in residences. Furthermore, they can be used as a reference to suggest more responsible and sustainable behaviours.

Even though this study presents results from several homes, a broader evaluation covering more users from more regions with similar climatological characteristics should be covered in the next steps of this study.

The application of predictive machine learning models will follow. Particularly, to model the house thermodynamics and predict the indoor temperature. This prediction will help to understand, in advance, when the house reaches or exceeds the comfort temperature. A lower setpoint shall be suggested when the predicted indoor temperature exceeds the desired indoor TC temperature. Moreover, to concretely understand the amount of energy that can be saved by adopting the suggested setpoints, a simulation stage could be explored. For that, the data-driven predictive models capable to simulate the thermal behaviour of the house must be employed and evaluated. As part of this step, new data such as local weather information, could be used to improve the predictive models.

References

1. Bosch thermotechnology. www.bosch-thermotechnology.com, Accessed 19 Jan 2022
2. Stepping up Europe's 2030 climate ambition. https://eur-lex.europa.eu/legal-content/en/ALL/?uri=CELEX:52020DC0562, Accessed 04 Dec 2020
3. CEN, E.: En 16798–1: 2019 energy performance of buildings-ventilation for buildings-part 1: Indoor environmental input parameters for design and assessment of energy performance of buildings addressing indoor air quality (2019)
4. De Dear, R., Brager, G.S.: Developing an adaptive model of thermal comfort and preference (1998)
5. Fanger, P.O.: Thermal comfort: analysis and applications in environmental engineering. Thermal comfort: analysis and applications in environmental engineering (PhD thesis) (1970)
6. Fanger, P.O., Toftum, J.: Extension of the PMV model to non-air-conditioned buildings in warm climates. Energy Build. **34**(6), 533–536 (2002)
7. Fanger, P.O.: Assessment of man's thermal comfort in practice. Occup. Environ. Med. **30**(4), 313–324 (1973)
8. Ioannou, A., Itard, L., Agarwal, T.: In-situ real time measurements of thermal comfort and comparison with the adaptive comfort theory in Dutch residential dwellings. Energy Build. **170**, 229–241 (2018)
9. ISO, I.: 7730: Ergonomics of the thermal environment analytical determination and interpretation of thermal comfort using calculation of the PMV and PPD indices and local thermal comfort criteria. Management **3**(605), e615 (2005)
10. McCartney, K.J., Nicol, J.F.: Developing an adaptive control algorithm for Europe. Energy Build. **34**(6), 623–635 (2002)
11. Nicol, F., Humphreys, M.: Derivation of the adaptive equations for thermal comfort in free-running buildings in European standard en15251. Build. Environ. **45**(1), 11–17 (2010)
12. Nicol, J.F., Humphreys, M.A.: Adaptive thermal comfort and sustainable thermal standards for buildings. Energy Build. **34**(6), 563–572 (2002)
13. Comite'Europe'en de Normalisation, C.: Indoor environmental input parameters for design and assessment of energy performance of buildings addressing indoor air quality, thermal environment, lighting, and acoustics. EN 15251 (2007)
14. Peeters, L., De Dear, R., Hensen, J., D'haeseleer, W.: Thermal comfort in residential buildings: comfort values and scales for building energy simulation. Appl. Energy **86**(5), 772–780 (2009)
15. Van Hoof, J.: Forty years of fanger's model of thermal comfort: comfort for all? Indoor Air **18**(3), 182–201 (2008)
16. Yang, L., Yan, H., Lam, J.C.: Thermal comfort and building energy consumption implications-a review. Appl. Energy **115**, 164–173 (2014)

Opt-SSL: An Enhanced Self-Supervised Framework for Food Recognition

Nil Ballús[1]([⊠]) [ID], Bhalaji Nagarajan[1] [ID], and Petia Radeva[1,2] [ID]

[1] Dept. de Matemàtiques i Informàtica, Universitat de Barcelona, Barcelona, Spain
nballuri7@alumnes.ub.edu, {bhalaji.nagarajan,petia.ivanova}@ub.edu
[2] Computer Vision Center, Cerdanyola, Barcelona, Spain

Abstract. Self-supervised Learning has been showing upbeat performance in several computer vision tasks. The popular contrastive methods make use of a Siamese architecture with different loss functions. In this work, we go deeper into two very recent state of the art frameworks, namely, SimSiam and Barlow Twins. Inspired by them, we propose a new self-supervised learning method we call Opt-SSL that combines both image and feature contrasting. We validate the proposed method on the food recognition task, showing that our proposed framework enables the self-learning networks to learn better visual representations.

Keywords: Self-supervised · Contrastive learning · Food recognition

1 Introduction

The advent of Deep Neural Networks (DNNs) has increased the performance of several challenging computer vision problems [24]. The success of this multi-fold advancements can be attributed to large amounts of annotated data [20]. However, building large labelled datasets has several challenges [21]. The labelling process is tedious, costly and time consuming. Moreover, it is not possible to create datasets for every possible scenario, and the difficulty in generalization is one of the main bottlenecks in these supervised models [23]. One of the recent possible alternatives to this approach is the Self-Supervised Learning (SSL). SSL focuses on models that adapt to learn different underlying representations of the data without use of supervised datasets [19]. SSL frameworks use a pretext learning task which requires only unlabeled data. This task enables the DNNs to construct data representations that are semantically meaningful and are later used to solve any downstream task such as image recognition [17].

Food Recognition is a fast growing, but challenging area of study [11]. However, food recognition is a very complex computer vision task, due to the nature of images [22]. Food images are very hard to study as they are often mixed or composed of many items in a single plate. The high intra-class and low inter-class variance makes it difficult to work for any food recognition model. The images

P. Radeva—IAPR Fellow.

© Springer Nature Switzerland AG 2022
A. J. Pinho et al. (Eds.): IbPRIA 2022, LNCS 13256, pp. 655–666, 2022.
https://doi.org/10.1007/978-3-031-04881-4_52

are very diverse in visual appearances and the complexity of background scenes add additional obstacles to train the models [27]. It is therefore very important for the DNNs to learn the underlying information of food images and not just fit with the given label information. Hence, the success of SSL framework has given a new way on the food recognition research direction.

The key challenge in any SSL framework is to learn effective image representation [31]. Siamese architectures [1] offer a very efficient architectures for learning data representations. In the Contrastive Learning algorithms, the models learn to reduce the distance between representations of the same image by using different augmentations of the same image. These pairs of images are known as 'positive pairs'. The earlier studies incorporated loss terms to increase the distance between negative pairs. However, it hugely increased the complexity of the learning algorithms. Recent studies have shown to achieve state of the art performance using only the positive pairs [13]. SimSiam [8] and Barlow Twins [33] are contrastive algorithms that were able to learn better visual representations and achieved higher performance on the downstream tasks. SimSiam works on learning representations specific to the pair of positive pairs whereas Barlow Twins uses features across different sets of images to learn the visual representation. Inspired in the SimSiam and Barlow Twins, we propose a new method called Opt-SSL, which modifies the contrastive learning loss function incorporating both image and feature comparison. Our main contributions are:

- First, we propose a new SSL which loss function incorporates cross-correlation matrices in the dimension of features along with the dimension of batches.
- Second, we do an extensive analysis of the new loss function applied to food recognition and show that our proposed method increases the performance of the baseline being at the same time computationally effective.

The rest of the paper is organized as follows. We discuss the related work in Sect. 2. We provide the rationale behind the proposed technique in Sect. 3 and explain the proposed method in Sect. 4. The results used to validate the proposed method are explained in Sect. 5 followed by conclusions at the end.

2   Related Work

SSL frameworks can be categorized into three classes of algorithms based on the objective of the SSL methods - Generative [18], Contrastive [33] and Generative-contrastive methods [23]. Generative SSL methods define pretext tasks that aim to reconstruct parts of the original input while learning effective latent representation for the data [18]. Contrastive SSL methods use contrastive pretext tasks that compare the similarity of inputs' representations and aim to get similar latent representations for similar inputs and dissimilar latent representations for non-similar inputs [33]. Generative-Contrastive methods define pretext tasks to generate fake samples from the input data and the objective is to distinguish them from real samples. These last methods arose from the idea of generative SSL, but instead of relying on sample reconstruction, they learn to reconstruct the original data distribution by minimizing the distributional divergence [12]. A detailed summary of the algorithms are presented in [23].

The fundamental principle behind the SSL methods relies on the cross-view prediction framework [3], where different representations of the same image are used to learn the representations. Contrastive learning algorithms focuses on learning to compare between different representations through a Noise Contrastive Estimation (NCE) [14] objective function, where the distance between different augmentations of the same image are brought close compared to the negative samples. One of the earlier approaches in Contrastive methods was to use clustering algorithms [5]. Deep Cluster [5] used clustering to yield pseudo-labels and used a discriminator to predict the labels of the images. Local aggregation [34] used an objective function that directly optimized a local soft-clustering metric instead of the cross entropy loss used in Deep Cluster. VQ-VAE [28] substituted each 1-dimensional vector in the matrix to the nearest one in an embedding dictionary. SwAV [6] used online clustering under balanced partition constraint to compute the assignment from one view and predicted it from another view. The main disadvantage of the cluster based approach is the two-stage training process [23], which gave rise to instance discrimination-based methods.

Contrastive Multiview Coding [29] used multiple different views of an image as positive samples and used another image as negative to make positive pairs near in the embedding space. InfoMin [30] studied the importance in selecting less mutual information between views of the same image for better-augmented views in contrastive learning. SimCLR [7] used a nonlinear transformation between different data augmentations and combined with a contrastive loss to improve the representations. However, SimCLR requires a large batch size to work well. MoCo [15] used different learning updates for the main network and the encoder network. BYOL [13] combined asymmetric learning with two networks - online and target, where the online network was trained using an augmentation of the image to predict the representation of the target with a different augmentation of the image. BYOL used only positive pairs and was able to learn faster and more efficient data representations.

SimSiam [8], similar to BYOL, modified both the network architecture and updated the parameters to introduce asymmetry. SimSiam used a predictor network to update the model parameters using one view of the network and used stop-gradient to prevent the model from collapsing. Barlow Twins [33] used a cross-correlation matrix between the two networks which were passed with different representations of the same image. The objective here is to make the distance as closer to the identity matrix as possible.

3 Rationale

In this section, we introduce the rationale behind our proposed method contrasting both SimSiam and Barlow Twins and introducing the Opt-SSL framework.

SimSiam [8] used a Siamese network where the two sub-networks of the Siamese network consist of an encoder and a predictor. The encoder consists of a backbone followed by a projection MLP head. The outputs of the encoder and the predictor are used to compute the loss function, which is a negative cosine similarity loss. The training is done using batches of images and a loss function is computed over each batch in order to optimize the goal. One of the main contributions of the SimSiam is the 'stop-gradient', which was proposed to avoid model collapsing.

On the other hand, Barlow Twins [33] used a loss function based on Horace Barlow's redundancy-reduction principle [2] applied to a Siamese network. This model also consists of an encoder, followed by a projector. The outputs of the projector are used in the loss function. The objective function is based on a cross-correlation matrix between the outputs of both projectors and is comprised of two terms - an *invariance term* and a *redundancy reduction term*. The *invariance term* aims to make the representations invariant to the distortions applied while the *redundancy reduction term* aims to decorrelate as much as possible the features of the embeddings, so as to reduce the redundancy between features.

The main conceptual difference between SimSiam and Barlow Twins is that former learns by directly contrasting the representations of different images, whereas the later does the contrasting along the features, trying to minimize their redundancy.

4   A New Self-Supervised Learning Method (Opt-SSL)

In this paper, we use an asymmetric Siamese network, inspired in the excellent results achieved by the SimSiam [8] and BYOL [13]. The proposed framework can be seen in Fig. 1. Here, we use a process T, wherein for each image x_i in a batch $X = \{x_i\}_{i=1}^N$, we generate two augmented views $x_i^A \sim T(x_i)$ and $x_i^B \sim T(x_i)$. This process leads to the creation of two batches of augmented images, $X^A = \{x_i^A\}_{i=1}^N$ and $X^B = \{x_i^B\}_{i=1}^N$. The two distorted batches are processed by an encoder f, which shares weights between the two batches, obtaining $Z^A = f(X^A)$ and $Z^B = f(X^B)$. Next, we use a predictor h to transform the output of one branch, obtaining $P^A = h(Z^A)$. Finally, the output of the first branch P^A is contrasted with the output of the second branch Z^B, by means of the proposed loss function explained bellow.

Another important component included in our training pipeline is the stop-gradient operation [8]. We implement it by treating Z^A and Z^B as constants in the computation of the loss function. This means that the backpropagation method is only carried out through the variables P^A and P^B. The encoder f will not receive gradients from Z^A and Z^B, but it will receive them from P^A and P^B to update the weights.

During the training phase, we also interchange the inputs so as to obtain $P^B = h(f(X^B))$ and $Z^A = f(X^A)$. These terms are added to the loss function, making it symmetric. Following the general conventions, we assume that $Z^A = (z_{ij}^{(A)})$, $Z^B = (z_{ij}^{(B)})$, $P^A = (p_{ij}^{(A)})$, $P^B = (p_{ij}^{(B)})$ where $\forall i \in \{1,...,N\}$ and $\forall j \in \{1,...,D\}$ $f(x_i^A) = (z_{i1}^{(A)}, z_{i2}^{(A)}, ..., z_{iD}^{(A)})$, $f(x_i^B) = (z_{i1}^{(B)}, z_{i2}^{(B)}, ..., z_{iD}^{(B)})$, $h(f(x_i^A)) = (p_{i1}^{(A)}, p_{i2}^{(A)}, ..., p_{iD}^{(A)})$ and $h(f(x_i^{(B)})) = (p_{i1}^{(B)}, p_{i2}^{(B)}, ..., p_{iD}^{(B)})$ are the

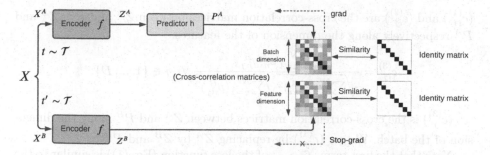

Fig. 1. Proposed Opt-SSL method.

representations of the two distorted views of the image x_i, N is the batch size, and D is the dimension of the representations. The symmetric loss function $\mathcal{L}_{Opt\text{-}SSL}$, is defined as follows:

$$\mathcal{L}_{Opt\text{-}SSL} = \mathcal{L}_{i\text{-}diag} + \lambda_1 \cdot \mathcal{L}_{i\text{-}off\text{-}diag} + \mathcal{L}_{f\text{-}diag} + \lambda_2 \cdot \mathcal{L}_{f\text{-}off\text{-}diag} \,, \tag{1}$$

where $\lambda_1, \lambda_2 \in \mathbb{R}_{>0}$ are constants that balance the terms importance, the values of λ_1 and λ_2 were empirically determined and set to be equal (e.g. 0.5).

$$\mathcal{L}_{i\text{-}diag} = \sqrt{\frac{1}{2N}\Big(\sum_{i=1}^{N}(1 - c_{ii}^{(1)})^2 + \sum_{i=1}^{N}(1 - c_{ii}^{(2)})^2\Big)} \,, \tag{2}$$

$$\mathcal{L}_{i\text{-}off\text{-}diag} = \sqrt{\frac{1}{2N(N-1)}\Big(\sum_{i=1}^{N}\sum_{j=1,j\neq i}^{N}(c_{ij}^{(1)})^2 + \sum_{i=1}^{N}\sum_{j=1,j\neq i}^{N}(c_{ij}^{(2)})^2\Big)} \,, \tag{3}$$

$$\mathcal{L}_{f\text{-}diag} = \sqrt{\frac{1}{2D}\Big(\sum_{i=1}^{D}(1 - c_{ii}^{(3)})^2 + \sum_{i=1}^{D}(1 - c_{ii}^{(4)})^2\Big)} \,,$$

$$\mathcal{L}_{f\text{-}off\text{-}diag} = \sqrt{\frac{1}{2D(D-1)}\Big(\sum_{i=1}^{D}\sum_{j=1,j\neq i}^{D}(c_{ij}^{(3)})^2 + \sum_{i=1}^{D}\sum_{j=1,j\neq i}^{D}(c_{ij}^{(4)})^2\Big)} \,.$$

where we define each term of the loss function as follows:

$$c_{ij}^{(1)} = \frac{\sum_{d=1}^{D} z_{id}^{A} p_{jd}^{B}}{\sqrt{\sum_{d=1}^{D}(z_{id}^{A})^2}\sqrt{\sum_{d=1}^{D}(p_{jd}^{B})^2}} \quad \forall i,j \in \{1,...,N\},$$

$$c_{ij}^{(2)} = \frac{\sum_{d=1}^{D} z_{id}^{B} p_{jd}^{A}}{\sqrt{\sum_{d=1}^{D}(z_{id}^{B})^2}\sqrt{\sum_{d=1}^{D}(p_{jd}^{A})^2}} \quad \forall i,j \in \{1,...,N\}.$$

$(c_{ij}^{(1)})$ and $(c_{ij}^{(2)})$ are the cross-correlation matrices between Z^A and P^B, Z^B and P^A respectively along the dimension of the features.

$$c_{ij}^{(3)} = \frac{\sum_{n=1}^{N} z_{ni}^A p_{nj}^B}{\sqrt{\sum_{n=1}^{N} (z_{ni}^A)^2} \sqrt{\sum_{n=1}^{N} (p_{nj}^B)^2}} \quad \forall i,j \in \{1,...,D\},$$

$(c_{ij}^{(3)})$ is the cross-correlation matrices between Z^A and P^B along the dimension of the batch. We get $(c_{ij}^{(4)})$ by replacing Z^A by Z^B and P^B by P^A.

Note that the first term, $\mathcal{L}_{i\text{-}diag}$, of the loss function (Eq. (1)) is similar to the loss function used in SimSiam. However, the second term, $\mathcal{L}_{i\text{-}off\text{-}diag}$, is added to compare the negative pairs of images. The third term, $\mathcal{L}_{f\text{-}diag}$, is comparable to the *invariance term* of Barlow Twins while the last term, $\mathcal{L}_{f\text{-}off\text{-}diag}$, is analogous to the *redundancy reduction term* of Barlow Twins. However, $\mathcal{L}_{Opt\text{-}SSL}$ is not the sum of SimSiam's and Barlow Twins' loss functions. We treat the image pairs differently compared to both the methods (Eq. 1).

$\mathcal{L}_{i\text{-}diag}$ includes the diagonal elements of matrices $(c_{ij}^{(1)})$ and $(c_{ij}^{(2)})$. These diagonal elements correspond to the cosine similarity between the predictor and the encoder representations of both distorted views of a same image. In this term, the positive pairs of views are contrasted. As each element is the cosine similarity between a positive pair of images, we tend to go towards unity (perfect correlation). $\mathcal{L}_{i\text{-}off\text{-}diag}$, whereas, includes the off-diagonal elements of matrices $(c_{ij}^{(1)})$ and $(c_{ij}^{(2)})$. These elements correspond to the cosine similarity between the predictor and the encoder representations of both distorted views of different images. Here, the negative pairs are contrasted and we tend this term to go towards 0 (no correlation).

$\mathcal{L}_{f\text{-}diag}$ includes the diagonal elements of matrices $(c_{ij}^{(3)})$ and $(c_{ij}^{(4)})$. The ith diagonal element of these matrices is the cosine similarity between the vector formed by the ith feature of the predictor of all the views in a batch and the vector formed by the ith feature of the encoder. This term makes the representations invariant to the distortions applied, making them to go towards unity. $\mathcal{L}_{f\text{-}off\text{-}diag}$ includes the off-diagonal elements of matrices $(c_{ij}^{(3)})$ and $(c_{ij}^{(4)})$. The element $c_{ij}^{(k)}, i \neq j$, of these matrices is the cosine similarity between the vector formed by the ith feature of the predictor of all the views in a batch and the vector formed by the jth feature of the encoder. This term aims to decorrelate as much as possible the features of the embeddings, so to reduce the redundancy between the features and is made to go towards 0.

Regarding the complexity of our proposal, Barlow Twins has a complexity of $O(m^2)$, where m is the number of features used in the latent space. SimSiam has a complexity of $O(n)$, where n is the batch size. In contrast, Opt-SSL has a complexity of $O(m^2) + O(n^2)$, since we are comparing both the positive and negative samples.

Alternatively to the Opt-SSL, we propose a reduced version, which we call Reduced Opt-SSL consisting of its first two terms (Eq. 1). Note that the complexity of the Reduced Opt-SSL is $O(n^2)$.

5   Validation

In this section, we describe the dataset used, we present the implementation setting and the evaluation metrics, and finally we discuss the obtained results.

5.1   Dataset

We use a popular food dataset, called Food-101 [4], to validate our proposed framework. Food-101 consists of 101 classes, with 1000 images (750 training, rest test) per class. We use the standard train-test split [26] for our experiments. The test set is used as validation set.

5.2   Implementation Details

Inspired in [33], the image augmentation process $\mathcal{T}$ used in Opt-SSL to generate the two distorted image views consists of the following transformations, applied sequentially: random resize cropping to 224 × 224, random horizontal flipping, color jittering, grayscale conversion, Gaussian blurring, and solarization. Random resize cropping is applied by default, whereas the rest are applied based on a probability score similar to the ones used in BYOL [13] and Barlow Twins [33].

The encoder f of the SSL network consists of a backbone, which is also a ResNet-50 [16], but without the final classification layer, followed by a projection MLP head. The MLP head has 3 fully-connected (fc) layers, each one with batch normalization applied. In the first two layers (hidden layers), a ReLU layer is applied as activation layer, but the last layer does not have any activation function. All the three layers have 2048 output units. The predictor h has 2 fc layers. Only the first layer has batch normalization applied and uses a ReLU as activation function. The input and output dimensions of h are both 2048 and the output of the hidden layer is 512. We use default PyTorch initialization for the projection MLP head and the predictor. The weights of each fc layer are initialized by a uniform distribution $\mathcal{U}(\sqrt{-k}, \sqrt{k})$, where $k = 1/\#(\text{input units})$.

For the SSL pretext tasks, we use a ResNet-50 pre-trained on ImageNet [10] with an input image of size 224 × 224 × 3. We use a batchsize of 32 and SGD optimizer with a weight decay of 0.0001, momentum of 0.9, and a learning rate of lr × batch size/256, starting with $lr = 0.05$. We make use of the cosine decay schedule. For reproducing the results of SimSiam and Barlow Twins, we use the default settings defined in the original papers, except the batch size, where we use 32 batch size. For all the studies, we train the SSL models for 150 epochs.

For the downstream tasks, we use the classifier architecture formed using the encoder of the SSL models which acts as backbone (ResNet-50 without the final classification layer), followed by a dropout layer (with a dropout rate of 0.4), and a classification layer with softmax activation. We use RandAugment [9] and standard data augmentations to reduce overfitting. We also used Stochastic Gradient Descent (SGD) with Warm Restarts [25] with a learning rate of 0.001, weight decay of 0.0005, momentum as 0.9 and T_{mul} of 2. We applied a batch size of 64. We used PyTorch framework in Nvidia RTX 20i or all our experiments.

Table 1. Validation accuracy of different SSL frameworks.

SSL model	λ	k-NN acc.		Linear classifier
		Top-1 acc. (%)	Top-5 acc. (%)	Val. Acc. (%)
Barlow twins [33]	0.0051	49.89	79.07	87.39
SimSiam [8]	–	54.33	81.62	88.60
Reduced Opt-SSL	0.5	**63.52**	**87.48**	88.53
Opt-SSL	0.5	62.06	86.43	**88.63**

5.3   Results and Analysis

We used two evaluation strategies to validate our proposed method. For the first evaluation, we computed the overall accuracy using a k-Nearest Neighbors (k-NN) classifier similar to the one by Wu et al. [32]. We used $k = 200$ and $\tau = 10$ for all the experiments. Table 1 compares the k-NN performance and the accuracy of the linear classifier using the three SSL methods. We can see that SimSiam has better learned representations compared to the Barlow Twins. Apart from the architecture and configuration setting variations between them, the most important difference is that SimSiam learns by directly contrasting the representations of different images, while the Barlow Twins method does the contrasting along the features, trying to minimize their redundancy. However, we can see that a hybrid combination of both approaches, as with the proposed Opt-SSL, is much better to learn the visual representations, which shows a 7.73% increase in the top-1 accuracy compared to SimSiam in the k-NN classifiers and a little higher on the linear classifier performance. The second optimal solution is the Reduced Opt-SSL that contains only the first two terms in Eq. (1). In fact, one can see that the Reduced Opt-SSL gave the best results when using the k-NN classifier but the best final linear classifier was for the Opt-SSL.

5.4   Empirical Study

Next we show an empirical study where we considered mainly the reduced version of Opt-SSL (Reduced Opt-SSL) in order to optimize the computation. The results can be shown to be easily extrapolated to the Opt-SSL loss function, as it is an extension of the previous one.

Normalizing the Opt-SSL's Loss Function. Each term of the loss function behaves differently and it is important to normalize the terms accordingly. We take the square root of the average of the atomic values of each term (Eq. (1)). We empirically show the importance of this normalization by comparing the performance of the first two terms of Eq. (1) with the non-normalized loss terms (Eq. (4)).

Table 2. Effect of normalizing the loss terms.

Experiment	λ	Top-1 acc. (%)	Top-5 acc. (%)
Non-normalized Reduced Opt-SSL	0.0	37.61	67.59
Reduced Opt-SSL	0.0	58.48	84.53
Non-normalized Reduced Opt-SSL	1.0	53.02	80.52
Reduced Opt-SSL	1.0	58.94	84.78

$$\mathcal{L}^{non-norm} = \Big(\sum_{i=1}^{N}(1 - c_{ii}^{(1)})^2 + \sum_{i=1}^{N}(1 - c_{ii}^{(2)})^2 \Big)$$
$$+\lambda \cdot \Big(\sum_{i=1}^{N}\sum_{j=1,j\neq i}^{N}(c_{ij}^{(1)})^2 + \sum_{i=1}^{N}\sum_{j=1,j\neq i}^{N}(c_{ij}^{(2)})^2 \Big). \tag{4}$$

From Table 2, we see that the performance improves with normalization of the loss terms. We also study the behaviour with different values of λ and find the improvement consistent.

Negative Sample Contrasting. As discussed in Sect. 4, the first term of Opt-SSL's loss function is similar to that of SimSiam. As an intermediate definition of the loss functions of SimSiam and the Reduced Opt-SSL, we extended the original loss function of SimSiam to include a second term to compare the negative samples represented by c_{ij}. To this aim, we defined an extended loss function of SimSiam as follows:

$$\mathcal{L}^{ext}_{SimSiam} = \mathcal{L}_{SimSiam} + \lambda \cdot \frac{1}{2N(N-1)} \Big(\sum_{i=1}^{N}\sum_{j=1,j\neq i}^{N}|c_{ij}^{(1)}| + \sum_{i=1}^{N}\sum_{j=1,j\neq i}^{N}|c_{ij}^{(2)}| \Big),$$

where $\mathcal{L}_{SimSiam}$ is the original SimSiam loss function, $\lambda \in \mathbb{R}_{>0}$ is a constant to balance the importance between the terms, and $(c_{ij}^{(1)})$ and $(c_{ij}^{(2)})$ are the cross-correlation matrices defined in Sect. 4.

In Table 3, we show the results with different values of λ of both the SimSiam with negative sample contrasting and the Reduced Opt-SSL loss function. In all the experiments, we observe that the loss terms of Opt-SSL behave better than the original SimSiam model. It can further be seen through the experiment with $\lambda = 0$, where the loss of Reduced Opt-SSL behaves in the same way as SimSiam, but shows a higher performance. We observe that the loss function of Reduced Opt-SSL (Eq. 2), which differs from SimSiam in the way the cosine similarity of samples is treated, gives a better performance.

Selection of λ. In Table 4, we show the experiments with respect to the selection of λ using Reduced Opt-SSL. We see that 0.5 shows the highest performance.

Table 3. Study of the behaviour of Loss Terms.

Experiment	λ	Top-1 acc. (%)	Top-5 acc. (%)
Original SimSiam	–	54.33	81.62
SimSiam with negative	0.5	57.47	83.71
Sample contrasting	1.0	57.88	83.92
Reduced Opt-SSL	0.0	58.48	84.53
	0.5	59.00	84.91
	1.0	58.94	84.78

Table 4. Selection of λ.

λ	Top-1 acc. (%)	Top-5 acc. (%)
0.0	58.48	84.53
0.5	59.00	84.91
0.75	58.82	84.81
1.0	58.94	84.78

Table 5. Study of different data augmentation techniques in SSL.

Experiment	λ	Top-1 acc. (%)	Top-5 acc. (%)
SimSiam (SimSiam DA)	–	54.33	81.62
SimSiam (Proposed DA)	–	61.76	86.48
Reduced Opt-SSL (SimSiam DA)	0.5	59.00	84.91
Reduced Opt-SSL (Proposed DA)	0.5	63.52	87.48

Data Augmentation Techniques. In this experiment, we check the importance of data augmentation techniques used in Opt-SSL. We compare the performance of data augmentation techniques used in SimSiam with the data augmentation techniques defined for Sect. 5.2. Just with changing the strategies, the original SimSiam model performed better. We also compare the performance of the image correlation terms (Eqs. (2) and (3)) with the proposed loss term and the best performance is achieved with the proposed method. From Table 5, we see that the combination of our proposed method and data augmentation techniques is optimal.

6   Conclusions

In this paper, we study the behaviour of the state of the art methods, specially, SimSiam and Barlow Twins and proposed a new modified SSL method, called Opt-SSL. In this work, we combine the feature representation learned by Barlow Twins with the image representation learned by SimSiam. We show the importance of each component of the proposed loss function using different experiments. We validated our method using a challenging food recognition task and shew to improve the results compared to the original state of the art SSL methods. One of the main advantages of our proposal is that it considers positive and negative pairs in an efficient way assuring that the resulting self-supervised approach is more beneficial. Next steps involve including annotations from supervised learning datasets to apply the self-learning heuristics to treat differently images from the same class and different classes.

Acknowledgements. This work was partially funded from the European Union's Horizon 2020 Research and Innovation programme under Open Call budget of the grant agreement No. 857159 (SHAPES), TIN2018-095232-B-C21, SGR-2017 1742 and CERCA Programme/Generalitat de Catalunya. B. Nagarajan acknowledges the support of FPI Becas, MICINN, Spain. We acknowledge the support of NVIDIA Corporation with the donation of the Titan Xp GPUs.

References

1. Bachman, P., Hjelm, R.D., Buchwalter, W.: Learning representations by maximizing mutual information across views. arXiv preprint arXiv:1906.00910 (2019)
2. Barlow, H.B., et al.: Possible principles underlying the transformation of sensory messages. Sensory Commun. **1**(01), 1–18 (1961)
3. Becker, S., Hinton, G.E.: Self-organizing neural network that discovers surfaces in random-dot stereograms. Nature **355**(6356), 161–163 (1992)
4. Bossard, L., Guillaumin, M., Van Gool, L.: Food-101 – mining discriminative components with random forests. In: Fleet, D., Pajdla, T., Schiele, B., Tuytelaars, T. (eds.) ECCV 2014. LNCS, vol. 8694, pp. 446–461. Springer, Cham (2014). https://doi.org/10.1007/978-3-319-10599-4_29
5. Caron, M., Bojanowski, P., Joulin, A., Douze, M.: Deep clustering for unsupervised learning of visual features. In: ECCV, pp. 132–149 (2018)
6. Caron, M., Misra, I., Mairal, J., Goyal, P., Bojanowski, P., Joulin, A.: Unsupervised learning of visual features by contrasting cluster assignments. arXiv preprint arXiv:2006.09882 (2020)
7. Chen, T., Kornblith, S., Norouzi, M., Hinton, G.: A simple framework for contrastive learning of visual representations. In: ICML, pp. 1597–1607. PMLR (2020)
8. Chen, X., He, K.: Exploring simple siamese representation learning. In: CVPR, pp. 15750–15758 (2021)
9. Cubuk, E., Zoph, B., Shlens, J., Le, Q.V.: Randaugment: practical automated data augmentation with a reduced search space. In: CVPR Workshops, pp. 702–703 (2020)
10. Deng, J., Dong, W., Socher, R., Li, L.J., Li, K., Fei-Fei, L.: Imagenet: a large-scale hierarchical image database. In: 2009 IEEE CVPR, pp. 248–255 (2009)
11. El Khoury, C.F., Karavetian, M., Halfens, R.J., Crutzen, R., Khoja, L., Schols, J.M.: The effects of dietary mobile apps on nutritional outcomes in adults with chronic diseases: a systematic review and meta-analysis. J. Acad. Nutr. Diet. **119**(4), 626–651 (2019)
12. Goodfellow, I., et al.: Generative adversarial nets. In: NIPS, vol. 27 (2014)
13. Grill, J.B., et al.: Bootstrap your own latent: a new approach to self-supervised learning. arXiv preprint arXiv:2006.07733 (2020)

14. Gutmann, M., Hyvärinen, A.: Noise-contrastive estimation: a new estimation principle for unnormalized statistical models. In: 13th ICAIS, pp. 297–304. JMLR Workshop (2010)
15. He, K., Fan, H., Wu, Y., Xie, S., Girshick, R.: Momentum contrast for unsupervised visual representation learning. In: CVPR, pp. 9729–9738 (2020)
16. He, K., Zhang, X., Ren, S., Sun, J.: Deep residual learning for image recognition. In: CVPR, pp. 770–778 (2016)
17. Jing, L., Tian, Y.: Self-supervised visual feature learning with deep neural networks: a survey. IEEE Trans. PAMI **43**, 4037–4058 (2020)
18. Kingma, D.P., Dhariwal, P.: Glow: generative flow with invertible 1×1 convolutions. In: 32nd NIPS, pp. 10236–10245 (2018)
19. Kolesnikov, A., Zhai, X., Beyer, L.: Revisiting self-supervised visual representation learning. In: Proceedings of the IEEE/CVF Conference on Computer Vision and Pattern Recognition, pp. 1920–1929 (2019)
20. Krizhevsky, A., Sutskever, I., Hinton, G.E.: Imagenet classification with deep convolutional neural networks. Adv. Neural Inf. Process. Syst. **25**, 1097–1105 (2012)
21. Liao, Y.H., Kar, A., Fidler, S.: Towards good practices for efficiently annotating large-scale image classification datasets. In: Proceedings of the IEEE/CVF Conference on Computer Vision and Pattern Recognition, pp. 4350–4359 (2021)
22. Liu, C., Liang, Y., Xue, Y., Qian, X., Fu, J.: Food and ingredient joint learning for fine-grained recognition. IEEE Trans. Circ. Syst. Video Technol. **31**, 2480–2493 (2020)
23. Liu, X., et al.: Self-supervised learning: generative or contrastive. IEEE Trans. Knowl. Data Eng. (2021)
24. Liu, X., Deng, Z., Yang, Y.: Recent progress in semantic image segmentation. Artif. Intell. Rev. **52**(2), 1089–1106 (2018). https://doi.org/10.1007/s10462-018-9641-3
25. Loshchilov, I., Hutter, F.: SGDR: stochastic gradient descent with warm restarts. arXiv preprint arXiv:1608.03983 (2016)
26. Martinel, N., Foresti, G.L., Micheloni, C.: Wide-slice residual networks for food recognition. In: 2018 WACV, pp. 567–576. IEEE (2018)
27. Meng, L., et al.: Learning using privileged information for food recognition. In: 27th ACM ICMM, pp. 557–565 (2019)
28. Razavi, A., van den Oord, A., Vinyals, O.: Generating diverse high-fidelity images with vq-vae-2. In: NIPS, pp. 14866–14876 (2019)
29. Tian, Y., Krishnan, D., Isola, P.: Contrastive multiview coding. In: Vedaldi, A., Bischof, H., Brox, T., Frahm, J.-M. (eds.) ECCV 2020. LNCS, vol. 12356, pp. 776–794. Springer, Cham (2020). https://doi.org/10.1007/978-3-030-58621-8_45
30. Tian, Y., Sun, C., Poole, B., Krishnan, D., Schmid, C., Isola, P.: What makes for good views for contrastive learning? arXiv preprint arXiv:2005.10243 (2020)
31. Wiskott, L., Sejnowski, T.J.: Slow feature analysis: unsupervised learning of invariances. Neural Comput. **14**(4), 715–770 (2002)
32. Wu, Z., Xiong, Y., Yu, S., Lin, D.: Unsupervised feature learning via non-parametric instance-level discrimination. arXiv preprint arXiv:1805.01978 (2018)
33. Zbontar, J., Jing, L., Misra, I., LeCun, Y., Deny, S.: Barlow twins: self-supervised learning via redundancy reduction. arXiv preprint arXiv:2103.03230 (2021)
34. Zhuang, C., Zhai, A.L., Yamins, D.: Local aggregation for unsupervised learning of visual embeddings. In: IEEE CVPR, pp. 6002–6012 (2019)

Using Bus Tracking Data to Detect Potential Hazard Driving Zones

Ana Almeida[1,2(✉)] ⓘ, Susana Brás[1,3] ⓘ, Susana Sargento[1,2] ⓘ,
and Ilídio Oliveira[1,3] ⓘ

[1] Departamento de Eletrónica, Telecomunicações e Informática, Universidade de
Aveiro, 3810-193 Aveiro, Portugal
{anaa,susana.bras,susana,ico}@ua.pt
[2] Instituto de Telecomunicações de Aveiro, 3810-193 Aveiro, Portugal
[3] Instituto de Engenharia Electrónica e Telemática de Aveiro, Universidade de
Aveiro, 3810-193 Aveiro, Portugal

Abstract. Urban mobility studies are critical to acquiring insights
about traffic management, urban mobility evolution, planning of urban
spaces, and many other applications. However, the study and mitigation
of recurrent problems have been postponed in real cases, given the difficulty of gathering mobility data. This paper presents a method to detect
potential hazard driving zones. We mapped bus GPS positions into road
segments, and we used bus speed, bus maximum allowed speed, and bus
acceleration to classify the driving behavior. We develop an interactive
web application that maps tracking data and provides rich visual insights
on potentially problematic areas. Side-by-side visualizations help with
the comparison of the traffic behavior in selected periods. We show that
we can map the most problematic zones in the city and the time of the
day. From the developed analysis, it is observed that some roads in the
city present a daily seasonality (problems occur in the same period of
the day); however, other roads present circulation issues independently
of the time period or day.

Keywords: Smart urban mobility · Hazard driving zones · Intelligent
Transportation Systems

This work is supported by FEDER, through POR LISBOA 2020 and COMPETE
2020 of the Portugal 2020 Project CityCatalyst POCI-01-0247-FEDER-046119, and by
the Urban Innovation Action EU/H2020 Aveiro STEAM City. Ana Almeida acknowledges the Doctoral Grant from Fundação para a Ciência e Tecnologia, with reference
2021.06222.BD. Susana Brás is funded by national funds, European Regional Development Fund, FSE, through COMPETE2020 and FCT, in the scope of the framework
contract foreseen in the numbers 4, 5 and 6 of the article 23, of the Decree-Law 57/2016,
of August 29, changed by Law 57/2017, of July 19. Ilídio Oliveira is funded by National
Funds through the FCT - Foundation for Science and Technology, in the context of the
project UIDB/00127/2020.

© Springer Nature Switzerland AG 2022
A. J. Pinho et al. (Eds.): IbPRIA 2022, LNCS 13256, pp. 667–679, 2022.
https://doi.org/10.1007/978-3-031-04881-4_53

1 Introduction

Humans are a key element in urban spaces; therefore, their characteristics, emotional and physical state can influence road safety. The study of driving behavior is valuable for law enforcement entities, city managers, and the insurance industry. Some of the factors that can contribute to safe or unsafe driving behavior are the safety of the zones, and the potential hazard driving zones. The objective of this paper is the automatic identification of those zones.

The work in [3] analyzed the driver's emotional state. Anxious driving behavior can be problematic, causing performance deficits (increases the frequency of driving errors), exaggerated caution, or even aggressive behavior. Besides anxious driving, cognitive distraction is one of the main reasons for traffic accidents. The authors in [11] applied a *Support Vector Machine* (SVM) (machine-learning technique) to data collected from a simulator (drivers' eye movement data and driving performance data) to detect if the driver was distracted. The work in [8] used driver's electroencephalogram signals and driving context awareness (environmental variables, driving behavior, and vehicle variables). They proposed a classification and prediction model based on SVM to create a feedback system that gives a warning when needed. The authors in [13] used data from sensors attached to a simulator, having 13 features that included drivers' behavior and vehicle information. They used a deep-learning method to generate real-time alerts regarding drivers' distractions. In [12] it is used semi-supervised learning to detect drivers distraction.

In [7] the authors performed their analysis based on controller area network bus signals (vehicles information). They achieved near-real-time analysis and classification using an unsupervised technique to distinguish between different groups of drivers. The authors in [9] took advantage of the existing car sensors to learn about the drivers. They built a classifier to separate the driving behavior of each driver. With this, they could predict the identity of the driver with a single turn. In [14], the authors extracted the behavioral patterns of the drivers and characterized those patterns by creating a driving skill level. They had sensors to monitor the environment, the vehicle, and the driver.

From the user's perspective, most of the works presented above monitor and evaluate the driver. In this paper, we aim to detect hazard driving areas automatically. We consider hazard driving areas, the locations where drivers do not respect speed limits and perform sudden accelerations and decelerations. We also develop a web application to help city decision-makers improve safe driving. Our method is focused on detecting problematic zones and areas, rather than performing a driver's evaluation. With this work, we can study driving behavior using inexpensive sensors. Our contributions can be outlined as:

- The extension of a method [6] proposed in the literature to detect unsafe driving behavior and areas.
- The creation of a tool to assist decision-makers to improve safe driving.
- The detection of unsafe driving patterns.

The remainder of this paper is organized as follows. Section 2 contains the related work about driving behavior, focusing on works that use similar data. Section 3 presents the *Aveiro Tech City Living Lab* (ATCLL) and the defined use cases. Section 4 describes the method to study the driving behavior and Sect. 5 presents its study. Finally, Sect. 6 contains the main conclusions and the proposed future work.

2 Related Work

One of the ways to perform a study of driving behavior is through the study of speed and acceleration to understand if the driver presents a safe or unsafe driving behavior. Studies based on speed and acceleration usually are based on the G-G diagram [5]. The G-G diagram gives the maximum acceleration that is possible to achieve for a given speed. It takes into consideration the longitudinal and lateral accelerations. However, there are other types of models.

Derbel et al. [5] proposed a system to evaluate the driving risk. This system is based on three types of factors. It is made a fusion of information about the driver, the vehicle, and the environment. The fusion happens at two levels. The first level is related to the type of factor and is made using Dempster-Shafer Theory [4]. The second level of fusion is a global fusion and is based on a Fuzzy theory [2] to designate basic degrees of membership assignment functions. The information about the driver that is considered relevant is the age and genre of the driver, since statistical studies reveal that they influence the probability of having an accident. They call vehicle information, the information about speed and acceleration, and use it as an indicator of the aggressiveness level. The environment information is based on statistical studies about the place, the time of the day, and the day of the week [5]. The vehicle factors are studied at two different levels. The first one is based on the Euclidean acceleration norm [10] and the speed of the vehicle. The second one is based on the G-G diagram, with lateral and longitudinal acceleration to identify danger left and right turns [5].

Eboli et al. [6] based their work on speed, acceleration, and road friction. The starting point was the G-G diagram. This model is more ethical than the previous model, since statistics cannot define a person. Equation (1) gives the relationship between the a (acceleration) and v (speed) to classify the driving behavior.

$$|\vec{a}| = g \times \left[0.198 \times \left(\frac{v}{100} \right)^2 - 0.592 \times \left(\frac{v}{100} \right) + 0.569 \right] \tag{1}$$

where g is the gravitational weight, and it is assumed a dry pavement and rural road. Using this equation, the authors can draw the limits for acceleration given the speed, to classify a driving behavior safe or unsafe.

Driving behavior studies can be very complex; since they can use a fusion of information about the driver, the vehicle, the road, and the weather in their analysis. We focus on those that use the same type of information we have access to.

3   Living Lab Scenario

The ATCLL was developed in the context of the Aveiro STEAM City project[1]. This project aims at the development of an urban platform based on a 5G infrastructure and the development of services that enable the mobility management of the city [1]. Sensors and collecting units were installed in 10 city buses. This information is gathered through the communication infrastructure, and is persisted in the ATCLL database.

The main use cases addressed in this work are:

- Classification of safe driving behavior versus unsafe driving behavior;
- Identification of road segments and zones that can be problematic;
- Comparison of temporal snapshots:
  - Compare different hours of the day of the same day;
  - Compare different periods like earlier in the day, midday, and end of the day;
  - Compare different days and different sets of days.
- Study the impact of periodicity in collecting data.

4   Method for Characterizing Traffic

Each bus communicates its speed, the heading, *Global Position System* (GPS) coordinates, and the timestamp. The data was collected every second, and was anonymized. It is not possible to identify the driver, as long as we consider multiple buses in the analysis.

The first step was data cleaning to deal with negative timestamps, outliers, and missing data. This was followed by the conversion of GPS coordinates into road segments. *OpenStreetMap* (OSM) API allows to get data about road infrastructure, and the bus network (including bus stops). We choose four weeks to study the driving behavior, from March 2 of 2020 to March 29 of 2020.

Our study of the driving behavior is described in Sect. 5, and it had in consideration the speed of the bus, the acceleration of the bus, the road friction coefficient (constant), and the maximum bus speed allowed. The following subsections explain the feature engineering performed.

4.1   Feature Engineering

The first challenge was to deal with imprecise GPS positions present in the bus data. Besides that, it is not practical to work with GPS coordinates because we can have a very large number of distinct GPS coordinates. The solution adopted was to associate the GPS positions with the closest road segment. The roads' information was obtained by using an OSM library.

---

[1] https://www.it.pt/Projects/Index/4613.

The information obtained from OSM contained some useful information about the maximum speed for cars. Since we are working with data from buses, it was necessary to adapt some of the data to get the maximum speed for buses.

For some of the road segments, there is not a value of maximum speed. In that case, the maximum bus speed was determined by the type of road. When information about the type of road is missing, then the maximum bus speed was 50km/h, the legal speed limit in residential areas in Portugal.

Figure 1 contains the road segments with the maximum bus speed associated. This image was obtained after performing the several corrections mentioned above. Note that, the figure can have some errors associated with our assumptions.

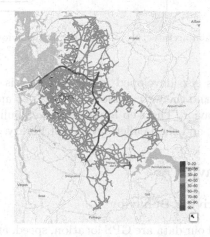

Fig. 1. Roads of Aveiro.

For the driving behavior study, we decided to calculate the traveled distance (using Vincenty's formula) and the acceleration of the bus (2)

$$a = \frac{\Delta v}{\Delta t},\qquad(2)$$

being Δv the difference between speeds, and Δt the difference between time.

4.2 Elements of the Driving Behavior Pipeline

The system architecture is depicted in Fig. 2. The information from buses is persisted in the Fiware infrastructure. The OSM library provides geographical data. The preprocessing module is responsible to process the collected data, and associate the buses GPS positions to road segments by using a library developed to get information from OSM. Besides that, the module will calculate the metrics, such as traveled distance, acceleration, and if the driving behavior is safe or not. The processed information is persisted in the AveiroBus database, since the process to associate GPS coordinates with road segments is computationally expensive. The modules were developed in Python.

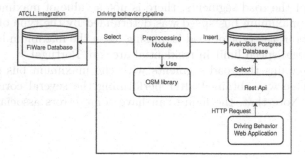

Fig. 2. System architecture of the driving behavior pipeline.

The Rest API was also developed in Python, and is the connection point between the database and the web application. The web application helps in the analysis of traffic behavior. The web application was built using HTML, CSS, JavaScript, Bootstrap, and the Leaflet JavaScript library.

5   Driving Behavior

5.1   Classifying Driving Behavior

The main attributes of our data are GPS location, speed, and timestamp. Therefore, it was necessary to choose a model based on this type of features, or features that could be calculated like acceleration, or traveled distance.

To study driving behavior, we developed a model inspired by the model proposed in [6]. This model considers the relationship between speed and acceleration, and the relationship between speed and road friction to determine if the driving behavior is safe or unsafe. The relationship between speed and friction is simplified, since the road material present in the city of Aveiro is mostly the same. Besides, the month when we collected the data had little precipitation. Because of this, the friction coefficient used is constant and we obtain the Eq. (1).

It was necessary to adapt the model to the reality of Aveiro. In Aveiro, even inside the city, there can be several maximum bus speeds. It is necessary to evaluate if the maximum allowed speed that the bus can achieve is respected. The created model evaluates if the driving is safe and has some conditions associated based on the maximum allowed bus speed beyond those described in the work [6].

We added an intermediate state between safe condition and unsafe condition based on the fact that, in Portugal there is a tolerance in the measurements surveillance systems (RADAR devices), which have an associated error component. The evaluation between safe driving behavior versus unsafe driving behavior is subjective and depends on the criteria chosen for the evaluation. Anyway, since the concepts of safe and unsafe driving behavior were very limited, we created six categories to evaluate driving behavior.

Table 1 contains different categories regarding driving behavior. For each one of the categories, we assigned a color in the representation of the driving behavior classification associated. The colors chosen for encoding this information on the map are based on the colors used in traffic signs. The green represents that everything is ok; the yellow represents that the driver should be careful; and the red indicates danger. The darker green, the orange, and the burgundy have the same connotations associated, but represent an additional danger because it is being performed significant variations in speed.

Table 1. Classification of the driving behavior.

Speed	Safety domain	
	Within	Outside
speed ≤ max. speed		
(speed > max. speed) and (speed ≤ max. speed + tolerance const.)		
speed > max. speed + tolerance const.		

The first line corresponds to safe speeds; the second line corresponds to the speeds that are in the threshold area between safe and unsafe (speeds that are bigger than the maximum bus speed, but are smaller or equal to the maximum bus speed plus a tolerance constant). All the other speeds belong to the unsafe driving behavior.

Classifying driving behavior can be a difficult task. Speed and acceleration can be useful. The GPS position allows to perform a more complete classification, because we can use the GPS position to associate the bus with road segments and we can obtain the maximum speed for those segments. This section details how the classification of driving behavior has been performed and presents a web application that is created to allow a more dynamic study.

Figure 3 contains the driving behavior for the first day of study for one of the buses. The blue lines are the limits that separate a safe driving behavior from a non-safe driving behavior. If the driver performs a safe driving, then all points or the vast majority of these belong inside the limits. In the figure, we observe that very few points are outside the limits. When the acceleration is larger than 0, it means that the bus driver is accelerating. If the acceleration is positive and is outside the bounds, then the driver performed a sudden acceleration. When the acceleration is smaller than 0, it means that the bus driver is decelerating and the bus driver can even be performing a braking. A negative acceleration outside the limits means that the driver performed a sudden braking. If we look carefully at the data, it is possible to observe that the majority of the points are centered in the graphic, but there are two distinct lines formed by the points. One of those lines has a positive trend, and the other is very close to the straight segment where the speed is 0. The line with the positive trend corresponds to the initial start when the driver turns on the bus. The other line corresponds when the driver is stopping the bus.

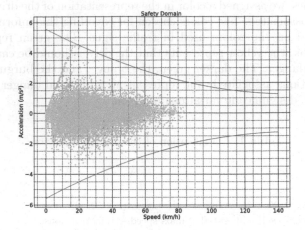

Fig. 3. Driving behavior - Relationship between speed and acceleration. (Color figure online)

The red lines symbolize the several limits of speed that the bus driver can find in the route. The first three are red dashed lines because they can only be applied if the bus is on a certain road. The last red line is the maximum limit of the maximum speed that the bus can achieve. There can be other limits for specific roads.

In order to achieve a more accurate classification, it was initially planned to create a code color for the points, but we could have an overlap of points and that could lead to wrong interpretations. With this in mind, it was planned the creation of a web application as a tool to study driving behavior and unsafe zones.

5.2   Traffic Behavior Web Dashboard

The goal of the web application is not to study the behavior of a specific driver, but the overall behavior in an area, considering aggregated data. Thus, it is possible to identify what are the situations that most contribute to more dangerous behavior. Some of the main goals in the study of driving behavior were the ability to: (1) visualize driving dynamics; (2) compare different temporal snapshots; (3) compare different days of the week, different weeks, etc.; (4) compare time periods (for example: the morning period); (5) focus the study in one street or region.

To achieve the desired goal, one of the possibilities was the creation of a tool to simplify user interaction and allow a more dynamic study. Therefore, it was created a web application. Since the goal is to perform comparisons, the web application is divided into two sides, each one of the sides contains a map. Figure 4 presents one of the sides of the web application. Note that both maps are synchronized, and performing zoom in one, will affect the other. The same happens when we change the visualization area of the map.

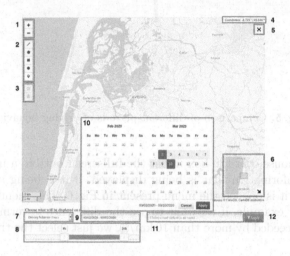

Fig. 4. Dashboard elements (**1**) Zoom in and zoom out (**2**) Add elements (**3**) Edit and delete elements (**4**) Mouse GPS position (**5**) Fullscreen (**6**) Minimap (it can be minimized) (**7**) Select information (**8**) Select hour interval (**9**) Pop-ups a calendar (**10**) Select day or an interval of days (**11**) Select road (with autocomplete functionality) (**12**) Apply changes.

In the type of information, the user can choose to analyze the driving behavior, the maximum bus speed, or the number of buses, the average speed, and the average acceleration. For the driving behavior there are two options: the user can choose the periodicity of the dataset of 1 s or 1 min. These two options are given to study the impact of increasing dataset periodicity. With that in mind, it was created the same functionality for all other options except the maximum bus speed. It was also necessary to recalculate the metrics as traveled distance, acceleration, and if the behavior is safe or unsafe.

Figure 5 contains the comparison with 1 s period between the same days interval in different hour intervals. It is possible to observe some differences and some patterns in the images. For example, there is one road segment that is red in both cases. Besides that, it is perceptible that the buses performed different paths in the morning shift versus in the afternoon shift, because there are some differences between the colored lines in both maps.

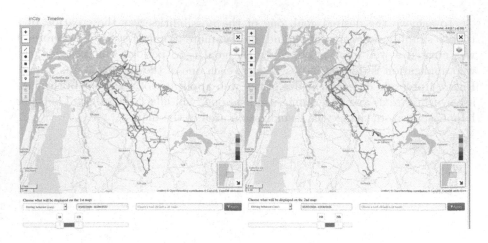

Fig. 5. Comparison of different periods of driving behavior.

With the mouse over the box under the fullscreen button, a menu is opened to select the information that is being displayed. If it is being studied driving behavior, then it is opened the menu present in Fig. 6. This menu is built based on Table 1. For example, if we want to see the difference in terms of maximum speed being exceeded by more than 10 km/h, we just select the third and fourth elements.

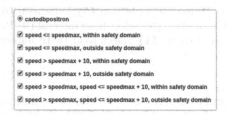

Fig. 6. Driving quality behavior menu with a tolerance of 10 km/h.

Through observation, it is possible to detect some roads in both maps in which the speed is exceeded by more than 10 km/h. If we want a more detailed analysis, we can focus the map in one specific area, as can be observed in Fig. 7. Some of the lines are more transparent than others; this is due to the percentage of unsafe driving behavior being different. For each segment, it was calculated the percentage of the several driving classification labels. A more opacity line means that the drivers performed more unsafe driving. The web application also allows the comparison of the driving behavior with the maximum bus speed, among others.

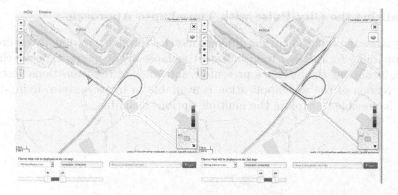

Fig. 7. Side-by-side comparison of different periods of driving behavior, maximum speed being exceeded by more than 10 km/h, and zoom on the area.

Impact of Data Frequency in Driving Behavior Analysis

One important aspect is the impact of having information every second versus every minute, Fig. 8a. As can be observed, there is a loss of information that results in some differences in Fig. 8a. The difference between the information that contributes to the creation of the maps is big. This happens because, instead of having 60 points per bus, per minute, we will have just 1. Figure 8b presents a more simplified example with just 5 points instead of 60. In the figure on the left, $t0$ represents when we start to measure the values, and $t5$ happens after we measure the other 4 values. If we ignore those, we will only have $t0$ and $t5$, or, as it is represented in the figure on the right, $t'0$ and $t'1$. This will alter the traveled distance, acceleration, and driving behavior classification.

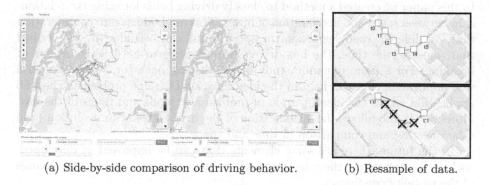

(a) Side-by-side comparison of driving behavior.   (b) Resample of data.

Fig. 8. The effects of periodicity in data - 1 s versus 1 min.

Visualizing the City Pulse with Time-Lapse Approach

To visualize the evolution of the different metrics through a period, it was created the possibility for the user to visualize a timelapse of the metric timeline chosen. Figure 9 contains the interface presented, and some of the transitions observed.

A version of the web application is available in https://aveiro-living-lab.it. pt. It is possible to explore the multiple options available.

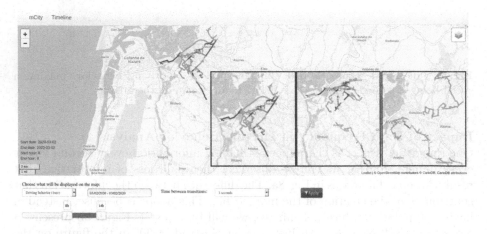

Fig. 9. Timelapse interface.

6   Conclusions

In this paper we created a method to classify driving behavior using the relationship between speed and acceleration of buses. This method has into consideration that, on different road segments, the speed limits can be different. We verified that, most of the time, the bus drivers do not perform hard accelerations or breaking. For a more dynamic study of the driving behavior, we developed a web application that can help in the identification of traffic problems. The web application allows the identification of problematic roads and times of the day. Moreover, it also allows the visualization of the driving behavior evolution (and other metrics) and comparison in time and space.

In the future, there are some elements that can be improved or developed, such as, understanding the causes for problematic zones, and enabling the change of the mobility conditions.

References

1. Aveiro Steam City: Aveiro Tech City, an Urban Innovative Actions project. www. aveirotechcity.pt/en/Projects/AVEIRO-STEAM-CITY, Accessed 15 Dec 2021
2. Boudraa, A.O., Bentabet, L., Salzenstein, F.: Dempster-shafer's basic probability assignment based on fuzzy membership functions. ELCVIA Electron. Lett. Comput. Vision Image Anal. **4**, 1–4 (2004). https://doi.org/10.1142/9789812834461_0006
3. Clapp, J.D., et al.: The driving behavior survey: Scale construction and validation. J. Anxiety Disord. **25**(1), 96–105 (2011)
4. Dempster, A.: A generalization of bayesian inference. J. Roy. Stat. Soc. Ser. B **30**, 205–247 (1968). https://doi.org/10.1111/j.2517-6161.1968.tb00722.x
5. Derbel, O., Landry], R.J.: Driver behavior assessment based on the g-g diagram in the dve system. IFAC-PapersOnLine **49**(11), 89–94 (2016). https://doi.org/10.1016/j.ifacol.2016.08.014, http://www.sciencedirect.com/science/article/pii/S2405896316313350
6. Eboli, L., Mazzulla, G., Pungillo, G.: Combining speed and acceleration to define car users' safe or unsafe driving behaviour. Transp. Res. Part C Emerg. Technol. **68**, 113–125 (2016). https://doi.org/10.1016/j.trc.2016.04.002
7. Fugiglando, U., et al.: Driving behavior analysis through can bus data in an uncontrolled environment. IEEE Trans. Intell. Transp. Syst. **20**(2), 737–748 (2019). https://doi.org/10.1109/TITS.2018.2836308
8. Guo, Z., Pan, Y., Zhao, G., Cao, S., Zhang, J.: Detection of driver vigilance level using EEG signals and driving contexts. IEEE Trans. Reliabil. **67**, 370–380 (2018)
9. Hallac, D., et al.: Driver identification using automobile sensor data from a single turn. CoRR http://arxiv.org/abs/1708.04636 (2017)
10. Jeffrey, A., Zwillinger, D., Gradshteyn, I., Ryzhik, I.: 15 - norms. In: Table of Integrals, Series, and Products (Seventh Edition), 7th edn., pp. 1081–1091. Academic Press, Boston (2007). https://doi.org/10.1016/B978-0-08-047111-2.50022-4, http://www.sciencedirect.com/science/article/pii/B9780080471112500224
11. Liang, Y., Reyes, M.L., Lee, J.D.: Real-time detection of driver cognitive distraction using support vector machines. IEEE Trans. Intell. Transp. Syst. **8**(2), 340–350 (2007). https://doi.org/10.1109/TITS.2007.895298
12. Liu, T., Yang, Y., Huang, G., Yeo, Y.K., Lin, Z.: Driver distraction detection using semi-supervised machine learning. IEEE Trans. Intell. Transp. Syst. **17**(4), 1108–1120 (2016)
13. Nakano, K., Chakraborty, B.: Real-time distraction detection from driving data based personal driving model using deep learning. Int. J. Intell. Transp. Syst. Res. (2022). https://doi.org/10.1007/s13177-021-00288-9
14. Zhang, Y., Lin, W.C., Chin, Y.K.S.: A pattern-recognition approach for driving skill characterization. IEEE Trans. Intell. Transp. Syst. **11**(4), 905–916 (2010). https://doi.org/10.1109/TITS.2010.2055239

Dynamic PCA Based Statistical Monitoring of Air Pollutant Concentrations in Wildfire Scenarios

Tobias Osswald[1]([✉]), Ana Patrícia Fernandes[2], Ana Isabel Miranda[2], and Sónia Gouveia[3]

[1] Department of Environment and Planning, University of Aveiro, Aveiro, Portugal
tobiasosswald@ua.pt
[2] Department of Environment and Planning and CESAM, University of Aveiro, Aveiro, Portugal
[3] Department of Electronics, Telecommunications and Informatics (DETI) and Institute of Electronics and Informatics Engineering of Aveiro (IEETA), University of Aveiro, Aveiro, Portugal
sonia.gouveia@ua.pt

Abstract. In 2021 the media reported multiple scenarios of devastation due to wildfires throughout the Mediterranean Basin. Wildfires have short- and long-term impacts on air quality due to the substantial release of air pollutants that can spread over large distances and may constitute a threat to human health. In this context, it is widely beneficial to develop suitable methods for continuous monitoring of air pollutants, able to produce quantitative time-variant snapshots of air quality. This work borrows some ideas of Statistical Process Monitoring (SPM) based on Dynamic Principal Component Analysis (DPCA). Shortly, the multivariate collected data are condensed into time-variant evaluations of the Hotelling's T^2 and Q statistics, which allow the quantification of the deviance of the process with respect to a reference behavior. The aim of this method is the identification of temporal episodes of large deviations from typical (generally safe) levels of concentrations. The analysis focuses on hourly time series of CO and PM concentrations, collected during 2021 at a monitoring station located in the populated area of Athens (Greece). Despite some methodological limitations and the reduced sample size, DPCA based SPM provides a promising framework to identify temporal episodes of large deviations from typical levels of air pollutants.

Keywords: Air pollutants · Multivariate monitoring · Statistical process control · Dynamic PCA

1 Introduction

In August of 2021 multiple wildfires took place around Athens, Greece. The temperatures reached 55 °C in the city center, relative humidity was low and the winds were strong. Large clouds of smoke covered the city and the region,

© Springer Nature Switzerland AG 2022
A. J. Pinho et al. (Eds.): IbPRIA 2022, LNCS 13256, pp. 680–692, 2022.
https://doi.org/10.1007/978-3-031-04881-4_54

while people struggled to protect their properties and evacuate houses [15]. One particular danger and consequence of this kind of events is the presence of air pollutants over populated areas which can lead to instantaneous respiratory problems or increase the incidence of long-term diseases, e.g. lung cancer [2]. In fact, literature studies support that air pollution has a relevant influence on short- and long-term mortality and morbidity. Martins et al. (2021) [10] analyzed the impact of short-term exposure to air pollution on daily hospital admissions in Portugal and showed it was a (statistically) significant factor, even after controlling for temperature and the history of hospital admissions.

In this context, Schneider et al. (2021) [13] analyzed data from air quality monitoring stations (i.e. continuous measurements of air pollutant concentrations) in tandem with smoke plume trajectories to characterize the impact of forest fires on the air quality at a given location. They showed that the levels of certain pollutants exhibited a consistent correlation in the presence of smoke. However, different absolute values of air pollutant concentrations were measured due to e.g. fluctuations in human-made background air pollution, site-specific features or the distance from the fire.

Aiming to find a way of detecting the presence of forest fire smoke from air quality monitoring stations alone, this work explores the use of Statistical Process Monitoring (SPM) based on Dynamic Principal Component Analysis (DPCA) to monitor air pollutants concentrations and, in particular, to identify temporal episodes of large deviations with respect to typical (and hopefully safe) levels of concentration. Shortly, DPCA considers the observed data and its time-lagged versions (up to a temporal lag l) to deal with the existing (temporal) correlations in the time series [9]. The number of observed variables is reduced to a (smaller) number of non-correlated DPCA components which are plugged-in to SPM. This work considers multivariate techniques of quality control to further reduce the number of DPCA components. The emphasis is given to Hotelling's T^2 and Q statistics to quantify the deviance with respect to a reference behavior [5]. The temporal tracking of the observed statistics against an upper control limit (UCL) allows the identification of exceedances which correspond to observations with unusually high values in some (or all) of the original variables, and may constitute (potential) episodes of disturbance on the process. The methods are illustrated with real data from the Air Quality e-Reporting database [3] for the year 2021, when several wildfires occurred in Greece. The analysis focused on data monitored at Piraeus station located in the metropolitan area of Athens, with more than 3.1 million inhabitants reported in 2021.

This paper is organized as follows. Section 2 presents the methods for multivariate statistical process monitoring, including the definition of the reference model (from DPCA), statistics of interest and UCLs. Section 3 presents the experimental data and a description of some occurring wildfire events. Results and discussion are presented in Sect. 4 and Sect. 5 is devoted to the conclusions.

2   Multivariate Statistical Process Monitoring

The T^2 and Q statistics require the definition of a (reference) model describing the target behavior of the process. This model is constructed with dynamic PCA and fit with a training sample representative of this behavior [17]. The same data are also used to determine UCLs (at a certain confidence level) based on the sampling distributions of T^2 and Q. Finally, during real-time monitoring, T^2 and Q are used to quantify deviations of the process against the reference.

The methods here described were implemented in the R (3.6.2) programming language for statistics [11]. The DPCA implementation was based on the PCA function available in the inbuilt "stats" package. Furthermore, the "openair" package was used to manage the air pollutant data and for visualization.

2.1   Reference Model for the Process

In a time series context, each observed variable is a time indexed variable $X(t)$ with $t = t_1, ...t_n$ being the observation time points. Assuming evenly spaced temporal data, $t = 1, \cdots, n$ is considered for simplicity in the notation [14].

For the proper application of any PCA-based transformation, the observed variables are first standardized, to be transformed into the same scale (i.e. zero mean, unitary standard deviation and no physical units) which guarantees that all variables contribute equally to the analysis. This standardization is accomplished by subtracting the mean and dividing it by the standard deviation for each value of each variable. Then, the standardized values $X_1(t), X_2(t), ... X_m(t)$ at a given time t can be organized into a $1 \times m$ vector

$$\mathbf{X}(t) = [X_1(t) \ X_2(t) \ \cdots \ X_m(t)] . \tag{1}$$

and, for the entire period, the information can be structured into a $n \times m$ matrix

$$\mathbf{X} := \begin{bmatrix} \mathbf{X}(1) \\ \cdots \\ \mathbf{X}(n) \end{bmatrix}, \tag{2}$$

where the x_{tj} entry corresponds to the standardized value of the variable $j = 1, \cdots, m$ observed at the time $t = 1, \cdots, n$. The DPCA requires the construction of an augmented matrix $\tilde{\mathbf{X}}^l$ by repeating the m columns of $\mathbf{X}$ with an increasing time-lag up to a maximum $l \geq 0$. The resulting matrix is defined as

$$\tilde{\mathbf{X}}^l = \begin{bmatrix} \mathbf{X}(1) & \mathbf{X}(0) & \cdots & \mathbf{X}(1-l) \\ \mathbf{X}(2) & \mathbf{X}(1) & \cdots & \mathbf{X}(2-l) \\ \vdots & \vdots & \ddots & \vdots \\ \mathbf{X}(n) & \mathbf{X}(n-1) & \cdots & \mathbf{X}(n-l) \end{bmatrix} . \tag{3}$$

The augmented matrix $\tilde{\mathbf{X}}^l$ has the special form of a Hankel matrix. Also, it increases the number of columns to $w = (l+1) \times m$ and maintains the number of rows as n by setting $\mathbf{X}(t) = \mathbf{X}(1)$ for $t \leq 0$.

The classical PCA algorithm is then applied to $\tilde{\mathbf{X}}^l$, by seeking for linear combinations of its columns with maximum variance [7]. The PCA ingredients can be obtained from the eigenvalue decomposition of the $w \times w$ covariance matrix of the augmented data, $\mathbf{S}$, i.e.

$$\mathbf{S} = \mathbf{U}\Lambda\mathbf{U}^T, \tag{4}$$

where $\mathbf{U} = [U_1 \cdots U_w]$ is an orthogonal matrix containing the w eigenvectors of $\mathbf{S}$ along its columns, $\mathbf{U}^T$ is the transpose of $\mathbf{U}$ and $\Lambda = \text{diag}(\lambda_1, \cdots, \lambda_w)$ is a matrix containing the corresponding eigenvalues along its diagonal, and zeros elsewhere, considering $\lambda_1 \geq \cdots \geq \lambda_w$. The principal components (PC) are then obtained from the orthogonal linear transformation

$$\mathbf{Y}^l = \tilde{\mathbf{X}}^l\mathbf{U}, \tag{5}$$

where $\mathbf{Y}^l$ is a matrix with the w PC scores along columns. From Eq. (5),

$$\tilde{\mathbf{X}}^l = \mathbf{Y}^l\mathbf{U}^{-1} \Leftrightarrow \tilde{\mathbf{X}}^l = \mathbf{Y}^l\mathbf{U}^T, \tag{6}$$

which highlights a convenient way for data reconstruction, due to the property $\mathbf{U}^{-1} = \mathbf{U}^T$ of the orthogonal matrix $\mathbf{U}$. Note that,

$$\tilde{\mathbf{X}}^l = \mathbf{Y}_k\mathbf{U}_k^T + \mathbf{e}_k^l, \tag{7}$$

with $\mathbf{Y}_k$ and $\mathbf{U}_k$ being the matrices composed of the first k columns in $\mathbf{Y}$ and $\mathbf{U}$, respectively. Moreover, $\mathbf{e}_k^l$ is a matrix of residuals expected to exhibit no auto- or crosscorrelation. Therefore,

$$\hat{\mathbf{X}}_k^l = \mathbf{Y}_k\mathbf{U}_k^T \tag{8}$$

can be used as a reconstruction (approximation) of the data and its dynamics up to lag l, considering the first k principal components. The variance explained by these k PC scores is given by

$$c_{l,k} = \frac{\sum_{i=1}^k \lambda_i^l}{\text{tr}(\Lambda)}. \tag{9}$$

where $\text{tr}()$ is the trace function of a matrix and λ_i^l represents the variance explained by the ith PC score out of the total of $w = (l+1) \times m$ scores.

The choice of l and k is a critical step in DPCA as, for certain choices of l and k, the DPCA principal components may still exhibit some correlation [8]. In agreement with previous studies [9], this paper considered the optimal (l,k) values as those minimizing the amount of autocorrelation structure in the residuals of the DPCA model. To accomplish that, the sum of squared residuals

$$r_{l,k}(t) = \mathbf{e}_k^l(t)\,\mathbf{e}_k^l(t)^T \tag{10}$$

was evaluated for each $t = 1, \cdots, n$ and for each (l,k) pair up to $l = 10$ (recall that k is constrained to be lower than $w = (l+1) \times m$). Then, the amount of autocorrelation in the $r_{l,k}(t)$ sequence was quantified through

$$\bar{\gamma}_{l,k} = \frac{1}{40} \sum_{i=1}^{40} \hat{\gamma}_{l,k}(i) .$$

(11)

i.e. the sample autocorrelation $\hat{\gamma}_{l,k}(i)$ averaged over 40 lags [14].

2.2   Expected Behavior of the Process During Monitoring

For any time interval to monitor, the expected behavior of the process assumes that the reference model is true. Recall that this model is characterized by the data standardization (mean and standard deviation calculated with training data) and the dynamic PCA settings, namely the orthogonal linear transformation provided through $\mathbf{U}_k$ with the chosen (l, k) values. Thus, for the new data to monitor, a new augmented matrix $\tilde{\mathbf{X}}^*$ is constructed by following the same procedures. In particular, the expected behavior of the process is given by reevaluating Eqs. (5) and (8) as follows

$$\mathbf{Y}^* = \tilde{\mathbf{X}}^* \mathbf{U}_k \text{ and } \hat{\mathbf{X}}^* = \mathbf{Y}^* \mathbf{U}_k^T,$$

(12)

where $\mathbf{Y}^*$ holds the PC scores for the new data and $\hat{\mathbf{X}}^*$ is the expected behavior of the process in the new time period. Note that $\mathbf{Y}^*$ and $\hat{\mathbf{X}}^*$ are $n \times k$ sized matrices and $\mathbf{Y}^*(t)$ is a $1 \times k$ vector with the PC scores evaluated for time t.

2.3   Hotelling's T^2 and Q Statistics Applied to SPM

As PC scores have zero mean and are uncorrelated, the Hotelling's T^2 statistic applied to the PC scores can be expressed simply by

$$T^2(t) = \mathbf{Y}^*(t) \, \Lambda^{-1} \, \mathbf{Y}^*(t)^T,$$

(13)

where the normalization by Λ ensures that all PC scores have the same contribution for the statistic. On the other hand, the Q statistic is defined as

$$Q(t) = \left(\tilde{\mathbf{X}}^*(t) - \hat{\mathbf{X}}^*(t) \right) \left(\tilde{\mathbf{X}}^*(t) - \hat{\mathbf{X}}^*(t) \right)^T,$$

(14)

where each factor expresses the difference between the observed $\tilde{\mathbf{X}}^*(t)$ and the predicted $\hat{\mathbf{X}}^*(t)$ given the reference model. Thus, $Q(t)$ corresponds to the sum of squared errors with respect to the reference, where larger values indicate larger deviation and imply that the correlation structure in the reference period is different from that observed in the new time period. In other words, $Q(t)$ is a goodness-of-fit measure between the new data and the reference model.

Note that $T^2(t)$ and $Q(t)$ together contain complementary information concerning the process dynamics as a function of time t. While $T^2(t)$ quantifies the "in control" component of the process (which is predicted by the reference model), $Q(t)$ expresses its "out of control" component (deviation from the reference model which includes the PC scores not considered in the model).

2.4   Upper Control Limit for Hotelling's T^2 and Q

Unusual high values in $T^2(t)$ or $Q(t)$ statistics are identified as exceedances of a given (large) threshold value named as upper control limit (UCL). There are theoretical results providing the sampling distribution (or approximations of it) for T^2 or Q statistics which allow an UCL for a given confidence level $1 - \alpha$ (e.g. 99%) to be derived. Here α represents the probability of the process exceeding the UCL given that the process behavior is in accordance with that of the reference model (type I error). Namely, the UCL for the T^2 statistic is based on the reference model and the $F-$distribution [17] and is given by

$$\text{UCL }_{T^2} = \frac{k(n+1)(n-1)}{n(n-k)} F_{(1-\alpha,k,n-k)}, \tag{15}$$

where k is the number of retained PCs, n is the number of observations used to fit the reference model and $F_{(1-\alpha,k,n-k)}$ is the $1-\alpha$ percentile of the $F-$distribution with k and $n - k$ degrees of freedom. On the other hand, an approximation of the UCL for the Q statistic is based on the normal distribution [6] being

$$\text{UCL }_Q = \theta_1 \left[\frac{z_{1-\alpha}\sqrt{2\theta_2 h_0^2}}{\theta_1} + 1 + \frac{\theta_2 h_0(h_0 - 1)}{\theta_1^2} \right]^{1/h_0} \tag{16}$$

where $z_{1-\alpha}$ is the $1 - \alpha$ percentile of the standard normal distribution $N(0,1)$ and $\theta_i = \sum_{j=k+1}^{w}(\lambda_j)^i$ for $i = 1, 2, 3$. Finally, h_0 is such that

$$h_0 = 1 - \frac{2\theta_1\theta_3}{3\theta_2^2}. \tag{17}$$

There are two aspects concerning the UCL for T^2 presented in Eq. (15) that are worth mentioning. First, its construction assumes time independent and multivariate normality data [9]. Although the literature has already prescribed some possible solutions to deal with the violation of such assumptions (e.g. by proposing autoregressive T^2 control charts for univariate autocorrelated processes [1], implementing robust estimators or Box-Cox transformations to correct deviations to normality [8]), this paper considers evaluating the use of the presented UCLs in the context of air pollutant data. Second, UCL $_{T^2}$ is known to be conservative when evaluated for the training sample [16], thus being expected to produce a percentage of exceedances lower than α in that sample.

3   Experimental Data

Several air pollutants emitted during a wildfire, including particulate matter with size lower than 2.5 um or than 10 um ($PM_{2.5}$ or PM_{10}) and carbon monoxide (CO), are often measured in air quality monitoring stations. As such, those air pollutants were chosen to illustrate the presented method. Hourly data for 2021, from 19 monitoring stations in Greece was made available through the Air Quality e-Reporting database [3]. It is common for air quality monitoring stations to become unavailable for several months, due to a number of issues associated with e.g. communication, maintenance operations or component failures. Of the 19 available stations, the "PIRAEUS-1" (EU-code STA-GR0030A) was the most adequate for this illustration since it had ample hourly data availability during the fire season in the three pollutants of interest. The "PIRAEUS-1" station is at latitude 37.94329N and longitude 23.64751E (see Fig. 1) and is located 20m above sea level, in the city center of Piraeus (a port city located 7km southwest of Athens' city center). The location is classified as urban and the station is optimally positioned to measure the direct effect of traffic emissions.

Information about wildfires in Greece during August 2021 was searched in publicly available reports, media and satellite products [4,12]. Figure 1 shows two examples of fires detected by satellite in which smoke plumes are visible. The fire in Evia (04-08-2021) produced air pollutants that were transported by the wind to Athens and created a smoke plume over the city. This behavior is not always observed and in certain cases air pollutants could be led away from the city. Such is the case of the wildfires in Evia and near Piraeus (06-08-2021) that did not originate, at the time of the satellite image acquisition, an accumulation of smoke over Athens. Therefore, it is not expected for all fires in Greece to have an impact on the air quality as measured in the "PIRAEUS-1" station.

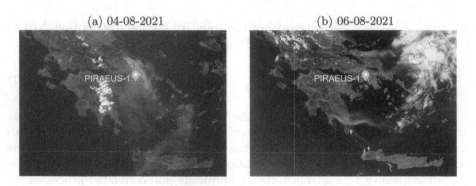

(a) 04-08-2021   (b) 06-08-2021

Fig. 1. Satellite images of Greece where clouds show as white puffs and smoke as gray fuzzy shades. Locations of the fires detected in the VIIRS Fire and Thermal Anomalies satellite product are overlaid in red (NASA Worldview website). (Color figure online)

In August, 13 days had at least one registered fire. Multiple fires occurred in Attica (Athens region) including one on the 3rd, a seven-day-fire starting on the 5th, a four-day-fire starting on the 16th and another on the 23rd. A single large fire in Evia (island to the North of Athens) on the 4th lasted for seven days.

4   Results and Discussion

The reference model was fit with training data representative of the "in control" behavior of air pollutants. The spring season (from 01-04-2021 to 30-06-2021) was considered to best fulfill this criterion since it typically presents low fire risk and exhibits anthropogenic emissions lower than those expected during winter.

Figure 2 shows the results concerning the choice of (l, k) for the reference model. On one hand, Fig. 2a clears out that the pair minimizing $\overline{\gamma}_{l,k}$ is $(l, k) = (1, 5)$. On the other hand, Fig. 2b shows that $c_{1,5} \gg 90\%$. Thus, $(l, k) = (1, 5)$ simultaneously minimizes $\overline{\gamma}_{l,k}$ and leads to a high percentage of explained variance by the DPCA model. Figure 2 also displays the results for $l = 0$ (and $k \leq 3$) and $k = 0$. The $l = 0$ setting leads to $\hat{\mathbf{X}}^0(t) = \mathbf{X}(t)$ i.e. the augmented data matrix equals the original (standardized) data. In this case, no lags are considered in the DPCA model reverting the analysis to the classical PCA. The $k = 0$ case (i.e. no PC scores in the reconstruction) will result in the approximation $\hat{\mathbf{X}}(t) = 0$. Thus, $\overline{\gamma}_{l,0}$ corresponds to the average autocorrelation of the sum of squared data in $\tilde{\mathbf{X}}^l(t)$. These two cases can be used as benchmarks for the interpretation of DPCA results, for several (l, k) combinations, with respect to a model based on classical PCA ($l = 0$) and a null model ($k = 0$). Figure 2 shows that the classical PCA ($l = 0$) retains $k = 2$ PC scores to achieve the minimal $\overline{\gamma}_{0,k}$. However, $\overline{\gamma}_{0,2}$ and $c_{0,2}$ are (both) less favorable with respect to $\overline{\gamma}_{1,5}$ and $c_{1,5}$, respectively, thus supporting that the dynamic PCA approach is more advantageous than the classical one for the analysis of these data. Finally, Fig. 2 also shows that combinations with $l > 1$ and appropriate values of k still yield high $c_{l,k}$ values but $\overline{\gamma}_{l,k}$ becomes progressively higher for increasing l.

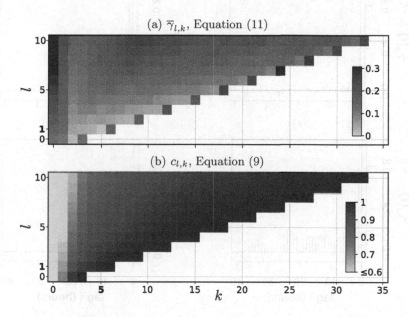

Fig. 2. Mosaic plot for (a) $\overline{\gamma}_{l,k}$ and (b) $c_{l,k}$, evaluated for $l \leq 10$ and $k \leq (l + 1) \times 3$. Results for the benchmark cases $l = 0$ and $k = 0$ are also shown.

For a more detailed analysis, Fig. 3 displays $\hat{\gamma}_{1,k}$ for several values of k. Figure 3a shows that the autocorrelation for $(l,k) = (1,5)$ is statistically significant solely for lag $i = 1$ hour (at 5% significance level). Note that, at 5% level, the sample autocorrelation is expected to surpass the band of no autocorrelation in 2 out of 40 lags and still support that $r_{l,k}(t)$ is a realization of a white noise process. Following the results in Fig. 2a, it can be seem that $\bar{\gamma}_{1,k}$ fairly decreases for increasing k which points out that the autocorrelation structure of the data is progressively retained in the PC scores. As shown in Fig. 3b, the $\hat{\gamma}_{1,0}(i)$ profile highlights that the data exhibit both short-term correlations (expressed by the exponential decaying values up to $i = 10$) and a strong daily periodic pattern (showed with the more pronounced values around $i = 24$). With $k = 1$ and $k = 2$ (Fig. 3) both of these features seem to be progressively removed from the $\hat{\gamma}_{1,k}(i)$ profile. This indicates that the first dynamic PC scores are retaining these features of the data, implying that they are shared simultaneously by CO and PM time series. Therefore, the DPCA model with $(l,k) = (1,5)$ adequately captures the temporal/dynamic structure of the original data. Note that even in the presence of a strong daily periodic pattern in the data, there is no need to set $l > 1$ because the pattern takes place simultaneously in all time series.

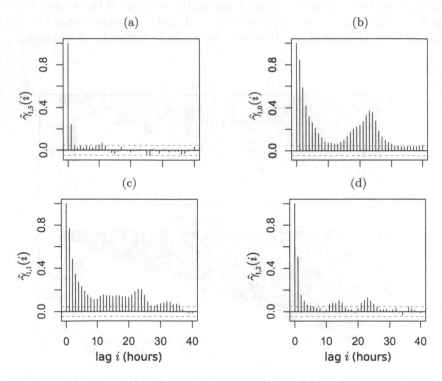

Fig. 3. Autocorrelation of the sum of squared residuals, $\hat{\gamma}_{1,k}(i)$ for $k = \{5, 0, 1, 2\}$ in subplots {a,b,c,d}, respectively. The blue dashed lines represent the limits of the approx. 95% confidence interval for no autocorrelation i.e. $\pm 1.96/\sqrt{n}$ [14]. (Color figure online)

After defining the reference model, Hotelling's T^2 and Q statistics were calculated along with the corresponding UCLs set at $\alpha = 0.01$. Recall that by setting $\alpha = 0.01$, 1% of exceedances are expected if the data are in accordance with the reference model. Figure 4 shows that the percentage of exceedances above the UCL is much higher than 1%, for both T^2 and Q. During the reference period, the percentage of exceedances reaches 5.6% for T^2 and 1.3% for Q. However, during the monitoring period, this percentage increases abruptly to 13.5% and to 3.7% respectively. This clearly indicates that, from the reference period to the monitoring period, the environmental variables change substantially (increased T^2) and are less in accordance with the reference model (increased Q). Thus, the "in control" component of the process decreases while the "out of control" one increases during the monitoring period.

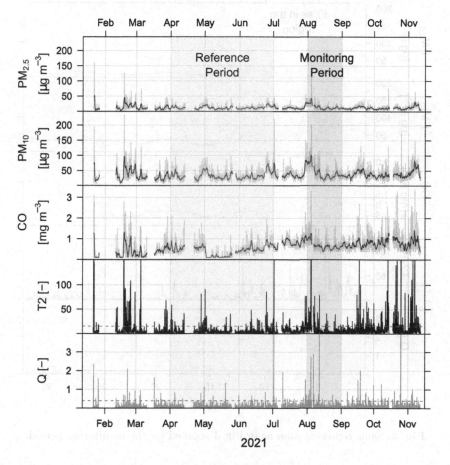

Fig. 4. Air pollutants monitored at PIRAEUS-1 station and daily moving averages (darker lines) for the sake of trend visualization. The T^2 and Q scale is limited at 10·UCL where UCLs are represented in dashed lines.

Figure 5 zooms into the monitoring period to better highlight changes in air pollutants and exceedances on T^2 and Q. There was one day with no available data (out of the 13 with registered fires). Thus, out of the 12 days, T^2 and Q exceed (individually) the UCL in 9 and 5 of those days, respectively. Although exceedances in both statistics were only observed in 3/12 days, in 11/12 days at least one statistic surpassed its UCL. On the 16th there were no exceedances in both statistics since the wind was blowing the smoke away from Athens. Finally, Fig. 5 also shows UCL crossings out of the period with registered fires. These exceedances are attributed to visible changes in air pollutants possibly linked to unreported fires or other events.

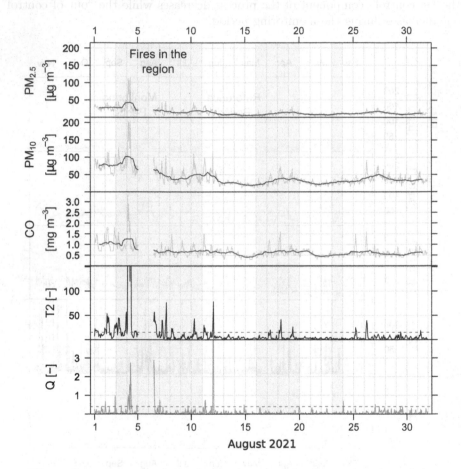

Fig. 5. Same representation as in Fig. 4 zoomed for the monitoring period.

5   Conclusions

A statistical process monitoring approach (developed for industrial processes) was applied to air pollutants and showed potential for the detection of large deviations from a reference behavior. Smoke from wildfires near Athens in 2021 was often detected in the control charts, suggesting that the method is a good basis for other similar cases. The percentage of exceedances was larger than the significance level set for the upper control limits. This may be an indication that further improvements should consider coping with deviations from independence and normality in the data. Nonetheless, the percentage of exceedances during the monitoring period is much larger than during the reference period.

Future work includes the use of an air pollution model to simulate the events presented in this paper with higher spatio-temporal resolution. This will provide a better understanding of the circulation of smoke around the monitoring stations and further assess the accuracy of this methodology.

Acknowledgments. This work was partially funded by the Foundation for Science and Technology, I.P. (FCT), through national (MEC) and European Structural (FEDER) funds, in the scope of UIDB/00127/2020 (IEETA, www.ieeta.pt) and UIDB/50017/2020 & UIDP/50017/2020 (CESAM, https://www.cesam.ua.pt/) projects. The authors also acknowledge the FCT funding through national funds under the projects SmokeStorm (PCIF/MPG/0147/2019) and FirEUrisk (EU's Horizon 2020 R&I programme, grant agreement No 101003890).

References

1. Apley, D.W., Tsung, F.: The autoregressive T2 chart for monitoring univariate autocorrelated processes. J. Qual. Technol. **34**(1), 80–96 (2002)
2. Cohen, A.: Air pollution and lung cancer: what more do we need to know? Thorax **58**, 1010–1012 (2003)
3. EEA: European Environmental Agency: Air quality e-reporting (2021). https://www.eea.europa.eu/data-and-maps/data/aqereporting-2, Accessed 30 Jan 2022
4. GDACS: Global disaster alert and coordination system (2021). https://www.gdacs.org/Alerts/default.asp, Accessed 30 Jan 2022
5. Jackson, J.E.: Multivariate quality control. Commun. Stat. Theory Methods **14**(11), 2657–2688 (1985)
6. Jackson, J.E., Mudholkar, G.S.: Control procedures for residuals associated with principal component analysis. Technometrics **21**(3), 341–349 (1979)
7. Jolliffe, I.T., Cadima, J.: Principal component analysis: a review and recent developments. Phil. Trans. Royal Soc. A **374**(2065) (2016)
8. Ketelaere, B.D., Hubert, M., Schmitt, E.: Overview of PCA-based statistical process-monitoring methods for time-dependent, high-dimensional data. J. Qual. Technol. **47**(4), 318–335 (2015)
9. Ku, W., Storer, R., Georgakis, C.: Disturbance detection and isolation by dynamic principal component analysis. Chemometr. Intell. Lab. Syst. **30**(1), 179–196 (1995)
10. Martins, A., Scotto, M., Deus, R., Monteiro, A., Gouveia, S.: Association between respiratory hospital admissions and air quality in Portugal: A count time series approach. PLoS ONE **16**(7), e0253455 (2021)

11. R Core Team: R: A Language and Environment for Statistical Computing. R Foundation for Statistical Computing, Vienna, Austria (2021)
12. Roesli, H.P., Fierli, F., Lancaster, S.: Smoke and burned areas from greek fires. EUMETSAT (2021). https://www.eumetsat.int/smoke-and-burned-areas-greek-fires, Accessed 30 Jan 2022
13. Schneider, S.R., Lee, K., Santos, G., Abbatt, J.P.D.: Air quality data approach for defining wildfire influence: Impacts on PM2.5, NO2, CO, and O3 in Western Canadian cities. Environ. Sci. Technol. **55**(20), 13709–13717 (2021)
14. Shumway, R., Stoffer, D.: Time Series Analysis and its Applications with R examples. Springer Texts in Statistics, Springer, Heidelberg (2011). https://doi.org/10.1007/978-3-319-52452-8
15. Smith, H.: 'Apocalyptic' scenes hit Greece as Athens besieged by fire (2021). https://www.theguardian.com/world/2021/aug/07, Accessed 09 Nov 2021
16. Tracy, N.D., Young, J.C., Mason, R.L.: Multivariate control charts for individual observations. J. Qual. Technol. **24**(2), 88–95 (1992)
17. Vanhatalo, E., Kulahci, M., Bergquist, B.: On the structure of dynamic principal component analysis used in statistical process monitoring. Chemometr. Intell. Lab. Syst. **167**, 1–11 (2017)

Author Index

Printed in the United States
by Baker & Taylor Publisher Services

Printed in the United States
by Baker & Taylor Publisher Services